PROBABILITY AND MEASURE

Probability and Measure

Patrick Billingsley

The University of Chicago

John Wiley & Sons, New York • Chichester • Brisbane • Toronto

Library of Congress Cataloging in Publication Data:

Billingsley, Patrick.
    Probability and measure.

    (Wiley series in probability and mathematical statistics)
    Bibliography: p.
    Includes index.
    1. Probabilities.   2. Measure theory.   I. Title.

QA273.B575   519.2   78-25632
ISBN 0-471-03173-9

Printed in the United States of America

10 9 8 7 6 5 4 3 2

Preface

Edward Davenant said he "would have a man knockt in the head that should write anything in Mathematiques that had been written of before." So reports John Aubrey in his *Brief Lives*. What is new here then?

To introduce the idea of measure the book opens with Borel's normal number theorem, proved by calculus alone, and there follow short sections establishing the existence and fundamental properties of probability measures, including Lebesgue measure on the unit interval. For simple random variables—ones with finite range—the expected value is a sum instead of an integral. Measure theory, without integration, therefore suffices for a completely rigorous study of infinite sequences of simple random variables, and this is carried out in the remainder of Chapter 1, which treats laws of large numbers, the optimality of bold play in gambling, Markov chains, large deviations, the law of the iterated logarithm. These developments in their turn motivate the general theory of measure and integration in Chapters 2 and 3.

Measure and integral are used together in Chapters 4 and 5 for the study of random sums, the Poisson process, queues, convergence of measures, characteristic functions, central limit theory. Chapter 6 begins with derivatives according to Lebesgue and Radon-Nikodym—a return to measure theory—then applies them to conditional expected values and martingales. Chapter 7 treats such topics in the theory of stochastic processes as Kolmogorov's existence theorem and separability, all illustrated by Brownian motion.

What is new, then, is the alternation of probability and measure, probability motivating measure theory and measure theory generating further probability. The book presupposes a knowledge of combinatorial and discrete probability, of rigorous calculus, in particular infinite series, and of elementary set theory.

Chapters 1 through 4 (starred passages excepted) are designed to be taken up in sequence. Apart from some examples and two starred topics, Chapters 5, 6, and 7 are independent of one another and can be read in any order.

My goal has been a book I would myself have liked when I first took up the subject, and the needs of students have been given precedence over the requirements of logical economy. For instance, Kolmogorov's existence theorem appears not in the first chapter but in the last, stochastic processes needed earlier having been constructed by special arguments which, although technically redundant, motivate the general result. And the general result is in the last chapter given two proofs at that. It is instructive, I think, to see the show in rehearsal as well as in performance.

Chicago, December 1978   *Patrick Billingsley*

Acknowledgments

Much of the work on this book was supported by the National Science Foundation. Michael Wichura taught from a preliminary version, and I am very grateful to him for his excellent suggestions concerning both detail and overall organization, which have greatly improved the book.

Contents

* Asterisk indicates topics that may be omitted on a first reading.

CHAPTER 2. MEASURE 131

PROBABILITY AND MEASURE

CHAPTER 1

Probability

SECTION 1. BOREL'S NORMAL NUMBER THEOREM

Although sufficient for the development of many interesting topics in mathematical probability, the theory of discrete probability spaces* does not go far enough for the rigorous treatment of problems of two kinds: those involving an infinitely repeated operation, as an infinite sequence of tosses of a coin, and those involving an infinitely fine operation, as the random drawing of a point from a segment. A mathematically complete development of probability, based on the theory of measure, puts these two classes of problem on the same footing, and as an introduction to measure-theoretic probability it is the purpose of the present section to show by example why this should be so.

The Unit Interval

The project is to construct simultaneously a model for the random drawing of a point from a segment and a model for an infinite sequence of tosses of a coin. The notions of independence and expected value, familiar in the discrete theory, will have analogues here, and some of the terminology of the discrete theory will be used in an informal way to motivate the development. The formal mathematics, which involves only such notions as the length of an interval and the Riemann integral of a step function, will be entirely rigorous. All the ideas will recur later in more general form.

Let Ω denote the unit interval $(0, 1]$; to be definite, take intervals open on the left and closed on the right. Let ω denote the generic point of Ω. If

---

* For the discrete theory, presupposed here, see the first half of FELLER, Volume I. (Names in capital letters refer to the bibliography on page 507.)

(1.1)
$$A = \bigcup_{i=1}^{n} (a_i, b_i],$$

where the intervals $(a_i, b_i]$ are disjoint* and are contained in Ω, assign to A the probability

(1.2)
$$P(A) = \sum_{i=1}^{n} (b_i - a_i).$$

If A and B are two such finite disjoint unions of intervals, then so is $A \cup B$, and if $A \cap B = 0$, then

(1.3)
$$P(A \cup B) = P(A) + P(B).$$

This relation, which is certainly obvious intuitively, is a consequence of the additivity of the Riemann integral:

(1.4)
$$\int_0^1 (f(\omega) + g(\omega)) \, d\omega = \int_0^1 f(\omega) \, d\omega + \int_0^1 g(\omega) \, d\omega.$$

If $f(\omega)$ is a step function taking value c_j in the interval $(x_{j-1}, x_j]$, where $0 = x_0 < x_1 < \cdots < x_k = 1$, then its integral in the sense of Riemann has the value

(1.5)
$$\int_0^1 f(\omega) \, d\omega = \sum_{j=1}^{k} c_j(x_j - x_{j-1}).$$

If $f = I_A$ and $g = I_B$ are the indicators[†] of A and B and if $A \cap B = 0$, then (1.3) follows from (1.4) and (1.5). This also shows that the definition (1.2) is unique—note that A will have many representations of the form (1.1) because $(a, b] \cup (b, c] = (a, c]$. Later these facts will be derived anew from the general theory of Lebesgue integration.[‡]

According to the usual models, if a radioactive substance has emitted a single α-particle during a unit interval of time, or if a single telephone call has arrived at an exchange during a unit interval of time, then the instant at which the emission or the arrival occurred is random in the sense that it lies in (1.1) with probability (1.2). Thus (1.2) is the starting place for the description of a point drawn at random from the unit interval: Ω is regarded as a sample space and the set (1.1) is identified with the event that the random point lies in it.

The definition (1.2) is also the starting point for a mathematical representation of an infinite sequence of tosses of a coin. With each ω associate its nonterminating dyadic expansion

---

* *Disjoint* will always mean *pairwise* disjoint: $\{B_i\}$ is disjoint if $B_i \cap B_j = 0$ for $i \neq j$, where 0 denotes the empty set.
† The *indicator* function I_A takes value 1 on A and 0 elsewhere. The term *characteristic function* will be reserved for the Fourier transform.
‡ Passages in small type concern side issues and technical matters, but their contents are sometimes required later.

(1.6) $$\omega = \sum_{n=1}^{\infty} \frac{d_n(\omega)}{2^n} = .d_1(\omega)\, d_2(\omega) \cdots ,$$

each $d_n(\omega)$ being 0 or 1. Thus

(1.7) $$(d_1(\omega), d_2(\omega), \cdots)$$

is the sequence of binary digits in the expansion of ω. For definiteness, a point such as $\frac{1}{2} = .1000 \ldots = .0111 \ldots$, which has two expansions, takes the non-terminating one; 1 takes the expansion $.111 \ldots$.

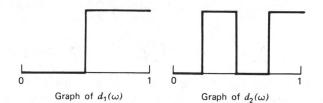

Graph of $d_1(\omega)$ Graph of $d_2(\omega)$

Imagine now a coin with faces labeled 1 and 0 instead of the usual heads and tails. If ω is drawn at random, then (1.7) behaves as if it resulted from an infinite sequence of tosses of the coin. Indeed, if $u_1, \ldots , u_n$ is any n-long sequence of 0's and 1's, then

(1.8) $$[\omega: d_i(\omega) = u_i,\, i = 1, \ldots , n] = \left(\sum_{i=1}^{n} \frac{u_i}{2^i}, \sum_{i=1}^{n} \frac{u_i}{2^i} + \frac{1}{2^n} \right].$$

The interval here is open on the left and closed on the right because the expansion (1.6) is the nonterminating one. In the model for coin tossing the set (1.8) represents the event that the first n tosses give the outcomes $u_1, \ldots , u_n$ in sequence. By (1.2) and (1.8),

(1.9) $$P[\omega: d_i(\omega) = u_i,\, i = 1, \ldots , n] = \frac{1}{2^n},$$

which is what probabilistic intuition requires.

The intervals (1.8) are called *dyadic* intervals, the endpoints being adjacent dyadic rationals with the same denominator, and n is the *rank* or *order* of the interval. For each n the 2^n dyadic intervals of rank n decompose or partition the unit interval. In the passage from the partition for n to that for $n + 1$, each interval (1.8) is split into two parts of equal length, a left half on which $d_{n+1}(\omega)$ is 0 and a right half on which $d_{n+1}(\omega)$ is 1. Hence $P[\omega: d_n(\omega) = 1] = \frac{1}{2}$ for every n. Note that $d_k(\omega)$ is constant on each dyadic interval of rank n if $k \leq n$.

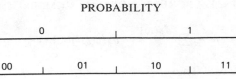

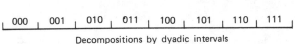

Decompositions by dyadic intervals

The probabilities of various familiar events can be written down immediately. The sum $\sum_{i=1}^{n} d_i(\omega)$ is the number of 1's among $d_1(\omega), \ldots, d_n(\omega)$, to be thought of as the number of heads in n tosses of a fair coin. The usual binomial formula is

$$(1.10) \qquad P\left[\omega : \sum_{i=1}^{n} d_i(\omega) = k\right] = \binom{n}{k} \frac{1}{2^n}, \qquad 0 \le k \le n.$$

This follows from the definitions: The set on the left in (1.10) is the union of those intervals (1.8) corresponding to sequences $u_1, \ldots, u_n$ containing k 1's and $n - k$ 0's; each such interval has length $1/2^n$ by (1.9) and there are $\binom{n}{k}$ of them, and so (1.10) follows from (1.2).

The functions $d_n(\omega)$ can be looked at in two ways. Fixing n and letting ω vary gives a real function $d_n = d_n(\cdot)$ on the unit interval. Fixing ω and letting n vary gives the sequence (1.7) of 0's and 1's. The probabilities (1.9) and (1.10) involve only finitely many of the components $d_i(\omega)$. The interest here, however, will center mainly on properties of the entire sequence (1.7). It will be seen that the mathematical properties of this sequence mirror the properties to be expected of a coin-tossing process that continues forever.

As the expansion (1.6) is the nonterminating one, there is the defect that for no ω is (1.7) the sequence $(1, 0, 0, 0, \ldots)$, for example. It seems clear that the chance should be 0 for the coin to turn up heads on the first toss and tails forever after, so that the absence of $(1, 0, 0, 0, \ldots)$—or of any other single sequence—should not matter. See on this point the additional remarks immediately preceding Theorem 1.2.

The Weak Law of Large Numbers

In studying the connection with coin tossing it is instructive to begin with a result that can, in fact, be treated within the framework of discrete probability, namely, the *weak law of large numbers*:

Theorem 1.1. *For each* $\epsilon,$*

---

* The standard ϵ and δ of analysis will always be understood to be positive. For the analysis presupposed here, see HARDY, for example, or RUDIN.

$$(1.11) \qquad \lim_{n \to \infty} P\left[\omega: \left|\frac{1}{n}\sum_{i=1}^{n} d_i(\omega) - \frac{1}{2}\right| \geq \epsilon\right] = 0.$$

Interpreted probabilistically, (1.11) says that if n is large, then there is small probability that the fraction or relative frequency of heads in n tosses will deviate much from $\frac{1}{2}$, an idea lying at the base of the frequency conception of probability. As a statement about the structure of the real numbers, (1.11) is also interesting arithmetically.

As will become clear in the course of the proof, the set in (1.11) is the union of certain of the intervals (1.8), and so its probability is well defined by (1.2).

PROOF OF THEOREM 1.1.   With the Riemann integral in the role of expected value, the usual application of Chebyshev's inequality will lead to a proof of (1.11). The argument becomes simpler if the $d_n(\omega)$ are replaced by the *Rademacher functions*,

$$(1.12) \qquad r_n(\omega) = 2d_n(\omega) - 1 = \begin{cases} +1 & \text{if } d_n(\omega) = 1, \\ -1 & \text{if } d_n(\omega) = 0. \end{cases}$$

Consider the partial sums

$$(1.13) \qquad s_n(\omega) = \sum_{i=1}^{n} r_i(\omega).$$

Since $\sum_{i=1}^{n} d_i(\omega) = (s_n(\omega) + n)/2$, (1.11) with $\epsilon/2$ in place of ϵ is the same thing as

$$(1.14) \qquad \lim_{n \to \infty} P\left[\omega: \left|\frac{1}{n}s_n(\omega)\right| \geq \epsilon\right] = 0.$$

This is the form in which the theorem will be proved.

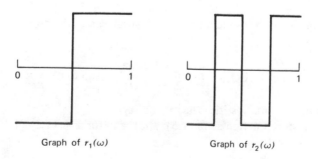

Graph of $r_1(\omega)$ Graph of $r_2(\omega)$

The Rademacher functions have themselves a direct probabilistic meaning.

If a coin is tossed successively, and if a particle starting from the origin performs a random walk on the real line by successively moving one unit in the positive or negative direction according as the coin falls heads or tails, then $r_i(\omega)$ represents the distance it moves on the ith step and $s_n(\omega)$ represents its position after n steps. There is also the gambling interpretation: If a gambler bets one dollar, say, on each toss of the coin, $r_i(\omega)$ represents his gain or loss on the ith play and $s_n(\omega)$ represents his gain or loss in n plays.

Each dyadic interval of rank $i-1$ splits into two dyadic intervals of rank i; $r_i(\omega)$ has value -1 on one of these and value $+1$ on the other. Thus $r_i(\omega)$ is -1 on a set of intervals of total length $\frac{1}{2}$ and $+1$ on a set of total length $\frac{1}{2}$. Hence $\int_0^1 r_i(\omega)\,d\omega = 0$ by (1.5), and

$$(1.15) \qquad \int_0^1 s_n(\omega)\,d\omega = 0$$

by (1.4). If the integral is viewed as an expected value, then (1.15) says that the mean position after n steps of a random walk is 0.

Suppose that $i < j \le n$. On a dyadic interval of rank $j-1$, $r_i(\omega)$ is constant and $r_j(\omega)$ has value -1 on the left half and $+1$ on the right. The product $r_i(\omega) \cdot r_j(\omega)$ therefore integrates to 0 over each of the dyadic intervals of rank $j-1$, and so

$$(1.16) \qquad \int_0^1 r_i(\omega)r_j(\omega)\,d\omega = 0, \qquad i \ne j.$$

This corresponds to the fact that independent random variables are uncorrelated. Since $r_i^2(\omega) = 1$, expanding the square of the sum (1.13) shows that

$$(1.17) \qquad \int_0^1 s_n^2(\omega)\,d\omega = n.$$

This corresponds to the fact that the variances of independent random variables add.

Applying Chebyshev's inequality in a formal way to the probability in (1.14) now leads to

$$(1.18) \qquad P[\omega: |s_n(\omega)| \ge n\epsilon] \le \frac{1}{n^2\epsilon^2}\int_0^1 s_n^2(\omega)\,d\omega = \frac{1}{n\epsilon^2}.$$

The following lemma justifies the inequality.

Let f be a step function as in (1.5): $f(\omega) = c_j$ for $\omega \in (x_{j-1}, x_j]$, where $0 = x_0 < \cdots < x_k = 1$.

Lemma. *If f is a nonnegative step function, then $[\omega: f(\omega) \ge \alpha]$ is for $\alpha > 0$ a finite union of intervals and*

(1.19) $$P[\omega: f(\omega) \geq \alpha] \leq \frac{1}{\alpha} \int_0^1 f(\omega) \, d\omega.$$

PROOF.   The set in question is the union of the intervals $(x_{j-1}, x_j]$ for which $c_j \geq \alpha$. If Σ' denotes summation over those j satisfying $c_j \geq \alpha$, then $P[\omega: f(\omega) \geq \alpha] = \Sigma'(x_j - x_{j-1})$ by the definition (1.2). On the other hand, since the c_j are all nonnegative by hypothesis, (1.5) gives

$$\int_0^1 f(\omega) \, d\omega = \sum_{j=1}^k c_j(x_j - x_{j-1}) \geq \Sigma' c_j(x_j - x_{j-1})$$

$$\geq \Sigma' \alpha(x_j - x_{j-1}).$$

Hence (1.19).   ■

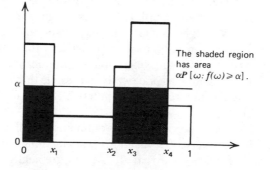

The shaded region has area
$\alpha P[\omega: f(\omega) \geq \alpha]$.

Taking $\alpha = n^2 \epsilon^2$ and $f(\omega) = s_n^2(\omega)$ in (1.19) gives (1.18). Clearly, (1.18) implies (1.14), and as already observed, this in turn implies (1.11).

The Strong Law of Large Numbers

It is possible with a minimum of technical apparatus to prove a stronger result that cannot even be formulated in the discrete theory of probability. Consider the set

(1.20) $$N = \left[\omega: \lim_{n \to \infty} \frac{1}{n} \sum_{i=1}^n d_i(\omega) = \frac{1}{2} \right]$$

consisting of those ω for which the asymptotic relative frequency of 1 in the sequence (1.7) is $\frac{1}{2}$. The points in (1.20) are called *normal numbers*. The idea is to show that a real number ω drawn at random from the unit interval is "practically certain" to be normal, or that there is "practical certainty" that 1 occurs in the sequence (1.7) of tosses with asymptotic relative frequency $1/2$. It is impossible at this stage to prove that $P(N) = 1$, because N is not a finite union of intervals and so has been assigned no probability. But the notion of "practical certainty" can be formalized in the following way.

Define a subset A of Ω to be *negligible** if for each positive ϵ there exists a finite or countable[†] collection $I_1, I_2, \ldots$ of intervals (they may overlap) satisfying

(1.21) $$A \subset \bigcup_k I_k$$

and

(1.22) $$\sum_k P(I_k) < \epsilon.$$

A negligible set is one that can be covered by intervals the total sum of whose lengths can be made arbitrarily small. If $P(A)$ is assigned to such an A in any reasonable way, $P(A) < \epsilon$ ought to hold for each positive ϵ, and hence $P(A)$ ought to be 0. Even without such an assignment of probability the definition of negligibility can serve as it stands as an explication of "practical impossibility" and "practical certainty": Regard it as practically impossible that the random ω will lie in A if A is negligible, and regard it as practically certain that ω will lie in A if A^c is negligible.[‡]

Although the fact plays no role in the arguments to follow, for an understanding of negligibility observe first that *a finite or countable union of negligible sets is negligible.* Indeed, suppose that $A_1, A_2, \ldots$ are negligible. Given ϵ, for each n choose intervals $I_{n1}, I_{n2}, \ldots$ such that $A_n \subset \bigcup_k I_{nk}$ and $\Sigma_k P(I_{nk}) < \epsilon/2^n$. All the intervals I_{nk} taken together form a countable collection covering $\bigcup_n A_n$, and their total length is $\Sigma_n \Sigma_k P(I_{nk}) < \Sigma_n \epsilon/2^n = \epsilon$. Therefore, $\bigcup_n A_n$ is negligible.

A set consisting of a single point is clearly negligible, and so every countable set is also negligible. The rationals for example form a negligible set. In the coin-tossing model, a single point of the unit interval has the role of a single sequence of 0's and 1's, or of a single sequence of heads and tails. It corresponds with intuition that it should be "practically impossible" to toss a coin infinitely

---

* The term *negligible* is introduced for the purposes of this section only. The negligible sets will reappear later as the sets of Lebesgue measure 0.
† *Countably infinite* is unambiguous. *Countable* will mean finite or countably infinite, although it will sometimes for emphasis be expanded as here to *finite or countable.*
‡ The *complement* $\Omega - A$ is denoted A^c. Other set-theoretic notation: the *difference* $A - B = A \cap B^c$ is well defined even if B is not a subset of A; the *symmetric difference* is $A \triangle B = (A - B) \cup (B - A)$; in accordance with standard usage, $A \subset B$ does not preclude $A = B$.

often and realize any one particular sequence set down in advance. It is for this reason not a real shortcoming of the model that for no ω is (1.7) the sequence $(1, 0, 0, 0, \ldots)$. In fact, since a countable set is negligible, it is not a shortcoming that (1.7) is never one of the countably many sequences that end in 0's.

Theorem 1.2. *The set of normal numbers has negligible complement.*

This is *Borel's normal number theorem*,* a special case of the *strong law of large numbers*. Like Theorem 1.1, it is of arithmetic as well as probabilistic interest.

The set N^c is not countable: Consider a point ω for which $(d_1(\omega), d_2(\omega), \ldots)$ $= (1, 1, u_3, 1, 1, u_6, \ldots)$—that is, a point for which $d_i(\omega) = 1$ unless i is a multiple of 3. Since $n^{-1} \sum_{i=1}^{n} d_i(\omega) \geq \frac{2}{3}$, such a point cannot be normal. But there are uncountably many such points, one for each infinite sequence $(u_3, u_6, \ldots)$ of 0's and 1's. Thus one cannot prove N^c negligible by proving it countable, and a deeper argument is required.

PROOF OF THEOREM 1.2. Clearly (1.20) and

$$(1.23) \qquad N = \left[\omega: \lim_{n \to \infty} \frac{1}{n} s_n(\omega) = 0 \right]$$

define the same set (see (1.13)). To prove N^c negligible requires constructing coverings that satisfy (1.21) and (1.22). The construction makes use of the inequality

$$(1.24) \qquad P[\omega: |s_n(\omega)| \geq n\epsilon] \leq \frac{1}{n^4 \epsilon^4} \int_0^1 s_n^4(\omega) \, d\omega.$$

This follows by the same argument that leads to the inequality in (1.18)—it is only necessary to take $f(\omega) = s_n^4(\omega)$ and $\alpha = n^4 \epsilon^4$ in (1.19). As the integral in (1.24) will be shown to have order n^2, the inequality is stronger than (1.18).

The integrand on the right in (1.24) is

$$(1.25) \qquad s_n^4(\omega) = \sum r_\alpha(\omega) r_\beta(\omega) r_\gamma(\omega) r_\delta(\omega),$$

where the four indices range independently from 1 to n. Depending on how the indices match up, each term in this sum reduces to one of the following five forms, where in each case the indices are now *distinct*:

---

* Émile Borel: Sur les probabilités dénombrables et leurs applications arithmétiques, *Circ. Mat. d. Palermo,* **29** (1909), 247–271.

$$(1.26) \qquad \begin{cases} r_i^4(\omega) = 1, \\ r_i^2(\omega)\, r_j^2(\omega) = 1, \\ r_i^2(\omega)\, r_j(\omega) r_k(\omega) = r_j(\omega) r_k(\omega), \\ r_i^3(\omega) r_j(\omega) = r_i(\omega) r_j(\omega), \\ r_i(\omega) r_j(\omega) r_k(\omega) r_l(\omega). \end{cases}$$

If, for example, k exceeds i, j, and l, then the last product in (1.26) integrates to 0 over each dyadic interval of rank $k - 1$ because $r_i(\omega) r_j(\omega) r_l(\omega)$ is constant there while $r_k(\omega)$ is -1 on the left half and $+1$ on the right. Adding over the dyadic intervals of rank $k - 1$ gives

$$\int_0^1 r_i(\omega) r_j(\omega) r_k(\omega) r_l(\omega)\, d\omega = 0.$$

This holds whenever the four indices are distinct. From this and (1.16) it follows that the last three forms in (1.26) integrate to 0 over the unit interval; of course, the first two forms integrate to 1.

The number of occurrences in the sum (1.25) of the first form in (1.26) is n. The number of occurrences of the second form is $3n(n - 1)$, because there are n choices for the α in (1.25), three ways to match it with β, γ, or δ, and $n - 1$ choices for the value common to the remaining two indices. A term-by-term integration of (1.25) therefore gives

$$(1.27) \qquad \int_0^1 s_n^4(\omega)\, d\omega = n + 3n(n - 1) \le 3n^2,$$

and it follows by (1.24) that

$$(1.28) \qquad P\!\left[\omega\colon \left| \frac{1}{n} s_n(\omega) \right| \ge \epsilon \right] \le \frac{3}{n^2 \epsilon^4}.$$

Fix a positive sequence $\{\epsilon_n\}$ going to 0 slowly enough that the series $\Sigma_n\, \epsilon_n^{-4} n^{-2}$ converges (take $\epsilon_n = n^{-1/8}$, for example). If $A_n = [\omega\colon |n^{-1} s_n(\omega)| \ge \epsilon_n]$, then $P(A_n) \le 3\epsilon_n^{-4} n^{-2}$ by (1.28), and so $\Sigma_n\, P(A_n) < \infty$.

If, for some m, ω lies in A_n^c for all the integers n exceeding m, then $|n^{-1} s_n(\omega)| < \epsilon_n$ for $n > m$ and ω is normal because $\epsilon_n \to 0$ (see (1.23)). In other words, for each m, $\bigcap_{n=m}^\infty A_n^c \subset N$, which is the same thing as $N^c \subset \bigcup_{n=m}^\infty A_n$. This last relation leads to the required covering: Given ϵ, choose m so that $\Sigma_{n=m}^\infty\, P(A_n) < \epsilon$. Now A_n is a finite disjoint union of intervals (1.8) whose lengths total $P(A_n)$, and therefore $\bigcup_{n=m}^\infty A_n$ is a countable union (not necessarily disjoint) of intervals whose lengths total less than ϵ. The intervals in this union provide a covering of N^c of the kind the definition of negligibility calls for. ∎

Strong Law versus Weak

Theorem 1.2 is stronger than Theorem 1.1. Since both are theorems—that is, are true propositions—it makes no sense to say that Theorem 1.2 implies Theorem 1.1 or that the reverse implication is false. A consideration of the *forms* of the two propositions will, however, show that the strong law does indeed go far beyond the weak law.

For each n let $f_n(\omega)$ be a step function on the unit interval and consider the relation

$$(1.29) \qquad\qquad \lim_{n \to \infty} P[\omega: |f_n(\omega)| \geq \epsilon] = 0$$

together with the set

$$(1.30) \qquad\qquad \left[\omega: \lim_{n \to \infty} f_n(\omega) = 0 \right].$$

If $f_n(\omega) = n^{-1}s_n(\omega)$, (1.29) reduces to the weak law (1.14) and (1.30) coincides with the set (1.23) of normal numbers. The general theory to follow (see the end of Section 4) will show that, whatever the step function $f_n(\omega)$ may be, if the set (1.30) has negligible complement, then (1.29) holds for each positive ϵ. For this reason, a proof of Theorem 1.2 is automatically a proof of Theorem 1.1.

But the converse fails. Define an infinite sequence $I_1, I_2, \ldots$ of intervals in the following way. Define the first two by

$$(1.31) \qquad\qquad I_1 = \left(0, \frac{1}{2}\right], \qquad I_2 = \left(\frac{1}{2}, 1\right].$$

Define the next four by

$$(1.32) \quad I_3 = \left(0, \frac{1}{4}\right], \qquad I_4 = \left(\frac{1}{4}, \frac{1}{2}\right], \qquad I_5 = \left(\frac{1}{2}, \frac{3}{4}\right], \qquad I_6 = \left(\frac{3}{4}, 1\right].$$

Define the next eight as the eight intervals (1.8) with $n = 3$. And so on. Now let $f_n(\omega)$ be the indicator function of I_n. The set in (1.29) is I_n itself (unless $\epsilon > 1$, in which case it is empty); since the length of I_n goes to 0, (1.29) holds. On the other hand, I_n moves repeatedly over $(0, 1]$, so that each ω lies in I_n for infinitely many values of n; hence $f_n(\omega)$ = 1 for infinitely many values of n, and the set (1.30) is empty. The complement of (1.30), the entire unit interval, is *not* negligible—an "obvious" fact whose proof is deferred (see Lemma 2 to Theorem 2.2).

Thus there exist step functions $f_n(\omega)$ which satisfy (1.29) for each positive ϵ but for which (1.30) fails to have a negligible complement. For this reason a proof of Theorem 1.1 is not automatically a proof of Theorem 1.2; the latter lies deeper and its proof is correspondingly more complex.

Extending the Probabilities

It is natural to try and go beyond the definition (1.2) and assign probabilities in a systematic way to sets other than finite unions of intervals. Since the set of nonnormal numbers is negligible, one feels that it ought to have probability 0.

For another probabilistically interesting set that is not a finite union of intervals, consider

$$(1.33) \qquad \bigcup_{n=1}^{\infty} [\omega: -a < s_1(\omega), \ldots, s_{n-1}(\omega) < b, s_n(\omega) = -a],$$

where a and b are positive integers. This is the event that the gambler's fortune $s_n(\omega)$ reaches $-a$ before it reaches $+b$; it represents ruin for a gambler with a dollars playing against an adversary with b dollars, the rule being that they play until one or the other runs out of capital.

The union in (1.33) is countable and disjoint, and for each n the set inside the union is itself a disjoint union of certain of the intervals (1.8). Thus (1.33) is a countably infinite disjoint union of intervals, and it is natural to take as its probability the sum of the lengths of these constituent intervals. Since the set of normal numbers is not a countable union of intervals, however, this extension of the definition of probability would still not cover all cases of interest.

It is, in fact, not fruitful to try to anticipate just which sets probabilistic analysis will require and then assign probabilities to them in some *ad hoc* way. The successful procedure is to develop a general theory that assigns probabilities at once to the sets of a class so extensive that most of its members never actually arise in probability theory. That being so, why not ask for a theory that goes all the way and applies to *every* set in the unit interval? For an arbitrary subset A of Ω, should there not exist a well-defined probability that the random point ω lies in A? The answer turns out to be no (see the end of Section 3), and it is necessary to work within certain subclasses of the class of all subsets of the unit interval. The classes of the appropriate kinds—the fields and σ-fields—are defined and studied in the next section. The theory there covers the unit interval as treated in this section, as well as discrete spaces, Euclidean spaces, and many others.

PROBLEMS

Some problems involve concepts not required for an understanding of the text, or concepts treated only in later sections; there are no problems whose solutions are used in the text itself. An arrow ↑ points back to a problem (the one immediately preceding if no number is given) the solution and terminology of which are assumed. See Notes on the Problems, page 478.

1.1. (a) Show that a *discrete* probability space (see Example 2.9 for the formal definition) cannot contain an infinite sequence $A_1, A_2, \ldots$ of independent events each of probability $\frac{1}{2}$. Since A_n could be identified with heads on the nth toss of a coin, the existence of such a sequence would make this section superfluous.

(b)   Suppose that $0 \le p_n \le 1$ and put $\alpha_n = \min\{p_n, 1 - p_n\}$. Show that, if $\Sigma_n \alpha_n$ diverges, then no discrete probability space can contain independent events A_1, A_2, . . . such that A_n has probability p_n.

1.2.   Use Lemma 2 to Theorem 2.2 to show that a nonempty interval is not negligible. Show that the set of algebraic numbers is negligible.

1.3.   Show that N and N^c (see (1.20)) are dense in (0, 1].

1.4.   Define a set A in (0, 1] to be *trifling** if for each ϵ there exists a *finite* sequence of intervals $I_1, I_2, \ldots$ satisfying (1.21) and (1.22).

(a)   Show that a trifling set is negligible.

(b)   Using the finite form of Lemma 2 to Theorem 2.2 show that a trifling set is nowhere dense.†

(c)   Show that finite unions of trifling sets are trifling but that this can fail for countable unions.

(d)   Put $B = \bigcup_n (r_n - 2^{-n-2}, r_n + 2^{-n-2}]$, where $r_1, r_2, \ldots$ is an enumeration of the rationals; show that (0, 1] $- B$ is nowhere dense but not trifling or even negligible.

1.5. ↑ **(a)**   For $\beta = 0, 1, \ldots, b - 1$, let $A_b(\beta)$ be the set of numbers in the unit interval whose representations in the base b do not contain the digit β. Show that $A_b(\beta)$ is trifling.

(b)   Find a trifling set A such that every point of the unit interval can be represented in the form $x + y$ with x and y in A.

(c)   Let $A_b(\beta_1, \ldots, \beta_k)$ consist of the numbers in whose expansions the digits $\beta_1, \ldots, \beta_k$ nowhere appear in that sequence. Show that it is trifling. What does this imply about the monkey that types at random?

1.6. ↑   From [0, 1] remove the open middle third ($\frac{1}{3}, \frac{2}{3}$); from the remainder, a union of two closed intervals, remove the two open middle thirds ($\frac{1}{9}, \frac{2}{9}$) and ($\frac{7}{9}, \frac{8}{9}$); the *Cantor set* is what remains when this process is continued ad infinitum. Arithmetically, C can be defined as the set of ω in [0, 1] that can be written in base 3 without any 1's. For example, C contains $\frac{1}{3} = .1000 \ldots = .0222 \ldots$ because it is possible to avoid 1 in the expansion. Consider $A_3(1)$ as in Problem 1.5 and show that C is trifling.

1.7.   Put $M(t) = \int_0^1 e^{ts_n(\omega)} \, d\omega$ and show by successive differentiations under the integral that

$$(1.34) \qquad M^{(k)}(0) = \int_0^1 s_n^k(\omega) \, d\omega.$$

Over each dyadic interval of rank n, $s_n(\omega)$ has a constant value of the form $\pm 1 \pm 1 \pm \cdots \pm 1$, and therefore $M(t) = 2^{-n} \Sigma \exp t(\pm 1 \pm 1 \pm \cdots \pm 1)$, where the

---

* Like *negligible, trifling* is a nonce word used only here.
† A set A is nowwhere dense if every interval contains some subinterval that does not meet A.

sum extends over all 2^n n-long sequences of $+1$'s and -1's. Thus

$$(1.35) \qquad M(t) = \left(\frac{e^t + e^{-t}}{2} \right)^n = (\cosh t)^n.$$

Use this and (1.34) to give new proofs of (1.15), (1.17), and (1.27). (This, the method of moment generating functions, will be investigated systematically in later sections.)

1.8. ↑ By an argument similar to that leading to (1.35) show that the Rademacher functions satisfy

$$\int_0^1 \exp \left[i \sum_{k=1}^n a_k r_k(\omega) \right] d\omega = \prod_{k=1}^n \frac{e^{ia_k} + e^{-ia_k}}{2}$$

$$= \prod_{k=1}^n \cos a_k.$$

Take $a_k = t2^{-k}$ and from $\Sigma_{k=1}^\infty r_k(\omega)2^{-k} = 2\omega - 1$ deduce

$$(1.36) \qquad \frac{\sin t}{t} = \prod_{k=1}^\infty \cos \frac{t}{2^k}$$

by letting $n \to \infty$ inside the integral above. Derive Vieta's remarkable formula

$$\frac{2}{\pi} = \frac{\sqrt{2}}{2} \frac{\sqrt{2 + \sqrt{2}}}{2} \frac{\sqrt{2 + \sqrt{2 + \sqrt{2}}}}{2} \cdots$$

1.9. In this problem let $d_n(\omega)$ stand for the nth digit in the nonterminating expansion of ω in the base b. Fix β, $0 \le \beta \le b - 1$, and redefine $r_n(\omega)$ as

$$r_n(\omega) = \begin{cases} b - 1 & \text{if } d_n(\omega) = \beta, \\ -1 & \text{if } d_n(\omega) \ne \beta. \end{cases}$$

Then (1.12) is the case $b = 2$, $\beta = 1$. Define $s_n(\omega)$ by (1.13) for the new $r_k(\omega)$ and prove that $\int_0^1 s_n(\omega) \, d\omega = 0$, $\int_0^1 s_n^2(\omega) \, d\omega = n(b - 1)$, and $\int_0^1 s_n^4(\omega) \, d\omega = n(b - 1)(b^2 - 3b + 3) + 3n(n - 1)(b - 1)^2$. Let $N_b(\beta)$ be the ω-set where b^{-1} is the asymptotic relative frequency $\lim_n n^{-1} \Sigma_{k=1}^n I_{\{\beta\}}(d_k(\omega))$ of β in the base-b expansion of ω. Let $N_b = \bigcap_{\beta=0}^{b-1} N_b(\beta)$; the numbers in N_b are *normal to the base* b. Show that N_b^c is negligible. That $N_3(1)$ has negligible complement proves the Cantor set negligible.

1.10. For a positive function $f(q)$ of integers q, let A_f be the set of ω in $(0, 1]$ that have close rational approximations in the sense that there are infinitely many irreducible fractions p/q satisfying $|\omega - p/q| < f(q)$. Show that A_f is negligible if $\Sigma_q qf(q) < \infty$.

According to a theorem of Diophantine approximation, $|\omega - p/q| < 1/q^2$ has infinitely many solutions for irrational ω. Except for a negligible set of ω, however, $|\omega - p/q| < 1/q^2 \log^2 q$ has only finitely many solutions.

1.11. A number ω is normal in the base 2 if and only if for each positive ϵ there exists an $n_0(\epsilon, \omega)$ such that $|n^{-1} \Sigma_{i=1}^n d_i(\omega) - \frac{1}{2}| < \epsilon$ for all n exceeding $n_0(\epsilon, \omega)$. Theorem 1.2 concerns the entire dyadic expansion, whereas Theorem 1.1 concerns only the beginning segment. Point up the difference by showing that for $\epsilon < \frac{1}{2}$ the $n_0(\epsilon, \omega)$ above cannot be the same for all ω in N—in other words, $n^{-1} \Sigma_{i=1}^n d_i(\omega)$ converges to $\frac{1}{2}$ for all ω in N, but not uniformly.

SECTION 2. PROBABILITY MEASURES

Spaces

Let Ω be an arbitrary space or set of points ω. In probability theory Ω consists of all the possible results or outcomes ω of an experiment or observation. For observing the number of heads in n tosses of a coin the space Ω is $\{0, 1, \ldots, n\}$; for describing the complete history of the n tosses Ω is the space of all 2^n n-long sequences of H's and T's; for an infinite sequence of tosses Ω can be taken as the unit interval as in the preceding section; for the number of α-particles emitted by a substance during a unit interval of time or for the number of telephone calls arriving at an exchange Ω is $\{0, 1, 2, \ldots\}$; for the position of a particle Ω is three-dimensional Euclidean space; for describing the motion of the particle Ω is an appropriate space of functions; and so on. Most Ω's to be considered are interesting from the point of view of geometry and analysis as well as that of probability.

Classes of Sets

Viewed probabilistically, a subset of Ω is an *event* and an element ω of Ω is a *sample point*. As pointed out at the end of Section 1, it is necessary to single out for treatment special classes of subsets of Ω. To be useful, such a class must be closed under various of the operations of set theory.

Example 2.1.*   Consider the set N of normal numbers in the form (1.23), where $s_n(\omega)$ is the sum of the first n Rademacher functions. Since a point ω lies in N if and only if $\lim_n n^{-1}s_n(\omega) = 0$, N can be put in the form

(2.1) $$N = \bigcap_{k=1}^{\infty} \bigcup_{m=1}^{\infty} \bigcap_{n=m}^{\infty} [\omega: |n^{-1}s_n(\omega)| < k^{-1}].$$

Indeed, because of the very meaning of union and of intersection, ω lies in the set on the right here if and only if for every k there exists an m such that $|n^{-1}s_n(\omega)| < k^{-1}$ holds for all $n \geq m$, and this is just the definition of convergence to 0—with the usual ϵ replaced by k^{-1} to avoid the formation of an uncountable intersection. Since $s_n(\omega)$ is constant over each dyadic interval of rank n, the set $[\omega: |n^{-1}s_n(\omega)| < k^{-1}]$ is a finite disjoint union of intervals. The formula (2.1) shows explicitly how N is constructed in steps from these simpler sets.∎

---

* Many of the examples in the book simply illustrate the concepts at hand, but others contain definitions and facts needed subsequently.

A systematic treatment of the ideas in Section 1 thus requires a class of sets that contains the intervals and is closed under the formation of countable unions and intersections. Note that a singleton* $\{x\}$ is a countable intersection $\bigcap_n (x - n^{-1}, x]$ of intervals. If a class contains all the singletons and is closed under the formation of *arbitrary* unions, then of course it contains all the subsets of Ω. As the theory of this section and the next does not apply to such extensive classes of sets, attention must be restricted to countable set-theoretic operations and in some cases just to finite ones.

A class $\mathcal{F}$ of subsets of a space Ω is called a *field*† if it contains Ω itself and is closed under the formation of complements and finite unions:

   (i) $\Omega \in \mathcal{F}$;
   (ii) $A \in \mathcal{F}$ implies $A^c \in \mathcal{F}$;
   (iii) $A, B \in \mathcal{F}$ implies $A \cup B \in \mathcal{F}$.

Since 0 and Ω are complementary, (i) is the same in the presence of (ii) as the assumption $0 \in \mathcal{F}$. In fact, (i) simply ensures that $\mathcal{F}$ is nonempty: If $A \in \mathcal{F}$, then $A^c \in \mathcal{F}$ by (ii) and $\Omega = A \cup A^c \in \mathcal{F}$ by (iii).

By DeMorgan's law, $A \cap B = (A^c \cup B^c)^c$ and $A \cup B = (A^c \cap B^c)^c$. If $\mathcal{F}$ is closed under complementation, therefore, it is closed under the formation of finite unions if and only if it is closed under the formation of finite intersections. Thus (iii) can be replaced by the requirement

   (iii′) $A, B \in \mathcal{F}$ implies $A \cap B \in \mathcal{F}$.

A class $\mathcal{F}$ of subsets of Ω is a *σ-field* if it is a field and if it is also closed under the formation of countable unions:

   (iv) $A_1, A_2, \ldots \in \mathcal{F}$ implies $A_1 \cup A_2 \cup \cdots \in \mathcal{F}$.

Note that (iv) implies (iii) in the presence of (i) and (ii): take $A_n = 0$ for $n \geq 3$. A field is sometimes called a *finitely additive* field to stress that it need not be a σ-field. A set in a given class $\mathcal{F}$ is said to be *measurable* $\mathcal{F}$ or to be an $\mathcal{F}$-*set*. A field or σ-field of subsets of Ω will sometimes be called a field or σ-field *in* Ω.

Example 2.2. Section 1 began with a consideration of the sets (1.1), the finite disjoint unions of subintervals of $\Omega = (0, 1]$. Augmented by the empty set, this class is a field $\mathcal{B}_0$: Suppose that $A = (a_1, a_1'] \cup \cdots \cup (a_m, a_m']$, where the notation is so chosen that $a_1 \leq \cdots \leq a_m$. If the $(a_i, a_i']$ are disjoint, then A^c is $(0, a_1] \cup (a_1', a_2] \cup \cdots \cup (a_{m-1}', a_m] \cup (a_m', 1]$ and so lies in $\mathcal{B}_0$ (some of these intervals may be empty, as a_i' and a_{i+1} may coincide). If $B = (b_1, b_1'] \cup \cdots \cup (b_n, b_n']$, the $(b_j, b_j']$ again disjoint, then $A \cap B = \bigcup_{i=1}^{m} \bigcup_{j=1}^{n} \{(a_i, a_i'] \cap (b_j, b_j']\}$;

---

* That is, a set consisting of a single point.
† The term *algebra* is sometimes used in place of *field*.

each intersection here is again an interval or else the empty set, and the union is disjoint, and hence $A \cap B$ is in $\mathcal{B}_0$. Thus $\mathcal{B}_0$ satisfies (i), (ii), and (iii′).

Although $\mathcal{B}_0$ is a field, it is not a σ-field, since it does not, for example, contain the set (1.33), a countable union of intervals that cannot be represented as a finite union of intervals. The set (2.1) of normal numbers is also outside $\mathcal{B}_0$. ∎

The definitions above involve distinctions perhaps most easily made clear by a pair of artificial examples.

Example 2.3. Let $\mathcal{F}$ consist of the finite and the cofinite sets (A being cofinite if A^c is finite). Then $\mathcal{F}$ is a field; it is a σ-field if and only if Ω is finite. ∎

Example 2.4. Let $\mathcal{F}$ consist of the countable and the co-countable sets (A being co-countable if A^c is countable). Then $\mathcal{F}$ is a σ-field. If Ω is uncountable, it contains an A such that A and A^c are both uncountable. This shows that a σ-field need not contain all subsets of Ω and need not be closed under the formation of arbitrary uncountable unions. ∎

The largest σ-field in Ω consists of all the subsets of Ω; the smallest consists only of the empty set and Ω itself.

The elementary facts about fields and σ-fields are easy to prove: If $\mathcal{F}$ is a field, then $A, B \in \mathcal{F}$ implies that $A - B = A \cap B^c \in \mathcal{F}$ and $A \triangle B = (A - B) \cup (B - A) \in \mathcal{F}$; further, $A_1, \ldots, A_n \in \mathcal{F}$ implies that $A_1 \cup \cdots \cup A_n \in \mathcal{F}$ and $A_1 \cap \cdots \cap A_n \in \mathcal{F}$. And if $\mathcal{F}$ is a σ-field, it follows by the infinite version of De-Morgan's law that $A_1, A_2, \ldots \in \mathcal{F}$ implies that $A_1 \cap A_2 \cap \cdots \in \mathcal{F}$.

Example 2.5. Suppose that $\mathcal{F}$ is a field and is closed under the formation of countable nondecreasing unions—that is, suppose that $\mathcal{F}$ contains $\bigcup_n A_n$ if it contains the A_n and if $A_n \subset A_{n+1}$. Then $\mathcal{F}$ is a σ-field, because an arbitrary union $\bigcup_n A_n$ can be represented as a nondecreasing union $\bigcup_n (\bigcup_{k \leq n} A_k)$, and $\mathcal{F}$ contains the $\bigcup_{k \leq n} A_k$ if it contains the A_n. ∎

A field is closed under the finite set-theoretic operations, and a σ-field is closed under the countable ones. The analysis of a probability problem usually begins with the sets of some rather small class $\mathcal{A}$, such as the class of subintervals of $(0, 1]$. As in Example 2.1, probabilistically natural constructions involving finite and countable operations can then lead to sets outside the initial class $\mathcal{A}$, but they can never lead outside the smallest σ-field over $\mathcal{A}$, namely, the intersection of all σ-fields containing $\mathcal{A}$. This class is called the σ-*field generated by* $\mathcal{A}$ and is denoted $\sigma(\mathcal{A})$; it is the intersection of all σ-fields containing $\mathcal{A}$.

There do exist σ-fields containing $\mathcal{A}$, the class of all subsets of Ω being one. Moreover, a completely arbitrary intersection of σ-fields (however many of them there may be) is itself a σ-field: Suppose that $\mathcal{F} = \bigcap_\theta \mathcal{F}_\theta$, where θ ranges over an arbitrary index set and each $\mathcal{F}_\theta$ is a σ-field. Then $A \in \mathcal{F}$ implies for each θ that $A \in \mathcal{F}_\theta$ and hence $A^c \in \mathcal{F}_\theta$, so that $A^c \in \mathcal{F}$. If $A_n \in \mathcal{F}$ for each n, then $A_n \in \mathcal{F}_\theta$ for each n and θ, so that $\bigcup_n A_n$ lies in each $\mathcal{F}_\theta$ and hence in $\mathcal{F}$.

Thus the intersection in the definition of $\sigma(\mathcal{A})$ is indeed a σ-field containing $\mathcal{A}$. It is minimal in the sense that it is contained in every σ-field that contains $\mathcal{A}$: *$\mathcal{A} \subset \sigma(\mathcal{A})$ and $\sigma(\mathcal{A})$ is a σ-field; if $\mathcal{A} \subset \mathcal{G}$ and $\mathcal{G}$ is a σ-field, then $\sigma(\mathcal{A}) \subset \mathcal{G}$.*

Example 2.6. If $\mathcal{F}$ is a σ-field, then obviously $\sigma(\mathcal{F}) = \mathcal{F}$. If $\mathcal{A}$ consists of the singletons, then $\sigma(\mathcal{A})$ is the σ-field in Example 2.4. If $\mathcal{A}$ is empty or $\mathcal{A} = \{0\}$ or $\mathcal{A} = \{\Omega\}$, then $\sigma(\mathcal{A}) = \{0, \Omega\}$. If $\mathcal{A} \subset \mathcal{A}' \subset \sigma(\mathcal{A})$, then $\sigma(\mathcal{A}') = \sigma(\mathcal{A})$. ∎

Example 2.7. Let $\mathcal{I}$ be the class of subintervals of $\Omega = (0, 1]$ and define $\mathcal{B} = \sigma(\mathcal{I})$. The elements of $\mathcal{B}$ are called the *Borel sets* of the unit interval. The field $\mathcal{B}_0$ of Example 2.2 satisfies $\mathcal{I} \subset \mathcal{B}_0 \subset \mathcal{B}$, and hence $\sigma(\mathcal{B}_0) = \mathcal{B}$.

Since $\mathcal{B}$ contains the intervals and is a σ-field, repeated finite and countable set-theoretic operations starting from intervals will never lead outside $\mathcal{B}$. Thus $\mathcal{B}$ contains the set (2.1) of normal numbers. It also contains for example the open sets in $(0, 1]$: If G is open and $x \in G$, then there exist rationals a_x and b_x such that $x \in (a_x, b_x] \subset G$. But then $G = \bigcup_{x \in G}(a_x, b_x]$; since there are only countably many intervals with rational endpoints, G is a *countable* union of elements of $\mathcal{I}$ and hence lies in $\mathcal{B}$.

In fact, $\mathcal{B}$ contains all the subsets of $(0, 1]$ actually encountered in ordinary analysis and probability. It is large enough for all "practical" purposes. It does not contain every subset of the unit interval, however; see the end of Section 3. The class $\mathcal{B}$ will play a fundamental role in all that follows. ∎

Probability Measures

A *set function* is a real-valued function defined on some class of subsets of Ω. A set function P on a field $\mathcal{F}$ is a *probability measure* if it satisfies these conditions:

(i) $0 \le P(A) \le 1$ for $A \in \mathcal{F}$;

(ii) $P(0) = 0, P(\Omega) = 1$;

(iii)   if $A_1, A_2, \ldots$ is a disjoint sequence of $\mathcal{F}$-sets and if $\bigcup_{k=1}^{\infty} A_k \in \mathcal{F}$, then*

---

* As the left side of (2.2) is invariant under permutations of the A_n, the same must be true of the right side. But in fact, according to *Dirichlet's theorem*, any nonnegative series has the same value whatever order the terms are summed in. Problem 2.23 outlines one proof of this basic theorem.

(2.2) $$P \left(\bigcup_{k=1}^{\infty} A_k \right) = \sum_{k=1}^{\infty} P(A_k).$$

The condition imposed on the set function P by (iii) is called *countable additivity*. Note that, since $\mathcal{F}$ is a field but perhaps not a σ-field, it is necessary in (iii) to assume that $\bigcup_{k=1}^{\infty} A_k$ lies in $\mathcal{F}$. If $A_1, \ldots, A_n$ are disjoint $\mathcal{F}$-sets, then $\bigcup_{k=1}^{n} A_k$ is also in $\mathcal{F}$ and (2.2) with $A_{n+1} = A_{n+2} = \cdots = 0$ gives

(2.3) $$P \left(\bigcup_{k=1}^{n} A_k \right) = \sum_{k=1}^{n} P(A_k).$$

The condition that (2.3) hold for disjoint $\mathcal{F}$-sets is *finite additivity*; it is a consequence of countable additivity.

The conditions above are redundant, because (i) can be replaced by $P(A) \geq 0$ and (ii) by $P(\Omega) = 1$. Indeed, the weakened forms (together with (iii)) imply that $P(\Omega) = P(\Omega) + P(0) + P(0) + \cdots$, so that $P(0) = 0$, and $1 = P(\Omega) = P(A) + P(A^c)$, so that $P(A) \leq 1$.

Example 2.8. Consider as in Example 2.2 the field $\mathcal{B}_0$ of finite disjoint unions of subintervals of $\Omega = (0, 1]$. The definition (1.2) assigns to each $\mathcal{B}_0$-set a number—the sum of the lengths of the constituent intervals—and hence specifies a set function P on $\mathcal{B}_0$. Extended inductively, (1.3) says that P is finitely additive. In Section 1 this property was deduced from the additivity of the Riemann integral (see (1.4)). At the end of this section the finite additivity of P will be proved from first principles, and it will be shown that P is, in fact, countably additive. Thus P is a probability measure on the field $\mathcal{B}_0$. ∎

If $\mathcal{F}$ is a σ-field in Ω and P is a probability measure on $\mathcal{F}$, the triple $(\Omega, \mathcal{F}, P)$ is called a *probability measure space,* or simply a *probability space*. A *support* of P is any $\mathcal{F}$-set A for which $P(A) = 1$.

Example 2.9. Let $\mathcal{F}$ be the σ-field of all subsets of a countable space Ω, and let $p(\omega)$ be a nonnegative function on Ω. Suppose that $\sum_{\omega \in \Omega} p(\omega) = 1$, and define $P(A) = \sum_{\omega \in A} p(\omega)$; since $p(\omega) \geq 0$, the order of summation is by Dirichlet's theorem irrelevant. Suppose that $A = \bigcup_{i=1}^{\infty} A_i$, where the A_i are disjoint, and let $\omega_{i1}, \omega_{i2}, \ldots$ be the points in A_i. By the theorem on nonnegative double series,* $P(A) = \sum_{ij} p(\omega_{ij}) = \sum_i \sum_j p(\omega_{ij}) = \sum_i P(A_i)$, and so P is countably additive. This $(\Omega, \mathcal{F}, P)$ is a *discrete* probability space. It is the formal basis for discrete probability theory.

More generally, a probability measure P on an arbitrary field or σ-field $\mathcal{F}$ in an arbitrary space Ω is discrete if there exist finitely or countably many points ω_k and masses m_k such that $P(A) = \sum_{\omega_k \in A} m_k$ for A in $\mathcal{F}$. This can be written

---

* For one proof of this standard analytic fact, see Problem 2.24.

in terms of indicator functions as $P(A) = \Sigma_k m_k I_A(\omega_k)$. Note that the set $\{\omega_1,$ $\omega_2, \ldots\}$ is a support of P (if it lies in $\mathscr{F}$).

If there is just one of these points, say ω_0, with mass $m_0 = 1$, then P is a *unit mass* at ω_0. In this case $P(A) = I_A(\omega_0)$ for $A \in \mathscr{F}$. ∎

Suppose that P is a probability measure on a field $\mathscr{F}$, and that $A, B \in \mathscr{F}$ and $A \subset B$. Since $P(A) + P(B - A) = P(B)$, P is *monotone*:

$$(2.4) \qquad\qquad P(A) \leq P(B) \qquad \text{if } A \subset B.$$

It follows further that $P(B - A) = P(B) - P(A)$, and as a special case,

$$(2.5) \qquad\qquad P(A^c) = 1 - P(A).$$

Other formulas familiar from the discrete theory are easily proved. For example,

$$(2.6) \qquad P(A \cup B) = P(A) + P(B) - P(A \cap B),$$

the common value of the two sides being $P(A \cap B^c) + P(A \cap B) + P(A^c \cap B)$. This is the case $n = 2$ of the general *inclusion-exclusion formula*:

$$(2.7) \qquad P\left(\bigcup_{k=1}^n A_k \right) = \sum_i P(A_i) - \sum_{i<j} P(A_i \cap A_j)$$

$$+ \sum_{i<j<k} P(A_i \cap A_j \cap A_k) + \cdots + (-1)^{n+1} P(A_1 \cap \cdots \cap A_n).$$

To deduce this inductively from (2.6), note that (2.6) gives

$$P\left(\bigcup_{k=1}^{n+1} A_k \right) = P\left(\bigcup_{k=1}^n A_k \right) + P(A_{n+1}) - P\left(\bigcup_{k=1}^n A_k \cap A_{n+1} \right).$$

Applying (2.7) to the first and third terms on the right gives (2.7) with $n + 1$ in place of n.

If $B_1 = A_1$ and $B_k = A_k \cap A_1^c \cap \cdots \cap A_{k-1}^c$, then the B_k are disjoint and $\bigcup_{k=1}^n A_k = \bigcup_{k=1}^n B_k$, so that $P(\bigcup_{k=1}^n A_k) = \Sigma_{k=1}^n P(B_k)$. Since $P(B_k) \leq P(A_k)$ by monotonicity, this establishes the *finite subadditivity* of P:

$$(2.8) \qquad\qquad P\left(\bigcup_{k=1}^n A_k \right) \leq \sum_{k=1}^n P(A_k).$$

Here, of course, the A_k need not be disjoint. Sometimes (2.8) is called *Boole's inequality*.

In these formulas all the sets are naturally assumed to lie in the field $\mathscr{F}$. The derivations above involve only the finite additivity of P. Countable additivity gives further properties:

Theorem 2.1. *Let P be a probability measure on a field $\mathcal{F}$.*

(i) *Continuity from below: If A_n and A lie in $\mathcal{F}$ and* $A_n\uparrow A$, then $P(A_n)$ $\uparrow P(A)$.*

(ii) *Continuity from above: If A_n and A lie in $\mathcal{F}$ and $A_n\downarrow A$, then $P(A_n)$ $\downarrow P(A)$.*

(iii) *Countable subadditivity: If $A_1, A_2, \ldots$ and $\bigcup_{k=1}^{\infty} A_k$ lie in $\mathcal{F}$, then*

$$(2.9) \qquad P\left(\bigcup_{k=1}^{\infty} A_k\right) \leq \sum_{k=1}^{\infty} P(A_k).$$

PROOF. For (i), put $B_1 = A_1$ and $B_k = A_k - A_{k-1}$. Then the B_k are disjoint, $A = \bigcup_{k=1}^{\infty} B_k$, and $A_n = \bigcup_{k=1}^{n} B_k$, so that by countable and finite additivity, $P(A)$ $= \sum_{k=1}^{\infty} P(B_k) = \lim_n \sum_{k=1}^{n} P(B_k) = \lim_n P(A_n)$. For (ii), observe that $A_n \downarrow A$ implies $A_n^c \uparrow A^c$, so that $1 - P(A_n) \uparrow 1 - P(A)$.

As for (iii), increase the right side of (2.8) to $\sum_{k=1}^{\infty} P(A_k)$ and then apply part (i) to the left side. ∎

Example 2.10. In this proof (ii) was deduced from (i), and a similar argument shows that (i) follows from (ii). In the presence of finite additivity these two equivalent conditions each imply countable additivity: If $A = \bigcup_n A_n$, the A_n disjoint, then $\bigcup_{k>n} A_k \downarrow 0$, and so (ii) and finite additivity give $P(A) - \sum_{k=1}^{n}$ $P(A_k) = P(\bigcup_{k>n} A_k) \to 0$. Note that here only a special case of (ii) is needed: $A_n \downarrow 0$ must imply that $P(A_n) \downarrow 0$. ∎

It is useful to note that the monotone passage to the limit in Theorem 2.1(i) applies if n is replaced by a continuous index. Suppose, for example, that $A_t \in \mathcal{F}$ for $t > 0$, that A_s $\subset A_t$ for $s < t$, and that $A = \bigcup_{t>0} A_t \in \mathcal{F}$. If $t_n \to \infty$, then $A_{t_n} \uparrow A$, and so $P(A_{t_n}) \uparrow P(A)$ by (i). But $P(A_t)$ is nondecreasing in t and hence has as $t \to \infty$, a limit that must coincide with $\lim_n P(A_{t_n})$. Thus $A_t \uparrow A$ implies that $P(A_t) \uparrow P(A)$. Moreover, if $\mathcal{F}$ is a σ-field, $A \in \mathcal{F}$ follows from $A_t \in \mathcal{F}$ because monotonicity implies $\bigcup_{t>0} A_t = \bigcup_n A_{t_n} \in \mathcal{F}$. Similar remarks apply to (ii).

Lebesgue Measure on the Unit Interval

To extend the probabilities in Section 1 and in Example 2.8 to a larger class of sets is to construct *Lebesgue measure*. This is a probability measure λ on the σ-field $\mathcal{B}$ of Borel sets in the unit interval. (In Chapter 2, λ will be defined for sets outside the unit interval as well.) It is specified by the requirement that

$$(2.10) \qquad \lambda(a, b] = b - a$$

for intervals. (In Chapter 1, λ was denoted P.) The following theorem is the first step in the construction.

---

* Here $A_n \uparrow A$ means that $A_n \subset A_{n+1}$ and $A = \bigcup_{n=1}^{\infty} A_n$; $A_n \downarrow A$ means that $A_n \supset A_{n+1}$ and $A = \bigcap_{n=1}^{\infty} A_n$.

Theorem 2.2. *As defined by* (2.10), λ *is a countably additive set function on the class of intervals.*

Note that allowing $a = b$ in (2.10) gives $\lambda(0) = 0$ for the empty interval. Of course, $\lambda(0, 1] = 1$. The proof will be given by two lemmas which apply to all finite intervals, not just to subintervals of $(0, 1]$.

Lemma 1. *If* $\bigcup_k (a_k, b_k] \subset (a, b]$ *for a finite or infinite sequence of disjoint intervals* $(a_k, b_k]$, *then* $\Sigma_k (b_k - a_k) \leq b - a$.

PROOF. Suppose first that there are only finitely many intervals, say n. The result being obvious for $n = 1$, assume that it holds for $n - 1$. If a_n is the largest among $a_1, \ldots, a_n$ (this is just a matter of notation), then $\bigcup_{k=1}^{n-1} (a_k, b_k] \subset (a, a_n]$, so that $\Sigma_{k=1}^{n-1} (b_k - a_k) \leq a_n - a$ by the induction hypothesis, and hence $\Sigma_{k=1}^{n} (b_k - a_k) \leq (a_n - a) + (b_n - a_n) \leq b - a$.

If there are infinitely many intervals, each finite subcollection satisfies the hypotheses of the lemma, and so $\Sigma_{k=1}^{n} (b_k - a_k) \leq b - a$ by the case just treated. But as n is arbitrary, the result follows.   ∎

Lemma 2. *If* $(a, b] \subset \bigcup_k (a_k, b_k]$ *for a finite or infinite sequence of intervals* (*not necessarily disjoint*), *then* $b - a \leq \Sigma_k (b_k - a_k)$.

PROOF. Assume that the result holds for the case of $n - 1$ intervals and that $(a, b] \subset \bigcup_{k=1}^{n} (a_k, b_k]$. Suppose that $a_n < b \leq b_n$ (notation again). If $a_n \leq a$, the result is obvious. Otherwise, $(a, a_n] \subset \bigcup_{k=1}^{n-1} (a_k, b_k]$, so that $\Sigma_{k=1}^{n-1} (b_k - a_k) \geq a_n - a$ by the induction hypothesis and hence $\Sigma_{k=1}^{n} (b_k - a_k) \geq (a_n - a) + (b_n - a_n) \geq b - a$. The finite case thus follows by induction.*

Now suppose that $(a, b] \subset \bigcup_{k=1}^{\infty} (a_k, b_k]$. If $0 < \epsilon < b - a$, the open intervals $(a_k, b_k + \epsilon 2^{-k})$ cover the closed interval $[a + \epsilon, b]$, and it follows by the Heine-Borel theorem that $[a + \epsilon, b] \subset \bigcup_{k=1}^{n} (a_k, b_k + \epsilon 2^{-k})$ for some n. But then $(a + \epsilon, b] \subset \bigcup_{k=1}^{n} (a_k, b_k + \epsilon 2^{-k}]$, and by the finite case, already established, $b - (a + \epsilon) \leq \Sigma_{k=1}^{n} (b_k + \epsilon 2^{-k} - a_k) \leq \Sigma_{k=1}^{\infty} (b_k - a_k) + \epsilon$. Since ϵ was arbitrary, the result follows.   ∎

Of course, Theorem 2.2 is an immediate consequence of these two lemmas.

---

* This and the finite case of Lemma 1 together imply (1.3). Like (1.3) they follow immediately from the additivity of the Riemann integral, but the point is to give an independent development of which the Riemann theory will be an eventual by-product.

Thus λ is countably additive on the class $\mathcal{J}$ of intervals in $(0, 1]$. If A is a finite disjoint union $\bigcup_{i=1}^{n} I_i$ of elements of $\mathcal{J}$, define

$$(2.11) \qquad \lambda(A) = \sum_{i=1}^{n} \lambda(I_i).$$

There is a question of uniqueness here, as A will have other representations as a finite disjoint union $\bigcup_{j=1}^{m} J_j$ of intervals. But $\mathcal{J}$ is closed under the formation of finite intersections, and so Theorem 2.2 gives

$$(2.12) \qquad \sum_{i=1}^{n} \lambda(I_i) = \sum_{i=1}^{n} \sum_{j=1}^{m} \lambda(I_i \cap J_j) = \sum_{j=1}^{m} \lambda(J_j).$$

(Some of the $I_i \cap J_j$ may be empty, but the corresponding λ-values are then 0.) Thus the definition is indeed consistent.

Thus (2.11) defines a set function on the class $\mathcal{B}_0$ of finite disjoint unions of intervals, the set function considered in Example 2.8. This set function is countably additive. To see this, suppose that $A = \bigcup_{k=1}^{\infty} A_k$, where $A \in \mathcal{B}_0$, $A_k \in \mathcal{B}_0$, and the A_k are disjoint. Then $A = \bigcup_{i=1}^{n} I_i$ and $A_k = \bigcup_{j=1}^{m_k} J_{kj}$ are finite disjoint unions of intervals, and the definition (2.11) and Theorem 2.2 give

$$(2.13) \qquad \lambda(A) = \sum_{i=1}^{n} \lambda(I_i) = \sum_{i=1}^{n} \sum_{k=1}^{\infty} \sum_{j=1}^{m_k} \lambda(I_i \cap J_{kj})$$

$$= \sum_{k=1}^{\infty} \sum_{j=1}^{m_k} \lambda(J_{kj}) = \sum_{k=1}^{\infty} \lambda(A_k).$$

Thus λ is a probability measure on the field $\mathcal{B}_0$. There remains the problem of extending λ to the larger class $\mathcal{B}$ of Borel sets, the σ-field generated by $\mathcal{B}_0$.

It is well to pause here and consider just what is involved in the construction of Lebesgue measure on the unit interval. That (2.10) defines a finitely additive set function on the class $\mathcal{J}$ of intervals in $(0, 1]$ is a consequence of Lemmas 1 and 2 for the case of only finitely many intervals. Finite additivity thus involves only the most elementary properties of the real number system. On the other hand, proving countable additivity involves compactness (the proof of Lemma 2 uses the Heine-Borel theorem), a profound idea underlying all of modern analysis. Once λ has been proved countably additive on the class $\mathcal{J}$, there is no real problem in extending it from $\mathcal{J}$ to the field $\mathcal{B}_0$ by the definition (2.11); the arguments involving (2.12) and (2.13) are easy. Difficulties again arise, however, in the further extension of λ from the field $\mathcal{B}_0$ to the σ-field $\mathcal{B} = \sigma(\mathcal{B}_0)$, and here new ideas are again required. These ideas are the subject of the next section, where it is shown that any probability measure on any field can be extended to the generated σ-field.

Constructing σ-Fields[†]

The σ-field $\sigma(\mathcal{A})$ generated by $\mathcal{A}$ was defined from above or from the outside, so to speak, by intersecting all the σ-fields that contain $\mathcal{A}$ (including the σ-field consisting of all the subsets of Ω). Can $\sigma(\mathcal{A})$ somehow be constructed from the inside by repeated finite and countable set-theoretic operations starting with sets in $\mathcal{A}$?

For any class $\mathcal{H}$ of sets in Ω let $\mathcal{H}^*$ consist of the sets in $\mathcal{H}$, the complements of sets in $\mathcal{H}$, and the finite and countable unions of sets in $\mathcal{H}$. Given a class $\mathcal{A}$, put $\mathcal{A}_1 = \mathcal{A}$ and define $\mathcal{A}_2, \mathcal{A}_3, \ldots$ inductively by

$$(2.14) \qquad\qquad \mathcal{A}_n = \mathcal{A}_{n-1}^*.$$

That each $\mathcal{A}_n$ is contained in $\sigma(\mathcal{A})$ follows by induction. One might hope that $\mathcal{A}_n = \sigma(\mathcal{A})$ for some n, or at least that $\bigcup_{n=1}^{\infty} \mathcal{A}_n = \sigma(\mathcal{A})$. But this process applied to the class of intervals fails to account for all the Borel sets.

Let $\mathcal{I}_0$ consist of the empty set and the intervals in $\Omega = (0, 1]$ with rational endpoints, and define $\mathcal{I}_n = \mathcal{I}_{n-1}^*$ for $n = 1, 2, \ldots$. *It will be shown that* $\bigcup_{n=0}^{\infty} \mathcal{I}_n$ *is strictly smaller than* $\mathcal{B} = \sigma(\mathcal{I}_0)$.[‡]

If a_n and b_n are rationals decreasing to a and b, then $(a, b] = \bigcup_m \bigcap_n (a_m, b_n] = \bigcup_m (\bigcup_n (a_m, b_n]^c)^c \in \mathcal{I}_4$. The result would therefore not be changed by including in $\mathcal{I}_0$ all the intervals in $(0, 1]$.

To prove $\bigcup_{n=0}^{\infty} \mathcal{I}_n$ smaller than $\mathcal{B}$, first put

$$(2.15) \qquad\qquad \Psi(A_1, A_2, \ldots) = A_1^c \cup A_2 \cup A_3 \cup A_4 \cup \cdots.$$

Since $\mathcal{I}_{n-1}$ contains $(0, 1]$ and the empty set, every element of $\mathcal{I}_n$ has the form (2.15) for some sequence $A_1, A_2, \ldots$ of sets in $\mathcal{I}_{n-1}$. Let every positive integer appear exactly once in the square array

$$m_{11}, m_{12}, \ldots$$

$$m_{21}, m_{22}, \ldots$$

$$\cdots\cdots\cdots$$

Inductively define

$$(2.16) \qquad \begin{cases} \Phi_0(A_1, A_2, \ldots) = A_1, \\ \cdot\Phi_n(A_1, A_2, \ldots) = \Psi(\Phi_{n-1}(A_{m_{11}}, A_{m_{12}}, \ldots), \Phi_{n-1}(A_{m_{21}}, A_{m_{22}}, \ldots), \ldots), \\ \qquad\qquad\qquad\qquad\qquad\qquad\qquad\qquad\qquad\qquad\qquad n = 1, 2, \ldots \end{cases}$$

It follows by induction that every element of $\mathcal{I}_n$ has the form $\Phi_n(A_1, A_2, \ldots)$ for some sequence of sets in $\mathcal{I}_0$. Finally, put

$$(2.17) \quad \Phi(A_1, A_2, \ldots) = \Phi_1(A_{m_{11}}, A_{m_{12}}, \ldots) \cup \Phi_2(A_{m_{21}}, A_{m_{22}}, \ldots) \cup \cdots.$$

[†] This topic may be omitted.

[‡] It might then be asked whether $(\bigcup_{n=0}^{\infty} \mathcal{I}_n)^* = \mathcal{B}$, but no countable repetition of the operation is sufficient to exhaust $\mathcal{B}$; see Problem 2.22.

Then every element of $\bigcup_{n=0}^{\infty} \mathcal{J}_n$ has the form (2.17) for some sequence $A_1, A_2, \ldots$ of sets in $\mathcal{J}_0$.

If $A_1, A_2, \ldots$ are in $\mathcal{B}$, then (2.15) is in $\mathcal{B}$; it follows by induction that each $\Phi_n(A_1, A_2, \ldots)$ is in $\mathcal{B}$ and therefore that (2.17) is in $\mathcal{B}$.

With each ω in $(0, 1]$ associate the sequence $(\omega_1, \omega_2, \ldots)$ of positive integers such that $\omega_1 + \cdots + \omega_k$ is the position of the kth 1 in the nonterminating dyadic expansion of ω (the smallest n for which $\Sigma_{j=1}^n d_j(\omega) = k$). Then $\omega \leftrightarrow (\omega_1, \omega_2, \ldots)$ is a one-to-one correspondence between $(0, 1]$ and the set of all sequences of positive integers. Let $I_1, I_2, \ldots$ be an enumeration of the sets in $\mathcal{J}_0$, put $\varphi(\omega) = \Phi(I_{\omega_1}, I_{\omega_2}, \ldots)$, and define $B = [\omega: \omega \notin \varphi(\omega)]$. It will be shown that B is a Borel set but is not contained in any of the $\mathcal{J}_n$.

Since ω lies in B if and only if ω lies outside $\varphi(\omega)$, $B \neq \varphi(\omega)$ for every ω. But every element of $\bigcup_{n=0}^{\infty} \mathcal{J}_n$ has the form (2.17) for some sequence in $\mathcal{J}_0$ and hence has the form $\varphi(\omega)$ for some ω. Therefore, B is not a member of $\bigcup_{n=0}^{\infty} \mathcal{J}_n$.

It remains to show that B is a Borel set. Let $D_k = [\omega: \omega \in I_{\omega_k}]$. Since $L_k(n) = [\omega: \omega_1 + \ldots + \omega_k = n] = [\omega: \Sigma_{j=1}^{n-1} d_j(\omega) < k = \Sigma_{j=1}^n d_j(\omega)]$ is a Borel set, so are $[\omega: \omega_k = n] = \bigcup_{m=1}^{\infty} L_{k-1}(m) \cap L_k(m + n)$ and $D_k = \bigcup_{n=1}^{\infty} [\omega: \omega_k = n] \cap I_n$. Suppose that it is shown that

$$(2.18) \qquad [\omega: \omega \in \Phi_n(I_{\omega_{u_1}}, I_{\omega_{u_2}} \ldots)] = \Phi_n(D_{u_1}, D_{u_2}, \ldots)$$

for every n and every sequence $u_1, u_2, \ldots$ of positive integers. It will then follow from the definition (2.17) that

$$B^c = [\omega: \omega \in \varphi(\omega)] = \bigcup_{n=1}^{\infty} [\omega: \omega \in \Phi_n(I_{\omega_{mn1}}, I_{\omega_{mn2}}, \ldots)]$$

$$= \bigcup_{n=1}^{\infty} \Phi_n(D_{mn1}, D_{mn2}, \ldots) = \Phi(D_1, D_2, \ldots).$$

But as remarked above, (2.17) is a Borel set if the A_n are. Therefore, (2.18) will imply that B^c and B are Borel sets.

If $n = 0$, (2.18) holds because it reduces by (2.16) to $[\omega: \omega \in I_{\omega_{u_1}}] = D_{u_1}$. Suppose that (2.18) holds with $n - 1$ in place of n. Consider the condition

$$(2.19) \qquad \omega \in \Phi_{n-1}(I_{\omega_{u m_{k1}}}, I_{\omega_{u m_{k2}}}, \ldots).$$

By (2.15) and (2.16), a necessary and sufficient condition for $\omega \in \Phi_n(I_{\omega_{u_1}}, I_{\omega_{u_2}}, \ldots)$ is that either (2.19) is false for $k = 1$ or else (2.19) is true for some k exceeding 1. But by the induction hypothesis, (2.19) and its negation can be replaced by $\omega \in \Phi_{n-1}(D_{u m_{k1}}, D_{u m_{k2}}, \ldots)$ and its negation. Therefore, $\omega \in \Phi_n(I_{\omega_{u_1}}, I_{\omega_{u_2}}, \ldots)$ if and only if $\omega \in \Phi_n(D_{u_1}, D_{u_2}, \ldots)$.

PROBLEMS

2.1. Define $x \vee y = \max\{x, y\}$ and for a collection $\{x_\alpha\}$ define $\bigvee_\alpha x_\alpha = \sup_\alpha x_\alpha$; define $x \wedge y = \min\{x, y\}$ and $\bigwedge_\alpha x_\alpha = \inf_\alpha x_\alpha$. Prove that $I_{A \cup B} = I_A \vee I_B$, $I_{A \cap B} = I_A \wedge I_B$, $I_{A^c} = 1 - I_A$, and $I_{A \triangle B} = |I_A - I_B|$, in the sense that there is equality at each point of Ω. Show that $A \subset B$ if and only if $I_A \leq I_B$ pointwise. Check the equation $x \wedge (y \vee z) = (x \wedge y) \vee (x \wedge z)$ and deduce the distributive law $A \cap (B \cup C) = (A \cap B) \cup (A \cap C)$. By similar arguments prove that

$$A \cup (B \cap C) = (A \cup B) \cap (A \cup C),$$

$$A \bigtriangleup C \subset (A \bigtriangleup B) \cup (B \bigtriangleup C),$$

$$\left(\bigcup_n A_n \right)^c = \bigcap_n A_n^c,$$

$$\left(\bigcap_n A_n \right)^c = \bigcup_n A_n^c.$$

2.2. Let $A_1, \ldots, A_n$ be arbitrary events and put $U_k = \bigcup (A_{i_1} \cap \cdots \cap A_{i_k})$ and $I_k = \bigcap (A_{i_1} \cup \cdots \cup A_{i_k})$, where the union and intersection extend over all the k-tuples satisfying $1 \leq i_1 < \cdots < i_k \leq n$. Show that $U_k = I_{n-k+1}$.

2.3. (a) Suppose that $\Omega \in \mathcal{F}$ and that $A, B \in \mathcal{F}$ implies $A - B = A \cap B^c \in \mathcal{F}$. Show that $\mathcal{F}$ is a field.

(b) Suppose that $\Omega \in \mathcal{F}$ and that $\mathcal{F}$ is closed under the formation of complements and finite disjoint unions. Show that $\mathcal{F}$ need not be a field.

2.4. (a) Suppose that $\mathcal{F}_n$ are fields satisfying $\mathcal{F}_n \subset \mathcal{F}_{n+1}$. Show that $\bigcup_{n=1}^{\infty} \mathcal{F}_n$ is a field.

(b) Suppose that $\mathcal{F}_n$ are σ-fields satisfying $\mathcal{F}_n \subset \mathcal{F}_{n+1}$. Show that $\bigcup_{n=1}^{\infty} \mathcal{F}_n$ need not be a σ-field.

2.5. Let $f(\mathcal{A})$ denote the field generated by $\mathcal{A}$; that is, $f(\mathcal{A})$ is the intersection of all fields containing $\mathcal{A}$.

(a) If $\mathcal{A}$ consists of the singletons, $f(\mathcal{A})$ is the field in Example 2.3. Show that $f(\mathcal{A}) \subset \sigma(\mathcal{A})$, that there is equality here if $\mathcal{A}$ is finite, and that $\sigma(f(\mathcal{A})) = \sigma(\mathcal{A})$.

(b) Show that $f(\mathcal{A})$ is the class of sets of the form $\bigcup_{i=1}^{m} \bigcap_{j=1}^{n_i} A_{ij}$, where for each i and j either $A_{ij} \in \mathcal{A}$ or $A_{ij}^c \in \mathcal{A}$, and where the m sets $\bigcap_{j=1}^{n_i} A_{ij}$, $1 \leq i \leq m$, are disjoint.

(c) Show that, if $\mathcal{A}$ is countable, then $f(\mathcal{A})$ is also countable.

2.6. Suppose for each A in $\mathcal{A}$ that A^c is a countable union of elements of $\mathcal{A}$. The class of intervals in $(0, 1]$ has this property. Show that $\sigma(\mathcal{A})$ coincides with the smallest class over $\mathcal{A}$ that is closed under the formation of countable unions and intersections.

2.7. Suppose that $\mathcal{A} = \{A_1, \ldots, A_n\}$ is a finite partition or decomposition of Ω and that $\mathcal{F}$ is a σ-field in Ω. Show that $\sigma(\mathcal{A} \cup \mathcal{F})$ consists of all the unions $\bigcup_{k=1}^{n} (A_k \cap B_k)$ with $B_k \in \mathcal{F}$.

2.8. Show that, if $B \in \sigma(\mathcal{A})$, then there exists a countable subclass $\mathcal{A}_B$ of $\mathcal{A}$ such that $B \in \sigma(\mathcal{A}_B)$.

2.9. Show that a class $\mathcal{A}$ of subsets of a countable Ω generates the σ-field of all subsets if and only if for each pair ω and ω' of distinct points there is in $\mathcal{A}$ an A such that $I_A(\omega) \neq I_A(\omega')$.

2.10. A σ-field is *separable* if it is generated by some countable class of sets.

(a) Show that the σ-field $\mathcal{B}$ of Borel sets is separable.

(b) Show that the σ-field of Example 2.4 is separable if and only if Ω is countable.

(c) Show that if $\mathcal{F}_1$ and $\mathcal{F}_2$ are σ-fields satisfying $\mathcal{F}_1 \subset \mathcal{F}_2$, separability of $\mathcal{F}_2$ does not imply separability of $\mathcal{F}_1$.

2.11. Show that a σ-field cannot be countably infinite—its cardinality must be finite or else at least that of the continuum.

2.12. (a) Let $\mathcal{F}$ be the field consisting of the finite and the cofinite sets in an infinite Ω and define P on $\mathcal{F}$ by taking $P(A)$ to be 0 or 1 as A is finite or cofinite. Show that P is finitely additive.

(b) Show that this P is not countably additive if Ω is countably infinite.

(c) Show that this P is countably additive if Ω is uncountable.

(d) Now let $\mathcal{F}$ be the σ-field consisting of the countable and the co-countable sets in an uncountable Ω and define P on $\mathcal{F}$ by taking $P(A)$ to be 0 or 1 as A is countable or co-countable. Show that P is countably additive.

2.13. Suppose that P is a probability measure on a field $\mathcal{F}$, that $A_1, A_2, \ldots$ and $A = \cup_n A_n$ lie in $\mathcal{F}$, and that the A_n are nearly disjoint, in the sense that $P(A_m \cap A_n) = 0$ for $m \neq n$. Show that $P(A) = \Sigma_n P(A_n)$.

2.14. On the field $\mathcal{B}_0$ of finite disjoint unions of intervals in $(0, 1]$ define $P(A)$ to be 1 or 0 as A does or does not for some positive ϵ contain the interval $(\frac{1}{2}, \frac{1}{2} + \epsilon]$. Show that P is finitely but not countably additive.

2.15. *Stochastic arithmetic.* Define a probability measure P_n on the class of all subsets of $\Omega = \{1, 2, \ldots\}$ by

(2.20) $$P_n(A) = \frac{1}{n} \#[m: 1 \leq m \leq n, m \in A];$$

among the first n integers the proportion that lie in A is just $P_n(A)$. The set A has *density*

(2.21) $$D(A) = \lim_n P_n(A),$$

provided this limit exists. Let $\mathcal{D}$ be the class of sets having density.

(a) Show that D is finitely but not countably additive on $\mathcal{D}$.

(b) Show that $\mathcal{D}$ contains the empty set and Ω and is closed under the formation of complements, proper differences, and finite disjoint unions but is not closed under the formation of countable disjoint unions or of finite unions that are not disjoint.

(c) Let $\mathcal{M}$ consist of the periodic sets $M_a = [ka: k = 1, 2, \ldots]$. Observe that

(2.22) $$P_n(M_a) = \frac{1}{n}\left[\frac{n}{a}\right] \to \frac{1}{a} = D(M_a).$$

Show that the field $f(\mathcal{M})$ generated by $\mathcal{M}$ (see Problem 2.5) is contained in $\mathcal{D}$. Show that D is completely determined on $f(\mathcal{M})$ by the value it gives for each a to the event that m is divisible by a.

(d) Assume that Σp^{-1} diverges (sum over all primes; see Problem 5.17 (e)) and prove that D, although finitely additive, is not countably additive on the field $f(\mathcal{M})$.

(e) Euler's function $\varphi(n)$ is the number of positive integers less than n and relatively prime to it. Let $p_1, \ldots, p_r$ be the distinct prime factors of n; from the inclusion-exclusion formula for the events $[m: p_i | m]$, (2.22), and the fact that the p_i divide n, deduce

$$(2.23) \qquad \frac{\varphi(n)}{n} = \prod_{p|n} \left(1 - \frac{1}{p}\right).$$

(f) Show for $0 \le x \le 1$ that $D(A) = x$ for some A.

2.16. Let Ω be *sequence space*, the set of infinite sequences of 0's and 1's. Let the nth component of ω be $a_n(\omega)$, so that $\omega = (a_1(\omega), a_2(\omega), \ldots)$. A set $[\omega: a_i(\omega) = u_i,$ $i = 1, \ldots, n]$, for a sequence $u_1, \ldots, u_n$ of 0's and 1's, is a *cylinder of rank n*; assign it P-measure 2^{-n}. Let $\mathcal{C}_0$ consist of the empty set and the finite disjoint unions $A = \bigcup_{i=1}^{k} C_i$ of cylinders C_i, and put $P(A) = \Sigma_{i=1}^{k} P(C_i)$.

(a) Show that the definition of P is consistent and that P is finitely additive on the field $\mathcal{C}_0$.

(b) Show that P is countably additive on $\mathcal{C}_0$.

(c) Compare sequence space with $(0, 1]$, cylinders with dyadic intervals, and P with Lebesgue measure.

2.17. A probability measure space $(\Omega, \mathcal{F}, P)$ is *nonatomic* if $P(A) > 0$ implies that there exists a B such that $B \subset A$ and $0 < P(B) < P(A)$ (A and B in $\mathcal{F}$, of course).

(a) Assuming the existence of Lebesgue measure λ on $\mathcal{B}$, prove that it is nonatomic.

(b) Show in the nonatomic case that $P(A) > 0$ and $\epsilon > 0$ imply that there exists a B such that $B \subset A$ and $0 < P(B) < \epsilon$.

(c) Show in the nonatomic case that $0 \le x \le P(A)$ implies that there exi·ts a B such that $B \subset A$ and $P(B) = x$.

Hint: Inductively define classes $\mathcal{H}_n$, numbers h_n, and sets H_n by $\mathcal{H}_0 = \{0\} = \{H_0\}$, $\mathcal{H}_n = [H: H \subset A - \bigcup_{k<n} H_k, P(\bigcup_{k<n} H_k) + P(H) \le x]$, $h_n = \sup [P(H): H \in \mathcal{H}_n]$, and $P(H_n) > h_n - n^{-1}$. Consider $\bigcup_k H_k$.

(d) Show in the nonatomic case that, if $p_1, p_2, \ldots$ are nonnegative and add to 1, then A can be decomposed into sets $B_1, B_2, \ldots$ such that $P(B_n) = p_n P(A)$.

2.18. ↑ **(a)** Show for nonatomic measures P_1 and P_2 on the same σ-field $\mathcal{F}$ that there exists an A such that $P_1(A) \ge 1/2$ and $P_2(A^c) \ge 1/2$.

(b) Extend to n measures.

2.19. For P a probability measure on a σ-field $\mathcal{F}$, metrize the elements of $\mathcal{F}$ by $d(A, B) = P(A \bigtriangleup B)$.

(a) Show that d is a metric except that $A = B$ may not follow from $d(A, B) = 0$. Identify sets at distance 0 and consider the resulting metric space.

(b) Suppose that $(\Omega, \mathcal{F}, P)$ is nonatomic and A, B, and $t, 0 \le t \le 1$, are given. Show that there exists an E_t such that $d(A, E_t) = td(A, B)$ and $d(E_t, B) = (1 - t)d(A, B)$. (A metric space with this property is called *convex*.)

2.20. (a) Suppose that $\mathcal{A} = \{A_1, A_2, \ldots\}$ is a countable partition of Ω. Show (see (2.14)) that $\mathcal{A}_2 = \mathcal{A}_1^* = \mathcal{A}^*$ coincides with $\sigma(\mathcal{A})$. This is a case where $\sigma(\mathcal{A})$ can be constructed "from the inside."

(b) Show that the set of normal numbers lies in $\mathcal{J}_8$.

2.21. Extend (2.14) to infinite ordinals α by defining $\mathcal{A}_\alpha = (\cup_{\beta<\alpha}\mathcal{A}_\beta)^*$. Show that, if Ω is the first uncountable ordinal, then $\cup_{\alpha<\Omega}\mathcal{A}_\alpha = \sigma(\mathcal{A})$. Show that, if the cardinality of $\mathcal{A}$ does not exceed that of the continuum, then the same is true of $\sigma(\mathcal{A})$. Thus $\mathcal{B}$ has the power of the continuum.

2.22. ↑   Extend (2.16) to ordinals $\alpha < \Omega$ as follows. Replace the right side of (2.15) by $\cup_{n=1}^\infty (A_{2n-1} \cup A_{2n}^c)$. Suppose that Φ_β is defined for $\beta < \alpha$. Let $\beta_\alpha(1), \beta_\alpha(2), \ldots$ be a sequence of ordinals such that $\beta_\alpha(n) < \alpha$ and such that if $\beta < \alpha$, then $\beta = \beta_\alpha(n)$ for infinitely many even n and for infinitely many odd n; define

(2.24) $\Phi_\alpha(A_1, A_2, \ldots)$

$$= \Psi(\Phi_{\beta_\alpha(1)}(A_{m_{11}}, A_{m_{12}}, \ldots), \Phi_{\beta_\alpha(2)}(A_{m_{21}}, A_{m_{22}}, \ldots), \ldots).$$

Prove by transfinite induction that (2.24) is in $\mathcal{B}$ if the A_n are, that every element of $\mathcal{J}_\alpha$ has the form (2.24) for sets A_n in $\mathcal{J}_0$, and that (2.18) holds with α in place of n. Define $\varphi_\alpha(\omega) = \Phi_\alpha(I_{\omega 1}, I_{\omega 2}, \ldots)$ and show that $B_\alpha = [\omega: \omega \notin \varphi_\alpha(\omega)]$ lies in $\mathcal{B} - \mathcal{J}_\alpha$ for $\alpha < \Omega$.

2.23. *Dirichlet's theorem.* Show that a nonnegative series $\Sigma_{n=1}^\infty x_n$ converges if and only if the sums $\Sigma_{k \in E} x_k$ for E finite are bounded, in which case the sum is the supremum: $\Sigma_{n=1}^\infty x_n = \sup_E \Sigma_{k \in E} x_k$. As this supremum is invariant under permutations, if $y_n = x_{f(n)}$ for a one-to-one map f of the positive integers onto themselves, then $\Sigma_n x_n$ and $\Sigma_n y_n$ diverge or converge together and in the latter case have the same sum.

2.24. ↑   *Nonnegative double series.* Suppose that $x_{ij}, i, j = 1, 2, \ldots,$ are nonnegative. The ith row gives a series $\Sigma_j x_{ij}$, and if each of these converges, one can form the series $\Sigma_i \Sigma_j x_{ij}$ of row sums. Let the terms x_{ij} be arranged in some order as a single infinite series $\Sigma_{ij} x_{ij}$ (by Dirichlet's theorem the order taken has no effect), and compare the results. Show in fact that $\Sigma_{ij} x_{ij}$ converges if and only if the $\Sigma_j x_{ij}$ all converge and $\Sigma_i \Sigma_j x_{ij}$ converges, in which case $\Sigma_{ij} x_{ij} = \Sigma_i \Sigma_j x_{ij}$.

SECTION 3.   EXISTENCE AND EXTENSION

The main theorem to be proved here may be compactly stated this way:

Theorem 3.1. *A probability measure on a field has a unique extension to the generated σ-field.*

In more detail the assertion is this: Suppose that P is a probability measure on a field $\mathcal{F}_0$ of subsets of Ω, and put $\mathcal{F} = \sigma(\mathcal{F}_0)$. Then there exists a probability measure Q on $\mathcal{F}$ such that $Q(A) = P(A)$ for $A \in \mathcal{F}_0$. Further, if Q' is another probability measure on $\mathcal{F}$ such that $Q'(A) = P(A)$ for $A \in \mathcal{F}_0$, then $Q'(A) = Q(A)$ for $A \in \mathcal{F}$.

Although the measure extended to $\mathcal{F}$ is usually denoted by the same letter as the original measure on $\mathcal{F}_0$, they are really different set functions, since they have different domains of definition. The class $\mathcal{F}_0$ is only assumed finitely additive in the theorem, but the set function P on it must be countably additive because this of course follows from the conclusion of the theorem.

As shown at the end of the last section, λ (initially defined for intervals by (2.10)) extends to a probability measure on the field $\mathcal{B}_0$ of finite disjoint unions of subintervals of $(0, 1]$. By Theorem 3.1, λ extends in a unique way from $\mathcal{B}_0$ to $\mathcal{B} = \sigma(\mathcal{B}_0)$, the class of Borel sets in $(0, 1]$. The extended λ is *Lebesgue measure* on the unit interval. Theorem 3.1 has many other applications as well.

The uniqueness in Theorem 3.1 will be proved later; see Theorem 3.3. The first project is to prove that an extension does exist.

Construction of the Extension

Let P be a probability measure on a field $\mathcal{F}_0$. The construction following extends P to a class that in general is much larger than $\sigma(\mathcal{F}_0)$ but nonetheless in general does not contain all the subsets of Ω.

For each subset A of Ω, define its *outer measure* by

$$(3.1) \qquad\qquad P^*(A) = \inf \sum_n P(A_n),$$

where the infimum extends over all finite and infinite sequences $A_1, A_2, \ldots$ of $\mathcal{F}_0$-sets satisfying $A \subset \bigcup_n A_n$. If the A_n form an efficient covering of A, in the sense that they do not overlap one another very much or extend much beyond A, then $\sum_n P(A_n)$ should be a good outer approximation to the measure of A if A is indeed to have a measure assigned it at all. Thus (3.1) represents a first attempt to assign a measure to A.

Because of the rule $P(A^c) = 1 - P(A)$ for complements (see (2.5)), it is natural in approximating A from the inside to approximate the complement A^c from the outside instead and then subtract from 1:

$$(3.2) \qquad\qquad P_*(A) = 1 - P^*(A^c).$$

This, the *inner measure* of A, is a second candidate for the measure of A.[†] A plausible procedure is to assign measure to those A for which (3.1) and (3.2)

[†] It will not do to define $P_*(A)$ as the supremum of the sums $\sum_n P(A_n)$ for disjoint $\mathcal{F}_0$-sequences contained in A because, for example, the set (2.1) of normal numbers contains no nonempty elements of $\mathcal{B}_0$ and so would have inner measure 0.

agree, and to take the common value $P^*(A) = P_*(A)$ as the measure. Since (3.1) and (3.2) agree if and only if

$$(3.3) \qquad\qquad P^*(A) + P^*(A^c) = 1,$$

the procedure would be to consider the class of A satisfying (3.3) and use $P^*(A)$ as the measure.

It turns out to be simpler to impose on A the more stringent requirement that

$$(3.4) \qquad\qquad P^*(A \cap E) + P^*(A^c \cap E) = P^*(E)$$

holds for every set E; (3.3) is the special case $E = \Omega$, because it will turn out that $P^*(\Omega) = 1$.[†] A set A is called P^*-*measurable* if (3.4) holds for all E; let $\mathcal{M}$ be the class of such sets. What will be shown is that $\mathcal{M}$ contains $\sigma(\mathcal{F}_0)$ and that the restriction of P^* to $\sigma(\mathcal{F}_0)$ is the required extension of P.

The set function P^* has four properties that will be needed:

   (i) $P^*(0) = 0$;
   (ii) P^* is nonnegative: $P^*(A) \geq 0$ for every $A \subset \Omega$;
   (iii) P^* is monotone: $A \subset B$ implies $P^*(A) \leq P^*(B)$;
   (iv) P^* is countably subadditive: $P^*(\bigcup_n A_n) \leq \Sigma_n P^*(A_n)$.

The others being obvious, only (iv) needs proof. For a given ϵ, choose $\mathcal{F}_0$-sets B_{nk} such that $A_n \subset \bigcup_k B_{nk}$ and $\Sigma_k P(B_{nk}) < P^*(A_n) + \epsilon 2^{-n}$, which is possible by the definition (3.1). Now $\bigcup_n A_n \subset \bigcup_{n,k} B_{nk}$, so that $P^*(\bigcup_n A_n) \leq \Sigma_{n,k} P(B_{nk}) < \Sigma_n P^*(A_n) + \epsilon$, and (iv) follows.[‡]

By definition A lies in the class $\mathcal{M}$ of P^*-measurable sets if it splits each E in such a way that P^* adds for the pieces—that is, if (3.4) holds. Because of subadditivity, (3.4) is equivalent to

$$(3.5) \qquad\qquad P^*(A \cap E) + P^*(A^c \cap E) \leq P^*(E).$$

Lemma 1. *The class $\mathcal{M}$ is a field.*

PROOF. It is clear that $\Omega \in \mathcal{M}$ and that $\mathcal{M}$ is closed under complementation. Suppose that $A, B \in \mathcal{M}$ and $E \subset \Omega$. Then

$$P^*(E) = P^*(B \cap E) + P^*(B^c \cap E)$$

$$= P^*(A \cap B \cap E) + P^*(A^c \cap B \cap E) + P^*(A \cap B^c \cap E)$$
$$+ P^*(A^c \cap B^c \cap E)$$

[†] It also turns out, after the fact, that (3.3) implies that (3.4) holds for all E anyway; see Problem 3.3.

[‡] Compare the proof on page 8 that a countable union of negligible sets is negligible.

$$\geq P^*(A \cap B \cap E) + P^*((A^c \cap B \cap E) \cup (A \cap B^c \cap E)$$
$$\cup (A^c \cap B^c \cap E))$$
$$= P^*((A \cap B) \cap E) + P^*((A \cap B)^c \cap E),$$

the inequality following by subadditivity. Hence[†] $A \cap B \in \mathcal{M}$, and $\mathcal{M}$ is a field. ∎

Lemma 2. *The set function* P^* *is finitely additive on* $\mathcal{M}$.

PROOF. If A and B lie in $\mathcal{M}$ and are disjoint, then $P^*(A \cup B) = P^*(A \cap (A \cup B)) + P^*(A^c \cap (A \cup B)) = P^*(A) + P^*(B)$. It follows by induction that $P^*(\bigcup_{k=1}^n A_k) = \sum_{k=1}^n P^*(A_k)$ for disjoint $\mathcal{M}$-sets $A_1, \ldots, A_k$. ∎

Lemma 3. *If* $A_1, A_2, \ldots$ *is a disjoint sequence of* $\mathcal{M}$-*sets, then* $\bigcup_n A_n \in \mathcal{M}$ *and* $P^*(\bigcup_n A_n) = \sum_n P^*(A_n)$.

PROOF. Put $A = \bigcup_n A_n$. By Lemma 2 and monotonicity, $\sum_{n \leq m} P^*(A_n) = P^*(\bigcup_{n \leq m} A_n) \leq P^*(A)$. Therefore, $\sum_n P^*(A_n) \leq P^*(A)$; the opposite inequality follows by subadditivity.

It remains to prove that A satisfies (3.5). Put $B_m = \bigcup_{n \leq m} A_n$ and consider the equation

$$(3.6) \qquad P^*(E \cap B_m) = \sum_{n \leq m} P^*(E \cap A_n).$$

This is certainly true for $m = 1$. Assume it is true for m, and split $E \cap B_{m+1}$ by the set B_m, which by Lemma 1 lies in $\mathcal{M}$: $P^*(E \cap B_{m+1}) = P^*(E \cap B_{m+1} \cap B_m) + P^*(E \cap B_{m+1} \cap B_m^c) = P^*(E \cap B_m) + P^*(E \cap A_{m+1}) = \sum_{n \leq m} P^*(E \cap A_n) + P^*(E \cap A_{m+1})$. Hence (3.6) holds for all m.

Now split E by B_m: $P^*(E) = P^*(E \cap B_m) + P^*(E \cap B_m^c) = \sum_{n \leq m} P^*(E \cap A_n) + P^*(E \cap B_m^c) \geq \sum_{n \leq m} P^*(E \cap A_n) + P^*(E \cap A^c)$. Let $n \to \infty$: $P^*(E) \geq \sum_n P^*(E \cap A_n) + P^*(E \cap A^c) \geq P^*(E \cap A) + P^*(E \cap A^c)$, which establishes (3.5). ∎

Lemma 4. *The class* $\mathcal{M}$ *is a* σ-*field, and* P^* *restricted to* $\mathcal{M}$ *is countably additive.*

PROOF. By Lemmas 1 and 2, P^* is finitely additive on the field $\mathcal{M}$. If $A_1, A_2, \ldots$ are in $\mathcal{M}$, put $B_1 = A_1$ and $B_n = A_n \cap A_1^c \cap \cdots \cap A_{n-1}^c$; then the B_n are

---

[†] This proof does not work if (3.4) is weakened to (3.3). Compare the fact that some inductive proofs go through only if the theorem in question is given a sufficiently strong formulation.

disjoint and by Lemma 1 lie in $\mathcal{M}$, so that $\bigcup_n A_n = \bigcup_n B_n$ lies in $\mathcal{M}$ by Lemma 3. Thus $\mathcal{M}$ is a σ-field. Lemma 3 also gives countable additivity. ∎

PROOF OF EXTENSION IN THEOREM 3.1.   Suppose that P^* is defined via (3.1) from a probability measure P on a field $\mathcal{F}_0$. The first step is to prove that $\mathcal{F}_0 \subset \mathcal{M}$.

Suppose that $A \in \mathcal{F}_0$. Given E and ϵ, choose $\mathcal{F}_0$-sets A_n such that $E \subset \bigcup_n A_n$ and $\Sigma_n P(A_n) \leq P^*(E) + \epsilon$. The sets $B_n = A_n \cap A$ and $C_n = A_n \cap A^c$ lie in $\mathcal{F}_0$ because it is a field. Also, $E \cap A \subset \bigcup_n B_n$ and $E \cap A^c \subset \bigcup_n C_n$; by the definition of P^* and the finite additivity of P, $P^*(E \cap A) + P^*(E \cap A^c) \leq \Sigma_n P(B_n) + \Sigma_n P(C_n) = \Sigma_n P(A_n) \leq P^*(E) + \epsilon$. Hence $A \in \mathcal{F}_0$ implies (3.5), and so $\mathcal{F}_0 \subset \mathcal{M}$.

It is obvious from the definition (3.1) that $P^*(A) \leq P(A)$ for A in $\mathcal{F}_0$. If $A \subset \bigcup_n A_n$, where A and the A_n are in $\mathcal{F}_0$, then by the countable subadditivity and monotonicity of P on $\mathcal{F}_0$, $P(A) \leq \Sigma_n P(A \cap A_n) \leq \Sigma_n P(A_n)$. Thus

$$(3.7) \qquad\qquad P(A) = P^*(A) \qquad \text{if } A \in \mathcal{F}_0.$$

Since $\Omega \in \mathcal{F}_0 \subset \mathcal{M}$, $P^*(\Omega) = P(\Omega) = 1$. By Lemma 4, P^* restricted to the σ-field $\mathcal{M}$ is thus a probability measure. Since the σ-field $\mathcal{M}$ contains $\mathcal{F}_0$, it also contains $\mathcal{F} = \sigma(\mathcal{F}_0)$, and P^* further restricted from $\mathcal{M}$ to $\mathcal{F}$ is also a probability measure. This measure on $\mathcal{F}$ is the required extension because by (3.7) it agrees with P on $\mathcal{F}_0$. ∎

Uniqueness and the π-λ Theorem

To prove the extension in Theorem 3.1 unique requires some auxiliary concepts. A class $\mathcal{P}$ of subsets of Ω is a *π-system* if it is closed under the formation of finite intersections:

(π) $A, B \in \mathcal{P}$ implies $A \cap B \in \mathcal{P}$.

A class $\mathcal{L}$ is a *λ-system* if it contains Ω and is closed under the formation of proper differences and countable, increasing unions:

(λ_1) $\Omega \in \mathcal{L}$;
(λ_2) $A, B \in \mathcal{L}$ and $A \subset B$ imply $B - A \in \mathcal{L}$;
(λ_3) $A_1, A_2, \ldots \in \mathcal{L}$ and $A_n \uparrow A$ imply $A \in \mathcal{L}$.

A class that is both a π-system and a λ-system is a field because it is closed under complementation (λ_1 and λ_2) and the formation of finite intersections (π); and because of (λ_3) it is, in fact, a σ-field (see Example 2.5).

Many uniqueness arguments depend on *Dynkin's π-λ theorem*. It is technical but extremely useful.

Theorem 3.2. *If P is a π-system and $\mathcal{L}$ is a λ-system, then $P \subset \mathcal{L}$ implies that $\sigma(P) \subset \mathcal{L}$.*

PROOF. Let $l(P)$ be the minimal λ-system over P—the intersection of all λ-systems containing P. If $l(P)$ is a π-system as well as a λ-system, then it will be a σ-field, as observed above, so that $\sigma(P) \subset l(P) \subset \mathcal{L}$. It therefore suffices to show that $l(P)$ is a π-system.

For each set A, let $\mathcal{G}_A$ be the class of sets B such that $A \cap B \in l(P)$. It is a simple matter to show that $\mathcal{G}_A$ is a λ-system if $A \in l(P)$. If $A \in P$ and $B \in P$, then $A \cap B \in P \subset l(P)$. Thus $A \in P$ implies that $P \subset \mathcal{G}_A$, and since $\mathcal{G}_A$ is a λ-system, minimality gives $l(P) \subset \mathcal{G}_A$.

Thus $A \in P$ implies that $l(P) \subset \mathcal{G}_A$, or, to put it another way, $A \in P$ and $B \in l(P)$ together imply that $A \cap B \in l(P)$. But this last implication means that $B \in l(P)$ implies that $P \subset \mathcal{G}_B$. Since $\mathcal{G}_B$ is a λ-system, it follows by minimality again that $B \in l(P)$ implies that $l(P) \subset \mathcal{G}_B$. Therefore, $B \in l(P)$ and $C \in l(P)$ together imply that $B \cap C \in l(P)$, and $l(P)$ is indeed a π-system. ∎

Since a field is certainly a π-system, the uniqueness asserted in Theorem 3.1 is a consequence of this result:

Theorem 3.3. *Suppose that P_1 and P_2 are probability measures on $\sigma(P)$, where P is a π-system. If P_1 and P_2 agree on P, then they agree on $\sigma(P)$.*

PROOF. Let $\mathcal{L}$ be the class of sets A in $\sigma(P)$ such that $P_1(A) = P_2(A)$. It is easy to check that $\mathcal{L}$ is a λ-system, and the hypothesis of the theorem is that $P \subset \mathcal{L}$. The π-λ theorem therefore gives $\sigma(P) \subset \mathcal{L}$, as required. ∎

If a probability measure P on a σ-field $\mathcal{F}$ is determined by its values on some smaller class $\mathcal{A}$, it is natural to ask whether $\mathcal{F}$-sets E can somehow be approximated by $\mathcal{A}$-sets A in such a way that $P(E)$ is approximated by $P(A)$. Theorem 11.4 is a result of this kind.

Monotone Classes

A class $\mathcal{M}$ of subsets of Ω is *monotone* if it is closed under the formation of monotone unions and intersections:

(i) $A_1, A_2, \ldots \in \mathcal{M}$ and $A_n \uparrow A$ imply $A \in \mathcal{M}$;
(ii) $A_1, A_2, \ldots \in \mathcal{M}$ and $A_n \downarrow A$ imply $A \in \mathcal{M}$.

The following *monotone class theorem* is a close relative of the π-λ theorem but will be less frequently used in this book.

Theorem 3.4. *If $\mathcal{F}_0$ is a field and $\mathcal{M}$ is a monotone class, then $\mathcal{F}_0 \subset \mathcal{M}$ implies that $\sigma(\mathcal{F}_0) \subset \mathcal{M}$.*

PROOF.   Let $m(\mathcal{F}_0)$ be the minimal monotone class over $\mathcal{F}_0$—the intersection of all monotone classes containing $\mathcal{F}_0$. It is enough to prove $\sigma(\mathcal{F}_0) \subset m(\mathcal{F}_0)$; this will follow if $m(\mathcal{F}_0)$ is shown to be a field because a monotone field is a σ-field.

Consider the class $\mathcal{G} = [A: A^c \in m(\mathcal{F}_0)]$. Since $m(\mathcal{F}_0)$ is monotone, so is $\mathcal{G}$. Since $\mathcal{F}_0$ is a field, $\mathcal{F}_0 \subset \mathcal{G}$, and so $m(\mathcal{F}_0) \subset \mathcal{G}$. Hence $m(\mathcal{F}_0)$ is closed under complementation.

Define $\mathcal{G}_1$ as the class of A such that $A \cup B \in m(\mathcal{F}_0)$ for all $B \in \mathcal{F}_0$. Then $\mathcal{G}_1$ is a monotone class and $\mathcal{F}_0 \subset \mathcal{G}_1$; by the minimality of $m(\mathcal{F}_0)$, $m(\mathcal{F}_0) \subset \mathcal{G}_1$. Define $\mathcal{G}_2$ as the class of B such that $A \cup B \in m(\mathcal{F}_0)$ for all $A \in m(\mathcal{F}_0)$. Then $\mathcal{G}_2$ is a monotone class. Now from $m(\mathcal{F}_0) \subset \mathcal{G}_1$ it follows that $A \in m(\mathcal{F}_0)$ and $B \in \mathcal{F}_0$ together imply that $A \cup B \in m(\mathcal{F}_0)$; in other words, $B \in \mathcal{F}_0$ implies that $B \in \mathcal{G}_2$. Thus $\mathcal{F}_0 \subset \mathcal{G}_2$; by minimality, $m(\mathcal{F}_0) \subset \mathcal{G}_2$, and hence $A, B \in m(\mathcal{F}_0)$ implies that $A \cup B \in m(\mathcal{F}_0)$.   ■

Completeness

This is the natural place to consider completeness, although it enters into probability theory in an essential way only in connection with the study of stochastic processes in continuous time; see Sections 37 and 38.

A probability measure space $(\Omega, \mathcal{F}, P)$ is *complete* if $A \subset B$, $B \in \mathcal{F}$, and $P(B) = 0$ together imply that $A \in \mathcal{F}$ (and hence that $P(A) = 0$). If $(\Omega, \mathcal{F}, P)$ is complete, then the conditions $A \in \mathcal{F}$, $A \Delta A' \subset B \in \mathcal{F}$, and $P(B) = 0$ together imply that $A' \in \mathcal{F}$ and $P(A') = P(A)$.

Suppose that $(\Omega, \mathcal{F}, P)$ is an arbitrary probability space. Define P^* by (3.1) with $\mathcal{F}_0 = \mathcal{F} = \sigma(\mathcal{F}_0)$, and consider the σ-field $\mathcal{M}$ of P^*-measurable sets. The arguments leading to Theorem 3.1 show that P^* restricted to $\mathcal{M}$ is a probability measure. If $P^*(B) = 0$ and $A \subset B$, then $P^*(A \cap E) + P^*(A^c \cap E) \leq P^*(B) + P^*(E) = P^*(E)$ by monotonicity, so that A satisfies (3.5) and hence lies in $\mathcal{M}$. Thus $(\Omega, \mathcal{M}, P^*)$ is a complete probability measure space. *In any probability space it is therefore possible to enlarge the σ-field and extend the measure in such a way as to get a complete space.*

Lebesgue Measure on the Unit Interval

Consider once again the unit interval $(0, 1]$ together with the field $\mathcal{B}_0$ of finite disjoint unions of subintervals (Example 2.2) and the σ-field $\mathcal{B} = \sigma(\mathcal{B}_0)$ of Borel sets in $(0, 1]$. According to Theorem 2.2, $\lambda(a, b] = b - a$ defines a countably additive set function on the class of subintervals, and by the argument following it, (2.10) and (2.11) define λ as a probability measure on $\mathcal{B}_0$. By Theorem 3.1, λ extends to $\mathcal{B}$, the extended λ being *Lebesgue measure*. The probability space $((0, 1], \mathcal{B}, \lambda)$ will be the basis for the probability theory in the remaining sections of this chapter. A few geometric properties of λ will be considered here. Since the intervals in $(0, 1]$ form a π-system generating $\mathcal{B}$, λ is the only probability measure on $\mathcal{B}$ that assigns to each interval its length as its measure.

Suppose that $((0, 1], \mathcal{B}, \lambda)$ is completed, as above. The sets in the completed σ-field $\mathcal{M}$ are called the *Lebesgue* sets and λ extended to $\mathcal{M}$ is still called Lebesgue measure.

Some Borel sets are difficult to visualize:

Example 3.1. Let $\{r_1, r_2, \ldots\}$ be an enumeration of the rationals in $(0, 1)$. Suppose that ϵ is small and choose an open interval $I_n = (a_n, b_n)$ such that $r_n \in I_n \subset (0, 1)$ and $\lambda(I_n) = b_n - a_n < \epsilon 2^{-n}$. Put $A = \bigcup_{n=1}^{\infty} I_n$. By subadditivity, $0 < \lambda(A) < \epsilon$.

Since A contains all the rationals in $(0, 1)$, it is dense there. Thus A is an open, dense set with measure near 0. If I is an open subinterval of $(0, 1)$, then I must intersect one of the I_n, and therefore $\lambda(A \cap I) > 0$.

If $B = (0, 1) - A$, then $1 - \epsilon < \lambda(B) < 1$. The set B contains no interval and is in fact nowhere dense (every interval contains a subinterval free of points of B). Despite this, B has measure nearly 1.   ∎

Example 3.2. There is a set defined in probability terms that has geometric properties similar to those in the preceding example. As in Section 1, let $d_n(\omega)$ be the nth digit in the dyadic expansion of ω; see (1.6). Let $A_n = [\omega \in (0, 1]$: $d_i(\omega) = d_{n+i}(\omega) = d_{2n+i}(\omega), i = 1, \ldots, n]$, and let $A = \bigcup_{n=1}^{\infty} A_n$. Probabilistically, A corresponds to the event that in an infinite sequence of tosses of a coin, some finite initial segment is immediately duplicated twice over. Since $\lambda(A_n) = 2^n \cdot 2^{-3n}$, $0 < \lambda(A) \leq \sum_{n=1}^{\infty} 2^{-2n} = 1/3$. Again A is dense in the unit interval; its measure, less than $1/3$, could be made less than ϵ by requiring that some initial segment be immediately duplicated k times over with k large.   ∎

The outer measure (3.1) corresponding to λ on $\mathcal{B}_0$ is the infimum of the sums $\sum_n \lambda(A_n)$ for which $A_n \in \mathcal{B}_0$ and $A \subset \bigcup_n A_n$. Since each A_n is a finite disjoint union of intervals, this outer measure is

$$(3.8) \qquad \lambda^*(A) = \inf \sum_n (b_n - a_n),$$

where the infimum extends over coverings of A by intervals $(a_n, b_n]$. The notion of negligibility in Section 1 can therefore be reformulated: A is negligible if and only if $\lambda^*(A) = 0$. For A in $\mathcal{B}$, this is the same thing as $\lambda(A) = 0$.

Since replacing $(a_n, b_n]$ by $(a_n, b_n + \epsilon 2^{-n})$ increases the sum in (3.8) only by ϵ, the infimum is unchanged if it extends over coverings of A by *open* intervals (a_n, b_n). Given ϵ choose the covering so that $\sum_n (b_n - a_n) < \lambda^*(A) + \epsilon$ and put $G = \bigcup_n (a_n, b_n)$. The construction proves this: *For $A \in \mathcal{B}$ and $\epsilon > 0$, there exists an open set G such that $A \subset G$ and $\lambda(G) < \lambda(A) + \epsilon$.* (Example 3.1 is this construction applied to the set of rationals.) A Borel set can thus in the sense of Lebesgue measure be approximated from above by the open sets; this property is called *regularity*.

Nonmeasurable Sets

There do exist sets outside $\mathcal{B}$. To construct one it is convenient to use addition modulo

1 in $(0, 1]$. For $x, y \in (0, 1]$ take $x \oplus y$ to be $x + y$ or $x + y - 1$ according as $x + y$ lies in $(0, 1]$ or not.* Put $A \oplus x = [a \oplus x: a \in A]$.

Let $\mathcal{L}$ be the class of Borel sets A such that $A \oplus x$ is a Borel set and $\lambda(A \oplus x) = \lambda(A)$. Then $\mathcal{L}$ is a λ-system containing the intervals, and so $\mathcal{B} \subset \mathcal{L}$ by the π-λ theorem. Thus $A \in \mathcal{B}$ implies that $A \oplus x \in \mathcal{B}$ and $\lambda(A \oplus x) = \lambda(A)$.

Define x and y to be equivalent $(x \sim y)$ if $x \oplus r = y$ for some rational r in $(0, 1]$. Let H be a subset of $(0, 1]$ consisting of exactly one representative point from each equivalence class; such a set exists under the assumption of the axiom of choice.† Consider now the countably many sets $H \oplus r$ for rational r.

These sets are disjoint because no two distinct points of H are equivalent. (If $H \oplus r_1$ and $H \oplus r_2$ share the point $h_1 \oplus r_1 = h_2 \oplus r_2$, then $h_1 \sim h_2$; this is impossible unless $h_1 = h_2$, in which case $r_1 = r_2$.) Each point of $(0, 1]$ lies in one of these sets because H has a representative from each equivalence class. (If $x \sim h \in H$, then $x = h \oplus r \in H \oplus r$ for some rational r.) Thus $(0, 1] = \bigcup_r (H \oplus r)$, a countable disjoint union.

If H were in $\mathcal{B}$, it would follow that $\lambda(0, 1] = \Sigma_r \lambda(H \oplus r)$. This is impossible: If the value common to the $\lambda(H \oplus r)$ is 0, it leads to $1 = 0$; if the common value is positive, it leads to a convergent infinite series of identical positive terms ($a + a + \cdots < \infty$ and $a \geqslant 0$). Thus H lies outside $\mathcal{B}$.

The argument can be recast in a straightforward way to prove (together with the axiom of choice) this result: There exists on the class of all subsets of $(0, 1]$ no probability measure P such that $P(A \oplus x) = P(A)$ for all A and x.

Under the assumption of the continuum hypothesis in addition to the axiom of choice it is possible to prove the stronger result that there exists on the class of all subsets of $(0, 1]$ no probability measure P such that the P-value of each singleton is 0.‡

PROBLEMS

3.1. For the following examples, describe P^* as defined by (3.1) and $\mathcal{M} = \mathcal{M}(P^*)$ as defined by the requirement (3.4). Sort out the cases in which P^* fails to agree with P on $\mathcal{F}_0$ and explain why.

(a) Let $\mathcal{F}_0$ consist of the sets $0, \{1\}, \{2, 3\},$ and $\Omega = \{1, 2, 3\}$, and define P_1 by $P_1\{1\} = 0$ and P_2 by $P_2\{2, 3\} = 0$. Note that $\mathcal{M}(P_1^*)$ and $\mathcal{M}(P_2^*)$ differ.

(b) Suppose that Ω is countably infinite, let $\mathcal{F}_0$ be the field of finite and cofinite sets, and take $P(A)$ to be 0 or 1 as A is finite or cofinite.

(c) The same, but suppose that Ω is uncountable.

(d) The probability in Problem 2.14.

(e) Let $P(A) = I_A(\omega_0)$ for $A \in \mathcal{F}_0$—a unit mass at ω_0 (assume $\{\omega\} \in \sigma(\mathcal{F}_0)$).

3.2. 2.15↑   Show that the σ-field generated by the class of periodic sets in $\Omega = \{1, 2, \ldots\}$ contains all subsets of the space. Conclude (this was proved differently in Problem 2.15 (d)) that D is not countably additive on the field generated by the periodic sets.

3.3. Let P be a probability measure on a field $\mathcal{F}_0$ and for every subset A of Ω define

---

* This amounts to working in the circle group.

† See ROYDEN, p. 18. In terms of the circle group, the rationals are a subgroup and the equivalence classes are the cosets.

‡ See BIRKHOFF, p. 87.

$P^*(A)$ by (3.1). Denote also by P the extension (Theorem 3.1) of P to $\mathcal{F} = \sigma(\mathcal{F}_0)$.

(a)   Show that

(3.9) $P^*(A) = \inf [P(B): A \subset B, B \in \mathcal{F}]$

and (see (3.2))

(3.10) $P_*(A) = \sup [P(C): C \subset A, C \in \mathcal{F}]$,

and show that the infimum and supremum are achieved.

(b)   Show that A is P^*-measurable if and only if $P_*(A) = P^*(A)$.

(c)   The outer and inner measures associated with a probability measure P on a σ-field $\mathcal{F}$ are usually *defined* by (3.9) and (3.10). Show that (3.9) and (3.10) are the same as (3.1) and (3.2) with $\mathcal{F}$ in the role of $\mathcal{F}_0$.

3.4.   Let Ω be the unit square $[(x, y): 0 < x, y \leq 1]$, let $\mathcal{F}$ be the class of sets of the form $[(x, y): x \in A, 0 < y \leq 1]$, where $A \in \mathcal{B}$, and let P have value $\lambda(A)$ at this set. Show that $(\Omega, \mathcal{F}, P)$ is a probability measure space. Show for $A = [(x, y): 0 < x \leq 1, y = \frac{1}{2}]$ that $P_*(A) = 0$ and $P^*(A) = 1$.

3.5.   As shown in the text, a probability measure space $(\Omega, \mathcal{F}, P)$ has a complete extension—that is, a complete probability measure space $(\Omega, \mathcal{F}_1, P_1)$ such that $\mathcal{F} \subset \mathcal{F}_1$ and P_1 agrees with P on $\mathcal{F}$.

(a)   Suppose that $(\Omega, \mathcal{F}_2, P_2)$ is a second complete extension. Show by an example in a space of two points that P_1 and P_2 need not agree on the σ-field $\mathcal{F}_1 \cap \mathcal{F}_2$.

(b)   There is, however, a unique minimal complete extension: Let $\mathcal{F}^+$ consist of the sets A for which there exist $\mathcal{F}$-sets B and C such that $A \triangle B \subset C$ and $P(C) = 0$. Show that $\mathcal{F}^+$ is a σ-field. For such a set A define $P^+(A) = P(B)$. Show that the definition is consistent, that P^+ is a probability measure on $\mathcal{F}^+$, and that $(\Omega, \mathcal{F}^+, P^+)$ is complete. Show that, if $(\Omega, \mathcal{F}_1, P_1)$ is any complete extension of $(\Omega, \mathcal{F}, P)$, then $\mathcal{F}^+ \subset \mathcal{F}_1$ and P_1 agrees with P^+ on $\mathcal{F}^+$; $(\Omega, \mathcal{F}^+, P^+)$ is the *completion* of $(\Omega, \mathcal{F}, P)$.

(c)   Show that $A \in \mathcal{F}^+$ if and only if $P_*(A) = P^*(A)$, where P_* and P^* are defined by (3.9) and (3.10), and that $P^+(A) = P_*(A) = P^*(A)$ in this case. Thus the complete extension constructed in the text is exactly *the* completion.

3.6.   Suppose that $(\Omega, \mathcal{F}, P)$ is a probability measure space and define P^* by (3.9). Suppose that $\Omega_0 \subset \Omega$ and let $\mathcal{F}_0 = [A \cap \Omega_0: A \in \mathcal{F}]$. Suppose finally that $P^*(\Omega_0) > 0$ and define P_0 on $\mathcal{F}_0$ by setting $P_0(A \cap \Omega_0) = P^*(A \cap \Omega_0)/P^*(\Omega_0)$ for $A \in \mathcal{F}$.

(a)   Show that $\mathcal{F}_0$ is a σ-field in Ω_0 and that P_0 is a probability measure on $\mathcal{F}_0$.

(b)   Let P_0^* be the outer measure in Ω_0 generated by P_0. Show that $P_0^*(A) = P^*(A)/P^*(\Omega_0)$ for $A \subset \Omega_0$.

3.7.   2.10   2.19↑   Show that the metric space in Problem 2.19 is separable if the σ-field $\mathcal{F}$ is separable in the sense of Problem 2.10.

3.8.   (a)   Show that a λ-system contains the empty set and is closed under complementation and under the formation of countable disjoint unions and countable decreasing intersections.

(b) Show that a λ-system need not be a field.

(c) Show that, if $\mathcal{L}$ satisfies (λ_1) and (λ_2) and is closed under the formation of countable disjoint unions, then $\mathcal{L}$ is a λ-system.

3.9. 2.5 3.8↑ Deduce the π-λ theorem from the monotone class theorem by showing directly that, if a λ-system $\mathcal{L}$ contains a π-system $\mathcal{P}$, then $\mathcal{L}$ also contains the field generated by $\mathcal{P}$.

3.10. Show that a λ-system is a monotone class. The large class in the monotone class theorem has less structure that the large class in the π-λ theorem, whereas the reverse is true of the small classes. Show that it follows from each of these two theorems that, if a field $\mathcal{F}_0$ is contained in a λ-system $\mathcal{L}$, then $\sigma(\mathcal{F}_0) \subset \mathcal{L}$. Show that each of these theorems would follow from a theorem asserting that, if a π-system $\mathcal{P}$ is contained in a monotone class $\mathcal{M}$, then $\sigma(\mathcal{P}) \subset \mathcal{M}$. Show, however, that there is no such theorem.

3.11. 2.5↑ **(a)** Suppose that $\mathcal{F}_0$ is a field and P_1 and P_2 are probability measures on $\sigma(\mathcal{F}_0)$. Show by the monotone class theorem that if P_1 and P_2 agree on $\mathcal{F}_0$, then they agree on $\sigma(\mathcal{F}_0)$.

(b) Let $\mathcal{F}_0$ be the smallest field over the π-system $\mathcal{P}$. Show by the inclusion-exclusion formula that probability measures agreeing on $\mathcal{P}$ must agree also on $\mathcal{F}_0$. Now deduce Theorem 3.3 from part (a).

3.12. Prove the π-λ theorem by establishing the following statements in sequence. Here $l(\mathcal{P})$ is the smallest λ-system over $\mathcal{P}$, $\mathcal{G}_A = [B: A \cap B \in l(\mathcal{P})]$, and the arrows denote implication.

1. $A \in l(\mathcal{P}) \to \mathcal{G}_A$ is a λ-system.
2. $A \in \mathcal{P} \,\&\, B \in \mathcal{P} \to A \cap B \in \mathcal{P} \subset l(\mathcal{P}) \to B \in \mathcal{G}_A$.
3. $A \in \mathcal{P} \to \mathcal{P} \subset \mathcal{G}_A \to l(\mathcal{P}) \subset \mathcal{G}_A$.
4. $A \in \mathcal{P} \,\&\, B \in l(\mathcal{P}) \to B \in \mathcal{G}_A \to A \cap B \in l(\mathcal{P})$.
5. $B \in l(\mathcal{P}) \,\&\, A \in \mathcal{P} \to B \cap A \in l(\mathcal{P}) \to A \in \mathcal{G}_B$.
6. $B \in l(\mathcal{P}) \to \mathcal{P} \subset \mathcal{G}_B \to l(\mathcal{P}) \subset \mathcal{G}_B$.
7. $B \in l(\mathcal{P}) \,\&\, A \in l(\mathcal{P}) \to A \in \mathcal{G}_B \to B \cap A \in l(\mathcal{P})$.
8. $l(\mathcal{P})$ is a π-system.

Lay out the proof of the monotone class theorem the same way.

3.13. Show that it is impossible in Theorem 3.3 to drop the assumption that $\mathcal{P}$ is a π-system.

3.14. 2.21↑ Prove the existence of a Lebesgue set that is not a Borel set.

3.15. The *outer content* of a set A in $(0, 1]$ is $c^*(A) = \inf \Sigma_n (b_n - a_n)$, where the infimum extends over *finite* coverings of A by intervals $(a_n, b_n]$. Thus A is trifling in the sense of Problem 1.4 if and only if $c^*(A) = 0$. Define *inner content* by $c_*(A) = 1 - c^*(A^c)$. Show that $c_*(A) \le c^*(A)$; if the two are equal, their common value is taken as the *content* $c(A)$ of A, which is then *Jordan measurable*. A trifling set for example has this property. For the set N of normal numbers, $0 = c_*(N) < c^*(N) = 1$, even though $\lambda_*(N) = \lambda^*(N) = 1$. Show that $c_*(A) \le \lambda_*(A) \le \lambda^*(A) \le c^*(A)$.

Show that $c_*(A) = \sup \Sigma_n (b_n - a_n)$, the supremum extending over finite disjoint unions of intervals $(a_n, b_n]$ contained in A. Of course, the analogue for λ_* fails.

3.16.   1.6↑   Deduce directly by countable additivity that the Cantor set has Lebesgue measure 0.

3.17.   From the fact that $\lambda(x \oplus A) = \lambda(A)$, deduce that sums and differences of normal numbers may be nonnormal.

3.18.   Let H be the nonmeasurable set constructed at the end of the section.
   (a)   Show that, if A is a Borel set and $A \subset H$, then $\lambda(A) = 0$—that is, $\lambda_*(H) = 0$.
   (b)   Show that, if $\lambda^*(E) > 0$, then E contains a nonmeasurable subset.

SECTION 4.   DENUMERABLE PROBABILITIES

Complex probability ideas can be made clear by the systematic use of measure theory, and probabilistic ideas of extramathematical origin, such as independence, can illuminate problems of purely mathematical interest. It is to this reciprocal exchange that measure-theoretic probability owes much of its interest.

The results of this section concern infinite sequences of events in a probability space.* They will be illustrated by examples in the *unit interval*. By this will always be meant the triple $(\Omega, \mathcal{F}, P)$ for which Ω is $(0, 1]$, $\mathcal{F}$ is the σ-field $\mathcal{B}$ of Borel sets there, and $P(A)$ is for A in $\mathcal{F}$ the Lebesgue measure $\lambda(A)$ of A. This is the space appropriate to the problems of Section 1, which will be pursued further. The definitions and theorems, as opposed to the examples, apply to all probability spaces. The unit interval will appear again and again in this chapter, and it is essential to keep in mind that there are many other important spaces to which the general theory will be applied later.

General Formulas

Formulas (2.4) through (2.8) will be used repeatedly. The sets involved in such formulas lie in the basic σ-field $\mathcal{F}$ by hypothesis. Any probability argument starts from given sets tacitly assumed to lie in $\mathcal{F}$; further sets constructed in the course of the argument must be shown to lie in $\mathcal{F}$ as well, but this is usually quite obvious.

If $P(A) > 0$, the *conditional probability* of B given A is defined in the usual way as

(4.1)
$$P(B|A) = \frac{P(A \cap B)}{P(A)}.$$

---

* They come under what Borel in his first paper on the subject (see the footnote on page 9) called *probabilités dénombrables;* hence the section heading.

There are the chain-rule formulas

$$(4.2) \quad \begin{cases} P(A \cap B) = P(A)P(B|A), \\ P(A \cap B \cap C) = P(A)P(B|A)P(C|A \cap B), \end{cases}$$

. . . .

If $A_1, A_2, \ldots$ partition Ω, then

$$(4.3) \qquad\qquad P(B) = \sum_n P(A_n)P(B|A_n).$$

Note that for fixed A the function $P(B|A)$ defines a probability measure as B varies over $\mathcal{F}$.

If $A_1, A_2, \ldots$ are sets of probability 0, then by subadditivity, so is $\bigcup_n A_n$. Passing to complements shows that, if $A_1, A_2, \ldots$ are sets of probability 1, then so is $\bigcap_n A_n$.

Limit Sets

For a sequence $A_1, A_2, \ldots$ of sets, define a set

$$(4.4) \qquad\qquad \limsup_n A_n = \bigcap_{n=1}^{\infty} \bigcup_{k=n}^{\infty} A_k$$

and a set

$$(4.5) \qquad\qquad \liminf_n A_n = \bigcup_{n=1}^{\infty} \bigcap_{k=n}^{\infty} A_k.$$

These sets* are the *limits superior and inferior* of the sequence $\{A_n\}$. They lie in $\mathcal{F}$ if all the A_n do. Now ω lies in (4.4) if and only if for each n there is some $k \geq n$ for which $\omega \in A_k$; in other words, ω lies in (4.4) if and only if it lies in *infinitely many* of the A_n. In the same way, ω lies in (4.5) if and only if it lies in *all but finitely many* of the A_n.

Of course, (4.4) contains (4.5). If there is inclusion in the other direction as well, write

$$(4.6) \qquad\qquad \lim_n A_n = \liminf_n A_n = \limsup_n A_n.$$

To say that A_n has limit A, written $A_n \to A$, means that the common value (4.6) exists and coincides with A. Since $\liminf_n A_n \subset \limsup_n A_n$ always holds, to check whether $A_n \to A$ is to check whether $\limsup_n A_n \subset A \subset \liminf_n A_n$. From $A_n \in \mathcal{F}$ and $A_n \to A$ follows $A \in \mathcal{F}$.

---

* See Problems 4.2 and 4.3 for the parallel between set-theoretic and numerical limits superior and inferior.

Example 4.1. Consider the functions $d_n(\omega)$ defined on the unit interval by the dyadic expansion (1.6), and let $l_n(\omega)$ be the length of the run of 0's starting at $d_n(\omega)$: $l_n(\omega) = k$ if $d_n(\omega) = \cdots = d_{n+k-1}(\omega) = 0$ and $d_{n+k}(\omega) = 1$; here $l_n(\omega) = 0$ if $d_n(\omega) = 1$. Probabilities can be computed by (1.9). Since $[\omega: l_n(\omega) = k]$ is a union of 2^{n-1} disjoint intervals of length 2^{-n-k}, it lies in $\mathcal{F}$ and has probability 2^{-k-1}. Therefore, $[\omega: l_n(\omega) \geq r] = [\omega: d_i(\omega) = 0, n \leq i < n + r]$ lies also in $\mathcal{F}$ and has probability $\sum_{k \geq r} 2^{-k-1}$:

$$(4.7) \qquad P[\omega: l_n(\omega) \geq r] = 2^{-r}.$$

If A_n is the event in (4.7), then (4.4) is the set of ω such that $l_n(\omega) \geq r$ for infinitely many n, or, n being regarded as a time index, such that $l_n(\omega) \geq r$ *infinitely often*. ∎

Because of the theory of Sections 2 and 3, statements like (4.7) are valid in the sense of ordinary mathematics, and using the traditional language of probability—"heads," "runs," and so on—does not change this.

When n has the role of time, (4.4) is frequently written

$$(4.8) \qquad \limsup_n A_n = [A_n \text{ i.o.}],$$

where "i.o." stands for "infinitely often." Similarly, (4.5) could be written

$$(4.9) \qquad \liminf_n A_n = [A_n \text{ a.a.}],$$

with "a.a." abbreviating "almost always," that is, "for all but finitely many values of n."

Theorem 4.1. (i) *For each sequence* $\{A_n\}$,

$$(4.10) \qquad P(\liminf_n A_n) \leq \liminf_n P(A_n)$$

$$\leq \limsup_n P(A_n) \leq P(\limsup_n A_n).$$

(ii) *If* $A_n \to A$, *then* $P(A_n) \to P(A)$.

PROOF. Clearly (ii) follows from (i). As for (i), if $B_n = \bigcap_{k=n}^{\infty} A_k$ and $C_n = \bigcup_{k=n}^{\infty} A_k$, then $B_n \uparrow \liminf_n A_n$ and $C_n \downarrow \limsup A_n$, so that, by parts (i) and (ii) of Theorem 2.1, $P(A_n) \geq P(B_n) \to P(\liminf_n A_n)$ and $P(A_n) \leq P(C_n) \to P(\limsup_n A_n)$. ∎

Example 4.2. Define $l_n(\omega)$ as in Example 4.1 and let $A_n = [\omega: l_n(\omega) \geq r]$ for fixed r. By (4.7) and (4.10), $P[\omega: l_n(\omega) \geq r \text{ i.o.}] \geq 2^{-r}$. Much stronger results will be proved later. ∎

Independent Events

Events A and B are *independent* if $P(A \cap B) = P(A)P(B)$. For events of positive probability, this is the same thing as requiring $P(B|A) = P(B)$ or $P(A|B) = P(A)$. More generally, a finite collection $A_1, \ldots, A_n$ of events is independent if

$$(4.11) \qquad P(A_{k_1} \cap \cdots \cap A_{k_j}) = P(A_{k_1}) \cdots P(A_{k_j})$$

for each set $k_1, \ldots, k_j$ of distinct indices between 1 and n. An infinite (perhaps uncountable) collection of events is independent if each of its finite subcollections is independent.

Example 4.3. On the unit interval the events $H_n = [\omega: d_n(\omega) = 0]$, $n = 1$, $2, \ldots$, are independent, the two sides of (4.11) having in this case value 2^{-j}. It seems intuitively clear that from this should follow the independence, for example, of $[\omega: d_2(\omega) = 0] = H_2$ and $[\omega: d_1(\omega) = 0, d_3(\omega) = 1] = H_1 \cap H_3^c$, since the two events involve disjoint sets of times. Further, any sets A and B depending, respectively, say, only on even and on odd times (like $[\omega: d_{2n}(\omega) = 0 \text{ i.o.}]$ and $[\omega: d_{2n+1}(\omega) = 1 \text{ i.o.}]$) ought also to be independent. This raises the general question of what it means for A to depend only on even times. Intuitively, it requires that knowing which ones among $H_2, H_4, \ldots$ occurred entails knowing whether or not A occurred—that is, it requires that the sets $H_2, H_4, \ldots$ "determine" A. The set-theoretic form of this requirement is that A is to lie in the σ-field generated by $H_2, H_4, \ldots$. From $A \in \sigma(H_2, H_4, \ldots)$ and $B \in \sigma(H_1, H_3, \ldots)$ it ought to be possible to deduce the independence of A and B. ∎

The next theorem and its corollaries make such deductions possible. Define *classes* $\mathcal{A}_1, \ldots, \mathcal{A}_n$ in $\mathcal{F}$ to be independent if for each choice of A_i from $\mathcal{A}_i$, $i = 1, \ldots; n$, the events $A_1, \ldots, A_n$ are independent. By the definition (4.11) this is exactly the same thing as requiring that $A_1, \ldots, A_n$ must satisfy $P(A_1 \cap A_2 \cap \cdots \cap A_n) = P(A_1)P(A_2) \cdots P(A_n)$ whenever for each i, $1 \leq i \leq n$, either $A_i \in \mathcal{A}_i$ or else $A_i = \Omega$; every A_i satisfying $A_i = \Omega$ can be suppressed on each side of the equation.

Theorem 4.2. *If $\mathcal{A}_1, \ldots, \mathcal{A}_n$ are independent and each $\mathcal{A}_i$ is a π-system, then $\sigma(\mathcal{A}_1), \ldots, \sigma(\mathcal{A}_n)$ are independent.*

PROOF. Fix sets $A_2, \ldots, A_n$ such that for $2 \leq i \leq n$ either $A_i \in \mathcal{A}_i$ or $A_i =$

Ω, and let $\mathcal{L}$ be the class of A in $\mathcal{F}$ for which $P(A \cap A_2 \cap \cdots \cap A_n) = P(A)$ $P(A_2) \cdots P(A_n)$. Then $\mathcal{L}$ is a λ-system containing the π-system $\mathcal{A}_1$ and hence (Theorem 3.2) containing $\sigma(\mathcal{A}_1)$. Therefore, $\sigma(\mathcal{A}_1)$, $\mathcal{A}_2$, $\ldots$, $\mathcal{A}_n$ are independent. Clearly, the argument applies if 1 is replaced by any of the indices 2, $\ldots$, n.

From the independence of $\sigma(\mathcal{A}_1)$, $\mathcal{A}_2$, $\ldots$, $\mathcal{A}_n$ now follows that of $\sigma(\mathcal{A}_1)$, $\sigma(\mathcal{A}_2)$, $\mathcal{A}_3$, $\ldots$, $\mathcal{A}_n$, and so on.   ■

If $\mathcal{A} = \{A_1, \ldots, A_k\}$ is finite, then each A in $\sigma(\mathcal{A})$ can be expressed by a "formula" such as $A = A_2 \cap A_5^c$ or $A = (A_2 \cap A_7) \cup (A_3 \cap A_7^c \cap A_8)$. If $\mathcal{A}$ is infinite, the sets in $\sigma(\mathcal{A})$ may be very complicated; the way to make precise the idea that the elements of $\mathcal{A}$ "determine" A is not to require formulas, but simply to require that A lie in $\sigma(\mathcal{A})$.

Independence for an infinite collection of classes can be defined just as in the finite case: $[\mathcal{A}_\theta : \theta \in \Theta]$ is independent if the collection $[A_\theta : \theta \in \Theta]$ of sets is independent for each choice of A_θ from $\mathcal{A}_\theta$. This is equivalent to the independence of each finite subcollection $\mathcal{A}_{\theta_1}, \ldots, \mathcal{A}_{\theta_n}$ of classes because of the way independence for infinite classes of sets is defined in terms of independence for finite classes. Hence Theorem 4.2 has an immediate consequence:

Corollary 1. *If $\mathcal{A}_\theta$, $\theta \in \Theta$, are independent and each $\mathcal{A}_\theta$ is a π-system, then $\sigma(\mathcal{A}_\theta)$, $\theta \in \Theta$, are independent.*

Corollary 2. *Suppose that the array*

(4.12)
$$A_{11}, A_{12}, \ldots$$
$$A_{21}, A_{22}, \ldots$$
$$\cdots \cdots \cdots$$

of events is independent; here each row is a finite or infinite sequence, and there are finitely or infinitely many rows. If $\mathcal{F}_i$ is the σ-field generated by the ith row, then $\mathcal{F}_1, \mathcal{F}_2, \ldots$ are independent.

PROOF.   If $\mathcal{A}_i$ is the class of finite intersections of elements of the ith row, then $\mathcal{A}_i$ is a π-system and $\sigma(\mathcal{A}_i) = \mathcal{F}_i$. For $1 \le i \le n$ let $B_i = \bigcap_{j=1}^{m_i} A_{i,k_i(j)}$ be an element of $\mathcal{A}_i$. Then $P(\bigcap_i B_i) = P(\bigcap_i \bigcap_j A_{i,k_i(j)}) = \Pi_i \Pi_j P(A_{i,k_i(j)}) = \Pi_i P(\bigcap_j A_{i,k_i(j)}) = \Pi_i P(B_i)$. It follows that the classes $\mathcal{A}_1, \mathcal{A}_2, \ldots$ are independent, so that Corollary 1 applies.   ■

Corollary 2 implies the independence of the events discussed in Example 4.3. The array (4.12) in this case has two rows:

$$H_2, H_4, H_6, \ldots$$
$$H_1, H_3, H_5, \ldots$$

Theorem 4.2 also implies, for example, that for independent $A_1, \ldots, A_n$,

(4.13)
$$P(A_1^c \cap \cdots \cap A_k^c \cap A_{k+1} \cap \cdots \cap A_n)$$
$$= P(A_1^c) \cdots P(A_k^c) P(A_{k+1}) \cdots P(A_n).$$

To prove this, let $\mathcal{A}_i$ consist of A_i alone; of course, $A_i^c \in \sigma(\mathcal{A}_i)$. In (4.13) any subcollection of the A_i could be replaced by their complements.

Subfields

Theorem 4.2 involves a number of σ-fields at once, which is characteristic of probability theory; measure theory not directed toward probability usually involves only one all-embracing σ-field $\mathcal{F}$. In probability, σ-fields in $\mathcal{F}$—that is, sub-σ-fields of $\mathcal{F}$—play an important role. To understand their function it helps to have an informal, intuitive way of looking at them.

A subclass $\mathcal{A}$ of $\mathcal{F}$ corresponds heuristically to *partial information.* Imagine that a point ω is drawn from Ω according to the probabilities given by P: ω lies in A with probability $P(A)$. Imagine also an observer who does not know which ω it is that has been drawn but who does know for each A in $\mathcal{A}$ whether $\omega \in A$ or $\omega \notin A$. Identifying this partial information with the class $\mathcal{A}$ itself will illuminate the connection between various measure-theoretic concepts and the premathematical ideas lying behind them.

The set B is by definition independent of the class $\mathcal{A}$ if $P(B|A) = P(B)$ for all sets A in $\mathcal{A}$ for which $P(A) > 0$. Thus if B is independent of $\mathcal{A}$, then the observer's probability for B is $P(B)$ even after he has received the information in $\mathcal{A}$; in this case $\mathcal{A}$ contains no information about B. The point of Theorem 4.2 is that this remains true even if the observer is given the information in $\sigma(\mathcal{A})$, provided that $\mathcal{A}$ is a π-system. It is to be stressed that here *information,* like *observer* and *know,* is an informal, extramathematical term (in particular, it is not information in the technical sense of entropy).

The notion of partial information can be looked at in terms of partitions. Say that points ω and ω' are $\mathcal{A}$-*equivalent* if, for every A in $\mathcal{A}$, ω and ω' lie either both in A or both in A^c—that is, if

(4.14)
$$I_A(\omega) = I_A(\omega'), \qquad A \in \mathcal{A}.$$

This relation partitions Ω into sets of equivalent points; call this the $\mathcal{A}$-*partition.*

Example 4.4. If ω and ω' are $\sigma(\mathcal{A})$-equivalent, then certainly they are $\mathcal{A}$-equivalent. For fixed ω and ω' the class of A such that $I_A(\omega) = I_A(\omega')$ is a σ-field (check that it is closed under the formation of complements, finite intersections, and countable increasing unions); if ω and ω' are $\mathcal{A}$-equivalent, then this σ-field contains $\mathcal{A}$ and hence $\sigma(\mathcal{A})$, so that ω and ω' are also $\sigma(\mathcal{A})$-equivalent. Thus $\mathcal{A}$-equivalence and $\sigma(\mathcal{A})$-equivalence are the same thing, and the $\mathcal{A}$-partition coincides with the $\sigma(\mathcal{A})$-partition. ∎

An observer with the information in $\sigma(\mathcal{A})$ knows, not the point ω drawn, but the equivalence class containing it.* That is the information he has.

Example 4.5. If $H_n = [\omega: d_n(\omega) = 0]$ as in Example 4.3, and if $\mathcal{A} = \{H_1, H_3, H_5, \ldots\}$, then ω and ω' satisfy (4.14) if and only if $d_n(\omega) = d_n(\omega')$ for all odd n. The information in $\sigma(\mathcal{A})$ is thus the set of values of $d_n(\omega)$ for n odd. ∎

The following example points up the informal nature of this interpretation of σ-fields as information.

Example 4.6. In the unit interval $(\Omega, \mathcal{F}, P)$ let $\mathcal{G}$ be the σ-field consisting of the countable and the co-countable sets. Since $P(G)$ is 0 or 1 for each G in $\mathcal{G}$, each set H in $\mathcal{F}$ is independent of $\mathcal{G}$. But in this case the $\mathcal{G}$-partition consists of the singletons, and so the information in $\mathcal{G}$ tells the observer exactly which ω in Ω has been drawn. In the sense that it is independent of H, the σ-field $\mathcal{G}$ contains no information at all about H; at the same time, in the sense that it tells the observer exactly which ω was drawn, the σ-field $\mathcal{G}$ contains all the information about H. ∎

The source of the difficulty or apparent paradox in this example lies in the unnatural structure of the σ-field $\mathcal{G}$ rather than in any deficiency in the notion of independence.† The heuristic equating of σ-fields and information is helpful even though it sometimes breaks down, and of course proofs are indifferent to whatever illusions and vagaries brought them into existence.

The Borel-Cantelli Lemmas

This is *the first Borel-Cantelli lemma*:

Theorem 4.3. *If* $\Sigma_n P(A_n)$ *converges, then* $P(\limsup_n A_n) = 0$.

PROOF. From $\limsup_n A_n \subset \bigcup_{k=m}^{\infty} A_k$ follows $P(\limsup_n A_n) \le P(\bigcup_{k=m}^{\infty} A_k) \le \Sigma_{k=m}^{\infty} P(A_k)$, and this sum tends to 0 as $m \to \infty$ if $\Sigma_n P(A_n)$ converges. ∎

By Theorem 4.1, $P(A_n) \to 0$ implies that $P(\liminf_n A_n) = 0$; in Theorem 4.3 hypothesis and conclusion are both stronger.

Example 4.7. For the run length $l_n(\omega)$ of Example 4.1, one can ask for conditions on a positive sequence $\{r_n\}$ under which

$$(4.15) \qquad P[\omega: l_n(\omega) \ge r_n \text{ i.o.}] = 0.$$

By (4.7), $P[\omega: l_n(\omega) \ge r_n] = 2^{-s_n}$, where s_n is r_n rounded upward to the next

---

* But see Problem 4.11.
† See Problem 4.13 for a more extreme example.

integer, and (4.15) will follow by the first Borel-Cantelli lemma if $\Sigma\, 2^{-s_n} < \infty$. Since $2^{-s_n} \le 2^{-r_n}$, $\Sigma\, 2^{-r_n} < \infty$ implies (4.15).

If $r_n = (1 + \epsilon)\log_2 n$ and ϵ is positive, there is convergence because $2^{-r_n} = 1/n^{1+\epsilon}$. Thus

$$(4.16) \qquad P[\omega: l_n(\omega) \ge (1 + \epsilon)\log_2 n \text{ i.o.}] = 0.$$

The limit superior of the ratio $l_n(\omega)/\log_2 n$ exceeds 1 if and only if ω belongs to the set in (4.16) for some positive rational ϵ. Since the union of this countable class of sets has probability 0,

$$(4.17) \qquad P[\omega: \limsup_n \frac{l_n(\omega)}{\log_2 n} > 1] = 0.$$

∎

In this example, whether the inequality $\limsup_n l_n(\omega)/\log_2 n \le 1$ holds or not is a property of ω, and the property in fact holds for ω in a set of probability 1. In such a case the property is said to hold *with probability 1*, or *almost surely*. In nonprobabilistic contexts, a property that holds for ω outside a set of measure 0 holds *almost everywhere,* or for *almost all* ω.

Example 4.8.   The preceding example has an interesting arithmetic consequence. Truncating the dyadic expansion at n gives the standard $(n - 1)$-place approximation $\sum_{k=1}^{n-1} d_k(\omega)2^{-k}$ to ω; the error is between 0 and 2^{-n+1}, and the error relative to the maximum is

$$(4.18) \qquad e_n(\omega) = \frac{\omega - \sum\limits_{k=1}^{n-1} d_k(\omega)2^{-k}}{2^{-n+1}} = \sum_{i=1}^{\infty} d_{n+i-1}(\omega)2^{-i},$$

which lies between 0 and 1. The binary expansion of $e_n(\omega)$ begins with $l_n(\omega)$ 0's, and then comes a 1. Hence $.0\dots01 \le e_n(\omega) \le .0\dots0111\dots$, where there are $l_n(\omega)$ 0's in the extreme terms. Therefore,

$$(4.19) \qquad \frac{1}{2^{l_n(\omega)+1}} \le e_n(\omega) \le \frac{1}{2^{l_n(\omega)}},$$

so that results on run length give information about the error of approximation.

By the left-hand inequality in (4.19), $e_n(\omega) \le x_n$ (assume that $0 < x_n \le 1$) implies that $l_n(\omega) \ge -\log_2 x_n - 1$; since $\Sigma 2^{-r_n} < \infty$ implies (4.15), $\Sigma x_n < \infty$ implies $P[\omega: e_n(\omega) \le x_n \text{ i.o.}] = 0$. (Clearly, $[\omega: e_n(\omega) \le x]$ is a Borel set.) In particular,

$$(4.20) \qquad P[\omega: e_n(\omega) \le 1/n^{1+\epsilon} \text{ i.o.}] = 0.$$

This probability refers to a point ω drawn at random from the unit interval.   ∎

Example 4.9. The final step in the proof of the normal number theorem was a disguised application of the first Borel-Cantelli lemma. If $A_n = [\omega: |n^{-1}s_n(\omega)| \geq n^{-1/8}]$, then $\Sigma P(A_n) < \infty$, as follows by (1.28), and so $P[A_n \text{ i.o.}] = 0$. But for ω in the set complementary to $[A_n \text{ i.o.}]$, $n^{-1}s_n(\omega) \to 0$.   ∎

This is *the second Borel-Cantelli lemma*:

Theorem 4.4. *If $\{A_n\}$ is an independent sequence of events and $\Sigma_n P(A_n)$ diverges, then $P(\limsup_n A_n) = 1$.*

PROOF.   It is enough to prove that $P(\bigcup_{n=1}^{\infty} \bigcap_{k=n}^{\infty} A_k^c) = 0$ and hence enough to prove that $P(\bigcap_{k=n}^{\infty} A_k^c) = 0$ for all n. Since $1 - x \leq e^{-x}$,

$$P\left(\bigcap_{k=n}^{n+j} A_k^c\right) = \prod_{k=n}^{n+j} (1 - P(A_k)) \leq \prod_{k=n}^{n+j} \exp [-P(A_k)].$$

Since $\Sigma_k P(A_k)$ diverges, the last expression tends to 0 as $j \to \infty$, and hence $P(\bigcap_{k=n}^{\infty} A_k^c) = \lim_j P(\bigcap_{k=n}^{n+j} A_k^c) = 0$.   ∎

By Theorem 4.1, $\limsup_n P(A_n) > 0$ implies that $P(\limsup_n A_n) > 0$; in Theorem 4.4 the hypothesis $\Sigma_n P(A_n) = \infty$ is weaker but the conclusion is stronger because of the additional hypothesis of independence.

Example 4.10.   Since the events $[\omega: l_n(\omega) = 0] = [\omega: d_n(\omega) = 1]$, $n = 1, 2, \ldots$, are independent and have probability $\frac{1}{2}$, $P[\omega: l_n(\omega) = 0 \text{ i.o.}] = 1$.
Since the events $A_n = [\omega: l_n(\omega) = 1] = [\omega: d_n(\omega) = 0, d_{n+1}(\omega) = 1]$, $n = 1, 2, \ldots$, are not independent, this argument is insufficient to prove that

(4.21) $P[\omega: l_n(\omega) = 1 \text{ i.o.}] = 1$.

But the events $A_2, A_4, A_6, \ldots$ are independent (Theorem 4.2) and their probabilities form a divergent series, and so $P[\omega: l_{2n}\omega) = 1 \text{ i.o.}] = 1$, which implies (4.21).   ∎

Significant applications of the second Borel-Cantelli lemma usually require some device of this kind to get around problems of dependence.

Example 4.11.   There is a complement to (4.16). First, if r_n is nondecreasing and $\Sigma 2^{-r_n}/r_n$ diverges, then

(4.22) $P[\omega: l_n(\omega) \geq r_n \text{ i.o.}] = 1$.

To prove this, note first that if r_n is rounded up to the next integer, then $\Sigma\, 2^{-r_n}/r_n$ certainly still diverges and (4.22) is unchanged. Assume then that $r_n = r(n)$ is integral and define $\{n_k\}$ inductively by $n_1 = 1$ and $n_{k+1} = n_k + r_{n_k}$, $k \geq 1$. Let $A_k = [\omega: l_{n_k}(\omega) \geq r_{n_k}] = [\omega: d_i(\omega) = 0,\, n_k \leq i < n_{k+1}]$; since the A_k involve nonoverlapping sequences of time indices, it follows by Corollary 2 to Theorem 4.2 that $A_1, A_2, \ldots$ are independent. By the second Borel-Cantelli lemma, $P[A_k$ i.o.$] = 1$ if $\Sigma_k P(A_k) = \Sigma_k\, 2^{-r(n_k)}$ diverges. But since r_n is nondecreasing,

$$\sum_{k \geq 1} 2^{-r(n_k)} = \sum_{k \geq 1} 2^{-r(n_k)} r^{-1}(n_k)(n_{k+1} - n_k)$$

$$\geq \sum_{k \geq 1} \sum_{n_k \leq n < n_{k+1}} 2^{-r_n} r_n^{-1} = \sum_{n \geq 1} 2^{-r_n} r_n^{-1}.$$

Thus the divergence of $\Sigma_n\, 2^{-r_n} r_n^{-1}$ implies that of $\Sigma_k\, 2^{-r(n_k)}$, and it follows that, with probability 1, $l_{n_k}(\omega) \geq r_{n_k}$ for infinitely many values of k. But this is stronger than (4.22).

The result in Example 4.2 follows if $r_n \equiv r$, but this is trivial. If $r_n = \log_2 n$, then $\Sigma\, 2^{-r_n}/r_n = \Sigma\, 1/(n \log_2 n)$ diverges, and therefore

(4.23) $$P[\omega: l_n(\omega) \geq \log_2 n \text{ i.o.}] = 1.$$

By (4.23) and (4.17),

(4.24) $$P[\omega: \limsup_n \frac{l_n(\omega)}{\log_2 n} = 1] = 1.$$

Thus for ω in a set of probability 1, $\log_2 n$ as a function of n is a kind of "upper envelope" for the function $l_n(\omega)$. ∎

Example 4.12.   By the right-hand inequality in (4.19), if $l_n(\omega) \geq \log_2 n$, then $e_n(\omega) \leq 1/n$. Hence (4.23) gives

(4.25) $$P[\omega: e_n(\omega) \leq \frac{1}{n} \text{ i.o.}] = 1.$$

This and (4.20) show that, with probability 1, $e_n(\omega)$ has $1/n$ as a "lower envelope." The discrepancy between ω and its $(n-1)$-place approximation $\Sigma_{k=1}^{n-1}$ $d_k(\omega)/2^k$ will fall infinitely often below $(n2^{n-1})^{-1}$ but not infinitely often below $(n^{1+\epsilon}2^{n-1})^{-1}$. ∎

The Zero-One Law

For a sequence $A_1, A_2, \ldots$ of events in a probability space $(\Omega, \mathcal{F}, P)$ consider the σ-fields $\sigma(A_n, A_{n+1}, \ldots)$ and their intersection

(4.26) $$\mathcal{T} = \bigcap_{n=1}^{\infty} \sigma(A_n, A_{n+1}, \ldots).$$

This is the *tail σ-field* associated with the sequence $\{A_n\}$, and its elements are called *tail events*. The idea is that a tail event is determined solely by the A_n for arbitrarily large n.

Example 4.13. Since $\limsup_m A_m = \bigcap_{k \geq n} \bigcup_{i \geq k} A_i$ and $\liminf_m A_m = \bigcup_{k \geq n}$ $\bigcap_{i \geq k} A_i$ are both in $\sigma(A_n, A_{n+1}, \ldots)$, the limits superior and inferior are tail events for the sequence $\{A_n\}$. ∎

Example 4.14. Let $l_n(\omega)$ be the run length, as before, and let $H_n = [\omega: d_n(\omega) = 0]$. For each n_0,

$$[\omega: l_n(\omega) \geq r_n \text{ i.o.}] = \bigcap_{n \geq n_0} \bigcup_{k \geq n} [\omega: l_k(\omega) \geq r_k]$$

$$= \bigcap_{n \geq n_0} \bigcup_{k \geq n} H_k \cap H_{k+1} \cap \cdots \cap H_{k+r_k-1}.$$

Thus $[\omega: l_n(\omega) \geq r_n \text{ i.o.}]$ is a tail event for the sequence $\{H_n\}$. ∎

The probabilities of tail events are governed by *Kolmogorov's zero-one law*:*

Theorem 4.5. *If $A_1, A_2, \ldots$ is an independent sequence of events, then for each event A in the tail σ-field (4.26), $P(A)$ is either 0 or 1.*

PROOF.   By Corollary 2 to Theorem 4.2, $\sigma(A_1), \ldots, \sigma(A_{n-1}), \sigma(A_n, A_{n+1}, \ldots)$ are independent. If $A \in \mathcal{T}$, then $A \in \sigma(A_n, A_{n+1}, \ldots)$ and therefore $A_1, \ldots, A_{n-1}, A$ are independent. Since independence of a collection of events is defined by independence of each finite subcollection, the sequence $A, A_1, A_2, \ldots$ is independent. By a second application of Corollary 2 to Theorem 4.2, $\sigma(A)$ and $\sigma(A_1, A_2, \ldots)$ are independent. But $A \in \mathcal{T} \subset \sigma(A_1, A_2, \ldots)$; from $A \in \sigma(A)$ and $A \in \sigma(A_1, A_2, \ldots)$ it follows that A is independent of itself: $P(A \cap A) = P(A)P(A)$. This is the same as $P(A) = (P(A))^2$ and can hold only if $P(A)$ is 0 or 1. ∎

Example 4.15. By the zero-one law and Example 4.13, $P(\limsup_n A_n)$ is 0 or 1 if the A_n are independent. The Borel-Cantelli lemmas in this case go further and give a specific criterion in terms of the convergence or divergence of $\Sigma P(A_n)$. ∎

---

* For a more general version, see Theorem 22.1.

Kolmogorov's result is surprisingly general, and it is in many cases quite easy to use it to show that the probability of some set must have one of the extreme values 0 and 1. It is perhaps curious that it should so often be very difficult to determine which of these extreme values is the right one.

Example 4.16. By Kolmogorov's theorem and Example 4.14, $[\omega: l_n(\omega) \geq r_n$ i.o.] has probability 0 or 1. Call the sequence $\{r_n\}$ an *outer boundary* or an *inner boundary* according as this probability is 0 or 1.

In Example 4.7 it was shown that $\{r_n\}$ is an outer boundary if $\Sigma\, 2^{-r_n} < \infty$. In Example 4.11 it was shown that $\{r_n\}$ is an inner boundary if r_n is nondecreasing and $\Sigma\, 2^{-r_n} r_n^{-1} = \infty$. By these criteria $r_n = \theta \log_2 n$ gives an outer boundary if $\theta > 1$ and an inner boundary if $\theta \leq 1$.

What about the sequence $r_n = \log_2 n + \theta \log_2 \log_2 n$? Here $\Sigma\, 2^{-r_n} = \Sigma\, 1/n(\log_2 n)^\theta$, and this converges for $\theta > 1$, which gives an outer boundary. Now $2^{-r_n} r_n^{-1}$ is of the order $1/n(\log_2 n)^{1+\theta}$, and this diverges if $\theta \leq 0$, which gives an inner boundary (this follows indeed from (4.23)). But this analysis leaves the range $0 < \theta \leq 1$ unresolved, although every sequence is either an inner or an outer boundary. This question is pursued further in Example 6.6.   ∎

Strong Laws versus Weak

If $A_n = [\omega: l_n(\omega) \geq \log_2 n]$, then by (4.7) and (4.23),

$$(4.27) \qquad P(A_n) \to 0, \qquad P(\limsup_n A_n) = 1.$$

Compare (4.10): if $P(A_n) \to 0$, then $\liminf_n A_n$ has probability 0, but $\limsup_n A_n$ may have probability 1.

Let I_1, I_2 be the dyadic intervals of rank 1 in $(0, 1]$, let I_3, I_4, I_5, I_6 be the dyadic intervals of rank 2, and so on; see (1.31) and (1.32). These intervals also have the property (4.27), as the length of I_n goes to 0 while $\limsup_n I_n$ is the entire interval $(0, 1]$. The point in Section 1 was that, if f_n is the indicator of I_n, then

$$(4.28) \qquad \lim_n P[\omega: |f_n(\omega)| \geq \epsilon] = 0$$

for each positive ϵ, even though

$$(4.29) \qquad P[\omega: \lim_n f_n(\omega) = 0] = 1$$

fails; in fact, the set in (4.29) is empty and hence has Lebesgue measure 0 (in the terminology of Section 1, it is negligible).

On the other hand, if each f_n is a step function on the unit interval, then (4.29) implies (4.28). Indeed, the set $[\omega: |f_n(\omega)| \geq \epsilon$ i.o.], being contained in $\Omega - [\omega:$

$\lim_n f_n(\omega) = 0$], must by (4.29) have measure 0; but then (4.28) follows by Theorem 4.1. As pointed out in Section 1', this implication (with $f_n(\omega) = n^{-1}s_n(\omega)$) shows that the strong law of large numbers, Theorem 1.2, is indeed stronger than the weak one, Theorem 1.1.

PROBLEMS

4.1. Suppose that sets $A_1, A_2, \ldots$ decompose Ω and that $P(A_n) > 0$, $P(B) > 0$. Derive Bayes' formula,

$$P(A_k|B) = \frac{P(A_k)P(B|A_k)}{\sum_n P(A_n)P(B|A_n)}.$$

4.2. 2.1 ↑ The limits superior and inferior of a numerical sequence $\{x_n\}$ can be defined as the supremum and infimum of the set of limit points—that is, the set of limits of convergent subsequences. This is the same thing as defining

$$\limsup_n x_n = \bigwedge_{n=1}^{\infty} \bigvee_{k=n}^{\infty} x_k$$

and

$$\liminf_n x_n = \bigvee_{n=1}^{\infty} \bigwedge_{k=n}^{\infty} x_k.$$

Compare these relations with (4.4) and (4.5) and prove that

$$I_{\limsup_n A_n} = \limsup_n I_{A_n}, \quad I_{\liminf_n A_n} = \liminf_n I_{A_n}.$$

Prove that $\lim_n A_n$ exists in the sense of (4.6) if and only if $\lim_n I_{A_n}(\omega)$ exists for each ω.

4.3. ↑ (a) Prove that

$$(\limsup_n A_n) \cap (\limsup_n B_n) \supset \limsup_n (A_n \cap B_n),$$

$$(\limsup_n A_n) \cup (\limsup_n B_n) = \limsup_n (A_n \cup B_n),$$

$$(\liminf_n A_n) \cap (\liminf_n B_n) = \liminf_n (A_n \cap B_n),$$

$$(\liminf_n A_n) \cup (\liminf_n B_n) \subset \liminf_n (A_n \cup B_n).$$

Show by example that the two inclusions can be strict.

(b) The numerical analogue of the first of the relations in part (a) is

$$(\limsup_n x_n) \wedge (\limsup_n y_n) \geq \limsup_n (x_n \wedge y_n).$$

Write out and verify the numerical analogues of the others.

(c) Show that

$$\limsup_n A_n^c = (\liminf A_n)^c,$$

$$\liminf_n A_n^c = (\limsup A_n)^c,$$

$$\limsup_n A_n - \liminf_n A_n = \limsup_n (A_n \cap A_{n+1}^c)$$

$$= \limsup_n (A_n^c \cap A_{n+1}).$$

(d) Show that $A_n \to A$ and $B_n \to B$ together imply that $A_n \cup B_n \to A \cup B$ and $A_n \cap B_n \to A \cap B$.

4.4. Let A_n be the square $[(x, y): |x| \leq 1, |y| \leq 1]$ rotated through the angle $2\pi n\theta$. Give geometric descriptions of $\limsup_n A_n$ and $\liminf A_n$ in case

(a) θ is rational;

(b) θ is irrational.

Hint: The $2\pi n\theta$ reduced modulo 2π are dense in $[0, 2\pi]$ if θ is irrational.

(c) When is there convergence in the sense of (4.6)?

4.5. (a) Find a sequence for which all three inequalities in (4.10) are strict.

(b) Show that $\lim_n P(\liminf_k A_n \cap A_k^c) = 0$.

Hint: Show that $\limsup_n \liminf_k A_n \cap A_k^c$ is empty.

Put $A^* = \limsup_n A_n$ and $A_* = \liminf_n A_n$.

(c) Show that $P(A_n - A^*) \to 0$ and $P(A_* - A_n) \to 0$.

(d) Show that $A_n \to A$ (in the sense that $A = A^* = A_*$) implies $P(A \triangle A_n) \to 0$.

(e) Suppose that A_n converges to A in the weaker sense that $P(A \triangle A^*) = P(A \triangle A_*) = 0$ (which implies that $P(A^* - A_*) = 0$). Show that $P(A \triangle A_n) \to 0$ (which implies that $P(A_n) \to P(A)$).

4.6. Assume that $P(\limsup_n A_n) = 1$ and $P(\liminf_n B_n) = 1$ and prove that $P(\limsup_n (A_n \cap B_n)) = 1$. What happens if the condition on $\{B_n\}$ is weakened to $P(\limsup_n B_n) = 1$?

4.7. 2.19 4.5 ↑   Show that the metric space of sets in Problem 2.19 is complete.

4.8. (a) Let A_1, A_2, A_3 be independent events of probability $\frac{1}{2}$ and put $A_{ij} = (A_i \triangle A_j)^c$. (If A_i is the event coin i falls heads, A_{ij} is the event coins i and j agree.) Show that A_{12}, A_{13}, A_{23} are not independent, even though they are independent in pairs.

(b) Find $n + 1$ events that are not independent even though every subcollection of n of them is independent.

4.9. In a space of six equally likely points (a die is rolled) find events A_1, A_2, A_3 that are not independent

(a)   even though $P(A_1 \cap A_2 \cap A_3) = P(A_1)P(A_2)P(A_3)$;

(b)   even though each is independent of the intersection of the other two.

4.10. For events $A_1, \ldots, A_n$, consider the 2^n equations $P(B_1 \cap \cdots \cap B_n) = P(B_1) \cdots P(B_n)$ with $B_i = A_i$ or $B_i = A_i^c$ for each i. Show that $A_1, \ldots, A_n$ are independent if all these equations hold.

4.11. Show that Theorem 4.2 fails if the $\mathcal{A}_i$ are not π-systems.

4.12. For each of the following classes $\mathcal{A}$, describe the $\mathcal{A}$-partition defined by (4.14).

(a)   The class of finite and cofinite sets.

(b)   The class of countable and co-countable sets.

(c)   A partition (of arbitrary cardinality) of Ω.

(d)   The level sets of $\sin x$ ($\Omega = R^1$).

(e)   The σ-field in Problem 3.4.

4.13. 3.3↑   There is in the unit interval a set of H that is nonmeasurable in the extreme sense that its inner and outer Lebesgue measures are 0 and 1 (see (3.9) and (3.10)): $\lambda_*(H) = 0$ and $\lambda^*(H) = 1$. See Problem 12.4 for the construction.

Let $\Omega = (0, 1]$, let $\mathcal{G}$ consist of the Borel sets in Ω, and let H be the set just described. Show that the class $\mathcal{F}$ of sets of the form $(H \cap G_1) \cup (H^c \cap G_2)$ for G_1 and G_2 in $\mathcal{G}$ is a σ-field and that $P[(H \cap G_1) \cup (H^c \cap G_2)] = \frac{1}{2}\lambda(G_1) + \frac{1}{2}\lambda(G_2)$ consistently defines a probability measure on $\mathcal{F}$. Show that $P(H) = \frac{1}{2}$ and that $P(G) = \lambda(G)$ for $G \in \mathcal{G}$. Show that $\mathcal{G}$ is separable—that is, $\mathcal{G}$ is generated by a countable subclass (see Problem 2.10). Show that $\mathcal{G}$ contains all the singletons and that H and $\mathcal{G}$ are independent.

The construction proves this: *There exists a probability space* $(\Omega, \mathcal{F}, P)$, *a σ-field $\mathcal{G}$ in $\mathcal{F}$, and a set H in $\mathcal{F}$, such that $P(H) = \frac{1}{2}$, H and $\mathcal{G}$ are independent, and $\mathcal{G}$ is separable and contains all the singletons.*

Example 4.6 is somewhat similar, but there the σ-field $\mathcal{G}$ is not separable and each set in it has probability either 0 or 1. In the present example $\mathcal{G}$ is separable and $P(G)$ assumes every value between 0 and 1 as G ranges over $\mathcal{G}$. Example 4.6 is to some extent unnatural because the $\mathcal{G}$ there is not separable. The present example, on the other hand, involves the pathological set H. This example is used in Section 33 in connection with conditional probability; see Problem 33.13.

4.14. (a)   If $A_1, A_2, \ldots$ are independent events, then $P(\bigcap_{n=1}^{\infty} A_n) = \prod_{n=1}^{\infty} P(A_n)$ and $P(\bigcup_{n=1}^{\infty} A_n) = 1 - \prod_{n=1}^{\infty} (1 - P(A_n))$. Prove these facts and from them derive the second Borel-Cantelli lemma by the well-known relation between infinite series and products.

(b)   Show that $P(\limsup_n A_n) = 1$ if for each k the series $\sum_{n>k} P(A_n | A_k^c \cap \cdots \cap A_{n-1}^c)$ diverges. From this deduce the second Borel-Cantelli lemma once again.

(c)   Show by example that $P(\limsup_n A_n) = 1$ does not follow from the divergence of $\sum_n P(A_n | A_1^c \cap \cdots \cap A_{n-1}^c)$ alone.

4.15. Suppose $A_1, A_2, \ldots$ are independent. There are four cases for the convergence or divergence of $\Sigma P(A_n)$ and $\Sigma P(A_n^c)$. Describe the pair $P(\lim \sup_n A_n)$ and $P(\lim \inf_n A_n)$ in each case.

4.16. (a) Show (see Example 4.16) that $\log_2 n + \log_2 \log_2 n + \theta \log_2 \log_2 \log_2 n$ is an outer boundary if $\theta > 1$. Generalize.

(b) Show that $\log_2 n + \log_2 \log_2 \log_2 n$ is an inner boundary.

4.17. 2.16↑ Extend to $\mathcal{C} = \sigma(\mathcal{C}_0)$ the measure P assigning probability 2^{-n} to each cylinder of rank n. Let $A_n = [\omega: a_n(\omega) = 1]$. Show that $A_1, A_2, \ldots$ are independent events and $P(A_n) = \frac{1}{2}$.

4.18. ↑ Suppose that $0 \le p_n \le 1$ and put $p_n(1) = p_n$ and $p_n(0) = 1 - p_n$. Generalize the construction in Problem 2.16 by defining P by $P[\omega: a_i(\omega) = u_i, i \le n] = p_1(u_1) \cdots p_n(u_n)$ for sequences $u_1, \ldots, u_n$ of 0's and 1's; Problem 2.16 covers the case $p_n \equiv \frac{1}{2}$. Show that P gives a probability measure on $\mathcal{C}_0$ and extend it to $\mathcal{C}$. Show that the events $A_n = [\omega: a_n(\omega) = 1]$ are independent and $P(A_n) = p_n$.

4.19. ↑ Suppose that $0 \le p_n \le 1$ and put $\alpha_n = \min \{p_n, 1 - p_n\}$. Show that, if $\Sigma \alpha_n$ converges, then in some discrete probability space there exist independent events A_n satisfying $P(A_n) = p_n$. Compare Problem 1.1(b).

4.20. 2.17↑ Suppose that there are in $(\Omega, \mathcal{F}, P)$ independent events $A_1, A_2, \ldots$ such that, if $\alpha_n = \min \{P(A_n), 1 - P(A_n)\}$, then $\Sigma \alpha_n = \infty$. Show that P is nonatomic.

4.21. 2.15↑ Let F be the set of square-free integers—those integers not divisible by any perfect square. Let F_l be the set of m such that $p^2 \mid m$ for no $p \le l$, and show that $D(F_l) = \prod_{p \le l}(1 - p^{-2})$. Show that $P_n(F_l - F) \le \Sigma_{p > l} p^{-2}$, and conclude that the square-free integers have density $\prod_p (1 - p^{-2}) = 6/\pi^2$.

4.22. Let $A_1, A_2, \ldots$ be independent events. Show that the event $[\omega: n^{-1} \Sigma_{k=1}^n I_{A_{nk}}(\omega) \to x]$ has probability either 0 or 1.

SECTION 5. SIMPLE RANDOM VARIABLES

Definition

Let $(\Omega, \mathcal{F}, P)$ be an arbitrary probability space and let X be a real-valued function on Ω; X is a *simple random variable* if it has finite range (assumes only finitely many values) and if

(5.1) $$[\omega: X(\omega) = x] \in \mathcal{F}$$

for each real x. (Of course, $[\omega: X(\omega) = x] = 0 \in \mathcal{F}$ for x outside the range of X.) Whether or not X satisfies this condition depends only on $\mathcal{F}$, not on P, but the point of the definition is to ensure that the probabilities $P[\omega: X(\omega) = x]$ are defined. Later sections will treat the theory of general random variables, of functions on Ω having arbitrary range; (5.1) will require modification in the general case.

The $d_n(\omega)$ of the preceding section (the digits of the dyadic expansion) are simple random variables on the unit interval. The run lengths $l_n(\omega)$ satisfy (5.1) but are not simple random variables because they have infinite range (they come under the general theory). In a discrete space, $\mathcal{F}$ consists of all subsets of Ω, so that (5.1) always holds. The indicator I_A of a set A is a simple random variable if and only if $A \in \mathcal{F}$.

It is customary in probability theory to omit the argument ω. Thus X stands for a general value $X(\omega)$ of the function as well as for the function itself, and $[X = x]$ is short for $[\omega: X(\omega) = x]$.

A finite sum

$$(5.2) \qquad\qquad X = \sum_i x_i I_{A_i}$$

is a random variable if the A_i form a finite partition of Ω into $\mathcal{F}$-sets. Moreover, any simple random variable can be represented in the form (5.2): for the x_i take the range of X, and put $A_i = [X = x_i]$. But X may have other such representations because $x_i I_{A_i}$ can be replaced by $\Sigma_j x_i I_{A_{ij}}$ if the A_{ij} form a finite decomposition of A_i into $\mathcal{F}$-sets.

If $\mathcal{G}$ is a sub-σ-field of $\mathcal{F}$, a simple random variable X is *measurable* $\mathcal{G}$ if $[X = x] \in \mathcal{G}$ for each x. Since $[X \in H] = \cup[X = x]$, where the union extends over the finitely many x lying both in H and in the range of X, $[X \in H] \in \mathcal{G}$ for every $H \subset R^1$ if X is a simple random variable measurable $\mathcal{G}$.

The σ-field $\sigma(X)$ *generated by* X is the smallest σ-field with respect to which X is measurable; that is, $\sigma(X)$ is the intersection of all σ-fields with respect to which X is measurable. For a finite or infinite sequence $X_1, X_2, \ldots$ of simple random variables, $\sigma(X_1, X_2, \ldots)$ is the smallest σ-field with respect to which *each* X_i is measurable. It can be described explicitly in the finite case:

Theorem 5.1. *Let $X_1, \ldots, X_n$ be simple random variables.*

(i) *The σ-field $\sigma(X_1, \ldots, X_n)$ consists of the sets*

$$(5.3) \qquad [(X_1, \ldots, X_n) \in H] = [\omega: (X_1(\omega), \ldots, X_n(\omega)) \in H]$$

for $H \subset R^n$; H in this representation may be taken finite.

(ii) *A simple random variable Y is measurable $\sigma(X_1, \ldots, X_n)$ if and only if*

$$(5.4) \qquad\qquad Y = f(X_1, \ldots, X_n)$$

for some $f: R^n \to R^1$.

PROOF. Sets of the form $\cap_{i=1}^n [X_i = x_i] = [(X_1, \ldots, X_n) = (x_1, \ldots, x_n)]$ must lie in $\sigma(X_1, \ldots, X_n)$; the sets (5.3) are finite unions of sets of this form because $(X_1, \ldots, X_n)$, as a mapping from Ω to R^n, has finite range. Thus each set (5.3) lies in $\sigma(X_1, \ldots, X_n)$. On the other hand, the sets (5.3) do constitute a σ-field $\mathcal{M}$ because $[(X_1, \ldots, X_n) \in H]^c = [(X_1, \ldots, X_n) \in H^c]$ and $\bigcup_j [(X_1, \ldots, X_n) \in H_j] = [(X_1, \ldots, X_n) \in \bigcup_j H_j]$. But each X_i is measurable with respect to this σ-field $\mathcal{M}$ because $[X_i = x]$ can be put in the form (5.3) by taking H to consist of those $(x_1, \ldots, x_n)$ in R^n for which $x_i = x$. As intersecting H with the range (finite) of $(X_1, \ldots, X_n)$ in R^n does not affect (5.3), H may be taken finite. This proves (i).

Assume that Y has the form (5.4)—that is, $Y(\omega) = f(X_1(\omega), \ldots, X_n(\omega))$ for every ω. Since $[Y = y]$ can be put in the form (5.3) by taking H to consist of those $x = (x_1, \ldots, x_n)$ for which $f(x) = y$, Y is measurable $\sigma(X_1, \ldots, X_n)$. Now assume that Y is measurable $\sigma(X_1, \ldots, X_n)$. Let $y_1, \ldots, y_r$ be the distinct values Y assumes. By part (i) there exist $H_1, \ldots, H_r$ such that $[\omega: Y(\omega) = y_i] = [\omega: (X_1(\omega), \ldots, X_n(\omega)) \in H_i] = [\omega: (X_1(\omega), \ldots, X_n(\omega)) \in H_i \cap H_1^c \cap \cdots \cap H_{i-1}^c]$, where the last equality holds because the $[\omega: Y(\omega) = y_i]$ are disjoint. It suffices to define $f(x_1, \ldots, x_n)$ as y_i on $H_i \cap H_1^c \cap \cdots \cap H_{i-1}^c$ and as 0, say, on $(\bigcup_{i=1}^r H_i)^c$. ∎

Since (5.4) implies that Y is measurable $\sigma(X_1, \ldots, X_n)$, it follows in particular that functions of simple random variables are again simple random variables. Thus X^2, e^{tX}, and so on are simple random variables along with X; and sums, products, and so on of simple random variables are simple random variables.

As explained on page 45, a sub-σ-field of $\mathcal{F}$ corresponds to partial information about ω. In this view, $\sigma(X_1, \ldots, X_n)$ corresponds to a knowledge of the values $X_1(\omega), \ldots, X_n(\omega)$. These values suffice to determine the value $Y(\omega)$ of Y if and only if (5.4) holds.*

Example 5.1.   For the dyadic digits $d_n(\omega)$ on the unit interval, d_3 is not measurable $\sigma(d_1, d_2)$; indeed, there exist ω' and ω'' such that $d_1(\omega') = d_1(\omega'')$ and $d_2(\omega') = d_2(\omega'')$ but $d_3(\omega') \neq d_3(\omega'')$, an impossibility if $d_3(\omega) = f(d_1(\omega), d_2(\omega))$ identically in ω. If such an f existed, one could unerringly predict the outcome $d_3(\omega)$ of the third toss from the outcomes $d_1(\omega)$ and $d_2(\omega)$ of the first two. ∎

Example 5.2.   Let $s_n(\omega) = \sum_{k=1}^n r_k(\omega)$ be the partial sums of the Rademacher functions—see (1.13). By Theorem 5.1(ii) s_k is measurable $\sigma(r_1, \ldots, r_n)$ for $k \leq n$, and $r_k = s_k - s_{k-1}$ is measurable $\sigma(s_1, \ldots, s_n)$ for $k \leq n$. Thus $\sigma(r_1, \ldots, r_n) = \sigma(s_1, \ldots, s_n)$. In random-walk terms, the first n positions contain the same

---

* The elements of the $\sigma(X_1, \ldots, X_n)$-partition (see (4.14)) are the sets $[X_1 = x_1, \ldots, X_n = x_n]$ with x_i in the range of X_i.

information as the first n distances moved. In gambling terms, to know the gambler's first n fortunes (relative to his initial fortune) is the same thing as to know his gains or losses on each of the first n plays.   ∎

Example 5.3. For sets A and A_i, $A \in \sigma(A_1, \ldots, A_n)$ if and only if $I_A = f(I_{A_1}, \ldots, I_{A_n})$ for some $f: R^n \to R^1$.

Independence

A sequence X_1, X_2, ... (finite or infinite) of simple random variables is by definition *independent* if the classes $\sigma(X_1)$, $\sigma(X_2)$, ... are independent in the sense of the preceding section. By Theorem 5.1(i), $\sigma(X_i)$ consists of the sets $[X_i \in H]$ for $H \subset R^1$. The condition for independence of $X_1, \ldots, X_n$ is therefore that

$$(5.5) \quad P[X_1 \in H_1, \ldots, X_n \in H_n] = P[X_1 \in H_1] \cdots P[X_n \in H_n]$$

for linear sets $H_1, \ldots, H_n$; for an infinite sequence $X_1, X_2, \ldots$, this must hold for each n. The sets $[X_i = x]$ and the empty set form a π-system generating $\sigma(X_i)$. By Theorem 4.2, (5.5) can therefore be replaced by the weaker condition that

$$(5.6) \quad P[X_1 = x_1, \ldots, X_n = x_n] = P[X_1 = x_1] \cdots P[X_n = x_n]$$

for real numbers $x_1, \ldots, x_n$.
    Suppose that

$$(5.7) \qquad \begin{array}{l} X_{11}, X_{12}, \ldots \\ X_{21}, X_{22}, \ldots \\ \cdots\cdots\cdots\cdots \end{array}$$

is an independent array of simple random variables. There may be finitely or infinitely many rows, each row finite or infinite. If $\mathcal{A}_i$ consists of the finite intersections of the sets $[X_{ij} = x]$, $j = 1, 2, \ldots$ and $x \in R^1$, an application of Theorem 4.2 shows that the σ-fields $\sigma(X_{i1}, X_{i2}, \ldots)$, $i = 1, 2, \ldots$ are independent. As a consequence, Y_1, Y_2, ... are independent if Y_i is measurable $\sigma(X_{i1}, X_{i2}, \ldots)$ for each i.

Example 5.4. The dyadic digits $d(\omega)$, $d_2(\omega)$, ... on the unit interval are an independent sequence of random variables for which

$$(5.8) \qquad\qquad P[d_n = 0] = P[d_n = 1] = \frac{1}{2}.$$

It is because of (5.8) and independence that the d_n give a model for tossing a fair coin.

The sequence $(d_1(\omega), d_2(\omega), \ldots)$ and the point ω determine one another. It can be imagined that ω is determined by the outcomes $d_n(\omega)$ of a sequence of tosses. It can also be imagined that ω is the result of drawing a point at random from the unit interval, and that ω determines the $d_n(\omega)$. In the second interpretation the $d_n(\omega)$ are all determined the instant ω is drawn, and so it should further be imagined that they are then revealed to the coin tosser or gambler one by one. For example, $\sigma(d_1, d_2)$ corresponds to knowing the outcomes of the first two tosses—to knowing not ω but only $d_1(\omega)$ and $d_2(\omega)$—and this does not help in predicting the value $d_3(\omega)$ because $\sigma(d_1, d_2)$ and $\sigma(d_3)$ are independent. See Example 5.1.   ∎

Example 5.5. Every permutation can be written as a product of cycles. For example,

$$\begin{pmatrix} 1 & 2 & 3 & 4 & 5 & 6 & 7 \\ 5 & 1 & 7 & 4 & 6 & 2 & 3 \end{pmatrix} = (1 \quad 5 \quad 6 \quad 2)(3 \quad 7)(4).$$

This permutation sends 1 to 5, 2 to 1, 3 to 7, and so on. The cyclic form on the right shows that 1 goes to 5, which goes to 6, which goes to 2, which goes back to 1; and so on. To standardize this cyclic representation, start the first cycle with 1 and each successive cycle with the smallest integer not yet encountered.

Let Ω consist of the $n!$ permutations of $1, 2, \ldots, n$, all equally probable; $\mathcal{F}$ contains all subsets of Ω and $P(A)$ is the fraction of points in A. Let $X_k(\omega)$ be 1 or 0 according as the element in the kth position in the cyclic representation of the permutation ω completes a cycle or not. Then $S(\omega) = \sum_{k=1}^{n} X_k(\omega)$ is the number of cycles in ω. In the example above, $n = 7$, $X_1 = X_2 = X_3 = X_5 = 0$, $X_4 = X_6 = X_7 = 1$, and $S = 3$. The following argument shows that $X_1, \ldots, X_n$ are independent and $P[X_k = 1] = (n - k + 1)^{-1}$.

Let $A_i(j)$ be the event that ω sends i to j. If $i_1, \ldots, i_k$ are distinct and $j_1, \ldots, j_k$ are distinct, then $P(\bigcap_{u=1}^{k} A_{i_u}(j_u)) = (n - k)!/n!$ For a sequence $\alpha = (a_1, \ldots, a_k)$, let B_α be the event that the first k positions in the cyclic representation of ω are occupied by $a_1, \ldots, a_k$. Now B_α has the form $\bigcap_{u=1}^{k-1} A_{a_u}(b_u)$, and a_u starts a new cycle if and only if $u = 1$ or $u > 1$ and a_u is distinct from the image b_{u-1} of a_{u-1} ($b_{u-1} \neq a_u$). Among $a_1, \ldots, a_k$, let a_{u_0} ($1 \le u_0 \le k$) be the rightmost one that starts a new cycle: $b_{u-1} = a_u$ for $u_0 < u \le k$, and either $u_0 = 1$ or $b_{u_0 - 1} \neq a_{u_0}$. If $\omega \in B_\alpha$, then $X_k(\omega) = 1$ if and only if ω sends a_k back to a_{u_0} to complete a cycle. Thus $B_\alpha \cap [X_k = 1] = \bigcap_{u=1}^{k-1} A_{a_u}(b_u) \cap A_{a_k}(a_{u_0})$, and it follows that $P(B_\alpha \cap [X_k = 1]) = (n - k)!/n! = P(B_\alpha)(n - k + 1)^{-1}$.

Let $\mathcal{B}_k$ consist of the empty set and the sets of this form B_α (k fixed). Since $\mathcal{B}_k$ is a π-system, $P(B \cap [X_k = 1]) = P(B)(n - k + 1)^{-1}$ for $B \in \sigma(\mathcal{B}_k)$. Since $X_1, \ldots, X_{k-1}$ are measurable $\sigma(\mathcal{B}_k)$, it follows by induction on k that $X_1, \ldots, X_k$ are independent and that $P[X_k = 1] = (n - k + 1)^{-1}$. This will lead later on to results on the probabilities $P[S \in H]$.   ∎

Existence of Independent Sequences

The *distribution* of a simple random variable X is the probability measure μ defined for all subsets A of the line by

(5.9) $\mu(A) = P[X \in A]$.

This does define a probability measure. It is discrete in the sense of Example 2.9: If $x_1, \ldots, x_l$ are the distinct points of the range of X, μ has mass $p_i = P[X = x_i] = \mu\{x_i\}$ at x_i, and $\mu(A) = \Sigma p_i$, the sum extending over those i for which $x_i \in A$. As $\mu(A) = 1$ if A is the range of X, not only is μ discrete, it has finite support.

Theorem 5.2. *Let $\{\mu_n\}$ be a sequence of probability measures on the class of all subsets of the line, each having finite support. There exists on some probability space $(\Omega, \mathcal{F}, P)$ an independent sequence $\{X_n\}$ of simple random variables such that X_n has distribution μ_n.*

What matters here is that there are finitely or countably many distributions μ_n. They need not be indexed by the integers; any countable index set will do.

PROOF.   The probability space will be the unit interval. Let $x_{n1}, \ldots, x_{nl_n}$ be the points on which μ_n concentrates its mass, and put $p_{ni} = \mu_n\{x_{ni}\}$, $1 \le i \le l_n$.

Decompose* $(0, 1]$ into l_1 subintervals $I_1^{(1)}, \ldots, I_{l_1}^{(1)}$ of respective lengths $p_{11}, \ldots, p_{1l_1}$. Define X_1 by setting $X_1(\omega) = x_{1i}$ for $\omega \in I_i^{(1)}$, $1 \le i \le l_1$. Then (P is Lebesgue measure) $P[\omega: X_1(\omega) = x_{1i}] = P(I_i^{(1)}) = p_{1i}$, $1 \le i \le l_1$. Thus X_1 is a simple random variable with distribution μ_1.

Next decompose each $I_i^{(1)}$ into l_2 subintervals $I_{i1}^{(2)}, \ldots, I_{il_2}^{(2)}$ of respective lengths $p_{1i}p_{21}, \ldots, p_{1i}p_{2l_2}$. Define $X_2(\omega) = x_{2j}$ for $\omega \in \bigcup_{i=1}^{l_1} I_{ij}^{(2)}$, $1 \le j \le l_2$. Then $P[\omega: X_1(\omega) = x_{1i}, X_2(\omega) = x_{2j}] = P(I_{ij}^{(2)}) = p_{1i}p_{2j}$. Adding out i shows that $P[\omega: X_2(\omega) = x_{2j}] = p_{2j}$, as required. Hence $P[X_1 = x_{1i}, X_2 = x_{2j}] = p_{1i}p_{2j} = P[X_1 = x_{1i}]P[X_2 = x_{2j}]$, and X_1 and X_2 are independent.

The construction proceeds inductively. Suppose that $(0, 1]$ has been decomposed into $l_1 \cdots l_n$ intervals

(5.10) $I_{i_1 \ldots i_n}^{(n)}$, $1 \le i_1 \le l_1, \ldots, 1 \le i_n \le l_n$,

of lengths

(5.11) $P(I_{i_1 \ldots i_n}^{(n)}) = p_{1,i_1} \cdots p_{n,i_n}$.

---

* If $b - a = \delta_1 + \cdots + \delta_l$ and $\delta_i \ge 0$, then $I_i = (a + \Sigma_{j<i}\delta_j, a + \Sigma_{j\le i}\delta_j]$ decomposes $(a, b]$ into subintervals $I_1, \ldots, I_l$ with lengths δ_i. Of course, I_i is empty if $\delta_i = 0$.

Decompose $I_{i_1...i_n}^{(n)}$ into l_{n+1} subintervals $I_{i_1...i_n1}^{(n+1)}, \ldots, I_{i_1...i_nl_{n+1}}^{(n+1)}$ of respective lengths $P(I_{i_1...i_n}^{(n)})p_{n+1,1}, \ldots, P(I_{i_1...i_n}^{(n)})p_{n+1,l_{n+1}}$. These are the intervals of the next decomposition. This construction gives a sequence of decompositions (5.10) of $(0,1]$ into subintervals; each decomposition satisfies (5.11) and each refines the preceding one. If μ_n is given for $1 \leq n \leq N$, the procedure terminates after N steps; for an infinite sequence it does not terminate at all.

For $1 \leq i \leq l_n$, put $X_n(\omega) = x_{ni}$ if $\omega \in \bigcup_{i_1...i_{n-1}} I_{i_1...i_{n-1}i}^{(n)}$. Since each decomposition (5.10) refines the preceding, $X_k(\omega) = x_{ki_k}$ for $\omega \in I_{i_1...i_k...i_n}^{(n)}$. Therefore, each element of (5.10) is contained in the element with the same label $i_1 \cdots i_n$ in the decomposition

$$A_{i_1 \ldots i_n} = [\omega: X_1(\omega) = x_{1i_1}, \ldots, X_n(\omega) = x_{ni_n}],$$
$$1 \leq i_1 \leq l_1, \ldots, 1 \leq i_n \leq l_n.$$

The two decompositions thus coincide, and it follows by (5.11) that $P[X_1 = x_{1i_1}, \ldots, X_n = x_{ni_n}] = p_{1,i_1} \cdots p_{n,i_n}$. Adding out the indices $i_1, \ldots, i_{n-1}$ shows that X_n has distribution μ_n and hence that $X_1, \ldots, X_n$ are independent. But n was arbitrary. ∎

Probability theorems such as those in the next sections concern independent sequences $\{X_n\}$ with specified distributions or with distributions having specified properties, and because of Theorem 5.2 these theorems are true not merely in the vacuous sense that their hypotheses are never fulfilled. Similar but more complicated existence theorems will come later. For most purposes the probability space on which the X_n are defined is largely irrelevant. Every independent sequence $\{X_n\}$ satisfying $P[X_n = 1] = p$ and $P[X_n = 0] = 1 - p$ is a model for Bernoulli trials, for example, and for an event like $\bigcup_{n=1}^{\infty}[\sum_{k=1}^{n}X_k > \alpha n]$, expressed in terms of the X_n alone, the calculation of its probability proceeds in the same way whatever the underlying space $(\Omega, \mathcal{F}, P)$ may be.

It is, of course, an advantage that such results apply not just to some canonical sequence $\{X_n\}$ (such as the one constructed in the proof above) but to every sequence with the appropriate distributions. In some applications of probability within mathematics itself, such as the arithmetic applications of run theory in Examples 4.8 and 4.12, the underlying Ω does play a role.

Expected Value

A simple random variable in the form (5.2) is assigned *expected value* or *mean value*

$$(5.12) \qquad E[X] = E\left[\sum_i x_i I_{A_i} \right] = \sum_i x_i P(A_i).$$

There is the alternative form

(5.13) $$E[X] = \sum_x xP[X=x],$$

the sum extending over the range of X; indeed, (5.12) and (5.13) both coincide with $\sum_x \sum_{i:x_i=x} x_i P(A_i)$. By (5.13) the definition (5.12) is consistent: different representations (5.2) give the same value to (5.12). From (5.13) it also follows that $E[X]$ depends only on the distribution of X; hence $E[X] = E[Y]$ if $P[X = Y] = 1$.

If X is a simple random variable on the unit interval and if the A_i in (5.2) happen to be subintervals, then (5.12) coincides with the Riemann integral as given by (1.5). More general notions of integral and expected value will be studied later. Simple random variables are easy to work with because the theory of their expected values is transparent and free of technical complications.

As a special case of (5.12) and (5.13),

(5.14) $$E[I_A] = P(A).$$

As another special case, if a constant α is identified with the random variable $X(\omega) \equiv \alpha$, then

(5.15) $$E[\alpha] = \alpha.$$

From (5.2) follows $f(X) = \sum_i f(x_i)I_{A_i}$, and hence

(5.16) $$E[f(X)] = \sum_i f(x_i)P(A_i) = \sum_x f(x)P[X = x],$$

the last sum extending over the range of X. For example, the kth *moment* $E[X^k]$ of X can be calculated by $E[X^k] = \sum_x x^k P[X = x]$.

If

(5.17) $$X = \sum_i x_i I_{A_i}, \qquad Y = \sum_j y_j I_{B_j}$$

are simple random variables, then $\alpha X + \beta Y = \sum_{ij} (\alpha x_i + \beta y_j)I_{A_i \cap B_j}$ has expected value $\sum_{ij}(\alpha x_i + \beta y_j)P(A_i \cap B_j) = \alpha\sum_i x_i P(A_i) + \beta\sum_j y_j P(B_j)$. Expected value is therefore *linear*:

(5.18) $$E[\alpha X + \beta Y] = \alpha E[X] + \beta E[Y].$$

If $X(\omega) \le Y(\omega)$ for all ω, then $x_i \le y_j$ if $A_i \cap B_j$ is nonempty, and hence $\sum_{ij} x_i P(A_i \cap B_j) \le \sum_{ij} y_j P(A_i \cap B_j)$. Expected value therefore *preserves order*:

(5.19) $$E[X] \le E[Y] \qquad \text{if } X \le Y.$$

(It is enough that $X \le Y$ on a set of probability 1.) Two applications of (5.19) give $E[-|X|] \le E[X] \le E[|X|]$, so that

(5.20) $$|E[X]| \le E[|X|].$$

The relations (5.14) through (5.20) will be used repeatedly, and so will the following theorem on expected values and limits. If there is a finite K such that $|X_n(\omega)| \leq K$ for all ω and all n, the X_n are *uniformly bounded*.

Theorem 5.3. *If $\{X_n\}$ is uniformly bounded, and if $X = \lim_n X_n$ on an $\mathcal{F}$-set of probability* 1*, then $E[X] = \lim_n E[X_n]$.*

PROOF. Let A be the $\mathcal{F}$-set of the hypothesis: $P(A) = 1$ and there is convergence inside A. Since redefining $X_n = X = 0$, say, outside A does not change the expected values, it can be assumed that $X = \lim_n X_n$ everywhere. Let K bound the X_n. Given $\epsilon > 0$, choose points $x_0 < x_1 < \cdots < x_k$ such that $x_0 < -K < K < x_k$ and $x_i - x_{i-1} < \epsilon, i = 1, \ldots, k$, and such that no x_i is the range of X. Since X cannot assume the values x_{i-1} and x_i, $A_{in} = [x_{i-1} < X_n \leq x_i]$ converges in the sense of (4.6) to $A_i = [x_{i-1} < X \leq x_i]$ and hence (Theorem 4.1) $P(A_{in}) \to P(A_i)$. Since (x_0, x_k) covers the range of each X_n, $X_n \leq \Sigma_{i=1}^k x_i I_{A_{in}}$; (5.19) implies that $E[X_n] \leq \Sigma_{i=1}^k x_i P(A_{ni}) \to \Sigma_{i=1}^k x_i P(A_i) \leq \Sigma_{i=1}^k (x_{i-1} + \epsilon) P(A_i)$. But $\Sigma_{i=1}^k x_{i-1} I_{A_i} \leq X$, and so $\Sigma_{i=1}^k x_{i-1} P(A_i) \leq E[X]$. Therefore, $\limsup_n E[X_n] \leq E[X] + \epsilon$. Since $\liminf_n E[X_n] \geq E[X] - \epsilon$ follows by a similar argument (or by the same argument applied to $-X_n$), and since ϵ was arbitrary, the result follows. ∎

Theorems of this kind are of constant use in probability and analysis. For the general version, Lebesgue's dominated convergence theorem, see Section 16.

Example 5.6. On the unit interval take $X(\omega)$ identically 0 and take $X_n(\omega)$ to be n^2 if $0 < \omega \leq n^{-1}$ and 0 if $n^{-1} < \omega \leq 1$. Then $X_n(\omega) \to X(\omega)$ for every ω, although $E[X_n] = n$ does not converge to $E[X] = 0$. Thus Theorem 5.3 fails without some hypothesis such as that of uniform boundedness. See also Example 7.7. ∎

An extension of (5.18) is an immediate consequence of Theorem 5.3:

Corollary. *If $X = \Sigma_n X_n$ on an $\mathcal{F}$-set of probability* 1*, and if the partial sums of $\Sigma_n X_n$ are uniformly bounded, then $E[X] = \Sigma_n E[X_n]$.*

Expected values for independent random variables satisfy the familiar product law. For X and Y as in (5.17), $XY = \Sigma_{ij} x_i y_j I_{A_i \cap B_j}$. If the x_i are distinct and the y_j are distinct, then $A_i = [X = x_i]$ and $B_j = [Y = y_j]$; for independent X and Y, $P(A_i \cap B_j) = P(A_i) P(B_j)$ by (5.6), and so $E[XY] = \Sigma_{ij} x_i y_j P(A_i) P(B_j) = E[X] E[Y]$. If X, Y, Z are independent, then XY and Z are independent by the argument involving (5.7), so that $E[XYZ] = E[XY] E[Z] = E[X] E[Y] E[Z]$. This obviously extends:

(5.21) $$E[X_i \cdots X_n] = E[X_1] \cdots E[X_n]$$

if $X_1, \ldots, X_n$ are independent.

Various concepts from discrete probability carry over to simple random variables. If $E[X] = m$, the *variance* of X is

(5.22) $$\text{Var } [X] = E[(X - m)^2] = E[X^2] - m^2;$$

the left-hand equality is a definition, the right-hand one a consequence of expanding the square. Since $\alpha X + \beta$ has mean $\alpha m + \beta$, its variance is $E[((\alpha X + \beta) - (\alpha m + \beta))^2] = E[\alpha^2(X - m)^2]$:

(5.23) $$\text{Var } [\alpha X + \beta] = \alpha^2 \text{ Var } [X].$$

If $X_1, \ldots, X_n$ have means $m_1, \ldots, m_n$, then $S = \Sigma_{i=1}^n X_i$ has mean $m = \Sigma_{i=1}^n m_i$, and $E[(S - m)^2] = E[(\Sigma_{i=1}^n (X_i - m_i))^2] = \Sigma_{i=1}^n E[(X_i - m_i)^2] + 2 \Sigma_{1 \le i < j \le n} E[(X_i - m_i)(X_j - m_j)]$. If the X_i are independent, then so are the $X_i - m_i$, and by (5.21) the last sum vanishes. This gives the familiar formula for the variance of a sum of independent random variables:

(5.24) $$\text{Var } \left[\sum_{i=1}^n X_i \right] = \sum_{i=1}^n \text{Var } [X_i].$$

Suppose that X is nonnegative; order its range: $0 \le x_1 < x_2 < \cdots < x_k$. Then

$$E[X] = \sum_{i=1}^k x_i P[X = x_i]$$

$$= \sum_{i=1}^{k-1} x_i(P[X \ge x_i] - P[X \ge x_{i+1}]) + x_k P[X \ge x_k]$$

$$= x_1 P[X \ge x_1] + \sum_{i=2}^k (x_i - x_{i-1})P[X \ge x_i].$$

Since $P[X \ge x] = P[X \ge x_1]$ for $0 \le x \le x_1$ and $P[X \ge x] = P[X \ge x_i]$ for $x_{i-1} < x \le x_i$, it is possible to write the final sum as the Riemann integral of a step function:

(5.25) $$E[X] = \int_0^\infty P[X \ge x] \, dx.$$

This holds if X is nonnegative. Since $P[X \ge x] = 0$ for $x > x_k$, the range of integration is really finite.

There is for (5.25) a simple geometric argument involving the "area over the curve." If $p_i = P[X = x_i]$, the area of the shaded region in the figure is the sum $p_1 x_1 + \cdots + p_k x_k = E[X]$ of the areas of the horizontal strips; it is also the integral of the height $P[X \ge x]$ of the region.

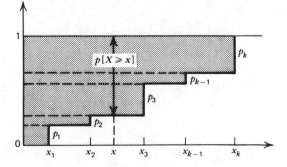

Inequalities

There are for expected values several standard inequalities that will be needed. If X is nonnegative, then for positive α (sum over the range of X) $E[X] = \Sigma_x xP[X = x] \geq \Sigma_{x:x \geq \alpha} xP[X = x] \geq \alpha\Sigma_{x:x \geq \alpha} P[X = x]$. Therefore,

$$(5.26) \qquad P[X \geq \alpha] \leq \frac{1}{\alpha} E[X].$$

This is the analogue of (1.19). Applied to $|X|^k$, it gives *Markov's inequality*,

$$(5.27) \qquad P[|X| \geq \alpha] \leq \frac{1}{\alpha^k} E[|X|^k].$$

If $k = 2$ and $m = E[X]$ is subtracted from X, this becomes the *Chebyshev* (or Chebyshev-Bienaymé) *inequality*:

$$(5.28) \qquad P[|X - m| \geq \alpha] \leq \frac{1}{\alpha^2} \operatorname{Var}[X].$$

A function φ on an interval is convex if $\varphi(px + (1 - p)y) \leq p\varphi(x) + (1 - p)\varphi(y)$ for $0 \leq p \leq 1$ and x and y in the interval. A sufficient condition for this is that φ have a nonnegative second derivative. It follows by induction that $\varphi(\Sigma_{i=1}^l p_i x_i) \leq \Sigma_{i=1}^l p_i \varphi(x_i)$ if the p_i are nonnegative and add to 1 and the x_i are in the domain of φ. If X assumes the value x_i with probability p_i, this becomes *Jensen's inequality*,

$$(5.29) \qquad \varphi(E[X]) \leq E[\varphi(X)],$$

valid if φ is convex on an interval containing the range of X.

Suppose that

$$(5.30) \qquad \frac{1}{p} + \frac{1}{q} = 1, \qquad p > 1, \quad q > 1.$$

Hölder's inequality is

$$(5.31) \qquad E[|XY|] \le E^{1/p}[|X|^p] \cdot E^{1/q}[|Y|^q].$$

If, say, the first factor on the right vanishes, then $X = 0$ with probability 1, hence $XY = 0$ with probability 1, and hence the left side vanishes also. Assume then that the right side of (5.31) is positive. If a and b are positive, there exist s and t such that $a = e^{p^{-1}s}$ and $b = e^{q^{-1}t}$. Since e^x is convex, $e^{p^{-1}s+q^{-1}t} \le p^{-1}e^s + q^{-1}e^t$, or

$$ab \le \frac{a^p}{p} + \frac{b^q}{q}.$$

This obviously holds for nonnegative as well as for positive a and b. Let u and v be the two factors on the right in (5.31). For each ω,

$$\left| \frac{X(\omega)Y(\omega)}{uv} \right| \le \frac{1}{p} \left| \frac{X(\omega)}{u} \right|^p + \frac{1}{q} \left| \frac{Y(\omega)}{v} \right|^q.$$

Taking expected values and applying (5.30) leads to (5.31).

If $p = q = 2$, Hölder's inequality becomes *Schwarz's inequality*:

$$(5.32) \qquad E[|XY|] \le E^{1/2}[X^2] \cdot E^{1/2}[Y^2].$$

Suppose that $0 < \alpha < \beta$. In (5.31) take $p = \beta/\alpha$, $q = \beta/(\beta - \alpha)$, and $Y(\omega) = 1$, and replace X by $|X|^\alpha$. The result is

$$(5.33) \qquad E^{1/\alpha}[|X|^\alpha] \le E^{1/\beta}[|X|^\beta], \qquad 0 < \alpha \le \beta.$$

PROBLEMS

5.1. (a) Show that X is measurable $\mathscr{G}$ if and only if $\sigma(X) \subset \mathscr{G}$.

(b) Show that, if $\mathscr{G} = \{0, \Omega\}$, then X is measurable $\mathscr{G}$ if and only if X is constant.

(c) Show that, if $P(A)$ is 0 or 1 for every A in $\mathscr{G}$ and X is measurable $\mathscr{G}$, then $P[X = c] = 1$ for a constant c.

(d) Complete Theorem 5.1(ii) by showing that Y is measurable $\sigma(X_1, \ldots, X_n)$ if and only if for all ω and ω', $X_i(\omega) = X_i(\omega')$, $i \le n$, implies that $Y(\omega) = Y(\omega')$.

5.2. 2.17 ↑ Show that the unit interval can be replaced by any nonatomic probability measure space in the proof of Theorem 5.2.

5.3. Show that $m = E[X]$ minimizes $[(X - m)^2]$.

5.4. Suppose that X assumes the values $m - \alpha$, m, $m + \alpha$ with probabilities $\frac{1}{2}\alpha^2$, $1 - 1/\alpha^2$, $\frac{1}{2}\alpha^2$, and show that there is equality in (5.28). Thus Chebyshev's inequality cannot be improved without special assumptions on X.

5.5. Suppose that X has mean m and variance σ^2.

(a) Prove *Cantelli's inequality*

$$P[X - m \geq \alpha] \leq \frac{\sigma^2}{\sigma^2 + \alpha^2}, \qquad \alpha \geq 0.$$

(b) Show that $P[|X - m| \geq \alpha] \leq 2\sigma^2/(\sigma^2 + \alpha^2)$. When is this better than Chebyshev's inequality?

(c) By considering a random variable assuming two values, show that Cantelli's inequality is sharp.

5.6. The polynomial $E[(t|X| + |Y|)^2]$ in t has at most one real zero. Deduce Schwarz's inequality once more.

5.7. (a) Write (5.33) in the form $E^{\beta/\alpha}[|X|^\alpha] \leq E[(|X|^\alpha)^{\beta/\alpha}]$ and deduce it directly from Jensen's inequality.

(b) Prove that $E[1/X^p] \geq 1/E^p[X]$ for $p > 0$ and X a positive random variable.

5.8. (a) Let f be a convex real function on a convex set C in the plane. Suppose that $(X(\omega), Y(\omega)) \in C$ for all ω and prove a two-dimensional Jensen's inequality:

(5.34) $$f(E[X], E[Y]) \leq E[f(X,Y)].$$

(b) Show that f is convex if it has continuous second derivatives that satisfy

(5.35) $$f_{11} \geq 0, \quad f_{22} \geq 0, \quad f_{11}f_{22} \geq f_{12}^2.$$

5.9. ↑ Hölder's inequality is equivalent to $E[X^{1/p}Y^{1/q}] \leq E^{1/p}[X] \cdot E^{1/q}[Y]$ ($p^{-1} + q^{-1} = 1$), where X and Y are nonnegative random variables. Derive this from (5.34).

5.10. ↑ *Minkowski's inequality* is

(5.36) $$E^{1/p}[|X + Y|^p] \leq E^{1/p}[|X|^p] + E^{1/p}[|Y|^p],$$

valid for $p \geq 1$. It is enough to prove that $E[(X^{1/p} + Y^{1/p})^p] \leq (E^{1/p}[X] + E^{1/p}[Y])^p$ for nonnegative X and Y. Use (5.34).

5.11. The *essential supremum* of $|X|$ is ess sup $|X| = \inf [x: P[|X| \leq x] = 1]$.

(a) Show that $E^{1/p}[|X|^p] \uparrow$ ess sup $|X|$ as $p \uparrow \infty$.

(b) Show that $E[|XY|] \leq E[|X|] \cdot$ ess sup $|Y|$ and ess sup $|X + Y| \leq$ ess sup $|X| +$ ess sup $|Y|$. If one formally writes ess sup $|X| = E^{1/\infty}[|X|^\infty]$, the first of these inequalities is Hölder's for $p = 1$ and $q = \infty$ (formally $1^{-1} + \infty^{-1} = 1$) and the second is Minkowski's inequality (5.36) for $p = \infty$.

5.12. For events $A_1, A_2, \ldots$, not necessarily independent, and let $N_n = \sum_{k=1}^n I_{A_k}$ be the number to occur among the first n. Let

(5.37) $$\alpha_n = \frac{1}{n} \sum_{k=1}^n P(A_k), \qquad \beta_n = \frac{2}{n(n-1)} \sum_{1 \leq j < k \leq n} P(A_j \cap A_k).$$

Show that

(5.38) $$E[n^{-1}N_n] = \alpha_n, \qquad \text{Var}\,[n^{-1}N_n] = \beta_n - \alpha_n^2 + \frac{\alpha_n - \beta_n}{n}.$$

Thus Var $[n^{-1}N_n] \to 0$ if and only if $\beta_n - \alpha_n^2 \to 0$, which holds if the A_n are independent and $P(A_n) = p$ (Bernoulli trials) because then $\alpha_n = p$ and $\beta_n = p^2 = \alpha_n^2$.

5.13. (a) Let S be the number of elements left fixed by a random permutation on n letters. Calculate $E[S]$ and Var $[S]$.

(b) A fair r-sided die is rolled n times. Let N be the number of pairs of trials on which the same face results. Calculate $E[N]$ and Var $[N]$.

Hint: In neither case consider the distribution of the random variable; instead, represent it as a sum of random variables assuming values 0 and 1.

5.14. Show that, if X has nonnegative integers as values, then $E[X] = \sum_{n=1}^{\infty} P[X \geq n]$.

5.15. Let $I_i = I_{A_i}$ be the indicators of n events having union A. Let $S_k = \Sigma I_{i_1} \cdots I_{i_k}$, where the summation extends over all k-tuples satisfying $1 \leq i_1 < \cdots < i_k \leq n$. Then $s_k = E[S_k]$ are the terms in the inclusion-exclusion formula $P(A) = s_1 - s_2 + \cdots \pm s_n$. Deduce the inclusion-exclusion formula from $I_A = S_1 - S_2 + \cdots \pm S_n$. Prove the latter formula by expanding the product $\Pi_{i=1}^n (1 - I_i)$. For a second proof, suppose that $\Sigma_{i=1}^n I_i = r$, calculate the S_k in terms of r, and use the identity $(1 - 1)^r = 0$.

5.16. 2.15 ↑ For integers m and primes p, let $\alpha_p(m)$ be the exact power of p in the prime factorization of m: $m = \Pi_p p^{\alpha_p(m)}$. Let $\delta_p(m)$ be 1 or 0 as p divides m or not. Under each P_n (see (2.20)) the α_p and δ_p are random variables. Show that for distinct primes $p_1, \ldots, p_u$,

(5.39) $P_n[\alpha_{p_i} \geq k_i, i \leq u] = \dfrac{1}{n}\left[\dfrac{n}{p_1^{k_1} \cdots p_u^{k_u}}\right] \to \dfrac{1}{p_1^{k_1} \cdots p_u^{k_u}}$

and

(5.40) $P_n[\alpha_{p_i} = k_i, i \leq u] \to \prod_{i=1}^{u}\left(\dfrac{1}{p_{p_i}^{k_i}} - \dfrac{1}{p_{p_i}^{k_i+1}}\right).$

Similarly,

(5.41) $P_n[\delta_{p_i} = 1, i \leq u] = \dfrac{1}{n}\left[\dfrac{n}{p_1 \cdots p_u}\right] \to \dfrac{1}{p_1 \cdots p_u}.$

According to (5.40), the α_p are for large n approximately independent under P_n, and according to (5.41), the same is true of the δ_p.

For a function f of positive integers, let

(5.42) $E_n[f] = \dfrac{1}{n} \sum_{m=1}^{n} f(m)$

be its expected value under the probability measure P_n. Show that

(5.43) $E_n[\alpha_p] = \sum_{k=1}^{\infty} \dfrac{1}{n}\left[\dfrac{n}{p^k}\right] \to \dfrac{1}{p - 1};$

this says roughly that $(p - 1)^{-1}$ is the average power of p in the factorization of large integers.

5.17. ↑ **(a)** From Stirling's formula, deduce

(5.44) $$E_n[\log] = \log n + O(1).$$

From this, the inequality $E_n[\alpha_p] \leq 2/p$, and the relation $\log m = \Sigma_p \alpha_p(m) \log p$, conclude that $\Sigma_p p^{-1} \log p$ diverges and that there are infinitely many primes.

(b) Let $\log^* m = \Sigma_p \delta_p(m) \log p$. Show that

(5.45) $$E_n[\log^*] = \sum_p \frac{1}{n}\left[\frac{n}{p}\right] \log p = \log n + O(1).$$

(c) Show that $[2n/p] - 2[n/p]$ is always nonegative and equals 1 in the range $n < p \leq 2n$. Deduce $E_{2n}[\log^*] - E_n[\log^*] = O(n)$ and conclude that

(5.46) $$\sum_{p \leq x} \log p = O(x).$$

Use this to estimate the error removing the integral-part brackets introduces into (5.45) and show that

(5.47) $$\sum_{p \leq x} p^{-1} \log p = \log x + O(1).$$

(d) Restrict the range of summation in (5.47) to $\theta x < p \leq x$ for an appropriate θ and conclude that

(5.48) $$\sum_{p \leq x} \log p \asymp x,$$

in the sense that the ratio of the two sides is bounded away from 0 and ∞.

(e) Use (5.48) and truncation arguments to prove for the number $\pi(x)$ of primes not exceeding x that

(5.49) $$\pi(x) \asymp \frac{x}{\log x}.$$

(By the prime number theorem the ratio of the two sides in fact goes to 1.) Conclude that the rth prime p_r satisfies $p_r \asymp r \log r$ and that

(5.50) $$\sum_p \frac{1}{p} = \infty.$$

5.18. Let $f_n(x)$ be $n^2 x$ or $2n - n^2 x$ or 0 according as $0 \leq x \leq n^{-1}$ or $n^{-1} \leq x \leq 2n^{-1}$ or $2n^{-1} \leq x \leq 1$. This gives a standard example of a sequence of continuous functions that converges to 0 but not uniformly. Note that $\int_0^1 f_n(x)\, dx$ does not converge to 0; relate to Example 5.6.

SECTION 6.   THE LAW OF LARGE NUMBERS

The Strong Law

Let $X_1, X_2, \ldots$ be a sequence of simple random variables on some probability space $(\Omega, \mathcal{F}, P)$. They are *identically distributed* if their distributions (in the sense of (5.9)) are all the same. Define $S_n = X_1 + \cdots + X_n$.

Theorem 6.1. *If the X_n are independent and identically distributed and $E[X_n]$ $= m$, then*

$$(6.1) \qquad\qquad P[\lim_n n^{-1}S_n = m] = 1.$$

PROOF.   The conclusion is that $n^{-1}S_n - m = n^{-1}\Sigma_{i=1}^n (X_i - m) \to 0$ except on an $\mathcal{F}$-set of probability 0. Replacing X_i by $X_i - m$ shows that there is no loss of generality in assuming $m = 0$.

Let $E[X_n^2] = \sigma^2$ and $E[X_n^4] = \xi^4$. The proof is like that for Theorem 1.2. In analogy with (1.25), $E[S_n^4] = \Sigma\, E[X_\alpha X_\beta X_\gamma X_\delta]$, the four indices ranging independently from 1 to n. Since $E[X_n] = 0$, it follows by the product rule (5.21) for independent random variables that the summand vanishes if there is one index different from the three others. This leaves terms of the form $E[X_i^4] = \xi^4$, of which there are n, and terms of the form $E[X_i^2 X_j^2] = E[X_i^2]E[X_j^2] = \sigma^4$ for $i \neq j$, of which there are $3n(n - 1)$. Hence

$$(6.2) \qquad\qquad E[S_n^4] = n\xi^4 + 3n(n - 1)\sigma^4 \leq Kn^2,$$

where K does not depend on n.

By Markov's inequality (5.27) for $k = 4$, $P[|S_n| \geq n\epsilon] \leq Kn^{-2}\epsilon^{-4}$, and so by the first Borel-Cantelli lemma, $P[|n^{-1}S_n| \geq \epsilon \text{ i.o.}] = 0$ for each positive ϵ. Now

$$(6.3) \qquad\quad \Omega - [\lim_n n^{-1}S_n = 0] = \bigcup [|n^{-1}S_n| \geq \epsilon \text{ i.o.}],$$

where the union extends over positive rational ϵ. This shows in the first place that the set on the left in (6.3) lies in $\mathcal{F}$ and in the second place that its probability is 0. Taking complements gives (6.1).   ■

Example 6.1.   The classical example is the strong law of large numbers for Bernoulli trials. Here $P[X_n = 1] = p$, $P[X_n = 0] = 1 - p$, $m = p$; S_n represents the number of successes in n trials, and $n^{-1}S_n \to p$ with probability 1. The idea of probability as frequency depends on the long-range stability of the success ratio S_n/n.   ■

Example 6.2. Theorem 1.2 is the case of Example 6.1 in which $(\Omega, \mathcal{F}, P)$ is the unit interval and the $X_n(\omega)$ are the digits $d_n(\omega)$ of the dyadic expansion of ω. Here $p = \frac{1}{2}$. The set (1.20) of normal numbers in the unit interval has by (6.1) Lebesgue measure 1; its complement has measure 0 (and so in the terminology of Section 1 is negligible). See Example 4.9. ∎

The Weak Law

If (6.3) has probability 0, so has $[|n^{-1}S_n| \geq \epsilon$ i.o.$]$ for each positive ϵ, and Theorem 4.1 implies that $P[|n^{-1}S_n| \geq \epsilon] \to 0$. (See the discussion centering on (4.28) and (4.29).) This follows directly from Chebyshev's inequality (5.28), which also leads to a more flexible result.

Suppose that $(\Omega_n, \mathcal{F}_n, P_n)$ is a probability space for each n and that $X_{n1}, \ldots, X_{nr_n}$ are simple random variables on this space. Let $S_n = X_{n1} + \cdots + X_{nr_n}$. If

(6.4) $$E[X_{nk}] = m_{nk}, \qquad \mathrm{Var}\,[X_{nk}] = \sigma_{nk}^2,$$

and if $X_{n1}, \ldots, X_{nr_n}$ are independent, then

(6.5) $$E[S_n] = m_n = \sum_{k=1}^{r_n} m_{nk}, \qquad \mathrm{Var}\,[S_n] = \sigma_n^2 = \sum_{k=1}^{r_n} \sigma_{nk}^2.$$

Chebyshev's inequality immediately gives this form of the weak law of large numbers:

Theorem 6.2. *Suppose that for each n, $X_{n1}, \ldots, X_{nr_n}$ are independent and $v_n > 0$. If $\sigma_n/v_n \to 0$, then*

(6.6) $$P_n\left[\left|\frac{S_n - m_n}{v_n}\right| \geq \epsilon\right] \to 0$$

for all positive ϵ.

Example 6.3. Suppose that the spaces $(\Omega_n, \mathcal{F}_n, P_n)$ are all the same, that $r_n = n$, and that $X_{nk} = X_k$, where $X_1, X_2, \ldots$ is an infinite sequence of independent random variables. If $v_n = n$, $E[X_k] = m$, and $\mathrm{Var}\,[X_k] = \sigma^2$, then $\sigma_n/v_n = \sigma/\sqrt{n} \to 0$. This gives the weak law corresponding to Theorem 6.1. ∎

Example 6.4. Let Ω_n consist of the $n!$ permutations of $1, 2, \ldots, n$, all equally probable, and let $X_{nk}(\omega)$ be 1 or 0 according as the kth element in the cyclic representation of $\omega \in \Omega_n$ completes a cycle or not. This is Example 5.5, although there the dependence on n was suppressed in the notation. Here $r_n = n$, $X_{n1}, \ldots, X_{nn}$ are independent, and $S_n = X_{n1} + \cdots + X_{nn}$ is the number of cycles; $m_{nk} = (n - k + 1)^{-1}$ and $\sigma_{nk}^2 = m_{nk} - m_{nk}^2$.

Take $v_n = m_n$. If $L_n = \Sigma_{k=1}^n k^{-1}$, then $v_n = m_n = L_n$ and $\sigma_n^2 = \Sigma_{k=1}^n (k^{-1} - k^{-2}) < L_n$. By Chebyshev's inequality, $P_n[|(S_n - L_n)/L_n| \geq \epsilon] \leq \sigma_n^2/L_n^2 \epsilon^2 < 1/L_n \epsilon^2 \to 0$. Of the $n!$ permutations on n letters, a proportion exceeding $1 - \epsilon^{-2}L_n^{-1}$ thus have their cycle number in the range $(1 \pm \epsilon)L_n$. Since $L_n = \log n + O(1)$, most permutations on n letters have about $\log n$ cycles. For a refinement, see Example 27.3.

Since Ω_n changes with n, it is in the nature of the case that there cannot be a strong law corresponding to this result.   ∎

Bernstein's Theorem

Some theorems that can be stated without reference to probability nonetheless have simple probabilistic proofs, as the last example shows. Bernstein's approach to the Weierstrass approximation theorem is another example.

Let f be a function on $[0, 1]$. The *Bernstein polynomial* of degree n associated with f is

$$(6.7) \qquad B_n(x) = \sum_{k=1}^n f\left(\frac{k}{n}\right) \binom{n}{k} x^k (1 - x)^{n-k}.$$

Theorem 6.3. *If f is continuous, $B_n(x)$ converges to $f(x)$ uniformly on $[0, 1]$.*

According to the Weierstrass approximation theorem, f can be uniformly approximated by polynomials; Bernstein's result goes further and specifies an approximating sequence.

PROOF.   Let $M = \sup_x |f(x)|$, and let $\delta(\epsilon) = \sup [|f(x) - f(y)|: |x - y| \leq \epsilon]$ be the modulus of continuity of f. It will be shown that

$$(6.8) \qquad \sup_x |f(x) - B_n(x)| \leq \delta(\epsilon) + \frac{2M}{n\epsilon^2}.$$

By the uniform continuity of f, $\lim_{\epsilon \to 0} \delta(\epsilon) = 0$, and so this inequality (for $\epsilon = n^{-1/3}$, say) will give the theorem.

Fix $n \geq 1$ and $x \in [0, 1]$ for the moment. Let $X_1, \ldots, X_n$ be independent random variables (on some probability space) such that $P[X_i = 1] = x$ and $P[X_i = 0] = 1 - x$; put $S = X_1 + \cdots + X_n$. Since $P[S = k] = \binom{n}{k}x^k(1 - x)^{n-k}$, the formula (5.16) for calculating expected values of functions of random variables gives $E[f(S/n)] = B_n(x)$. By the law of large numbers, there should be high probability that S/n is near x and hence (f being continuous) that $f(S/n)$ is near $f(x)$. This is the idea behind the proof and, indeed, behind the definition (6.7) itself.

Bound $|f(n^{-1}S) - f(x)|$ by $\delta(\epsilon)$ on the set $[|n^{-1}S - x| < \epsilon]$ and by $2M$ on the complementary set. Since $E[S] = nx$, Chebyshev's inequality gives

$$|B_n(x) - f(x)| \leq E[|f(n^{-1}S) - f(x)|]$$

$$\leq \delta(\epsilon)P[|n^{-1}S - x| < \epsilon] + 2MP[|n^{-1}S - x| \geq \epsilon]$$

$$\leq \delta(\epsilon) + 2M \operatorname{Var}[S]/n^2\epsilon^2;$$

since $\operatorname{Var}[S] = nx(1 - x) \leq n$, (6.8) follows. ∎

A Refinement of the Second Borel-Cantelli Lemma

For a sequence $A_1, A_2, \ldots$ of events, consider the number $N_n = I_{A_1} + \cdots + I_{A_n}$ of occurrences among $A_1, \ldots, A_n$. Since $[A_n \text{ i.o.}] = [\omega: \sup_n N_n(\omega) = \infty]$, $P[A_n \text{ i.o.}]$ can be studied by means of the random variables N_n.

Suppose that the A_n are independent. Put $p_k = P(A_k)$ and $m_n = p_1 + \cdots + p_n$. From $E[I_{A_k}] = p_k$ and $\operatorname{Var}[I_{A_k}] = p_k(1 - p_k) \leq p_k$ follow $E[N_n] = m_n$ and $\operatorname{Var}[N_n] = \Sigma_{k=1}^n \operatorname{Var}[I_{A_k}] \leq m_n$. If $m_n > x$, then

$$(6.9) \qquad P[N_n \leq x] \leq P[|N_n - m_n| \geq m_n - x]$$

$$\leq \frac{\operatorname{Var}[N_n]}{(m_n - x)^2} \leq \frac{m_n}{(m_n - x)^2}.$$

If $\Sigma p_n = \infty$, so that $m_n \to \infty$, it follows that $\lim_n P[N_n \leq x] = 0$ for each x. Since

$$(6.10) \qquad P[\sup_k N_k \leq x] \leq P[N_n \leq x],$$

$P[\sup_k N_k \leq x] = 0$ and hence (take the union over $x = 1, 2, \ldots$) $P[\sup_k N_k < \infty] = 0$. Thus $P[A_n \text{ i.o.}] = P[\sup_n N_n = \infty] = 1$ if the A_n are independent and $\Sigma p_n = \infty$, which proves the second Borel-Cantelli lemma once again.

Independence was used in this argument only to estimate $\operatorname{Var}[N_n]$. Even without independence, $E[N_n] = m_n$ and the first two inequalities in (6.9) hold.

Theorem 6.4.* *If $\Sigma P(A_n)$ diverges and*

$$(6.11) \qquad \liminf_n \frac{\sum\limits_{j,k \leq n} P(A_j \cap A_k)}{\left(\sum\limits_{k \leq n} P(A_k)\right)^2} \leq 1,$$

then $P[A_n \text{ i.o.}] = 1$.

* For an extension, see Problem 6.15.

PROOF.   Let θ_n denote the ratio in (6.11). In the notation above,

$$\text{Var}\,[N_n] = E[N_n^2] - m_n^2 = \sum_{j,k \leq n} E[I_{A_j}I_{A_k}] - m_n^2$$

$$= \sum_{j,k \leq n} P(A_j \cap A_k) - m_n^2 = (\theta_n - 1)m_n^2.$$

Hence (see (6.9)) $P[N_n \leq x] \leq (\theta_n - 1)m_n^2/(m_n - x)^2$ for $x < m_n$. Since $m_n^2/(m_n - x)^2 \to 1$, (6.11) implies that $\liminf_n P[N_n \leq x] = 0$. It still follows by (6.10) that $P[\sup_k N_k \leq x] = 0$, and the rest of the argument is as before.   ∎

Example 6.5.   If as in the second Borel-Cantelli lemma, the A_n are independent (or even if they are merely independent in pairs), the ratio in (6.11) is $1 + \sum_{k \leq n} (p_k - p_k^2)/m_n^2$, so that $\Sigma P(A_n) = \infty$ itself implies (6.11).   ∎

Example 6.6.   Return once again to the run lengths $l_n(\omega)$ of Section 4. It was shown in Example 4.16 that $\{r_n\}$ is an outer boundary ($P[l_n \geq r_n \text{ i.o.}] = 0$) if $\Sigma\ 2^{-r_n} < \infty$. It was also shown that $\{r_n\}$ is an inner boundary ($P[l_n \geq r_n \text{ i.o.}] = 1$) if r_n is nondecreasing and $\Sigma\ 2^{-r_n}r_n^{-1} = \infty$, but Theorem 6.4 can be used to prove this under the sole assumption that $\Sigma\ 2^{-r_n} = \infty$.

As usual, the r_n can be taken to be positive integers. Let $A_n = [l_n \geq r_n] = [d_n = \cdots = d_{n+r_n-1} = 0]$. If $j + r_j \leq k$, then A_j and A_k are independent. Suppose that $j < k < j + r_j$ and put $m = \max\{j + r_j, k + r_k\}$; then $A_j \cap A_k = [d_j = \cdots = d_{m-1} = 0]$, and so $P(A_j \cap A_k) = 2^{-(m-j)} \leq 2^{-(k+r_k-j)} = 2^{-(k-j)}P(A_k)$. Therefore,

$$\sum_{j,k \leq n} P(A_j \cap A_k) \leq \sum_{k \leq n} P(A_k) + 2 \sum_{\substack{j < k \leq n \\ j+r_j \leq k}} P(A_j)P(A_k)$$

$$+ 2 \sum_{\substack{j < k \leq n \\ k < j+r_j}} 2^{-(k-j)}P(A_k) \leq \sum_{k \leq n} P(A_k) + \left(\sum_{k \leq n} P(A_k)\right)^2 + 2 \sum_{k \leq n} P(A_k).$$

If $\Sigma P(A_n) = \Sigma\ 2^{-r_n}$ diverges, then (6.11) follows.

Thus $\{r_n\}$ is an outer or an inner boundary according as $\Sigma\ 2^{-r_n}$ converges or diverges, which completely settles the issue. In particular, $r_n = \log_2 n + \theta \log_2 \log_2 n$ gives an outer boundary for $\theta > 1$ and an inner boundary for $\theta \leq 1$.   ∎

Example 6.7.   It is now possible to complete the analysis in Examples 4.8 and 4.12 of the relative error $e_n(\omega)$ in the approximation of ω by $\Sigma_{k=1}^{n-1} d_k(\omega)2^{-k}$. If $l_n(\omega) \geq -\log_2 x_n$ $(0 < x_n < 1)$, then $e_n(\omega) \leq x_n$ by (4.19). By the preceding example for the case $r_n = -\log_2 x_n$, $\Sigma\ x_n = \infty$ implies that $P[\omega : e_n(\omega) \leq x_n \text{ i.o.}] = 1$. By this and Example 4.8, $[\omega : e_n(\omega) \leq x_n \text{ i.o.}]$ has Lebesgue measure 0 or 1 according as Σx_n converges or diverges.   ∎

PROBLEMS

6.1. Show that $Z_n \to Z$ with probability 1 if and only if for every positive ϵ there exists an n such that $P[|Z_k - Z| < \epsilon, n \le k \le m] > 1 - \epsilon$ for all m exceeding n. This describes convergence with probability 1 in "finite" terms.

6.2. Show in Example 6.4 that $P[|S_n - L_n| \ge L_n^{1/2+\epsilon}] \to 0$.

6.3. As in Examples 5.5 and 6.4, let ω be a random permutation of $1, 2, \ldots, n$. Each k, $1 \le k \le n$, occupies some position in the bottom row of the permutation ω; let $X_{nk}(\omega)$ be the number of smaller elements (between 1 and $k - 1$) lying to the right of k in the bottom row. The sum $S_n = X_{n1} + \cdots + X_{nn}$ is the total number of *inversions*—the number of pairs appearing in the bottom row in reverse order of size. For the permutation in Example 5.5 the values of $X_{71}, \ldots, X_{77}$ are 0, 0, 0, 2, 4, 2, 4, and $S_7 = 12$. Show that $X_{n1}, \ldots, X_{nn}$ are independent and $P[X_{nk} = i] = k^{-1}$ for $0 \le i < k$. Calculate $E[S_n]$ and Var $[S_n]$. Show that S_n is likely to be near $n^2/4$.

6.4. For a function f on $[0, 1]$ write $\|f\| = \sup_x |f(x)|$. Show that, if f has a continuous derivative f', then $\|f - B_n\| \le \epsilon\|f'\| + 2\|f\|/n\epsilon^2$. Conclude that $\|f - B_n\| = O(n^{-1/3})$.

6.5. From Theorem 6.2 deduce *Poisson's theorem*: If $A_1, A_2, \ldots$ are independent events and $P(A_n) = p_n$, if $\bar{p}_n = (p_1 + \cdots + p_n)/n$, and if $N_n = \Sigma_{k=1}^n I_{A_k}$, then $P[|n^{-1}N_n - \bar{p}_n| \ge \epsilon] \to 0$.

In the following problems $S_n = X_1 + \cdots + X_n$.

6.6. Prove *Cantelli's theorem*: If $X_1, X_2, \ldots$ are independent, $E[X_n] = 0$, and $E[X_n^4]$ is bounded, then $n^{-1}S_n \to 0$ with probability 1. The X_n need not be identically distributed.

6.7. Suppose that $n^{-2}S_{n^2} \to 0$ with probability 1 and that the X_n are uniformly bounded ($\sup_{n,\omega} |X_n(\omega)| < \infty$). Show that $n^{-1}S_n \to 0$ with probability 1. Here the X_n need not be identically distributed or even independent.

6.8. ↑   Suppose that $X_1, X_2, \ldots$ are independent and uniformly bounded and $E[X_n]$ = 0. Using only the preceding result, the first Borel-Cantelli lemma, and Chebyshev's inequality, prove that $n^{-1}S_n \to 0$ with probability 1.

6.9. ↑   Use the ideas of Problem 6.8 to give a new proof of Borel's normal number theorem, Theorem 1.2. The point is to return to first principles and use only negligibility and the other ideas of Section 1, not the apparatus of Sections 2 through 6.

6.10. 5.12 6.7 ↑   Suppose that (in the notation of (5.37)) $\beta_n - \alpha_n^2 = O(1/n)$. Show that $n^{-1}N_n - \alpha_n \to 0$ with probability 1. What condition on $\beta_n - \alpha_n^2$ will imply a weak law? Note that independence is not assumed here.

6.11. Suppose that $X_1, X_2, \ldots$ are *m-dependent* in the sense that random variables more than m apart in the sequence are independent. More precisely, let $\mathcal{A}_j^k = \sigma(X_j, \ldots, X_k)$, and assume that $\mathcal{A}_{j_1}^{k_1}, \ldots, \mathcal{A}_{j_l}^{k_l}$ are independent if $k_{i-1} + m < j_i$ for $i = 2, \ldots, l$. (Independent random variables are 0-dependent.) Suppose that the X_n have this property and are uniformly bounded and that $E[X_n] = 0$. Show that $n^{-1}S_n \to 0$.

Hint: Consider the subsequences $X_i, X_{i+m+1}, X_{i+2(m+1)}, \ldots$ for $1 \le i \le m +$ 1.

6.12. ↑ Suppose that the X_n are independent and assume the values $x_1, \ldots, x_l$ with probabilities $p(x_1), \ldots, p(x_l)$. For $u_1, \ldots, u_k$ a k-tuple of the x_i's, let $N_n(u_1, \ldots, u_k)$ be the frequency of the k-tuple in the first $n + k - 1$ trials, that is, the number of m such that $1 \le m \le n$ and $X_m = u_1, \ldots, X_{m+k-1} = u_k$. Show that with probability 1, all asymptotic relative frequencies are what they should be—that is, with probability 1, $n^{-1}N_n(u_1, \ldots, u_k) \to p(u_1) \cdots p(u_k)$ for every k and every k-tuple $u_1, \ldots, u_k$.

6.13. ↑ A number ω in the unit interval is *completely normal* if, for every base b and every k and every k-tuple of base-b digits, the k-tuple appears in the base-b expansion of ω with asymptotic relative frequency b^{-k}. Show that the set of completely normal numbers has Lebesgue measure 1.

6.14. *Shannon's theorem.* Suppose that $X_1, X_2, \ldots$ are independent, identically distributed random variables taking on the values $1, \ldots, r$ with positive probabilities $p_1, \ldots, p_r$. If $p_n(i_1, \ldots, i_n) = p_{i_1} \cdots p_{i_n}$, and if $p_n(\omega) = p_n(X_1(\omega), \ldots, X_n(\omega))$, then $p_n(\omega)$ is the probability that a new sequence of n trials would produce the particular sequence $X_1(\omega), \ldots, X_n(\omega)$ of outcomes that happens actually to have been observed. Show that

$$-\frac{1}{n} \log p_n(\omega) \to h = -\sum_{i=1}^{r} p_i \log p_i$$

with probability 1.

In information theory $1, \ldots, r$ are interpreted as the letters of an *alphabet*, $X_1, X_2, \ldots$ are the successive letters produced by an information *source*, and h is the *entropy* of the source. Prove the *asymptotic equipartition property*: For large n there is probability exceeding $1 - \epsilon$ that the probability $p_n(\omega)$ of the observed n-long sequence, or *message*, is in the range $e^{-n(h \pm \epsilon)}$.

6.15. *The Rényi-Lamperti lemma.* Let the limit inferior in (6.11) be c. Show that $P[A_n \text{ i.o.}] \ge 2 - c$. Note that c cannot be less than 1.

6.16. In the terminology of Example 6.6, show that $\log_2 n + \log_2 \log_2 n + \theta \log_2 \log_2 \log_2 n$ is an outer or inner boundary as $\theta > 1$ or $\theta \le 1$. Generalize. (Compare Problem 4.16.)

6.17. 5.17 ↑ Let $g(m) = \sum_p \delta_p(m)$ be the number of distinct prime devisors of m. For $a_n = E_n[g]$ (see (5.42)) show that $a_n \to \infty$. Show that

$$(6.12) \qquad E_n\left[\left(\delta_p - \frac{1}{n}\left[\frac{n}{p}\right]\right) \left(\delta_q - \frac{1}{n}\left[\frac{n}{q}\right]\right) \right] \le \frac{1}{np} + \frac{1}{nq}$$

for $p \ne q$ and hence that the variance of g under P_n satisfies

$$(6.13) \qquad \text{Var}_n[g] \le 3 \sum_{p \le n} \frac{1}{p}.$$

Prove the *Hardy-Ramanujan theorem*:

$$(6.14) \qquad \lim_n P_n\left[m: \left|\frac{g(m)}{a_n} - 1\right| \ge \epsilon \right] = 0.$$

Since $a_n \sim \log \log n$ (see Problem 18.14), an integer under n has around $\log \log n$ distinct prime devisors. Strengthen (6.14) by letting $\epsilon = \epsilon_n$ vary with n. What is the condition on ϵ_n?

SECTION 7.   GAMBLING SYSTEMS

Let $X_1, X_2, \ldots$ be an independent sequence of random variables (on some (Ω, $\mathscr{F}$, P)) taking on the two values $+1$ and -1 with probabilities $P[X_n = +1] = p$ and $P[X_n = -1] = q = 1 - p$. Throughout the section X_n will be viewed as the gambler's gain on the nth of a series of plays at unit stakes. The game is favorable to the gambler if $p > \frac{1}{2}$, fair if $p = \frac{1}{2}$, and unfavorable if $p < \frac{1}{2}$. The case $p \leq \frac{1}{2}$ will be called the *subfair case*.

After the classical gambler's ruin problem has been solved, it will be shown that every gambling system is in certain respects without effect and that some gambling systems are in other respects optimal. Gambling problems of the sort considered here have inspired many ideas in the mathematical theory of probability, ideas that carry far beyond their origin.

Red-and-black will provide numerical examples. Of the 38 spaces on a roulette wheel, 18 are red, 18 are black, and 2 are green. In betting either on red or on black the chance of winning is $\frac{18}{38}$.

Gambler's Ruin

Suppose that the gambler enters the casino with capital a and adopts the strategy of continuing to bet at unit stakes until his fortune increases to c or his funds are exhausted. What is the probability of ruin, the probability that he will lose his capital, a? What is the probability he will achieve his goal, c? Here a and c are integers.

Let

$$(7.1) \qquad S_n = X_1 + \cdots + X_n, \qquad S_0 = 0.$$

The gambler's fortune after n plays is $a + S_n$. The event

$$(7.2) \qquad A_{a,n} = [a + S_n = c, \, 0 < a + S_k < c, \, k < n]$$

represents success for the gambler at time n, and

$$(7.3) \qquad B_{a,n} = [a + S_n = 0, \, 0 < a + S_k < c, \, k < n]$$

represents ruin at time n. If $s_c(a)$ denotes the probability of ultimate success, then

$$(7.4) \qquad s_c(a) = P\left(\bigcup_{n=1}^{\infty} A_{a,n} \right) = \sum_{n=1}^{\infty} P(A_{a,n})$$

for $0 < a < c$.

Fix c and let a vary. For $n \geq 1$ and $0 < a < c$, define $A_{a,n}$ by (7.2), and adopt the conventions $A_{a,0} = 0$ for $0 \leq a < c$ and $A_{c,0} = \Omega$ (success is impossible at time 0 if $a < c$ and certain if $a = c$), as well as $A_{0,n} = A_{c,n} = 0$ for $n \geq 1$ (play never starts if a is 0 or c). By these conventions, $s_c(0) = 0$ and $s_c(c) = 1$.

Define $A'_{a,n}$ just as $A_{a,n}$ but with $S'_n = X_2 + \cdots + X_{n+1}$ in place of S_n in (7.2). Now $P[X_i = x_i, i \leq n] = P[X_{i+1} = x_i, i \leq n]$ for each sequence $x_1, \ldots, x_n$ of $+1$'s and -1's, and therefore $P[(X_1, \ldots, X_n) \in H] = P[(X_2, \ldots, X_{n+1}) \in H]$ for $H \subset R^n$. Take H to be the set of $x = (x_1, \ldots, x_n)$ in R^n satisfying $x_i = \pm 1$, $a + x_1 + \cdots + x_n = c$, and $0 < a + x_1 + \cdots + x_k < c$ for $k < n$. It follows then that

$$(7.5) \qquad P(A_{a,n}) = P(A'_{a,n}).$$

Moreover, $A_{a,n} = ([X_1 = +1] \cap A'_{a+1,n-1}) \cup ([X_1 = -1] \cap A'_{a-1,n-1})$ for $n \geq 1$ and $0 < a < c$. By independence and (7.5), $P(A_{a,n}) = pP(A_{a+1,n-1}) + qP(A_{a-1,n-1})$; adding over n now gives

$$(7.6) \qquad s_c(a) = ps_c(a + 1) + qs_c(a - 1), \qquad 0 < a < c.$$

The chance of success is thus the chance of winning the first wager times the chance of success for an initial fortune $a + 1$, plus the chance of losing the first wager times the chance of success for an initial fortune $a - 1$. Note that the argument leading to this equation involves the entire infinite sequence X_1, $X_2, \ldots$.

It remains to solve this difference equation with the side conditions $s_c(0) = 0$, $s_c(c) = 1$. Let $\rho = q/p$ be the odds against the gambler. Then (see (7.10)) there exist constants A and B such that, for $0 \leq a \leq c$, $s_c(a) = A + B\rho^a$ if $p \neq q$ and $s_c(a) = A + Ba$ if $p = q$. The requirements $s_c(0) = 0$ and $s_c(c) = 1$ determine A and B, which gives the solution:

The probability the gambler can before ruin attain his goal of c from an initial capital of a is

$$(7.7) \qquad s_c(a) = \begin{cases} \dfrac{\rho^a - 1}{\rho^c - 1}, & 0 \leq a \leq c, \quad \rho = q/p \neq 1, \\[2ex] \dfrac{a}{c}, & 0 \leq a \leq c, \quad \rho = q/p = 1. \end{cases}$$

Example 7.1. The gambler's initial capital is \$900 and his goal is \$1000. If $p = \frac{1}{2}$, his chance of success is very good: $s_{1000}(900) = .9$. At red-and-black, $p = \frac{18}{38}$ and hence $\rho = \frac{20}{18}$; in this case his chance of success as computed by (7.7) is only about .00003. ∎

Example 7.2. It is the gambler's desperate intention to convert his \$100 into \$20,000. For a game in which $p = \frac{1}{2}$ (no casino has one), his chance of success is $100/20,000 = .005$; at red-and-black it is minute—about 3×10^{-911}. ∎

In the analysis leading to (7.7), replace (7.2) by (7.3). It follows that (7.7) with p and q interchanged (ρ goes to ρ^{-1}) and a and $c - a$ interchanged gives the probability $r_c(a)$ of ruin for the gambler: $r_c(a) = (\rho^{-(c-a)} - 1)/(\rho^{-c} - 1)$ if $\rho \neq 1$ and $r_c(a) = (c - a)/c$ if $\rho = 1$. Hence $s_c(a) + r_c(c) = 1$ holds in all cases: *The probability is 0 that play continues forever.*

For positive integers a and b, let $H_{a,b} = \bigcup_{n=1}^{\infty}[S_n = b, -a < S_k < b, k < n]$ be the event that S_n reaches $+b$ before reaching $-a$. Its probability is simply (7.7) with $c = a + b$: $P(H_{a,b}) = s_{a+b}(a)$. Now let $H_b = \bigcup_{a=1}^{\infty} H_{a,b} = \bigcup_{n=1}^{\infty}[S_n = b] = [\sup_n S_n \geq b]$ be the event that S_n ever reaches $+b$. Since $H_{a,b} \uparrow H_b$ as $a \to \infty$, $P(H_b) = \lim_a s_{a+b}(a)$; this is 1 if $\rho = 1$ or $\rho < 1$, and it is $1/\rho^b$ if $\rho > 1$. Thus

$$(7.8) \qquad P\left[\sup_n S_n \geq b\right] = \begin{cases} 1 & \text{if } p \geq q, \\ (p/q)^b & \text{if } p < q. \end{cases}$$

This is the probability that a gambler with unlimited capital can ultimately gain b units.

Example 7.3. The gambler in Example 7.1 has capital 900 and the goal of winning $b = 100$; in Example 7.2 he has capital 100 and b is 19,900. Suppose, instead, that his capital is infinite. If $p = \frac{1}{2}$, the chance of achieving his goal increases from .9 to 1 in the first example and from .005 to 1 in the second. At red-and-black, however, the two probabilities $.9^{100}$ and $.9^{19900}$ remain essentially what they were before (.00003 and 3×10^{-911}). ■

The Difference Equation

Consider a finite sequence $x_0, \ldots, x_N$ or an infinite sequence $x_0, x_1, \ldots$, and suppose that the difference equation

$$(7.9) \qquad x_n = px_{n+1} + qx_{n-1}$$

holds for $n = 1, \ldots, N - 1$ or for $n = 1, 2, \ldots$, as the case may be. The general solution has the form

$$(7.10) \qquad x_n = \begin{cases} A + B(q/p)^n & \text{if } p \neq q, \\ A + Bn & \text{if } p = q. \end{cases}$$

That (7.10) always solves (7.9) is easily checked. Suppose that $\{x_n\}$ is given by (7.10) and $\{a_n\}$ is an arbitrary solution of (7.9). It will follow by induction that $a_n = x_n$ for all n if $a_0 = x_0$ and $a_1 = x_1$. But the latter requirements give a pair of linear equations for A and B, and the system is easily seen to be solvable in each case.

Thus each solution of (7.9) is given by (7.10) for some pair A and B. Specifying x_n for any two distinct values of n suffices to determine A and B completely, again because it always gives a nonsingular pair of linear equations in A and B.

Selection Systems

Players often try to improve their luck by betting only when in the preceding trials the wins and losses form an auspicious pattern. Perhaps the gambler bets on the nth trial only when among $X_1, \ldots, X_{n-1}$ there are many more $+1$'s than -1's, the idea being to ride winning streaks (he is "in the vein"). Or he may bet only when there are many more -1's than $+1$'s, the idea being it is then surely time a $+1$ came along (the "maturity of the chances"). There is a mathematical theorem which translated into gaming language says that all such systems are futile.

The gambler's strategy is described by random variables $B_1, B_2, \ldots$ taking the two values 0 and 1: If $B_n = 1$, the gambler places a bet on the nth trial; if $B_n = 0$, he skips that trial. If B_n were $(X_n + 1)/2$, so that $B_n = 1$ for $X_n = +1$ and $B_n = 0$ for $X_n = -1$, the gambler would win every time he bet, but of course such a system requires he be prescient—he must know the outcome X_n in advance. For this reason the value of B_n is assumed to depend only on the values of $X_1, \ldots, X_{n-1}$: there exists some function $b_n: R^{n-1} \to R^1$ such that

(7.11) $B_n = b_n(X_1, \ldots, X_{n-1}).$

(Here B_1 is constant.) Thus the mathematics avoids, as it must, the question of whether prescience is actually possible.

Define

(7.12) $\begin{cases} \mathcal{F}_n = \sigma(X_1, \ldots, X_n), & n = 1, 2, \ldots, \\ \mathcal{F}_0 = \{0, \Omega\}. \end{cases}$

The σ-field $\mathcal{F}_{n-1}$ generated by $X_1, \ldots, X_{n-1}$ corresponds to a knowledge of the outcomes of the first $n - 1$ trials. The requirement (7.11) ensures that B_n is measurable $\mathcal{F}_{n-1}$ (Theorem 5.1) and so depends only on the information actually available to the gambler just before the nth trial.

For $n = 1, 2, \ldots$, let N_n be the time at which the gambler places his nth bet. This nth bet is placed at time k or earlier if and only if the number $\sum_{i=1}^{k} B_i$ of bets placed up to and including time k is n or more; in fact, N_n is the smallest k for which $\sum_{i=1}^{k} B_i = n$. Thus the event $[N_n \leq k]$ coincides with $[\sum_{i=1}^{k} B_i \geq n]$; by (7.11) this latter event lies in $\sigma(B_1, \ldots, B_k) \subset \sigma(X_1, \ldots, X_{k-1}) = \mathcal{F}_{k-1}$. Therefore,

(7.13) $[N_n = k] = [N_n \leq k] - [N_n \leq k - 1] \in \mathcal{F}_{k-1}.$

(Even though $[N_n = k]$ lies in $\mathcal{F}_{k-1}$ and hence in $\mathcal{F}$, N_n is as a function on Ω generally not a simple random variable because it has infinite range. This makes no difference because expected values of the N_n will play no role; (7.13) is the essential property.)

To ensure that play continues forever (stopping rules will be considered later) and that the N_n have finite values with probability 1, make the further assumption that

(7.14) $P[B_n = 1 \text{ i.o.}] = 1.$

A sequence $\{B_n\}$ of random variables assuming the values 0 and 1, having the form (7.11), and satisfying (7.14) is a *selection system*.

Let Y_n be the gambler's gain on the nth of the trials at which he does bet: $Y_n = X_{N_n}$. It is only on the set $[B_n = 1 \text{ i.o.}]$ that all the N_n and hence all the Y_n are well defined. To complete the definition, set $Y_n = -1$, say, on $[B_n = 1 \text{ i.o.}]^c$; since this set has probability 0 by (7.14), it really makes no difference how Y_n is defined on it.

Now Y_n is a complicated function on Ω because $Y_n(\omega) = X_{N_n(\omega)}(\omega)$. Nonetheless,

$$[\omega: Y_n(\omega) = 1] = \bigcup_{k=1}^{\infty} ([\omega: N_n(\omega) = k] \cap [\omega: X_k(\omega) = 1])$$

lies in $\mathcal{F}$ and so does its complement $[\omega: Y_n(\omega) = -1]$. Hence Y_n is a simple random variable.

Example 7.4. An example will fix these ideas. Suppose that the rule is always to bet on the first trial, to bet on the second trial if and only if $X_1 = +1$, to bet on the third trial if and only if $X_1 = X_2$, and to bet on all subsequent trials. Here $B_1 = 1$, $[B_2 = 1] = [X_1 = +1]$, $[B_3 = 1] = [X_1 = X_2]$, and $B_4 = B_5 = \cdots = 1$. The table shows the ways the gambling can start out. A dot represents a value undetermined by X_1, X_2, X_3. Ignore the rightmost column for the moment.

X_1	X_2	X_3	B_1	B_2	B_3	N_1	N_2	N_3	N_4	Y_1	Y_2	Y_3	τ
-1	-1	-1	1	0	1	1	3	4	5	-1	-1	.	1
-1	-1	$+1$	1	0	1	1	3	4	5	-1	$+1$	.	1
-1	$+1$	-1	1	0	0	1	4	5	6	-1	.	.	1
-1	$+1$	$+1$	1	0	0	1	4	5	6	-1	.	.	1
$+1$	-1	-1	1	1	0	1	2	4	5	$+1$	-1	.	2
$+1$	-1	$+1$	1	1	0	1	2	4	5	$+1$	-1	.	2
$+1$	$+1$	-1	1	1	1	1	2	3	4	$+1$	$+1$	-1	3
$+1$	$+1$	$+1$	1	1	1	1	2	3	4	$+1$	$+1$	$+1$	.

In the evolution represented by the first line of the table, the second bet is placed on the third trial ($N_2 = 3$), which results in a loss because $Y_2 = X_{N_2} = X_3 = -1$. Since $X_3 = -1$, the gambler was "wrong" to bet. But remember that before the third trial he does not know $X_3(\omega)$ (much less ω itself); he knows only $X_1(\omega)$ and $X_2(\omega)$. See the discussion in Example 5.4. ∎

Selection systems achieve nothing because $\{Y_n\}$ has the same structure as $\{X_n\}$:

Theorem 7.1. *For any selection system,* $\{Y_n\}$ *is independent and* $P[Y_n = +1] = p$, $P[Y_n = -1] = q$.

PROOF.   Relabel p and q as $p(+1)$ and $p(-1)$, so that $P[X_k = x] = p(x)$ for $x = \pm 1$. If $A \in \mathcal{F}_{k-1}$, then A and $[X_k = x]$ are independent, and so $P(A \cap [X_k = x]) = P(A)p(x)$. Therefore, by (7.13), $P[Y_n = x] = \Sigma_{k=1}^{\infty} P[N_n = k, X_k = x] = \Sigma_{k=1}^{\infty} P[N_n = k]p(x) = p(x)$.

More generally, for any sequence $x_1, \ldots, x_n$ of ± 1's, $P[Y_i = x_i, i \leq n] = \Sigma_{k_1 < \ldots < k_n} P[N_i = k_i, X_{k_i} = x_i, i \leq n]$, where the sum extends over n-tuples of positive integers satisfying $k_1 < \cdots < k_n$. The event $[N_i = k_i, i \leq n] \cap [X_{k_i} = x_i, i < n]$ lies in $\mathcal{F}_{k_n-1}$ (note there is no condition on X_{k_n}), and therefore $P[Y_i = x_i, i \leq n] = \Sigma_{k_1 < \ldots < k_n} P([N_i = k_i, i \leq n] \cap [X_{k_i} = x_i, i < n])p(x_n)$. Summing k_n over $k_{n-1} + 1, k_{n-1} + 2, \ldots$ brings this last sum to $\Sigma_{k_1 < \ldots < k_{n-1}} P[N_i = k_i, X_{k_i} = x_i, i < n]p(x_n)$, and therefore $P[Y_i = x_i, i \leq n] = P[Y_i = x_i, i < n]p(x_n)$. It now follows by induction that $P[Y_i = x_i, i \leq n] = \prod_{i \leq n} p(x_i) = \prod_{i \leq n} P[Y_i = x_i]$, and so the Y_i are independent (see (5.6)).   ∎

Gambling Policies

There are schemes that go beyond selection systems and tell the gambler not only whether to bet but how much. Gamblers frequently contrive or adopt such schemes in the confident expectation that they can by pure force of arithmetic counter the most adverse workings of chance. If the wager specified for the nth trial is in the amount W_n and the gambler cannot see into the future, then W_n must depend only on $X_1, \ldots, X_{n-1}$. Assume therefore that W_n is a nonnegative function of these random variables: there is an $f_n : R^{n-1} \to R^1$ such that

$$(7.15) \qquad W_n = f_n(X_1, \ldots, X_{n-1}) \geq 0.$$

Apart from nonnegativity there are at the outset no constraints on the f_n, although in an actual casino their values must be integral multiples of a basic unit. Such a sequence $\{W_n\}$ is a *betting system*. Since $W_n = 0$ corresponds to a decision not to bet at all, betting systems in effect include selection systems. In the double-or-nothing system, $W_n = 2^{n-1}$ if $X_1 = \cdots = X_{n-1} = -1$ ($W_1 = 1$) and $W_n = 0$ otherwise.

The amount the gambler wins on the nth play is $W_n X_n$. If his fortune at time n is F_n, then

$$(7.16) \qquad F_n = F_{n-1} + W_n X_n.$$

This also holds for $n = 1$ if F_0 is taken as his initial (nonrandom) fortune. It is convenient to let W_n depend on F_0 as well as the past history of play and hence to generalize (7.15) to

$$(7.17) \qquad W_n = g_n(F_0, X_1, \ldots, X_{n-1}) \geq 0$$

for a function $g_n : R^n \to R^1$. In expanded notation, $W_n(\omega) = g_n(F_0, X_1(\omega), \ldots,$

$X_{n-1}(\omega)$). The symbol W_n does not show the dependence on ω or on F_0 either. For each *fixed* initial fortune F_0, W_n is a simple random variable; by (7.17) it is measurable $\mathcal{F}_{n-1}$. Similarly, F_n is a function of F_0 as well as of $X_1(\omega), \ldots, X_n(\omega)$: $F_n = F_n(F_0, \omega)$.

If $F_0 = 0$ and $g_n \equiv 1$, the F_n reduce to the partial sums (7.1).

Since $\mathcal{F}_{n-1}$ and $\sigma(X_n)$ are independent, and since W_n is measurable $\mathcal{F}_{n-1}$ (for each fixed F_0), W_n and X_n are independent. Therefore, $E[W_n X_n] = E[W_n] \cdot E[X_n]$. Now $E[X_n] = p - q \le 0$ in the subfair case ($p \le \frac{1}{2}$), with equality in the fair case ($p = \frac{1}{2}$). Since $E[W_n] \ge 0$, (7.16) implies that $E[F_n] \le E[F_{n-1}]$. Therefore,

$$(7.18) \qquad F_0 \ge E[F_1] \ge \cdots \ge E[F_n] \ge \cdots$$

in the subfair case, and

$$(7.19) \qquad F_0 = E[F_1] = \cdots = E[F_n] = \cdots$$

in the fair case. (If $p < q$ and $P[W_n > 0] > 0$, there is strict inequality in (7.18).) Thus no betting system can convert a subfair game into a profitable enterprise.

Suppose that in addition to a betting system, the gambler adopts some policy for quitting. Perhaps he stops when his fortune reaches a set target, or his funds are exhausted, or the auguries are in some way dissuasive. The decision to stop must depend only on the initial fortune and the history of play up to the present.

Let $\tau(F_0, \omega)$ be a nonnegative integer for each ω in Ω and each $F_0 \ge 0$. If $\tau = n$, the gambler plays on the nth trial (betting W_n) and then stops; if $\tau = 0$, he does not begin gambling in the first place. The event $[\omega: \tau(F_0, \omega) = n]$ represents the decision to stop just after the nth trial, and so, whatever value F_0 may have, it must depend only on $X_1, \ldots, X_n$. Therefore, assume that

$$(7.20) \qquad [\omega: \tau(F_0, \omega) = n] \in \mathcal{F}_n, \qquad n = 0, 1, 2, \ldots$$

A τ satisfying this requirement is a *stopping time*. (In general it has infinite range and hence is not a simple random variable; as expected values of τ play no role here, this does not matter.) It is technically necessary to let $\tau(F_0, \omega)$ be undefined or infinite on an ω-set of probability 0. This has no effect on the requirement (7.20), which must hold for each finite n. Since τ is finite with probability 1, play is certain to terminate.

A betting system together with a stopping time is a *gambling policy*. Let π denote such a policy.

Example 7.5. Suppose that the betting system is given by $W_n = B_n$, with B_n as in Example 7.4. Suppose that the stopping rule is to quit after the first loss of a wager. Then $[\tau = n] = \bigcup_{k=1}^{\infty} [N_k = n, Y_1 = \cdots = Y_{k-1} = +1, Y_k = -1]$.

For $j \leq k$, $[N_k = n, Y_j = x] = \bigcup_{m=1}^n [N_k = n, N_j = m, X_m = x]$ lies in $\mathcal{F}_n$ by (7.13); hence τ is a stopping time. The values of τ are shown in the right column of the table.   ∎

The sequence of fortunes is governed by (7.16) until play terminates, and then the fortune remains for all future time fixed at F_τ (with value $F_{\tau(F_0, \omega)}(\omega)$). Therefore, the gambler's fortune at time n is

$$(7.21) \qquad\qquad F_n^* = \begin{cases} F_n & \text{if } \tau \geq n, \\ F_\tau & \text{if } \tau \leq n. \end{cases}$$

Note that the case $\tau = n$ is covered by both clauses here. If $n - 1 < n \leq \tau$, then $F_n^* = F_n = F_{n-1} + W_n X_n = F_{n-1}^* + W_n X_n$; if $\tau \leq n - 1 < n$, then $F_n^* = F_\tau = F_{n-1}^*$. Therefore, if $W_n^* = I_{[\tau \geq n]} W_n$, then

$$(7.22) \qquad\qquad F_n^* = F_{n-1}^* + I_{[\tau \geq n]} W_n X_n = F_{n-1}^* + W_n^* X_n.$$

But this is the equation for a new betting system in which the wager placed at time n is W_n^*. If $\tau \geq n$ (play has not already terminated), W_n^* is the old amount W_n; if $\tau < n$ (play has terminated), W_n^* is 0. Now by (7.20), $[\tau \geq n] = [\tau < n]^c$ lies in $\mathcal{F}_{n-1}$. Thus $I_{[\tau \geq n]}$ is measurable $\mathcal{F}_{n-1}$, so that W_n^* as well as W_n is measurable $\mathcal{F}_{n-1}$, and $\{W_n^*\}$ represents a legitimate betting system. Therefore, (7.18) and (7.19) apply to the new system:

$$(7.23) \qquad\qquad F_0 = F_0^* \geq E[F_1^*] \geq \cdots \geq E[F_n^*] \geq \cdots$$

if $p \geq \frac{1}{2}$, and

$$(7.24) \qquad\qquad F_0 = F_0^* = E[F_1^*] = \cdots = E[F_n^*] = \cdots$$

if $p = \frac{1}{2}$.

The gambler's ultimate fortune is F_τ. Now $\lim_n F_n^* = F_\tau$ with probability 1 since in fact $F_n^* = F_\tau$ for $n \geq \tau$. If

$$(7.25) \qquad\qquad \lim_n E[F_n^*] = E[F_\tau],$$

then (7.23) and (7.24), respectively, imply that $E[F_\tau] \leq F_0$ and $E[F_\tau] = F_0$. According to Theorem 5.3, (7.25) does hold if the F_n^* are uniformly bounded.

Call the policy *bounded by M* (M nonrandom) if

$$(7.26) \qquad\qquad 0 \leq F_n^* \leq M, \qquad n = 0, 1, 2, \ldots .$$

If F_n^* is not bounded above, the gambler's adversary must have infinite capital. A negative F_n^* represents a debt, and if F_n^* is not bounded below, the gambler must have a patron of infinite wealth and generosity from whom to borrow and so must in effect have infinite capital. In case F_n^* is bounded below, 0 is the convenient lower bound—the gambler is assumed to have in hand all the capital to which he has access. In any real case, (7.26) holds and (7.25) follows. (There is a technical point that arises because the general theory of integration has been postponed: F_τ must be assumed to have finite range so that it will be a simple random variable and hence have an expected value in the sense of Section 5.) The argument has led to this result:

Theorem 7.2. *For any policy, (7.23) holds if $p \leq \frac{1}{2}$ and (7.24) holds if $p = \frac{1}{2}$. If the policy is bounded (and F_τ has finite range), then $E[F_\tau] \leq F_0$ for $p \leq \frac{1}{2}$ and $E[F_\tau] = F_0$ for $p = \frac{1}{2}$.*

Example 7.6. The gambler has initial capital a and plays at unit stakes until his capital increases to c $(0 \leq a \leq c)$ or he is ruined. Here $F_0 = a$ and $W_n = 1$, and so $F_n = a + S_n$. The policy is bounded by c and F_τ is c or 0 according as the gambler succeeds or fails. If $p = \frac{1}{2}$ and if s is the probability of success, then $a = F_0 = E[F_\tau] = sc$. Thus $s = a/c$. This gives a new derivation of (7.7) for the case $p = \frac{1}{2}$. The argument assumes however that play is certain to terminate. If $p \leq \frac{1}{2}$, Theorem 7.2 only gives $s \leq a/c$, which is weaker than (7.7). ∎

Example 7.7. Suppose as before that $F_0 = a$ and $W_n = 1$, so that $F_n = a + S_n$, but suppose the stopping rule is to quit as soon as F_n reaches $a + b$. Here F_n^* is bounded above by $a + b$ but is not bounded below. If $p = \frac{1}{2}$, the gambler is by (7.8) certain to achieve his goal, so that $F_\tau = a + b$. In this case $F_0 = a < a + b = E[F_\tau]$. This illustrates the effect of infinite capital. It also illustrates the need for uniform boundedness in Theorem 5.3 (compare Example 5.6). ∎

For some other systems (gamblers call them "martingales"), see the problems. For most such systems there is a large chance of a small gain and a small chance of a large loss.

Bold Play†

The formula (7.7) gives the chance that a gambler betting unit stakes can increase his fortune from a to c before being ruined. Suppose that a and c happen to be even and that at each trial the wager is two units instead of one. Since this has the effect of halving a and c, the chance of success is now

$$\frac{\rho^{a/2} - 1}{\rho^{c/2} - 1} = \frac{\rho^a - 1}{\rho^c - 1} \frac{\rho^{c/2} + 1}{\rho^{a/2} + 1}, \quad \frac{q}{p} = \rho \neq 1.$$

† This topic may be omitted.

If $\rho > 1$ ($p < \frac{1}{2}$), the second factor on the right exceeds 1: Doubling the stakes increases the probability of success in the unfavorable case $\rho > 1$. In case $\rho = 1$, the probability remains the same.

There is a sense in which large stakes are optimal. It will be convenient to rescale so that the initial fortune satisfies $0 \leq F_0 \leq 1$ and the goal is 1. The policy of *bold play* is this: At each stage the gambler bets his entire fortune, unless a win would carry him past his goal of 1, in which case he bets just enough that a win would exactly achieve that goal:

$$(7.27) \qquad W_n = \begin{cases} F_{n-1} & \text{if } 0 \leq F_{n-1} \leq \frac{1}{2}, \\ 1 - F_{n-1} & \text{if } \frac{1}{2} \leq F_{n-1} \leq 1. \end{cases}$$

(It is convenient to allow even irrational fortunes.) As for stopping, the policy is to quit as soon as F_n reaches 0 or 1.

For play to continue beyond time n under this policy, X_k must for $k \leq n$ be $+1$ or -1 according as $F_{k-1} \leq \frac{1}{2}$ or $F_{k-1} \geq \frac{1}{2}$. Since the probability of this is at most m^n, where m is the maximum of p and q, play is certain to terminate (τ is finite with probability 1).

It will be shown that in the subfair case, bold play maximizes the probability of successfully reaching the goal of 1. It will further be shown that there are other policies that are also optimal in this sense, and this maximum probability will be calculated. Bold play can be substantially better than betting at constant stakes. This contrasts with Theorems 7.1 and 7.2 concerning respects in which gambling systems are worthless.

From now on, consider only policies π that are bounded by 1 (see (7.26)). Suppose further that play stops as soon as F_n reaches 0 or 1 and that this is certain eventually to happen. Since F_τ assumes the values 0 and 1, and since $[F_\tau = x] = \bigcup_{n=0}^{\infty} [\tau = n] \cap [F_n = x]$ for $x = 0$ and $x = 1$, F_τ is a simple random variable. Bold play is one such policy π.

The policy π leads to success if $F_\tau = 1$. Let $Q_\pi(x)$ be the probability of this for an initial fortune $F_0 = x$:

$$(7.28) \qquad Q_\pi(x) = P[F_\tau = 1], \qquad F_0 = x.$$

Since F_n is a function $\psi_n(F_0, X_1(\omega), \ldots, X_n(\omega)) = \Psi_n(F_0, \omega)$, (7.28) in expanded notation is $Q_\pi(x) = P[\omega: \Psi_{\tau(x,\omega)}(x, \omega) = 1]$. As π specifies that play stops at the boundaries 0 and 1,

$$(7.29) \qquad \begin{cases} Q_\pi(0) = 0, & Q_\pi(1) = 1, \\ 0 \leq Q_\pi(x) \leq 1, & 0 \leq x \leq 1. \end{cases}$$

Let Q be the Q_π for bold play. (The notation does not show the dependence of Q and Q_π on p, which is fixed.)

Theorem 7.3. *In the subfair case, $Q_\pi(x) \leq Q(x)$ for all π and all x.*

PROOF. Under the assumption $p \le q$, it will be shown later that

$$(7.30) \qquad Q(x) \ge pQ(x+t) + qQ(x-t), \qquad 0 \le x - t \le x \le x + t \le 1.$$

This can be interpreted as saying that the chance of success under bold play starting at x is at least as great as the chance of success if the amount t is wagered and bold play then pursued from $x + t$ in case of a win and from $x - t$ in case of a loss. Under the assumption of (7.30), optimality can be proved as follows.

Consider a policy π, and let F_n and F_n^* be the simple random variables defined by (7.16) and (7.21) for this policy. Now $Q(x)$ is a real function, and so $Q(F_n^*)$ is also a simple random variable; it can be interpreted as the chance of success if π is replaced by bold play after time n. By (7.22), $F_n^* = x + tX_n$ if $F_{n-1}^* = x$ and $W_n^* = t$. Therefore,

$$Q(F_n^*) = \sum_{x,t} I_{[F_{n-1}^* = x, \; W_n^* = t]} Q(x + tX_n),$$

where x and t vary over the (finite) ranges of F_{n-1}^* and W_n^*, respectively.

For each x and t, the indicator above is measurable $\mathcal{F}_{n-1}$ and $Q(x + tX_n)$ is measurable $\sigma(X_n)$; since the X_n are independent, (5.21) and (5.14) give

$$(7.31) \qquad E[Q(F_n^*)] = \sum_{x,t} P[F_{n-1}^* = x, \; W_n^* = t] E[Q(x + tX_n)].$$

By (7.30), $E[Q(x + tX_n)] \le Q(x)$ if $0 \le x - t \le x \le x + t \le 1$. As it is assumed of π that F_n^* lies in $[0, 1]$ (that is, $W_n^* \le \min\{F_{n-1}^*, 1 - F_{n-1}^*\}$), the probability in (7.31) is 0 unless x and t satisfy this constraint. Therefore,

$$E[Q(F_n^*)] \le \sum_{x,t} P[F_{n-1}^* = x, \; W_n^* = t] Q(x)$$

$$= \sum_x P[F_{n-1}^* = x] Q(x) = E[Q(F_{n-1}^*)].$$

This is true for each n, and so $E[Q(F_n^*)] \le E[Q(F_0^*)] = Q(F_0)$. Since $Q(F_n^*) = Q(F_\tau)$ for $n \ge \tau$, Theorem 5.3 implies that $E[Q(F_\tau)] \le Q(F_0)$. Since $x = 1$ implies that $Q(x) = 1$, $P[F_\tau = 1] \le E[Q(F_\tau)] \le Q(F_0)$. Thus $Q_\pi(F_0) \le Q(F_0)$ for the policy π, whatever F_0 may be.

It remains to analyze Q and prove (7.30). Everything hinges on the functional equation

$$(7.32) \qquad Q(x) = \begin{cases} pQ(2x), & 0 \le x \le \frac{1}{2}, \\ p + qQ(2x - 1), & \frac{1}{2} \le x \le 1. \end{cases}$$

For $x = 0$ and $x = 1$ this is obvious because $Q(0) = 0$ and $Q(1) = 1$. The idea is this: Suppose that the initial fortune is x. If $x \le \frac{1}{2}$, the first stake under bold play is x; if the gambler is to succeed in reaching 1, he must win the first trial (probability p) and then from his new fortune $2x$ go on to succeed (probability $Q(2x)$). If $x \geq \frac{1}{2}$, the first stake is $1 - x$; the gambler can succeed either by winning the first trial (probability p) or by losing the first trial (probability q) and then going on from his new fortune $x - (1 - x) = 2x - 1$ to succeed (probability $Q(2x - 1)$).

A formal proof of (7.32) can be constructed as for the difference equation (7.6). If $\beta(x)$ is x for $x \leq \frac{1}{2}$ and $1 - x$ for $x \geq \frac{1}{2}$, then under bold play $W_n = \beta(F_{n-1})$. Starting from $f_0(x) = x$, recursively define

$$f_n(x; x_1, \ldots, x_n) = f_{n-1}(x; x_1, \ldots, x_{n-1}) + \beta(f_{n-1}(x; x_1, \ldots, x_{n-1}))x_n,$$

and put

$$g_n(x; x_1, \ldots, x_n) = \max_{0 \leq k \leq n} f_k(x; x_1, \ldots, x_k).$$

Then $F_n = f_n(F_0; X_1, \ldots, X_k)$, and if $F_0 = x$, then $T_n(x) = [g_n(x; X_1, \ldots, X_n) = 1]$ is the event that bold play will by time n successfully increase the gambler's fortune to 1. From the recursive definition it follows by induction on n that $f_n(x; x_1, \ldots, x_n) = f_{n-1}(x + \beta(x)x_1; x_2, \ldots, x_n)$ and hence that $g_n(x; x_1, \ldots, x_n) = \max\{x, g_{n-1}(x + \beta(x)x_1; x_2, \ldots, x_n)\}$ for $n \geq 1$. Therefore, $T_n(x) = [g_{n-1}(x + \beta(x)X_1; X_2, \ldots, X_n) = 1]$, and since the X_i are independent and identically distributed, $P(T_n(x)) = P([X_1 = +1] \cap T_n(x)) + P([X_1 = -1] \cap T_n(x)) = pP[g_{n-1}(x + \beta(x); X_2, \ldots, X_n) = 1] + qP[g_{n-1}(x - \beta(x); X_2, \ldots, X_n) = 1] = pP(T_{n-1}(x + \beta(x))) + qP(T_{n-1}(x - \beta(x)))$. Letting $n \to \infty$ now gives $Q(x) = pQ(x + \beta(x)) + qQ(x - \beta(x))$, which reduces to (7.32) because $Q(0) = 0$ and $Q(1) = 1$.

Suppose that $y = f_{n-1}(x; x_1, \ldots, x_{n-1})$ is nondecreasing in x. If $x_n = +1$, then $f_n(x; x_1, \ldots, x_n)$ is $2y$ if $0 \leq y \leq \frac{1}{2}$ and 1 if $\frac{1}{2} \leq y \leq 1$; if $x_n = -1$, then $f_n(x; x_1, \ldots, x_n)$ is 0 if $0 \leq y \leq \frac{1}{2}$ and $2y - 1$ if $\frac{1}{2} \leq y \leq 1$. In any case, $f_n(x; x_1, \ldots, x_n)$ is also nondecreasing in x, and by induction this is true for every n. It follows that the same is true of $g_n(x; x_1, \ldots, x_n)$, of $P(T_n(x))$, and of $Q(x)$. Thus $Q(x)$ is nondecreasing.

Since $Q(1) = 1$, (7.32) implies that $Q(\frac{1}{2}) = pQ(1) = p$, $Q(\frac{1}{4}) = pQ(\frac{1}{2}) = p^2$, $Q(\frac{3}{4}) = p + qQ(\frac{1}{2}) = p + pq$. More generally, if $p_0 = p$ and $p_1 = q$, then

$$(7.33) \quad Q\left(\frac{k}{2^n}\right) = \Sigma \left[p_{u_1} \ldots p_{u_n}: \sum_{i=1}^{n} \frac{u_i}{2^i} < \frac{k}{2^n}\right], \qquad 0 < k \leq 2^n,$$

the sum extending over n-tuples $(u_1, \ldots, u_n)$ of 0's and 1's satisfying the condition indicated. Indeed, it is easy to see that (7.33) is the same thing as

$$(7.34) \qquad Q(.u_1 \ldots u_n + 2^{-n}) - Q(.u_1 \ldots u_n) = p_{u_1} p_{u_2} \ldots p_{u_n}$$

for each dyadic rational $.u_1 \ldots u_n$ of rank n. If $.u_1 \ldots u_n + 2^{-n} \leq \frac{1}{2}$, then $u_1 = 0$ and by (7.32) the difference in (7.34) is $p_0[Q(.u_2 \ldots u_n + 2^{-n+1}) - Q(.u_2$

$\ldots u_n)$]. But (7.34) follows inductively from this and a similar relation for the case $.u_1 \ldots u_n \geq \frac{1}{2}$.

Therefore $Q(k2^{-n}) - Q((k-1)2^{-n})$ is bounded by max $\{p^n, q^n\}$, and so by monotonicity Q is continuous.

The inequality (7.30) is still to be proved. It is equivalent to the assertion that

$$\Delta(r, s) = Q(a) - pQ(s) - qQ(r) \geq 0$$

if $0 \leq r \leq s \leq 1$, where a stands for the average: $a = \frac{1}{2}(r + s)$. Since Q is continuous, it suffices to prove the inequality for r and s of the form $k/2^n$, and this will be done by induction on n. Checking all cases disposes of $n = 0$. Assume that the inequality holds for a particular n, and that r and s have the form $k/2^{n+1}$. There are four cases to consider.

Case 1. $s \leq \frac{1}{2}$. By the first part of (7.32), $\Delta(r, s) = p\Delta(2r, 2s)$. Since $2r$ and $2s$ have the form $k/2^n$, the induction hypothesis implies that $\Delta(2r, 2s) \geq 0$.

Case 2. $\frac{1}{2} \leq r$. The proof is as before but uses the second part of (7.32).

Case 3. $r \leq a \leq \frac{1}{2} \leq s$. By (7.32),

$$\Delta(r, s) = pQ(2a) - p[p + qQ(2s - 1)] - q[pQ(2r)].$$

Since $\frac{1}{2} \leq s \leq r + s = 2a \leq 1$, $Q(2a) = p + qQ(4a - 1)$; since $0 \leq 2a - \frac{1}{2} \leq \frac{1}{2}$, $Q(2a - \frac{1}{2}) = pQ(4a - 1)$. Therefore, $pQ(2a) = p^2 + qQ(2a - \frac{1}{2})$, and it follows that

$$\Delta(r, s) = q[Q(2a - \frac{1}{2}) - pQ(2s - 1) - pQ(2r)].$$

Since $p \leq q$, the right side does not increase if either of the two p's is changed to q. Hence

$$\Delta(r, s) \geq q \max [\Delta(2r, 2s - 1), \Delta(2s - 1, 2r)].$$

The induction hypothesis applies to $2r \leq 2s - 1$ or to $2s - 1 \leq 2r$, as the case may be, and so one of the two Δ's on the right is nonnegative.

Case 4. $r \leq \frac{1}{2} \leq a \leq s$. This case is symmetric to the preceding one.

This completes the proof of (7.30) and hence of Theorem 7.3.   ■

The equation (7.33) has an interesting interpretation. Let $Z_1, Z_2, \ldots$ be independent random variables satisfying $P[Z_n = 0] = p_0 = p$ and $P[Z_n = 1] = p_1 = q$. From $P[Z_n = 1 \text{ i.o.}] = 1$ and $\Sigma_{i>n} Z_i 2^{-i} \leq 2^{-n}$ it follows that $P[\Sigma_{i=1}^{\infty} Z_i 2^{-i} \leq k2^{-n}] \leq P[\Sigma_{i=1}^{n} Z_i 2^{-i} < k2^{-n}] \leq P[\Sigma_{i=1}^{\infty} Z_i 2^{-i} \leq k2^{-n}]$. Since by (7.33) the middle term is $Q(k2^{-n})$,

(7.35)
$$Q(x) = P\left[\sum_{i=1}^{\infty} Z_i 2^{-i} \le x \right]$$

holds for dyadic rational x and hence by continuity holds for all x. In Section 31, Q will reappear as a continuous, strictly increasing function singular in the sense of Lebesgue. On page 362 is a graph for the case $p_0 = .3$.

Note that $Q(x) \equiv x$ in the fair case $p = \frac{1}{2}$. In fact, for a bounded policy Theorem 7.2 implies that $E[F_\tau] = F_0$ in the fair case, and if the policy is to stop as soon as the fortune reaches 0 or 1, then the chance of successfully reaching 1 is $P[F_\tau = 1] = E[F_\tau] = F_0$. Thus in the fair case with initial fortune x, the chance of success is x for *any* policy that stops at the boundaries, and x is an upper bound even if stopping earlier is allowed.

Example 7.8. The gambler of Example 7.1 has capital $900 and goal $1000. For a fair game ($p = \frac{1}{2}$) his chance of success is .9 whether he bets unit stakes or adopts bold play. At red-and-black ($p = \frac{18}{38}$), his chance of success with unit stakes is .00003; an approximate calculation based on (7.33) shows that under bold play his chance $Q(.9)$ of success increases to about .88, which compares well with the fair case.   ∎

Example 7.9. In Example 7.2 the capital is $100 and the goal $20,000. At unit stakes the chance of successes is .005 for $p = \frac{1}{2}$ and 3×10^{-911} for $p = \frac{18}{38}$. Another approximate calculation shows that bold play at red-and-black gives the gambler probability about .003 of success, which again compares well with the fair case.

This example illustrates the point of Theorem 7.3. The gambler enters the casino knowing that he must by dawn convert his $100 into $20,000 or face certain death at the hands of criminals to whom he owes that amount. Only red-and-black is available to him. The question is not whether to gamble—he *must* gamble. The question is how to gamble so as to maximize the chance of survival, and bold play is the answer.   ∎

There are policies other than the bold one which achieve the maximum success probability $Q(x)$. Suppose that as long as the gambler's fortune x is less than $\frac{1}{2}$ he bets x for $x \le \frac{1}{4}$ and $\frac{1}{2} - x$ for $\frac{1}{4} \le x \le \frac{1}{2}$. This is, in effect, the bold-play strategy scaled down to the interval $[0, \frac{1}{2}]$, and so the chance he ever reaches $\frac{1}{2}$ is $Q(2x)$ for an initial fortune of x. Suppose further that if he does reach the goal of $\frac{1}{2}$, or if he starts with fortune at least $\frac{1}{2}$ in the first place, he continues with ordinary bold play. For an initial fortune $x \ge \frac{1}{2}$, the overall chance of success is of course $Q(x)$, and for an initial fortune $x < \frac{1}{2}$, it is $Q(2x)Q(\frac{1}{2}) = pQ(2x) = Q(x)$. The success probability is indeed $Q(x)$ as for bold play, although the policy is different. With this example in mind, one can generate a whole series of distinct optimal policies.

Timid Play*

The optimality of bold play seems reasonable when one considers the effect of its opposite, timid play. Let the ϵ-*timid* policy be to bet $W_n = \min\{\epsilon, F_{n-1}, 1 - F_{n-1}\}$ and stop when F_n reaches 0 or 1. Suppose that $p < q$, fix an initial fortune $x = F_0$ with $0 \le x < 1$, and consider what happens as $\epsilon \to 0$. By the strong law of large numbers, $\lim_n n^{-1} S_n = E[X_1] = p - q < 0$. There is therefore probability 1 that $\sup_k S_k < \infty$ and $\lim_n S_n = -\infty$. Given $\eta > 0$, choose ϵ so that $P[\sup_k (x + \epsilon S_k) < 1] > 1 - \eta$. Since $P(\bigcup_{n=1}^{\infty} [x + \epsilon S_n < 0]) = 1$, with probability at least $1 - \eta$ there exists an n such that $x + \epsilon S_n < 0$ and $\max_{k < n}(x + \epsilon S_k) < 1$. But under the ϵ-timid policy the gambler is in this circumstance ruined. If $Q_\epsilon(x)$ is the probability of success under the ϵ-timid policy, then $\lim_{\epsilon \to 0} Q_\epsilon(x) = 0$ for $0 \le x < 1$. The law of large numbers carries the timid player to his ruin.

PROBLEMS

7.1. A gambler with initial capital a plays until his fortune increases b units or he is ruined. Suppose that $\rho > 1$. The chance of success is multiplied by $1 + \theta$ if his initial capital is infinite instead of a. Show that $0 < \theta < (\rho^a - 1)^{-1} < (a(\rho - 1))^{-1}$; relate to Example 7.3.

7.2. As shown on page 79, there is probability 1 that the gambler either achieves his goal of c or is ruined. For $p \ne q$, deduce this directly from the strong law of large numbers. Deduce it (for all p) via the Borel-Cantelli lemma from the fact that if play never terminates, there can never occur c successive $+1$'s.

7.3. 6.12↑ If V_n is the set of n-long sequences of ± 1's, the function b_n in (7.11) maps V_{n-1} into $\{0, 1\}$. A selection system is a sequence of such maps. Although there are uncountably many selection systems, how many have an *effective* description in the sense of an algorithm or finite set of instructions by means of which a deputy (perhaps a machine) could operate the system in the gambler's stead? A full analysis of the question is a matter for mathematical logic, but one can see in an informal way that there must be only countably many algorithms or finite sets of rules expressed in finite alphabets.

Let $Y_1^{(\sigma)}, Y_2^{(\sigma)}, \ldots$ be the random variables of Theorem 7.1 for a particular system σ, and let C_σ be the ω-set where every k-tuple of ± 1's (k arbitrary) occurs in $Y_1^{(\sigma)}(\omega), Y_2^{(\sigma)}(\omega), \ldots$ with the right asymptotic relative frequency (in the sense of Problem 6.12). Let C be the intersection of C_σ over all effective selection systems σ. Show that C lies in $\mathcal{F}$ (the σ-field in the probability space $(\Omega, \mathcal{F}, P)$ on which the X_n are defined) and that $P(C) = 1$. A sequence $(X_1(\omega), X_2(\omega), \ldots)$ for ω in C is called a *collective*: a subsequence chosen by any of the effective rules σ contains all k-tuples in the correct proportions.

---

* This topic may be omitted.

7.4. Let D_n be 1 or 0 according as $X_{2n-1} \neq X_{2n}$ or not, and let M_k be the time of the kth 1—the smallest n such that $\sum_{i=1}^{n} D_i = k$. Let $Z_k = X_{2M_k}$. In other words, look at successive nonoverlapping pairs (X_{2n-1}, X_{2n}), discard accordant $(X_{2n-1} = X_{2n})$ pairs, and keep the second element of discordant $(X_{2n-1} \neq X_{2n})$ pairs. Show that this process simulates a fair coin: $Z_1, Z_2, \ldots$ are independent and identically distributed and $P[Z_k = +1] = P[Z_k = -1] = \frac{1}{2}$, whatever p may be.

7.5. Suppose that a gambler with initial fortune 1 stakes a proportion θ $(0 < \theta < 1)$ of his current fortune: $F_0 = 1$ and $W_n = \theta F_{n-1}$. Show that $F_n = \prod_{k=1}^{n} (1 + \theta X_k)$ and hence that

$$\log F_n = \frac{n}{2} \left[\frac{S_n}{n} \log \frac{1+\theta}{1-\theta} + \log (1 - \theta^2) \right].$$

Show that $F_n \to 0$ with probability 1 in the subfair case.

7.6. In "doubling," $W_1 = 1$, $W_n = 2W_{n-1}$, and the rule is to stop after the first win. For any positive p, play is certain to terminate. Here $F_\tau = F_0 + 1$, but of course infinite capital is required. If $F_0 = 2^k - 1$ and W_n cannot exceed F_{n-1}, the probability of $F_\tau = F_0 + 1$ in the fair case is $1 - 2^{-k}$. Prove this via Theorem 7.2 and also directly.

7.7. In "progress and pinch," the wager, initially some integer, is increased by 1 after a loss and decreased by 1 after a win, the stopping rule being to quit if the next bet is 0. Show that play is certain to terminate if and only if $p \geq \frac{1}{2}$. Show that $F_\tau = F_0 + \frac{1}{2} W_1^2 + \frac{1}{2}(\tau - 1)$. Again, infinite capital is required.

7.8. Here is a common martingale. At each stage the gambler has before him a pattern $x_1, \ldots, x_k$ of numbers. He bets $x_1 + x_k$, or x_1 in case $k = 1$. If he loses, at the next stage he uses the pattern $x_1, \ldots, x_k, x_1 + x_k$. If he wins, at the next stage he uses the pattern $x_2, \ldots, x_{k-1}$, unless k is 1 or 2, in which case he quits. Show that play is certain to terminate if $p > \frac{1}{3}$ and that the ultimate gain is the sum of the numbers in the initial pattern. Infinite capital is again required.

7.9. Suppose that $W_k = 1$, so that $F_k = F_0 + S_k$. Suppose that $p \geq q$ and τ is a stopping time such that $1 \leq \tau \leq n$ with probability 1. Show that $E[F_\tau] \leq E[F_n]$, with equality in case $p = q$. Interpret this result in terms of a stock option that must be exercised by time n, $F_0 + S_k$ representing the price of the stock at time k.

7.10. Let u be a real function on $[0, 1]$, $u(x)$ representing the *utility* of the fortune x. Consider policies bounded by 1; see (7.26). Let $Q_\pi(F_0) = E[u(F_\tau)]$; this represents the expected utility under the policy π of an initial fortune F_0. Suppose of a policy π_0 that

(7.36) $u(x) \leq Q_{\pi_0}(x),$ $0 \leq x \leq 1,$

and that

(7.37) $Q_{\pi_0}(x) \geq p Q_{\pi_0}(x + t) + q Q_{\pi_0}(x - t),$
$$0 \leq x - t \leq x \leq x + t \leq 1.$$

Show that $Q_\pi(x) \leq Q_{\pi_0}(x)$ for all x and all policies π. Such a π_0 is optimal.

Theorem 7.3 is the special case of this result for $p \leq \frac{1}{2}$, bold play in the role of π_0, and $u(x) = 1$ or $u(x) = 0$ according as $x = 1$ or $x < 1$.

Condition (7.36) says that gambling with policy π_0 is at least as good as not gambling at all; (7.37) says that, although the prospects even under π_0 become on the average less sanguine as time passes, it is better to use π_0 now than to use some other policy for one step and then change to π_0.

7.11. The functional equation (7.32) and the assumption that Q is bounded suffice to determine Q completely. First, $Q(0)$ and $Q(1)$ must be 0 and 1, respectively, and so (7.33) holds. Let $T_0 x = \frac{1}{2} x$ and $T_1 x = \frac{1}{2} x + \frac{1}{2}$; let $f_0 x = px$ and $f_1 x = p + qx$. Then $Q(T_{u_1} \cdots T_{u_n} x) = f_{u_1} \cdots f_{u_n} Q(x)$. If the binary expansions of x and y both begin with the digits $u_1, \ldots, u_n$, they have the form $x = T_{u_1} \cdots T_{u_n} x'$ and $y = T_{u_1} \cdots T_{u_n} y'$. If K bounds Q and if $m = \max\{p, q\}$, it follows that $|Q(x) - Q(y)| \le Km^n$. Therefore, Q is continuous and satisfies (7.33).

SECTION 8. MARKOV CHAINS

As Markov chains illustrate well the connection between probability and measure, their basic properties are here developed in a measure-theoretic setting.

Definitions

Let S be a finite or countable set. Suppose that to each pair i and j in S there is assigned a nonnegative number p_{ij} and that these numbers satisfy the constraint

$$(8.1) \qquad \sum_{j \in S} p_{ij} = 1, \qquad i \in S.$$

Let $X_0, X_1, X_2, \ldots$ be a sequence of random variables whose ranges are contained in S. The sequence is a *Markov chain* or *Markov process* if

$$(8.2) \qquad \begin{aligned} P[X_{n+1} = j \,|\, X_0 = i_0, \ldots, X_n = i_n] \\ = P[X_{n+1} = j \,|\, X_n = i_n] = p_{i_n j} \end{aligned}$$

for every n and every sequence $i_0, \ldots, i_n$ in S for which $P[X_0 = i_0, \ldots, X_n = i_n] > 0$. The set S is the *state space* or *phase space* of the process, and the p_{ij} are the *transition probabilities*. Part of the defining condition (8.2) is that the transition probability

$$(8.3) \qquad P[X_{n+1} = j \,|\, X_n = i] = p_{ij}$$

does not vary with n.*

The elements of S are thought of as the possible states of a *system*, X_n representing the state at *time n*. The sequence or process $X_0, X_1, X_2, \ldots$ then rep-

---

* Sometimes in the definition of Markov chain $P[X_{n+1} = j \,|\, X_n = i]$ is allowed to depend on n. A chain satisfying (8.3) is then said to have *stationary transition probabilities*, a phrase that will be omitted here because (8.3) will always be assumed.

resents the history of the system, which evolves in accordance with the probability law (8.2). The conditional distribution of the *next* state X_{n+1} given the *present* state X_n must not further depend on the *past* $X_0, \ldots, X_{n-1}$. This is what (8.2) requires, and it leads to a copious theory.

The *initial probabilities* are

(8.4) $$\pi_i = P[X_0 = i].$$

The definition of Markov chain places no restrictions on the π_i.

Example 8.1. *The Bernoulli-Laplace model of diffusion.* Imagine r black balls and r white balls distributed between two boxes with the constraint that each box contains r balls. The state of the system is specified by the number of white balls in the first box, so that the state space is $S = \{0, 1, \ldots, r\}$. The transition mechanism is this: at each stage one ball is chosen at random from each box and the two are interchanged. If the present state is i, the chance of a transition to $i - 1$ is the chance i/r of drawing one of the i white balls from the first box times the chance i/r of drawing one of the i black balls from the second box. Together with similar arguments for the other possibilities, this shows that the transition probabilities are

$$p_{i,i-1} = \left(\frac{i}{r}\right)^2, \qquad p_{i,i+1} = \left(\frac{r-i}{r}\right)^2, \qquad p_{ii} = 2\,\frac{i(r-i)}{r^2},$$

the others being 0. This is the probabilistic analogue of the model for the flow of two liquids between two containers.* ∎

The p_{ij} form the *transition matrix* $P = [p_{ij}]$ of the process. A *stochastic matrix* is one whose entries are nonnegative and satisfy (8.1); the transition matrix of course has this property.

Example 8.2. *Random walk with absorbing barriers.* Suppose that $S = \{0, 1, \ldots, r\}$ and

$$P = \begin{bmatrix} 1 & 0 & 0 & 0 & \ldots & 0 & 0 & 0 & 0 \\ q & 0 & p & 0 & \ldots & 0 & 0 & 0 & 0 \\ 0 & q & 0 & p & \ldots & 0 & 0 & 0 & 0 \\ & & & \ldots & & & & & \\ 0 & 0 & 0 & 0 & \ldots & q & 0 & p & 0 \\ 0 & 0 & 0 & 0 & \ldots & 0 & q & 0 & p \\ 0 & 0 & 0 & 0 & \ldots & 0 & 0 & 0 & 1 \end{bmatrix}$$

---

* For an excellent collection of examples from physics and biology, see FELLER, Volume 1, Chapter XV.

That is, $p_{i,i+1} = p$ and $p_{i,i-1} = q = 1 - p$ for $0 < i < r$ and $p_{00} = p_{rr} = 1$. The chain represents a particle in *random walk*. The particle moves one unit to the right or left, the respective probabilities being p and q, except that each of 0 and r is an *absorbing* state—once the particle enters, it cannot leave. The state can also be viewed as a gambler's fortune; absorption in 0 represents ruin for the gambler, absorption in r ruin for his adversary (see Section 7). The gambler's initial fortune is usually regarded as nonrandom, so that (see (8.4)) $\pi_i = 1$ for some i. ■

Example 8.3. *Unrestricted random walk.* Let S consist of all the integers $i = 0, \pm 1, \pm 2, \ldots$, and take $p_{i,i+1} = p$ and $p_{i,i-1} = q = 1 - p$. This chain represents a random walk without barriers, the particle being free to move anywhere on the integer lattice. The walk is *symmetric* if $p = q$. ■

The state space may, as in the preceding example, be countably infinite. If so, the Markov chain consists of functions X_n on a probability space $(\Omega, \mathcal{F}, P)$, but these will have infinite range and hence will not be random variables in the sense of the preceding sections. This will cause no difficulty however, because expected values of the X_n will not be considered. All that is required is that for each $i \in S$ the set $[\omega: X_n(\omega) = i]$ lie in $\mathcal{F}$ and hence have a probability.

Example 8.4. *Symmetric random walk in space.* Let S consist of the integer lattice points in k-dimensional Euclidean space R^k; $x = (x_1, \ldots, x_k)$ lies in S if the coordinates are all integers. Now x has $2k$ neighbors, points of the form $y = (x_1, \ldots, x_i \pm 1, \ldots, x_k)$; for each such y let $p_{xy} = (2k)^{-1}$. The chain represents a particle moving randomly in space; for $k = 1$ it reduces to Example 8.3 with $p = q = \frac{1}{2}$. The cases $k \leq 2$ and $k \geq 3$ exhibit an interesting difference. If $k \leq 2$, the particle is certain to return to its initial position, but this is not so if $k \geq 3$; see Example 8.6. ■

Since the state space in this example is not a subset of the line, the $X_0, X_1, \ldots$ do not assume real values. This is immaterial because expected values of the X_n play no role. All that is necessary is that X_n be a mapping from Ω into S (finite or countable) such that $[\omega: X_n(\omega) = i] \in \mathcal{F}$ for $i \in S$. There will be expected values $E[f(X_n)]$ for real functions f on S with finite range, but then $f(X_n(\omega))$ is a simple random variable as defined before.

Example 8.5.   A Selection Problem. A princess must choose from among r suitors. She is definite in her preferences and if presented with all r at once could choose her favorite and could even rank the whole group. They are ushered into her presence one by one in random order, however, and she must at each stage either stop and accept the suitor or else reject him and proceed in the hope that a better one will come along. What strategy will maximize her chance of stopping with the best suitor of all?

Shorn of some details, the analysis is this. Let $S_1, S_2, \ldots, S_r$ be the suitors in order of presentation; this sequence is a random permutation of the set of suitors. Let $X_1 = 1$, and let $X_2, X_3, \ldots$ be the successive positions of suitors who dominate (are preferable to) all their predecessors. Thus $X_2 = 4$ and $X_3 = 6$ means that S_1 dominates S_2 and S_3 but S_4 dominates S_1, S_2, S_3, and that S_4 dominates S_5 but S_6 dominates $S_1, \ldots, S_5$. There can be at most r of these dominant suitors; if there are exactly m, $X_{m+1} = X_{m+2} = \cdots = r + 1$ by convention.

As the suitors arrive in random order, the chance that S_i ranks highest among $S_1, \ldots, S_i$ is $(i - 1)!/i! = 1/i$. The chance that S_j ranks highest among $S_1, \ldots, S_j$ and S_i ranks next is $(j - 2)!/j! = 1/j(j - 1)$. This leads to a chain with transition probabilities

$$(8.5) \qquad P[X_{n+1} = j \mid X_n = i] = \frac{i}{j(j - 1)}, \qquad 1 \le i < j \le r.$$

If $X_n = i$, then $X_{n+1} = r + 1$ means that S_i dominates $S_{i+1}, \ldots, S_r$ as well as $S_1, \ldots, S_i$, and the conditional probability of this is

$$(8.6) \qquad P[X_{n+1} = r + 1 \mid X_n = i] = \frac{i}{r}, \qquad 1 \le i \le r.$$

As downward transitions are impossible and $r + 1$ is absorbing, this specifies a transition matrix for $S = \{1, 2, \ldots, r + 1\}$.

It is quite clear that in maximizing her chance of selecting the best suitor of all, the princess should reject those who do not dominate their predecessors. Her strategy therefore will be to stop with the suitor in position X_τ, where τ is a random variable representing her strategy. Since her decision to stop must depend only on the suitors she has seen thus far, the event $[\tau = n]$ must lie in $\sigma(X_1, \ldots, X_n)$. If $X_\tau = i$, then by (8.6) the conditional probability of success is $f(i) = i/r$. The probability of success is therefore $E[f(X_\tau)]$, and the problem is to choose the strategy τ so as to maximize it. For the solution, see Example 8.16.* ∎

Higher-Order Transitions

The properties of the Markov chain are entirely determined by the transition and initial probabilities. The chain rule (4.2) for conditional probabilities gives

$$P[X_0 = i_0, X_1 = i_1, X_2 = i_2]$$

$$= P[X_0 = i_0]P[X_1 = i_1 \mid X_0 = i_0]P[X_2 = i_2 \mid X_0 = i_0, X_1 = i_1]$$

$$= \pi_{i_0} p_{i_0 i_1} p_{i_1 i_2}.$$

Similarly,

$$(8.7) \qquad P[X_t = i_t, 0 \le t \le m] = \pi_{i_0} p_{i_0 i_1} \cdots p_{i_{m-1} i_m}$$

for any sequence $i_0, i_1, \ldots, i_m$ of states.

Further,

---

* With the princess replaced by an executive and the suitors by applicants for an office job, this is known as the *secretary problem*.

$P[X_2 = j_2, X_3 = j_3 | X_0 = i_0, X_1 = i_1]$

$$= P[X_2 = j_2 | X_0 = i_0, X_1 = i_1]P[X_3 = j_3 | X_0 = i_0, X_1 = i_1, X_2 = j_2]$$

$$= p_{i_1 j_2} p_{j_2 j_3}.$$

An inductive extension of this argument gives

(8.8) $P[X_{m+t} = j_t, \ 1 \leq t \leq n | X_s = i_s, \ 0 \leq s \leq m]$

$$= p_{i_m j_1} p_{j_1 j_2} \cdots p_{j_{n-1} j_n}.$$

Adding out the intermediate states now gives the formula

(8.9) $p_{ij}^{(n)} = P[X_{m+n} = j | X_m = i]$

$$= \sum_{k_1 \ldots k_{n-1}} p_{ik_1} p_{k_1 k_2} \cdots p_{k_{n-1} j},$$

(the k_l range over S) for the *nth order transition probabilities*.

Notice that $p_{ij}^{(n)}$ is the entry in position (i, j) of P^n, the nth power of the transition matrix P. If S is infinite, P is a matrix with infinitely many rows and columns; as the terms in (8.9) are nonnegative, there are no convergence problems. It is natural to put

$$p_{ij}^{(0)} = \delta_{ij} = \begin{cases} 1 & \text{if } i = j, \\ 0 & \text{if } i \neq j. \end{cases}$$

Thus P^0 is the identity, as it should be. From (8.1) and (8.9) follow

(8.10) $p_{ij}^{(n+1)} = \sum_{\nu} p_{i\nu}^{(n)} p_{\nu j} = \sum_{\nu} p_{i\nu} p_{\nu j}^{(n)}, \qquad \sum_j p_{ij}^{(n)} = 1.$

An Existence Theorem

Theorem 8.1. *Suppose that $P = [p_{ij}]$ is a stochastic matrix and that π_i are nonnegative numbers satisfying $\Sigma_{i \in S} \pi_i = 1$. There exists on some $(\Omega, \mathcal{F}, P)$ a Markov chain $X_0, X_1, X_2, \ldots$ with initial probabilities π_i and transition probabilities p_{ij}.*

PROOF.   Reconsider the proof of Theorem 5.2. There the space $(\Omega, \mathcal{F}, P)$ was the unit interval, and the central part of the argument was the construction of the decompositions (5.10). Suppose for the moment that $S = \{1, 2, \ldots\}$. First construct a partition $I_1^{(0)}, I_2^{(0)}, \ldots$ of $(0, 1]$ into countably many* subintervals of lengths (P is again Lebesgue measure) $P(I_i^{(0)}) = \pi_i$. Next decompose each $I_i^{(0)}$ into subintervals $I_{ij}^{(1)}$ of lengths $P(I_{ij}^{(1)}) = \pi_i p_{ij}$. Continuing inductively gives

* If $\delta_1 + \delta_2 + \cdots = b - a$ and $\delta_i \geq 0$, then $I_i = (b - \Sigma_{j \leq i} \delta_j, \ b - \Sigma_{j < i} \delta_j], \ i = 1, 2, \ldots$, decompose $(a, b]$ into intervals of lengths δ_i.

a sequence of partitions $\{I_{i_0\ldots i_n}^{(n)}:i_0, \ldots, i_n = 1, 2, \ldots\}$ such that each refines the preceding and $P(I_{i_0\ldots i_n}^{(n)}) = \pi_{i_0} p_{i_0 i_1} \cdots p_{i_{n-1} i_n}$.

Put $X_n(\omega) = i$ if $\omega \in \bigcup_{i_0\ldots i_{n-1}} I_{i_0\ldots i_{n-1} i}^{(n)}$. It follows just as in the proof of Theorem 5.2 that the set $[X_0 = i_0, \ldots, X_n = i_n]$ coincides with the interval $I_{i_0\ldots i_n}^{(n)}$. Thus $P[X_0 = i_0, \ldots, X_n = i_n] = \pi_{i_0} p_{i_0 i_1} \cdots p_{i_{n-1} i_n}$. From this, (8.2) and (8.4) follow immediately.

That completes the construction for the case $S = \{1, 2, \ldots\}$. For the general countably infinite S, let g be a one-to-one mapping of $\{1, 2, \ldots\}$ onto S and replace the X_n as already constructed by $g(X_n)$; the assumption $S = \{1, 2, \ldots\}$ was merely for notational convenience. The same argument obviously works if S is finite. ∎

Although strictly speaking the Markov chain *is* the sequence $X_0, X_1, \ldots$, one often speaks as though the chain were the matrix P together with the initial probabilities π_i or even P with some unspecified set of π_i. Theorem 8.1 justifies this attitude: For given P and π_i the corresponding X_n do exist, and the apparatus of probability theory—the Borel-Cantelli lemmas and so on—is available for the study of P and of systems evolving in accordance with the Markov rule.

From now on fix a chain $X_0, X_1, \ldots$ satisfying $\pi_i > 0$ for all i. Denote by P_i probabilities conditional on $[X_0 = i]$: $P_i(A) = P[A|X_0 = i]$. Thus

$$(8.11) \qquad P_i[X_t = i_t, 1 \le t \le n] = p_{ii_1} p_{i_1 i_2} \cdots p_{i_{n-1} i_n},$$

by (8.8). The interest centers on these conditional probabilities, and the actual initial probabilities are now largely irrelevant. From (8.11) follows

$$(8.12) \quad P_i[X_1 = i_1, \ldots, X_m = i_m, X_{m+1} = j_1, \ldots, X_{m+n} = j_n]$$
$$= P_i[X_1 = i_1, \ldots, X_m = i_m]P_{i_m}[X_1 = j_1, \ldots, X_n = j_n].$$

Suppose that I is a set (finite or infinite) of m-long sequences of states, J is a set of n-long sequences of states, and every sequence in I ends in j. Adding both sides of (8.12) for $(i_1, \ldots, i_m)$ ranging over I and $(j_1, \ldots, j_n)$ ranging over J gives

$$(8.13) \quad P_i[(X_1, \ldots, X_m) \in I, (X_{m+1}, \ldots, X_{m+n}) \in J]$$
$$= P_i[(X_1, \ldots, X_m) \in I]P_j[(X_1, \ldots, X_n) \in J].$$

For this to hold it is essential that each sequence in I end in j. Formulas (8.12) and (8.13) are of central importance.

Transience and Persistence

Let

$$(8.14) \qquad f_{ij}^{(n)} = P_i[X_1 \ne j, \ldots, X_{n-1} \ne j, X_n = j]$$

be the probability of a first visit to j at time n for a system that starts in i, and let

$$(8.15) \qquad f_{ij} = P_i \left(\bigcup_{n=1}^{\infty} [X_n = j] \right) = \sum_{n=1}^{\infty} f_{ij}^{(n)}$$

be the probability of an eventual visit. A state i is *persistent* if a system starting at i is certain sometime to return to i: $f_{ii} = 1$. The state is *transient* in the opposite case: $f_{ii} < 1$.

Suppose that $n_1, \ldots, n_k$ are integers satisfying $1 \le n_1 < \cdots < n_k$ and consider the event that the system visits j at times $n_1, \ldots, n_k$ but not in between; this event is determined by the conditions

$$(8.16) \qquad \begin{cases} X_1 \ne j, \ldots, X_{n_1-1} \ne j, \ X_{n_1} = j, \\ X_{n_1+1} \ne j, \ldots, X_{n_2-1} \ne j, X_{n_2} = j, \\ \quad\quad \cdots\cdots\cdots\cdots\cdots\cdots \\ X_{n_{k-1}+1} \ne j, \ldots, X_{n_k-1} \ne j, X_{n_k} = j. \end{cases}$$

Repeated application of (8.13) shows that under P_i the probability of (8.16) is $f_{ij}^{(n_1)} f_{jj}^{(n_2-n_1)} \cdots f_{jj}^{(n_k-n_{k-1})}$. Add this over the k-tuples $n_1, \ldots, n_k$: the P_i-probability that $X_n = j$ for at least k different values of n is $f_{ij} f_{jj}^{k-1}$. Letting $k \to \infty$ therefore gives

$$(8.17) \qquad P_i[X_n = j \text{ i.o.}] = \begin{cases} 0 & \text{if } f_{jj} < 1, \\ f_{ij} & \text{if } f_{jj} = 1. \end{cases}$$

Recall that *i.o.* means *infinitely often.*

Theorem 8.2. (i) *Persistence of i is equivalent to $P_i[X_n = i$ i.o.$] = 1$ and to $\Sigma_n p_{ii}^{(n)} = \infty$.*
(ii) *Transience of i is equivalent to $P_i[X_n = i$ i.o.$] = 0$ and to $\Sigma_n p_{ii}^{(n)} < \infty$.*

In particular, $P_i[X_n = i$ i.o.$]$ is either 0 or 1; compare the zero-one law— Theorem 4.5—but note that the events $[X_n = i]$ here are not in general independent.

PROOF. Take $i = j$ in (8.17): $P_i[X_n = i$ i.o.$]$ is 0 or 1 according as $f_{ii} < 1$ or $f_{ii} = 1$. If $\Sigma_n p_{ii}^{(n)} < \infty$, then $P_i[X_n = i$ i.o.$]$ is 0 by the first Borel-Cantelli lemma, so that i is transient. Therefore, the entire theorem will be proved if it is shown that $\Sigma_n p_{ii}^{(n)} = \infty$ implies that i is persistent.

The proof uses a first-passage argument: by (8.13),

$$p_{ij}^{(n)} = P_i[X_n = j] = \sum_{s=0}^{n-1} P_i[X_1 \ne j, \ldots, X_{n-s-1} \ne j, X_{n-s} = j, X_n = j]$$

$$= \sum_{s=0}^{n-1} P_i[X_1 \neq j, \ldots, X_{n-s-1} \neq j, X_{n-s} = j]P_j[X_s = j]$$

$$= \sum_{s=0}^{n-1} f_{ij}^{(n-s)} p_{jj}^{(s)}.$$

Therefore,

$$\sum_{t=1}^{n} p_{ii}^{(t)} = \sum_{t=1}^{n} \sum_{s=0}^{t-1} f_{ii}^{(t-s)} p_{ii}^{(s)}$$

$$= \sum_{s=0}^{n-1} p_{ii}^{(s)} \sum_{t=s+1}^{n} f_{ii}^{(t-s)} \leq \sum_{s=0}^{n} p_{ii}^{(s)} f_{ii}.$$

Thus $(1 - f_{ii}) \sum_{t=1}^{n} p_{ii}^{(t)} \leq f_{ii}$, and if $f_{ii} < 1$, this puts a bound on the partial sums $\sum_{t=1}^{n} p_{ii}^{(t)}$ ∎

Example 8.6. *Polya's Theorem.* For the symmetric k-dimensional random walk (Example 8.4) the probability $p_{ii}^{(n)}$ of return in n steps is obviously the same for all states i; denote this probability by $a_n^{(k)}$ to indicate the dependence on the dimension k. Clearly, $a_{2n+1}^{(k)} = 0$. Suppose that $k = 1$. Since return in $2n$ steps means n steps east and n steps west,

$$a_{2n}^{(1)} = \binom{2n}{n} \frac{1}{2^{2n}}.$$

By Stirling's formula, $a_{2n}^{(1)} \sim (\pi n)^{-1/2}$. Therefore, $\sum_n a_n^{(1)} = \infty$ and all states are persistent by Theorem 8.2.

In the plane, a return to the starting point in $2n$ steps means equal numbers of steps east and west as well as equal numbers north and south:

$$a_{2n}^{(2)} = \sum_{u=0}^{n} \frac{(2n)!}{u!u!(n-u)!(n-u)!} \frac{1}{4^{2n}}$$

$$= \frac{1}{4^{2n}} \binom{2n}{n} \sum_{u=0}^{n} \binom{n}{u} \binom{n}{n-u}.$$

It can be seen on combinatorial grounds that the last sum is $\binom{2n}{n}$, and so $a_{2n}^{(2)} = (a_{2n}^{(1)})^2 \sim (\pi n)^{-1}$. Again, $\sum_n a_n^{(2)} = \infty$ and every state is persistent.

For three dimensions,

$$a_{2n}^{(3)} = \sum \frac{(2n)!}{u!u!v!v!(n-u-v)!(n-u-v)!} \frac{1}{6^{2n}},$$

the sum extending over nonnegative u and v satisfying $u + v \leq n$. This reduces to

(8.18) $$a_{2n}^{(3)} = \sum_{l=0}^{n} \binom{2n}{2l} \left(\frac{1}{3}\right)^{2n-2l} \left(\frac{2}{3}\right)^{2l} a_{2n-2l}^{(1)} a_{2l}^{(2)},$$

as can be checked by substitution. (To see the probabilistic meaning of this formula, condition on there being $2n - 2l$ steps parallel to the vertical axis and $2l$ steps parallel to the horizontal plane.) It will be shown that $a_{2n}^{(3)} = O(n^{-3/2})$, which will imply that $\Sigma_n a_n^{(3)} < \infty$. The terms in (8.18) for $l = 0$ and $l = n$ are each $O(n^{-3/2})$ and hence can be omitted. Now $a_u^{(1)} \le Ku^{-1/2}$ and $a_u^{(2)} \le Ku^{-1}$, as already seen, and so the sum in question is at most

$$K^2 \sum_{l=1}^{n-1} \binom{2n}{2l} \left(\frac{1}{3}\right)^{2n-2l} \left(\frac{2}{3}\right)^{2l} (2n - 2l)^{-1/2}(2l)^{-1}.$$

Since $(2n - 2l)^{-1/2} \le 2n^{1/2} (2n - 2l)^{-1} \le 4n^{1/2}(2n - 2l + 1)^{-1}$ and $(2l)^{-1} \le 2(2l + 1)^{-1}$, this is at most a constant times

$$n^{1/2} \frac{(2n)!}{(2n + 2)!} \sum_{l=1}^{n-1} \binom{2n + 2}{2l + 1} \left(\frac{1}{3}\right)^{2n-2l-1} \left(\frac{2}{3}\right)^{2l+1} = O(n^{-3/2}).$$

Thus $\Sigma_n a_n^{(3)} < \infty$, and the states are transient. The same is true for $k = 4, 5, \ldots$, since an inductive extension of the argument shows that $a_n^{(k)} = O(n^{-k/2})$. ∎

It is possible for a system starting in i to reach j $(f_{ij} > 0)$ if and only if $p_{ij}^{(n)} > 0$ for some n. If this is true for all i and j, the Markov chain is *irreducible*.

Theorem 8.3. *If the Markov chain is irreducible, then one of the following two alternatives holds.*

(i) *All states are transient,* $P_i(\bigcup_j [X_n = j \text{ i.o.}]) = 0$ *for all* i, *and* $\Sigma_n p_{ij}^{(n)} < \infty$ *for all* i *and* j.

(ii) *All states are persistent,* $P_i (\bigcap_j [X_n = j \text{ i.o.}]) = 1$ *for all* i, *and* $\Sigma_n p_{ij}^{(n)} = \infty$ *for all* i *and* j.

The irreducible chain itself can accordingly be called persistent or transient. In the persistent case the system visits every state infinitely often. In the transient case it visits each state only finitely often, hence visits each finite set only finitely often, and so may be said to go to infinity.

PROOF.   For each i and j there exist r and s such that $p_{ij}^{(r)} > 0$ and $p_{ji}^{(s)} > 0$. Now

(8.19) $p_{ii}^{(r+s+n)} \ge p_{ij}^{(r)} p_{jj}^{(n)} p_{ji}^{(s)}$,

and since $p_{ij}^{(r)} p_{ji}^{(s)} > 0$, $\Sigma_n p_{ii}^{(n)} < \infty$ implies that $\Sigma_n p_{jj}^{(n)} < \infty$: if one state is transient, they all are. In this case (8.17) gives $P_i[X_n = j \text{ i.o.}] = 0$ for all i and j, so that $P_i(\bigcup_j [X_n = j \text{ i.o.}]) = 0$ for all i. Since $\Sigma_{n=1}^\infty p_{ij}^{(n)} = \Sigma_{n=1}^\infty \Sigma_{\nu=1}^n f_{ij}^{(\nu)} p_{jj}^{(n-\nu)} = \Sigma_{\nu=1}^\infty f_{ij}^{(\nu)} \Sigma_{m=0}^\infty p_{jj}^{(m)} \le \Sigma_{m=0}^\infty p_{jj}^{(m)}$, if j is transient, then (Theorem 8.2) $\Sigma_n p_{ij}^{(n)}$ converges for every i.

The other possibility is that all states are persistent. In this case $P_j[X_n = j \text{ i.o.}]$

= 1 by Theorem 8.2, and it follows by (8.13) that

$$p_{ji}^{(m)} = P_j([X_m = i] \cap [X_n = j \text{ i.o.}])$$

$$\leq \sum_{n>m} P_j[X_m = i, X_{m+1} \neq j, \ldots, X_{n-1} \neq j, X_n = j]$$

$$= \sum_{n>m} p_{ji}^{(m)} f_{ij}^{(n-m)} = p_{ji}^{(m)} f_{ij}.$$

There is an m for which $p_{ji}^{(m)} > 0$, and therefore $f_{ij} = 1$. By (8.17), $P_i[X_n = j \text{ i.o.}]$ $= f_{ij} = 1$. If $\Sigma_n p_{ij}^{(n)}$ were to converge for some i and j, it would follow by the first Borel-Cantelli lemma that $P_i[X_n = j \text{ i.o.}] = 0$.   ■

Example 8.7. Since $\Sigma_j p_{ij}^{(n)} = 1$, the first alternative in Theorem 8.3 is impossible if S is finite: In a finite, irreducible Markov chain, all states are persistent.   ■

Example 8.8. The chain in Polya's theorem is certainly irreducible. If the dimension is 1 or 2, there is probability 1 that a particle in symmetric random walk visits every state infinitely often. If the dimension is 3 or more, the particle goes to infinity.   ■

Example 8.9. Consider the unrestricted random walk on the line (Example 8.3). According to the ruin calculation (7.8), $f_{01} = p/q$ for $p < q$. Since the chain is irreducible, all states are transient. By symmetry, of course, the chain is also transient if $p > q$, although in this case (7.8) gives $f_{01} = 1$. Thus $f_{ij} = 1$ ($i \neq j$) is possible in the transient case.

If $p = q = \frac{1}{2}$, the chain is persistent by Polya's theorem. If n and $j - i$ have the same parity,

$$p_{ij}^{(n)} = \binom{n}{\dfrac{n+j-i}{2}} \frac{1}{2^n}, \qquad |j - i| \leq n.$$

This is maximal if $j = i$ or $j = i \pm 1$, and by Stirling's formula the maximal value is of order $n^{-1/2}$. Therefore, $\lim_n p_{ij}^{(n)} = 0$, which always holds in the transient case but is thus possible in the persistent case as well.   ■

Another Criterion for Persistence

Let $Q = [q_{ij}]$ be a matrix with rows and columns indexed by the elements of a finite or countable set U. Suppose it is *substochastic* in the sense that $q_{ij} \geq 0$ and $\Sigma_j q_{ij} \leq 1$. Let $Q^n = [q_{ij}^{(n)}]$ be the nth power, so that

(8.20) $q_{ij}^{(n+1)} = \sum_\nu q_{i\nu}q_{\nu j}^{(n)}, \qquad q_{ij}^{(0)} = \delta_{ij}.$

Consider the row sums

(8.21) $\sigma_i^{(n)} = \sum_j q_{ij}^{(n)}.$

From (8.20) follows

(8.22) $\sigma_i^{(n+1)} = \sum_j q_{ij}\sigma_j^{(n)}.$

Since Q is substochastic, $\sigma_i^{(1)} \le 1$, and hence $\sigma_i^{(n+1)} = \sum_j\sum_\nu q_{i\nu}^{(n)} q_{\nu j} = \sum_\nu q_{i\nu}^{(n)}\sigma_\nu^{(1)}$
$\le \sigma_i^{(n)}$. Therefore, the monotone limits

(8.23) $\sigma_i = \lim_n \sum_j q_{ij}^{(n)}$

exist. By (8.22) and the Weierstrass M-test,* $\sigma_i = \sum_j q_{ij}\sigma_j$. Thus the σ_i solve
the system

(8.24) $\begin{cases} x_i = \sum_{j\in U} q_{ij}x_j, & i \in U, \\ 0 \le x_i \le 1, & i \in U. \end{cases}$

For an arbitrary solution, $x_i = \sum_j q_{ij}x_j \le \sum_j q_{ij} = \sigma_i^{(1)}$, and $x_i \le \sigma_i^{(n)}$ for all
i implies that $x_i \le \sum_j q_{ij}\sigma_j^{(n)} = \sigma_i^{(n+1)}$ by (8.22). Thus $x_i \le \sigma_i^{(n)}$ for all n by induction, and so $x_i \le \sigma_i$. Thus the σ_i give the *maximal* solution to (8.24):

Lemma 1. *For a substochastic matrix Q the limits* (8.23) *are the maximal
solution of* (8.24).

Now suppose that U is a subset of the state space S. The p_{ij} for i and j in U
give a substochastic matrix Q. The row sums (8.21) are $\sigma_i^{(n)} = \sum p_{ij_1}p_{j_1j_2}\cdots$
$p_{j_{n-1}j_n}$, where the $j_1,\ldots,j_n$ range over U, and so $\sigma_i^{(n)} = P_i[X_t \in U, t \le n]$. Let
$n \to \infty$:

(8.25) $\sigma_i = P_i[X_t \in U, t = 1, 2, \ldots], \qquad i \in U.$

The following theorem is now an immediate consequence of Lemma 1.

Theorem 8.4. *For $U \subset S$ the probabilities* (8.25) *are the maximal solution of*

---

* The *Weierstrass M-test:* If $\lim_n x_{nk} = x_k$ and $|x_{nk}| \le M_k$, where $\sum_k M_k < \infty$, then $\lim_n \sum_k x_{nk}$
$= \sum_k x_k$. *Proof:* $|\sum_k x_{nk} - \sum_k x_k| \le \sum_{k\le k_0} |x_{nk} - x_k| + 2\sum_{k>k_0} M_k$; let $n \to \infty$ and then $k_0 \to$
∞. ∎ See Example 16.7.

the system

$$
(8.26) \qquad
\begin{cases}
x_i = \displaystyle\sum_{j \in U} p_{ij} x_j, & i \in U, \\[2mm]
0 \le x_i \le 1, & i \in U.
\end{cases}
$$

Example 8.10. For the random walk on the line consider the set $U = \{0, 1, 2, \ldots\}$. The system (8.26) is

$$
\begin{cases}
x_i = p x_{i+1} + q x_{i-1}, & i \ge 1, \\
x_0 = p x_1.
\end{cases}
$$

By (7.10) it follows that $x_n = A + An$ if $p = q$ and $x_n = A - A(q/p)^{n+1}$ if $p \ne q$. The only bounded solution is $x_n \equiv 0$ if $q \ge p$, and in this case there is probability 0 of staying forever among the nonnegative integers. If $q < p$, $A = 1$ gives the maximal solution $x_n = 1 - (q/p)^{n+1}$. Compare (7.8) and Example 8.9. ∎

Now consider the system (8.26) with $U = S - \{i_0\}$ for a single state i_0:

$$
(8.27) \qquad
\begin{cases}
x_i = \displaystyle\sum_{j \ne i_0} p_{ij} x_j, & i \ne i_0, \\[2mm]
0 \le x_i \le 1, & i \ne i_0.
\end{cases}
$$

There is always the trivial solution—the one for which $x_i \equiv 0$.

Theorem 8.5. *An irreducible chain is transient if and only if* (8.27) *has a nontrivial solution.*

PROOF. The probabilities

$$
(8.28) \qquad 1 - f_{ii_0} = P_i[X_n \ne i_0, n \le 1], \qquad i \ne i_0,
$$

are by Theorem 8.4 the maximal solution of (8.27) and hence vanish identically if and only if (8.27) has no nontrivial solution. If the chain is persistent, then by Theorem 8.3 the probabilities (8.28) are 0 for all i and i_0. Now

$$
f_{i_0 i_0} = P_{i_0}[X_1 = i_0] + \sum_{n=2}^{\infty} \sum_{i \ne i_0} P_{i_0}[X_1 = i, X_2 \ne i_0, \ldots, X_{n-1} \ne i_0, X_n = i_0]
$$

$$
= p_{i_0 i_0} + \sum_{i \ne i_0} p_{i_0 i} f_{i i_0}.
$$

If the chain is transient, then $f_{i_0 i_0} < 1$ and hence $f_{i i_0} < 1$ for some $i \ne i_0$. ∎

Since the equations (8.27) are homogeneous, the issue is whether they have a solution that is nonnegative, nontrivial, and *bounded*. If they do, $0 \leq x_i \leq 1$ can be arranged by rescaling.

Example 8.11.   In the simplest of *queueing models* the state space is $\{0, 1, 2, \ldots\}$ and the transition matrix has the form

$$\begin{bmatrix} \alpha_0 & \alpha_1 & \alpha_2 & 0 & 0 & 0 & \cdots \\ \alpha_0 & \alpha_1 & \alpha_2 & 0 & 0 & 0 & \cdots \\ 0 & \alpha_0 & \alpha_1 & \alpha_2 & 0 & 0 & \cdots \\ 0 & 0 & \alpha_0 & \alpha_1 & \alpha_2 & 0 & \cdots \\ 0 & 0 & 0 & \alpha_0 & \alpha_1 & \alpha_2 & \cdots \\ & & \cdots & & & & \end{bmatrix}$$

If there are i customers in the queue and $i \geq 1$, the customer at the head of the queue is served and leaves, and then 0, 1, or 2 new customers arrive (probabilities $\alpha_0, \alpha_1, \alpha_2$), which leaves a queue of length $i - 1$, i, or $i + 1$. If $i = 0$, no one is served, and the new customers bring the queue length to 0, 1, or 2. Assume that α_0 and α_2 are positive, so that the chain is irreducible.

For $i_0 = 0$ the system (8.27) is

$$(8.29) \qquad \begin{cases} x_1 = \alpha_1 x_1 + \alpha_2 x_2, \\ x_k = \alpha_0 x_{k-1} + \alpha_1 x_k + \alpha_2 x_{k+1}, \qquad k \geq 2. \end{cases}$$

Since $\alpha_0, \alpha_1, \alpha_2$ have the form $q(1 - \alpha), \alpha, p(1 - \alpha)$ for appropriate p, q, α, the second line of (8.29) has the form $x_k = px_{k+1} + qx_{k-1}, k \geq 2$. Now (7.10) gives the solution, and the A there can because of the first equation in (8.29) be expressed in terms of B. The result is

$$x_k = \begin{cases} B((q/p)^k - 1) & \text{if } p \neq q, \\ Bk & \text{if } p = q. \end{cases}$$

There is a nontrivial bounded solution if $q < p$ but not if $q \geq p$.

If $q < p$, the chain is thus transient and the queue size goes to infinity. If $q \geq p$, the queue is persistent. For a nonempty queue the expected increase in queue length in one step is $(1 - \alpha)(p - q)$, and the queue goes out of control if and only if this is positive.   ∎

Stationary Distributions

Suppose that the initial probabilities for the chain satisfy

$$(8.30) \qquad \sum_{i \in S} \pi_i p_{ij} = \pi_j, \qquad j \in S.$$

It then follows by induction that

$$(8.31) \qquad \sum_{i \in S} \pi_i p_{ij}^{(n)} = \pi_j, \qquad j \in S, \quad n = 0, 1, 2, \ldots .$$

If π_i is the probability that $X_0 = i$, then the left side of (8.31) is the probability that $X_n = j$, and thus (8.30) implies that the probability of $[X_n = j]$ is the same for all n. A set of probabilities satisfying (8.30) is for this reason called a *stationary distribution*. The existence of such a distribution implies that the chain is very stable.

To discuss this requires the notion of periodicity. The state i has *period t* if $p_{ii}^{(n)} > 0$ implies that t divides n and if t is the largest integer with this property. If the chain is irreducible, for each pair i and j there exist r and s for which (8.19) holds and $p_{ij}^{(r)} p_{ji}^{(s)} > 0$. If t is the period i, then (take $n = 0$) t divides $r + s$; therefore, $p_{jj}^{(n)} > 0$ implies that t divides n. Since i and j can be interchanged in this argument, i and j have the same period. One can thus speak of the period of the chain itself in the irreducible case. The random walk on the line has period 2, for example. If the period is 1, the chain is *aperiodic*.

Lemma 2. *In an irreducible, aperiodic chain, for each i and j, $p_{ij}^{(n)} > 0$ for all n exceeding some $n_0(i, j)$.*

PROOF. Since j has period 1, there are finitely many integers $n_1, \ldots, n_l$ which have 1 as their greatest common divisor and which satisfy $p_{jj}^{(n_\nu)} > 0$. But every sufficiently large n has the form $n = \Sigma_\nu a_\nu n_\nu$ for positive integers $a_1, \ldots, a_l$,[*] and $p_{jj}^{(n)} \geq \prod_\nu p_{jj}^{(a_\nu n_\nu)} \geq \prod_\nu (p_{jj}^{(n_\nu)})^{a_\nu} > 0$. Thus $p_{jj}^{(n)} > 0$ for all large enough n. Since $p_{ij}^{(s)} > 0$ for some s and $p_{ij}^{(s+n)} \geq p_{ij}^{(s)} p_{jj}^{(n)}$, $p_{ij}^{(m)} > 0$ for all large enough m. ∎

Theorem 8.6. *Suppose of an irreducible, aperiodic chain that there exists a stationary distribution—a solution of (8.30) satisfying $\pi_i \geq 0$ and $\Sigma_i \pi_i = 1$. Then the chain is persistent, $\lim_n p_{ij}^{(n)} = \pi_j$ for all i and j, and the stationary distribution is unique.*

The main point of the conclusion is that, since $p_{ij}^{(n)}$ approaches π_j, the effect of the initial state i wears off.

PROOF. If the chain is transient, then $p_{ij}^{(n)} \to 0$ for all i and j by Theorem 8.3, and it follows by (8.31) and the M-test that π_j is identically 0, which contradicts

---

[*] There are integers $b_1, \ldots, b_l$, not necessarily positive, such that $\Sigma_\nu b_\nu n_\nu = 1$ (see HARDY & WRIGHT, p. 21). Put $N = \Sigma_\nu n_\nu$, and express n as $n = qN + r$, $0 \leq r < N$, so that $n = \Sigma_\nu (q + b_\nu r) n_\nu$; since $q + b_\nu r \geq q - |b_\nu| N$, the coefficients of the n_ν are all positive if $[n/N] = q > N \max_\nu |b_\nu|$.

$\Sigma_i \pi_i = 1$. The existence of a stationary distribution therefore implies that the chain is persistent.

Consider now a Markov chain with state space $S \times S$ and transition probabilities $p(ij, kl) = p_{ik}p_{jl}$ (it is easy to verify that these form a stochastic matrix). Call this the *coupled* chain; it describes the joint behavior of a pair of independent systems, each evolving according to the laws of the original Markov chain. By Theorem 8.1 there exists a Markov chain (X_n, Y_n), $n = 0, 1, \ldots$, having positive initial probabilities and transition probabilities

$$P[(X_{n+1}, Y_{n+1}) = (k, l) | (X_n, Y_n) = (i, j)] = p(ij, kl).$$

For n exceeding some n_0 depending on i, j, k, l, $p^{(n)}(ij, kl) = p_{ik}^{(n)}p_{jl}^{(n)}$ is positive by Lemma 2. Therefore, the coupled chain is irreducible. (This is the only place where the hypothesis of aperiodicity is used.)

It is easy to check that $\pi(ij) = \pi_i\pi_j$ forms a set of stationary initial probabilities for the coupled chain, which, like the original one, must therefore be persistent. It follows that, for an arbitrary initial state (i, j) for the chain $\{(X_n, Y_n)\}$ and an arbitrary i_0 in S, $P_{ij}[(X_n, Y_n) = (i_0, i_0) \text{ i.o.}] = 1$. If τ is the smallest integer such that $X_\tau = Y_\tau = i_0$, then τ is finite with probability 1 under P_{ij}. The idea of the proof is now this: X_n starts in i and Y_n starts in j; once $X_n = Y_n = i_0$ occurs, X_n and Y_n follow identical probability laws, and hence the initial states i and j will lose their influence.

By (8.13) applied to the coupled chain, if $m \leq n$, then

$$P_{ij}[(X_n, Y_n) = (k, l), \tau = m]$$
$$= P_{ij}[(X_t, Y_t) \neq (i_0, i_0), t < m, (X_m, Y_m) = (i_0, i_0)]$$
$$\times P_{i_0 i_0}[(X_{n-m}, Y_{n-m}) = (k, l)]$$
$$= P_{ij}[\tau = m]p_{i_0 k}^{(n-m)}p_{i_0 l}^{(n-m)}.$$

Adding out l gives $P_{ij}[X_n = k, \tau = m] = P_{ij}[\tau = m]p_{i_0 k}^{(n-m)}$, and adding out k gives $P_{ij}[Y_n = l, \tau = m] = P_{ij}[\tau = m]p_{i_0 l}^{(n-m)}$. Take $k = l$, equate probabilities, and add over m:

$$P_{ij}[X_n = k, \tau \leq n] = P_{ij}[Y_n = k, \tau \leq n].$$

From this follows

$$P_{ij}[X_n = k] \leq P_{ij}[X_n = k, \tau \leq n] + P_{ij}[\tau > n]$$
$$= P_{ij}[Y_n = k, \tau \leq n] + P_{ij}[\tau > n]$$
$$\leq P_{ij}[Y_n = k] + P_{ij}[\tau > n].$$

This and the same inequality with X and Y interchanged give

$$|p_{ik}^{(n)} - p_{jk}^{(n)}| = |P_{ij}[X_n = k] - P_{ij}[Y_n = k]| \leq P_{ij}[\tau > n].$$

Since τ is finite with probability 1,

(8.32) $$\lim_n |p_{ik}^{(n)} - p_{jk}^{(n)}| = 0.$$

By (8.31), $\pi_k - p_{jk}^{(n)} = \Sigma_i \pi_i(p_{ik}^{(n)} - p_{jk}^{(n)})$, and this goes to 0 by the M-test if (8.32) holds. Thus $\lim_n p_{jk}^{(n)} = \pi_k$. As this holds for each stationary distribution, there can be only one of them. ∎

Example 8.12. For the queueing model in Example 8.11 the equations (8.30) are

(8.33)
$$\begin{cases} \pi_0 = \pi_0\alpha_0 + \pi_1\alpha_0, \\ \pi_1 = \pi_0\alpha_1 + \pi_1\alpha_1 + \pi_2\alpha_0, \\ \pi_2 = \pi_0\alpha_2 + \pi_1\alpha_2 + \pi_2\alpha_1 + \pi_3\alpha_0, \\ \pi_k = \pi_{k-1}\alpha_2 + \pi_k\alpha_1 + \pi_{k+1}\alpha_0, \qquad k \geq 3. \end{cases}$$

Again write $\alpha_0, \alpha_1, \alpha_2$, as $q(1 - \alpha), \alpha, p(1 - \alpha)$. Since the last line in (8.33) is $\pi_k = q\pi_{k+1} + p\pi_{k-1}$,

$$\pi_k = \begin{cases} A + B(p/q)^k & \text{if } p \neq q, \\ A + Bk & \text{if } p = q, \end{cases}$$

for $k \geq 2$. If $p > q$ and $\Sigma \pi_k$ converges, then $\pi_k \equiv 0$, and hence there is no stationary distribution; but this is not new, because it was shown in Example 8.11 that the chain is transient in this case. If $p = q$, there is again no stationary distribution, and this is new because the chain was in Example 8.11 shown to be persistent in this case.

If $p < q$, then $\Sigma \pi_k$ converges, provided that $A = 0$. Solving for π_0 and π_1 in the first two equations of (8.33) gives $\pi_0 = B\alpha_2$ and $\pi_1 = B\alpha_2(1 - \alpha_0)/\alpha_0$. From $\Sigma \pi_k = 1$ it now follows that $B = (q - p)/p$, and the π_k can be written down explicitly. Since $\pi_k = B(p/q)^k$ for $k \geq 2$, there is small chance of a large queue length. ∎

If $p = q$ in this queueing model, the chain is persistent (Example 8.11) but has no stationary distribution (Example 8.12). This phenomenon can also be seen in the following simpler example.

Example 8.13. Suppose that the states are 0, 1, 2, . . . and the transition matrix is

$$\begin{bmatrix} q_0 & p_0 & 0 & 0 & \cdots \\ q_1 & 0 & p_1 & 0 & \cdots \\ q_2 & 0 & 0 & p_2 & \cdots \\ \multicolumn{5}{c}{\cdots\cdots\cdots\cdots\cdots\cdots} \end{bmatrix}$$

where the p_i and q_i are positive. The state i represents the length of a success

run, the conditional chance of a further sucess being p_j. A solution of the system
(8.27) for checking transience must have the form $x_k = x_1/p_1 \cdots p_{k-1}$, and so
there is a bounded, nontrivial solution if and only if $\Pi_{k=1}^n (1 - q_k)$ is bounded
away from 0, which holds if and only if $\Sigma_k q_k < \infty$. Thus the chain is persistent
if and only if $\Sigma_k q_k = \infty$.

Any solution of the steady-state equations (8.30) has the form $\pi_k = \pi_0 p_0 \cdots$
p_{k-1}, and so there is a stationary distribution if and only if $\Sigma_k p_0 \cdots p_k$ con-
verges. If Σq_k and $\Sigma p_0 \cdots p_k$ both diverge, as happens, for example, if $q_k = 1/k$, then the chain is persistent but there is no stationary distribution. Now $f_{00}^{(n)}$
$= p_0 \cdots p_{n-2} q_{n-1}$, and so $\Sigma_{n=1}^{\infty} n f_{00}^{(n)} = \Sigma_{n=1}^{\infty} n(p_0 \cdots p_{n-2} - p_0 \cdots p_{n-1}) = 1$
$+ \Sigma_{k=0}^{\infty} p_0 \cdots p_k$. If $\Sigma_k p_0 \cdots p_k = \infty$, it therefore follows that $\Sigma_n n f_{00}^{(n)} = \infty$,
which means in the persistent case that, although the system is certain to return
to 0, the average return time is infinite. ∎

This example illustrates the general fact that for an irreducible, aperiodic,
persistent chain there fails to exist a stationary distribution if and only if the
average return times are infinite.*

Exponential Convergence†

In the finite case $p_{ij}^{(n)}$ converges to π_j at an exponential rate:

Theorem 8.7. *If the state space is finite and the chain is irreducible and aperiodic, then there is a stationary distribution* $\{\pi_i\}$, *and*

$$|p_{ij}^{(n)} - \pi_j| \leq A\rho^n,$$

where $A > 0$ and $0 < \rho < 1$.

PROOF. Let $m_j^{(n)} = \min_i p_{ij}^{(n)}$ and $M_j^{(n)} = \max_i p_{ij}^{(n)}$. By (8.10),

$$m_j^{(n+1)} = \min_i \sum_\nu p_{i\nu} p_{\nu j}^{(n)} \geq \min_i \sum_\nu p_{i\nu} m_j^{(n)} = m_j^{(n)},$$

$$M_j^{(n+1)} = \max_i \sum_\nu p_{i\nu} p_{\nu j}^{(n)} \leq \max_i \sum_\nu p_{i\nu} M_j^{(n)} = M_j^{(n)}.$$

Since obviously $m_j^{(n)} \leq M_j^{(n)}$,

(8.34) $$0 \leq m_j^{(1)} \leq m_j^{(2)} \leq \cdots \leq M_j^{(2)} \leq M_j^{(1)} \leq 1.$$

Suppose temporarily that all the p_{ij} are positive. Let r be the number of states
and let $\delta = \min_{ij} p_{ij}$. Since $\Sigma_j p_{ij} \geq r\delta$, $0 < \delta \leq r^{-1}$. Fix states u and v for the

---

* See FELLER, Volume 1, Chapter XV, and the Addendum on page 510.
† This topic may be omitted.

moment; let Σ' denote summation over j in S satisfying $p_{uj} \geq p_{vj}$ and let Σ'' denote summation over j satisfying $p_{uj} < p_{vj}$. Then

(8.35) $$\Sigma'(p_{uj} - p_{vj}) + \Sigma''(p_{uj} - p_{vj}) = 1 - 1 = 0.$$

Since $\Sigma' p_{vj} + \Sigma'' p_{uj} \geq r\delta$,

(8.36) $$\Sigma'(p_{uj} - p_{vj}) = 1 - \Sigma'' p_{uj} - \Sigma' p_{vj} \leq 1 - r\delta.$$

Apply (8.35) and then (8.36):

$$
\begin{aligned}
p_{uk}^{(n+1)} - p_{vk}^{(n+1)} &= \sum_j (p_{uj} - p_{vj}) p_{jk}^{(n)} \\
&\leq \Sigma'(p_{uj} - p_{vj}) M_k^{(n)} + \Sigma''(p_{uj} - p_{vj}) m_k^{(n)} \\
&= \Sigma'(p_{uj} - p_{vj})(M_k^{(n)} - m_k^{(n)}) \\
&\leq (1 - r\delta)(M_k^{(n)} - m_k^{(n)}).
\end{aligned}
$$

Since u and v are arbitrary,

$$M_k^{(n+1)} - m_k^{(n+1)} \leq (1 - r\delta)(M_k^{(n)} - m_k^{(n)}).$$

Therefore, $M_k^{(n)} - m_k^{(n)} \leq (1 - r\delta)^n$. It follows by (8.34) that $m_j^{(n)}$ and $M_j^{(n)}$ have a common limit π_j and that

(8.37) $$|p_{ij}^{(n)} - \pi_j| \leq (1 - r\delta)^n.$$

Take $A = 1$ and $\rho = 1 - r\delta$. Passing to the limit in (8.10) shows that the π_i are stationary probabilities. (Note that the proof thus far makes almost no use of the preceding theory.)

If the p_{ij} are not all positive, apply Lemma 2: since there are only finitely many states, there exists an m such that $p_{ij}^{(m)} > 0$ for all i and j. By the case just treated, $M_j^{(mt)} - m_j^{(mt)} \leq \rho^t$. Take $A = \rho^{-1}$ and then replace ρ by $\rho^{1/t}$. ∎

Example 8.14. Suppose that

$$
P = \begin{bmatrix}
p_0 & p_1 & \dots & p_{r-1} \\
p_{r-1} & p_0 & \dots & p_{r-2} \\
\multicolumn{5}{c}{\dots\dots\dots\dots} \\
p_1 & p_2 & \dots & p_0
\end{bmatrix}
$$

The rows of P are the cyclic permutations of the first row: $p_{ij} = p_{j-i}$, $j - i$ reduced modulo r. Since the columns of P add to 1 as well as the rows, the steady-state equations (8.30) have the solution $\pi_i \equiv r^{-1}$. If the p_i are all positive, the theorem implies that $p_{ij}^{(n)}$ converges to r^{-1} at an exponential rate. If $X_0, Y_1, Y_2, \dots$ are independent random variables with range $\{0, 1, \dots, r - 1\}$, if each Y_n has distribution $\{p_0, \dots, p_{r-1}\}$, and if $X_n = X_0 + Y_1 + \cdots + Y_n$, where the sum is reduced modulo r, then $P[X_n = j] \to r^{-1}$. The X_n describe a random walk on a circle of points, and whatever the initial distribution, the positions become equally likely in the limit. ∎

Optimal Stopping*

Assume throughout the rest of the section that S *is finite.* Consider a function τ on Ω for which $\tau(\omega)$ is a nonnegative integer for each ω. Let $\mathcal{F}_n = \sigma(X_0, X_1, \ldots, X_n)$; τ is a *stopping time* or a *Markov time* if

$$(8.38) \qquad\qquad [\omega: \tau(\omega) = n] \in \mathcal{F}_n$$

for $n = 0, 1, \ldots$. This is analogous to the condition (7.20) on the gambler's stopping time. It will be necessary to allow $\tau(\omega)$ to assume the special value ∞, but only on a set of probability 0. This has no effect on the requirement (8.38), which concerns finite n only.

If f is a real function on the state space, then $f(X_0), f(X_1), \ldots$ are simple random variables. Imagine an observer who follows the successive states $X_0, X_1, \ldots$ of the system. He stops at time τ, when the state is X_τ (or $X_{\tau(\omega)}(\omega)$), and receives a reward or payoff $f(X_\tau)$. The condition (8.38) prevents prevision on the part of the observer. This is a kind of game, the stopping time is a strategy, and the problem is to find a strategy that maximizes the expected payoff $E[f(X_\tau)]$. The problem in Example 8.5 had this form; there $S = \{1, 2, \ldots, r + 1\}$, and the payoff function is $f(i) = i/r$ for $i \le r$ (set $f(r + 1) = 0$).

If $P(A) > 0$ and $Y = \sum_j y_j I_{B_j}$ is a simple random variable, the B_j forming a finite decomposition of Ω into $\mathcal{F}$-sets, the conditional expected value of Y given A is defined by

$$E[Y|A] = \sum_j y_j P(B_j|A).$$

Denote by E_i conditional expected values for the case $A = [X_0 = i]$:

$$E_i[Y] = E[Y|X_0 = i] = \sum_j y_j P_i(B_j).$$

The stopping-time problem is to choose τ so as simultaneously to maximize $E_i[f(X_\tau)]$ for all initial states i. If x lies in the range of f, which is finite, and if τ is everywhere finite, then $[\omega: f(X_{\tau(\omega)}(\omega)) = x] = \bigcup_{n=0}^\infty [\omega: \tau(\omega) = n, f(X_n(\omega)) = x]$ lies in $\mathcal{F}$, and so $f(X_\tau)$ is a simple random variable. In order that this always hold, put $f(X_{\tau(\omega)}(\omega)) = 0$, say, if $\tau(\omega) = \infty$ (which happens only on a set of probability 0).

The game with payoff function f has at i the *value*

---

* This topic may be omitted.

$$v(i) = \sup E_i[f(X_\tau)],$$

the supremum extending over all Markov times τ. (It is possible a priori that there is a different optimal τ for each i.) The problem is to calculate $v(i)$ and find the best τ. If the chain is irreducible, the system must pass through every state, and the best strategy is obviously to wait until the system enters a state for which f is maximal. This describes an optimal τ, and $v(i) = \max f$ for all i. For this reason the interesting cases are those in which some states are transient and others are absorbing ($p_{ii} = 1$).

A function φ on S is *excessive*, or *superharmonic*, if*

$$(8.39) \qquad\qquad \varphi(i) \geq \sum_j p_{ij}\varphi(j), \qquad i \in S.$$

In terms of conditional expectation the requirement is $\varphi(i) \geq E_i[\varphi(X_1)]$.

Lemma 3. *The value function v is excessive.*

PROOF. Given ϵ choose for each j in S a "good" stopping time τ_j satisfying $E_j[f(X_{\tau_j})] > v(j) - \epsilon$. By (8.38), $[\tau_j = n] = [(X_0, \ldots, X_n) \in I_{jn}]$ for some set I_{jn} of $(n + 1)$-long sequences of states. Set $\tau = n + 1$ ($n \geq 0$) on the set $[X_1 = j] \cap [(X_1, \ldots, X_{n+1}) \in I_{jn}]$; that is, take one step and then from the new state X_1 add on the "good" stopping time for that state. Then τ is a stopping time and

$$E_i[f(X_\tau)] = \sum_{n=0}^{\infty} \sum_j \sum_k P_i[X_1 = j, (X_1, \ldots, X_{n+1}) \in I_{jn}, X_{n+1} = k]f(k)$$

$$= \sum_{n=0}^{\infty} \sum_j \sum_k p_{ij}P_j[(X_0, \ldots, X_n) \in I_{jn}, X_n = k]f(k)$$

$$= \sum_j p_{ij}E_j[f(X_{\tau_j})].$$

Therefore, $v(i) \geq E_i[f(X_\tau)] \geq \sum_j p_{ij}(v(j) - \epsilon) = \sum_j p_{ij}v(j) - \epsilon$. Since ϵ was arbitrary, v is excessive. ∎

Lemma 4. *Suppose that φ is excessive.*
  (i) *For all stopping times τ, $\varphi(i) \geq E_i[\varphi(X_\tau)]$.*
  (ii) *For all pairs of stopping times satisfying $\tau \leq \tau'$, $E_i[\varphi(X_\tau)] \geq E_i[\varphi(X_{\tau'})]$.*

Part (i) says that for an excessive payoff function $\tau \equiv 0$ represents an optimal strategy.

* Compare the conditions (7.30) and (7.37).

PROOF.   To prove (i), put $\tau_N = \min\{\tau, N\}$. Then τ_N is a stopping time, and

$$(8.40) \quad E_i[\varphi(X_{\tau_N})] = \sum_{n=0}^{N-1} \sum_k P_i[\tau = n, X_n = k]\varphi(k)$$

$$+ \sum_k P_i[\tau \geq N, X_N = k]\varphi(k).$$

Since $[\tau \geq N] = [\tau < N]^c \in \mathcal{F}_{N-1}$, the final sum here is by (8.13)

$$\sum_k \sum_j P_i[\tau \geq N, X_{N-1} = j, X_N = k]\varphi(k)$$

$$= \sum_k \sum_j P_i[\tau \geq N, X_{N-1} = j]p_{jk}\varphi(k) \leq \sum_j P_i[\tau \geq N, X_{N-1} = j]\varphi(j).$$

Substituting this into (8.40) leads to $E_i[\varphi(X_{\tau_N})] \leq E_i[\varphi(X_{\tau_{N-1}})]$. Since $\tau_0 = 0$ and $E_i[\varphi(X_0)] = \varphi(i)$, $E_i[\varphi(X_{\tau_N})] \leq \varphi(i)$ for all N. But for $\tau(\omega)$ finite, $\varphi(X_{\tau_N(\omega)}(\omega)) \to \varphi(X_{\tau(\omega)}(\omega))$ (there is equality for large N), and so $E_i[\varphi(X_{\tau_N})] \to E[\varphi(X_\tau)]$ by Theorem 5.3.

The proof of (ii) is essentially the same. If $\tau'_N = \min\{\tau', \tau + N\}$, then τ'_N is a stopping time, and

$$E_i[\varphi(X'_{\tau_N})] = \sum_{m=0}^{\infty} \sum_{n=0}^{N-1} \sum_k P_i[\tau = m, \tau' = m + n, X_{m+n} = k]\varphi(k)$$

$$+ \sum_{m=0}^{\infty} \sum_k P_i[\tau = m, \tau' \geq m + N, X_{m+N} = k]\varphi(k).$$

Since $[\tau = m, \tau' \geq m + N] = [\tau = m] - [\tau = m, \tau' < m + N] \in \mathcal{F}_{m+N-1}$, again $E_i[\varphi(X'_{\tau_N})] \leq E_i[\varphi(X'_{\tau_{N-1}})] \leq E_i[\varphi(X'_{\tau_0})]$. Since $\tau'_0 = \tau$, part (ii) follows from part (i) by another passage to the limit.   ∎

Lemma 5.   *If an excessive function φ dominates the payoff function f, then it dominates the value function v as well.*

Of course, by definition g dominates h if and only if $g(i) \geq h(i)$ for all i.

PROOF.   By Lemma 4, $\varphi(i) \geq E_i[\varphi(X_\tau)] \geq E_i[f(X_\tau)]$ for all Markov times τ, and so $\varphi(i) \geq v(i)$ for all i.   ∎

Since $\tau \equiv 0$ is a stopping time, v dominates f. Lemmas 3 and 5 immediately characterize v:

Theorem 8.8.   *The value function v is the minimal excessive function dominating f.*

There remains the problem of constructing the optimal strategy τ. Let M be the set of states i for which $v(i) = f(i)$; M, the *support set,* is nonempty since it at least contains those i that maximize f. Let $A = \bigcap_{n=0}^{\infty} [X_n \notin M]$ be the event that the system never enters M. The following argument shows that $P_i(A) = 0$ for each i. As this is trivial if $M = S$, assume that $M \neq S$. Choose $\delta > 0$ so that $f(i) \leq v(i) - \delta$ for $i \in S - M$. Now $E_i[f(X_\tau)] = \sum_{n=0}^{\infty} \sum_k P_i[\tau = n, X_n = k]f(k)$; replacing the $f(k)$ by $v(k)$ or $v(k) - \delta$ according as $k \in M$ or $k \in S - M$ gives $E_i[f(X_\tau)] \leq E_i[v(X_\tau)] - \delta P_i[X_\tau \in S - M] \leq E_i[v(X_\tau)] - \delta P_i(A) \leq v(i) - \delta P_i(A)$, the last inequality by Lemmas 3 and 4. Since this holds for every Markov time, taking the supremum over τ gives $P_i(A) = 0$. Whatever the initial state, the system is thus certain to enter the support set M.

Let $\tau_0(\omega) = \min[n: X_n \in M]$ be the *hitting time* for M. Then τ_0 is a Markov time and $\tau_0 = 0$ if $X_0 \in M$. It may be that $X_n(\omega) \notin M$ for all n, in which case $\tau_0(\omega) = \infty$, but as just shown, the probability of this is 0.

Theorem 8.9. *The hitting time τ_0 is optimal: $E_i[f(X_{\tau_0})] = v(i)$ for all i.*

PROOF.   By the definition of τ_0, $f(X_{\tau_0}) = v(X_{\tau_0})$. Put $\varphi(i) = E_i[f(X_{\tau_0})] = E_i[v(X_{\tau_0})]$. The first step is to show that φ is excessive. If $\tau_1 = \min[n: n \geq 1, X_n \in M]$, then τ_1 is a Markov time and

$$E_i[v(X_{\tau_1})] = \sum_{n=1}^{\infty} \sum_{k \in M} P_i[X_1 \notin M, \ldots, X_{n-1} \notin M, X_n = k]v(k)$$

$$= \sum_{n=1}^{\infty} \sum_{k \in M} \sum_{j \in S} p_{ij}P_j[X_0 \notin M, \ldots, X_{n-2} \notin M, X_{n-1} = k]v(k)$$

$$= \sum_j p_{ij}E_j[v(X_{\tau_0})].$$

Since $\tau_0 \leq \tau_1$, $E_i[v(X_{\tau_0})] \geq E_i[v(X_{\tau_1})]$ by Lemmas 3 and 4.

This shows that φ is excessive. By Lemmas 3 and 4, $v(i) \geq E_i[v(X_{\tau_0})] = \varphi(i)$. If $\varphi(i) \geq f(i)$ is proved, it will follow by Theorem 8.8 that $\varphi(i) \geq v(i)$ and hence that $\varphi(i) = v(i)$. Since $\tau_0 = 0$ for $X_0 \in M$, $i \in M$ implies that $\varphi(i) = E_i[f(X_0)] = f(i)$. Suppose that $\varphi(i) < f(i)$ for some values of i in $S - M$, and choose i_0 to maximize $f(i) - \varphi(i)$. Then $\psi(i) = \varphi(i) + f(i_0) - \varphi(i_0)$ dominates f and is excessive, being the sum of a constant and an excessive function. By Theorem 8.8, ψ must dominate v, so that $\psi(i_0) \geq v(i_0)$, or $f(i_0) \geq v(i_0)$. But this implies that $i_0 \in M$, a contradiction.   ∎

The optimal strategy need not be unique. If f is constant, for example, all strategies have the same value.

Example 8.15.   For the symmetric random walk with absorbing barriers at

0 and r (Example 8.2) a function φ on $S = \{0, 1, \ldots, r\}$ is excessive if $\varphi(i) \geq \frac{1}{2}\varphi(i - 1) + \frac{1}{2}\varphi(i + 1)$ for $1 \leq i \leq r - 1$. The requirement is that φ give a concave function when extended by linear interpolation from S to the entire interval $[0, r]$. Hence v thus extended is the minimal concave function dominating f. The figure shows the geometry: the ordinates of the dots are the values of f and the polygonal line describes v. The optimal strategy is to stop at a state for which the dot lies on the polygon.

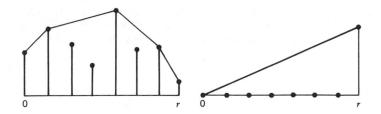

If $f(r) = 1$ and $f(i) = 0$ for $i < r$, v is a straight line; $v(i) = i/r$. The optimal Markov time τ_0 is the hitting time for $M = \{0, r\}$, and $v(i) = E_i[f(X_{\tau_0})]$ is the probability of absorption in the state r. This gives another solution of the gambler's ruin problem for the symmetric case.   ∎

Example 8.16.   For the selection problem in Example 8.5, the p_{ij} are given by (8.5) and (8.6) for $1 \leq i \leq r$, while $p_{r+1,r+1} = 1$. The payoff is $f(i) = i/r$ for $i \leq r$ and $f(r + 1) = 0$. Thus $v(r + 1) = 0$, and since v is excessive,

$$(8.41) \qquad v(i) \geq g(i) = \sum_{j=i+1}^{r} \frac{i}{j(j - 1)} v(j), \qquad 1 \leq i < r.$$

By Theorem 8.8, v is the smallest function satisfying (8.41) and $v(i) \geq f(i) = i/r$, $1 \leq i \leq r$. Since (8.41) puts no lower limit on $v(r)$, $v(r) = f(r) = 1$, and r lies in the support set M. By minimality,

$$(8.42) \qquad v(i) = \max\{f(i), g(i)\}, \qquad 1 \leq i < r.$$

If $i \in M$, then $f(i) = v(i) \geq g(i) \geq \sum_{j=i+1}^{r} ij^{-1}(j - 1)^{-1}f(j) = f(i)\sum_{j=i+1}^{r} (j - 1)^{-1}$, and hence $\sum_{j=i+1}^{r} (j - 1)^{-1} \leq 1$. On the other hand, if this inequality holds and $i + 1$, $\ldots, r$ all lie in M, then $g(i) = \sum_{j=i+1}^{r} ij^{-1} (j - 1)^{-1}f(j) = f(i) \sum_{j=i+1}^{r} (j - 1)^{-1} \leq f(i)$, so that $i \in M$ by (8.42). Therefore, $M = \{i_r, i_r+1, \ldots, r, r + 1\}$, where i_r is determined by

$$(8.43) \qquad \frac{1}{i_r} + \frac{1}{i_r+1} + \cdots + \frac{1}{r} \leq 1 < \frac{1}{i_r - 1} + \frac{1}{i_r} + \cdots + \frac{1}{r}.$$

If $i < i_r$, so that $i \notin M$, then $v(i) > f(i)$ and so, by (8.42),

$$v(i) = g(i) = \sum_{j=i+1}^{i_r-1} \frac{i}{j(j - 1)} v(j) + \sum_{j=i_r}^{r} \frac{i}{j(j - 1)} f(j)$$

$$= \sum_{j=i+1}^{i_r-1} \frac{i}{j(j - 1)} v(j) + \frac{i}{r} \left(\frac{1}{i_r - 1} + \cdots + \frac{1}{r - 1} \right).$$

It follows by backward induction starting with $i = i_r - 1$ that

(8.44) $$v(i) = p_r = \frac{i_r - 1}{r}\left(\frac{1}{i_r - 1} + \cdots + \frac{1}{r - 1}\right)$$

is constant for $1 \leq i < i_r$.

In the selection problem as originally posed, $X_1 = 1$. The optimal strategy is to stop with the first X_n that lies in M. The princess should therefore reject the first $i_r - 1$ suitors and accept the next one who is preferable to all his predecessors (is dominant). The probability of success is p_r as given by (8.44). Failure can happen two ways. Perhaps the first dominant suitor after i_r is not the best of all suitors; in this case the princess will be unaware of failure. Perhaps no dominant suitor comes after i_r; in this case the princess is obliged to take the last suitor of all and may be well aware of failure. Recall that the problem was to maximize the chance of getting the best suitor of all rather than, say, the chance of getting a suitor in the top half.

If r is large, (8.43) essentially requires that $\log r - \log i_r$ be near 1, so that $i_r \approx r/e$. In this case, $p_r \approx 1/e$.

Note that although the system starts in state 1 in the original problem, its resolution by means of the preceding theory requires consideration of all possible initial states.

■

PROBLEMS

8.1. Show that (8.2) holds if the first and third terms are equal for all $n, i_0, \ldots, i_n$, and j.

8.2. Let $Y_0, Y_1, \ldots$ be independent and identically distributed with $P[Y_n = 1] = p$, $P[Y_n = 0] = q, p \neq q$. Put $X_n = Y_n + Y_{n+1}$ (mod 2). Show that $X_0, X_1, \ldots$ is not a Markov chain even though $P[X_{n+1} = j | X_{n-1} = i] = P[X_{n+1} = j]$. Does this last relation hold for all Markov chains? Why?

8.3. Show by example that a function $f(X_0), f(X_1), \ldots$ of a Markov chain need not be a Markov chain.

8.4. 2.16   4.18↑   Given a 2×2 stochastic matrix and initial probabilities π_0 and π_1, define P for cylinders in sequence space by $P[\omega: a_i(\omega) = u_i, 1 \leq i \leq n] = \pi_{u_1}p_{u_1u_2}\cdots p_{u_{n-1}u_n}$ for sequences $u_1, \ldots, u_n$ of 0's and 1's (Problem 4.18 is the case $p_{ij} = \pi_j = p_j$). Extend P as a probability measure to $\mathcal{C}_0$ and then to $\mathcal{C}$. Show that $a_1(\omega), a_2(\omega), \ldots$ is a Markov chain with transition probabilities p_{ij}.

8.5. A set C in S is *closed* if $\Sigma_{j \in C} p_{ij} = 1$ for $i \in C$: Once the system enters C it cannot leave. Show that a chain is irreducible if and only if S has no proper closed subset.

8.6. Show by solving (8.27) that the unrestricted random walk on the line (Example 8.3) is persistent if and only if $p = \frac{1}{2}$.

8.7. Suppose that $U_0 \subset U$. Show that $x_i = P_i([X_t \in U, t \geq 1] \cap [X_n \in U_0 \text{ i.o.}])$, $i \in U$, solves (8.26) but may not be maximal.

8.8. Suppose that an irreducible chain has period $t > 1$. Show that S decomposes into sets $S_0, \ldots, S_{t-1}$ such that $p_{ij} > 0$ only if $i \in S_\nu$ and $j \in S_{\nu+1}$ for some ν ($\nu +$

1 reduced modulo t). Thus the system passes through the S_ν in cyclic succession.

8.9. Show that $\sup_{ij} n_0(i, j) = \infty$ is possible in Lemma 2.

8.10. Suppose that $\{\pi_i\}$ solves (8.30), where it is assumed that $\Sigma_i |\pi_i| < \infty$, so that the left side is well defined. Show in the irreducible case that the π_i are either all positive or all negative or all 0. Stationary probabilities thus exist in the irreducible case if and only if (8.30) has a nontrivial solution $\{\pi_i\}$ ($\Sigma_i \pi_i$ absolutely convergent), in which case each state has positive stationary probability.

8.11. Show by example that the coupled chain in the proof of Theorem 8.6 need not be irreducible if the original chain is not aperiodic.

8.12. 8.8↑ Suppose that an irreducible chain of period $t > 1$ has a stationary distribution $\{\pi_j\}$. Show that, if $i \in S_\nu$ and $j \in S_{\nu+\alpha}$ ($\nu + \alpha$ reduced modulo t), then $\lim_n p_{ij}^{(nt+\alpha)} = \pi_j$. Show that $\lim_n n^{-1} \Sigma_{m=1}^n p_{ij}^{(m)} = \pi_j/t$ for all i and j.

8.13. Show by example that the coupled chain may be persistent even though there is no stationary distribution.

8.14. Suppose that S consists of all the integers and

$$\begin{cases} p_{0,-1} = p_{0,0} = p_{0,+1} = \dfrac{1}{3}, \\ p_{k,k-1} = q, \qquad p_{k,k+1} = p, \qquad k \le -1, \\ p_{k,k-1} = p, \qquad p_{k,k+1} = q, \qquad k \ge 1. \end{cases}$$

Show that the chain is irreducible and aperiodic. For which p's is the chain persistent? For which p's are there stationary probabilities?

8.15. Analyze the chain with transition matrix (all p_i positive)

$$\begin{bmatrix} p_0 & p_1 & p_2 & \cdots \\ 1 & 0 & 0 & \cdots \\ 0 & 1 & 0 & \cdots \\ 0 & 0 & 1 & \cdots \\ \cdots\cdots\cdots\cdots\cdots \end{bmatrix}$$

8.16. Show for a two-state chain with $S = \{0, 1\}$ that $p_{00}^{(n)} = p_{10} + (p_{00} - p_{10})p_{00}^{(n-1)}$. Derive an explicit formula for $p_{00}^{(n)}$ and show that it converges to $p_{10}/(p_{01} + p_{10})$ if all p_{ij} are positive. Calculate the other $p_{ij}^{(n)}$ and verify that the limits satisfy (8.30). Verify that the rate of convergence is exponential.

8.17. A thinker who owns r umbrellas travels back and forth between home and office, taking along an umbrella (if there is one at hand) in rain (probability p) but not in shine (probability q). Let the state be the number of umbrellas at hand, irrespective of whether the thinker is at home or at work. Set up the transition matrix and find the stationary probabilities. Find the steady-state probability of his getting wet, and show that five umbrellas will protect him at the 5% level against any climate (any p).

8.18. (a) A transition matrix is *doubly stochastic* if $\Sigma_i p_{ij} = 1$ for each j. For a finite, irreducible, aperiodic chain with doubly stochastic transition matrix, show that the stationary probabilities are all equal.

(b)   Generalize example 8.14: Let S be a finite group, let $p(i)$ be probabilities, and put $p_{ij} = p(j \cdot i^{-1})$, where product and inverse refer to the group operation. Show that, if all $p(i)$ are positive, the states are all equally likely in the limit.

(c)   Let S be the symmetric group on 52 elements. What has (b) to say about card shuffling?

8.19.   Suppose that $\{X_n\}$ is a Markov chain with state space S and put $Y_n = (X_n, X_{n+1})$. Let T be the set of pairs (i, j) such that $p_{ij} > 0$ and show that Y_n is a Markov chain with state space T. Write down the transition probabilities. Show that, if $\{X_n\}$ is irreducible and aperiodic, so is $\{Y_n\}$. Show that, if π_i are stationary probabilities for $\{X_n\}$, then $\pi_i p_{ij}$ are stationary probabilities for $\{Y_n\}$.

8.20.   6.10   8.19↑   Suppose that the chain is finite, irreducible, and aperiodic and that the initial probabilities are the stationary ones. Fix a state i, let $A_n = [X_n = i]$, and let N_n be the number of passages through i in the first n steps. Calculate α_n and β_n as defined by (5.37). Show that $\beta_n - \alpha_n^2 = O(1/n)$, so that $n^{-1} N_n \to \pi_i$ with probability 1. Show for a function f on the state space that $n^{-1} \Sigma_{k=1}^n f(X_k) \to \Sigma_i \pi_i f(i)$ with probability 1. Show that $n^{-1} \Sigma_{k=1}^n g(X_k, X_{k+1}) \to \Sigma_{ij} \pi_i p_{ij} g(i, j)$ for functions g on $S \times S$.

8.21.   6.14   8.20↑   If $X_0(\omega) = i_0, \ldots, X_n(\omega) = i_n$ for states $i_0, \ldots, i_n$, put $p_n(\omega) = \pi_{i_0} p_{i_0 i_1} \cdots p_{i_{n-1} i_n}$, so that $p_n(\omega)$ is the probability of the observation observed. Show that $-n^{-1} \log p_n(\omega) \to h = -\Sigma_{ij} \pi_i p_{ij} \log p_{ij}$ with probability 1 if the chain is finite, irreducible, and aperiodic. Extend to this case the notions of source, entropy, and asymptotic equipartition.

8.22.   A sequence $\{X_n\}$ is a Markov chain *of second order* if $P[X_{n+1} = j | X_0 = i_0, \ldots, X_n = i_n] = P[X_{n+1} = j | X_{n-1} = i_{n-1}, X_n = i_n] = p_{i_{n-1}, i_n; j}$. Show that nothing really new is involved because the sequence of pairs (X_n, X_{n+1}) is an ordinary Markov chain (of first order). Compare Problem 8.19. Generalize this idea to chains of order r.

8.23.   By the method of Example 8.15, solve the gambler's ruin problem for $p \neq q$.

8.24.   Apply Theorems 8.8 and 8.9 to the indicator f of a set V in S. Show that $1 - v(i)$ is the maximal solution of

$$\begin{cases} x_i \leq \sum_{j \in S} p_{ij} x_j, & i \in S, \\ x_i = 0, & i \in V, \\ 0 \leq x_i \leq 1, & i \in S - V. \end{cases}$$

Show that the x_i for $i \in U = S - V$ are the maximal solution of (8.26) and deduce Theorem 8.4 for the finite case.

8.25.   Suppose that a chain is finite and irreducible. Show by Theorem 8.7 that an excessive function must be constant.

8.26.   Suppose that $S = \{0, 1, 2, \ldots\}$, $p_{00} = 1$, and $f_{i0} > 0$ for all i.
(a)   Show that $P_i(\bigcup_{j=1}^{\infty} [X_n = j \text{ i.o.}]) = 0$ for all i.
(b)   Regard the state as the size of a population and interpret the conditions $p_{00} = 1$ and $f_{i0} > 0$ and the conclusion in part (a).

SECTION 9.   LARGE DEVIATIONS AND THE LAW OF THE ITERATED LOGARITHM*

It is interesting in connection with the strong law of large numbers to estimate the rate at which S_n/n converges to the mean m. The proof of the strong law used upper bounds for the probabilities $P[|S_n - m| \geq \alpha]$ for large α. Accurate upper and lower bounds for these probabilities will lead to the law of the iterated logarithm, a theorem giving very precise rates for $S_n/n \to m$.

The first concern will be to estimate the probability of large deviations from the mean, which will require the method of moment generating functions. The estimates will be applied first to a problem in statistics and·then to the law of the iterated logarithm.

Moment Generating Functions

Let X be a simple random variable assuming values $x_1, \ldots, x_l$ with respective probabilities $p_1, \ldots, p_l$. Its *moment generating function* is

$$(9.1) \qquad M(t) = E[e^{tX}] = \sum_i p_i e^{tx_i}.$$

(See (5.16) for expected values of functions of random variables.) This function, defined for all real t, can be regarded as associated with X itself or as associated with its distribution—that is, with the measure on the line having mass p_i at x_i (see (5.9)).

If $c = \max_i |x_i|$, the partial sums of the series $e^{tX} = \sum_{k=0}^{\infty} t^k X^k/k!$ are bounded by $e^{|t|c}$, and so the corollary to Theorem 5.3 applies:

$$(9.2) \qquad M(t) = \sum_{k=0}^{\infty} \frac{t^k}{k!} E[X^k].$$

Thus $M(t)$ has a Taylor expansion, and as follows from the general theory, the coefficient of t^k must be $M^{(k)}(0)/k!$ Thus

$$(9.3) \qquad E[X^k] = M^{(k)}(0).$$

Furthermore, term-by-term differentiation in (9.1) gives

$$M^{(k)}(t) = \sum_i p_i x_i^k e^{tx_i} = E[X^k e^{tX}];$$

taking $t = 0$ here gives (9.3) again. Thus the moments of X can be calculated by successive differentiation, whence $M(t)$ gets its name. Note that $M(0) = 1$.

___
* This section may be omitted.

Example 9.1. If X assumes the values 1 and 0 with probabilities p and $q = 1 - p$, as in Bernoulli trials, its moment generating function is $M(t) = pe^t + q$. The first two moments are $M'(0) = p$ and $M''(0) = p$, and the variance is $p - p^2 = pq$. ∎

If $X_1, \ldots, X_n$ are independent, then for each t (see the argument following (5.7)), $e^{tX_1}, \ldots, e^{tX_n}$ are also independent. Let M and $M_1, \ldots, M_n$ be the respective moment generating functions of $S = X_1 + \cdots + X_n$ and of $X_1, \ldots, X_n$; of course, $e^{tS} = \Pi_i e^{tX_i}$. Since by (5.21) expected values multiply for independent random variables, there results the fundamental relation

$$(9.4) \qquad\qquad M(t) = M_1(t) \cdots M_n(t).$$

This is an effective way of calculating the moment generating function of the sum S. The real interest, however, centers on the distribution of S, and so it is important to know that distributions can in principle be recovered from their moment generating functions.

Consider along with (9.1) another finite exponential sum $N(t) = \Sigma_j q_j e^{ty_j}$, and suppose that $M(t) = N(t)$ for all t. If $x_{i_0} = \max x_i$ and $y_{j_0} = \max y_j$, then $M(t) \sim p_{i_0} e^{tx_{i_0}}$ and $N(t) \sim q_{j_0} e^{ty_{j_0}}$ as $t \to \infty$, and so $x_{i_0} = y_{j_0}$ and $p_{i_0} = q_{i_0}$. The same argument now applies to $\Sigma_{i \neq i_0} p_i e^{tx_i} = \Sigma_{j \neq j_0} q_j e^{ty_j}$, and it follows inductively that with appropriate relabeling, $x_i = y_i$ and $p_i = q_i$ for each i. *Thus the function* (9.1) *does uniquely determine the* x_i *and* p_i.

Example 9.2. If $X_1, \ldots, X_n$ are independent, each assuming values 1 and 0 with probabilities p and q, then $S = X_1 + \cdots + X_n$ is the number of successes in n Bernoulli trials. By (9.4) and Example 9.1, S has the moment generating function

$$E[e^{tS}] = (pe^t + q)^n = \sum_{k=0}^{n} \binom{n}{k} p^k q^{n-k} e^{tk}.$$

The right-hand form shows this to be the moment generating function of a distribution with mass $\binom{n}{k} p^k q^{n-k}$ at the integer k, $0 \leq k \leq n$. The uniqueness just established therefore yields the standard fact that $P[S = k] = \binom{n}{k} p^k q^{n-k}$. ∎

The *cumulant generating function* of X (or of its distribution) is

$$(9.5) \qquad\qquad C(t) = \log M(t) = \log E[e^{tX}].$$

(Note that $M(t)$ is strictly positive.) Since $C' = M'/M$ and $C'' = (MM'' - (M')^2)/M^2$, and since $M(0) = 1$,

$$(9.6) \qquad C(0) = 0, \qquad C'(0) = E[X], \qquad C''(0) = \text{Var } [X].$$

Let $m_k = E[X^k]$. The leading term in (9.2) is $m_0 = 1$, and so a formal expansion of the logarithm in (9.5) gives

$$(9.7) \qquad C(t) = \sum_{v=1}^{\infty} \frac{(-1)^{v+1}}{v} \left(\sum_{k=1}^{\infty} \frac{m_k}{k!} t^k \right)^v.$$

Since $M(t) \to 1$ as $t \to 0$, this expansion is valid for t in some neighborhood of 0. By the theory of series, the powers on the right can be expanded and terms with a common factor t^i collected together. This gives an expansion

$$(9.8) \qquad C(t) = \sum_{i=1}^{\infty} \frac{c_i}{i!} t^i,$$

valid in some neighborhood of 0.

The c_i are the *cumulants* of X. Equating coefficients in the expansions (9.7) and (9.8) leads to $c_1 = m_1$ and $c_2 = m_2 - m_1^2$, which checks with (9.6). Each c_i can be expressed as a polynomial in $m_1, \dots, m_i$ and conversely, although the calculations soon become tedious. If $E[X] = 0$, however, so that $m_1 = c_1 = 0$, it is not hard to check that

$$(9.9) \qquad c_3 = m_3, \qquad c_4 = m_4 - 3m_2^2.$$

Taking logarithms converts the multiplicative relation (9.4) into the additive relation

$$(9.10) \qquad C(t) = C_1(t) + \cdots + C_n(t)$$

for the corresponding cumulant generating functions; it is valid in the presence of independence. By this and the definition (9.8), it follows that cumulants add for independent random variables.

Clearly, $M''(t) = E[X^2 e^{tX}] \geq 0$. Since $(M'(t))^2 = E^2[Xe^{tX}] \leq E[e^{tX}] \cdot E[X^2 e^{tX}] = M(t)M''(t)$ by Schwarz's inequality (5.32), $C''(t) \geq 0$. *Thus the moment generating function and the cumulant generating function are both convex.*

Large Deviations

Let Y be a simple random variable assuming values y_j with probabilities p_j. The problem is to estimate $P[Y \geq \alpha]$ when Y has mean 0 and α is positive. It is notationally convenient to subtract α away from Y and instead estimate $P[Y \geq 0]$ when Y has negative mean.

Assume then that

$$(9.11) \qquad E[Y] < 0, \qquad P[Y > 0] > 0,$$

the second assumption to avoid trivialities. Let $M(t) = \Sigma_j p_j e^{ty_j}$ be the moment generating function of Y. Then $M'(0) < 0$ by the first assumption in (9.11), and $M(t) \to \infty$ as $t \to \infty$ by the second. Since $M(t)$ is convex, it has its minimum

ρ at a positive argument τ:

(9.12) $\qquad \inf_t M(t) = M(\tau) = \rho, \qquad 0 < \rho < 1, \quad \tau > 0.$

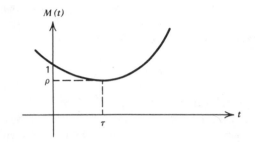

Construct (on an entirely irrelevant probability space) an auxiliary random variable Z such that

(9.13) $$P[Z = y_j] = \frac{e^{\tau y_j}}{\rho} P[Y = y_j]$$

for each y_j in the range of Y. Note that the probabilities on the right do add to 1. The moment generating function of Z is

(9.14) $$E[e^{tZ}] = \sum_j \frac{e^{\tau y_j}}{\rho} p_j e^{t y_j} = \frac{M(\tau + t)}{\rho}$$

and therefore

(9.15) $\qquad E[Z] = \dfrac{M'(\tau)}{\rho} = 0, \qquad s^2 = E[Z^2] = \dfrac{M''(\tau)}{\rho} > 0.$

For all positive t, $P[Y \geq 0] = P[e^{tY} \geq 1] \leq M(t)$ by Markov's inequality (5.27), and hence

(9.16) $\qquad\qquad\qquad\qquad P[Y \geq 0] \leq \rho.$

Inequalities in the other direction are harder to obtain. If Σ' denotes summation over those indices j for which $y_j \geq 0$, then

(9.17) $\qquad P[Y \geq 0] = \Sigma' p_j = \rho \Sigma' e^{-\tau y_j} P[Z = y_j].$

Put the final sum here in the form $e^{-\theta}$, and let $p = P[Z \geq 0]$. By (9.16), $\theta \geq 0$. Since $\log x$ is concave, Jensen's inequality (5.29) gives

$$-\theta = \log \Sigma' e^{-\tau y_j} p^{-1} P[Z = y_j] + \log p$$
$$\geq \Sigma'(-\tau y_j) p^{-1} P[Z = y_j] + \log p$$

$$= -\tau s p^{-1} \sum' \frac{y_j}{s} P[Z = y_j] + \log p.$$

By (9.15) and (5.33),

$$\sum' \frac{y_j}{s} P[Z = y_j] \le \frac{1}{s} E[|Z|] \le \frac{1}{s} E^{1/2}[Z^2] = 1.$$

The last two inequalities give

(9.18) $$0 \le \theta \le \frac{\tau s}{P[Z \ge 0]} - \log P[Z \ge 0].$$

This proves the following result.

Theorem 9.1. *Suppose that Y satisfies* (9.11). *Define ρ and τ by* (9.12), *let Z be a random variable with distribution* (9.13), *and define s^2 by* (9.15). *Then $P[Y \le 0] = \rho e^{-\theta}$, where θ satisfies* (9.18).

To use (9.18) requires a lower bound for $P[Z \ge 0]$.

Theorem 9.2. *If $E[Z] = 0$, $E[Z^2] = s^2$, and $E[Z^4] = \xi^4 > 0$, then $P[Z \ge 0] \ge s^4/4\xi^4$.*

PROOF. Let $Z^+ = ZI_{[Z \ge 0]}$ and $Z^- = -ZI_{[Z < 0]}$. Then Z^+ and Z^- are non-negative, $Z = Z^+ - Z^-$, $Z^2 = (Z^+)^2 + (Z^-)^2$, and

(9.19) $$s^2 = E[(Z^+)^2] + E[(Z^-)^2].$$

Let $p = P[Z \ge 0]$. By Schwarz's inequality (5.32),

$$E[(Z^+)^2] = E[I_{[Z \ge 0]}Z^2]$$
$$\le E^{1/2}[I_{[Z \ge 0]}^2]E^{1/2}[Z^4] = p^{1/2}\xi^2.$$

By Hölder's inequality (5.31) (for $p = \frac{3}{2}$ and $q = 3$)

$$E[(Z^-)^2] = E[(Z^-)^{2/3}(Z^-)^{4/3}]$$
$$\le E^{2/3}[Z^-]E^{1/3}[(Z^-)^4] \le E^{2/3}[Z^-]\xi^{4/3}.$$

Since $E[Z] = 0$, another application of Hölder's inequality (for $p = 4$ and $q = \frac{4}{3}$) gives

$$E[Z^-] = E[Z^+] = E[ZI_{[Z \ge 0]}]$$
$$\le E^{1/4}[Z^4]E^{3/4}[I_{[Z \ge 0]}^{4/3}] = \xi p^{3/4}.$$

Combining these three inequalities with (9.19) gives $s^2 \le p^{1/2}\xi^2 + (\xi p^{3/4})^{2/3}\xi^{4/3} = 2p^{1/2}\xi^2$. ∎

Chernoff's Theorem*

Theorem 9.3. *Let $X_1, X_2, \ldots$ be independent, identically distributed simple random variables satisfying $E[X_n] < 0$ and $P[X_n > 0] > 0$, let $M(t)$ be their common moment generating function, and put $\rho = \inf_t M(t)$. Then*

(9.20) $$\lim_{n \to \infty} \frac{1}{n} \log P[X_1 + \cdots + X_n \ge 0] = \log \rho.$$

PROOF.   Put $Y_n = X_1 + \cdots + X_n$. Then $E[Y_n] < 0$ and $P[Y_n > 0] \ge P^n[X_1 > 0] > 0$, and so the hypotheses of Theorem 9.1 are satisfied. Define ρ_n and τ_n by $\inf_t M_n(t) = M_n(\tau_n) = \rho_n$, where $M_n(t)$ is the moment generating function of Y_n. Since $M_n(t) = M^n(t)$, it follows that $\rho_n = \rho^n$ and $\tau_n = \tau$, where $M(\tau) = \rho$.

Let Z_n be the analogue for Y_n of the Z described by (9.13). Its moment generating function (see (9.14)) is $M_n(\tau + t)/\rho^n = (M(\tau + t)/\rho)^n$. This is also the moment generating function of $Z'_1 + \cdots + Z'_n$ for independent random variables $Z'_1, \ldots, Z'_n$ each having moment generating function $M(\tau + t)/\rho$. Now each Z'_i has (see (9.15)) mean 0 and some positive variance σ^2 and fourth moment ξ^4 independent of i. Since Z_n must have the same moments as $Z'_1 + \cdots + Z'_n$, it has mean 0, variance $s_n^2 = n\sigma^2$, and fourth moment $\xi_n^4 = n\xi^4 + 3n(n - 1)\sigma^4 = O(n^2)$ (see (6.2)). By Theorem 9.2, $P[Z_n \ge 0] \ge s_n^4/4\xi_n^4 \ge \alpha$ for some positive α independent of n. By Theorem 9.1 then, $P[Y_n \ge 0] = \rho^n e^{-\theta_n}$, where $0 \le \theta_n \le \tau_n s_n \alpha^{-1} - \log \alpha = \tau \alpha^{-1} \sigma \sqrt{n} - \log \alpha$. This gives (9.20), and shows, in fact, that the rate of convergence is $O(n^{-1/2})$.   ∎

This result is important in the theory of statistical hypothesis testing. An informal treatment of the Bernoulli case will illustrate the connection.

Suppose $S_n = X_1 + \cdots + X_n$, where the X_i are independent and assume the values 1 and 0 with probabilities p and q. Now $P[S_n \ge na] = P[\sum_{k=1}^{n} (X_k - a) \ge 0]$, and Chernoff's theorem applies if $p < a < 1$. In this case $M(t) = E[e^{t(X_1 - a)}] = e^{-ta}(pe^t + q)$. Minimizing this shows that the ρ of Chernoff's theorem satisfies

$$-\log \rho = K(a, p) = a \log \frac{a}{p} + b \log \frac{b}{q},$$

where $b = 1 - a$. By (9.20), $n^{-1} \log P[S_n \ge na] \to -K(a, p)$; express this as

(9.21) $$P[S_n \ge na] \approx e^{-nK(a,p)}.$$

Suppose now that p is unknown and that there are two competing hypotheses concerning its value, the hypothesis H_1 that $p = p_1$ and the hypothesis H_2 that $p = p_2$, where $p_1 < p_2$. Given the observed results $X_1, \ldots, X_n$ of n Bernoulli trials, one decides in favor of H_2 if $S_n \ge na$ and in favor of H_1 if $S_n < na$, where a is some number satisfying $p_1 < a < p_2$. The problem is to find an advantageous value for the threshold a.

---

* This theorem is not needed for the law of the iterated logarithm, Theorem 9.5.

By (9.21),

(9.22) $$P[S_n \geq na \,|\, H_1] \approx e^{-nK(a,p_1)},$$

where the notation indicates that the probability is calculated for $p = p_1$—that is, under the assumption of H_1. By symmetry,

(9.23) $$P[S_n < na \,|\, H_2] \approx e^{-nK(a,p_2)}.$$

The left sides of (9.22) and (9.23) are the probabilities of erroneously deciding in favor of H_2 when H_1 is, in fact, true and of erroneously deciding in favor of H_1 when H_2 is, in fact, true—the probabilities describing the level and power of the test.

Suppose a is chosen so that $K(a, p_1) = K(a, p_2)$, which makes the two error probabilities approximately equal. This constraint gives for a a linear equation with solution

(9.24) $$a = a(p_1, p_2) = \frac{\log (q_1/q_2)}{\log (p_2/p_1) + \log (q_1/q_2)},$$

where $q_i = 1 - p_i$. The common error probability is approximately $e^{-nK(a,p_1)}$ for this value of a, and so the larger $K(a, p_1)$ is, the easier it is to distinguish statistically between p_1 and p_2.

Although $K(a(p_1, p_2), p_1)$ is a complicated function, it has a simple approximation for p_1 near p_2. As $x \to 0$, $\log(1 + x) = x - \frac{1}{2}x^2 + O(x^3)$. Using this in the definition of K and collecting terms gives

(9.25) $$K(p + x, p) = \frac{x^2}{2pq} + O(x^3), \qquad x \to 0.$$

Fix $p_1 = p$, and let $p_2 = p + t$; (9.24) becomes a function $\psi(t)$ of t, and expanding the logarithms gives

(9.26) $$\psi(t) = p + \frac{1}{2} t + O(t^2), \qquad t \to 0$$

after some reductions. Finally, (9.25) and (9.26) together imply that

(9.27) $$K(\psi(t), p) = \frac{t^2}{8pq} + O(t^3), \qquad t \to 0.$$

In distinguishing $p_1 = p$ from $p_2 = p + t$ for small t, if a is chosen to equalize the two error probabilities, then their common value is about $e^{-nt^2/8pq}$. For t fixed, the nearer p is to $\frac{1}{2}$, the larger this probability is and the more difficult it is to distinguish p from $p + t$. As an example, compare $p = .1$ with $p = .5$. Now $.36nt^2/8(.1)(.9) = nt^2/8(.5)(.5)$; with a sample only .36 as large, .1 can therefore be distinguished from $.1 + t$ with the same precision as .5 can be distinguished from $.5 + t$.

The Law of the Iterated Logarithm

The analysis of the rate at which S_n/n approaches the mean depends on the following variant of the theorem on large deviations.

Theorem 9.4. *Let $S_n = X_1 + \cdots X_n$, where the X_n are independent and identically distributed simple random variables with mean 0 and variance 1. If a_n are constants satisfying*

(9.28) $$a_n \to \infty, \qquad \frac{a_n}{\sqrt{n}} \to 0,$$

then

(9.29) $$P[S_n \geq a_n\sqrt{n}] = e^{-a_n^2(1+\zeta_n)/2}$$

for a sequence ζ_n going to 0.

PROOF.   Put $Y_n = S_n - a_n\sqrt{n} = \sum_{k=1}^{n}(X_k - a_n/\sqrt{n})$. Then $E[Y_n] < 0$. Since X_1 has mean 0 and variance 1, $P[X_1 > 0] > 0$, and it follows by (9.28) that $P[X_1 > a_n/\sqrt{n}] > 0$ for n sufficiently large, in which case $P[Y_n > 0] \geq P^n[X_1 - a_n/\sqrt{n} > 0] > 0$. Thus Theorem 9.1 applies to Y_n for all large enough n.

Let $M_n(t)$, ρ_n, τ_n, and Z_n be associated with Y_n as in the theorem. If $m(t)$ and $c(t)$ are the moment and cumulant generating functions of the X_n, then $M_n(t)$ is the nth power of the moment generating function $e^{-ta_n/\sqrt{n}}m(t)$ of $X_1 - a_n/\sqrt{n}$, and so Y_n has cumulant generating function

(9.30) $$C_n(t) = -ta_n\sqrt{n} + nc(t).$$

Since τ_n is the unique minimum of $C_n(t)$, and since $C_n'(t) = -a_n\sqrt{n} + nc'(t)$, τ_n is determined by the equation $c'(\tau_n) = a_n/\sqrt{n}$. Since X_1 has mean 0 and variance 1, it follows by (9.6) that

(9.31) $$c(0) = c'(0) = 0, \qquad c''(0) = 1.$$

Now $c'(t)$ is nondecreasing because $c(t)$ is convex, and since $c'(\tau_n) = a_n/\sqrt{n}$ goes to 0, τ_n must therefore go to 0 as well and must in fact be $O(a_n/\sqrt{n})$. By the second-order mean-value theorem for $c'(t)$, $a_n/\sqrt{n} = c'(\tau_n) = \tau_n + O(\tau_n^2)$, from which follows

(9.32) $$\tau_n = \frac{a_n}{\sqrt{n}} + O\left(\frac{a_n^2}{n}\right).$$

By the third-order mean-value theorem for $c(t)$,

$$\log \rho_n = C_n(\tau_n) = -\tau_n a_n\sqrt{n} + nc(\tau_n)$$

$$= -\tau_n a_n\sqrt{n} + n\left[\frac{1}{2}\tau_n^2 + O(\tau_n^3)\right].$$

Applying (9.32) gives

(9.33) $$\log \rho_n = -\frac{1}{2}a_n^2 + o(a_n^2).$$

Now (see (9.14)) Z_n has moment generating function $M_n(\tau_n + t)/\rho_n$ and (see (9.30)) cumulant generating function $D_n(t) = C_n(\tau_n + t) - \log \rho_n = -(\tau_n + t)a_n\sqrt{n} + nc(t + \tau_n) - \log \rho_n$. The mean of Z_n is $D_n'(0) = 0$. Its variance s_n^2 is $D_n''(0)$; by (9.31), this is

(9.34) $\qquad s_n^2 = nc''(\tau_n) = n(c''(0) + O(\tau_n)) = n(1 + o(1)).$

The fourth cumulant of Z_n is $D_n''''(0) = nc''''(\tau_n) = O(n)$. By formula (9.9) relating moments and cumulants (applicable because $E[Z_n] = 0$), $E[Z_n^4] = 3s_n^4 + D_n''''(0)$. Therefore, $E[Z_n^4]/s_n^4 \to 3$, and it follows by Theorem 9.2 that there exists an α such that $P[Z_n \geq 0] \geq \alpha > 0$ for all sufficiently large n.

By Theorem 9.1, $P[Y_n \geq 0] = \rho_n e^{-\theta_n}$ with $0 \leq \theta_n \leq \tau_n s_n \alpha^{-1} + \log \alpha$. By (9.28), (9.32), and (9.34), $\theta_n = O(a_n) = o(a_n^2)$, and it follows by (9.33) that $P[Y_n \geq 0] = e^{-a_n^2(1+o(1))/2}$. ∎

The law of the iterated logarithm is this:

Theorem 9.5. *Let $S_n = X_1 + \cdots + X_n$, where the X_n are independent, identically distributed simple random variables with mean 0 and variance 1. Then*

(9.35) $\qquad P\left[\limsup_n \frac{S_n}{\sqrt{2n \log \log n}} = 1 \right] = 1.$

Equivalent to (9.35) is the assertion that for positive ϵ

(9.36) $\qquad P[S_n \geq (1 + \epsilon)\sqrt{2n \log \log n} \text{ i.o.}] = 0$

and

(9.37) $\qquad P[S_n \geq (1 - \epsilon)\sqrt{2n \log \log n} \text{ i.o.}] = 1.$

The set in (9.35) is, in fact, the intersection over positive rational ϵ of the sets in (9.37) minus the union over positive rational ϵ of the sets in (9.36).

The idea of the proof is this. Write

(9.38) $\qquad \phi(n) = \sqrt{2n \log \log n}\,.$

If $A_n^{\pm} = [S_n \geq (1 \pm \epsilon)\phi(n)]$, then by (9.29), $P(A_n^{\pm})$ is near $(\log n)^{-(1\pm\epsilon)^2}$. If n_k increases exponentially, say $n_k \sim \theta^k$ for $\theta > 1$, then $P(A_{n_k}^{\pm})$ is of the order $k^{-(1\pm\epsilon)^2}$. Now $\Sigma_k k^{-(1\pm\epsilon)^2}$ converges if the sign is $+$ and diverges if the sign is $-$. It will follow by the first Borel-Cantelli lemma that there is probability 0 that $A_{n_k}^+$ occurs for infinitely many k. In proving (9.36), an extra argument is required to get around the fact that the A_n^+ for $n \neq n_k$ must also be accounted for (this involves choosing θ near 1). If the A_n^- were independent, it would follow by the second Borel-Cantelli lemma that with probability 1, $A_{n_k}^-$ occurs for infinitely many k, which would in turn imply (9.37). An extra argument is required to get around the fact that the $A_{n_k}^-$ are dependent (this involves choosing θ large).

For the proof of (9.36) a preliminary result is needed. Put $M_k = \max\{S_0, S_1, \ldots, S_k\}$, where $S_0 = 0$.

Theorem 9.6. *If the X_k are independent with mean 0 and variance 1, then for $\alpha \geq \sqrt{2}$,*

(9.39) $$P\left[\frac{M_n}{\sqrt{n}} \geq \alpha\right] \leq 2P\left[\frac{S_n}{\sqrt{n}} \geq \alpha - \sqrt{2}\right].$$

PROOF. If $A_j = [M_{j-1} < \alpha\sqrt{n} \leq M_j]$, then

$$P\left[\frac{M_n}{\sqrt{n}} \geq \alpha\right] \leq P\left[\frac{S_n}{\sqrt{n}} \geq \alpha - \sqrt{2}\right] + \sum_{j=1}^{n-1} P\left(A_j \cap \left[\frac{S_n}{\sqrt{n}} \leq \alpha - \sqrt{2}\right]\right).$$

Since $S_n - S_j$ has variance $n - j$, it follows by independence and Chebyshev's inequality that the probability in the sum is at most

$$P\left(A_j \cap \left[\frac{|S_n - S_j|}{\sqrt{n}} > \sqrt{2}\right]\right) = P(A_j)P\left[\frac{|S_n - S_j|}{\sqrt{n}} > \sqrt{2}\right]$$

$$\leq P(A_j)\frac{n - j}{2n} \leq \frac{1}{2}P(A_j).$$

Since $\bigcup_{j=1}^{n-1} A_j \subset [M_n \geq \alpha\sqrt{n}]$,

$$P\left[\frac{M_n}{\sqrt{n}} \geq \alpha\right] \leq P\left[\frac{S_n}{\sqrt{n}} \geq \alpha - \sqrt{2}\right] + \frac{1}{2}P\left[\frac{M_n}{\sqrt{n}} \geq \alpha\right]. \quad \blacksquare$$

PROOF OF (9.36). Given ϵ, choose θ so that $\theta > 1$ but $\theta^2 < 1 + \epsilon$. Let $n_k = [\theta^k]$ and $x_k = \theta(2 \log \log n_k)^{1/2}$. By (9.29) and (9.39),

$$P\left[\frac{M_{n_k}}{\sqrt{n_k}} \geq x_k\right] \leq 2 \exp\left\{-\frac{1}{2}(x_k - \sqrt{2})^2(1 + \xi_k)\right\},$$

where $\xi_k \to 0$. The negative of the exponent is asymptotically $\theta^2 \log k$ and hence for large k exceeds $\theta \log k$, so that

$$P\left[\frac{M_{n_k}}{\sqrt{n_k}} \geq x_k\right] \leq \frac{2}{k^\theta}.$$

Since $\theta > 1$, it follows by the first Borel-Cantelli lemma that there is probability 0 that (see (9.38))

(9.40) $$M_{n_k} \geq \theta\phi(n_k)$$

for infinitely many k. Suppose that $n_{k-1} < n \leq n_k$ and that

(9.41) $$S_n > (1 + \epsilon)\phi(n).$$

Now $\phi(n) \geq \phi(n_{k-1}) \sim \theta^{-1/2}\phi(n_k)$; hence, by the choice of θ, $(1 + \epsilon)\phi(n) > \theta\phi(n_k)$ if k is large enough. Thus for sufficiently large k, (9.41) implies (9.40) (if $n_{k-1} < n \leq n_k$), and there is therefore probability 0 that (9.41) holds for infinitely many n. $\quad \blacksquare$

PROOF OF (9.37). Given ϵ, choose an integer θ so large that $3\theta^{-1/2} < \epsilon$. Take $n_k = \theta^k$. Now $n_k - n_{k-1} \to \infty$, and (9.29) applies with $n = n_k - n_{k-1}$ and $a_n = x_k/\sqrt{n_k - n_{k-1}}$, where $x_k = (1 - \theta^{-1})\phi(n_k)$. It follows that

$$P[S_{n_k} - S_{n_{k-1}} \geq x_k] = P[S_{n_k-n_{k-1}} \geq x_k] = \exp\left[-\frac{1}{2}\frac{x_k^2}{n_k - n_{k-1}}(1 + \xi_k) \right],$$

where $\xi_k \to 0$. The negative of the exponent is asymptotically $(1 - \theta^{-1}) \log k$ and so for large k is less than $\log k$, in which case $P[S_{n_k} - S_{n_{k-1}} \geq x_k] \geq k^{-1}$. The events here being independent, it follows by the second Borel-Cantelli lemma that with probability 1, $S_{n_k} - S_{n_{k-1}} \geq x_k$ for infinitely many k. On the other hand, by (9.36) applied to $\{-X_n\}$, there is probability 1 that $-S_{n_{k-1}} \leq 2\phi(n_{k-1})$ $\leq 2\theta^{-1/2}\phi(n_k)$ for all but finitely many k. These two inequalities give $S_{n_k} \geq x_k - 2\theta^{-1/2}\phi(n_k) > (1 - \epsilon)\phi(n_k)$, the last inequality because of the choice of θ. ∎

That completes the proof of Theorem 9.5.

PROBLEMS

9.1. Prove (6.2) by using (9.9) and the fact that cumulants add in the presence of independence.

9.2. Let Y be a simple random variable with moment generating function $M(t)$. Show that $P[Y > 0] = 0$ implies that $P[Y \geq 0] = \inf_t M(t)$. Is the infimum achieved in this case?

9.3. In the Bernoulli case, (9.21) gives

$$P[S_n \geq np + x_n] = \exp\left[-nK\left(p + \frac{x_n}{n}, p\right)(1 + o(1)) \right],$$

where $p < a < 1$ and $x_n = n(a - p)$. Theorem 9.4 gives

$$P[S_n \geq np + x_n] = \exp\left[-\frac{x_n^2}{2npq}(1 + o(1)) \right],$$

where $x_n = a_n\sqrt{npq}$. Resolve the apparent discrepancy. Use (9.25) to compare the two expressions in case x_n/n is small. See Problem 27.17.

9.4. Relabeled the binomial parameter p as $\theta = f(p)$, where f is increasing and continuously differentiable. Show by (9.27) that the distinguishability of θ from $\theta + \Delta\theta$, as measured by K, is $(\Delta\theta)^2/8p(1 - p)(f'(p))^2 + O(\Delta\theta)^3$. The leading coefficient is independent of θ if $f(p) = \arcsin\sqrt{p}$.

9.5. From (9.35) and the same result for $\{-X_n\}$, together with the uniform boundedness of the X_n, deduce that with the probability 1 the set of limit points of the sequence $\{S_n(2n \log \log n)^{-1/2}\}$ is the closed interval from -1 to $+1$.

9.6. ↑ Suppose X_n takes the values ± 1 with probability $\frac{1}{2}$ each and show that $P[S_n$

$= 0$ i.o.$] = 1$. (This gives still another proof of the persistence of symmetric random walk on the line (Example 8.3).) Show more generally that, if the X_n are bounded by M, then $P[|S_n| \leq M$ i.o.$] = 1$.

9.7. Weakened versions of (9.36) are quite easy to prove. By a fourth moment argument (see 6.2)), show that $P[S_n > n^{3/4}(\log n)^{(1+\epsilon)/4}$ i.o.$] = 0$. Use (9.29) to give a simple proof that $P[S_n > (3n \log n)^{1/2}$ i.o.$] = 0$.

9.8. What happens in Theorem 9.6 if $\sqrt{2}$ is replaced by θ ($0 < \theta < \alpha$)?

9.9. Show that (9.35) is true if S_n is replaced by $|S_n|$ or $\max_{k \leq n} S_k$ or $\max_{k \leq n} |S_k|$.

CHAPTER 2

Measure

SECTION 10. GENERAL MEASURES

Lebesgue measure on the unit interval was central to the ideas in Chapter 1. Lebesgue measure on the entire real line is important in probability as well as in analysis generally, and a uniform treatment of this and other examples requires a notion of measure for which infinite values are possible. The present chapter extends the ideas of Sections 2 and 3 to this more general case.

Classes of Sets

The σ-field of Borel sets in $(0, 1]$ played an essential role in Chapter 1, and it is necessary to construct the analogous classes for the entire real line and for k-dimensional Euclidean space.

Example 10.1.   Let $x = (x_1, \ldots, x_k)$ be the generic point of Euclidean k-space R^k. The bounded rectangles

$$(10.1) \qquad [x = (x_1, \ldots, x_k): a_i < x_i \le b_i, i = 1, \ldots, k]$$

will play in R^k the role the intervals $(a, b]$ played in $(0, 1]$. Let $\mathcal{R}^k$ be the σ-field generated by these rectangles. This is the analogue of the class $\mathcal{B}$ of Borel sets in $(0, 1]$; see Example 2.7. The elements of $\mathcal{R}^k$ are the *k-dimensional Borel sets*. For $k = 1$ they are also called the *linear Borel sets*.

Call the rectangle (10.1) *rational* if the a_i and b_i are all rational. If G is an open set in R^k and $y \in G$, then there is a rational rectangle A_y such that $y \in A_y \subset G$. But then $G = \bigcup_{y \in G} A_y$, and since there are only countably many rational rectangles, this is a countable union. Thus $\mathcal{R}^k$ *contains the open sets*. Since a closed set has open complement, $\mathcal{R}^k$ also contains the *closed* sets. Just as $\mathcal{B}$ contains all the tangible sets in $(0, 1]$, $\mathcal{R}^k$ contains all the tangible sets in R^k.

The σ-field $\mathcal{R}^k$ is generated by subclasses other than the class of rectangles. If A_n is the x-set where $a_i < x_i < b_i + n^{-1}, i = 1, \ldots, k$, then A_n is open and (10.1) is $\bigcap_n A_n$. Thus $\mathcal{R}^k$ *is generated by the open sets*. Similarly, it is generated by the closed sets. Now an open set is a countable union of rational rectangles. *Therefore, the (countable) class of rational rectangles generates* $\mathcal{R}^k$. ∎

The σ-field $\mathcal{R}^1$ on the line R^1 is by definition generated by the finite intervals. The σ-field $\mathcal{B}$ in $(0, 1]$ is generated by the subintervals of $(0, 1]$. The question naturally arises whether the elements of $\mathcal{B}$ are the elements of $\mathcal{R}^1$ that happen to lie inside $(0, 1]$. The answer is yes, as shown by the following technical but frequently useful result.

Theorem 10.1. *Suppose that* $\Omega_0 \subset \Omega$.

(i) *If $\mathcal{F}$ is a σ-field in Ω, then $\mathcal{F}_0 = [A \cap \Omega_0: A \in \mathcal{F}]$ is a σ-field in Ω_0.*

(ii) *If $\mathcal{A}$ generates the σ-field $\mathcal{F}$ in Ω, then $\mathcal{A}_0 = [A \cap \Omega_0: A \in \mathcal{A}]$ generates the σ-field $\mathcal{F}_0 = [A \cap \Omega_0: A \in \mathcal{F}]$ in Ω_0.*

PROOF OF (i).   Of course, $\Omega_0 = \Omega \cap \Omega_0 \in \mathcal{F}_0$. If $B = A \cap \Omega_0$ and $A \in \mathcal{F}$, then $\Omega_0 - B = (\Omega - A) \cap \Omega_0$ and $\Omega - A \in \mathcal{F}$. If $B_n = A_n \cap \Omega_0$ and $A_n \in \mathcal{F}$, then $\bigcup_n B_n = (\bigcup_n A_n) \cap \Omega_0$ and $\bigcup_n A_n \in \mathcal{F}$. Thus $\mathcal{F}_0$ is a σ-field in Ω_0. ∎

PROOF OF (ii).   Let $\sigma_0(\mathcal{A}_0)$ be the σ-field $\mathcal{A}_0$ generates in Ω_0. Since $\mathcal{F}_0$ is a σ-field in Ω_0 by part (i), and since $\mathcal{A}_0 \subset \mathcal{F}_0$, certainly $\sigma_0(\mathcal{A}_0) \subset \mathcal{F}_0$.

Now $\mathcal{F}_0 \subset \sigma_0(\mathcal{A}_0)$ will follow if it is shown that $A \in \mathcal{F}$ implies that $A \cap \Omega_0 \in \sigma_0(\mathcal{A}_0)$, or, to put it another way, if it is shown that $\mathcal{F}$ is contained in $\mathcal{G} = [A: A \cap \Omega_0 \in \sigma_0(\mathcal{A}_0)]$. Since $A \in \mathcal{A}$ implies that $A \cap \Omega_0 \in \mathcal{A}_0 \subset \sigma_0(\mathcal{A}_0)$, it follows that $\mathcal{A} \subset \mathcal{G}$. It is therefore enough to show that $\mathcal{G}$ is a σ-field in Ω. But if $A \in \mathcal{G}$, then $(\Omega - A) \cap \Omega_0 = \Omega_0 - (A \cap \Omega_0) \in \sigma_0(\mathcal{A})$ and hence $\Omega - A \in \mathcal{G}$. If $A_n \in \mathcal{G}$ for all n, then $(\bigcup_n A_n) \cap \Omega_0 = \bigcup_n (A_n \cap \Omega_0) \in \sigma_0(\mathcal{A}_0)$ and hence $\bigcup_n A_n \in \mathcal{G}$. ∎

In the case $\Omega_0 \in \mathcal{F}$, $\mathcal{F}_0 = [A: A \subset \Omega_0, A \in \mathcal{F}]$. If $\Omega = R^1$, $\Omega_0 = (0, 1]$, and $\mathcal{F} = \mathcal{R}^1$, and if $\mathcal{A}$ is the class of finite intervals, then $\mathcal{A}_0$ is the class of subintervals of $(0, 1]$ and $\mathcal{B} = \sigma_0(\mathcal{A}_0)$ is given by

$$(10.2) \qquad\qquad \mathcal{B} = [A: A \subset (0, 1], A \in \mathcal{R}^1].$$

A subset of $(0, 1]$ is thus a Borel set (lies in $\mathcal{B}$) if and only if it is a linear Borel set (lies in $\mathcal{R}^1$), and the distinction in terminology can be dropped.

Conventions Involving ∞

Measures assume values in the set $[0, \infty]$ consisting of the ordinary nonnegative reals and the special value ∞, and some arithmetic conventions are called for.

For $x, y \in [0, \infty]$, $x \leq y$ means that $y = \infty$ or else x and y are finite (that is, are ordinary real numbers) and $x \leq y$ holds in the usual sense. Similarly, $x < y$ means that $y = \infty$ and x is finite or else x and y are both finite and $x < y$ holds in the usual sense.

For a finite or infinite sequence $x, x_1, x_2, \ldots$ in $[0, \infty]$,

(10.3) $$x = \sum_k x_k$$

means that either (i) $x = \infty$ and $x_k = \infty$ for some k, or (ii) $x = \infty$ and $x_k < \infty$ for all k and $\Sigma_k x_k$ is an ordinary divergent infinite series, or (iii) $x < \infty$ and $x_k < \infty$ for all k and (10.3) holds in the usual sense for $\Sigma_k x_k$ an ordinary finite sum or convergent infinite series. By these conventions and Dirichlet's theorem the order of summation in (10.3) makes no difference.

For an infinite sequence $x, x_1, x_2, \ldots$ in $[0, \infty]$,

(10.4) $$x_k \uparrow x$$

means in the first place that $x_k \leq x_{k+1} \leq x$ and in the second place that either (i) $x < \infty$ and there is convergence in the usual sense, or (ii) $x_k = \infty$ for some k, or (iii) $x = \infty$ and the x_k are finite reals converging to infinity in the usual sense.

Measures

A set function μ on a field $\mathcal{F}$ in Ω is a *measure* if it satisfies these conditions:

(i) $\mu(A) \in [0, \infty]$ for $A \in \mathcal{F}$;
(ii) $\mu(0) = 0$;
(iii) if $A_1, A_2, \ldots$ is a disjoint sequence of $\mathcal{F}$-sets and if $\bigcup_{k=1}^{\infty} A_k \in \mathcal{F}$, then (see (10.3))

$$\mu \left(\bigcup_{k=1}^{\infty} A_k \right) = \sum_{k=1}^{\infty} \mu(A_k).$$

The measure μ is *finite* or *infinite* as $\mu(\Omega) < \infty$ or $\mu(\Omega) = \infty$; it is a *probability* measure if $\mu(\Omega) = 1$ as in Chapter 1. If $\Omega = A_1 \cup A_2 \cup \ldots$ for a finite or countable collection of $\mathcal{F}$-sets satisfying $\mu(A_k) < \infty$, then μ is *σ-finite*. The significance of this concept will be seen later. A finite measure is by definition σ-finite; a σ-finite measure may be either finite or infinite. If $\mathcal{A}$ is a subclass of $\mathcal{F}$, μ is *σ-finite on $\mathcal{A}$* if $\Omega = \bigcup_k A_k$ for $\mathcal{A}$-sets satisfying $\mu(A_k) < \infty$.

If μ is a measure on a σ-field $\mathcal{F}$ in Ω, the triple $(\Omega, \mathcal{F}, \mu)$ is a *measure space*. (This term is not used if $\mathcal{F}$ is merely a field.) It is an infinite, a σ-finite, a finite, or a probability measure space according as μ has the corresponding properties. If $\mu(A^c) = 0$ for an $\mathcal{F}$-set A, then A is a *support* of μ, and μ is *concentrated* on A. For a finite measure, A is a support if and only if $\mu(A) = \mu(\Omega)$.

The pair $(\Omega, \mathcal{F})$ itself is a *measurable space* if $\mathcal{F}$ is a σ-field in Ω. To say that μ is a measure on $(\Omega, \mathcal{F})$ indicates clearly both the space and the class of sets involved.

As in the case of probability measures, (iii) above is the condition of *countable additivity*, and it implies *finite additivity*: If $A_1, \ldots, A_n$ are disjoint $\mathcal{F}$-sets, then

$$\mu \left(\bigcup_{k=1}^{n} A_k \right) = \sum_{k=1}^{n} \mu(A_k).$$

Example 10.2. A measure μ on $(\Omega, \mathcal{F})$ is *discrete* if there are finitely or countably many points ω_i in Ω and masses m_i in $[0, \infty]$ such that $\mu(A) = \Sigma_{\omega_i \in A} \, m_i$ for $A \in \mathcal{F}$. It is an infinite, a finite, or a probability measure as $\Sigma_i m_i$ diverges, or converges, or converges to 1; the last case was treated in Example 2.9. If $\mathcal{F}$ contains each singleton $\{\omega_i\}$, then μ is σ-finite if and only if $m_i < \infty$ for all i. ∎

Example 10.3. Let $\mathcal{F}$ be the σ-field of all subsets of an arbitrary Ω, and let $\mu(A)$ be the number of points in A, where $\mu(A) = \infty$ if A is not finite. This μ is *counting measure*; it is finite if and only if Ω is finite and is σ-finite if and only if Ω is countable. Even if $\mathcal{F}$ does not contain every subset of Ω, counting measure is well defined on $\mathcal{F}$. ∎

Example 10.4. Specifying a measure includes specifying its domain. If μ is a measure on a field $\mathcal{F}$ and $\mathcal{F}_0$ is a field contained in $\mathcal{F}$, then the restriction μ_0 of μ to $\mathcal{F}_0$ is also a measure. Although often denoted by the same symbol, μ_0 is really a different measure from μ unless $\mathcal{F}_0 = \mathcal{F}$. Its properties may be different: If μ is counting measure on the σ-field $\mathcal{F}$ of all subsets of a countably infinite Ω, μ is σ-finite, but its restriction to the σ-field $\mathcal{F}_0 = \{0 \; \Omega\}$ is not σ-finite. ∎

Certain properties of probability measures carry over immediately to the general case. First, μ is *monotone*: $\mu(A) \leq \mu(B)$ if $A \subset B$. This is derived, just as its special case (2.4), from $\mu(A) + \mu(B - A) = \mu(B)$. But it is possible to go on and write $\mu(B - A) = \mu(B) - \mu(A)$ only if $\mu(B) < \infty$ (because $\infty - x$ has not been defined). The inclusion-exclusion formula (2.7) also carries over without change to $\mathcal{F}$-sets of finite measure:

$$(10.5) \quad \mu\left(\bigcup_{k=1}^{n} A_k\right) = \sum_i \mu(A_i) - \sum_{i<j} \mu(A_i \cap A_j) + \cdots$$
$$+ (-1)^{n+1}\mu(A_1 \cap \cdots \cap A_n).$$

The proof of *finite subadditivity* also goes through just as before:

$$\mu\left(\bigcup_{k=1}^{n} A_k\right) \leq \sum_{k=1}^{n} \mu(A_k);$$

here the A_k need not have finite measure.

Theorem 10.2. *Let μ be a measure on a field $\mathcal{F}$.*

(i) *Continuity from below: If A_n and A lie in $\mathcal{F}$ and $A_n \uparrow A$, then* $\mu(A_n)$ $\uparrow \mu(A)$.*

(ii) *Continuity from above: If A_n and A lie in $\mathcal{F}$ and $A_n \downarrow A$, and if $\mu(A_1)$ $< \infty$, then $\mu(A_n) \downarrow \mu(A)$.*

---

* See (10.4).

(iii) *Countable subadditivity: If $A_1, A_2, \ldots$ and $\bigcup_{k=1}^{\infty} A_k$ lie in $\mathcal{F}$, then*

$$\mu\left(\bigcup_{k=1}^{\infty} A_k\right) \le \sum_{k=1}^{\infty} \mu(A_k).$$

(iv) *If μ is σ-finite on $\mathcal{F}$, then $\mathcal{F}$ cannot contain an uncountable, disjoint collection of sets of positive μ-measure.*

PROOF. The proofs of (i) and (iii) are exactly as for the corresponding parts of Theorem 2.1. The same is essentially true of (ii): If $\mu(A_1) < \infty$, subtraction is possible and $A_1 - A_n \uparrow A_1 - A$ implies that $\mu(A_1) - \mu(A_n) = \mu(A_1 - A_n) \uparrow \mu(A_1 - A) = \mu(A_1) - \mu(A)$.

There remains (iv). Let $[B_\theta: \theta \in \Theta]$ be a disjoint collection of $\mathcal{F}$-sets satisfying $\mu(B_\theta) > 0$. Consider an $\mathcal{F}$-set A for which $\mu(A) < \infty$. If $\theta_1, \ldots, \theta_n$ are distinct indices satisfying $\mu(A \cap B_{\theta_i}) \ge \epsilon > 0$, then $n\epsilon \le \sum_{i=1}^{n} \mu(A \cap B_{\theta_i}) \le \mu(A)$, and so $n \le \mu(A)/\epsilon$. Thus the index set $[\theta: \mu(A \cap B_\theta) \ge \epsilon]$ is finite, and hence (take the union over positive rational ϵ) $[\theta: \mu(A \cap B_\theta) > 0]$ is countable. Since μ is σ-finite, $\Omega = \bigcup_k A_k$ for some finite or countable sequence of $\mathcal{F}$-sets A_k satisfying $\mu(A_k) < \infty$. But then $\Theta_k = [\theta: \mu(A_k \cap B_\theta) > 0]$ is countable for each k. If $\mu(A_k \cap B_\theta) = 0$ for all k, then $\mu(B_\theta) \le \Sigma_k \mu(A_k \cap B_\theta) = 0$. Therefore, $\Theta = \bigcup_k \Theta_k$, and so Θ is countable. ∎

Uniqueness

According to Theorem 3.3, probability measures agreeing on a π-system $\mathcal{P}$ agree on $\sigma(\mathcal{P})$. There is an extension to the general case.

Theorem 10.3. *Suppose that μ_1 and μ_2 are measures on $\sigma(\mathcal{P})$, where $\mathcal{P}$ is a π-system, and suppose they are σ-finite* on $\mathcal{P}$. If μ_1 and μ_2 agree on $\mathcal{P}$, then they agree on $\sigma(\mathcal{P})$.*

PROOF. If $B \in \mathcal{P}$ and $\mu_1(B) = \mu_2(B) < \infty$, then the argument for the probability case (Theorem 3.3) shows without change that $\mu_1(B \cap A) = \mu_2(B \cap A)$ for all $A \in \sigma(\mathcal{P})$. By σ-finiteness there exist $\mathcal{P}$-sets B_k satisfying $\Omega = \bigcup_k B_k$ and $\mu_1(B_k) = \mu_2(B_k) < \infty$. By the inclusion-exclusion formula (10.5),

$$\mu_\alpha\left(\bigcup_{i=1}^{n}(B_i \cap A)\right) = \sum_{1 \le i \le n} \mu_\alpha(B_i \cap A) - \sum_{1 \le i < j \le n} \mu_\alpha(B_i \cap B_j \cap A) + \cdots$$

for $\alpha = 1,2$ and for all n. Since $\mathcal{P}$ is a π-system containing the B_i, it contains the $B_i \cap B_j$, and so on. Whatever $\sigma(\mathcal{P})$-set A may be, the terms on the right above are therefore the same for $\alpha = 1$ as for $\alpha = 2$. The left side is then the same for $\alpha = 1$ as for $\alpha = 2$; letting $n \to \infty$ gives $\mu_1(A) = \mu_2(A)$.† ∎

---

* A measure can be σ-finite on $\sigma(\mathcal{P})$ without being σ-finite on $\mathcal{P}$; that μ_1 and μ_2 are σ-finite on $\mathcal{P}$ implies in particular that Ω is a countable union of $\mathcal{P}$-sets. See Problem 10.4.

† If the B_i are disjoint (which can always be arranged if $\mathcal{P}$ is a field), the argument simplifies.

As the half-infinite intervals $(-\infty, x]$ constitute a π-system generating $\mathcal{R}^1$, measures agreeing for such sets agree for all sets in $\mathcal{R}^1$, provided that they are σ-finite on the class of half-infinite intervals.

PROBLEMS

10.1. Let $(\Omega_n, \mathcal{F}_n, \mu_n)$ be a measure space, $n = 1, 2, \ldots$, and suppose that the sets Ω_n are disjoint. Let $\Omega = \bigcup_n \Omega_n$, let $\mathcal{F}$ consist of the sets of the form $A = \bigcup_n A_n$ with $A_n \in \mathcal{F}_n$, and for such an A let $\mu(A) = \Sigma_n \mu_n(A_n)$. Show that $(\Omega, \mathcal{F}, \mu)$ is a measure space. Under what conditions is it σ-finite? Finite?

10.2. On the σ-field of all subsets of $\Omega = \{1, 2, \ldots\}$, put $\mu(A) = \Sigma_{k \in A} 2^{-k}$ if A is finite and $\mu(A) = \infty$ otherwise. Is μ finitely additive? Countably additive?

10.3. 3.5↑   A measure space $(\Omega, \mathcal{F}, \mu)$ is *complete* if $A \subset B$, $B \in \mathcal{F}$, and $\mu(B) = 0$ together imply that $A \in \mathcal{F}$—the definition is just as in the probability case. Use the ideas of Problem 3.5 to construct a complete measure space $(\Omega, \mathcal{F}^+, \mu^+)$ such that $\mathcal{F} \subset \mathcal{F}^+$ and μ and μ^+ agree on $\mathcal{F}$.

10.4. **(a)**   Show that Theorem 10.3 fails without the σ-finiteness condition.

(b)   Show by example that a measure can be σ-finite on $\sigma(\mathcal{P})$ without being σ-finite on $\mathcal{P}$.

10.5. Suppose that μ_1 and μ_2 are measures on a σ-field $\mathcal{F}$ and are σ-finite on a field $\mathcal{F}_0$ generating $\mathcal{F}$.

(a)   Show that, if $\mu_1(A) \le \mu_2(A)$ holds for all A in $\mathcal{F}_0$, then it holds for all A in $\mathcal{F}$.

(b)   Show that this fails if $\mathcal{F}_0$ is only a π-system.

SECTION 11.   OUTER MEASURE

Outer Measure

An *outer measure* is a set function μ^* that is defined for all subsets of a space Ω and has these four properties:

(i) $\mu^*(A) \in [0, \infty]$ for every $A \subset \Omega$;

(ii) $\mu^*(0) = 0$;

(iii) μ^* is monotone: $A \subset B$ implies $\mu^*(A) \le \mu^*(B)$;

(iv) μ^* is countably subadditive: $\mu^*(\bigcup_n A_n) \le \Sigma_n \mu^*(A_n)$.

The set function P^* defined by (3.1) is an example, one that generalizes:

Example 11.1.   Let ρ be a set function on a class $\mathcal{A}$ in Ω. Assume that $0 \in \mathcal{A}$ and $\rho(0) = 0$, and that $\rho(A) \in [0, \infty]$ for $A \in \mathcal{A}$; ρ and $\mathcal{A}$ are otherwise arbitrary. Put

(11.1) $$\mu^*(A) = \inf \sum_n \rho(A_n),$$

where the infimum extends over all finite and countable coverings of A by $\mathcal{A}$-sets A_n. If no such covering exists, take $\mu^*(A) = \infty$ in accordance with the convention that the infimum over an empty set is ∞.

That μ^* satisfies (i), (ii), and (iii) is clear. If $\mu^*(A_n) = \infty$ for some n, then obviously $\mu^*(\bigcup_n A_n) \le \Sigma_n \mu^*(A_n)$. Otherwise, cover each A_n by $\mathcal{A}$-sets B_{nk} satisfying $\Sigma_k \rho(B_{nk}) < \mu^*(A_n) + \epsilon/2^n$; then $\mu^*(\bigcup_n A_n) \le \Sigma_{n,k} \rho(B_{nk}) < \Sigma_n \mu^*(A_n) + \epsilon$. Thus μ^* is an outer measure. ∎

Define A to be μ^*-*measurable* if

(11.2) $$\mu^*(A \cap E) + \mu^*(A^c \cap E) = \mu^*(E)$$

for every E. This is the general version of the definition (3.4) used in Section 3. It is by subadditivity equivalent to

(11.3) $$\mu^*(A \cap E) + \mu^*(A^c \cap E) \le \mu^*(E).$$

Denote by $\mathcal{M}(\mu^*)$ the class of μ^*-measurable sets.

The extension property for probability measures in Theorem 3.1 was proved via a sequence of four lemmas. These lemmas carry over directly to the case of the general outer measure: if P^* is replaced by μ^* and $\mathcal{M}$ by $\mathcal{M}(\mu^*)$ at each occurrence, the proofs hold verbatum. The fourth lemma in Section 3 becomes this:

Theorem 11.1.   *If μ^* is an outer measure, then $\mathcal{M}(\mu^*)$ is a σ-field, and μ^* restricted to $\mathcal{M}(\mu^*)$ is a measure.*

This will be used to prove an extension theorem, but it has other applications as well; see Sections 19 and 29.

Extension

Theorem 11.2.   *A measure on a field has an extension to the generated σ-field.*

For the question of the extent to which the extension is unique, see Theorem 10.3.

Theorem 11.2 can be deduced from Theorem 11.1 by the arguments used in the proof of Theorem 3.1.* It is unnecessary to retrace the steps, however, because the ideas will appear in stronger form in the proof of the next result, which generalizes Theorem 11.2.

---

* See also Problem 11.1.

Define a class $\mathcal{A}$ of subsets of Ω to be a *semiring* if

(i) $0 \in \mathcal{A}$;

(ii) $A, B \in \mathcal{A}$ implies $A \cap B \in \mathcal{A}$;

(iii)· if $A, B \in \mathcal{A}$ and $A \subset B$, then there exist disjoint $\mathcal{A}$-sets $C_1, \ldots, C_n$ such that $B - A = \bigcup_{k=1}^{n} C_k$.

The class of finite intervals in $\Omega = R^1$ and the class of subintervals of $\Omega = (0, 1]$ are the simplest examples of semirings.

Theorem 11.3. *Suppose that μ is a set function on a semiring $\mathcal{A}$. Suppose that μ has values in $[0, \infty]$, that $\mu(0) = 0$, and that μ is finitely additive and countably subadditive. Then μ extends to a measure on $\sigma(\mathcal{A})$.*[†]

This contains Theorem 11.2, because the conditions are all satisfied if $\mathcal{A}$ is a field and μ is a measure on it.

PROOF.   If A, B, and the C_k are related as in condition (iii) above, then by finite additivity $\mu(B) = \mu(A) + \sum_{k=1}^{n} \mu(C_k) \geq \mu(A)$. Thus μ is monotone.

Define an outer measure μ^* by (11.1) for $\rho = \mu$:

(11.4) $$\mu^*(A) = \inf \sum_n \mu(A_n),$$

the infimum extending over coverings of A by $\mathcal{A}$-sets. Suppose that $A \in \mathcal{A}$. If $\mu^*(E) = \infty$, then (11.3) holds trivially. If $\mu^*(E) < \infty$, for given ϵ choose $\mathcal{A}$-sets A_n such that $E \subset \bigcup_n A_n$ and $\sum_n \mu(A_n) < \mu^*(E) + \epsilon$. Since $\mathcal{A}$ is a semiring, $B_n = A \cap A_n$ lies in $\mathcal{A}$ and $A^c \cap A_n = A_n - B_n$ has the form $\bigcup_{k=1}^{m_n} C_{nk}$ for disjoint $\mathcal{A}$-sets C_{nk}. Note that $A_n = B_n \cup \bigcup_{k=1}^{m_n} C_{nk}$, where the union is disjoint, and that $A \cap E \subset \bigcup_n B_n$ and $A^c \cap E \subset \bigcup_n \bigcup_{k=1}^{m_n} C_{nk}$. By the definition of μ^* and the assumed finite additivity of μ,

$$\mu^*(A \cap E) + \mu^*(A^c \cap E) \leq \sum_n \mu(B_n) + \sum_n \sum_{k=1}^{m_n} \mu(C_{nk})$$

$$= \sum_n \mu(A_n) < \mu^*(E) + \epsilon.$$

Since ϵ is arbitrary, (11.3) follows. Thus $\mathcal{A} \subset \mathcal{M}(\mu^*)$.

If $A \subset \bigcup_n A_n$ for $\mathcal{A}$-sets A and A_n, then by the assumed countable subadditivity of μ and the monotonicity established above, $\mu(A) \leq \sum_n \mu(A \cap A_n) \leq \sum_n \mu(A_n)$. Therefore, $A \in \mathcal{A}$ implies that $\mu(A) \leq \mu^*(A)$ and hence, since the reverse inequality is an immediate consequence of (11.4), $\mu(A) = \mu^*(A)$. Thus μ^* agrees with μ on $\mathcal{A}$. As $\mathcal{A} \subset \mathcal{M}(\mu^*)$ and μ^* is countably additive on the

[†] And so μ must have been countably additive on $\mathcal{A}$ in the first place; see Problem 11.4.

σ-field $\mathcal{M}(\mu^*)$ by Theorem 11.1, μ^* restricted to $\sigma(\mathcal{A})$ is the required extension of μ on $\mathcal{A}$. ∎

Example 11.2. For $\mathcal{A}$ take the semiring of subintervals of $\Omega = (0, 1]$ together with the empty set. For μ take length λ: $\lambda(a, b] = b - a$. The finite additivity of λ follows by Theorem 2.2 and the countable subadditivity by Lemma 2 to that theorem.[†] By Theorem 11.3, λ extends to a measure on the class $\sigma(\mathcal{A}) = \mathcal{B}$ of Borel sets in $(0, 1]$. ∎

This gives a second construction of Lebesgue measure in the unit interval. In the first construction λ was extended first from the intervals to the field $\mathcal{B}_0$ of finite disjoint unions of intervals (see (2.11)) and then by Theorem 11.2 (in its special form Theorem 3.1) from $\mathcal{B}_0$ to $\mathcal{B} = \sigma(\mathcal{B}_0)$. Using Theorem 11.3 instead of Theorem 11.2 effects a slight economy, since the extension then goes from $\mathcal{A}$ directly to $\mathcal{B}$ without the intermediate stop at $\mathcal{B}_0$, and the arguments involving (2.12) and (2.13) become unnecessary.

Example 11.3. In Theorem 11.3 take for $\mathcal{A}$ the semiring of finite intervals on the real line R^1, and consider $\lambda_1(a, b] = b - a$. The arguments for Theorem 2.2 and its two lemmas in no way require that the (finite) intervals in question be contained in $(0, 1]$, and so λ_1 is finitely additive and countably subadditive on this class $\mathcal{A}$. Hence λ_1 extends to the σ-field $\mathcal{R}^1$ of linear Borel sets, which is by definition generated by $\mathcal{A}$. This defines Lebesgue measure λ_1 over the whole real line. ∎

A subset of $(0, 1]$ lies in $\mathcal{B}$ if and only if it lies in $\mathcal{R}^1$ (see (10.2)). Now $\lambda_1(A) = \lambda(A)$ for subintervals A of $(0, 1]$, and it follows by uniqueness (Theorem 3.3) that $\lambda_1(A) = \lambda(A)$ for all A in $\mathcal{B}$. Thus there is no inconsistency in dropping λ_1 and using λ to denote Lebesgue measure on $\mathcal{R}^1$ as well as on $\mathcal{B}$.

Example 11.4. The class of bounded rectangles in R^k is a semiring, a fact needed in the next section. Suppose $A = [x: x_i \in I_i, i \le k]$ and $B = [x: x_i \in J_i, i \le k]$ are nonempty rectangles, the I_i and J_i being finite intervals. If $A \subset B$, then $I_i \subset J_i$, so that $J_i - I_i$ is a disjoint union $I_i' \cup I_i''$ of intervals (possibly empty). Consider the 3^k disjoint rectangles $[x: x_i \in U_i, i \le k]$, where for each i, U_i is I_i or I_i' or I_i''. One of these rectangles is A itself, and $B - A$ is the union of the others. The rectangles thus form a semiring. ∎

An Approximation Theorem

If $\mathcal{A}$ is semiring, then by Theorem 10.3 a measure on $\sigma(\mathcal{A})$ is determined by its values

---

[†] On a field, countable additivity implies countable subadditivity, but $\mathcal{A}$ is merely a semiring. Hence the necessity of this observation; but see Problem 11.4.

on $\mathcal{A}$. The following theorem shows more explicitly how the measure of a $\sigma(\mathcal{A})$-set can be approximated by the measures of $\mathcal{A}$-sets.

Theorem 11.4. *Suppose that $\mathcal{A}$ is a semiring, μ is a measure on $\mathcal{F} = \sigma(\mathcal{A})$, and μ is σ-finite on $\mathcal{A}$.*

(i) *If $B \in \mathcal{F}$ and $\epsilon > 0$, there exists a finite or infinite disjoint sequence $A_1, A_2, \ldots$ of $\mathcal{A}$-sets such that $B \subset \bigcup_k A_k$ and $\mu((\bigcup_k A_k) - B) < \epsilon$.*

(ii) *If $B \in \mathcal{F}$ and $\epsilon > 0$, and if $\mu(B) < \infty$, then there exists a finite disjoint sequence $A_1, \ldots, A_n$ of $\mathcal{A}$-sets such that $\mu(B \, \Delta (\bigcup_{k=1}^n A_k)) < \epsilon$.*

PROOF. Return to the proof of Theorem 11.3. If μ^* is the outer measure defined by (11.4), then $\mathcal{F} \subset \mathcal{M}(\mu^*)$ and μ^* agrees with μ on $\mathcal{A}$, as was shown. Since μ^* restricted to $\mathcal{F}$ is a measure, it follows by Theorem 10.3 that μ^* agrees with μ on $\mathcal{F}$ as well. If $B \in \mathcal{F}$ and $\mu(B) = \mu^*(B) < \infty$, there exist $\mathcal{A}$-sets A_k such that $B \subset \bigcup_k A_k$ and $\mu(\bigcup_k A_k) \leq \Sigma_k \mu(A_k) < \mu(B) + \epsilon$; but then $\mu((\bigcup_k A_k) - B) < \epsilon$.

By σ-finiteness there are $\mathcal{A}$-sets C_m such that $\Omega = \bigcup_m C_m$ and $\mu(C_m) < \infty$. For the general $\mathcal{F}$-set B there are $\mathcal{A}$-sets A_{mk} such that $B \cap C_m \subset \bigcup_k A_{mk}$ and $\mu((\bigcup_k A_{mk}) - (B \cap C_m)) < \epsilon/2^m$. The sets A_{mk} taken all together provide a sequence $A_1, A_2, \ldots$ of $\mathcal{A}$-sets for which $B \subset \bigcup_k A_k$ and $\mu((\bigcup_k A_k) - B) < \epsilon$. To make the sequence disjoint, replace A_k by $A_k \cap A_1^c \cap \cdots \cap A_{k-1}^c$; as follows by the semiring property and induction, this last set is a finite disjoint union of $\mathcal{A}$-sets.

To prove part (ii), consider the A_k of part (i). If B has finite measure, so has $A = \bigcup_k A_k$, and hence by continuity from above (Theorem 10.2(ii)), $\mu(A - \bigcup_{k \leq n} A_k) < \epsilon$ for some n. But then $\mu(B \, \Delta (\bigcup_{k=1}^n A_k)) < 2 \, \epsilon$. ∎

If, for example, B is a linear Borel set of finite Lebesgue measure, then $\lambda(B \, \Delta (\bigcup_{k=1}^n A_k)) < \epsilon$ for some finite intervals $A_1, \ldots, A_n$.

Carathéodory's Condition[†]

Suppose that μ^* is an outer measure on the class of all subsets of R^k. It satisfies *Carathéodory's condition* if

$$(11.5) \qquad \qquad \mu^*(A \cup B) = \mu^*(A) + \mu^*(B)$$

whenever A and B are at positive distance in the sense that dist $(A, B) = \inf [\,|x - y| : x \in A, y \in B]$ is positive.

Theorem 11.5. *If μ^* is an outer measure on R^k satisfying Carathéodory's condition, then $\mathcal{R}^k \subset \mathcal{M}(\mu^*)$.*

PROOF. Since $\mathcal{M}(\mu^*)$ is a σ-field (Theorem 11.1) and the closed sets generate $\mathcal{R}^k$ (Example 10.1), it is enough to show that each closed set is μ^*-measurable. The problem is thus to prove (11.3) for A closed and E arbitrary.

Let $B = A \cap E$ and $C = A^c \cap E$. Let $C_n = [x \in C : \text{dist} (x, A) \geq n^{-1}]$, so that

---

† This topic may be omitted; it is needed only in Section 19 for the construction of Hausdorff measure.

$C_n \uparrow C$ and dist $(C_n, A) \geq n^{-1}$. By Carathéodory's condition, $\mu^*(E) = \mu^*(B \cup C) \geq \mu^*(B \cup C_n) = \mu^*(B) + \mu^*(C_n)$. Hence it suffices to prove that $\mu^*(C_n) \to \mu^*(C)$, or, since $C_n \uparrow C$, that $\mu^*(C) \leq \lim_n \mu^*(C_n)$.

Let $D_n = C_{n+1} - C_n$; if $x \in D_{n+1}$ and $|x - y| < n^{-1}(n + 1)^{-1}$, then dist $(y, A) \leq |y - x| + $ dist $(x, A) < n^{-1}$, so that $y \notin C_n$. Thus $|x - y| \geq n^{-1}(n + 1)^{-1}$ if $x \in D_{n+1}$ and $y \in C_n$, and so D_{n+1} and C_n are at positive distance if they are nonempty. By the Carathéodory condition extended inductively,

$$(11.6) \qquad \mu^*(C_{2n+1}) \geq \mu^* \left(\bigcup_{k=1}^{n} D_{2k} \right) = \sum_{k=1}^{n} \mu^*(D_{2k}),$$

and

$$(11.7) \qquad \mu^*(C_{2n}) \geq \mu^* \left(\bigcup_{k=1}^{n} D_{2k-1} \right) = \sum_{k=1}^{n} \mu^*(D_{2k-1}).$$

By subadditivity,

$$(11.8) \qquad \mu^*(C) \leq \mu^*(C_{2n}) + \sum_{k=n}^{\infty} \mu^*(D_{2k}) + \sum_{k=n+1}^{\infty} \mu^*(D_{2k-1}).$$

If $\sum \mu^*(D_{2k})$ or $\sum \mu^*(D_{2k-1})$ diverges, $\mu^*(C) \leq \lim_n \mu^*(C_n)$ follows from (11.6) or (11.7); otherwise, it follows from (11.8). ∎

Example 11.5. For $A \subset R^1$ define $\lambda^*(A) = \inf \sum_n (b_n - a_n)$, where the infimum extends over countable coverings of A by intervals $(a_n, b_n]$. Then λ^* is an outer measure (Example 11.1). Suppose that A and B are at positive distance and consider any covering of $A \cup B$ by intervals $(a_n, b_n]$; (11.5) will follow if $\sum(b_n - a_n) \geq \lambda^*(A) + \lambda^*(B)$ necessarily holds. The sum here is unchanged if $(a_n, b_n]$ is replaced by a disjoint union of subintervals each of length less than dist (A, B). But if $b_n - a_n < $ dist (A, B), then $(a_n, b_n]$ cannot meet both A and B. Let M and N be the sets of n for which $(a_n, b_n]$ meets A and B, respectively. Then M and N are disjoint, $A \subset \bigcup_{n \in M} (a_n, b_n]$, and $B \subset \bigcup_{n \in N} (a_n, b_n]$; hence $\sum(b_n - a_n) \geq \sum_{n \in M} (b_n - a_n) + \sum_{n \in N} (b_n - a_n) \geq \lambda^*(A) + \lambda^*(B)$.

Thus λ^* satisfies Carathéodory's condition. By Theorem 11.5, $\mathcal{R}^1 \subset \mathcal{M}(\lambda^*)$, and so λ^* restricted to $\mathcal{R}^1$ is a measure by Theorem 11.1. By Lemma 2 in Section 2 (page 22), $\lambda^*(a, b] = b - a$. Thus λ^* restricted to $\mathcal{R}^1$ is λ, which gives still another construction of Lebesgue measure. ∎

PROBLEMS

11.1. The proof of Theorem 3.1 obviously applies if the probability measure is replaced by a finite measure, since this is only a matter of rescaling. Take as a starting point then the fact that a finite measure on a field extends uniquely to the generated σ-field. By the following steps prove Theorem 11.1—that is, remove the assumption of finiteness.

(a) Let μ be a measure (not necessarily even σ-finite) on a field $\mathcal{F}_0$ and let $\mathcal{F}$ $= \sigma(\mathcal{F}_0)$. If $A \in \mathcal{F}_0$ and $\mu(A) < \infty$, restrict μ to a finite measure μ_A on the field $\mathcal{F}_0(A) = [B: B \subset A, B \in \mathcal{F}_0]$ in A, and extend μ_A to a finite measure (still denoted μ_A) on the σ-field $\mathcal{F}(A) = [B: B \subset A, B \in \mathcal{F}]$ generated in A by $\mathcal{F}_0(A)$.

(b) Suppose that $E \in \mathcal{F}$. If there exist disjoint $\mathcal{F}_0$-sets A_n such that $\mu(A_n) < \infty$ and $E \subset \bigcup_n A_n$, put $\mu(E) = \Sigma_n \mu_{A_n}(E \cap A_n)$ and prove consistency; otherwise, put $\mu(E) = \infty$.

(c) Show that μ is a measure on $\mathcal{F}$ and agrees with the original μ on $\mathcal{F}_0$.

11.2. (a) If μ^* is an outer measure and $\nu^*(E) = \mu^*(E \cap A)$, then ν^* is an outer measure.

(b) If μ_n^* are outer measures and $\mu^*(E) = \Sigma_n \mu_n^*(E)$ or $\mu^*(E) = \sup \mu_n^*(E)$ then μ^* is an outer measure.

11.3. Suppose in Example 11.1 that $\rho(A) = 1$ if $A \neq 0$. Describe μ^*.

11.4. Suppose that $\mathcal{A}$ is a semiring and μ is a finitely additive set function on $\mathcal{A}$ having values in $[0, \infty]$ and satisfying $\mu(0) = 0$.

(a) Show by induction on l that, if $A, B_1, \ldots, B_l$ lie in $\mathcal{A}$, then $A \cap B_1^c \cap \cdots \cap B_l^c$ has the form $\bigcup_{k=1}^n C_k$ for disjoint $\mathcal{A}$-sets C_k.

(b) Suppose that $A, A_1, \ldots, A_m$ are $\mathcal{A}$-sets. Show that $\Sigma_{k=1}^m \mu(A_k) \leq \mu(A)$ if $\bigcup_{k=1}^m A_k \subset A$ and the A_k are disjoint. Show that $\mu(A) \leq \Sigma_{k=1}^m \mu(A_k)$ if $A \subset \bigcup_{k=1}^m A_k$.

(c) Show that μ is countably additive on $\mathcal{A}$ if and only if it is countably sub-additive.

11.5. Measure theory is sometimes based on the notion of a *ring*. A class $\mathcal{A}$ is a ring if it contains the empty set and is closed under the formation of differences and finite unions; it is a σ-ring if it is also closed under the formation of countable unions.

(a) Show that a ring (σ-ring) containing Ω is a field (σ-field). Find a ring that is not a field and a σ-ring that is not a σ-field.

(b) Show that a ring is closed under the formation of symmetric differences and finite intersections. Show that a σ-ring is closed under the formation of countable intersections.

(c) Show that the field $f(\mathcal{A})$ generated (see Problem 2.5) by a ring $\mathcal{A}$ consists of the sets A for which either $A \in \mathcal{A}$ or $A^c \in \mathcal{A}$. Show that $f(\mathcal{A}) = \sigma(\mathcal{A})$ if $\mathcal{A}$ is a σ-ring.

(d) Show that a (countably additive) measure on a ring $\mathcal{A}$ extends to $f(\mathcal{A})$. Of course this follows by Theorem 11.3, but prove it directly.

(e) Let $\mathcal{A}$ be a ring and let $\mathcal{A}'$ be the σ-ring generated (define) by $\mathcal{A}$. Show that a measure on $\mathcal{A}$ extends to a measure on $\mathcal{A}'$ and that the extension is unique if the original measure is σ-finite on $\mathcal{A}$.

Hint: To prove uniqueness, first show by adapting the proof of Theorem 3.4 that a monotone class containing $\mathcal{A}$ must contain $\mathcal{A}'$.

11.6. Call $\mathcal{A}$ a *strong* semiring if it contains the empty set and is closed under the formation of finite intersections and if $A, B \in \mathcal{A}$ and $A \subset B$ imply that there exist $\mathcal{A}$-sets $A_0, \ldots, A_n$ such that $A = A_0 \subset A_1 \subset \cdots \subset A_n = B$ and $A_k - A_{k-1} \in \mathcal{A}$, $k = 1, \ldots, n$.

(a) Show that the class of bounded rectangles in R^k is a strong semiring.

(b) Show that a semiring need not be a strong semiring.

(c) Call μ on $\mathcal{A}$ *pairwise* additive if $A, B, A \cup B \in \mathcal{A}$ and $A \cap B = 0$ imply that $\mu(A \cup B) = \mu(A) + \mu(B)$. It can be shown that, if μ is pairwise additive on a strong semiring, then it is finitely additive there. Show that this is false if $\mathcal{A}$ is only a semiring.

11.7. Suppose that μ is a measure on the σ-field $\mathcal{F}$ generated by the field $\mathcal{F}_0$. Suppose that $A \in \mathcal{F}$ and $\mu(A) < \infty$ and that $A_1, \ldots, A_n$ is a decomposition of A into $\mathcal{F}_0$-sets. Show that there exists an $\mathcal{F}_0$-set B and a decomposition $B_1, \ldots, B_n$ of B into $\mathcal{F}_0$-sets such that $\mu(A_k \triangle B_k) < \epsilon$, $k = 1, \ldots, n$. Show that B may be taken as A if $A \in \mathcal{F}_0$.

11.8. Suppose that μ is σ-finite on a field $\mathcal{F}_0$ generating the σ-field $\mathcal{F}$. Show that for $A \in \mathcal{F}$ there exist $\mathcal{F}_0$-sets C_{ni} and D_{ni} such that, if $C = \bigcup_n \bigcap_i C_{ni}$ and $D = \bigcap_n \bigcup_i D_{ni}$, then $C \subset A \subset D$ and $\mu(D - C) = 0$.

11.9. Here is another approach to Theorem 11.4. Suppose for simplicity that μ is a finite measure on the σ-field $\mathcal{F}$ generated by a field $\mathcal{F}_0$. Let $\mathcal{G}$ be the class of $\mathcal{F}$-sets B for which for each ϵ there exist $\mathcal{F}_0$-sets A_k and A'_k such that $\bigcap_k A_k \subset B \subset \bigcup_k A'_k$ and $\mu((\bigcup_k A'_k) - (\bigcap_k A_k)) < \epsilon$. Show that $\mathcal{G}$ is a σ-field containing $\mathcal{F}_0$.

11.10. Show that σ-finiteness is essential in Theorem 11.4. Show that part (ii) can fail if $\mu(B) = \infty$.

11.11. Assume that volume is finitely additive and countably subadditive for bounded rectangles and extend the construction in Example 11.5 to R^k.

SECTION 12. MEASURES IN EUCLIDEAN SPACE

Lebesgue Measure

In Example 11.3 Lebesgue measure λ was constructed on the class $\mathcal{R}^1$ of linear Borel sets. By Theorem 10.3, λ is the only measure on $\mathcal{R}^1$ satisfying $\lambda(a, b] = b - a$ for all intervals. There is in k-space an analogous *k-dimensional Lebesgue measure* λ_k on the class $\mathcal{R}^k$ of k-dimensional Borel sets (Example 10.1). It is specified by the requirement that bounded rectangles have measure

$$(12.1) \qquad \lambda_k [x: a_i < x_i \leq b_i, i = 1, \ldots, k] = \prod_{i=1}^{k} (b_i - a_i).$$

This is ordinary volume—that is, length ($k = 1$), area ($k = 2$), volume ($k = 3$), or hypervolume ($k \geq 4$).

Since an intersection of rectangles is again a rectangle, the uniqueness theorem shows that (12.1) completely determines λ_k. That there does exist such a measure on $\mathcal{R}^k$ can be proved in several ways. One is to use the ideas involved in the case

$k = 1$. A second construction is given in Theorem 12.5. A third, independent, construction uses the general theory of product measures; this is carried out in Section 18. For the moment, assume the existence on $\mathcal{R}^k$ of a measure λ_k satisfying (12.1). Of course, λ_k is σ-finite.

A basic property of λ_k is *translation invariance.**

Theorem 12.1. *If $A \in \mathcal{R}^k$, then $A + x = [a + x: a \in A] \in \mathcal{R}^k$ and $\lambda_k(A) = \lambda_k(A + x)$ for all x.*

PROOF. If $\mathcal{G}$ is the class of A such that $A + x$ is in $\mathcal{R}^k$ for all x, then $\mathcal{G}$ is a σ-field containing the bounded rectangles, and so $\mathcal{G} \supset \mathcal{R}^k$. Thus $A + x \in \mathcal{R}^k$ for $A \in \mathcal{R}^k$.

For fixed x define a measure μ on $\mathcal{R}^k$ by $\mu(A) = \lambda(A + x)$. Then μ and λ agree on the π-system of bounded rectangles and so agree for all Borel sets. ∎

If A is a $(k - 1)$-dimensional subspace and x lies outside A, the hyperplanes $A + tx$ for real t are disjoint and by Theorem 12.1, all have the same measure. Since only countably many disjoint sets can have positive measure (Theorem 10.2 (iv)), the measure common to the $A + tx$ must be 0. *Every $(k - 1)$-dimensional hyperplane has k-dimensional Lebesgue measure 0.*

The Lebesgue measure of a rectangle is its ordinary volume. The following theorem makes it possible to calculate the measures of simple figures.

Theorem 12.2. *If $T:R^k \to R^k$ is linear and nonsingular, then $A \in \mathcal{R}^k$ implies that $TA \in \mathcal{R}^k$ and*

$$(12.2) \qquad\qquad \lambda_k(TA) = |\det T| \cdot \lambda_k(A).$$

As a parallelepiped is the image of a rectangle under a linear transformation, (12.2) can be used to compute its volume. If T is a rotation or a reflection—an orthogonal or a unitary transformation—det $T = \pm 1$, and so $\lambda_k(TA) = \lambda_k(A)$. Hence every rigid transformation (a rotation followed by a translation) preserves Lebesgue measure.

PROOF OF THE THEOREM. Since $T \bigcup_n A_n = \bigcup_n TA_n$ and $TA^c = (TA)^c$ because of the assumed nonsingularity of T, the class $\mathcal{G} = [A: TA \in \mathcal{R}^k]$ is a σ-field. Since TA is open for open A, again by the assumed nonsingularity of T, $\mathcal{G}$ contains all the open sets and hence (Example 10.1) all the Borel sets. Therefore, $A \in \mathcal{R}^k$ implies that $TA \in \mathcal{R}^k$.

For $A \in \mathcal{R}^k$, set $\mu_1(A) = \lambda_k(TA)$ and $\mu_2(A) = |\det T| \cdot \lambda_k(A)$. Then μ_1 and μ_2 are measures, and by Theorem 10.3 they will agree on $\mathcal{R}^k$ (which is the assertion (12.2)) if they agree on the π-system consisting of the rectangles $[x: a_i < x_i \le b_i, i = 1, \ldots, k]$ for which the a_i and the b_i are all rational (Example 10.1). It suffices therefore to prove

---

* An analogous fact was used in the construction of a nonmeasurable set at the end of Section 3.

(12.2) for rectangles with sides of rational length. Since such a rectangle is a finite disjoint union of cubes and λ_k is translation-invariant, it is enough to check (12.2) for cubes

$$(12.3) \qquad A = [x: 0 < x_i \leq c, i = 1, \ldots, k]$$

that have their lower corner at the origin.

Now the general T can by elementary row and column operations* be represented as a product of linear transformations of these three special forms:

(1°) $T(x_1, \ldots, x_k) = (x_{\pi 1}, \ldots, x_{\pi k})$, where π is a permutation of the set $\{1, 2, \ldots, k\}$;

(2°) $T(x_1, \ldots, x_k) = (\alpha x_1, x_2, \ldots, x_k)$;

(3°) $T(x_1, \ldots, x_k) = (x_1 + x_2, x_2, \ldots, x_k)$.

Because of the rule for multiplying determinants, it suffices to check (12.2) for T of these three forms. And, as observed, for each such T it suffices to consider cubes (12.3).

(1°) Such a T is a permutation matrix, and so $\det T = \pm 1$. Since (12.3) is invariant under T, (12.2) is in this case obvious.

(2°) Here $\det T = \alpha$, and $TA = [x: x_1 \in H, 0 < x_i \leq c, i = 2, \ldots, k]$, where $H = (0, \alpha c]$ if $\alpha > 0$, $H = \{0\}$ if $\alpha = 0$, and $H = [\alpha c, 0)$ if $\alpha < 0$. In each case, $\lambda_k(TA) = |\alpha| \cdot c^k = |\alpha| \cdot \lambda_k(A)$.

(3°) Here $\det T = 1$. Let $B = [x: 0 < x_i \leq c, i = 3, \ldots, k]$, where $B = R^k$ if $k < 3$, and define

$$B_1 = [x: 0 < x_1 \leq x_2 \leq c] \cap B,$$

$$B_2 = [x: 0 < x_2 < x_1 \leq c] \cap B,$$

$$B_3 = [x: c < x_1 \leq c + x_2, 0 < x_2 \leq c] \cap B.$$

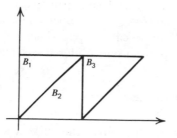

Then $A = B_1 \cup B_2$, $TA = B_2 \cup B_3$, and $B_1 + (c, 0, \ldots, 0) = B_3$. Since $\lambda_k(B_1) = \lambda_k(B_3)$ by translation invariance, (12.2) follows by additivity. ∎

If T is singular, then $\det T = 0$ and TA lies in a $(k - 1)$-dimensional subspace. Since such a subspace has measure 0, (12.2) holds if A and TA lie in R^k. The surprising thing is that $A \in R^k$ need not imply that $TA \in R^k$ if T is singular. For even a very simple transformation such as the projection $T(x_1, x_2) = (x_1, 0)$ in the plane, there exist Borel sets A for which TA is not a Borel set.[†] But if A is a rectangle, TA is a Borel set by the argument involving (1°), (2°), and (3°)—note that α can be 0 in (2°).

* BIRKHOFF & MAC LANE, Section 8.9.
[†] See HAUSDORFF, p. 241.

Regularity

Important among measures on $\mathcal{R}^k$ are those assigning finite measure to bounded sets. They share with λ_k the property of *regularity*:

Theorem 12.3. *Suppose that μ is a measure on $\mathcal{R}^k$ such that $\mu(A) < \infty$ if A is bounded.*

(i) *For $A \in \mathcal{R}^k$ and $\epsilon > 0$, there exist a closed C and an open G such that $C \subset A \subset G$ and $\mu(G - F) < \epsilon$.*

(ii) *If $\mu(A) < \infty$, then $\mu(A) = \sup \mu(K)$, the supremum extending over the compact subsets K of A.*

PROOF. The second part of the theorem follows from the first: $\mu(A) < \infty$ implies that $\mu(A - A_0) < \epsilon$ for a bounded subset A_0 of A, and it then follows from the first part that $\mu(A_0 - K) < \epsilon$ for a closed and hence compact subset K of A_0.

To prove (i) consider first a bounded rectangle $A = [x: a_i < x_i \le b_i, i \le k]$. The set $G_n = [x: a_i < x_i < b_i + n^{-1}, i \le k]$ is open and $G_n \downarrow A$. Since $\mu(G_1)$ is finite by hypothesis, it follows by continuity from above that $\mu(G_n - A) < \epsilon$ for large n. A bounded rectangle can therefore be approximated from the outside by open sets.

The rectangles form a semiring (Example 11.4). For an arbitrary set A in $\mathcal{R}^k$, by Theorem 11.4 there therefore exist bounded rectangles A_k such that $A \subset \bigcup_k A_k$ and $\mu((\bigcup_k A_k) - A) < \epsilon$. Choose open sets G_k such that $A_k \subset G_k$ and $\mu(G_k - A_k) < \epsilon/2^k$. Then $G = \bigcup_k G_k$ is open and $\mu(G - A) < 2\epsilon$. Thus the general k-dimensional Borel set can be approximated from the outside by open sets. To approximate from the inside by closed sets, pass to complements. ∎

Specifying Measures on the Line

There are on the line many measures other than λ which are important for probability theory. There is a useful way to describe the collection of all measures on $\mathcal{R}^1$ that assign finite measure to each bounded set.

If μ is such a measure, define a real function F by

$$(12.4) \qquad F(x) = \begin{cases} \mu(0, x] & \text{if } x \ge 0, \\ -\mu(x, 0] & \text{if } x \le 0. \end{cases}$$

It is because $\mu(A) < \infty$ for bounded A that F is a finite function. Clearly, F is nondecreasing. Suppose that $x_n \downarrow x$. If $x \ge 0$, apply part (ii) of Theorem 10.2, and if $x < 0$, apply part (i); in either case, $F(x_n) \downarrow F(x)$ follows. Thus F is continuous from the right. Finally,

(12.5) $$\mu(a, b] = F(b) - F(a)$$

for every bounded interval $(a, b]$. If μ is Lebesgue measure, then (12.4) gives $F(x) = x$.

The finite intervals form a π-system generating $\mathcal{R}^1$, and therefore by Theorem 10.3 the function F completely determines μ through the relation (12.5). But (12.5) and μ do not determine F: if $F(x)$ satisfies (12.5), then so does $F(x) + c$.

On the other hand, for a given μ, (12.5) certainly determines F to within such an additive constant. For finite μ, it is customary to standardize F by defining it not by (12.4) but by

(12.6) $$F(x) = \mu(-\infty, x];$$

then $\lim_{x \to -\infty} F(x) = 0$ and $\lim_{x \to \infty} F(x) = \mu(R^1)$. If μ is a probability measure, F is called a *distribution function* (the adjective *cumulative* is sometimes added).

Measures μ are often specified by means of the function F. The following theorem ensures that to each F there does exist a μ.

Theorem 12.4. *If F is a nondecreasing, right-continuous real function on the line, there exists on $\mathcal{R}^1$ a unique measure μ satisfying (12.5) for all a and b.*

As noted above, uniqueness is a simple consequence of Theorem 10.3. The proof of existence is almost the same as the construction of Lebesgue measure, the case $F(x) = x$. This proof is not carried through at this point, because it is contained in a parallel, more general construction for k-dimensional space in the next theorem. For a very simple argument establishing Theorem 12.4, see the second proof of Theorem 14.1.

Specifying Measures in R^k

The σ-field $\mathcal{R}^k$ of k-dimensional Borel sets is generated by the class of bounded rectangles

(12.7) $$A = [x : a_i < x_i \le b_i, i = 1, \ldots, k]$$

(Example 10.1). If $I_i = (a_i, b_i]$, A has the form of a cartesian product

(12.8) $$A = I_1 \times \cdots \times I_n.$$

Consider the sets of the special form

(12.9) $S_x = [y: y_i \le x_i, i = 1, \ldots, k];$

S_x consists of the points "southwest" of $x = (x_1, \ldots, x_k)$; in the case $k = 1$ it is the half-infinite interval $(-\infty, x]$. Now S_x is closed, and (12.7) has the form

(12.10)

$$A = S_{(b_1 \ldots b_k)} - [S_{(a_1 b_2 \ldots b_k)} \cup S_{(b_1 a_2 \ldots b_k)} \cup \cdots \cup S_{(b_1 b_2 \ldots a_k)}].$$

Therefore, the class of sets (12.9) generates $\mathcal{R}^k$. This class is a π-system.

The objective is to find a version of Theorem 12.4 for k-space. This will in particular give k-dimensional Lebesgue measure. The first problem is to find the analogue of (12.5).

A bounded rectangle (12.7) has 2^k vertices—the points $x = (x_1, \ldots, x_k)$ for which each x_i is either a_i or b_i. Let $\mathrm{sgn}_A x$, the signum of the vertex, be $+1$ or -1, according as the number of i ($1 \le i \le k$) satisfying $x_i = a_i$ is even or odd. For a real function F on R^k, the difference of F around the vertices of A is $\Delta_A F = \Sigma \ \mathrm{sgn}_A x \cdot F(x)$, the sum extending over the 2^k vertices x of A. In the case $k = 1$, $A = (a, b]$ and $\Delta_A F = F(b) - F(a)$. In the case $k = 2$, $\Delta_A F = F(b_1, b_2) - F(b_1; a_2) - F(a_1, b_2) + F(a_1, a_2)$.

Since the k-dimensional analogue of (12.4) is complicated, suppose at first that μ is a finite measure on $\mathcal{R}^k$ and consider instead the analogue of (12.6), namely

(12.11) $F(x) = \mu[y: y_i \le x_i, i = 1, \ldots, k].$

Suppose that S_x is defined by (12.9) and A is a bounded rectangle (12.7). Then

(12.12) $\mu(A) = \Delta_A F.$

To see this, apply to the union on the right in (12.10) the inclusion-exclusion formula (10.5). The k sets in the union give $2^k - 1$ intersections, and these are the sets S_x for x ranging over the vertices of A other than $(b_1, \ldots, b_k)$. Taking into account the signs in (10.5) leads to (12.12).

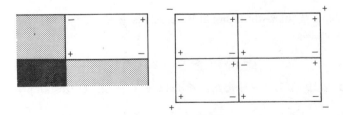

Suppose $x^{(n)} \downarrow x$ in the sense that $x_i^{(n)} \downarrow x_i$ as $n \to \infty$ for each $i = 1, \ldots, k$.

Then $S_{x^{(n)}} \downarrow S_x$, and hence $F(x^{(n)}) \to F(x)$ by Theorem 10.2(ii). In this sense, F is *continuous from above*.

Theorem 12.5. *Suppose that the real function F on R^k is continuous from above and satisfies $\Delta_A F \geq 0$ for bounded rectangles A. Then there exists a unique measure μ on $\mathcal{R}^k$ satisfying (12.12) for bounded rectangles A.*

The empty set can be taken as a bounded rectangle (12.7) for which $a_i = b_i$ for some i, and for such a set A, $\Delta_A F = 0$. Thus (12.12) defines a finite-valued set function μ on the class of bounded rectangles. The point of the theorem is that μ extends uniquely to a measure on $\mathcal{R}^k$. The uniqueness is an immediate consequence of Theorem 10.3, since the bounded rectangles form a π-system generating $\mathcal{R}^k$.

If F is bounded, then μ will be a finite measure. But the theorem does not require that F be bounded. The most important unbounded F is $F(x) = x_1 \cdots x_k$. Here $\Delta_A F = (b_1 - a_1) \cdots (b_k - a_k)$ for A given by (12.7). This is the ordinary volume of A as specified by (12.1). The corresponding measure extended to $\mathcal{R}^k$ is k-dimensional Lebesgue measure as described at the beginning of this section.

PROOF OF THEOREM 12.5.   As already observed, the uniqueness of the extension is easy to prove. To prove its existence it will first be shown that μ as defined by (12.12) is finitely additive on the class of bounded rectangles. Suppose that each side $I_i = (a_i, b_i]$ of a bounded rectangle (12.7) is partitioned into n_i subintervals $J_{ij} = (t_{i,j-1}, t_{ij}], j = 1, \ldots, n_i$, where $a_i = t_{i0} < t_{i1} < \cdots < t_{in_i} = b_i$. The $n_1 n_2 \cdots n_k$ rectangles

$$(12.13) \qquad B_{j_1 \ldots j_k} = J_{1j_1} \times \cdots \times J_{kj_k}, \qquad 1 \leq j_1 \leq n_1, \ldots, 1 \leq j_k \leq n_k,$$

then partition A. Call such a partition *regular*. It will first be shown that μ adds for regular partitions:

$$(12.14) \qquad \mu(A) = \sum_{j_1 \ldots j_k} \mu(B_{j_1 \ldots j_k}).$$

The right side of (12.14) is $\sum_B \sum_x \text{sgn}_B x \cdot F(x)$, where the outer sum extends over the rectangles B of the form (12.13) and the inner sum extends over the vertices x of B. Now

$$(12.15) \qquad \sum_B \sum_x \text{sgn}_B x \cdot F(x) = \sum_x F(x) \sum_B \text{sgn}_B x,$$

where on the right the outer sum extends over each x that is a vertex of one or more of the B's, and for fixed x the inner sum extends over the B's of which it is a vertex. Suppose that x is a vertex of one or more of the B's but is not a vertex of A. Then there must be an i such that x_i is neither a_i nor b_i. There may be several such i, but fix on one of them and suppose for notational convenience that it is $i = 1$. Then $x_1 = t_{1j}$ with $0 < j < n_1$. The rectangles (12.13) of which x is a vertex therefore come in pairs $B' = B_{jj_2 \ldots j_k}$ and $B'' = B_{j+1,j_2 \ldots j_k}$, and $\text{sgn}_{B'} x = -\text{sgn}_{B''} x$. Thus the inner sum on the right in (12.15) is 0 if x is not a vertex of A.

On the other hand, if x is a vertex of A as well as of at least one B, then for each i either $x_i = a_i = t_{i0}$ or $x_i = b_i = t_{in_i}$. In this case x is vertex of only one B of the form (12.13)—the one for which $j_i = 1$ or $j_i = n_i$, according as $x_i = a_i$ or $x_i = b_i$—and $\operatorname{sgn}_B x = \operatorname{sgn}_A x$. Thus the right side of (12.15) reduces to $\Delta_A F$, which proves (12.14).

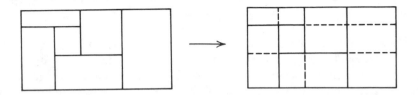

Now suppose that $A = \bigcup_{u=1}^{n} A_u$, where A is the bounded rectangle (12.8), $A_u = I_{1u} \times \cdots \times I_{ku}$ for $u = 1, \ldots, n$, and the A_u are disjoint. For each i ($1 \leq i \leq k$), the intervals $I_{i1}, \ldots, I_{in}$ have I_i as their union, although they need not be disjoint. But their endpoints split I_i into disjoint subintervals $J_{i1}, \ldots, J_{in_i}$ such that each I_{iu} is the union of certain of the J_{ij}. The rectangles B of the form (12.13) are a regular partition of A as before; furthermore, the B's contained in a single A_u form a regular partition of A_u. Since the A_u are disjoint, it follows by (12.14) that

$$\mu(A) = \sum_B \mu(B) = \sum_{u=1}^{n} \sum_{B \subset A_u} \mu(B) = \sum_{u=1}^{n} \mu(A_u).$$

Therefore, μ is finitely additive on the class $\mathcal{J}^k$ of bounded k-dimensional rectangles.

As shown in Example 11.4, $\mathcal{J}^k$ is a semiring, and so Theorem 11.3 applies.

Now suppose that $\bigcup_{u=1}^{n} A_u \subset A$, where A and the A_u are in $\mathcal{J}^k$ and the latter are disjoint. If follows by induction from the semiring property that $A - \bigcup_{u=1}^{n} A_u$ is a finite disjoint union $\bigcup_{v=1}^{m} B_v$ of sets in $\mathcal{J}^k$. Therefore, $\Sigma_{u=1}^{n} \mu(A_j) + \Sigma_{v=1}^{m} \mu(B_v) = \mu(A)$, and so

$$(12.16) \qquad \sum_{u=1}^{n} \mu(A_u) \leq \mu(A) \quad \text{if} \quad \bigcup_{u=1}^{n} A_u \subset A, \qquad A_u \cap A_v = 0.$$

Now suppose that $A \subset \bigcup_{u=1}^{n} A_u$, where A and the A_u are in $\mathcal{J}^k$. Let $B_1 = A \cap A_1$ and $B_u = A \cap A_u - (A_1 \cup \cdots \cup A_{u-1})$. As observed above, each B_u is a finite disjoint union $\bigcup_v B_{uv}$ of elements of $\mathcal{J}^k$. The B_u are disjoint, and so the B_{uv} taken all together are also disjoint. They have union A, so that $\mu(A) = \Sigma_u \Sigma_v \mu(B_{uv})$. Since $\bigcup_v B_{uv} = B_u \subset A_u$, an application of (12.16) gives

$$(12.17) \qquad \mu(A) \leq \sum_{u=1}^{n} \mu(A_u) \quad \text{if} \quad A \subset \bigcup_{u=1}^{n} A_u.$$

Of course, (12.16) and (12.17) give back finite additivity again.

To apply Theorem 11.3 requires showing that μ is countably subadditive on $\mathcal{J}^k$. Suppose then that $A \subset \bigcup_{u=1}^{\infty} A_u$, where A and the A_u are in $\mathcal{J}^k$. The problem is to prove that

$$(12.18) \qquad \mu(A) \leq \sum_{u=1}^{\infty} \mu(A_u).$$

Suppose that $\epsilon > 0$. If A is given by (12.7) and $B = [x : a_i + \delta < x_i \leq b_i, i \leq k]$, then $\mu(B) > \mu(A) - \epsilon$ for small enough positive δ because μ is defined by (12.12) and F is continuous from above. Note that A contains the closure $B^- = [x: a_i + \delta \leq x_i \leq b_i, i \leq k]$ of B. Similarly, for each u there is in $\mathcal{J}^k$ a set $B_u = [x: a_{iu} < x_i \leq b_{iu} + \delta_u, i \leq k]$ such that $\mu(B_u) < \mu(A_u) + \epsilon/2^u$ and A_u is in the interior $B_u^\circ = [x: a_{iu} < x_i < b_{iu} + \delta_u, i \leq k]$ of B_u.

Since $B^- \subset A \subset \bigcup_{u=1}^\infty A_u \subset \bigcup_{u=1}^\infty B_u^\circ$, it follows by the Heine-Borel theorem that $B \subset B^- \subset \bigcup_{u=1}^n B_u^\circ \subset \bigcup_{u=1}^n B_u$ for some n. Now (12.17) applies, and so $\mu(A) - \epsilon < \mu(B) \leq \Sigma_{u=1}^n \mu(B_u) < \Sigma_{u=1}^\infty \mu(A_u) + \epsilon$. Since ϵ was arbitrary, the proof of (12.18) is complete.

Thus μ as defined by (12.12) is finitely additive and countably subadditive on the semiring $\mathcal{J}^k$. By Theorem 11.3, μ extends to a measure on $\mathcal{R}^k = \sigma(\mathcal{J}^k)$. ∎

PROBLEMS

12.1. Suppose that μ is a measure on $\mathcal{R}^1$ which is finite for bounded sets and is translation-invariant: $\mu(A + x) = \mu(A)$. Show that $\mu(A) = \alpha\lambda(A)$ for some $\alpha \geq 0$. Extend to R^k.

12.2. Suppose that $A \in \mathcal{R}^1$, $\lambda(A) > 0$, and $0 < \theta < 1$. Show that there is a bounded open interval I such that $\lambda(A \cap I) \geq \theta \lambda(I)$.

Hint: Show that $\lambda(A)$ may be assumed finite and choose an open G such that $A \subset G$ and $\lambda(A) \geq \theta \lambda(G)$. Now $G = \bigcup_n I_n$ for disjoint open intervals I_n, and $\Sigma_n \lambda(A \cap I_n) \geq \theta \Sigma_n \lambda(I_n)$; use an I_n.

12.3. ↑ If $A \in \mathcal{R}^1$ and $\lambda(A) > 0$, then the origin is interior to the difference set $D(A) = [x - y: x, y \in A]$.

Hint: Choose a bounded open interval I as in Problem 12.2 for $\theta = \frac{3}{4}$. Suppose that $|z| < \lambda(I)/2$; since $A \cap I$ and $(A \cap I) + z$ are contained in an interval of length less than $3 \lambda(I)/2$ and hence cannot be disjoint, $z \in D(A)$.

12.4. 3.3   4.13   12.3 ↑   The following construction leads to a subset H of $(0, 1]$ which is nonmeasurable in the extreme sense that (see (3.9) and (3.10)) $\lambda_*(H) = 0$ and $\lambda^*(H) = 1$. Complete the details. The notation is as in the construction of the nonmeasurable set at the end of Section 3.

(a) Fix an irrational θ. For $n = 0, \pm1, \pm2, \ldots$, let θ_n be $n\theta$ reduced modulo 1: $\theta_n = n\theta - N_n$, where $0 < \theta_n \leq 1$ and N_n is an integer. Split the unit interval into finitely many intervals of lengths less than ϵ; since one of them must contain infinitely many θ_{2n}, $0 < \theta_{2m} - \theta_{2n} < \epsilon$ for some m and n. Show that $\{\theta_{2n}: n = 0, \pm1, \ldots\}$ and $\{\theta_{2n+1}: n = 0, \pm1, \ldots\}$ are dense in $(0, 1]$.

(b) Take x and y to be equivalent if $x \oplus \theta_n = y$ for some n and let S contain one representative from each equivalence class. Show that $(0, 1] = \bigcup_n (S \oplus \theta_n)$, the union disjoint. Put $H = \bigcup_n (S \oplus \theta_{2n})$ and prove that $(0, 1] - H = H \oplus \theta$.

(c) Suppose that $A \subset H$ for a Borel set A. If $\lambda(A) > 0$, then $D = [x - y: x, y \in A]$ contains some interval about 0; hence for some integers k, m, n and elements s_1 and s_2 of H, D contains $\theta_{2k+1} = (s_1 \oplus \theta_{2m}) - (s_2 \oplus \theta_{2n})$. Deduce that $s_1 = s_2$ and obtain a contradiction. Conclude that $\lambda_*(H) = 0$. Prove similarly that $\lambda^*(H) = 1 - \lambda_*(H + \theta) = 1$.

12.5. Suppose that μ is nonnegative and finitely additive on $\mathcal{R}^k$ and $\mu(R^k) < \infty$. Suppose further that $\mu(A) = \sup \mu(K)$, where K ranges over the compact subsets of A. Show that μ is countably additive. (Compare Theorem 12.3(ii).)

12.6. Suppose μ is a measure on $\mathcal{R}^k$ such that bounded sets have finite measure. Given A, show that there exist an F_σ set U (a countable union of closed sets) and a G_δ set V (a countable intersection of open sets) such that $U \subset A \subset V$ and $\mu(V - U) = 0$.

12.7. 2.17 ↑ Suppose that μ is a nonatomic probability measure on $(R^k, \mathcal{R}^k)$ and that $\mu(A) > 0$. Show that there is an uncountable compact set K such that $K \subset A$ and $\mu(K) = 0$.

12.8. The *closed support* of a measure μ on $\mathcal{R}^k$ is a closed set C_0 such that $C_0 \subset C$ for closed C if and only if C supports μ. Prove its existence and uniqueness. Characterize the points of C_0 as those x such that $\mu(U) > 0$ for every neighborhood U of x. If $k = 1$ and if μ and the function $F(x)$ are related by (12.5), the condition is $F(x - \epsilon) < F(x + \epsilon)$ for all ϵ; x is in this case called a *point of increase* of F.

12.9. Let $\mathcal{G}$ be the class of sets in $\mathcal{R}^2$ of the form $[(x_1, x_2): x_1 \in H]$ for some H in $\mathcal{R}^1$. Show that $\mathcal{G}$ is a σ-field in $\mathcal{R}^2$ and that two-dimensional Lebesgue measure is not σ-finite on $\mathcal{G}$ even though it is σ-finite on $\mathcal{R}^2$.

12.10. Of minor interest is the k-dimensional analogue of (12.4). Let I_t be $(0, t]$ for $t \geq 0$ and $(t, 0]$ for $t \leq 0$, and let $A_x = I_{x_1} \times \cdots \times I_{x_k}$. Let $\varphi(x)$ be $+1$ or -1 according as the number of i, $1 \leq i \leq k$, for which $x_i < 0$ is even or odd. Show that, if $F(x) = \varphi(x) \mu(A_x)$, then (12.12) holds for bounded rectangles A.

Call F degenerate if it is a function of some $k - 1$ of the coordinates, the requirement in the case $k = 1$ being that F is constant. Show that $\Delta_A F = 0$ for every bounded rectangle if and only if F is a finite sum of degenerate functions; (12.12) determines F to within addition of a function of this sort.

12.11. Let G be a nondecreasing, right-continuous function on the line and put $F(x, y) = \min \{G(x), y\}$. Show that F satisfies the conditions of Theorem 12.5 and that the curve $[(x, G(x)): x \in R^1]$ supports the corresponding measure.

12.12. Let F_1 and F_2 be bounded, nondecreasing, right-continuous functions on the line and put $F(x_1, x_2) = F_1(x_1) F_2(x_2)$. Show that F satisfies the conditions of Theorem 12.5. Let μ, μ_1, μ_2 be the measures corresponding to F, F_1, F_2, and prove that $\mu(A_1 \times A_1) = \mu_1(A_1) \mu_2(A_2)$ for linear Borel sets A_1 and A_2. This μ is the *product* of μ_1 and μ_2; products are studied in a general setting in Section 18.

12.13. Let $\mathcal{G}$ be the class of A such that for each ϵ, there exist an open G and a closed C such that $C \subset A \subset G$ and $\mu(G - C) < \epsilon$. Prove Theorem 12.3 for the case of a finite μ by showing that $\mathcal{G}$ is a σ-field containing the closed sets.

SECTION 13. MEASURABLE FUNCTIONS AND MAPPINGS

If a real function X on Ω has finite range, it is by the definition in Section 5 a simple random variable if $[\omega: X(\omega) = x]$ lies in the basic σ-field $\mathcal{F}$ for each x. The requirement appropriate for the general real function X is stronger; namely, $[\omega: X(\omega) \in H]$ must lie in $\mathcal{F}$ for each linear Borel set H. An abstract version of this definition greatly simplifies the theory of such functions.

Measurable Mappings

Let $(\Omega, \mathcal{F})$ and $(\Omega', \mathcal{F}')$ be two measurable spaces. For a mapping $T: \Omega \to \Omega'$, consider the inverse images $T^{-1}A' = [\omega \in \Omega: T\omega \in A']$ for $A' \subset \Omega'$. The mapping T is *measurable* $\mathcal{F}/\mathcal{F}'$ if $T^{-1}A' \in \mathcal{F}$ for each $A' \in \mathcal{F}'$.

For a real function f, the image space Ω' is the line R^1, and in this case $\mathcal{R}^1$ is always tacitly understood to play the role of $\mathcal{F}'$. A real function f on Ω is thus measurable $\mathcal{F}$ (or simply measurable if it is clear from the context what $\mathcal{F}$ is involved) if it is measurable $\mathcal{F}/\mathcal{R}^1$—that is, if $f^{-1}H = [\omega: f(\omega) \in H] \in \mathcal{F}$ for every $H \in \mathcal{R}^1$. In probability contexts, a real measurable function is called a *random variable*. The point of the definition is to ensure that $[\omega: f(\omega) \in H]$ has a measure or probability for all sufficiently regular sets H of real numbers—that is, for all Borel sets H.

A real function f with finite range is measurable if $f^{-1}\{x\} \in \mathcal{F}$ for each singleton $\{x\}$, but this is too weak a condition to impose on the general f. (It is satisfied if $(\Omega, \mathcal{F}) = (R^1, \mathcal{R}^1)$ and f is any one-to-one map of the line into itself, but in this case $f^{-1}H$ even for so simple a set H as an interval can for an appropriately chosen f be any uncountable set, say the non-Borel set constructed in Section 3.) On the other hand, for a measurable f with finite range, $f^{-1}H \in \mathcal{F}$ for every $H \subset R^1$, but this is too strong a condition to impose on the general f. (For $(\Omega, \mathcal{F}) = (R^1, \mathcal{R}^1)$, even $f(x) \equiv x$ fails to satisfy it.) Notice that nothing is required of fA; it need not lie in $\mathcal{R}^1$ for A in $\mathcal{F}$.

If in addition to $(\Omega, \mathcal{F})$, $(\Omega', \mathcal{F}')$, and the map $T: \Omega \to \Omega'$ there is a third measurable space $(\Omega'', \mathcal{F}'')$ and a map $T': \Omega' \to \Omega''$, the composition $T'T = T' \circ T$ is the mapping $\Omega \to \Omega''$ that carries ω to $T'(T(\omega))$.

Theorem 13.1. (i) *If $T^{-1}A' \in \mathcal{F}$ for each $A' \in \mathcal{A}'$ and $\mathcal{A}'$ generates $\mathcal{F}'$, then T is measurable $\mathcal{F}/\mathcal{F}'$.*

(ii) *If T is measurable $\mathcal{F}/\mathcal{F}'$ and T' is measurable $\mathcal{F}'/\mathcal{F}''$, then $T'T$ is measurable $\mathcal{F}/\mathcal{F}''$.*

PROOF. Since $T^{-1}(\Omega' - A') = \Omega - T^{-1}A'$ and $T^{-1}(\bigcup_n A'_n) = \bigcup_n T^{-1}A'_n$, and since $\mathcal{F}$ is a σ-field in Ω, the class $[A': T^{-1}A' \in \mathcal{F}]$ is a σ-field in Ω'. If this σ-field contains $\mathcal{A}'$, it must also contain $\sigma(\mathcal{A}')$, and (i) follows.

As for (ii), it follows by the hypotheses that $A'' \in \mathcal{F}''$ implies that $(T')^{-1}A''$

$\in \mathcal{F}'$, which in turn implies that $(T'T)^{-1}A'' = [\omega: T'T\omega \in A''] = [\omega: T\omega \in (T')^{-1}A''] = T^{-1}((T')^{-1}A'') \in \mathcal{F}$.   ∎

By part (i), if f is a real function such that $[\omega: f(\omega) \le x]$ lies in $\mathcal{F}$ for all x, then f is measurable $\mathcal{F}$. This condition is usually easy to check.

Mappings into R^k

For a mapping $f: \Omega \to R^k$ carrying Ω into k-space, $\mathcal{R}^k$ is always understood to be the σ-field in the image space. In probabilistic contexts, a measurable mapping into R^k is called a *random vector*. Now f must have the form

(13.1) $f(\omega) = (f_1(\omega), \ldots, f_k(\omega))$

for real functions $f_j(\omega)$. Since the sets (12.9) (the "southwest regions") generate $\mathcal{R}^k$, Theorem 13.1(i) implies that f is measurable $\mathcal{F}$ if and only if the set

(13.2) $[\omega: f_1(\omega) \le x_1, \ldots, f_k(\omega) \le x_k] = \bigcap_{j=1}^{k} [\omega: f_j(\omega) \le x_j]$

lies in $\mathcal{F}$ for each $(x_1, \ldots, x_k)$. This condition holds if each f_j is measurable $\mathcal{F}$. On the other hand, if $x_j = x$ is fixed and $x_1 = \cdots = x_{j-1} = x_{j+1} = \cdots = x_k = n \to \infty$, the sets (13.2) increase to $[\omega: f_j(\omega) \le x]$; the condition thus implies that each f_j is measurable. Therefore, *f is measurable $\mathcal{F}$ if and only if each component function f_j is measurable $\mathcal{F}$.* This provides a practical criterion for mappings into R^k.

A mapping $f: R^i \to R^k$ is defined to be measurable if it is measurable $\mathcal{R}^i/\mathcal{R}^k$. Such functions are often called *Borel functions*. To sum up, $T: \Omega \to \Omega'$ is measurable $\mathcal{F}/\mathcal{F}'$ if $T^{-1}A' \in \mathcal{F}$ for all $A' \in \mathcal{F}'$; $f: \Omega \to R^k$ is measurable $\mathcal{F}$ if it is measurable $\mathcal{F}/\mathcal{R}^k$; and $f: R^i \to R^k$ is measurable (a Borel function) if it is measurable $\mathcal{R}^i/\mathcal{R}^k$. If $H \in \mathcal{R}^1$ ($i = k = 1$), I_H is not a Borel function.

Theorem 13.2.   *If $f: R^i \to R^k$ is continuous, then it is measurable.*

PROOF.   As noted above, it suffices to check that each set (13.2) lies in $\mathcal{R}^i$. But each is closed because of continuity.   ∎

Theorem 13.3.   *If $f_j: \Omega \to R^1$ is measurable $\mathcal{F}$, $j = 1, \ldots, k$, then $g(f_1(\omega), \ldots, f_k(\omega))$ is measurable $\mathcal{F}$ if $g: R^k \to R^1$ is measurable—in particular, if it is continuous.*

PROOF.   If the f_j are measurable, then so is (13.1), so that the result follows by Theorem 13.1(ii).   ∎

Taking $g(x_1, \ldots, x_k)$ to be $\Sigma_{i=1}^k x_i$, $\Pi_{i=1}^k x_i$, and $\max\{x_1, \ldots, x_k\}$ in turn shows that *sums, products, and maxima of measurable functions are measurable.* If $f(\omega)$ is real and measurable, then so are $\sin f(\omega)$, $e^{tf(\omega)}$, and so on, and if $f(\omega)$ never vanishes, then $1/f(\omega)$ is measurable as well.

Limits and Measurability

For a real function f it is often convenient to admit the artificial values ∞ and $-\infty$—to work with the *extended real line* $[-\infty, \infty]$. Such an f is by definition measurable $\mathcal{F}$ if $[\omega: f(\omega) \in H]$ lies in $\mathcal{F}$ for each Borel set H of (finite) real numbers and if $[\omega: f(\omega) = \infty]$ and $[\omega: f(\omega) = -\infty]$ both lie in $\mathcal{F}$. This extension of the notion of measurability is convenient in connection with limits and suprema, which need not be finite.

Theorem 13.4.   *Suppose that $f_1, f_2, \ldots$ are real functions measurable $\mathcal{F}$.*

(i)   *The functions $\sup_n f_n$, $\inf_n f_n$, $\limsup_n f_n$, and $\liminf_n f_n$ are measurable $\mathcal{F}$.*

(ii)   *If $\lim_n f_n$ exists everywhere, then it is measurable $\mathcal{F}$.*

(iii)   *The ω-set where $\{f_n(\omega)\}$ converges lies in $\mathcal{F}$.*

(iv)   *If f is measurable $\mathcal{F}$, then the ω-set where $f_n(\omega) \to f(\omega)$ lies in $\mathcal{F}$.*

PROOF.   Clearly, $[\sup_n f_n \le x] = \bigcap_n [f_n \le x]$ lies in $\mathcal{F}$ even for $x = \infty$ and $x = -\infty$, and so $\sup_n f_n$ is measurable. The measurability of $\inf_n f_n$ follows the same way, and hence $\limsup_n f_n = \inf_n \sup_{k \ge n} f_k$ and $\liminf_n f_n = \sup_n \inf_{k \ge n} f_k$ are measurable. If $\lim_n f_n$ exists, it coincides with these last two functions and hence is measurable. Finally, the set in (iii) is the set where $\limsup_n f_n(\omega) = \liminf_n f_n(\omega)$, and that in (iv) is the set where this common value is $f(\omega)$.   ■

Special cases of this theorem have been encountered before—part (iv), for example, in connection with the strong law of large numbers. The last three parts of the theorem obviously carry over to mappings into R^k.

A *simple* real function is one with finite range; it can be put in the form

$$(13.3) \qquad f = \sum_i x_i I_{A_i},$$

where the A_i form a finite decomposition of Ω. It is measurable $\mathcal{F}$ if each A_i lies in $\mathcal{F}$. The simple random variables of Section 5 have this form.

Many results concerning measurable functions are most easily proved first for simple functions and then, by an appeal to the next theorem and a passage to the limit, for the general measurable function.

Theorem 13.5.   *If f is real and measurable $\mathcal{F}$, there exists a sequence $\{f_n\}$ of* *simple functions, each measurable $\mathcal{F}$, such that*

$$(13.4) \qquad\qquad 0 \leq f_n(\omega) \uparrow f(\omega) \qquad if f(\omega) \geq 0$$

and

$$(13.5) \qquad\qquad 0 \geq f_n(\omega) \downarrow f(\omega) \qquad if f(\omega) \leq 0.$$

PROOF.   Define

$$(13.6) \quad f_n(\omega) = \begin{cases} -n & \text{if } -\infty \leq f(\omega) \leq -n, \\ -(k-1)2^{-n} & \text{if } -k2^{-n} < f(\omega) \leq -(k-1)2^{-n}, \\ & \qquad\qquad 1 \leq k \leq n2^n, \\ (k-1)2^{-n} & \text{if } (k-1)2^{-n} \leq f(\omega) < k2^{-n}, \\ & \qquad\qquad 1 \leq k \leq n2^n, \\ n & \text{if } n \leq f(\omega) \leq \infty. \end{cases}$$

This sequence has the required properties.   ∎

Note that (13.6) covers the possibilities $f(\omega) = \infty$ and $f(\omega) = -\infty$.

If $A \in \mathcal{F}$, a function f defined only on A is by definition measurable if $[\omega \in A: f(\omega) \in H]$ lies in $\mathcal{F}$ for $H \in \mathcal{R}^1$ and for $H = \{\infty\}$ and $H = \{-\infty\}$.

Transformations of Measures

Let $(\Omega, \mathcal{F})$ and $(\Omega', \mathcal{F}')$ be measurable spaces and suppose that the mapping T: $\Omega \to \Omega'$ is measurable $\mathcal{F}/\mathcal{F}'$. Given a measure μ on $\mathcal{F}$, define a set function μT^{-1} on $\mathcal{F}'$ by

$$(13.7) \qquad\qquad \mu T^{-1}(A') = \mu(T^{-1}A'), \qquad A' \in \mathcal{F}'.$$

That is, μT^{-1} assigns value $\mu(T^{-1}A')$ to the set A'. If $A' \in \mathcal{F}'$, then $T^{-1}A' \in \mathcal{F}$ because T is measurable, and hence the set function μT^{-1} is well defined on $\mathcal{F}'$. Since $T^{-1}\bigcup_n A'_n = \bigcup_n T^{-1}A'_n$ and the $T^{-1}A'_n$ are disjoint sets in Ω if the A'_n are disjoint sets in Ω', the countable additivity of μT^{-1} follows from that of μ. *Thus μT^{-1} is a measure.* This way of transferring a measure from Ω to Ω' will prove useful in a number of ways.

If μ is σ-finite, or finite, or a probability measure, μT^{-1} obviously has the same property.

PROBLEMS

13.1.   Show by examples in which $\Omega = \Omega'$ is a space of two points that measurability $\mathcal{F}/\mathcal{F}'$ is compatible with all but one of the eight statements

$$A \in \mathcal{F} \text{ and } TA \in \mathcal{F}', \qquad A' \in \mathcal{F}' \text{ and } T^{-1}A' \in \mathcal{F},$$

$$A \in \mathcal{F} \text{ and } TA \notin \mathcal{F}', \qquad A' \in \mathcal{F}' \text{ and } T^{-1}A' \notin \mathcal{F},$$

$$A \notin \mathcal{F} \text{ and } TA \in \mathcal{F}', \qquad A' \notin \mathcal{F}' \text{ and } T^{-1}A' \in \mathcal{F},$$

$$A \notin \mathcal{F} \text{ and } TA \notin \mathcal{F}', \qquad A' \notin \mathcal{F}' \text{ and } T^{-1}A' \notin \mathcal{F}.$$

13.2. Although an indicator I_A is measurable $\mathcal{F}$ if and only if $A \in \mathcal{F}$, the simple function (13.3) can be measurable $\mathcal{F}$ even if some of the A_i lie outside $\mathcal{F}$.

13.3. Show that a monotone real function is measurable.

13.4. Functions are often defined in pieces (for example, let $f(x)$ be x^3 or x^{-1} as $x \geq 0$ or $x < 0$), and the following result shows that the function is measurable if the pieces are.

Consider measurable spaces $(\Omega, \mathcal{F})$ and $(\Omega', \mathcal{F}')$ and a map $T\colon \Omega \to \Omega'$. Let $A_1, A_2, \ldots$ be a countable covering of Ω by $\mathcal{F}$-sets. Consider the σ-field $\mathcal{F}_n = [A\colon A \subset A_n, A \in \mathcal{F}]$ in A_n and the restriction T_n of T to A_n. Show that T is measurable $\mathcal{F}/\mathcal{F}'$ if and only if T_n is measurable $\mathcal{F}_n/\mathcal{F}'$ for each n.

13.5. (a)   For a map T and σ-fields $\mathcal{F}$ and $\mathcal{F}'$, define $T^{-1}\mathcal{F}' = [T^{-1}A'\colon A' \in \mathcal{F}']$ and $T\mathcal{F} = [A'\colon T^{-1}A' \in \mathcal{F}]$. Show that $T^{-1}\mathcal{F}'$ and $T\mathcal{F}$ are sigma fields and that measurability $\mathcal{F}/\mathcal{F}'$ is equivalent to $T^{-1}\mathcal{F}' \subset \mathcal{F}$ and to $\mathcal{F}' \subset T\mathcal{F}$.
(b)   For given $\mathcal{F}'$, $T^{-1}\mathcal{F}'$, which is the smallest σ-field for which T is measurable $\mathcal{F}/\mathcal{F}'$, is the σ-field *generated* by T. For simple random variables describe $\sigma(X_1, \ldots, X_n)$ in these terms.

13.6. ↑   Suppose that $f\colon \Omega \to R^1$. Show that f is measurable $T^{-1}\mathcal{F}'$ if and only if there exists a map $\varphi\colon \Omega' \to R^1$ such that φ is measurable $\mathcal{F}'$ and $f = \varphi T$.

Hint: First consider simple functions and then use Theorem 13.5.

13.7. ↑   Relate the result in Problem 13.6 to Theorem 5.1(ii).

13.8. Show of real functions f and g that $f(\omega) + g(\omega) < x$ if and only if there exist rationals r and s such that $r + s < x$, $f(\omega) < r$, and $g(\omega) < s$. Prove directly that $f + g$ is measurable $\mathcal{F}$ if f and g are.

13.9. Let $\mathcal{F}$ be a σ-field in R^1. Show that $\mathcal{R}^1 \subset \mathcal{F}$ if and only if every continuous function is measurable $\mathcal{F}$. Thus $\mathcal{R}^1$ is the smallest σ-field with respect to which all the continuous functions are measurable.

Hint: If $f_n(x)$ is 1 or $1 - n(x - \alpha)$ or 0 as $x \leq \alpha$ or $\alpha \leq x \leq \alpha + n^{-1}$ or $\alpha + n^{-1} \leq x$, then $f_n(x) \to I_{(-\infty, \alpha]}(x)$.

13.10. ↑   Consider on R^1 the smallest class $\mathcal{K}$ (that is, the intersection of all classes) of real functions containing all the continuous functions and closed under pointwise passages to the limit. The elements of $\mathcal{K}$ are called *Baire* functions. Show that Baire functions and Borel functions are the same thing.

13.11. A real function f on the line is upper semicontinuous at x if for each ϵ there is a δ such that $|x - y| < \delta$ implies that $f(y) < f(x) + \epsilon$. Show that, if f is everywhere upper semicontinuous, then it is measurable.

13.12. If f is bounded and measurable $\mathcal{F}$, then there exist simple functions, each measurable $\mathcal{F}$, such that $f_n(\omega) \to f(\omega)$ uniformly in ω.

13.13. Suppose that f_n and f are finite-valued, $\mathcal{F}$-measurable functions such that $f_n(\omega) \to f(\omega)$ for $\omega \in A$, where $\mu(A) < \infty$ (μ a measure on $\mathcal{F}$). Prove *Egoroff's theorem*: For each ϵ there exists a subset B of A such that $\mu(B) < \epsilon$ and $f_n(\omega) \to f(\omega)$ uniformly on $A - B$.

> *Hint*:   Let $B_n^{(k)}$ be the set of ω in A such that $|f(\omega) - f_i(\omega)| > k^{-1}$ for some $i \geq n$. Show that $B_n^{(k)} \downarrow 0$ as $n \uparrow \infty$, choose n_k so that $\mu(B_{n_k}^{(k)}) < \epsilon/2^k$, and put $B = \bigcup_{k=1}^{\infty} B_{n_k}^{(k)}$.

13.14. ↑   Show that Egoroff's theorem is false without the hypothesis $\mu(A) < \infty$.

13.15. Suppose that $(\Omega, \mathcal{F}, P)$ is a probability measure space, $T: \Omega \to \Omega'$ is measurable $\mathcal{F}/\mathcal{F}'$, and $Q = T^{-1}P$ on $\mathcal{F}'$. Consider the outer and inner measures as defined by (3.9) and (3.10). Show that $Q_*(A') \leq P_*(T^{-1}A') \leq P^*(T^{-1}A') \leq Q^*(A')$ and show by example that the inequalities can be strict. Show that $P^*(A) \leq Q^*(TA)$ and that, if T is one-to-one, $Q_*(TA) \leq P_*(A)$.

13.16. 11.7↑   Show that, if $\mathcal{F}_0$ is a field generating $\mathcal{F}$, if f is measurable $\mathcal{F}$, and if $\mu(\Omega) < \infty$, then for positive ϵ there exists a simple function $g = \Sigma_i x_i I_{A_i}$ such that $A_i \in \mathcal{F}_0$ and $\mu[\omega: |f(\omega) - g(\omega)| > \epsilon] < \epsilon$.

> *Hint*:   First approximate f by a simple function measurable $\mathcal{F}$, and then use Problem 11.7.

13.17. 2.8.↑   Show that, if f is measurable $\sigma(\mathcal{A})$, then there exists a countable subclass $\mathcal{A}_f$ of $\mathcal{A}$ such that f is measurable $\sigma(\mathcal{A}_f)$.

13.18. *Circular Lebesgue measure* Let C be the unit circle in the complex plane and define $T: [0, 1) \to C$ by $T\omega = e^{2\pi i\omega}$. Let $\mathcal{B}$ consist of the Borel subsets of $[0, 1)$ and let λ be Lebesgue measure on $\mathcal{B}$. Show that $\mathcal{C} = [A: T^{-1}A \in \mathcal{B}]$ consists of the sets in $\mathcal{R}^2$ (identify with the complex plane) that are contained in C. Show that $\mathcal{C}$ is generated by the arcs in C. Show that λT^{-1}, circular Lebesgue measure, is invariant under rotations: $\lambda T^{-1} [\theta z: z \in A] = T^{-1}A$ for $A \in \mathcal{C}$ and $\theta \in C$.

SECTION 14.   DISTRIBUTION FUNCTIONS

Distribution Functions

A random variable as defined in Section 13 is a measurable real function X on a probability measure space $(\Omega, \mathcal{F}, P)$. The *distribution* or *law* of the random variable is the probability measure μ on $(R^1, \mathcal{R}^1)$ defined by

(14.1) $\mu(A) = P[X \in A], \qquad A \in \mathcal{R}^1.$

As in the case of the simple random variables in Chapter 1, the argument ω is usually omitted: $P[X \in A]$ is short for $P[\omega: X(\omega) \in A]$. In the notation (13.7), the distribution is PX^{-1}.

For simple random variables the distribution was defined in Section 5—see (5.9). There μ was defined for every subset of the line, however; from now on μ will be defined only for Borel sets, because unless X is simple, one cannot in general be sure that $[X \in A]$ has a probability for A outside $\mathcal{R}^1$.

The *distribution function* of X is defined by

$$(14.2) \qquad F(x) = \mu(-\infty, x] = P[X \le x]$$

for real x. By continuity from above (Theorem 10.2(ii)) for μ, F is right-continuous. Since F is nondecreasing, the left-hand limit $F(x-) = \lim_{y \uparrow x} F(y)$ exists, and by continuity from below (Theorem 10.2(i)) for μ,

$$(14.3) \qquad F(x-) = \mu(-\infty, x) = P[X < x].$$

Thus the jump or saltus $F(x) - F(x-)$ is $\mu\{x\}$, and hence (Theorem 10.2 (iv)) F can have at most countably many points of discontinuity. Clearly,

$$(14.4) \qquad \lim_{x \to -\infty} F(x) = 0, \qquad \lim_{x \to \infty} F(x) = 1.$$

A function with these properties must, in fact, be the distribution function of some random variable:

Theorem 14.1. *If F is a nondecreasing, right-continuous function satisfying (14.4), then there exists on some probability space a random variable X for which $F(x) = P[X \le x]$.*

FIRST PROOF. By Theorem 12.4, if F is nondecreasing and right-continuous, there is on $(R^1, \mathcal{R}^1)$ a measure μ for which $\mu(a, b] = F(b) - F(a)$. But $\lim_{x \to -\infty} F(x) = 0$ implies that $\mu(-\infty, x] = F(x)$, and $\lim_{x \to \infty} F(x) = 1$ implies that $\mu(R^1) = 1$. For the probability space take $(\Omega, \mathcal{F}, P) = (R^1, \mathcal{R}^1, \mu)$ and for X take the identity function: $X(\omega) \equiv \omega$. Then $P[X \le x] = \mu[\omega \in R^1: \omega \le x] = F(x)$. ∎

SECOND PROOF. There is a proof that uses only the existence of Lebesgue measure on the unit interval and does not require Theorem 12.4. For the probability space take the open unit interval: Ω is $(0, 1)$, $\mathcal{F}$ consists of the Borel subsets of $(0, 1)$, and $P(A)$ is the Lebesgue measure of A.

To understand the method, suppose at first that F is continuous and strictly increasing. Then F is a one-to-one mapping of R^1 onto $(0, 1)$; let $\varphi: (0, 1) \to R^1$ be the inverse mapping. For $0 < \omega < 1$, let $X(\omega) = \varphi(\omega)$. Since φ is increasing, certainly X is measurable $\mathcal{F}$. If $0 < u < 1$, then $\varphi(u) \le x$ if and only if $u \le F(x)$. Since P is Lebesgue measure, $P[X \le x] = P[\omega \in (0, 1): \varphi(\omega) \le x] = P[\omega \in (0, 1): \omega \le F(x)] = F(x)$, as required.

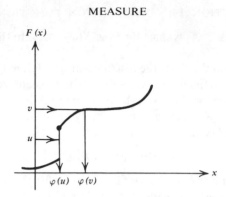

If F has discontinuities or is not strictly increasing, define

(14.5) $\varphi(u) = \inf[x\colon u \leq F(x)]$

for $0 < u < 1$. Since F is nondecreasing, $[x\colon u \leq F(x)]$ is an interval stretching to ∞; since F is right-continuous, this interval is closed on the left. For $0 < u < 1$, therefore, $[x\colon u \leq F(x)] = [\varphi(u), \infty)$, and so $\varphi(u) \leq x$ if and only if $u \leq F(x)$. If $X(\omega) = \varphi(\omega)$ for $0 < \omega < 1$, then by the same reasoning as before, X is a random variable with $P[X \leq x] = F(x)$.   ∎

This last argument actually provides a second proof of Theorem 12.4 for a probability distribution* F: the distribution μ (as defined by (14.1)) of the random variable just constructed satisfies $\mu(-\infty, x] = F(x)$ and hence $\mu(a, b] = F(b) - F(a)$.

Exponential Distributions

There are a number of results which for their interpretation require random variables, independence, and other probabilistic concepts, but which can be discussed technically in terms of distribution functions alone and do not require the apparatus of measure theory.

Suppose as an example that F is the distribution function of the waiting time to the occurrence of some event—say the arrival of the next customer at a queue or the next call at a telephone exchange. As the waiting time must be positive, assume that $F(0) = 0$. Suppose that $F(x) < 1$ for all x and furthermore suppose that

(14.6) $\dfrac{1 - F(x + y)}{1 - F(x)} = 1 - F(y), \qquad x, y \geq 0.$

---

* For the general case, see Problem 14.2.

The right side of this equation is the probability that the waiting time exceeds y; by the definition (4.1) of conditional probability, the left side is the probability that the waiting time exceeds $x + y$ given that it exceeds x. Thus (14.6) attributes to the waiting-time mechanism a kind of lack of memory or aftereffect: If after a time lapse of x the event has not yet occurred, the waiting time still remaining is conditionally distributed just as the entire waiting time from the beginning. For reasons that will emerge later (see Section 23), waiting times often have this property.

The condition (14.6) completely determines the form of F. If $U(x) = 1 - F(x)$, (14.6) is $U(x + y) = U(x)U(y)$. This is a form of Cauchy's equation, and since U is bounded, the corollary to Theorem 14.4 (page 168) implies that $U(x) = e^{-\alpha x}$ for some α. Since $\lim_{x \to \infty} U(x) = 0$, α must be positive. Thus (14.6) implies that F has the exponential form

$$(14.7) \qquad F(x) = \begin{cases} 0 & \text{if } x \le 0, \\ 1 - e^{-\alpha x} & \text{if } x \ge 0. \end{cases}$$

Weak Convergence

Random variables $X_1, \ldots, X_n$ are defined to be independent if the events $[X_1 \in A_1], \ldots, [X_n \in A_n]$ are independent for all Borel sets $A_1, \ldots, A_n$, so that $P[X_i \in A_i, i = 1, \ldots, n] = \Pi_{i=1}^{n} P[X_i \in A_i]$. To find the distribution function of the maximum $M_n = \max \{X_1, \ldots, X_n\}$, take $A_1 = \cdots = A_n = (-\infty, x]$. This gives $P[M_n \le x] = \Pi_{i=1}^{n} P[X_i \le x]$. If the X_i are independent and have common distribution function G and M_n has distribution function F_n, then

$$(14.8) \qquad F_n(x) = G^n(x).$$

It is possible without any appeal to measure theory to study the real function F_n solely by means of the relation (14.8), which can indeed be taken as defining F_n. It is possible in particular to study the asymptotic properties of F_n:

Example 14.1. Consider a stream or sequence of events, say arrivals of calls at a telephone exchange. Suppose that the times between successive events, the interarrival times, are independent and that each has the exponential form (14.7) with a common value of α. By (14.8) the maximum M_n among the first n interarrival times has distribution function $F_n(x) = (1 - e^{-\alpha x})^n$, $x \ge 0$. For each x, $\lim_n F_n(x) = 0$, which means that M_n tends to be large for n large. But $P[M_n - \alpha^{-1} \log n \le x] = F_n(x + \alpha^{-1} \log n)$. This is the distribution function of $M_n - \alpha^{-1} \log n$, and it satisfies

$$(14.9) \qquad F_n(x + \alpha^{-1} \log n) = (1 - e^{-(\alpha x + \log n)})^n \to e^{-e^{-\alpha x}}$$

as $n \to \infty$; the equality here holds if $\log n \ge -\alpha x$, and so the limit holds for all

x. This gives for large n the approximate distribution of the normalized random variable $M_n - \alpha^{-1} \log n$.   ∎

If F_n and F are distribution functions, F_n *converges weakly to* F, written $F_n \Rightarrow F$, if

$$(14.10) \qquad\qquad \lim_n F_n(x) = F(x)$$

for each x at which F is continuous.* To study the approximate distribution of a random variable Y_n it is often necessary to study instead the normalized or rescaled random variable $(Y_n - b_n)/a_n$ for appropriate constants a_n and b_n. If Y_n has distribution function F_n and if $a_n > 0$, then $P[(Y_n - b_n)/a_n \le x] = P[Y_n \le a_n x + b_n]$, and therefore $(Y_n - b_n)/a_n$ has distribution function $F_n(a_n x + b_n)$. For this reason weak convergence often appears in the form[†]

$$(14.11) \qquad\qquad F_n(a_n x + b_n) \Rightarrow F(x).$$

An example of this is (14.9): there $a_n = 1$, $b_n = \alpha^{-1} \log n$, and $F(x) = e^{-e^{-\alpha x}}$.

Example 14.2. Consider again the distribution function (14.8) of the maximum, but suppose that G has the form

$$G(x) = \begin{cases} 0 & \text{if } x \le 1, \\ 1 - x^{-\alpha} & \text{if } x \ge 1, \end{cases}$$

where $\alpha > 0$. Here $F_n(n^{1/\alpha}x) = (1 - n^{-1}x^{-\alpha})^n$ for $x \ge n^{-1/\alpha}$, and therefore

$$\lim_n F_n(n^{1/\alpha}x) = \begin{cases} 0 & \text{if } x \le 0, \\ e^{-x^{-\alpha}} & \text{if } x > 0. \end{cases}$$

This is an example of (14.11) in which $a_n = n^{1/\alpha}$ and $b_n = 0$.   ∎

Example 14.3. Consider (14.8) once more, but for

$$G(x) = \begin{cases} 0 & \text{if } x \le 0, \\ 1 - (1 - x)^\alpha & \text{if } 0 \le x \le 1, \\ 1 & \text{if } x \ge 1, \end{cases}$$

where $\alpha > 0$. This time $F_n(n^{-1/\alpha}x + 1) = (1 - n^{-1}(-x)^\alpha)^n$ if $-n^{1/\alpha} \le x \le 0$.

---

* For the role of continuity, see Example 14.4.
† To write $F_n(a_n x + b_n) \Rightarrow F(ax + b)$ ignores the distinction between a function and its value at an unspecified value of its argument, but the meaning of course is that $F_n(a_n x + b_n) \to F(x)$ at continuity points x of F.

Therefore,

$$\lim_n F_n(n^{-1/\alpha}x + 1) = \begin{cases} e^{-(-x)^\alpha} & \text{if } x \le 0, \\ 1 & \text{if } x > 0, \end{cases}$$

a case of (14.11) in which $a_n = n^{-1/\alpha}$ and $b_n = 1$. ∎

Let Δ be the distribution function with a unit jump at the origin:

$$(14.12) \qquad \Delta(x) = \begin{cases} 0 & \text{if } x < 0, \\ 1 & \text{if } x \ge 0. \end{cases}$$

If $X(\omega) \equiv 0$, then X has distribution function Δ.

Example 14.4. Let $X_1, X_2, \ldots$ be independent random variables for which $P[X_k = 1] = P[X_k = -1] = \frac{1}{2}$ and put $S_n = X_1 + \cdots + X_n$. By the weak law of large numbers,

$$(14.13) \qquad P[|n^{-1}S_n| > \epsilon] \to 0$$

for $\epsilon > 0$. Let F_n be the distribution function of $n^{-1}S_n$. If $x > 0$, then $F_n(x) = 1 - P[n^{-1}S_n > x] \to 1$; if $x < 0$, then $F_n(x) \le P[|n^{-1}S_n| \ge |x|] \to 0$. As this accounts for all the continuity points of Δ, $F_n \Rightarrow \Delta$. It is easy to turn the argument around and deduce (14.13) from $F_n \Rightarrow \Delta$. Thus the weak law of large numbers is equivalent to the assertion that the distribution function of $n^{-1}S_n$ converges weakly to Δ.

If n is odd, so that $S_n = 0$ is impossible, then by symmetry the events $[S_n \le 0]$ and $[S_n \ge 0]$ each have probability $\frac{1}{2}$ and hence $F_n(0) = \frac{1}{2}$. Thus $F_n(0)$ does not converge to $\Delta(0) = 1$, but because Δ is discontinuous at 0, the definition of weak convergence does not require this. ∎

Allowing (14.10) to fail at discontinuity points x of F thus makes it possible to bring the weak law of large numbers under the theory of weak convergence. But if (14.10) need hold only for certain values of x, there arises the question of whether weak limits are unique. Suppose that $F_n \Rightarrow F$ and $F_n \Rightarrow G$. Then $F(x) = \lim_n F_n(x) = G(x)$ if F and G are both continuous at x. Since F and G each have only countably many points of discontinuity,* the set of common continuity points is dense, and it follows by right continuity that F and G are identical. A sequence can thus have at most one weak limit.

Convergence of distribution functions is studied in detail in Chapter 5. The remainder of this section is devoted to some weak-convergence theorems which are interesting both for themselves and for the reason that they require so little technical machinery.

---

* The proof following (14.3) uses measure theory, but this is not necessary: If the saltus $\sigma(x) = F(x) - F(x-)$ exceeds ϵ at $x_1 < \cdots < x_n$, then $F(x_i) - F(x_{i-1}) > \epsilon$ (take $x_0 < x_1$), and so $n\epsilon \le F(x_n) - F(x_0) \le 1$; hence $[x: \sigma(x) > \epsilon]$ is finite and $[x: \sigma(x) > 0]$ is countable.

Convergence of Types*

Distribution functions F and G are of the same *type* if there exist constants a and b, $a > 0$, such that $F(ax + b) = G(x)$ for all x. A distribution function is *degenerate* if it has the form $\Delta(x - x_0)$ (see (14.12)) for some x_0; otherwise, it is *nondegenerate*.

Theorem 14.2. *Suppose that $F_n(x) \Rightarrow F(x)$ and $F_n(a_nx + b_n) \Rightarrow G(x)$, where $a_n > 0$ and F and G are nondegenerate. Then there exist a and b, $a > 0$, such that $a_n \rightarrow a$, $b_n \rightarrow b$, and $F(ax + b) = G(x)$.*

It follows easily from this that if $F_n(u_nx + v_n) \Rightarrow F(x)$ and $F_n(a_nx + b_n) \Rightarrow G(x)$ ($u_n, a_n > 0$), where F and G are nondegenerate, then a_n/u_n and $(b_n - v_n)/u_n$ converge, and F and G are of the same type. Thus there can be only one possible limit type and essentially only one possible sequence of norming constants.

The proof of the theorem is for clarity set out in a sequence of lemmas. In all of them, a and the a_n are assumed to be positive.

Lemma 1. *If $F_n \Rightarrow F$, $a_n \rightarrow a$, and $b_n \rightarrow b$, then $F_n(a_nx + b_n) \Rightarrow F(ax + b)$.*

PROOF. If x is a continuity point of $F(ax + b)$ and $\epsilon > 0$, choose continuity points u and v of F so that $u < ax + b < v$ and $F(v) - F(u) < \epsilon$; this is possible because F has only countably many discontinuities. For large enough n, $u < a_nx + b_n < v$, $|F_n(u) - F(u)| < \epsilon$, and $|F_n(v) - F(v)| < \epsilon$; but then $F(ax + b) - 2\epsilon < F(u) - \epsilon < F_n(u) \le F_n(a_nx + b_n) \le F_n(v) < F(v) + \epsilon < F(ax + b) + 2\epsilon$.   ∎

Lemma 2. *If $F_n \Rightarrow F$ and $a_n \rightarrow \infty$, then $F_n(a_nx) \Rightarrow \Delta(x)$.*

PROOF. Given ϵ choose a continuity point u of F so large that $F(u) > 1 - \epsilon$. If $x > 0$, then for all large enough n, $a_nx > u$ and $|F_n(u) - F(u)| < \epsilon$, so that $F_n(a_nx) \ge F_n(u) > F(u) - \epsilon > 1 - 2\epsilon$. Thus $\lim_n F_n(a_nx) = 1$ for $x \gtrsim 0$; similarly, $\lim_n F_n(a_nx) = 0$ for $x < 0$.   ∎

Lemma 3. *If $F_n \Rightarrow F$ and b_n is unbounded, then $F_n(x + b_n)$ cannot converge weakly.*

PROOF. Suppose that b_n is unbounded and that $b_n \rightarrow \infty$ along some subsequence (the case $b_n \rightarrow -\infty$ is similar). Suppose that $F_n(x + b_n) \Rightarrow G(x)$. Given

* This topic may be omitted.

ϵ choose a continuity point u of F so that $F(u) > 1 - \epsilon$. Whatever x may be, for n far enough out in the subsequence, $x + b_n > u$ and $F_n(u) > 1 - 2\epsilon$, so that $F_n(x + b_n) > 1 - 2\epsilon$. Thus $G(x) = \lim_n F_n(x + b_n) = 1$ for all continuity points x of G, which is impossible.   ∎

Lemma 4. *If $F_n(x) \Rightarrow F(x)$ and $F_n(a_n x + b_n) \Rightarrow G(x)$, where F and G are nondegenerate, then*

(14.14) $$0 < \inf_n a_n \le \sup_n a_n < \infty, \qquad \sup_n |b_n| < \infty.$$

PROOF. Suppose that a_n is not bounded above. Arrange by passing to a subsequence that $a_n \to \infty$. Then by Lemma 2, $F_n(a_n x) \Rightarrow \Delta(x)$, and by Lemma 3 and the hypothesis, b_n is bounded along this subsequence. Arrange by passing to a further subsequence that b_n converges to some b. Then by Lemma 1, $F_n(a_n x + b_n) \Rightarrow \Delta(x + b)$, and so by hypothesis $G(x) = \Delta(x + b)$, contrary to the assumption that G is nondegenerate.

Thus a_n is bounded above. If $G_n(x) = F_n(a_n x + b_n)$, then $G_n(x) \Rightarrow G(x)$ and $G_n(a_n^{-1} x - a_n^{-1} b_n) = F_n(x) \Rightarrow F(x)$. The result just proved shows that a_n^{-1} is bounded.

Thus a_n is bounded away from 0 and ∞. If b_n is not bounded, pass to a subsequence along which $b_n \to \pm\infty$ and a_n converges to some a. Since $F_n(a_n x) \Rightarrow F(ax)$ along the subsequent by Lemma 1, $b_n \to \pm\infty$ is by Lemma 3 incompatible with the hypothesis that $F_n(a_n x + b_n) \Rightarrow G(x)$.   ∎

Lemma 5. *If $F(x) = F(ax + b)$ for all x and F is nondegenerate, then $a = 1$ and $b = 0$.*

PROOF. Since $F(x) = F(a^n x + (a^{n-1} + \cdots + a + 1)b)$, it follows by Lemma 4 that a^n is bounded away from 0 and ∞, so that $a = 1$, and it then follows that nb is bounded, so that $b = 0$.   ∎

PROOF OF THEOREM 14.2. The hypothesis implies (14.14). Fix any subsequence along which a_n converges to some positive a and b_n converges to some b. By Lemma 1, $F_n(a_n x + b_n) \Rightarrow F(ax + b)$ along this subsequence, and the hypothesis gives $F(ax + b) = G(x)$.

Suppose that along some other sequence, $a_n \to u > 0$ and $b_n \to v$. Then $F(ux + v) = G(x)$ and $F(ax + b) = G(x)$ both hold, so that $u = a$ and $v = b$ by Lemma 5. Every convergent subsequence of $\{(a_n, b_n)\}$ thus converges to (a, b), and so the entire sequence does.   ∎

Extremal Distributions*

A distribution function F is *extremal* if it is nondegenerate and if, for some distribution function G and constants a_n ($a_n > 0$) and b_n,

$$(14.15) \qquad\qquad G^n(a_n x + b_n) \Rightarrow F(x).$$

These are the possible limiting distributions of normalized maxima (see (14.8)), and Examples 14.1, 14.2, and 14.3 give three specimens. The following analysis shows that these three examples exhaust the possible types.

Assume that F is extremal. From (14.15) follow $G^{nk}(a_n x + b_n) \Rightarrow F^k(x)$ and $G^{nk}(a_{nk} x + b_{nk}) \Rightarrow F(x)$, and so by Theorem 14.2 there exist constants c_k and d_k such that c_k is positive and

$$(14.16) \qquad\qquad F^k(x) = F(c_k x + d_k).$$

From $F(c_{jk} x + d_{jk}) = F^{jk}(x) = F^j(c_k x + d_k) = F(c_j(c_k x + d_k) + d_j)$ follow (Lemma 5) the relations

$$(14.17) \qquad c_{jk} = c_j c_k, \qquad d_{jk} = c_j d_k + d_j = c_k d_j + d_k.$$

Of course, $c_1 = 1$ and $d_1 = 0$. There are three cases to be considered separately.

Case 1. Suppose that $c_k = 1$ for all k. Then

$$(14.18) \qquad F^k(x) = F(x + d_k), \qquad F^{1/k}(x) = F(x - d_k).$$

This implies that $F^{j/k}(x) = F(x + d_j - d_k)$. For positive rational $r = j/k$, put $\delta_r = d_j - d_k$; (14.17) implies that the definition is consistent, and $F^r(x) = F(x + \delta_r)$. Since F is nondegenerate, there is an x such that $0 < F(x) < 1$, and it follows by (14.18) that d_k is decreasing in k, so that δ_r is strictly decreasing in r.

For positive real t let $\varphi(t) = \inf_{0 < r < t} \delta_r$ (r rational in the infimum). Then $\varphi(t)$ is decreasing in t and

$$(14.19) \qquad\qquad F^t(x) = F(x + \varphi(t))$$

for all x and all positive t. Further, (14.17) implies that $\varphi(st) = \varphi(s) + \varphi(t)$, so that by Theorem 14.4 applied to $\varphi(e^x)$, $\varphi(t) = -\beta \log t$, where $\beta > 0$ because $\varphi(t)$ is strictly decreasing. Now (14.19) with $t = e^{x/\beta}$ gives $F(x) = \exp\{e^{-x/\beta} \log F(0)\}$, and so F must be of the same type as

$$(14.20) \qquad\qquad F_1(x) = e^{-e^{-x}}.$$

---

* This topic may be omitted.

Example 14.1 shows that this distribution function can arise as a limit of distributions of maxima—that is, F_1 is indeed extremal. ∎

Case 2. Suppose that $c_{k_0} \neq 1$ for some k_0, which necessarily exceeds 1. Then there exists an x' such that $c_{k_0}x' + d_{k_0} = x'$; but (14.16) then gives $F^{k_0}(x') = F(x')$, so that $F(x')$ is 0 or 1. (In Case 1, F has the type (14.20) and so never assumes the values 0 and 1.)

Now suppose further that, in fact, $F(x') = 0$. Let x_0 be the supremum of those x for which $F(x) = 0$. By passing to a new F of the same type one can arrange that $x_0 = 0$; then $F(x) = 0$ for $x < 0$ and $F(x) > 0$ for $x > 0$. The new F will satisfy (14.16), but with new constants d_k.

If a (new) d_k is distinct from 0, then there is an x near 0 for which the arguments on the two sides of (14.16) have opposite signs. Therefore, $d_k = 0$ for all k, and

$$(14.21) \qquad F^k(x) = F(c_k x), \qquad F^{1/k}(x) = F\left(\frac{x}{c_k}\right)$$

for all k and x. This implies that $F^{j/k}(x) = F(xc_j/c_k)$. For positive rational $r = j/k$, put $\gamma_r = c_j/c_k$. The definition is by (14.17) again consistent, and $F^r(x) = F(\gamma_r x)$. Since $0 < F(x) < 1$ for some x, necessarily positive, it follows by (14.21) that c_k is decreasing in k, so that γ_r is strictly decreasing in r. Put $\psi(t) = \inf_{0 < r < t} \gamma_r$ for positive real t. From (14.17) follows $\psi(st) = \psi(s)\psi(t)$, and by the corollary to Theorem 14.4 applied to $\psi(e^x)$, $\psi(t) = t^{-\xi}$ for some $\xi > 0$. Since $F^t(x) = F(\psi(t)x)$ for all x and for t positive, $F(x) = \exp\{x^{-1/\xi}\log F(1)\}$ for $x > 0$. Thus (take $\alpha = 1/\xi$) F is of the same type as

$$(14.22) \qquad F_{2,\alpha}(x) = \begin{cases} 0 & \text{if } x < 0, \\ e^{-x^{-\alpha}} & \text{if } x \geq 0. \end{cases}$$

Example 14.2 shows that this case can arise. ∎

Case 3. Suppose as in Case 2 that $c_{k_0} \neq 1$ for some k_0, so that $F(x')$ is 0 or 1 for some x', but this time suppose that $F(x') = 1$. Let x_1 be the infimum of those x for which $F(x) = 1$. By passing to a new F of the same type, arrange that $x_1 = 0$: $F(x) < 1$ for $x < 0$ and $F(x) = 1$ for $x \geq 0$. If $d_k \neq 0$, then for some x near 0, one side of (14.16) is 1 and the other is not. Thus $d_k = 0$ for all k, and (14.21) again holds. And again $\gamma_{j/k} = c_j/c_k$ consistently defines a function satisfying $F^r(x) = F(\gamma_r x)$. Since F is nondegenerate, $0 < F(x) < 1$ for some x, but this time x is necessarily negative, so that c_k is increasing.

The same analysis as before shows that there is a positive ξ such that $F^t(x) = F(t^\xi x)$ for all x and for t positive. Thus $F(x) = \exp\{(-x)^{1/\xi}\log F(-1)\}$ for $x < 0$, and F is of the type

(14.23) $F_{3,\alpha}(x) = \begin{cases} e^{-(-x)^\alpha} & \text{if } x \le 0, \\ 1 & \text{if } x \ge 0. \end{cases}$

Example 14.3 shows that this distribution function is indeed extremal.   ∎

This completely characterizes the class of extremal distributions:

Theorem 14.3.   *The class of extremal distribution functions consists exactly of the distribution functions of the types* (14.20), (14.22), *and* (14.23).

It is possible to go on and characterize the *domains of attraction*. That is, it is possible for each extremal distribution function F to describe the class of G satisfying (14.15) for some constants a_n and b_n—the class of G *attracted* to F.*

Cauchy's Equation

Theorem 14.4.   *Let f be a real function on $(0, \infty)$ and suppose that f satisfies Cauchy's equation: $f(x + y) = f(x) + f(y)$ for $x, y > 0$. If there is some interval on which f is bounded above, then $f(x) = xf(1)$ for $x > 0$.*

PROOF.   The problem is to prove that $g(x) = f(x) - xf(1)$ vanishes identically. Clearly, $g(1) = 0$, and g satisfies Cauchy's equation and on some interval is bounded above. By induction, $g(nx) = ng(x)$; hence $ng(m/n) = g(m) = mg(1) = 0$, so that $g(r) = 0$ for positive rational r. Suppose that $g(x_0) \ne 0$ for some x_0. If $g(x_0) < 0$, then $g(r_0 - x_0)$ $= -g(x_0) > 0$ for rational $r_0 > x_0$. It is thus no restriction to assume that $g(x_0) > 0$. Let I be an open interval in which g is bounded above. Given a number M, choose n so that $ng(x_0) > M$ and then choose a rational r so that $nx_0 + r$ lies in I; since $g(nx_0 + r) =$ $g(nx_0) = ng(x_0) > M$ and M was arbitrary, g is not bounded above in I, a contradiction.
                                                                            ∎

Obviously, the same proof works if f is bounded below in some interval.

Corollary.   *Let U be a real function on $(0, \infty)$ and suppose that $U(x + y) = U(x)U(y)$ for $x, y > 0$. Suppose further that there is some interval on which U is bounded above. Then either $U(x) = 0$ for $x > 0$, or else there is an A such that $U(x) = e^{Ax}$ for $x > 0$.*

PROOF.   Since $U(x) = U^2(x/2)$, U is nonnegative. If $U(x) = 0$, then $U(x/2^n) = 0$ and so U vanishes at points arbitrarily near 0. If U vanishes at a point, it must by the functional equation vanish everywhere to the right of that point. Hence U is identically 0 or else everywhere positive.
   In the latter case, the theorem applies to $f(x) = \log U(x)$, this function being bounded above in some interval, and so $f(x) = Ax$ for $A = \log U(1)$.   ∎

---

* This theory is associated with the names of Fisher, Fréchet, Gnedenko, and Tippet. For further information, see GALAMBOS.

PROBLEMS

14.1. Prove that a distribution function has only countably many jumps by considering the intervals $(F(x-), F(x)]$—each nonempty one contains a rational.

14.2. For distribution functions F, the second proof of Theorem 14.1 shows how to construct a measure μ on $(R^1, \mathcal{R}^1)$ such that $\mu(a, b] = F(b) - F(a)$.

    (a)   Extend to the case F bounded.

    (b)   Extend to the general case.

    Hint: *Let* $F_n(x)$ *be* $-n$ *or* $F(x)$ *or* n *as* $F(x) < -n$ *or* $-n \leq F(x) < n$ *or* $n \leq F(x)$. *Construct the corresponding* μ_n *and define* $\mu(A) = \lim_n \mu_n(A)$.

14.3. Suppose that X assumes positive integers as values. Show that $P[X \geq n + m \mid X \geq n] = P[X \geq m]$ if and only if X has a Pascal distribution: $P[X = n] = q^{n-1}p, n = 1, 2, \ldots$. This is the discrete analogue of the exponential distribution.

14.4. **(a)**   Suppose that X has a continuous, strictly increasing distribution function F. Show that the random variable $F(X)$ is uniformly distributed over the unit interval in the sense that $P[F(X) \leq u] = u$ for $0 \leq u \leq 1$. Passing from X to $F(X)$ is called the *probability transformation*.

    (b)   Show that the function $\varphi(u)$ defined by (14.5) satisfies $F(\varphi(u) -) \leq u \leq F(\varphi(u))$ and that, if F is continuous (but not necessarily strictly increasing), then $F(\varphi(u)) = u$ for $0 < u < 1$.

    (c)   Show that $P[F(X) < u] = F(\varphi(u)-)$ and hence that the result in part (a) holds as long as F is continuous.

    (d)   Show that, if A is a Borel set and F is continuous at each point of A, then $P[F(X) \in A]$ is at most the Lebesgue measure of A.

14.5. A distribution function F can be described by its derivative or density $f = F'$, provided that f exists and can be integrated back to F. One thinks in terms of the approximation

$$(14.24) \qquad P[x < X \leq x + dx] \approx f(x)\, dx.$$

    (a)   Show that, if X has distribution function with density f and g is increasing and differentiable, then the distribution function of $g(X)$ has density $f(g^{-1}(x))/g'(g^{-1}(x))$.

    (b)   Check part (a) of Problem 14.4 under the assumption that F is differentiable.

14.6. **(a)**   Suppose that $X_1, \ldots, X_n$ are independent and that each has distribution function F. Let $X_{(k)}$ be the kth-*order statistic*, the kth largest among the X_i. Show that $X_{(k)}$ has distribution function

$$G_k(x) = \sum_{u=k}^{n} \binom{n}{u} (F(x))^u (1 - F(x))^{n-u}.$$

    (b)   Assume that F has density f and show that G_k has density

$$\frac{n!}{(k - 1)!(n - k)!} (F(x))^{k-1} f(x)(1 - F(x))^{n-k}.$$

Interpret this in terms of (14.24).

Hint: Do not differentiate directly. Let A be the event that $k - 1$ of the X_i lie in $(-\infty, x]$, one lies in $(x, x + h]$, and $n - k$ lie in (x, ∞). Show that $P[x < X_{(k)} \leq x + h] - P(A)$ is nonnegative and does not exceed the probability that two of the X_i lie in $(x, x + h]$. Estimate the latter probability, calculate $P(A)$ exactly, and let h go to 0.

14.7. Show for every distribution function F that there exist distribution functions F_n such that $F_n \Rightarrow F$ and

(a) F_n is continuous;

(b) F_n is constant over each interval $((k - 1)n^{-1}, kn^{-1}]$, $k = 0, \pm 1, \pm 2,$
. . . .

14.8. The *Lévy distance* $d(F, G)$ between two distribution functions is the infimum of those ϵ such that $G(x - \epsilon) - \epsilon \leq F(x) \leq G(x + \epsilon) + \epsilon$ for all x. Verify that this is a metric on the set of distribution functions. Show that a necessary and sufficient condition for $F_n \Rightarrow F$ is that $d(F_n, F) \to 0$.

14.9. Simplify the proof of Theorem 14.4 under the assumption that f is continuous, or right continuous. Simplify it further under the assumption that f is differentiable.

14.10. 12.3↑ A Borel function satisfying Cauchy's equation is automatically bounded in some interval and hence satisfies $f(x) = xf(1)$.

Hint: Take K large enough that $\lambda[x: x > 0, |f(x)| \leq K] > 0$. Apply Problem 12.3 and conclude that f is bounded in some interval to the right of 0.

14.11. ↑ Consider sets S of reals that are linearly independent over the field of rationals in the sense that $n_1 x_1 + \cdots + n_k x_k = 0$ for points x_i in S and integers n_i (positive or negative) is impossible unless $n_i \equiv 0$.

(a) By Zorn's lemma find a maximal such S. Show that it is a *Hamel basis*. That is, show that each real x can be written uniquely as $x = n_1 x_1 + \cdots + n_k x_k$ for points x_i in S and integers n_i.

(b) Define f arbitrarily on S and define it elsewhere by $f(n_1 x_1 + \cdots + n_k x_k) = n_1 f(x_1) + \cdots + n_k f(x_k)$. Show that f satisfies Cauchy's equation but need not satisfy $f(x) = xf(1)$. Relate to Theorem 14.4.

(c) By means of Problem 14.10 give a new construction of a nonmeasurable set.

14.12. 14.8. ↑ (a) Show that if a distribution function F is everywhere continuous, then it is uniformly continuous.

(b) Let $\delta_F(\epsilon) = \sup [F(x) - F(y): |x - y| \leq \epsilon]$ be the modulus of coninuity of F. Show that $d(F,G) < \epsilon$ implies that $\sup_x |F(x) - G(x)| \leq \epsilon + \delta_F(\epsilon)$.

(c) Show that, if $F_n \Rightarrow F$ and F is everywhere continuous, then $F_n(x) \to F(x)$ uniformly in x. What if F is continuous over a closed interval?

14.13. Show that (14.22) and (14.23) are everywhere infinitely differentiable, although not analytic.

CHAPTER 3

Integration

SECTION 15. THE INTEGRAL

Expected values of simple random variables and Riemann integrals of continuous functions can be brought together with other related concepts under a general theory of integration, and this theory is the subject of the present chapter.

Definition

Throughout this section, f, g, and so on will denote real measurable functions, the values $\pm\infty$ allowed, on a measure space $(\Omega, \mathcal{F}, \mu)$. The object is to define and study the definite integral

$$\int f \, d\mu = \int_\Omega f(\omega) \, d\mu(\omega) = \int_\Omega f(\omega)\mu \, (d\omega).$$

Suppose first f that is nonnegative. For each finite decomposition $\{A_i\}$ of Ω into $\mathcal{F}$-sets, consider the sum

(15.1) $$\sum_i \left[\inf_{\omega \in A_i} f(\omega) \right] \mu(A_i).$$

In computing the products here, the conventions about infinity are

(15.2) $$\begin{cases} 0 \cdot \infty = \infty \cdot 0 = 0, \\ x \cdot \infty = \infty \cdot x = \infty \qquad \text{if } 0 < x < \infty, \\ \infty \cdot \infty = \infty. \end{cases}$$

The reasons for these conventions will become clear later. Also in force are the conventions of Section 10 for sums and limits involving infinity; see (10.3) and (10.4). The integral of f is defined as the supremum of the sums (15.1):

171

(15.3) $$\int f \, d\mu = \sup_i \sum_i \left[\inf_{\omega \in A_i} f(\omega) \right] \mu(A_i).$$

The supremum here extends over all finite decompositions $\{A_i\}$ of Ω into $\mathcal{F}$-sets.

For the general f, consider its *positive part*,

(15.4) $$f^+(\omega) = \begin{cases} f(\omega) & \text{if } 0 \le f(\omega) \le \infty, \\ 0 & \text{if } -\infty \le f(\omega) \le 0 \end{cases}$$

and its *negative part*,

(15.5) $$f^-(\omega) = \begin{cases} -f(\omega) & \text{if } -\infty \le f(\omega) \le 0, \\ 0 & \text{if } 0 \le f(\omega) \le \infty. \end{cases}$$

These functions are nonnegative and measurable, and $f = f^+ - f^-$. The general integral is defined by

(15.6) $$\int f \, d\mu = \int f^+ \, d\mu - \int f^- \, d\mu,$$

unless $\int f^+ \, d\mu = \int f^- \, d\mu = \infty$, in which case f has no integral.

If $\int f^+ \, d\mu$ and $\int f^- \, d\mu$ are both finite, f is *integrable*, or integrable μ, or summable, and has (15.6) as its *definite integral*. If $\int f^+ \, d\mu = \infty$ and $\int f^- \, d\mu < \infty$, f is not integrable but is in accordance with (15.6) assigned ∞ as its definite integral. Similarly, if $\int f^+ \, d\mu < \infty$ and $\int f^- \, d\mu = \infty$, f is not integrable but has definite integral $-\infty$. Note that f can have a definite integral without being integrable; it fails to have a definite integral if and only if its positive and negative parts both have infinite integrals.

The really important case of (15.6) is that in which $\int f^+ \, d\mu$ and $\int f^- \, d\mu$ are both finite. Allowing infinite integrals is a convention that simplifies the statements of various theorems, especially theorems involving nonnegative functions. Note that (15.6) is defined unless it involves "$\infty - \infty$;" if one term on the right is ∞ and the other is a finite real x, the difference is defined by the conventions $\infty - x = \infty$ and $x - \infty = -\infty$.

The extension of the integral from the nonnegative case to the general case is consistent: (15.6) agrees with (15.3) if f is nonnegative because then $f^- \equiv 0$.

Nonnegative Functions

It is convenient first to analyze nonnegative functions.

Theorem 15.1. (i) *If $f = \sum_i x_i I_{A_i}$ is a nonnegative simple function, $\{A_i\}$ being a finite decomposition of Ω into $\mathcal{F}$-sets. then $\int f \, d\mu = \sum_i x_i \mu(A_i)$.*

(ii) *If $0 \le f(\omega) \le g(\omega)$ for all ω, then $\int f \, d\mu \le \int g \, d\mu$.*

(iii) *If* $0 \le f_n(\omega) \uparrow f(\omega)$ *for all* ω, *then* $0 \le \int f_n \, d\mu \uparrow \int f \, d\mu$.

(iv) *For nonnegative functions* f *and* g *and nonnegative constants* α *and* β, $\int (\alpha f + \beta g) \, d\mu = \alpha \int f \, d\mu + \beta \int g \, d\mu$.

PROOF OF (i). Let $\{B_j\}$ be a finite decomposition of Ω and let β_j be the infimum of f over B_j. If $A_i \cap B_j \ne 0$, then $\beta_j \le x_i$; therefore, $\Sigma_j \, \beta_j \mu(B_j) = \Sigma_{ij} \, \beta_j \mu(A_i \cap B_j) \le \Sigma_{ij} \, x_i \mu(A_i \cap B_j) = \Sigma_i \, x_i \mu(A_i)$. On the other hand, there is equality here if $\{B_j\}$ coincides with $\{A_i\}$.

PROOF OF (ii). The sums (15.1) obviously do not decrease if f is replaced by g.

PROOF OF (iii). By (ii) the sequence $\int f_n \, d\mu$ is nondecreasing and bounded above by $\int f \, d\mu$. It therefore suffices to show that $\int f \, d\mu \le \lim_n \int f_n \, d\mu$, or that

(15.7) $$\sum_i \left[\inf_{\omega \in A_i} f(\omega) \right] \mu(A_i) \le \lim_n \int f_n \, d\mu$$

for each finite decomposition $\{A_i\}$.

Fix $\{A_i\}$ and let v_i be the infimum in (15.7). It is enough to show that $x < \Sigma_i \, v_i \mu(A_i)$ implies that $x < \lim_n \int f_n \, d\mu$. Choose real numbers u_i such that $x < \Sigma_i \, u_i \mu(A_i)$ and such that for each i either $0 < u_i < v_i$ or else $u_i = v_i = 0$. (Let $u_i = (1 - k^{-1})v_i$ if $0 \le v_i < \infty$ and $u_i = k$ if $v_i = \infty$; these u_i will work if k is taken sufficiently large.) If $\omega \in A_i$, then $f_n(\omega) \uparrow f(\omega) \ge v_i$ and hence $f_n(\omega) \ge u_i$ for all sufficiently large n. If $A_{in} = [\omega \in A_i : f_n(\omega) \ge u_i]$, it follows that $A_{in} \uparrow A_i$ and hence (Theorem 10.2(i)) that $\mu(A_{in}) \uparrow \mu(A_i)$. Therefore,

$$\int f_n \, d\mu \ge \sum_i \left[\inf_{\omega \in A_{in}} f_n(\omega) \right] \mu(A_{in}) + \sum_i \left[\inf_{\omega \in A_i - A_{in}} f_n(\omega) \right] \mu(A_i - A_{in})$$

$$\ge \sum_i u_i \mu(A_{in}) \to \sum_i u_i \mu(A_i) > x.$$

Technicalities apart, the ideas in this argument are the same as those in the proof of Theorem 5.3.

PROOF OF (iv). Suppose at first that $f = \Sigma_i \, x_i I_{A_i}$ and $g = \Sigma_j \, y_j I_{B_j}$ are simple. Then $\alpha f + \beta g = \Sigma_{ij} \, (\alpha x_i + \beta y_j) I_{A_i \cap B_j}$, and so

$$\int (\alpha f + \beta g) \, d\mu = \sum_{ij} (\alpha x_i + \beta y_j) \mu(A_i \cap B_j)$$

$$= \alpha \sum_i x_i \mu(A_i) + \beta \sum_j y_j \mu(B_j) = \alpha \int f \, d\mu + \beta \int g \, d\mu.$$

Note that the argument is valid if some of α, β, x_i, y_j are infinite. Apart from this possibility, the ideas are as in the proof of (5.18).

For general nonnegative f and g, there exist by Theorem 13.5 simple functions f_n and g_n such that $0 \leq f_n \uparrow f$ and $0 \leq g_n \uparrow g$. But then $0 \leq \alpha f_n + \beta g_n \uparrow \alpha f + \beta g$ and $\int (\alpha f_n + \beta g_n) \, d\mu = \alpha \int f_n \, d\mu + \beta \int g_n \, d\mu$, so that (iv) follows via (iii). ∎

By part (i) of Theorem 15.1, the expected values of simple random variables in Chapter 1 are integrals: $E[X] = \int X(\omega)P(d\omega)$. This also covers the step functions in Section 1 (see (1.5)). The relation between the Riemann integral and the integral as defined here will be studied in Section 17.

Example 15.1. Consider the line $(R^1, \mathcal{R}^1, \lambda)$ with Lebesgue measure. Suppose that $-\infty < a_0 \leq a_1 \leq \cdots \leq a_m < \infty$ and let f be the function with nonnegative value x_i on $(a_{i-1}, a_i]$, $i = 1, \ldots, m$, and value 0 on $(-\infty, a_0]$ and (a_m, ∞). By part (i) of Theorem 15.1, $\int f \, d\lambda = \sum_{i=1}^{m} x_i(a_i - a_{i-1})$ because of the convention $0 \cdot \infty = 0$ — see (15.2). If the "area under the curve" to the left of a_0 and the right of a_m is to be 0, this convention is inevitable. It must then work the other way as well: $\infty \cdot 0 = 0$, so that $\int f \, d\lambda = 0$ if f is ∞ at a single point (say) and 0 elsewhere.

If $f = I_{(a,\infty)}$, the area-under-the-curve point of view makes $\int f \, d\mu = \infty$ natural. Hence the second convention in (15.2), which also requires that the integral be infinite if f is ∞ on a nonempty interval and 0 elsewhere. ∎

Recall that *almost everywhere* means outside a set of measure 0.

Theorem 15.2. *Suppose that f and g are nonnegative.*
   (i)   *If $f = 0$ almost everywhere, then $\int f \, d\mu = 0$.*
   (ii) *If $\mu[\omega: f(\omega) > 0] > 0$, then $\int f \, d\mu > 0$.*
   (iii) *If $\int f \, d\mu < \infty$, then $f < \infty$ almost everywhere.*
   (iv) *If $f = g$ almost everywhere, then $\int f \, d\mu = \int f \, d\mu$.*

PROOF. Suppose that $f = 0$ almost everywhere. If A_i meets $[\omega: f(\omega) = 0]$, then the infimum in (15.1) is 0; otherwise, $\mu(A_i) = 0$. Hence each sum (15.1) is 0, and (i) follows.

If $A_\epsilon = [\omega: f(\omega) \geq \epsilon]$, then $A_\epsilon \uparrow [\omega: f(\omega) > 0]$ as $\epsilon \downarrow 0$, so that under the hypothesis of (ii) there is a positive ϵ for which $\mu(A_\epsilon) > 0$. But $f \geq \epsilon I_{A_\epsilon}$ and so (Theorem 15.1, parts (i) and (ii)) $\int f \, d\mu \geq \epsilon\mu(A_\epsilon) > 0$.

If $\mu[f = \infty] > 0$ and g is ∞ on $[f = \infty]$ and 0 elsewhere, then $f \geq g$ and $\int f \, d\mu \geq \int g \, d\mu = \infty$. Hence (iii).

To prove (iv), let $h(\omega)$ be 0 if $f(\omega) = g(\omega)$ and ∞ otherwise; then $h = 0$ almost everywhere and $f(\omega) \leq g(\omega) + h(\omega)$ for all ω. By Theorem 15.1 and part (i) here,

$\int f \, d\mu \leq \int (g + h) \, d\mu = \int g \, d\mu + \int h \, d\mu = \int g \, d\mu$. Reversing the roles of f and g completes the proof.

Uniqueness

Although there are various ways to frame the definition of the integral, they are all equivalent—they all assign the same value to $\int f \, d\mu$. This is because the integral is uniquely determined by certain simple properties it is natural to require of it.

It is natural to want the integral to have properties (i) and (iii) of Theorem 15.1. But these uniquely determine the integral for nonnegative functions: For f nonnegative, there exist by Theorem 13.5 simple functions f_n such that $0 \leq f_n \uparrow f$; by (iii), $\int f \, d\mu$ must be $\lim_n \int f_n \, d\mu$, and (i) determines the value of each $\int f_n \, d\mu$.

Property (i) can itself be derived from (iv) (linearity) together with the assumption that $\int I_A \, d\mu = \mu(A)$ for indicators I_A: $\int (\Sigma_i \, x_i I_{A_i}) \, d\mu = \Sigma_i \, x_i \int I_{A_i} \, d\mu = \Sigma_i \, x_i \mu(A_i)$.

If (iv) of Theorem 15.1 is to persist when the integral is extended beyond the class of nonnegative functions, $\int f \, d\mu$ must be $\int (f^+ - f^-) \, d\mu = \int f^+ \, d\mu - \int f^- \, d\mu$, which makes the definition (15.6) inevitable.

PROBLEMS

These problems outline alternative definitions of the integral. In all of them f is assumed measurable and *nonnegative*. Call (15.3) the *lower integral* and write it as

$$(15.8) \qquad \int_* f \, d\mu = \sup \sum_i \left[\inf_{\omega \in A_i} f(\omega) \right] \mu(A_i)$$

to distinguish it from the *upper integral*

$$(15.9) \qquad \int^* f \, d\mu = \inf \sum_i \left[\sup_{\omega \in A_i} f(\omega) \right] \mu(A_i).$$

The infimum in (15.9), like the supremum in (15.8), extends over all finite partitions $\{A_i\}$ of Ω into $\mathcal{F}$-sets.

15.1. Show that $\int^* f \, d\mu = \infty$ if $\mu[\omega: f(\omega) > 0] = \infty$ or if $\mu[\omega: f(\omega) \geq x] > 0$ for all x.

As there are many fs of this last kind that ought to be integrable, (15.9) is inappropriate as a definition of $\int f \, d\mu$. The only trouble with (15.9), however, is that it treats infinity incorrectly. To see this, and to focus on essentials, assume in the following problems that $\mu(\Omega) < \infty$ and that f is bounded and nonnegative.

15.2. ↑ (a) Show that

$$\sum_i \left[\inf_{\omega \in A_i} f(\omega) \right] \mu(A_i) \le \sum_j \left[\inf_{\omega \in B_j} f(\omega) \right] \mu(B_j)$$

if $\{B_j\}$ refines $\{A_i\}$. Prove a dual relation for the sums in (15.9) and conclude that $\int_* f \, d\mu \le \int^* f \, d\mu$.
(b) Suppose that M bounds f and consider the partition $A_i = [\omega: i\epsilon \le f(\omega) < (i + 1)\epsilon]$, $i = 1, \ldots, N$, where N is large enough that $N\epsilon > M$. Show that

$$\sum_i \left[\sup_{\omega \in A_i} f(\omega) \right] \mu(A_i) - \sum_i \left[\inf_{\omega \in A_i} f(\omega) \right] \mu(A_i) \le \epsilon \mu(\Omega).$$

Conclude that

(15.10) $$\int_* f \, d\mu = \int^* f \, d\mu.$$

To define the integral as the common value in (15.10) is the *Darboux-Young* approach. The advantage of (15.3) as a definition is that it applies at once to unbounded f and infinite μ.

15.3.   3.3   15.2 ↑ For $A \subset \Omega$, define $\mu^*(A)$ and $\mu_*(A)$ by (3.9) and (3.10) with μ in place of P. Suspend for the moment the assumption that f is measurable and show that $\int^* I_A \, d\mu = \mu^*(A)$ and $\int_* I_A \, d\mu = \mu_*(A)$ for every A. Hence (15.10) can fail if f is not measurable. (Where was measurability used in the proof of (15.10)?) Even though the definition (15.3) makes formal sense for all nonnegative functions, it is thus idle to apply it to nonmeasurable ones.

Assume again that f is measurable (as well as nonnegative and bounded, and that μ is finite).

15.4. ↑ Show that for positive ϵ there exists a finite partition $\{A_i\}$ such that, if $\{B_j\}$ is any finer partition and $\omega_j \in B_j$, then

$$\left| \int f \, d\mu - \sum_j f(\omega_j) \mu(B_j) \right| < \epsilon.$$

15.5. ↑ Show that

$$\int f \, d\mu = \lim_n \sum_{k=1}^{n2^n} \frac{k-1}{2^n} \mu\left[\omega: \frac{k-1}{2^n} \le f(\omega) < \frac{k}{2^n} \right].$$

The limit on the right here is *Lebesgue's* definition of the integral.

15.6. ↑ Suppose that the integral is *defined* for simple functions by $\int (\sum_i x_i I_{A_i}) \, d\mu = \sum_i x_i \mu(A_i)$. Suppose that f_n and g_n are simple and nondecreasing and have a common limit: $0 \le f_n \uparrow f$ and $0 \le g_n \uparrow f$. Adapt the arguments used to prove Theorem 15.1(iii) and show that $\lim_n \int f_n \, d\mu = \lim_n \int g_n \, d\mu$. Thus $\int f \, d\mu$ can (Theorem 13.5) consistently be defined as $\lim_n \int f_n \, d\mu$ for simple functions for which $0 \le f_n \uparrow f$.

SECTION 16.   PROPERTIES OF THE INTEGRAL

Equalities and Inequalities

By definition, the requirement for integrability of f is that $\int f^+ \, d\mu$ and $\int f^-\, d\mu$ both be finite, which is the same as the requirement that $\int f^+ \, d\mu + \int f^- \, d\mu < \infty$ and hence the same as the requirement that $\int (f^+ + f^-) \, d\mu < \infty$ (Theorem 15.1(iv)). Since $f^+ + f^- = |f|$, f is integrable if and only if

$$(16.1) \qquad\qquad \int |f| \, d\mu < \infty.$$

It follows that if $|f| \leq |g|$ almost everywhere and g is integrable, then f is integrable as well. If $\mu(\Omega) < \infty$, a bounded f is integrable.

Theorem 16.1.   (i)   *Monotonicity: If f and g are integrable and $f \leq g$ almost everywhere, then*

$$(16.2) \qquad\qquad \int f \, d\mu \leq \int g \, d\mu.$$

(ii)   *Linearity: If f and g are integrable and α, β are finite real numbers, then $\alpha f + \beta g$ is integrable and*

$$(16.3) \qquad\qquad \int (\alpha f + \beta g) \, d\mu = \alpha \int f \, d\mu + \beta \int g \, d\mu.$$

PROOF OF (i).   For nonnegative f and g such that $f(\omega) \leq g(\omega)$ for all ω, (16.2) follows by Theorem 15.1; if $f(\omega) \leq g(\omega)$ merely holds almost everywhere, (16.2) still holds because of Theorem 15.2(iv). And for general f and g, if $f \leq g$ almost everywhere, then $f^+ \leq g^+$ almost everywhere and $f^- \geq g^-$ almost everywhere, and so (16.2) follows via the definition (15.6).

PROOF OF (ii).   First, $\alpha f + \beta g$ is integrable because, by Theorem 15.1, $\int |\alpha f + \beta g| \, d\mu \leq \int (|\alpha| \cdot |f| + |\beta| \cdot |g|) \, d\mu = |\alpha| \int |f| \, d\mu + |\beta| \int |g| \, d\mu < \infty$. By Theorem 15.1(iv) and the definition (15.6), $\int (\alpha f) \, d\mu = \alpha \int f \, d\mu$—consider separately the cases $\alpha \geq 0$ and $\alpha \leq 0$. Therefore, it is enough to check (16.3) for the case $\alpha = \beta = 1$. By definition, $(f + g)^+ - (f + g)^- = f + g = f^+ - f^- + g^+ - g^-$ and therefore $(f + g)^+ + f^- + g^- = (f + g)^- + f^+ + g^+$. All these functions being nonnegative, $\int (f + g)^+ \, d\mu + \int f^- \, d\mu + \int g^- \, d\mu = \int (f + g)^- \, d\mu + \int f^+ \, d\mu + \int g^+ \, d\mu$, which can be rearranged to give $\int (f + g)^+ \, d\mu - \int (f + g)^- \, d\mu = \int f^+ \, d\mu - \int f^- \, d\mu + \int g^+ \, d\mu - \int g^- \, d\mu$. But this reduces to (16.3).   ∎

Since $-|f| \leq f \leq |f|$, it follows by Theorem 16.1 that

(16.4)
$$\left| \int f \, d\mu \right| \leq \int |f| \, d\mu$$

for integrable f. Applying this to integrable f and g gives

(16.5)
$$\left| \int f \, d\mu - \int g \, d\mu \right| \leq \int |f - g| \, d\mu.$$

Example 16.1. Suppose that Ω is countable, that $\mathcal{F}$ consists of all the subsets of Ω, and that μ is counting measure: each one-point set has measure 1. To be definite, take $\Omega = \{1, 2, \ldots\}$. A function is then a sequence $x_1, x_2, \ldots$. If x_{nm} is x_m or 0 as $m \leq n$ or $m > n$, the function corresponding to $x_{n1}, x_{n2}, \ldots$ has integral $\Sigma_{m=1}^{n} x_m$ by Theorem 15.1(i) (consider the decomposition $\{1\}, \ldots, \{n\}$, $\{n + 1, n + 2, \ldots\}$). It follows by Theorem 15.1(iii) that in the nonnegative case the integral of the function given by $\{x_m\}$ is the sum $\Sigma_m x_m$ (finite or infinite) of the corresponding infinite series. In the general case the function is integrable if and only if $\Sigma_{m=1}^{\infty} |x_m|$ is a convergent infinite series, in which case the integral is $\Sigma_{m=1}^{\infty} x_m^{+} - \Sigma_{m=1}^{\infty} x_m^{-}$.

The function $x_m = (-1)^{m+1} m^{-1}$ is not integrable by this definition and even fails to have a definite integral, since $\Sigma_{m=1}^{\infty} x_m^{+} = \Sigma_{m=1}^{\infty} x_m^{-} = \infty$. This invites comparison with the ordinary theory of infinite series, according to which the alternating harmonic series does converge in the sense that $\lim_M \Sigma_{m=1}^{M}$ $(-1)^{m+1} m^{-1} = \log 2$. But since this says that the sum of the *first* M terms has a limit, it requires that the elements of the space Ω be ordered. If Ω consists not of the positive integers but, say, of the integer lattice points in 3-space, it has no canonical linear ordering. And if $\Sigma_m x_m$ is to have the same finite value no matter what the order of summation, the series must be absolutely convergent.* This helps to explain why f is defined to be integrable only if $\int f^{+} \, d\mu$ and $\int f^{-} \, d\mu$ are both finite. ∎

Example 16.2.   In connection with Example 15.1, consider the function $f = 3I_{(a,\infty)} - 2I_{(-\infty,a)}$. There is no natural value for $\int f \, d\lambda$ (it is "$\infty - \infty$"), and none is assigned by the definition.

If f is bounded on bounded intervals, then each function $f_n = fI_{(-n,n)}$ is integrable with respect to λ. Since $f = \lim_n f_n$, the limit of $\int f_n \, d\lambda$, if it exists, is sometimes called the "principal value" of the integral of f. Although it is natural for some purposes to integrate symmetrically about the origin, this is not the right definition of the integral in the context of general measure theory. The functions $g_n = fI_{(-n,n+1)}$ for example also converge to f, and $\int g_n \, d\lambda$ may have some other limit, or none at all; $f(x) = x$ is a case in point. There is no general reason why f_n should take precedence over g_n.

* RUDIN, p. 76.

As in the preceding example, $f = \sum_{k=1}^{\infty} (-1)^k k^{-1} I_{(k,k+1)}$ has no integral, even though the $\int f_n \, d\lambda$ above converge. ∎

Integration to the Limit

The first result, the *monotone convergence theorem,* merely restates Theorem 15.1(iii).

Theorem 16.2 *If $0 \le f_n \uparrow f$ almost everywhere, then $\int f_n \, d\mu \uparrow \int f \, d\mu$.*

As the functions are nonnegative, all the integrals exist. The conclusion of the theorem is that $\lim_n \int f_n \, d\mu$ and $\int f \, d\mu$ are both infinite or both finite and in the latter case are equal.

Example 16.3. Consider the space $\{1, 2, \ldots\}$ together with counting measure, as in Example 16.1. If for each m, $0 \le x_{nm} \uparrow x_m$ as $n \to \infty$, then $\lim_n \sum_m x_{nm} = \sum_m x_m$. ∎

Example 16.4. If μ is a measure on $\mathscr{F}$ and $\mathscr{F}_0$ is a σ-field contained in $\mathscr{F}$, then the restriction μ_0 of μ to $\mathscr{F}_0$ is another measure (Example 10.4). If $f = I_A$ and $A \in \mathscr{F}_0$, then

$$\int f \, d\mu = \int f \, d\mu_0,$$

the common value being $\mu(A) = \mu_0(A)$. The same is by linearity true for non-negative simple functions measurable $\mathscr{F}_0$. It holds by Theorem 16.2 for all nonnegative f measurable $\mathscr{F}_0$ because (Theorem 13.5) $0 \le f_n \uparrow f$ for simple functions f_n measurable $\mathscr{F}_0$. For functions measurable $\mathscr{F}_0$, integration with respect to μ is thus the same thing as integration with respect to μ_0. ∎

In this example a property was extended by linearity from indicators to nonnegative simple functions and thence to the general nonnegative function by a monotone passage to the limit. This is a technique of very frequent application.

Example 16.5. For a finite or infinite sequence of measures μ_n on $\mathscr{F}$, $\mu(A) = \sum_n \mu_n(A)$ defines another measure (countably additive because sums can be reversed in a nonnegative double series). For indicators f,

$$\int f \, d\mu = \sum_n \int f \, d\mu_n,$$

and by linearity the same holds for simple $f \ge 0$. If $0 \le f_k \uparrow f$ for simple f_k, then by Theorem 16.2 and Example 16.3, $\int f \, d\mu = \lim_k \int f_k \, d\mu = \lim_k \sum_n \int f_k \, d\mu_n$

$= \Sigma_n \lim_k \int f_k \, d\mu_n = \Sigma_n \int f \, d\mu_n$. The relation in question thus holds for all nonnegative f.　　　　　　　　　　　　　　　　　　　　　■

An important consequence of the monotone convergence theorem is *Fatou's lemma*:

Theorem 16.3.　*For nonnegative f_n,*

(16.6)　　　　　$$\int \liminf_n f_n \, d\mu \le \liminf_n \int f_n \, d\mu.$$

PROOF.　If $g_n = \inf_{k \ge n} f_k$, then $0 \le g_n \uparrow g = \liminf_n f_n$, and the preceding two theorems give $\int f_n \, d\mu \ge \int g_n \, d\mu \to \int g \, d\mu$.

Example 16.6.　On $(R^1, \mathcal{R}^1, \lambda)$, $f_n = n^2 I_{(0, n^{-1})}$ and $f \equiv 0$ satisfy $f_n(x) \to f(x)$ for each x, but $\int f \, d\lambda = 0$ and $\int f_n \, d\lambda = n \to \infty$. This shows that the inequality in (16.6) can be strict and that it is not always possible to integrate to the limit. This phenomenon has been encountered before; see Examples 5.6 and 7.7.　■

Fatou's lemma leads to *Lebesgue's dominated convergence theorem:*

Theorem 16.4.　*If $|f_n| \le g$ almost everywhere, where g is integrable, and if $f_n \to f$ almost everywhere, then f and the f_n are integrable and $\int f_n \, d\mu \to \int f \, d\mu$.*

PROOF.　Since $|f| \le g$ almost everywhere, f and all the f_n are integrable. Put $h_n = |f_n - f|$ and $h = 2g$. Then $h_n \to 0$ and $0 \le h_n \le h$ almost everywhere, h is integrable, and by (16.5) it suffices to prove that $\int h_n \, d\mu \to 0$. By Fatou's lemma, $\int h \, d\mu = \int \liminf_n (h - h_n) \, d\mu \le \liminf_n \int (h - h_n) \, d\mu = \int h \, d\mu - \limsup_n \int h_n \, d\mu$. Therefore, $\limsup_n \int h_n \, d\mu = 0$.　　　　　　■

Example 16.6 shows that this theorem can fail if no dominating g exists.

Example 16.7.　*The Weierstrass M-test for series.** Consider the space $\{1, 2, \ldots\}$ together with counting measure, as in Example 16.1. If $|x_{nm}| \le M_m$ and $\Sigma_m M_m < \infty$, and if $\lim_n x_{nm} = x_m$ for each m, then $\lim_n \Sigma_m x_{nm} = \Sigma_m x_m$. The function given by the sequence $M_1, M_2, \ldots$ plays the role of g.　■

The next result, the *bounded convergence theorem,* is a special case of Theorem 16.4. It contains Theorem 5.3 as a further special case.

Theorem 16.5.　*If $\mu(\Omega) < \infty$ and the f_n are uniformly bounded, then $f_n \to f$ almost everywhere implies that $\int f_n \, d\mu \to \int f \, d\mu$.*

* See the footnote on page 103.

The next two theorems are simply the series versions of the monotone and the dominated convergence theorems.

Theorem 16.6. *If $f_n \geq 0$, then $\int \Sigma_n f_n \, d\mu = \Sigma_n \int f_n \, d\mu$.*

The members of this last equation are both equal either to ∞ or to the same finite, nonnegative real number.

Theorem 16.7. *If $\Sigma_n f_n$ converges almost everywhere and $|\Sigma_{k=1}^n f_k| \leq g$ almost everywhere, where g is integrable, then $\Sigma_n f_n$ is integrable and $\int \Sigma_n f_n \, d\mu = \Sigma_n \int f_n \, d\mu$.*

Corollary. *If $\Sigma_n \int |f_n| \, d\mu < \infty$, then $\Sigma_n f_n$ converges almost everywhere and is integrable, and $\int \Sigma_n f_n \, d\mu = \Sigma_n \int f_n \, d\mu$.*

PROOF. Since $g = \Sigma_n |f_n|$ is integrable by Theorem 16.6, $\Sigma_n f_n$ converges almost everywhere and Theorem 16.7 applies. ∎

In place of a sequence $\{f_n\}$ of real measurable functions on $(\Omega, \mathcal{F}, \mu)$, consider a family $[f_t : t > 0]$ indexed by a continuous parameter t. Suppose of a measurable f that

$$(16.7) \qquad\qquad \lim_{t \to \infty} f_t(\omega) = f(\omega)$$

on a set A, where

$$(16.8) \qquad\qquad A \in \mathcal{F}, \qquad \mu(\Omega - A) = 0.$$

A technical point arises here, since $\mathcal{F}$ need not contain the ω-set where (16.7) holds:

Example 16.8. Let $\mathcal{F}$ consist of the Borel subsets of $\Omega = (0, 1]$, and let H be a nonmeasurable set—a subset of Ω that does not lie in $\mathcal{F}$ (see the end of Section 3). Define $f_t(\omega) = 1$ if ω equals the fractional part $t - [t]$ of t and their common value lies in H^c; define $f_t(\omega) = 0$ otherwise. Each f_t is measurable $\mathcal{F}$, but if $f(\omega) \equiv 0$, then the ω-set where (16.7) holds is exactly H. ∎

Because of such examples, the set A above must be assumed to lie in $\mathcal{F}$. (Because of Theorem 13.4, no such assumption is necessary in the case of sequences.) The assumption that (16.7) holds on a set A satisfying (16.8) can still be expressed as the assumption that (16.7) holds almost everywhere.

If $I_t = \int f_t \, d\mu$ converges to $I = \int f \, d\mu$ as $t \to \infty$, then certainly $I_{t_n} \to I$ for each sequence $\{t_n\}$ going to infinity. But the converse holds as well: if I_t does not converge to I, then there is a positive ϵ such that $|I_{t_n} - I| > \epsilon$ for a sequence $\{t_n\}$ going to infinity. To the question of whether I_{t_n} converges to I the previous theorems apply.

Suppose that (16.7) holds almost everywhere and that $|f_t| \leq g$ almost everywhere, where g is integrable. By the dominated convergence theorem, f must then be integrable and $I_{t_n} \to I$ for each sequence $\{t_n\}$ going to infinity. It follows that $\int f_t \, d\mu \to \int f \, d\mu$. In this result t could go continuously to 0 or to some other value instead of to infinity.

Theorem 16.8. *Suppose that $f(\omega, t)$ is a measurable function of ω for each t in (a, b).*

(i) *Suppose that $f(\omega, t)$ is almost everywhere continuous in t at t_0; suppose further that, for each t in some neighborhood of t_0, $|f(\omega, t)| \le g(\omega)$ almost everywhere, where g is integrable. Then $\int f(\omega, t)\mu(d\omega)$ is continuous in t at t_0.*

(ii) *Suppose that for $\omega \in A$, where A satisfies (16.8), $f(\omega, t)$ has in (a, b) a derivative $f'(\omega, t)$ with respect to t; suppose further that $|f'(\omega, t)| \le g(\omega)$ for $\omega \in A$ and $t \in (a, b)$, where g is integrable. Then $\int f(\omega, t)\mu(d\omega)$ has derivative $\int f'(\omega, t)\mu(d\omega)$ on (a, b).*

PROOF.   Part (i) is an immediate consequence of the preceding discussion. To prove part (ii), consider a fixed t. By the mean value theorem,

$$\frac{f(\omega, t + h) - f(\omega, t)}{h} = f'(\omega, s),$$

where s lies between t and $t + h$. The left side goes to $f'(\omega, t)$ as $h \to 0$ and is by the hypothesis dominated by the integrable function $g(\omega)$. Therefore,

$$\frac{1}{h}\left[\int f(\omega, t + h)\mu(d\omega) - \int f(\omega, t)\mu(d\omega) \right] \to \int f'(\omega, t)\mu(d\omega). \qquad \blacksquare$$

The condition involving g can be localized. It suffices to assume that for each t there is an integrable $g(\omega, t)$ such that $|f'(\omega, s)| \le g(\omega, t)$ for all s in some neighborhood of t.

Densities

The integral of f over a set A in $\mathcal{F}$ is defined by

$$(16.9) \qquad\qquad \int_A f \, d\mu = \int I_A f \, d\mu.$$

The definition applies if f is defined only on A in the first place (set $f = 0$ outside A). Notice that $\int_A f \, d\mu = 0$ if $\mu(A) = 0$.

All the concepts and theorems above carry over in an obvious way to integrals over A. Theorems 16.6 and 16.7 yield this result:

Theorem 16.9.   *If $A_1, A_2, \ldots$ are disjoint, and if f is either nonnegative or integrable, then $\int_{\cup_n A_n} f \, d\mu = \Sigma_n \int_{A_n} f \, d\mu$.*

Example 16.9.   The integrals (16.9) suffice to determine f. Suppose that f and g are integrable and $\int_A f \, d\mu = \int_A g \, d\mu$ for all $A \in \mathcal{F}$. From $\int_{f>g} (f - g) \, d\mu = 0$, it follows by Theorem 15.2 (i) that $\mu[f > g] = 0$. Similarly, $\mu[f < g] = 0$, and so $f = g$ almost everywhere.

Suppose that f and g are integrable and $\int_A f \, d\mu = \int_A g \, d\mu$ for A in a π-system $\mathcal{P}$ that generates $\mathcal{F}$. Then $\int_A (f^+ + g^-) \, d\mu = \int_A (f^- + g^+) \, d\mu$ for $A \in \mathcal{P}$ and hence (Theorem 10.3) for $A \in \mathcal{F}$. Therefore, $f^+ + g^- = f^- + g^+$ almost everywhere, and again $f = g$ almost everywhere.

The class of A for which $\int_A f \, d\mu \le \int_A g \, d\mu$ holds is a monotone class, and so if it holds for A in a field generating $\mathcal{F}$, then (Theorem 3.4) it holds for all A in $\mathcal{F}$, from which it follows that $f \le g$ almost everywhere. $\blacksquare$

Suppose that δ is a nonnegative function and define a measure ν by (Theorem 16.9)

$$(16.10) \qquad \nu(A) = \int_A \delta \, d\mu, \qquad A \in \mathcal{F};$$

δ is not assumed integrable with respect to μ. Many measures arise this way. Note that $\mu(A) = 0$ implies that $\nu(A) = 0$. Clearly, ν is finite if and only if δ is integrable μ. Another function δ' gives rise to the same ν if $\delta = \delta'$ almost everywhere. On the other hand, $\nu(A) = \int_A \delta' \, d\mu$ and (16.10) together imply that $\delta = \delta'$ almost everywhere if ν is finite, as was shown in Example 16.9. This is true also for σ-finite ν, as follows by decomposition of Ω into sets of finite ν-measure.

The measure ν defined by (16.10) is said to have *density* δ with respect to μ. A density is by definition nonnegative.

Formal substitution $d\nu = \delta \, d\mu$ gives formulas (16.11) and (16.12).

Theorem 16.10. *If ν has density δ with respect to μ, then*

$$(16.11) \qquad \int f \, d\nu = \int f\delta \, d\mu$$

holds for nonnegative f. Moreover, f (not necessarily nonnegative) is integrable with respect to ν if and only if $f\delta$ is integrable with respect to μ, in which case (16.11) and

$$(16.12) \qquad \int_A f \, d\nu = \int_A f\delta \, d\mu$$

both hold.

PROOF.   If $f = I_A$, then $\int f \, d\nu = \nu(A)$, so that (16.11) reduces to the definition (16.10). If f is a simple nonnegative function, (16.11) then follows by linearity. If f is nonnegative, $\int f_n \, d\nu = \int f_n \delta \, d\mu$ for the simple functions f_n of Theorem 13.5, and (16.11) follows by a monotone passage to the limit—that is, by Theorem 16.2. Note that both sides of (16.11) may be infinite.

Even if f is not nonnegative, (16.11) applies to $|f|$, whence it follows that f is integrable with respect to ν if and only if $f\delta$ is integrable with respect to μ. And if f is integrable, (16.11) follows from differencing the same result for f^+ and f^-. Replacing f by fI_A leads from (16.11) to (16.12).   ∎

This result has two features in common with a number of theorems about integration. (i) The relation in question, (16.11) in this case, in addition to holding for integrable functions, holds for all nonnegative functions—the point being that if one side of the equation is infinite, then so is the other, and if both are finite, then they have the same value. This is useful in checking for integrability in the first place. (ii) The result is proved first for indicator functions, then for simple functions, then for nonnegative functions, then for integrable functions. In this connection, see Examples 16.4 and 16.5.

The next result is *Scheffé's theorem.*

Theorem 16.11. *Suppose that* $\nu_n(A) = \int_A \delta_n \, d\mu$ *and* $\nu(A) = \int_A \delta \, d\mu$ *for densities* δ_n *and* δ. *If*

$$(16.13) \qquad \nu_n(\Omega) = \nu(\Omega) < \infty, \qquad n = 1, 2, \ldots,$$

and if $\delta_n \rightarrow \delta$ *except on a set of* μ-*measure* 0, *then*

$$(16.14) \qquad \sup_{A \in \mathcal{F}} |\nu(A) - \nu_n(A)| \le \int_\Omega |\delta - \delta_n| \, d\mu \rightarrow 0.$$

PROOF.   The inequality in (16.14) of course follows from (16.5). Let $g_n = \delta - \delta_n$. The positive part g_n^+ of g_n converges to 0 except on a set of μ-measure 0. Moreover, $0 \le g_n^+ \le \delta$ and δ is integrable, and so the dominated convergence theorem applies: $\int g_n^+ \, d\mu \rightarrow 0$. But $\int g_n \, d\mu = 0$ by (16.13), and therefore

$$\int_\Omega |g_n| \, d\mu = \int_{[g_n \ge 0]} g_n \, d\mu - \int_{[g_n < 0]} g_n \, d\mu$$

$$= 2 \int_{[g_n \ge 0]} g_n \, d\mu = 2 \int_\Omega g_n^+ \, d\mu \rightarrow 0. \qquad \blacksquare$$

A corollary concerning infinite series follows immediately—take μ as counting measure on $\Omega = \{1, 2, \ldots\}$.

Corollary.   *If* $\Sigma_m x_{nm} = \Sigma_m x_m < \infty$, *the terms being nonnegative, and if* $\lim_n x_{nm} = x_m$ *for each* m, *then* $\lim_n \Sigma_m |x_{nm} - x_m| = 0$. *If* y_m *is bounded, then* $\lim_n \Sigma_m y_m x_{nm} = \Sigma_m y_m x_m$.

Change of Variable

Let $(\Omega, \mathcal{F})$ and $(\Omega', \mathcal{F}')$ be measurable spaces, and suppose that the mapping $T: \Omega \rightarrow \Omega'$ is measurable $\mathcal{F} / \mathcal{F}'$. For a measure μ on $\mathcal{F}$, define a measure μT^{-1} on $\mathcal{F}'$ by

$$(16.15) \qquad \mu T^{-1}(A') = \mu(T^{-1}A'), \qquad A' \in \mathcal{F}',$$

as at the end of Section 13.

Suppose f is a real function on Ω' that is measurable $\mathcal{F}'$, so that the composition fT is a real function on Ω that is measurable $\mathcal{F}$ (Theorem 13.1(ii)). The change-of-variable formulas are (16.16) and (16.17). If $A' = \Omega'$, the second reduces to the first.

Theorem 16.12   *If f is nonnegative, then*

$$(16.16) \qquad \int_\Omega f(T\omega)\mu(d\omega) = \int_{\Omega'} f(\omega')\mu T^{-1}(d\omega').$$

A function f (not necessarily nonnegative) is integrable with respect to μT^{-1} if and only if fT is integrable with respect to μ, in which case (16.16) *and*

$$(16.17) \qquad \int_{T^{-1}A'} f(T\omega)\mu(d\omega) = \int_{A'} f(\omega')\mu T^{-1}(d\omega')$$

hold.

PROOF.   If $f = I_{A'}$, then $fT = I_{T^{-1}A'}$, and so (16.16) reduces to the definition (16.15). By linearity, (16.16) holds for nonnegative simple functions. If f_n are simple functions for which $0 \le f_n \uparrow f$, then $0 \le f_n T \uparrow fT$, and (16.16) follows by the monotone convergence theorem.

An application of (16.16) to $|f|$ establishes the assertion about integrability, and for integrable f, (16.16) follows by decomposition into positive and negative parts. Finally, if f is replaced by $fI_{A'}$, (16.16) reduces to (16.17).   ∎

Example 16.10.   Suppose that $(\Omega', \mathcal{F}') = (R^1, \mathcal{R}^1)$ and $T = \varphi$ is an ordinary real function, measurable $\mathcal{F}$. If $f(x) = x$, (16.16) becomes

$$(16.18) \qquad \int_\Omega \varphi(\omega)\mu(d\omega) = \int_{R^1} x\, \mu\varphi^{-1}(dx).$$

If $\varphi = \Sigma_i\, x_i I_{A_i}$ is simple, then $\mu\varphi^{-1}$ has mass $\mu(A_i)$ at x_i, and each side of (16.18) reduces to $\Sigma_i\, x_i\mu(A_i)$.   ∎

Uniform Integrability

If f is integrable, then $|f|I_{[|f|\ge \alpha]}$ goes to 0 almost everywhere as $\alpha \to \infty$ and is dominated by $|f|$, and hence

$$(16.19) \qquad \lim_{\alpha\to\infty} \int_{[|f|\ge\alpha]} |f|\, d\mu = 0.$$

A sequence $\{f_n\}$ is *uniformly integrable* if (16.19) holds uniformly in n:

$$(16.20) \qquad \lim_{\alpha\to\infty} \sup_n \int_{[|f_n|\ge\alpha]} |f_n|\, d\mu = 0.$$

Theorem 16.13.   *If $\mu(\Omega) < \infty$ and $f_n \to f$ almost everywhere for uniformly integrable f_n, then f is integrable and $\int f_n\, d\mu \to \int f\, d\mu$.*

PROOF. Since $\int |f_n| \, d\mu \leq \alpha\mu(\Omega) + \int_{[|f_n| \geq \alpha]} |f| \, d\mu$, it follows by (16.20) that $\int |f_n| \, d\mu$ is bounded. By Fatou's lemma f is thus integrable.

Put $h_n = |f_n - f|$. The h_n are uniformly integrable (use (16.19)) and $h_n \to 0$ almost everywhere. If $h_n^{(\alpha)} = h_n I_{[h_n < \alpha]}$, then for each α, $\lim_n h_n^{(\alpha)} = 0$ and so $\lim_n \int h_n^{(\alpha)} \, d\mu = 0$ by the bounded convergence theorem. Now

$$\int h_n \, d\mu = \int_{[h_n \geq \alpha]} h_n \, d\mu + \int h_n^{(\alpha)} \, d\mu.$$

Given ϵ, choose α so that the first term on the right is less than ϵ for all n; then choose n_0 so that the second term on the right is less than ϵ for $n \geq n_0$. This shows that $\int h_n \, d\mu \to 0$. ■

Suppose that

(16.21)
$$\sup_n \int |f_n|^{1+\epsilon} \, d\mu < \infty.$$

If K is the supremum here, then

$$\int_{[|f_n| \geq \alpha]} |f_n| \, d\mu \leq \frac{1}{\alpha^\epsilon} \int_{[|f_n| \geq \alpha]} |f_n|^{1+\epsilon} \, d\mu \leq \frac{K}{\alpha^\epsilon},$$

and so $\{f_n\}$ is uniformly integrable.

Complex Functions

A complex-valued function on Ω has the form $f(\omega) = g(\omega) + ih(\omega)$, where g and h are ordinary finite-valued real functions on Ω. It is, by definition, measurable $\mathscr{F}$ if g and h are. If g and h are integrable, then f is by definition integrable, and its integral is of course taken as

(16.22)
$$\int (g + ih) \, d\mu = \int g \, d\mu + i \int h \, d\mu.$$

Since $\max\{|g|, |h|\} \leq |f| \leq |g| + |h|$, f is integrable if and only if $\int |f| \, d\mu < \infty$, just as in the real case.

The linearity equation (16.3) extends to complex functions and coefficients—the proof requires only that everything be decomposed into real and imaginary parts. Consider the inequality (16.4): $|\int f \, d\mu| \leq \int |f| \, d\mu$. If $f = g + ih$ and g and h are simple, the corresponding partitions can be taken to be the same ($g = \Sigma_k x_k I_{A_k}$ and $h = \Sigma_k y_k I_{A_k}$), and (16.4) follows by the triangle inequality. For the general integrable f, represent g and h as limits of simple functions dominated by $|f|$ and pass to the limit.

The results on integration to the limit extend as well. Suppose that $f_k = g_k + ih_k$ are complex functions satisfying $\Sigma_k \int |f_k| \, d\mu < \infty$. Then $\Sigma_k \int |g_k| \, d\mu < \infty$, and so by the corollary to Theorem 16.7, $\Sigma_k g_k$ is integrable and integrates to $\Sigma_k \int g_k \, d\mu$. The same is true of the imaginary parts, and hence $\Sigma_k f_k$ is integrable and

(16.23)
$$\int \sum_k f_k \, d\mu = \sum_k \int f_k \, d\mu.$$

PROBLEMS

16.1. Suppose that $\mu(A) < \infty$ and $\alpha \le f \le \beta$ almost everywhere on A. Show that $\alpha\mu(A) \le \int_A f \, d\mu \le \beta\mu(A)$.

16.2. Deduce parts (i) and (ii) of Theorem 10.2 from Theorems 16.2 and 16.4, respectively. Deduce the left-hand inequality in (4.10) from Theorem 16.3.

16.3. If $\mu(\Omega) < \infty$ and $f_n \to f$ uniformly for integrable f_n and f, then $\int f_n \, d\mu \to \int f \, d\mu$ by Theorem 16.5. Deduce this directly from (16.5).

16.4. Suppose that μ is Lebesgue measure on the unit interval Ω and that $(a, b) = (0, 1)$ in Theorem 16.8. If $f(\omega, t)$ is 1 or 0 according as $\omega \le t$ or $\omega > t$, then for each t, $f'(\omega, t) = 0$ almost everywhere. But $\int f(\omega, t)\mu(d\omega)$ does not differentiate to 0. Why does this not contradict the theorem? Examine the proof for this case.

16.5. If $f_n = nI_{(0, n^{-1})} - nI_{(-n^{-1}, 0)}$, the f_n can be integrated over $(-1, 1)$ (Lebesgue measure) to the limit. Show that there is no integrable g satisfying $|f_n| \le g$ for all n and that the f_n are not uniformly integrable.

16.6. Suppose that functions a_n, b_n, f_n converge almost everywhere to functions a, b, f, respectively. Suppose that the first two sequences may be integrated to the limit—that is, the functions are all integrable and $\int a_n \, d\mu \to \int a \, d\mu$, $\int b_n \, d\mu \to \int b \, d\mu$. Suppose, finally, that the first two sequences enclose the third: $a_n \le f_n \le b_n$ almost everywhere. Show that the third may be integrated to the limit.

Hint: The dominated convergence theorem is no help because of Problem 16.5.

16.7. Show that, for f integrable and ϵ positive, there exists an integrable simple g such that $\int |f - g| \, d\mu < \epsilon$. If $\mathcal{A}$ is a field generating $\mathcal{F}$, g can be taken as $\Sigma_i \, x_i I_{A_i}$ with $A_i \in \mathcal{A}$.

Hint: Use Theorems 13.5, 16.4, and 11.4.

16.8. Suppose that $f(\omega, \cdot)$ is, for each ω, a function on an open set W in the complex plane and that $f(\cdot, z)$ is for z in W measurable $\mathcal{F}$. Suppose that A satisfies (16.8), that $f(\omega, \cdot)$ is analytic in W for ω in A, and that for each z_0 in W there is an integrable $g(\cdot, z_0)$ such that $|f(\omega, z)| \le g(\omega, z_0)$ for all $\omega \in A$ and all z in some neighborhood of z_0. Show that $\int f(\omega, z)\mu(d\omega)$ is analytic in W.

16.9. Suppose that f_n are integrable and $\sup_n \int f_n \, d\mu < \infty$. Show that, if $f_n \uparrow f$, then f is integrable and $\int f_n \, d\mu \to \int f \, d\mu$. This is *Beppo Levi's theorem*.

16.10. Suppose of integrable f_n that $\alpha_{mn} = \int |f_m - f_n| \, d\mu \to 0$ as $m, n \to \infty$. Then there exists an integrable f such that $\beta_n = \int |f - f_n| \, d\mu \to 0$. Prove this by the following steps.

   (a) Choose n_k so that $m, n \ge n_k$ implies that $\alpha_{mn} < 2^{-k}$. Use the corollary to Theorem 16.7 to show that $f = \lim_k f_{n_k}$ exists and is integrable and $\lim_k \int |f - f_{n_k}| \, d\mu = 0$.

   (b) For $n \ge n_k$, write $\int |f - f_n| \, d\mu \le \int |f - f_{n_k}| \, d\mu + \int |f_{n_k} - f_n| \, d\mu$.

16.11. Show that, if μ is σ-finite and δ is finite except on a set of μ-measure 0, then the ν defined by (16.10) is σ-finite. Show that if ν is not σ-finite, then δ may not be unique even up to sets of μ-measure 0.

16.12. Show that, if $\{f_n\}$ is uniformly integrable and $\mu(\Omega) < \infty$, then for each ϵ there is a δ such that $\mu(A) < \delta$ implies that $\int_A |f_n|\, d\mu < \epsilon$ for all n.

16.13. Suppose that $\mu(\Omega) < \infty$, f and f_n are integrable, $0 \leq f_n \to f$, and $\int f_n\, d\mu \to \int f\, d\mu$. Show that the f_n are uniformly integrable.

16.14. Show that, if $\mu(\Omega) < \infty$ and $|f_n| \leq g$ for an integrable g, then the f_n are uniformly integrable. Compare the hypotheses of Theorems 16.4 and 16.13.

16.15. $\uparrow$ On the unit interval with Lebesgue measure, let $f_n = I_{(0,n^{-1})}n/\log n$. Then $f_n \to 0$. Show that the f_n are uniformly integrable and $\int f_n\, d\mu \to 0$, although no integrable function dominates the f_n.

16.16. Generalize the condition (16.21): If $\lim_{x\to\infty} \varphi(x)/x = \infty$ and $\int \varphi(|f_n|)\, d\mu$ is bounded, then $\{f_n\}$ is uniformly integrable.

16.17. If $\{f_n\}$ and $\{g_n\}$ are uniformly integrable, then so is $\{\alpha f_n + \beta g_n\}$.

16.18. (a) Suppose that $f_1, f_2, \ldots$ are nonnegative and put $g_n = \max\{f_1, \ldots, f_n\}$. Show that $\int_{g_n \geq \alpha} g_n\, d\mu \leq \sum_{k=1}^{n} \int_{f_k \geq \alpha} f_k\, d\mu$.
(b) Suppose further that $\{f_n\}$ is uniformly integrable and $\mu(\Omega) < \infty$. Show that $\int g_n\, d\mu = o(n)$.

16.19. Show by example that Theorem 16.13 is false without the hypothesis $\mu(\Omega) < \infty$.

16.20. Let f be a complex-valued function integrating to $re^{i\theta}$, $r \geq 0$. From $\int (|f(\omega)| - e^{-i\theta}f(\omega))\mu(d\omega) = \int |f|\, d\mu - r$, deduce (16.4).

SECTION 17.   INTEGRAL WITH RESPECT TO LEBESGUE MEASURE

The Lebesgue Integral on the Line

A real measurable function on the line is *Lebesgue integrable* if it is integrable with respect to Lebesgue measure λ, and its *Lebesgue integral* $\int f\, d\lambda$ is denoted by $\int f(x)\, dx$ or, in case of integration over an interval, by $\int_a^b f(x)\, dx$. It is instructive to compare it with the Riemann integral.

The Riemann Integral

A real function f on an interval $(a, b]$ is by definition *Riemann integrable*, with integral r, if this condition holds: For each ϵ there exists a δ with the property that

$$(17.1) \qquad\qquad \left| r - \sum_j f(x_j)\lambda(J_j) \right| < \epsilon$$

if $\{J_j\}$ is any finite partition of $(a, b]$ into subintervals satisfying $\lambda(J_j) < \delta$ and if $x_j \in J_j$ for each j. The Riemann integral for step functions was used in Section 1.

Suppose that f is Borel measurable, and suppose that f is bounded, so that it is Lebesgue integrable. If f is also Riemann integrable, the r of (17.1) must coincide with the Lebesgue integral $\int_a^b f(x)\,dx$. To see this, first note that letting x_j vary over J_j leads from (17.1) to

$$(17.2) \qquad \left| r - \sum_j \sup_{x \in J_j} f(x) \cdot \lambda(J_j) \right| \leq \epsilon.$$

Consider the simple function g with value $\sup_{x \in J_j} f(x)$ on J_j. Now $f \leq g$, and the sum in (17.2) is the Lebesgue integral of g. By monotonicity of the Lebesgue integral, $\int_a^b f(x)\,dx \leq \int_a^b g(x)\,dx \leq r + \epsilon$. The reverse inequality follows in the same way, and so $\int_a^b f(x)\,dx = r$. *Therefore, the Riemann integral when it exists coincides with the Lebesgue integral.*

Suppose that f is continuous on $[a, b]$. By uniform continuity, for each ϵ there exists a δ such that $|f(x) - f(y)| < \epsilon/(b - a)$ if $|x - y| < \delta$. If $\lambda(J_j) < \delta$ and $x_j \in J_j$, then $g = \Sigma_j f(x_j) I_{J_j}$ satisfies $|f - g| < \epsilon/(b - a)$ and hence $|\int_a^b f\,dx - \int_a^b g\,dx| < \epsilon$. But this is (17.1) with r replaced (as it must be) by the Lebesgue integral $\int_a^b f\,dx$: *A continuous function is Riemann integrable.*

Example 17.1. If f is the indicator of the set of rationals in $(0, 1]$, then the Lebesgue integral $\int_0^1 f(x)\,dx$ is 0 because $f = 0$ almost everywhere. But for an arbitrary partition $\{J_j\}$ of $(0, 1]$ into intervals, $\Sigma_j f(x_j)\lambda(J_j)$ for $x_j \in J_j$ is 1 if each x_j is taken from the rationals and 0 if each x_j is taken from the irrationals. Thus f is not Riemann integrable. ∎

Example 17.2. For the f of Example 17.1, there exists a g (namely, $g \equiv 0$) such that $f = g$ almost everywhere and g is Riemann integrable. To show that the Lebesgue theory is not reducible to the Riemann theory by the casting out of sets of measure 0, it is of interest to produce an f (bounded and Borel measurable) for which no such g exists.

In Examples 3.1 and 3.2 there were constructed Borel subsets A of $(0, 1]$ such that $0 < \lambda(A) < 1$ and such that $\lambda(A \cap J) > 0$ for each subinterval J of $(0, 1]$. Take $f = I_A$. Suppose that $f = g$ almost everywhere and that $\{J_j\}$ is a decomposition of $(0, 1]$ into subintervals. That g is not Riemann integrable will follow if the J_j are shown to contain points x_j and y_j such that

$$(17.3) \qquad \sum_j g(x_j)\lambda(J_j) \leq \lambda(A) < 1 = \sum_j g(y_j)\lambda(J_j).$$

In case $\lambda(J_j - A) = 0$, choose from $J_j \cap [f = g]$ (a set of measure $\lambda(J_j) > 0$) any point x_j; then $g(x_j) = f(x_j) \leq 1$. In the opposite case, choose from $[f = g] \cap (J_j - A)$ (a set of measure $\lambda(J_j - A) > 0$) any point x_j; then $g(x_j) = f(x_j) = 0$. If Σ' denotes summation over those j for which $\lambda(J_j - A) = 0$, then the left-hand sum in (17.3) is $\Sigma' g(x_j)\lambda(J_j) \leq \Sigma' \lambda(J_j) = \Sigma' \lambda(J_j \cap A) \leq \lambda(A)$. (The argument thus far applies to any A.)

To construct the y_j, note first that $\lambda(J_j \cap A \cap [f = g]) = \lambda(J_j \cap A) > 0$ by the construction of A. Thus $g(y_j) = f(y_j) = 1$ for some y_j in $J_j \cap A$, from which the right-hand equality in (17.3) follows. (It is actually enough just to establish this last equality, because the Riemann integral of g, if it had one, would be $\lambda(A)$.)   ∎

It is because of their extreme oscillations that the functions in Examples 17.1 and 17.2 fail to be Riemann integrable. (It can be shown that a function is Riemann integrable if and only if the set of its discontinuities has Lebesgue measure 0.*) This cannot happen in the case of the Lebesgue integral of a measurable function: if f fails to be Lebesgue integrable, it is because its positive part or its negative part is too large, not because one or the other is too irregular.

Example 17.3.   It is an important analytic fact that

(17.4) $$\lim_{t \to \infty} \int_0^t \frac{\sin x}{x}\, dx = \frac{\pi}{2}.$$

The existence of the limit is simple to prove because $\int_{(n-1)\pi}^{n\pi} x^{-1} \sin x\, dx$ converges to 0 and alternates in sign; the value of the limit will be identified in the next section (Example 18.3). On the other hand, $x^{-1} \sin x$ is not Lebesgue integrable over $(0, \infty)$ because its positive and negative parts integrate to ∞. Within the conventions of the Lebesgue theory (17.4) thus cannot be written $\int_0^\infty x^{-1} \sin x\, dx = \pi/2$—although such "improper" integrals appear in calculus texts. It is, of course, just a question of choosing the terminology most convenient for the subject at hand.   ∎

The function in Example 17.2 is not equal almost everywhere to any Riemann integrable function. Every Lebesgue integrable function can, however, be approximated in a certain sense by Riemann integrable functions of two kinds.

Theorem 17.1.   *Suppose that $\int |f|\,dx < \infty$ and $\epsilon > 0$.*

(i)   *There is a step function $g = \Sigma_{i=1}^k x_i I_{A_i}$, with bounded intervals as the A_i, such that $\int |f - g|\,dx < \epsilon$.*

(ii)   *There is a continuous h such that $\int |f - h|\,dx < \epsilon$.*

PROOF.   By the construction (13.6) and the dominated convergence theorem, (i) holds if the A_i are not required to be intervals; moreover, $\lambda(A_i) < \infty$ for each i for which $x_i \neq 0$. By Theorem 11.4 there exists a finite disjoint union B_i of intervals such that $\lambda(A_i \Delta B_i) < \epsilon/k|x_i|$. But then $\Sigma_i x_i I_{B_i}$ satisfies the requirements of (i) with 2ϵ in place of ϵ.

* See Problems 17.4 and 25.15.

To prove (ii) it is only necessary to show that for the g of (i) there is a continuous h such that $\int |g - h| \, dx < \epsilon$. Suppose that $A_i = (a_i, b_i]$; let $h_i(x)$ be 1 on $(a_i, b_i]$ and 0 outside $(a_i - \delta, b_i + \delta]$, and let it increase linearly from 0 to 1 over $(a_i - \delta, a_i]$ and decrease linearly from 1 to 0 over $(b_i, b_i + \delta]$. Since $\int |I_{A_i} - h_i| \, dx \to 0$ as $\delta \to 0$, $h = \Sigma_i \, x_i h_i$ for sufficiently small δ will satisfy the requirements. ∎

The Lebesgue integral is thus determined by its values for continuous functions.*

The Fundamental Theorem of Calculus

For positive h,

$$\left| \frac{1}{h} \int_x^{x+h} f(y) \, dy - f(x) \right| \leq \frac{1}{h} \int_x^{x+h} |f(y) - f(x)| \, dx$$

$$\leq \sup \, [|f(y) - f(x)| : x \leq y \leq x + h],$$

and the right side goes to 0 with h if f is continuous at x. The same thing holds for negative h,[†] and therefore $\int_a^x f(y) \, dy$ has derivative $f(x)$:

(17.5) $$\frac{d}{dx} \int_a^x f(y) \, dy = f(x)$$

if f is continuous at x.

Suppose that F is a function with continuous derivative $F' = f$; that is, suppose that F is a *primitive* of the continuous function f. Then

(17.6) $$\int_a^b f(x) \, dx = \int_a^b F'(x) \, dx = F(b) - F(a),$$

as follows from the fact that $F(x)$ and $\int_a^x f(y) \, dy$ have by (17.5) identical derivatives. For continuous f, (17.5) and (17.6) are two ways of stating the fundamental theorem of calculus. To the calculation of Lebesgue integrals the methods of elementary calculus thus apply.

As will follow from the general theory of derivatives in Section 31, (17.5) holds almost everywhere if f is integrable—it need not be continuous. As the following example shows, however, (17.6) can fail for discontinuous f.

Example 17.4. Define $F(x) = x^2 \sin x^{-2}$ for $0 \leq x \leq \frac{1}{2}$ and $F(x) = 0$ for $x \leq 0$ and for $x \geq 1$. Now for $\frac{1}{2} < x < 1$ define $F(x)$ in such a way that F is continuously differentiable over $(0, \infty)$. Then F is everywhere differentiable, but $F'(0) = 0$ and $F'(x) = 2x \sin x^{-2} - 2x^{-1} \cos x^{-2}$ for $0 < x < \frac{1}{2}$. Thus F' is discontinuous at 0; F' is, in fact, not even integrable over $(0, 1]$, which makes (17.6) impossible for $a = 0$.

For a more extreme example, decompose $(0, 1]$ into countably many subintervals $(a_n,$

---

* This provides another way of *defining* the Lebesgue integral on the line. See RIESZ & SZ.-NAGY, Chapter 2.

† This requires defining $\int_u^v = - \int_v^u$ for $u > v$.

b_n]. Define $G(x) = 0$ for $x \leq 0$ and $x \geq 1$, and on $(a_n, b_n]$ define $G(x) = F((x - a_n)/(b_n - a_n))$. Then G is everywhere differentiable but (17.6) is impossible for G if $(a, b]$ contains any of the $(a_n, b_n]$ because G is not integrable over any of them. ∎

Change of Variable

Suppose that T is a continuous, increasing mapping of (a, b) onto (Ta, Tb). If T has a continuous derivative T', then the formal substitutions $y = T(x)$ and $dy = T'(x) \, dx$ give the usual formula for changing variables:

$$(17.7) \qquad \int_a^b f(T(x)) T'(x) \, dx = \int_{Ta}^{Tb} f(y) \, dy.$$

There is a more general formula. If T has a continuous derivative on (a, b) and is either increasing or decreasing, then

$$(17.8) \qquad \int_A f(T(x)) |T'(x)| \, dx = \int_{TA} f(y) \, dy$$

for Borel subsets A of (a, b). To prove this, define $\mu(A) = \int_A |T'(x)| \, dx$ for $A \subset (a, b)$. If B is a subinterval of $T(a, b)$, then

$$(17.9) \qquad \mu(T^{-1}B) = \int_{T^{-1}B} |T'(x)| \, dx = \lambda(B),$$

as follows by the fundamental theorem of calculus (consider increasing and decreasing T separately). By Theorem 10.3, (17.9) holds for all Borel sets B in $T(a, b)$. By the general change-of-variable formula (16.17), $\int_{T^{-1}B} f(Tx) \mu(dx) = \int_B f(y) \, dy$. Putting $B = TA$ (T is one-to-one) and transforming the left side by the formula (16.12) for integration with respect to a density gives (17.8).

Note that T' may vanish and T^{-1} may be nondifferentiable at places, for example if $T(x) = x^3$. The intervals (a, b) and $T(a, b)$ may be infinite. Finally, (17.8) holds for all nonnegative f as well as for integrable f.

Example 17.5. Put $T(x) = \sin x / \cos x$ on $(-\pi/2, \pi/2)$. Then $T'(x) = 1 + T^2(x)$, and (17.7) for $f(y) = (1 + y^2)^{-1}$ gives

$$(17.10) \qquad \int_{-\infty}^{\infty} \frac{dy}{1 + y^2} = \pi.$$ ∎

The Lebesgue Integral in R^k

The k-dimensional Lebesgue integral, the integral in $(R^k, \mathcal{R}^k, \lambda_k)$, is denoted $\int f(x) \, dx$, it being understood that $x = (x_1, \ldots, x_k)$. In low-dimensional cases it is also denoted $\iint_A f(x_1, x_2) \, dx_1 \, dx_2$, and so on.

A mapping $T: V \to R^k$ carrying a set V in R^k into R^k has the form $T(x) =$

$(t_1(x), \ldots, t_k(x))$. Suppose that the partial derivatives $t_{ij} = \partial t_i/\partial x_j$ exist, let D_x be the $k \times k$ matrix with entries $t_{ij}(x)$, and let $J(x)$ denote the *Jacobian*:

$$(17.11) \qquad J(x) = \det D_x = \det [t_{ij}(x)].$$

Theorem 17.2. *Let $T: V \to TV$ be a one-to-one mapping of an open set V onto an open set TV. Suppose that T is continuous and that T has continuous partial derivatives t_{ij}. Then*

$$(17.12) \qquad \int_V f(Tx)|J(x)| \, dx = \int_{TV} f(y) \, dy.$$

As usual this holds for nonnegative f and for f integrable over TV. Replacing f by fI_{TA} for $A \subset V$ gives

$$(17.13) \qquad \int_A f(Tx)|J(x)| \, dx = \int_{TA} f(y) \, dy.$$

In the case $k = 1$, this reduces to (17.8).

If T is linear, D_x is for each x the matrix of the transformation. If T is identified with this matrix, (17.13) reduces to

$$(17.14) \qquad |\det T| \cdot \int_A f(Tx) \, dx = \int_{TA} f(y) \, dy.$$

Only this special case will be proved here, because Theorem 19.3 contains Theorem 17.2.* Suppose that T is linear and nonsingular and $V = R^k$. It is enough to prove (17.14) for $A = R^k$. If f is an indicator, this reduces to (12.2), and it follows in the usual way that it holds for simple f, then for nonnegative f, and finally for the general f.

Example 17.6. If $T(r, \theta) = (r \cos \theta, r \sin \theta)$, then the Jacobian (17.11) is $J(r, \theta) = r$, which gives the formula for integrating in polar coordinates:

$$(17.15) \qquad \iint_{\substack{r>0 \\ 0<\theta<2\pi}} f(r \cos \theta, r \sin \theta) r \, dr \, d\theta = \iint_{R^2} f(x, y) \, dx \, dy.$$

Stieltjes Integrals

Suppose that F is a function on R^k satisfying the hypotheses of Theorem 12.5, so that there exists a measure μ such that $\mu(A) = \Delta_A F$ for bounded rectangles A. In integrals with respect to μ, $\mu(dx)$ is often replaced by $dF(x)$:

---

* The proof in Section 19 uses properties of Hausdorff measure. For an alternative argument, see, for example, RUDIN', p. 184.

(17.16) $$\int_A f(x)\, dF(x) = \int_A f(x)\mu(dx).$$

The left side of this equation is the *Stieltjes integral* of f with respect to F; since it is defined by the right side of the equation, nothing new is involved.

Suppose that f is uniformly continuous on a rectangle A, and suppose that A is decomposed into rectangles A_m small enough that $|f(x) - f(y)| < \epsilon/\mu(A)$ for $x, y \in A_m$. Then

$$\left| \int_A f(x)\, dF(x) - \sum_m f(x_m)\Delta_{A_m}F \right| < \epsilon$$

for $x_m \in A_m$. In this case (17.16) can be defined as the limit of these approximating sums without any reference to the general theory of measure and for historical reasons is sometimes called the *Riemann-Stieltjes* integral; (17.16) for the general f is then called the *Lebesgue-Stieltjes* integral. Since these distinctions are unimportant in the context of general measure theory, $\int f(x)\, dF(x)$ and $\int f\, dF$ are best regarded as merely notational variants for $\int f(x)\mu(dx)$ and $\int f\, d\mu$.

PROBLEMS

17.1. Extend Theorem 17.1 to R^k.

17.2. The Lebesgue integral as well as Lebesgue measure is translation-invariant. State this precisely and prove it.

17.3. Suppose that μ is a finite measure on $\mathscr{R}^k$ and A is closed. Show that $\mu(x + A)$ is upper semicontinuous in x and hence measurable.

17.4. Let A_ϵ be the set of x in the unit interval for which

$$\sup_{|y-x|<\delta} f(y) - \inf_{|y-x|<\delta} f(y) \ge \epsilon$$

for every positive δ. Show that A_ϵ is closed relative to $(0, 1]$ and that f is discontinuous at x if and only if x lies in A_ϵ for some positive ϵ. Suppose that $(0, 1]$ is decomposed into intervals $(a_i, b_i]$. If (a_i, b_i) meets A_ϵ, choose x_i and x_i' in (a_i, b_i) in such a way that $f(x_i) - f(x_i') \ge \epsilon/2$; otherwise, take $x_i = x_i'$. Show that $\Sigma(f(x_i) - f(x_i'))(b_i - a_i) \ge \epsilon\lambda(A_\epsilon)/2$. Show that, if f is Riemann integrable, then it is continuous almost everywhere. See Problem 25.15.

17.5. A function f on $[0, 1]$ has *Kurzweil integral* I if for each ϵ there exists a positive function $\delta(\cdot)$ over $[0, 1]$ such that, if $x_0, \dots, x_n$ and $\xi_1, \dots, \xi_n$ satisfy

(17.17) $$\begin{cases} 0 = x_0 < x_1 < \cdots < x_n = 1, \\ x_{i-1} \le \xi_i \le x_i,\; x_i - x_{i-1} < \delta(\xi_i), \qquad i = 1, \dots, n, \end{cases}$$

then

(17.18) $$\left| \sum_{k=1}^n f(\xi_i)(x_i - x_{i-1}) - I \right| \le \epsilon.$$

Riemann's definition is narrower because it requires the function $\delta(\cdot)$ to be constant. Show in fact by the following steps that, if a Borel function over $[0, 1]$ is Lebesgue integrable, then it has a Kurzweil integral, namely its Lebesgue integral.

(a) The Kurzweil integral being obviously additive, assume that $f \geq 0$. Choose k_0 so that $\int_{f \geq k_0} f\, dx < \epsilon$, and for $k = 1, 2, \ldots$ choose open sets G_k such that $G_k \supset A_k = [(k-1)\epsilon \leq f < k\epsilon]$ and $\lambda(G_k - A_k) < 1/kk_0^2 2^k$. Define $\delta(\xi) = \mathrm{dist}\,(\xi, G_k^c)$ for $\xi \in A_k$.

(b) Suppose that (17.17) holds. Note that $\xi_i \in A_k$ implies that $[x_{i-1}, x_i] \subset G_k$ and show that $\sum_{i=1}^n f(\xi_i)(x_i - x_{i-1}) \leq \int f\, dx + 2\epsilon$ by splitting the sum according to which A_k it is that ξ_i lies in.

(c) Let $\Sigma^{(k)}$ denote summation over the i for which $\xi_i \in A_k$, and show that $\Sigma^{(k)}(x_i - x_{i-1}) = 1 - \Sigma_{l \neq k}\Sigma^{(l)}(x_i - x_{i-1}) \geq \lambda(A_k) - k_0^{-2}$. Conclude that $\sum_{i=1}^n f(\xi_i)(x_i - x_{i-1}) \geq \int_{f < k_0} f\, dx - 2\epsilon \geq \int f\, dx - 3\epsilon$. Thus (17.17) implies (17.18) for $I = \int f\, dx$ (with 3ϵ in place of ϵ).

17.6. $\uparrow$ (a) Suppose that f is continuous on $(0, 1]$ and

$$(17.19) \qquad \lim_{t \to 0} \int_t^1 f(x)\, dx = I.$$

Choose η so that $t \leq \eta$ implies that $|I - \int_t^1 f\, dx| < \epsilon$; define $\delta_0(\cdot)$ over $(0, 1]$ in such a way that $0 < x, y \leq 1$ and $|x - y| < \delta_0(x)$ imply that $|f(x) - f(y)| < \epsilon$, and put

$$\delta(x) = \begin{cases} \min\left\{\dfrac{1}{2}x,\, \delta_0(x)\right\} & \text{if } 0 < x \leq 1, \\[2mm] \min\left\{\eta,\, \dfrac{\epsilon}{1 + |f(0)|}\right\} & \text{if } x = 0. \end{cases}$$

From (17.17) deduce that $\xi_1 = 0$ and then that (17.18) holds with 3ϵ in place of ϵ.

(b) Let F be the function in Example 17.4, and put $f = F'$ on $[0, 1]$. Show that f has a Kurzweil integral even though it is not Lebesgue integrable.

17.7. 17.5$\uparrow$ Use the Kurzweil integral to prove Theorem 17.1(i) for functions on $[0, 1]$.

17.8. 13.13$\uparrow$ Let f be a finite-valued Borel function over $(0, 1)$. By the following steps, prove *Lusin's theorem*: For each ϵ there exists a continuous function g such that $\lambda[x \in (0, 1): f(x) \neq g(x)] < \epsilon$.

(a) Show that f may be assumed integrable, or even bounded.

(b) By Theorem 17.1(ii), find continuous functions g_n such that $\int_0^1 |f - g_n|\, dx < n^{-2}$. Show that $g_n(x) \to f(x)$ except on a set of measure 0.

(c) Combine Egoroff's theorem and Theorem 12.3 to show that convergence is uniform on a compact set K such that $\lambda((0, 1) - K) < \epsilon$. The limit $\lim_n g_n(x) = f(x)$ must be continuous when restricted to K.

(d) Exhibit $(0, 1) - K$ as a disjoint union of open intervals I_k; define g as f on K and define it by linear interpolation on each I_k.

17.9. Let $f_n(x) = x^{n-1} - 2x^{2n-1}$. Calculate and compare $\int_0^1 \Sigma_{n=1}^\infty f_n(x)\, dx$ and $\Sigma_{n=1}^\infty$

$\int_0^1 f_n(x)\, dx$. Relate this to Theorem 16.6 and to Theorem 16.7 and its corollary.

17.10. Suppose that f_n are differentiable functions on (a, b) and $f_n(x) \to f(x)$ for all x. One asks for conditions under which f is differentiable and $f_n' \to f'$. If the f_n' are continuous and $f_n'(x) \to g(x)$ uniformly, the standard conditions, then g is continuous and Theorem 16.5 implies that $\int_{x_0}^x g(t)\, dt = \lim_n \int_{x_0}^x f_n'(t)\, dt = f(x) - f(x_0)$, so that $f'(x) = g(x)$ by the fundamental theorem of the calculus. Devise a more general theorem with Theorem 16.13, say, in place of Theorem 16.5.

17.11. Show that $(1 + y^2)^{-1}$ has equal integrals over $(-\infty, -1)$, $(-1, 0)$, $(0, 1)$, $(1, \infty)$. Conclude from (17.10) that $\int_0^1 (1 + y^2)^{-1}\, dy = \pi/4$. Expand the integrand in a geometric series and deduce Leibniz's formula

$$\frac{\pi}{4} = 1 - \frac{1}{3} + \frac{1}{5} - \frac{1}{7} + \cdots$$

by Theorem 16.7 (note that its corollary does not apply).

17.12. Take $T(x) = \sin x$ and $f(y) = (1 - y^2)^{-1/2}$ in (17.7) and show that $t = \int_0^{\sin t} (1 - y^2)^{-1/2}\, dy$ for $0 \le t \le \pi/2$. From Newton's binomial formula and two applications of Theorem 16.6, conclude that

$$(17.20) \qquad \frac{\pi^2}{8} = \sum_{n=0}^{\infty} \binom{-\frac{1}{2}}{n} (-1)^n \frac{1}{2n+1} \int_0^{\pi/2} (\sin t)^{2n+1}\, dt.$$

17.13. If T has a continuous derivative on $[a, b]$ and f is a bounded Borel function on an interval containing the range of T, then (17.7) holds.

(a) Prove this for continuous f by replacing b by a variable upper limit and differentiating.

(b) Prove it for bounded f by exhibiting f as a limit (almost everywhere) of uniformly bounded continuous functions (see part (b) of Problem 17.8).

SECTION 18. PRODUCT MEASURE AND FUBINI'S THEOREM

Let $(X, \mathcal{X})$ and $(Y, \mathcal{Y})$ be measurable spaces. For given measures μ and ν on these spaces, the problem is to construct on the cartesian product $X \times Y$ a *product measure* π such that $\pi(A \times B) = \mu(A)\nu(B)$ for $A \subset X$ and $B \subset Y$. In the case where μ and ν are Lebesgue measure on the line, π will be Lebesgue measure in the plane. The main result is *Fubini's theorem*, according to which double integrals can be calculated as iterated integrals.

Product Spaces

It is notationally convenient in this section to change from $(\Omega, \mathcal{F})$ to $(X, \mathcal{X})$ and $(Y, \mathcal{Y})$. In the product space $X \times Y$ a *measurable rectangle* is a product $A \times B$ for which $A \in \mathcal{X}$ and $B \in \mathcal{Y}$. The natural class of sets in $X \times Y$ to consider is the σ-field $\mathfrak{X} \times \mathfrak{Y}$ generated by the measurable rectangles. (Of course, $\mathfrak{X} \times \mathfrak{Y}$ is not a cartesian product in the usual sense.)

Example 18.1.   Suppose that $X = Y = R^1$ and $\mathfrak{X} = \mathfrak{Y} = \mathcal{R}^1$. Then a measurable rectangle is a cartesian product $A \times B$ in which A and B are linear Borel sets. The term *rectangle* has up to this point been reserved for cartesian products of intervals, and so a measurable rectangle is more general. As the measurable rectangles do include the ordinary ones and the latter generate $\mathcal{R}^2$, $\mathcal{R}^2 \subset \mathcal{R}^1 \times \mathcal{R}^1$. On the other hand, if A is an interval, $[B: A \times B \in \mathcal{R}^2]$ contains R^1 ($A \times R^1 = \bigcup_n A \times (-n, n] \in \mathcal{R}^2$) and is closed under the formation of proper differences and countable unions; thus it is a σ-field containing the intervals and hence the Borel sets. Therefore, if B is a Borel set, $[A: A \times B \in \mathcal{R}^2]$ contains the intervals and hence, being a σ-field, contains the Borel sets. Thus all the measurable rectangles are in $\mathcal{R}^2$, and so $\mathcal{R}^1 \times \mathcal{R}^1 = \mathcal{R}^2$ consists exactly of the two-dimensional Borel sets.   ∎

As this example shows, $\mathfrak{X} \times \mathfrak{Y}$ is in general much larger than the class of measurable rectangles.

Theorem 18.1.   (i)   *If $E \in \mathfrak{X} \times \mathfrak{Y}$, then for each x the set $[y: (x, y) \in E]$ lies in $\mathfrak{Y}$ and for each y the set $[x: (x, y) \in E]$ lies in $\mathfrak{X}$.*

(ii)   *If f is measurable $\mathfrak{X} \times \mathfrak{Y}$, then for each fixed x the function $f(x, y)$ with y varying is measurable $\mathfrak{Y}$, and for each fixed y the function $f(x, y)$ with x varying is measurable $\mathfrak{X}$.*

The set $[y: (x, y) \in E]$ is the *section of E* determined by x, and $f(x, y)$ as a function of y is the *section of f* determined by x.

PROOF.   Fix x and consider the mapping $T_x: Y \to X \times Y$ defined by $T_x y = (x, y)$. If $E = A \times B$ is a measurable rectangle, $T_x^{-1} E$ is B or $\emptyset$ according as A contains x or not, and in either case $T_x^{-1} E \in \mathfrak{Y}$. By Theorem 13.1(i), T_x is measurable $\mathfrak{Y} / \mathfrak{X} \times \mathfrak{Y}$. Hence $[y: (x, y) \in E] = T_x^{-1} E \in \mathfrak{Y}$ for $E \in \mathfrak{X} \times \mathfrak{Y}$. By Theorem 13.1(ii), if f is measurable $\mathfrak{X} \times \mathfrak{Y} / \mathcal{R}^1$, then $f T_x$ is measurable $\mathfrak{Y} / \mathcal{R}^1$. Hence $f(x, y) = f T_x(y)$ as a function of y is measurable $\mathfrak{Y}$. The symmetric statements for fixed y are proved the same way.   ∎

Product Measure

Now suppose that $(X, \mathfrak{X}, \mu)$ and $(Y, \mathfrak{Y}, \nu)$ are measure spaces, and suppose for the moment that μ and ν are *finite*. By the theorem just proved $\nu[y: (x, y) \in E]$ is a well-defined function of x. If $\mathcal{L}$ is the class of E in $\mathfrak{X} \times \mathfrak{Y}$ for which this function is measurable $\mathfrak{X}$, it is not hard to show that $\mathcal{L}$ is a λ-system. Since the function is $I_A(x)\nu(B)$ for $E = A \times B$, $\mathcal{L}$ contains the π-system consisting of the measurable rectangles. Hence $\mathcal{L}$ coincides with $\mathcal{X} \times \mathcal{Y}$ by the π-λ theorem. It follows without difficulty that

$$(18.1) \qquad \pi'(E) = \int_X \nu[y: (x, y) \in E]\mu(dx), \qquad E \in \mathcal{X} \times \mathcal{Y},$$

is a finite measure on $\mathcal{X} \times \mathcal{Y}$, and similarly for

$$(18.2) \qquad \pi''(E) = \int_Y \mu[x: (x, y) \in E] \nu(dy), \qquad E \in \mathcal{X} \times \mathcal{Y}.$$

For measurable rectangles,

$$(18.3) \qquad\qquad \pi'(A \times B) = \pi''(A \times B) = \mu(A) \cdot \nu(B).$$

The class of E in $\mathcal{X} \times \mathcal{Y}$ for which $\pi'(E) = \pi''(E)$ thus contains the measurable rectangles; since this class is a λ-system, it contains $\mathcal{X} \times \mathcal{Y}$. The common value $\pi'(E) = \pi''(E)$ is the product measure sought.

To show that (18.1) and (18.2) also agree for σ-finite μ and ν, let $\{A_m\}$ and $\{B_n\}$ be decompositions of X and Y into sets of finite measure, and put $\mu_m(A) = \mu(A \cap A_m)$ and $\nu_n(B) = \nu(B \cap B_n)$. Since $\nu(B) = \Sigma_m \nu_m(B)$, the integrand in (18.1) is measurable $\mathcal{X}$ in the σ-finite as well as in the finite case; hence π' is a well-defined measure on $\mathcal{X} \times \mathcal{Y}$ and so is π''. If π'_{mn} and π''_{mn} are (18.1) and (18.2) for μ_m and ν_n, then by the finite case, already treated, $\pi'(E) = \Sigma_{mn} \pi'_{mn}(E) = \Sigma_{mn} \pi''_{mn}(E) = \pi''(E)$. Thus (18.1) and (18.2) coincide in the σ-finite case as well. Moreover, $\pi'(A \times B) = \Sigma_{mn} \mu_m(A) \nu_n(B) = \mu(A)\nu(B)$.

Theorem 18.2. *If $(X, \mathcal{X}, \mu)$ and $(Y, \mathcal{Y}, \nu)$ are σ-finite measure spaces, $\pi(E) = \pi'(E) = \pi''(E)$ defines a σ-finite measure on $\mathcal{X} \times \mathcal{Y}$; it is the only measure such that $\pi (A \times B) = \mu(A) \cdot \nu(B)$ for measurable rectangles.*

PROOF. Only σ-finiteness and uniqueness remain to be proved. The products $A_m \times B_n$ for $\{A_m\}$ and $\{B_n\}$ as above decompose $X \times Y$ into measurable rectangles of finite π-measure. This proves both σ-finiteness and uniqueness, since the measurable rectangles form a π-system generating $\mathcal{X} \times \mathcal{Y}$ (Theorem 10.3). ■

The π thus defined is called *product measure*; it is usually denoted $\mu \times \nu$. Note that the integrands in (18.1) and (18.2) may be infinite for certain x and y, which is one reason for introducing functions with infinite values. Note also that (18.3) in some cases requires the conventions (15.2).

Fubini's Theorem

Integrals with respect to π are usually computed via the formulas

(18.4) $$\int_{X \times Y} f(x, y)\, \pi\,(d(x, y)) = \int_X \left[\int_Y f(x, y)\, \nu(dy) \right] \mu(dx)$$

and

(18.5) $$\int_{X \times Y} f(x, y)\pi(d(x, y)) = \int_Y \left[\int_X f(x, y)\mu(dx) \right] \nu(dy).$$

The left side here is a *double* integral, and the right sides are *iterated* integrals. The formulas hold very generally, as the following argument shows.

Consider (18.4). The inner integral on the right is

(18.6) $$\int_Y f(x, y)\nu(dy).$$

Because of Theorem 18.1(ii), for f measurable $\mathcal{X} \times \mathcal{Y}$ the integrand here is measurable $\mathcal{Y}$; the question is whether the integral exists, whether (18.6) is measurable $\mathcal{X}$ as a function of x, and whether it integrates to the left side of (18.4).

First consider nonnegative f. If $f = I_E$, everything follows from Theorem 18.2: (18.6) is $\nu[y: (x, y) \in E]$ and (18.4) reduces to $\pi(E) = \pi'(E)$. Because of linearity (Theorem 15.1(iv)), if f is a nonnegative simple function, then (18.6) is a linear combination of functions measurable $\mathcal{X}$ and hence is itself measurable $\mathcal{X}$; further application of linearity to the two sides of (18.4) shows that (18.4) again holds. The general nonnegative f is the monotone limit of nonnegative simple functions; applying the monotone convergence theorem to (18.6) and then to each side of (18.4) shows that again f has the properties required.

Thus for nonnegative f, (18.6) is a well-defined function of x (the value ∞ is not excluded), measurable $\mathcal{X}$, whose integral satisfies (18.4). If one side of (18.4) is infinite, so is the other; if both are finite, they have the same finite value.

Now suppose that f, not necessarily nonnegative, is integrable with respect to π. Then the two sides of (18.4) are finite if f is replaced by $|f|$. Suppose further that

(18.7) $$\int_Y |f(x, y)|\, \nu(dy) < \infty$$

for all x. Then

(18.8) $$\int_Y f(x, y)\nu(dy) = \int_Y f^+(x, y)\nu(dy) - \int_Y f^-(x, y)\nu(dy).$$

The functions on the right here are measurable $\mathcal{X}$ and (since $f^+, f^- \leq |f|$) integrable with respect to μ, and so the same is true of the function on the left. Integrating out the x and applying (18.4) to f^+ and to f^- gives (18.4) for f itself.

The set A_0 of x satisfying (18.7) need not coincide with X, but $\mu(X - A_0) = 0$ if f is integrable with respect to π because the function in (18.7) integrates to $\int |f| d\pi$ (Theorem 15.2(iii)). Now (18.8) holds on A_0, (18.6) is measurable $\mathcal{X}$ on A_0, and (18.4) again follows if the inner integral on the right is given some arbitrary constant value on $X - A_0$.

The same analysis applies to (18.5):

Theorem 18.3. *Under the hypotheses of Theorem* 18.2, *for nonnegative f the functions*

$$(18.9) \qquad \int_Y f(x, y)\nu(dy), \quad \int_X f(x, y)\mu(dx)$$

are measurable $\mathcal{X}$ and $\mathcal{Y}$, respectively, and (18.4) *and* (18.5) *hold. If f (not necessarily nonnegative) is integrable with respect to π, then the two functions* (18.9) *are finite and measurable on A_0 and on B_0, respectively, where $\mu(X - A_0) = \nu(Y - B_0) = 0$, and again* (18.4) *and* (18.5) *hold.*

It is understood here that the inner integrals on the right in (18.4) and (18.5) are set equal to 0 (say) outside A_0 and B_0.

This is *Fubini's theorem*; the part concerning nonnegative f is sometimes called *Tonelli's theorem.* Application of the theorem usually follows a two-step procedure that parallels its proof. First, one of the iterated integrals is computed (or estimated above) with $|f|$ in place of f. If the result is finite, then the double integral (integral with respect to π) of $|f|$ must be finite, so that f is integrable with respect to π; then the value of the double integral of f is found by computing one of the iterated integrals of f. If the iterated integral of $|f|$ is infinite, f is not integrable π.

Example 18.2. Let $I = \int_{-\infty}^{\infty} e^{-x^2} dx$. By Fubini's theorem applied in the plane and by the polar-coordinate formula (see (17.15)),

$$I^2 = \iint_{R^2} e^{-(x^2+y^2)} dx\, dy = \iint_{\substack{r>0 \\ \theta<\theta<2\pi}} e^{-r^2} r\, dr\, d\theta.$$

The double integral on the right can be evaluated as an iterated integral by another application of Fubini's theorem, which leads to the famous formula

$$(18.10) \qquad \int_{-\infty}^{\infty} e^{-x^2} dx = \sqrt{\pi}.$$

As the integrand in this example is nonnegative, the question of integrability does not arise.   ∎

Example 18.3.   It is possible by means of Fubini's theorem to identify the limit in (17.4). First,

$$\int_0^t e^{-ux} \sin x \, dx = \frac{1}{1 + u^2} [1 - e^{-ut}(u \sin t + \cos t)],$$

as follows by differentiation with respect to t. Since

$$\int_0^t \left[\int_0^\infty |e^{-ux} \sin x| \, du \right] dx = \int_0^t |\sin x| \cdot x^{-1} \, dx \le t < \infty,$$

Fubini's theorem applies to the integration of $e^{-ux} \sin x$ over $(0, t) \times (0, \infty)$:

$$\int_0^t \frac{\sin x}{x} \, dx = \int_0^t \sin x \left[\int_0^\infty e^{-ux} \, du \right] dx$$

$$= \int_0^\infty \left[\int_0^t e^{-ux} \sin x \, dx \right] du$$

$$= \int_0^\infty \frac{du}{1 + u^2} - \int_0^\infty \frac{e^{-ut}}{1 + u^2} (u \sin t + \cos t) \, du.$$

The next-to-last integral is $\pi/2$ (see (17.10)), and a change of variable $ut = s$ shows that the final integral goes to 0 as $t \to \infty$. Therefore,

(18.11) $$\lim_{t \to \infty} \int_0^t \frac{\sin x}{x} \, dx = \frac{\pi}{2}.$$ ∎

Integration by Parts

Let F and G be two nondecreasing, right-continuous functions on an interval $[a, b]$, and let μ and ν be the corresponding measures:

$$\mu(x, y] = F(y) - F(x), \qquad \nu(x, y] = G(y) - G(x), \qquad a \le x \le y \le b.$$

In accordance with the convention (17.16), write $dF(x)$ and $dG(x)$ in place of $\mu(dx)$ and $\nu(dx)$.

Theorem 18.4.   *If F and G have no common points of discontinuity in $(a, b]$, then*

(18.12) $$\int_{(a,b]} G(x) \, dF(x)$$

$$= F(b)G(b) - F(a)G(a) - \int_{(a,b]} F(x) \, dG(x).$$

In brief: $\int G\, dF = \Delta FG - \int F\, dG$. This is one version of the partial integration formula.

PROOF. Note first that replacing $F(x)$ by $F(x) - C$ leaves (18.12) unchanged—it merely adds and subtracts $Cv(a, b]$ on the right. Hence (take $C = F(a)$) it is no restriction to assume that $F(x) = \mu(a, x]$ and no restriction to assume that $G(x) = v(a, x]$. If $\pi = \mu \times v$ is product measure in the plane, then by Fubini's theorem,

(18.13) $\pi[(x, y): a < y \le x \le b]$

$$= \int_{(a,b]} v(a, x]\mu(dx) = \int_{(a,b]} G(x)\, dF(x)$$

and

(18.14) $\pi[(x, y): a < x \le y \le b]$

$$= \int_{(a,b]} \mu(a, y]v(dy) = \int_{(a,b]} F(y)\, dG(y).$$

The two sets on the left have as their union the square $S = (a, b] \times (a, b]$. The diagonal of S has π-measure

$$\pi[(x, y): a < x = y \le b] = \int_{(a,\ b]} v\{x\}\mu(dx) = 0$$

because of the assumption that μ and v share no points of positive measure. Thus the left sides of (18.13) and (18.14) add to $\pi(S) = \mu(a, b]v(a, b] = F(b)G(b)$. ∎

Suppose that v has a density g with respect to Lebesgue measure and let $G(x) = c + \int_a^x g(t)\, dt$. Transform the right side of (18.12) by the formula (16.12) for integration with respect to a density; the result is

(18.15) $$\int_{(a,b]} G(x)\, dF(x) = F(b)G(b) - F(a)G(a) - \int_a^b F(x)g(x)\, dx.$$

A consideration of positive and negative parts shows that this holds for any g integrable over $(a, b]$.

Suppose further that μ has a density f with respect to Lebesgue measure and let $F(x) = c' + \int_a^x f(t)\, dt$. Then (18.15) further reduces to

(18.16) $$\int_a^b G(x)f(x)\, dx = F(b)G(b) - F(a)G(a) - \int_a^b F(x)g(x)\, dx.$$

Again, f can be any integrable function. This is the classical formula for integration by parts.

Under the appropriate integrability conditions, $(a, b]$ can be replaced by an unbounded interval.

Products of Higher Order

Suppose that $(X, \mathcal{X}, \mu)$, $(Y, \mathcal{Y}, \nu)$, and $(Z, \mathcal{Z}, \eta)$ are three σ-finite measure spaces. In the usual way, identify the products $X \times Y \times Z$ and $(X \times Y) \times Z$. Let $\mathcal{X} \times \mathcal{Y} \times Z$ be the σ-field in $X \times Y \times Z$ generated by the $A \times B \times C$ with A, B, C in $\mathcal{X}, \mathcal{Y}, \mathcal{Z}$, respectively. For C in $\mathcal{Z}$, let $\mathcal{G}_C$ be the class of $E \in \mathcal{X} \times \mathcal{Y}$ for which $E \times C \in \mathcal{X} \times \mathcal{Y} \times Z$. Then $\mathcal{G}_C$ is a σ-field containing the measurable rectangles in $X \times Y$, and so $\mathcal{G}_C = \mathcal{X} \times \mathcal{Y}$. Therefore, $(\mathcal{X} \times \mathcal{Y}) \times \mathcal{Z} \subset \mathcal{X} \times \mathcal{Y} \times Z$. But the reverse relation is obvious, and so $(\mathcal{X} \times Z) \times \mathcal{Y} = \mathcal{X} \times \mathcal{Y} \times Z$.

Define the product $\mu \times \nu \times \eta$ on $\mathcal{X} \times \mathcal{Y} \times Z$ as $(\mu \times \nu) \times \eta$. It gives to $A \times B \times C$ the value $(\mu \times \nu)(A \times B) \cdot \eta(C) = \mu(A)\nu(B)\eta(C)$, and it is the only measure that does. The formulas (18.4) and (18.5) extend in the obvious way.

Products of four or more components can clearly be treated in the same way. This leads in particular to another construction of Lebesgue measure in $R^k = R^1 \times \cdots \times R^1$ (see Example 18.1) as the product $\lambda \times \cdots \times \lambda$ (k factors) on $\mathcal{R}^k = \mathcal{R}^1 \times \cdots \times \mathcal{R}^1$. Fubini's theorem of course gives a practical way to calculate volumes:

Example 18.4. Let V_k be the volume of the sphere of radius 1 in R^k; by Theorem 12.2, a sphere in R^k with radius r has volume $r^k V_k$. Let A be the unit sphere in R^k, let $B = [(x_1, x_2): x_1^2 + x_2^2 \leq 1]$, and let $C(x_1, x_2) = [(x_3, \ldots, x_k): \Sigma_{i=3}^k x_i^2 \leq 1 - x_1^2 - x_2^2]$. By Fubini's theorem,

$$V_k = \int_A dx_1 \cdots dx_k = \int_B dx_1 \, dx_2 \int_{C(x_1, x_2)} dx_3 \cdots dx_k$$

$$= \int_B dx_1 \, dx_2 \, V_{k-2}(1 - x_1^2 - x_2^2)^{(k-2)/2}$$

$$= V_{k-2} \iint_{\substack{0 < \theta < 2\pi \\ 0 < \rho < 1}} (1 - \rho^2)^{(k-2)/2} \rho \, d\rho \, d\theta$$

$$= \pi V_{k-2} \int_0^1 t^{(k-2)/2} \, dt = \frac{2\pi V_{k-2}}{k}.$$

If V_0 is taken as 1, this holds for $k = 2$ as well as for $k \geq 3$. Since $V_1 = 2$, it follows by induction that

$$V_{2i-1} = \frac{2(2\pi)^{i-1}}{1 \cdot 3 \cdot 5 \cdots (2i-1)}, \qquad V_{2i} = \frac{(2\pi)^i}{2 \cdot 4 \cdots (2i)}$$

for $i = 1, 2, \ldots$.   ∎

PROBLEMS

18.1. The assumption of σ-finiteness in Theorem 18.2 is essential: Let μ be Lebesgue measure on the line, let ν be counting measure on the line, and take $E = [(x, y): x = y]$. Then (18.1) and (18.2) do not agree.

18.2. Suppose Y is the real line and ν is Lebesgue measure. Suppose that, for each x in $X, f(x, y)$ has with respect to y a derivative $f'(x, y)$. Formally,

$$\int_a^y \left[\int_X f'(x, \xi)\mu(dx) \right] d\xi = \int_X [f(x, y) - f(x, a)]\mu(dx)$$

and hence $\int_X f(x, y)\mu(dx)$ has derivative $\int_X f'(x, y)\mu(dx)$. Use this idea to prove Theorem 16.8 (ii).

18.3. Suppose that f is nonnegative on a σ-finite measure space $(\Omega, \mathcal{F}, \mu)$. Show that

$$\int_\Omega f \, d\mu = (\mu \times \lambda)[(\omega, y) \in \Omega \times R^1 : 0 \leq y \leq f(\omega)].$$

Prove that the set on the right is measurable. This gives the "area under the curve." Given the existence of $\mu \times \lambda$ on $\Omega \times R^1$, one can use the right side of this equation as an alternative definition of the integral.

18.4. Reconsider Problem 12.12.

18.5. Suppose that $\nu[y: (x, y) \in E] = \nu[y: (x, y) \in F]$ for all x and show that $(\mu \times \nu)(E) = (\mu \times \nu)(F)$. This is a general version of *Cavalieri's principle*.

18.6. 2.8↑ Suppose that $X = Y$ is uncountable and $\mathcal{X} = \mathcal{Y}$ consists of the countable and the co-countable sets. Show that the diagonal $E = [(x, y): x = y]$ does not lie in $\mathcal{X} \times \mathcal{Y}$, even though $[y: (x, y) \in E] \in \mathcal{Y}$ and $[x: (x, y) \in E] \in \mathcal{X}$ for all x and y.

18.7. (a) Prove the corollary to Theorem 16.7 by Fubini's theorem in $\Omega \times \{1, 2, \ldots\}$.

(b) Relate the series in Problem 17.9 to Fubini's theorem.

18.8. (a) Let $\mu = \nu$ be counting measure on $X = Y = \{1, 2, \ldots\}$. If

$$f(x, y) = \begin{cases} 2 - 2^{-x} & \text{if } x = y, \\ -2 + 2^{-x} & \text{if } x = y + 1, \\ 0 & \text{otherwise,} \end{cases}$$

then the iterated integrals exist but are unequal. Why does this not contradict Fubini's theorem?

(b) Show that $xy/(x^2 + y^2)^2$ is not integrable over the square $[(x, y): |x|, |y| \leq 1]$ even though the iterated integrals exist and are equal.

18.9. For an integrable function f on $(0, 1]$, consider a Riemann sum in the form $R = \sum_{j=1}^n f(x_j)(x_j - x_{j-1})$, where $0 = x_0 < \cdots < x_n = 1$. Extend f to $(0, 2]$ by setting $f(x) = f(x - 1)$ for $1 < x \leq 2$, and define

$$R(t) = \sum_{j=1}^n f(x_j + t)(x_j - x_{j-1}).$$

For $0 < t < 1$, the points $x_j + t$ reduced modulo 1 give a partition of $(0, 1]$, and $R(t)$ is essentially the corresponding Riemann sum, differing from it by at most three terms. If $\max_j (x_j - x_{j-1})$ is small, then $R(t)$ is for most values of t a good approximation to $\int_0^1 f(x)dx$, even though $R = R(0)$ may be a poor one. Show in fact that

$$
\begin{aligned}
&\int_0^1 \left| R(t) - \int_0^1 f(x)\, dx \right| dt \\
\text{(18.17)} \quad &= \int_0^1 \left| \sum_{j=1}^n \int_{x_{j-1}}^{x_j} [f(x_j + t) - f(x + t)]\, dx \right| dt \\
&\leq \sum_{j=1}^n \int_{x_{j-1}}^{x_j} \int_0^1 |f(x_j + t) - f(x + t)|\, dt\, dx.
\end{aligned}
$$

Given ϵ choose a continuous g such that $\int_0^1 |f(x) - g(x)|\, dx < \epsilon^2/3$ and then choose δ so that $|x - y| < \delta$ implies that $|g(x) - g(y)| < \epsilon^2/3$. Show that max $(x_j - x_{j-1}) < \delta$ implies that the first integral in (18.17) is at most ϵ^2 and hence that $|R(t) - \int_0^1 f(x)\, dx| < \epsilon$ on a set of t (in $(0, 1]$) of Lebesgue measure exceeding $1 - \epsilon$.

18.10. Prove for distribution functions F that $\int_{-\infty}^{\infty} (F(x + c) - F(x))\, dx = c$.

18.11. Prove for continuous distribution functions that $\int_{-\infty}^{\infty} F(x)\, dF(x) = 1/2$.

18.12. Suppose that a number f_n is defined for each $n \geq n_0$ and put $F(x) = \sum_{n_0 \leq n \leq x} f_n$. Deduce from (18.15) that

$$
\text{(18.18)} \qquad \sum_{n_0 \leq n \leq x} G(n)f_n = F(x)G(x) - \int_{n_0}^x F(t)g(t)\, dt
$$

if G has continuous derivative g. (First assume that the f_n are nonnegative.)

18.13. ↑ Take $n_0 = 1, f_n = 1$, and $G(x) = 1/x$ and derive $\sum_{n \leq x} n^{-1} = \log x + \gamma + O(1/x)$, where $\gamma = 1 - \int_1^\infty (t - [t])t^{-2}\, dt$ is Euler's constant.

18.14. 5.17 18.12↑ Use (18.18) and (5.47) to prove that there exists a constant c such that

$$
\text{(18.19)} \qquad \sum_{p \leq x} \frac{1}{p} = \log \log x + c + O\left(\frac{1}{\log x}\right).
$$

18.15. (a) Let $I_n = \int_0^{\pi/2} (\sin x)^n\, dx$. Show by partial integration that $I_n = (n - 1)(I_{n-2} - I_n)$ and hence that $I_n = (n - 1)n^{-1}I_{n-2}$.
(b) Show by induction that

$$
I_{2n+1}^{-1} = (-1)^n (2n + 1) \binom{-1/2}{n}.
$$

From (17.20) deduce that

$$
\frac{\pi^2}{8} = 1 + \frac{1}{3^2} + \frac{1}{5^2} + \frac{1}{7^2} + \cdots
$$

and hence that

$$\frac{\pi^2}{6} = 1 + \frac{1}{2^2} + \frac{1}{3^2} + \frac{1}{4^2} + \cdots.$$

(c)   Show by induction that

$$I_{2n} = \frac{2n-1}{2n} \frac{2n-3}{2n-2} \cdots \frac{3}{4} \frac{1}{2} \frac{\pi}{2}, \qquad I_{2n+1} = \frac{2n}{2n+1} \frac{2n-2}{2n-1} \cdots \frac{4}{5} \frac{2}{3}.$$

From $I_{2n-1} > I_{2n} > I_{2n+1}$ deduce that

$$\frac{\pi}{2} \frac{2n}{2n+1} < \frac{2^2 \cdot 4^2 \cdots (2n)^2}{3^2 \cdot 5^2 \cdots (2n-1)^2(2n+1)} < \frac{\pi}{2},$$

and from this derive *Wallis's formula*,

(18.20) $$\frac{\pi}{2} = \frac{2}{1} \frac{2}{3} \frac{4}{3} \frac{4}{5} \frac{6}{5} \frac{6}{7} \cdots$$

18.16.   *Stirling's formula.* The area under the curve $y = \log x$ between the abscissas 1 and n is $A_n = n \log n - n + 1$. The area under the corresponding inscribed polygon is $B_n = \log n! - \frac{1}{2}\log n$.

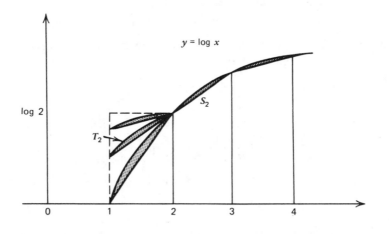

(a)   Let S_k be the sliver-shaped region bounded by curve and polygon between the abscissas k and $k + 1$, and let T_k be this region translated so that its right-hand vertex is at $(2, \log 2)$. Show that the T_k do not overlap and that $T_n \cup T_{n+1} \cup \cdots$ is contained in a triangle of area $\frac{1}{2} \log (1 + 1/n)$. Conclude that

(18.21) $$\log n! = \left(n + \frac{1}{2}\right)\log n - n + c + \alpha n,$$

where c is 1 minus the area of $T_1 \cup T_2 \cup \cdots$ and

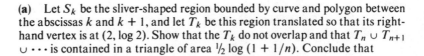

$$0 < \alpha_n < \frac{1}{2}\log \left(1 + \frac{1}{n}\right) < \frac{1}{n}.$$

(b) By Wallis's formula (18.20) show that

$$\log\sqrt{\frac{\pi}{2}} = \lim_n [2n \log 2 + 2 \log n! - \log (2n)! - \frac{1}{2} \log (2n + 1)].$$

Substitute (18.21) and show that $c = \log \sqrt{2\pi}$. This gives Stirling's formula,

$$n! \sim \sqrt{2\pi} \, n^{n+(1/2)}e^{-n}.$$

18.17. Euler's *gamma function* is defined for positive t by $\Gamma (t) = \int_0^\infty x^{t-1}e^{-x} \, dx$.
   (a) Prove that $\Gamma^{(k)} (t) = \int_0^\infty x^{t-1}(\log x)^k e^{-x} \, dx$.
   (b) Show by partial integration that $\Gamma(t + 1) = t \, \Gamma(t)$ and hence that $\Gamma(n + 1) = n!$ for integral n.
   (c) From (18.10) deduce $\Gamma(\frac{1}{2}) = \sqrt{\pi}$.
   (d) Show that the unit sphere in R^k has volume (see Example 18.4)

   (18.22)
   $$V_k = \frac{\pi^{k/2}}{\Gamma\left(\dfrac{k}{2} + 1\right)}.$$

18.18. By partial integration prove that $\int_0^\infty ((\sin x)/x)^2 \, dx = \pi/2$ and $\int_{-\infty}^\infty (1 - \cos x)x^{-2} \, dx = \pi$.

18.19. Suppose that μ is a probability measure on $(X, \mathcal{X})$ and that, for each x in X, ν_x is a probability measure on $(Y, \mathcal{Y})$. Suppose further that, for each B in $\mathcal{Y}$, $\nu_x(B)$ is, as a function of x, measurable $\mathcal{X}$. Regard the $\mu(A)$ as initial probabilities and the $\nu_x(B)$ as transition probabilities.
   (a) Show that, if $E \in \mathcal{X} \times \mathcal{Y}$, then $\nu_x[y: (x, y) \in E]$ in measurable $\mathcal{X}$.
   (b) Show that $\pi(E) = \int_X \nu_x[y: (x, y) \in E] \, \mu(dx)$ defines a probability measure on $\mathcal{X} \times \mathcal{Y}$. If $\nu_x = \nu$ does not depend on x, this is just (18.1).
   (c) Show that if f is measurable $\mathcal{X} \times \mathcal{Y}$ and nonnegative, then $\int_Y f(x, y) \nu_x(dy)$ is measurable $\mathcal{X}$. Show further that

   $$\int_{X \times Y} f(x, y)\pi(d(x, y)) = \int_X \left[\int_Y f(x, y) \, \nu_x(dy) \right]\nu(dx),$$

   which extends Fubini's theorem (in the probability case). Consider also f's that may be negative.
   (d) Let $\nu(B) = \int_X \nu_x(B) \, \mu (dx)$. Show that $\pi(X \times B) = \nu(B)$ and

   $$\int_Y f(y) \, \nu(dy) = \int_X \left[\int_Y f(y)\nu_x(dy) \right]\mu(dx).$$

SECTION 19.   HAUSDORFF MEASURE*

The theory of Lebesgue measure gives the area of a figure in the plane and the volume of a solid in 3-space, but how can one define the area of a curved surface in 3-space? This section is devoted to such geometric questions.

The Definition

Suppose that A is a set in R^k. For positive m and ϵ put

$$(19.1) \qquad h_{m,\epsilon}(A) = \inf c_m \Sigma_n (\operatorname{diam} B_n)^m,$$

where the infimum extends over countable coverings of A by sets B_n with diameters $\operatorname{diam} B_n = \sup [|x - y| : x, y \in B_n]$ less than ϵ. Here c_m is a positive constant to be assigned later. As ϵ decreases, the infimum extends over smaller classes, and so $h_{m,\epsilon}(A)$ does not decrease. Thus $h_{m,\epsilon}(A)$ has a limit $h_m(A)$, finite or infinite:

$$(19.2) \qquad h_{m,\epsilon}(A) \uparrow h_m(A) \qquad \text{as } \epsilon \downarrow 0.$$

This limit $h_m(A)$ is the *m-dimensional outer Hausdorff measure* of A. Although h_m on R^k is technically a different set function for each k, no confusion results from suppressing k in the notation.

Since $h_{m,\epsilon}$ is an outer measure (Example 11.1), $h_{m,\epsilon}(\cup_n A_n) \leq \Sigma_n h_{m,\epsilon}(A_n) \leq \Sigma_n h_m(A_n)$, and hence $h_m(\cup_n A_n) \leq \Sigma_n h_m(A_n)$. Since it is obviously monotone, h_m is thus an outer measure. If A and B are at positive distance, choose ϵ so that $\epsilon < \operatorname{dist}(A, B)$. If $A \cup B \subset \cup_n C_n$ and $\operatorname{diam} C_n < \epsilon$ for all n, then no C_n can meet both A and B. Splitting the sum $c_m \Sigma_n (\operatorname{diam} C_n)^m$ according as C_n meets A or not shows that it is at least $h_{m,\epsilon}(A) + h_{m,\epsilon}(B)$. Therefore, $h_m(A \cup B) \geq h_m(A) + h_m(B)$, and h_m satisfies the Carathéodory condition (11.5).

It follows by Theorem 11.5 that *every Borel set is h_m-measurable*. Therefore, by Theorem 11.1, h_m restricted to $\mathcal{R}^k$ is a measure. It is immediately obvious from the definitions that h_m *is invariant under isometries*: If $A \subset R^k$, $\varphi: A \to R^j$, and $|\varphi(x) - \varphi(y)| = |x - y|$ for $x, y \in A$, then $h_m(\varphi A) = h_m(A)$.

Example 19.1.   Suppose that $m = k = 1$. If $a_n = \inf B_n$ and $b_n = \sup B_n$, then $[a_n, b_n]$ and B_n have the same diameter $b_n - a_n$. Thus (19.1) is in this case unchanged if the B_n are taken to be intervals. Further, any sum $\Sigma_n \operatorname{diam} B_n$ remains unchanged if each interval B_n is split into subintervals of length less than ϵ. Thus $h_{1,\epsilon}(A)$ does not depend on ϵ, and $h_1(A) = h_{1,\epsilon}(A)$ is for $A \subset R^1$ the ordinary outer Lebesgue measure $\lambda^*(A)$ of A, provided that c_1 is taken to be 1, as it will.

* This section may be omitted.

A planar segment $S_t = [(x, t) \, 0 \leq x \leq l]$, as an isometric copy of the interval $[0, l]$, satisfies $h_1(S_t) = l$. This is a first indication that Hausdorff measure has the correct geometric properties. Since the segments S_t are disjoint, one-dimensional Hausdorff measure in the plane is not σ-finite. ∎

Example 19.2. Consider the planar segments $A = [(x, 0): 0 < x \leq 1]$ and $B = [(0, y): 0 < y \leq 1]$ and their union C. Since $h_1(A) = h_1(B) = 1$, $h_1(C) = 2$, as intuition requires. If $D_\epsilon = [(x, y): x, y \geq 0, x + y \leq \epsilon]$, then $C - D_\epsilon$ can be split into disjoint segments of diameters less than ϵ, and it follows that $h_{1,\epsilon}(C) \leq 2(1 - \epsilon) + \epsilon\sqrt{2} < 2$. This shows the role of ϵ: if $h_m(A)$ were defined, not by (19.2), but as the infimum in (19.1) without the requirement diam $B_n < \epsilon$, then C would have the wrong one-dimensional measure ($\sqrt{2}$ instead of 2). ∎

The Normalizing Constant

In the definition m can be any positive number, but suppose from now on that it is an integer between 1 and the dimension of the space R^k: $m = 1, \ldots, k$. The problem is to choose c_m so as to satisfy the geometric intention of the definition.

Let V_m be the volume of a sphere of radius 1 in R^m; the value of V_m is explicitly calculated in Example 18.4. And now take the c_m in (19.1) as the volume of a sphere of diameter 1:

$$(19.3) \qquad\qquad c_m = \frac{V_m}{2^m}.$$

A sphere in R^m of diameter d has volume $c_m d^m$. It will be proved below that for this choice of c_m the k-dimensional Hausdorff and Lebesgue measures agree for k-dimensional Borel sets:

Theorem 19.1. *If $A \in \mathcal{R}^k$, then $h_k(A) = \lambda_k(A)$.*

If $A \in \mathcal{R}^m$ and B is a set in R^k that is an isometric copy of A, it will follow by this theorem that $h_m(B) = h_m(A) = \lambda_m(A)$.

The unit cube C in R^k can be covered by n^k cubes of side n^{-1} and diameter $k^{1/2}n^{-1}$; taking $n > k^{1/2}\epsilon^{-1}$ shows that $h_{k,\epsilon}(C) \leq c_k n^k (k^{1/2}n^{-1})^k = c_k k^{k/2}$, and so $h_k(C) < \infty$. If $C \subset \bigcup_n B_n$ and diam $B_n = d_n$, enclose B_n in a sphere S_n of radius d_n. Then $C \subset \bigcup_n S_n$, and so $1 = \lambda_k(C) \leq \Sigma_n \lambda_k(S_n) = \Sigma_n V_k d_n^k$; hence $h_k(C) \geq c_k/V_k$. Thus $h_k(C)$ is finite and positive.

Put $h_k(C) = K$, where again C is the unit cube in R^k. Then of course

$$(19.4) \qquad\qquad h_k(A) = K\lambda_k(A)$$

for $A = C$. Consider the linear transformation carrying x to θx, where $\theta > 0$. By

Theorem 12.2, $\lambda_k(\theta A) = \theta^k \lambda_k(A)$ for $A \in \mathcal{R}^k$. Since diam $\theta A = \theta^k$ diam A, the definition (19.2) gives

$$(19.5) \qquad\qquad h_k(\theta A) = \theta^k h_k(A).$$

Thus (19.4) holds for all cubes and hence by additivity holds for rectangles whose vertices have rational coordinates. The latter sets form a π-system generating $\mathcal{R}^k$, and so (19.4) holds for all $A \in \mathcal{R}^k$, by Theorem 10.3.

Thus Theorem 19.1 will follow once it has been shown that the h_k-measure of the unit cube in R^k, the K in (19.4), is 1. To establish this requires two lemmas connecting measure with geometry. The first is a version of the *Vitali covering theorem*.

Lemma 1. *Suppose that G is a bounded open set in R^k and that $\epsilon > 0$. Then there exists in G a disjoint sequence $S_1, S_2, \ldots$ of closed spheres such that $\lambda_k(G - \bigcup_n S_n) = 0$ and $0 < $ diam $S_n < \epsilon$.*

PROOF.   Let S_1 be any closed sphere in G satisfying $0 < $ diam $S_1 < \epsilon$. Suppose that $S_1, \ldots, S_n$ have already been constructed. Let $\mathcal{S}_n$ be the class of closed spheres in G that have diameter less than ϵ and do not meet $S_1 \cup \cdots \cup S_n$. Let s_n be the supremum of the radii of the spheres in $\mathcal{S}_n$, and let S_{n+1} be an element of $\mathcal{S}_n$ whose radius r_{n+1} exceeds $s_n/2$. Since the S_n are disjoint subsets of G, $\sum_n V_k r_n^k = \sum_n \lambda_k(S_n) \le \lambda_k(G) < \infty$, and hence

$$(19.6) \qquad\qquad r_n \to 0.$$

Put $B = G - \bigcup_n S_n$. It is to be shown that $\lambda_k(B) = 0$. Assume the opposite. Let S_n' be the sphere having the same center as S_n and having radius $4r_n$. Then $\sum_n \lambda_k(S_n') = 4^k \sum_n \lambda_k(S_n) \le 4^k \lambda_k(G)$. Choose N so that $\sum_{n>N} \lambda_k(S_n') < \lambda_k(B)$. Then there is an x_0 such that

$$(19.7) \qquad\qquad x_0 \in B, \qquad x_0 \notin \bigcup_{n>N} S_n'.$$

As the S_n are closed, there is about x_0 a closed sphere S with radius r small enough that

$$(19.8) \qquad\qquad S \cap (S_1 \cup \cdots \cup S_N) = 0.$$

If S does not meet $S_1 \cup \cdots \cup S_n$, then $r \le s_n < 2r_{n+1}$; this cannot hold for all n because of (19.6). Thus S meets some S_n; let S_{n_0} be the first of these:

$$(19.9) \qquad S \cap S_{n_0} \ne 0, \qquad S \cap (S_1 \cup \cdots \cup S_{n_0-1}) = 0.$$

Because of (19.8), $n_0 > N$, and by (19.7), $x_0 \notin S_{n_0}'$. But since $x_0 \notin S_{n_0}'$ and $S \cap S_{n_0} \ne 0$, S has radius $r \ge 3r_{n_0} > \frac{3}{2}s_{n_0-1}$. By (19.9), $r \le s_{n_0-1}$. Thus $s_{n_0-1} \ge r > \frac{3}{2}s_{n_0-1}$, a contradiction. ∎

Lemma 2.   *If $A \in \mathcal{R}^k$, then $\lambda_k(A) \leq c_k(\text{diam } A)^k$.*

In other words, the volume of a set cannot exceed that of a sphere with the same diameter.

PROOF.   Represent the general point of R^k as (t, y), where $t \in R^1, y \in R^{k-1}$. Let $A_y = [t: (t, y) \in A]$. Since A is bounded, each A_y is bounded and $A_y = 0$ for y outside some bounded set. By the results of Section 18, $A_y \in \mathcal{R}^1$ and $\lambda(A_y)$ is measurable $\mathcal{R}^{k-1}$. Let I_y be the open (perhaps empty) interval $(-\lambda(A_y)/2, + \lambda(A_y)/2)$, and set $S_1A = \bigcup_y [(t, y): t \in I_y]$. If $0 \leq f_n(y) \uparrow \lambda(A_y)/2$ and f_n is simple, then $[(t, y): |t| < f_n(y)]$ is in $\mathcal{R}^k$ and increases to S_1A; thus $S_1A \in \mathcal{R}^k$. By Fubini's theorem, $\lambda_k(A) = \lambda_k(S_1A)$. The passage from A to S_1A is *Steiner symmetrization* with respect to the hyperplane $H_1 = [x \in R^k: x_1 = 0]$.

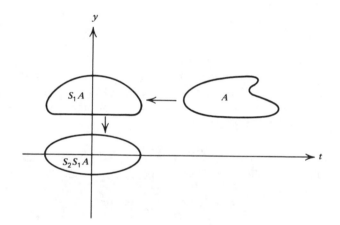

Let J_y be the closed interval from inf A_y to sup A_y; it has length $\lambda(J_y) \geq \lambda(A_y)$. Let m_y be the midpoint of J_y. If $s \in I_y$ and $s' \in I_{y'}$, then $|s - s'| \leq \frac{1}{2}\lambda(A_y) + \frac{1}{2}\lambda(A_{y'}) \leq |m_y - m_{y'}| + \frac{1}{2}\lambda(J_y) + \frac{1}{2}\lambda(J_{y'})$, and this last sum is $|t - t'|$ for appropriate endpoints t of J_y and t' of $J_{y'}$. Thus for any points (s, y) and (s', y') of S_1A, there are points (t, y) and (t', y') that are at least as far apart and are limits of points in A. Hence diam $S_1A \leq$ diam A.

For each i there is a Steiner symmetrization S_i with respect to the hyperplane $H_i = [x: x_i = 0]$; $\lambda_k(S_iA) = \lambda_k(A)$ and diam $S_iA \leq$ diam A. It is obvious by the construction that S_iA is symmetric with respect to H_i: if x' is x with its ith coordinate reversed in sign, then x and x' both lie in S_iA or neither does. If A is symmetric with respect to H_i ($i > 1$) and y' is y with its $(i-1)$st coordinate reversed in sign, then, in the notation above, $A_y = A_{y'}$, and so S_1A is symmetric with respect to H_i. More generally, if A is symmetric with respect to H_i and $j \neq i$, then $S_j A$ is also symmetric with respect to H_j.

Pass from A to $SA = S_k \cdots S_1 A$. This is *central symmetrization*. Now $\lambda_k(SA) = \lambda_k(A)$, diam $SA \leq$ diam A, and SA is symmetric with respect to each H_i, so that $x \in SA$ implies $-x \in SA$. If $|x| > \frac{1}{2}$ diam A, then $|x - (-x)| >$ diam $A \geq$ diam SA, so that SA cannot contain x and $-x$. Thus SA is contained in the closed sphere with center 0 and diameter diam A. $\blacksquare$

PROOF OF THEOREM 19.1.   It is only necessary to show that $K = 1$ in (19.4). By Lemma 1 the interior of the unit cube C in R^k contains disjoint spheres S_1, $S_2, \ldots$ such that diam $S_n < \epsilon$ and $h_{k,\epsilon}(C - \cup_n S_n) \leq h_k(C - \cup_n S_n) = K\lambda_k(C - \cup_n S_n) = 0$. By (19.3), $h_{k,\epsilon}(C) = h_{k,\epsilon}(\cup_n S_n) \leq c_k \Sigma_n$ (diam $S_n)^k = c_k \Sigma_n c_k^{-1} \lambda_k(S_n) \leq \lambda_k(C) = 1$. Hence $h_k(C) \leq 1$.

If $C \subset \cup_n B_n$, then by Lemma 2, $1 \leq \Sigma_n \lambda_k(B_n) \leq \Sigma_n c_k$ (diam $B_n)^k$, and hence $h_k(C) \geq 1$. $\blacksquare$

Change of Variable

If $x_1, \ldots, x_m$ are points in R^k, let $P(x_1, \ldots, x_m)$ be the parallelepiped $[\Sigma_{i=1}^m \alpha_i x_i : 0 < \alpha_1, \ldots, \alpha_m \leq 1]$. Suppose that D is a $k \times m$ matrix, and let $x_1, \ldots, x_m$ be its columns, viewed as points of R^k; define $|D| = h_m(P(x_1, \ldots, x_m))$. If C is the unit cube in R^m and D is viewed as a linear map from R^m to R^k, then $P(x_1, \ldots, x_m) = DC$, so that

$$(19.10) \qquad\qquad |D| = h_m(DC).$$

In the case $m = k$ it follows by Theorem 12.2 that

$$(19.11) \qquad\qquad |D| = |\det D|, \qquad m = k.$$

Theorem 12.2 has an extension:

Theorem 19.2.   *Suppose that D is univalent.* If $A \in \mathcal{R}^m$, then $DA \in \mathcal{R}^k$ and*

$$(19.12) \qquad\qquad h_m(DA) = |D| \cdot h_m(A).$$

PROOF.   Since D is univalent, $D \cup_n A_n = \cup_n DA_n$ and $D(R^m - A) = DR^m - DA$; since DR^m is closed, it is in $\mathcal{R}^k$. Thus $[A : DA \in \mathcal{R}^k]$ is a σ-field. Since D is continuous, DA is compact if A is. By Theorem 13.1(i), D is measurable $\mathcal{R}^m / \mathcal{R}^k$.

Each side of (19.12) is a measure as A varies over $\mathcal{R}^m$. They agree by defi-

_____
* That is, one-to-one: $Du = 0$ implies that $u = 0$. In this case $m \leq k$.

nition if A is the unit cube in R^m, and so by (19.5) they agree if A is any cube. By the usual argument they agree for all A in $\mathcal{R}^m$.   ∎

If D is not univalent, $A \in \mathcal{R}^m$ need not imply that $DA \in \mathcal{R}^k$, but (19.12) still holds because each side vanishes.

The Jacobian formula (17.12) for change of variable also generalizes. Suppose that V is an open set in R^m and T is a continuous, one-to-one mapping of V onto the set TV in R^k, where $m \leq k$. Let $t_i(x)$ be the ith coordinate of Tx and suppose that $t_{ij}(x) = \partial t_i(x)/\partial x_j$ is continuous on V. Put

$$(19.13) \qquad D_x = \begin{bmatrix} t_{11}(x) \dots t_{1m}(x) \\ \cdots\cdots\cdots\cdots \\ t_{k1}(x) \dots t_{km}(x) \end{bmatrix};$$

$|D_x|$ as defined by (19.10) plays the role of the modulus of the Jacobian (17.11) and by (19.11) coincides with it in case $m = k$.

Theorem 19.3.   *Suppose that T is a continuous, one-to-one map of an open set V in R^m into R^k. If T has continuous partial derivatives and $A \subset V$, then*

$$(19.14) \qquad \int_A |D_x| h_m(dx) = h_m(TA)$$

and

$$(19.15) \qquad \int_A f(Tx)|D_x| h_m(dx) = \int_{TA} f(y) h_m(dy).$$

By Theorem 19.1, h_m on the left in (19.14) and (19.15) can be replaced by λ_m.

This generalizes Theorem 17.2. Several explanations are required. Since V is open, it is a countable union of compact sets K_n. If A is closed, $T(A \cap K_n)$ is compact because T is continuous, and so $T(A \cap V) = \bigcup_n T(A \cap K_n) \in \mathcal{R}^k$. In particular, $TV \in \mathcal{R}^k$, and it follows by the usual argument that $TA \in \mathcal{R}^k$ for Borel subsets A of V.

During the course of the proof of the theorem it will be seen that $|D_x|$ is continuous in x. Thus the formulas (19.14) and (19.15) make sense if A and f are measurable. As usual, (19.15) holds if f is nonnegative and also if the two sides are finite when f is replaced by $|f|$. And (19.15) follows from (19.14) by the standard argument starting with indicators. Hence only (19.14) need be proved.

To see why (19.14) ought to hold, imagine that A is a rectangle split into many small rectangles A_n; choose x_n in A_n and approximate the integral in (19.14) by $\Sigma |D_{x_n}| h_m(A_n)$. For y in A_n, Ty has the linear approximation $Tx_n + D_{x_n}(y - x_n)$, and so TA_n is approximated by $Tx_n + D_{x_n}(A_n - x_n)$, an m-dimensional parallelepiped for which h_m is $|D_{x_n}|h_m(A_n - x_n) = |D_{x_n}|h_m(A_n)$ by (19.12). Hence $\Sigma|D_{x_n}|h_m(A_n) \approx \Sigma h_m(TA_n) = h_m(TA)$.

The first step in a proof based on these ideas is to split the region of integration into small sets (not necessarily rectangles) on each of which the D_x are all approximately equal to a linear transformation that in turn gives a good local approximation to T. Let V_0 be the set of x in V for which D_x is univalent.

Lemma 3.   *If $\theta > 1$, then there exists a countable decomposition of V_0 into Borel sets B_l such that, for some linear maps $M_l:R^m \to R^k$,*

$$(19.16) \quad \theta^{-1}|M_l v| \leq |D_x v| \leq \theta|M_l v| \qquad \text{for } x \in B_l, v \in R^m,$$

and

$$(19.17) \quad \theta^{-1}|M_l x - M_l y| \leq |Tx - Ty| \leq \theta|M_l x - M_l y|$$
$$\text{for } x, y \in B_l.$$

PROOF.   Choose ϵ and θ_0 in such a way that $\theta^{-1} + \epsilon < \theta_0^{-1} < 1 < \theta_0 < \theta - \epsilon$. Let $\mathcal{M}$ be the set of $k \times m$ matrices with rational coefficients. For M in $\mathcal{M}$ and p a positive integer let $E(M, p)$ be the set of x in V such that

$$(19.18) \qquad \theta_0^{-1}|Mv| \leq |D_x v| \leq \theta_0|Mv| \qquad \text{for } v \in R^m$$

and

$$(19.19) \quad \theta^{-1}|Mx - My| \leq |Tx - Ty| \leq \theta|Mx - My|$$
$$\text{for } |y - x| < p^{-1}, \quad y \in V.$$

Imposing these conditions for countable, dense sets of v's and y's shows that $E(M, p)$ is a Borel set.

The essential step is to show that the $E(M, p)$ cover V_0. If D_x is univalent, then $\delta_0 = \inf \left[|D_x v| : |v| = 1 \right]$ is positive. Take δ so that $\delta \leq (\theta_0 - 1)\delta_0$ and $\delta \leq (1 - \theta_0^{-1})\delta_0$, and then choose in $\mathcal{M}$ an M such that $|Mv - D_x v| \leq \delta|v|$ for all $v \in R^m$. Then $|Mv| \leq |D_x v| + |Mv - D_x v| \leq |D_x v| + \delta|v| \leq \theta_0|D_x v|$, and (19.18) follows from this and a similar inequality going the other way.

As for (19.19), it follows from (19.18) that M is univalent, and so there is a positive η such that $\eta|v| \leq |Mv|$ for all v. Since T is differentiable at x, there exists a p such that $|y - x| < p^{-1}$ implies that $|Ty - Tx - D_x(y - x)| \leq \epsilon\eta|y - x| \leq \epsilon|M(y - x)|$. This and (19.18) imply that $|Ty - Tx| \leq |D_x(x - y)| + |Tx - Ty - D_x(x - y)| \leq \theta_0|M(x - y)| + \epsilon|M(x - y)| \leq \theta|M(x - y)|$. This inequality and a similar one in the other direction give (19.19).

Thus the $E(M, p)$ cover V_0. Now $E(M, p)$ is a countable union of sets B such that diam $B < p^{-1}$. These sets B (for all M in $\mathcal{M}$ and all p), made disjoint in the usual way, together with the corresponding maps M, satisfy the requirements of the theorem. ∎

Also needed is an extension of the fact that isometries preserve Hausdorff measure.

Lemma 4. *Suppose that $A \subset R^k$, $\varphi \colon A \to R^i$, and $\psi \colon A \to R^j$. If $|\varphi x - \varphi y| \le \theta |\psi x - \psi y|$ for $x, y \in A$, then $h_m(\varphi A) \le \theta^m h_m(\psi A)$.*

PROOF. Given ϵ and δ, cover ψA by sets B_n such that $\theta \operatorname{diam} B_n < \epsilon$ and $c_m \Sigma_n (\operatorname{diam} B_n)^m \le h_m(\psi A) + \delta$. The sets $\varphi \psi^{-1} B_n$ cover φA and diam $\varphi \psi^{-1} B_n \le \theta$ diam $B_n < \epsilon$, whence follows $h_{m,\epsilon}(\varphi A) \le \theta^m (h_m(\psi A) + \delta)$. ∎

PROOF OF THEOREM 19.3: FIRST CASE. Suppose first that A is contained in the set V_0 where D_x is univalent. Given $\theta > 1$, choose sets B_l as in Lemma 3 and put $A_l = A \cap B_l$. If C is the unit cube in R^m, it follows by (19.16) and Lemma 4 that $\theta^{-m} h_m(M_l C) \le h_m(D_x C) \le \theta^m h_m(M_l C)$ for $x \in A_l$; by (19.10), $\theta^{-m} |M_l| \le |D_x| \le \theta^m |M_l|$ for $x \in A_l$.* By (19.17) and Lemma 4, $\theta^{-m} h_m(M_l A_l) \le h_m(T A_l) \le \theta^m h_m(M_l A_l)$. Now $h_m(M_l A_l) = |M_l| h_m(A_l)$ by Theorem 19.2, and so $\theta^{-2m} h_m(T A_l) \le \int_{A_l} |D_x| h_m(dx) \le \theta^{2m} h_m(T A_l)$. Adding over A_l and letting $\theta \to 1$ gives (19.14).

SECOND CASE. Suppose that A is contained in the set $V - V_0$ where D_x is not univalent. It is to be shown that each side of (19.14) vanishes. Since the entries of (19.13) are continuous in x and V is a countable union of compact sets, it is no restriction to assume that there exists a constant K such that $|D_x v| \le K|v|$ for $x \in A$, $v \in R^m$. It is also no restriction to assume that $h_m(A) = \lambda_m(A)$ is finite.

Suppose that $0 < \epsilon < K$. Define $T' \colon R^m \to R^k \times R^m$ by $T'x = (Tx, \epsilon x)$. For the corresponding matrix D'_x of derivatives, $D'_x v = (D_x v, \epsilon v)$, and so D'_x is univalent. Fixing x in A for the moment, choose in R^m an orthonormal set $v_1, \dots, v_m$ such that $D_x v_1 = 0$, which is possible because $x \in V - V_0$. Then $|D'_x v_1| = |(0, \epsilon v_1)| = \epsilon$ and $|D'_x v_i| \le |D_x v_i| + \epsilon |v_i| \le 2K$. Since $h_m(P(v_1, \dots, v_m)) = 1$, it follows by Theorem 19.2 that $|D'_x| = h_m(D'_x P(v_1, \dots, v_m))$. But the parallelepiped $D'_x P(v_1, \dots, v_m)$ of dimension m in R^{k+m} has one edge of length ϵ and $m - 1$ edges of length at most $2K$ and hence is isometric with a subset of $[x \in R^m \colon |x_1| \le \epsilon, |x_i| \le 2mK, i = 2, \dots, m]$; therefore, $|D'_x| \le \epsilon(2mK)^{m-1}$. Now (19.14) for the univalent case already treated gives $h_m(TA) \le \int_A \epsilon(2mK)^{m-1} h_m(dx)$. If $\pi \colon R^k \times R^m \to R^k$ is defined by $\pi(z, v) = z$, then $TA = \pi T'A$, and Lemma 4 gives $h_m(TA) \le h_m(T'A) \le \epsilon(2mK)^{m-1} h_m(A)$. Hence $h_m(TA) = 0$.

If D_x is not univalent, $D_x R^m$ is contained in an $(m - 1)$-dimensional subspace

---

* Since the entries of D_x are continuous in x, if y is close enough to x, then $\theta^{-1}|D_x v| \le |D_y v| \le \theta |D_x v|$ for $v \in R^m$, so that by the same argument, $\theta^{-m}|D_x| \le |D_y| \le \theta^m |D_x|$; hence $|D_x|$ is continuous in x.

of R^k and hence is isometric to a set in R^m of Lebesgue measure 0. Thus $|D_x|$ = 0 for $x \in V - V_0$, and both sides of (19.14) do vanish. ∎

Calculations

The h_m on the left in (19.14) and (19.15) can be replaced by λ_m. To evaluate the integrals, however, still requires calculating $|D|$ as defined by (19.10). Here only the case $m = k - 1$ will be considered.*

If D is a $k \times (k - 1)$ matrix, let z be the point in R^k whose ith component is $(-1)^{i+1}$ times the determinant of D with its ith row removed. Then

(19.20) $$|D| = |z|.$$

To see this, note first that for each x in R^k the inner product $x \cdot z$ is the determinant of D augmented by adding x as its first column (expand by cofactors). Hence z is orthogonal to the columns $x_1, \ldots, k_{k-1}$ of D. If $z \neq 0$, let $z_0 = z/|z|$; otherwise, let z_0 be any unit vector orthogonal to $x_1, \ldots, x'_{k-1}$. In either case a rotation and an application of Fubini's theorem shows that $|D| = h_{k-1}(P(x_1, \ldots, x_{k-1})) = \lambda_k(P(z_0, x_1, \ldots, x_{k-1}))$; this last quantity is by Theorem 12.2 the absolute value of the determinant of the matrix with columns $z_0, x_1, \ldots, x_{k-1}$. But, as above, this is $z_0 \cdot z = |z|$.

Example 19.3.   Let V be the unit sphere in R^{k-1}. If $t_i(x) = x_i$ for $i < k$ and $t_k(x) = [1 - \Sigma_{i=1}^{k-1} x_i^2]^{1/2}$, then TV is the top half of the surface of the unit sphere in R^k. The determinants of the $(k - 1) \times (k - 1)$ submatrices of D_x are easily calculated, and $|D_x|$ comes out to $t_k^{-1}(x)$. If A_k is the surface area of the unit sphere in R^k, then by (19.14), $A_k = 2h_{k-1}(TV) = 2 \int_V t_k^{-1}(x)\, dx$. Integrating out one of the variables shows that $A_k = 2\pi V_{k-2}$, where V_{k-2} is the volume of the unit sphere in R^{k-2}. The calculation in Example 18.4 now gives $A_2 = 2\pi$ and

$$A_{2i-1} = \frac{2(2\pi)^{i-1}}{1 \cdot 3 \cdot 5 \cdots (2i - 3)}, \qquad A_{2i} = \frac{(2\pi)^i}{2 \cdot 4 \cdot 6 \cdots (2i - 2)}$$

for $i = 2, 3, \ldots$. Note that $A_k = kV_k$ for $k = 1, 2, \ldots$. By (19.5) the sphere in R^k with radius r has surface area $r^{k-1}A_k$. ∎

PROBLEMS

19.1.   Describe $S_2 S_1 A$ for the general triangle A in the plane.

19.2.   Show that in (19.1) finite coverings by open sets suffice if A is compact: for

---
* For the general case, see SPIVAK.

positive ϵ there exist finitely many open sets B_n such that diam $B_n < \epsilon$ and $c_m \Sigma_n$ (diam $B_n)^m < h_{m,\epsilon}(A) + \epsilon$.

19.3. ↑ Let f be an *arc* or *curve*—a continuous mapping of an interval $[a, b]$ into R^k—and let $f[a, b] = [f(t): a \le t \le b]$ be its *trace* or graph. The *arc length* $L(f)$ of f is the supremum of the lengths of inscribed polygons:

$$L(f) = \sup \sum_{i=1}^{n} |f(t_i) - f(t_{i-1})|,$$

where the supremum extends over sets $\{t_i\}$ for which $a = t_0 \le t_1 \le \cdots \le t_n = b$. The curve is *rectifiable* if $L(f)$ is finite.

(a) Let $l[u, v]$ be the length of f restricted to a subinterval $[u, v]$. Show that $l[u, v] = l[u, t] + l[t, v]$ for $u \le t \le v$.

(b) Let $l(t)$ be the length of f restricted to $[a, t]$. Show for $s \le t$ that $|f(t) - f(s)| \le l[s, t] = l(t) - l(s)$ and deduce that $h_1(f[a, b]) \le L(f)$: the one-dimensional measure of the trace is at most the length of the arc.

(c) Show that $h_1(f[a, b]) < L(f)$ is possible.

(d) Show that $|f(b) - f(a)| \le h_1(f[a, b])$.

Hint: Use a compactness argument—cover $f[a, b]$ by finitely many open sets $B_1, \ldots, B_N$ such that $\Sigma_{n=1}^{N}$ diam $B_n < h_{1,\epsilon}(f[a, b]) + \epsilon$.

(e) Show that $h_1(f[a, b]) = L(f)$ if the arc has no multiple points (f is one-to-one).

19.4. ↑ Suppose the arc is smooth in the sense that $f(t) = (f_1(t), \ldots, f_k(t))$ has continuous derivative $f'(t) = (f'_1(t), \ldots, f'_k(t))$. Show that $L(f) = \int_a^b |f'(t)| \, dt$. The curve is said to be parameterized by arc length if $|f'(t)| \equiv 1$, because in that case $f[s, t]$ has length $t - s$.

19.5. ↑ Suppose that $f(t) = (x(t), y(t))$, $a \le t \le b$, is parametrized by arc length. For $a \le \xi \le b$, define $\psi(\xi, \eta) = (x(\xi), y(\xi), \eta)$. Interpret the transformation geometrically and show that it preserves area and arc length.

19.6. ↑ (a) Calculate the length of the helix $(\sin \theta, \cos \theta, \theta)$, $0 \le \theta \le 2\pi$.

(b) Calculate the area of the helical surface $(t \sin \theta, t \cos \theta, \theta)$, $0 \le \theta \le 2\pi$, $0 \le t \le 1$.

19.7. Since arc length can be approximated by the lengths of incribed polygons, it might be thought that surface area could be approximated by the areas of inscribed polyhedra. The matter is very complicated, however, as an example will show. Split a 1 by 2π rectangle into mn rectangles, each m^{-1} by $2\pi n^{-1}$, and split each of these into four triangles by means of the two diagonals. Bend the large rectangle into a cylinder of height 1 and circumference 2π, and use the vertices of the $4mn$ triangles as the vertices of an inscribed polyhedron with $4mn$ flat triangular faces. Show that the resulting polyhedron has area

$$A_{mn} = 2n \sin \frac{\pi}{2n} + n \sin \frac{\pi}{n} \sqrt{1 + 16 \, m^2 \sin^4 \frac{\pi}{2n}}.$$

Show that as $m, n \to \infty$, the limit points of A_{mn} fill out the interval $[2\pi, \infty]$.

19.8.   A point on the unit sphere in R^3 is specified by its longitude θ, $0 \le \theta < 2\pi$, and its latitude ϕ, $-\pi/2 \le \phi \le \pi/2$. Calculate the surface area of the sector $\theta_1 < \theta < \theta_2$, $\phi_1 < \phi < \phi_2$.

19.9.   By (18.22), (19.3) can be written $c_m = 2^{-m}\pi^{m/2}/\Gamma(1 + m/2)$, and this is defined for all $m \ge 0$. Thus h_m is well defined for all m. Show that h_0 is counting measure.

19.10.   ↑ **(a)**   Show that, if $h_m(A) < \infty$ and $\delta > 0$, then $h_{m+\delta}(A) = 0$.
(b)   Because of this, there exists a nonnegative number dim A, called the *Hausdorff dimension* of A, such that $h_m(A) = \infty$ if $m < \dim A$ and $h_m(A) = 0$ if $m > \dim A$. Thus $0 < h_m(A) < \infty$ implies that dim $A = m$. Show that "most" sets in R^k specified by m smooth parameters have (integral) dimension m.
(c)   Show that dim $A \le \dim B$ if $A \subset B$ and that dim $\cup_n A_n = \sup_n \dim A_n$.

19.11.   ↑ Show that $h_m(A)$ is well defined by (19.1) and (19.2) for sets A in an arbitrary metric space. Show that h_m is an outer measure, whatever the space. Show that part (a) of the preceding problem goes through, so that Hausdorff dimension is defined for all sets in all metric spaces, and show that part (c) again holds.

19.12.   Let $x = (x_1, \dots, x_k)$ range over the surface of the unit sphere in R^k and consider the set where $\alpha_u < x_u \le \beta_u$, $1 \le u \le t$; here $t \le k - 2$ and $-1 \le \alpha_u \le \beta_u \le 1$. Show that its area is

$$(19.21) \qquad A_{k-t} \int_S \left[1 - \sum_{i=1}^t x_i^2 \right]^{(k-t-2)/2} dx_1 \cdots dx_t,$$

where A_{k-t} is as in Example 19.3 and S is the set where $\sum_{i=1}^t x_i^2 \le 1$ and $\alpha_u < x_u \le \beta_u$, $1 \le u \le t$.

CHAPTER 4

Random Variables
and Expected Values

SECTION 20. RANDOM VARIABLES AND DISTRIBUTIONS

This section and the next cover random variables and the machinery for dealing with them—expected values, distributions, moment generating functions, independence, convolution.

Random Variables and Vectors

A *random variable* on a probability space $(\Omega, \mathcal{F}, P)$ is a real-valued function $X = X(\omega)$ measurable $\mathcal{F}$. Sections 5 through 9 dealt with random variables of a special kind, namely simple random variables, those with finite range. All concepts and facts concerning real measurable functions carry over to random variables; any changes are matters of viewpoint, notation, and terminology only.

The positive and negative parts X^+ and X^- of X are defined as in (15.4) and (15.5). Theorem 13.5 also applies: Define

$$(20.1) \quad \psi_n(x) = \begin{cases} (k-1)2^{-n} & \text{if } (k-1)2^{-n} \le x < k2^{-n}, \quad 1 \le k \le n2^n, \\ n & \text{if } x \ge n. \end{cases}$$

If X is nonnegative and $X_n = \psi_n(X)$, then $0 \le X_n \uparrow X$. If X is not necessarily nonnegative, define

$$(20.2) \qquad\qquad X_n = \begin{cases} \psi_n(X) & \text{if } X \ge 0, \\ -\psi_n(-X) & \text{if } X \le 0. \end{cases}$$

(This is the same as (13.6).) Then $0 \le X_n(\omega) \uparrow X(\omega)$ if $X(\omega) \ge 0$ and $0 \ge X_n(\omega)$

219

$\downarrow X(\omega)$ if $X(\omega) \le 0$; and $|X_n(\omega)| \uparrow |X(\omega)|$ for every ω. The random variable X_n is in each case simple.

A *random vector* is a mapping from Ω to R^k measurable $\mathcal{F}$. Any mapping from Ω to R^k must have the form $\omega \to X(\omega) = (X_1(\omega), \dots, X_k(\omega))$, where each $X_i(\omega)$ real; as shown in Section 13 (see (13.2)), X is measurable if and only if each X_i is. Thus a random vector is simply a k-tuple $X = (X_1, \dots, X_k)$ of random variables.

Subfields

If $\mathcal{G}$ is a σ-field for which $\mathcal{G} \subset \mathcal{F}$, a k-dimensional random vector X is of course measurable $\mathcal{G}$ if $[\omega: X(\omega) \in H] \in \mathcal{G}$ for every H in $\mathcal{R}^k$. The σ-field $\sigma(X)$ generated by X is the smallest σ-field with respect to which it is measurable.

As explained in Sections 4 and 5, a sub-σ-field corresponds to partial information about ω. The information contained in $\sigma(X) = \sigma(X_1, \dots, X_k)$ consists of the k numbers $X_1(\omega), \dots, X_k(\omega)$.* The following theorem is instructive in this connection, although only part (i) is used much. It is the analogue of Theorem 5.1, but there are technical complications in its proof.

Theorem 20.1. *Let $X = (X_1, \dots, X_k)$ be a random vector.*

(i)   *The σ-field $\sigma(X_1, \dots, X_k)$ consists exactly of the sets $[X \in H]$ for $H \in \mathcal{R}^k$.*

(ii)   *In order that a random variable Y be measurable $\sigma(X_1, \dots, X_k)$ it is necessary and sufficient that there exist a measurable map $f: R^k \to R^1$ such that $Y(\omega) = f(X_1(\omega), \dots, X_k(\omega))$ for all ω.*

PROOF OF (i).   The class $\mathcal{G}$ of sets of the form $[X \in H]$ for $H \in \mathcal{R}^k$ is a σ-field. Since X is measurable $\sigma(X)$, $\mathcal{G} \subset \sigma(X)$. Since X is measurable $\mathcal{G}$, $\sigma(X) \subset \mathcal{G}$. ∎

PROOF OF (ii).   Measurability of f refers of course to measurability $\mathcal{R}^k/\mathcal{R}^1$. The sufficiency is easy: if such an f exists, Theorem 13.1 (ii) implies that Y is measurable $\sigma(X_1, \dots, X_k)$.

To prove necessity,[†] suppose at first that Y is a simple random variable and let $y_1, \dots, y_m$ be its different possible values. Since $A_i = [\omega: Y(\omega) = y_i]$ lies in $\sigma(X_1, \dots, X_k)$, it must by part (i) have the form $[\omega: (X_1(\omega), \dots, X_k(\omega)) \in H_i]$ for some H_i in $\mathcal{R}^k$. Put $f = \Sigma_i y_i I_{H_i}$; certainly f is measurable. Since the A_i are disjoint, no $X(\omega)$ can lie in more than one H_i (even though the latter need not be disjoint), and hence $f(X(\omega)) = Y(\omega)$.

To treat the general case, consider simple random variables Y_n such that $Y_n(\omega) \to Y(\omega)$ for each ω. For each n, there is a measurable function $f_n: R^k \to R^1$ such that $Y_n(\omega) = f_n(X_1(\omega), \dots, X_k(\omega))$ for all ω. Let M be the set of x in R^k for which $\{f_n(x)\}$ converges;

---

* The partition defined by (4.14) consists of the sets $[\omega: X(\omega) = x]$ for $x \in R^k$.

† For a general version of this argument, see Problem 13.6.

by Theorem 13.4(iii), M lies in $\mathcal{R}^k$. For x in M, let $f(x) = \lim_n f_n(x)$, and let $f(x) = 0$ for x in $R^k - M$ (in case it is nonempty); redefine $f_n(x) = 0$ for x in $R^k - M$. Then f is measurable by Theorem 13.4(ii). For each ω, $Y(\omega) = \lim_n Y_n(\omega) = \lim_n f_n(X_1(\omega),$ $\ldots, X_k(\omega))$; this implies in the first place that $(X_1(\omega), \ldots, X_k(\omega))$ lies in M and in the second place that $\lim_n f_n(X_1(\omega), \ldots, X_k(\omega)) = f(X_1(\omega), \ldots, X_k(\omega))$.   ■

Distributions

The distribution or law of a random variable X was in Section 14 defined as the probability measure on the line given by $\mu = PX^{-1}$ (see (13.7)), or

(20.3) $\mu(A) = P[X \in A], \quad A \in \mathcal{R}^1.$

The distribution function of X was defined by

(20.4) $F(x) = \mu(-\infty, x] = P[X \le x]$

for real x. The left-hand limit satisfies

(20.5) $\begin{cases} F(x-) = \mu(-\infty, x) = P[X < x], \\ F(x) - F(x-) = \mu\{x\} = P[X = x], \end{cases}$

and F has at most countably many discontinuities. Further, F is nondecreasing and right-continuous, and $\lim_{x\to-\infty} F(x) = 0$, $\lim_{x\to\infty} F(x) = 1$. By Theorem 14.1, for each F with these properties there exists on some probability space a random variable having F as its distribution function.

A support for μ is a Borel set S for which $\mu(S) = 1$. A random variable, its distribution, and its distribution function are *discrete* if μ has a countable support $S = \{x_1, x_2, \ldots\}$. In this case μ is completely determined by the values $\mu\{x_1\}$, $\mu\{x_2\}, \ldots$.

A familiar discrete distribution is the *binomial:*

(20.6) $P[X = r] = \mu\{r\} = \binom{n}{r} p^r(1 - p)^{n-r}, \quad r = 0, 1, \ldots, n.$

There are many random variables, on many spaces, with this distribution: If $\{X_k\}$ is an independent sequence such that $P[X_k = 1] = p$ and $P[X_k = 0] = 1 - p$ (see Theorem 5.2), then X could be $\sum_{i=1}^n X_i$, or $\sum_{i=9}^{8+n} X_i$, or the sum of any n of the X_i. Or Ω could be $\{0, 1, \ldots, n\}$ if $\mathcal{F}$ consists of all subsets, $P\{r\} = \mu\{r\}$, $r = 0, 1, \ldots, n$, and $X(r) \equiv r$. Or again the space and random variable could be those given by the construction in either of the two proofs of Theorem 14.1. These examples show that, although the distribution of a random variable X contains all the information about the probabilistic behavior of X itself, it contains beyond this no further information about the underlying probability space $(\Omega, \mathcal{F}, P)$ or about the interaction of X with other random variables on the space.

Another common discrete distribution is the *Poisson* distribution with parameter $\lambda > 0$:

$$(20.7) \qquad P[X = r] = \mu\{r\} = e^{-\lambda}\frac{\lambda^r}{r!}, \qquad r = 0, 1, \dots .$$

A *constant* c can be regarded as a discrete random variable with $X(\omega) \equiv c$. In this case $P[X = c] = \mu\{c\} = 1$. For an artificial discrete example, let $\{x_1, x_2, \dots\}$ be an enumeration of the rationals and put

$$(20.8) \qquad\qquad \mu\{x_r\} = 2^{-r};$$

the point of the example is that the support need not be contained in a lattice.

A random variable and its distribution have *density* f with respect to Lebesgue measure if f is a nonnegative function on R^1 and

$$(20.9) \qquad P[X \in A] = \mu(A) = \int_A f(x)\, dx, \qquad A \in \mathcal{R}^1.$$

In other words, the requirement is that μ have a density with respect to λ in the sense of (16.10). The density is assumed to be with respect to λ if no other measure is specified.

Taking $A = R^1$ in (20.9) shows that f must integrate to 1. Note that f is determined only to within a set of Lebesgue measure 0: if $f = g$ except on a set of Lebesgue measure 0, then g can also serve as a density for X and μ.

It follows by Theorem 3.3 that (20.9) holds for every Borel set A if it holds for every interval—that is, if

$$F(b) - F(a) = \int_a^b f(x)\, dx$$

holds for every a and b. Note that F need not differentiate to f everywhere (see (20.13), for example); all that is required is that f integrate properly—that (20.9) hold. On the other hand, if F does differentiate to f and f is continuous, it follows by the fundamental theorem of calculus that f is indeed a density for F. The general question of the relation between differentiation and integration is taken up in Section 31.

For the *exponential distribution* with parameter $\alpha > 0$, the density is

$$(20.10) \qquad\qquad f(x) = \begin{cases} 0 & \text{if } x < 0, \\ \alpha e^{-\alpha x} & \text{if } x \geq 0. \end{cases}$$

The corresponding distribution function

$$(20.11) \qquad\qquad F(x) = \begin{cases} 0 & \text{if } x \leq 0, \\ 1 - e^{-\alpha x} & \text{if } x \geq 0 \end{cases}$$

was studied in Section 14.

For the *normal distribution* with parameters m and σ, $\sigma > 0$,

$$(20.12) \qquad f(x) = \frac{1}{\sqrt{2\pi}\sigma} \exp\left[-\frac{(x-m)^2}{2\sigma^2}\right], \qquad -\infty < x < \infty;$$

a change of variable together with (18.10) shows that f does integrate to 1. For the *standard* normal distribution, $m = 0$ and $\sigma = 1$.

For the *uniform* distribution over an interval $(a, b]$,

$$(20.13) \qquad f(x) = \begin{cases} \dfrac{1}{b-a} & \text{if } a < x \le b, \\ 0 & \text{otherwise.} \end{cases}$$

The distribution function F is useful if it has a simple expression, as in (20.11). It is ordinarily simpler to describe μ via the density $f(x)$ or the discrete probabilities $\mu\{x_r\}$.

If F comes from a density, it is continuous. In the discrete case, F increases in jumps; the example (20.8), in which the points of discontinuity are dense, shows that it may nonetheless be very irregular. There exist distributions that are not discrete but are not continuous either. An example is $\mu(A) = \frac{1}{2}\mu_1(A) + \frac{1}{2}\mu_2(A)$ for μ_1 discrete and μ_2 coming from a density; such mixed cases arise, but they are few. Section 31 has examples of a more interesting kind, namely functions F that are continuous but do not come from any density. These are the functions singular in the sense of Lebesgue; the $Q(x)$ describing bold play in gambling (see (7.35)) turns out to be one of them.

If X has distribution μ and g is a real function of a real variable,

$$(20.14) \qquad P[g(X) \in A] = P[X \in g^{-1}A] = \mu(g^{-1}A).$$

Thus the distribution of $g(X)$ is μg^{-1} in the notation (13.7).

In the case where there is a density, f and F are related by

$$(20.15) \qquad F(x) = \int_{-\infty}^{x} f(t)\, dt.$$

Hence f at its continuity points must be the derivative of F. As noted above, if F has a continuous derivative, this derivative can serve as the density f. Suppose that f is continuous and g is increasing, and let $T = g^{-1}$. The distribution function of $g(X)$ is

$$P[g(X) \le x] = P[X \le T(x)] = F(T(x)).$$

If T is differentiable, this differentiates to

$$(20.16) \qquad \frac{d}{dx} P[g(X) \le x] = f(T(x))T'(x),$$

which is thus the density for $g(X)$ (as follows also by (17.8)).

If X has the normal density (20.12) and $a > 0$, (20.16) shows that $aX + b$ has the normal density with parameters $am + b$ and $a\sigma$. Rather than to calculate (20.16), it is generally simpler to calculate the distribution function of $g(X)$ from first principles and then differentiate; this works even if g is many-to-one:

Example 20.1. If X has the standard normal distribution, then

$$P[X^2 \leq x] = \frac{1}{\sqrt{2\pi}} \int_{-\sqrt{x}}^{\sqrt{x}} e^{-t^2/2} \, dt = \frac{2}{\sqrt{2\pi}} \int_0^{\sqrt{x}} e^{-t^2/2} \, dt$$

for $x > 0$. Hence X^2 has density

$$f(x) = \begin{cases} 0 & \text{if } x \leq 0, \\ \dfrac{1}{\sqrt{2\pi}} x^{-1/2} e^{-x/2} & \text{if } x > 0. \end{cases}$$ ∎

The distribution of a k-dimensional random vector $X = (X_1, \ldots, X_k)$ is the probability measure in R^k defined by

$$(20.17) \qquad \mu(A) = P[(X_1, \ldots, X_k) \in A], \qquad A \in \mathcal{R}^k.$$

Often μ is called the *joint* distribution of the random variables $X_1, \ldots, X_k$. The distribution function is defined by

$$(20.18) \qquad F(x_1, \ldots, x_k) = P[X_1 \leq x_1, \ldots, X_k \leq x_k] = \mu(S_{x_1 \ldots x_k}),$$

where S_x consists of the points "southwest" of x (see (12.9)). The distribution function satisfies the hypotheses of Theorem 12.5. Moreover, F is nondecreasing in each varaible, $F(x_1, \ldots, x_k) \to 0$ if $x_i \to -\infty$ (the other coordinates held fixed), $F(x_1, \ldots, x_k) \to 1$ if all $x_i \to \infty$, and $\Delta_A F \geq 0$ for bounded rectangles A (see (12.12)). For any such F there is by Theorem 12.5 a unique probability measure μ on $\mathcal{R}^k$ such that $\mu(A) = \Delta_A F$ for bounded rectangles A.

There is always a random vector having a given distribution and distribution function: Take $(\Omega, \mathcal{F}, P) = (R^k, \mathcal{R}^k, \mu)$ and $X(\omega) \equiv \omega$. This is the obvious extension of the construction in the first proof of Theorem 14.1.

The distribution may as for the line be discrete in the sense of having countable support. It may have density f with respect to k-dimensional Lebesgue measure: $\mu(A) = \int_A f(x) \, dx$. As in the case $k = 1$, the distribution μ is more fundamental than the distribution function F, and usually μ is described not by F but by a density or by discrete probabilities.

If X is a k-dimensional random vector and $g: R^k \to R^i$ is measurable, then $g(X)$ is an i-dimensional random vector; if the distribution of X is μ, the distribution of $g(X)$ is μg^{-1}, just as in the case $k = 1$—see (20.14). If $g_j: R^k \to R^1$ is defined by $g_j(x_1, \ldots, x_k) = x_j$, then $g_j(X)$ is X_j, and its distribution $\mu_j = \mu g_j^{-1}$

is given by $\mu_j(A) = \mu[(x_1, \ldots, x_k): x_j \in A] = P[X_j \in A]$ for $A \in \mathcal{R}^1$. The μ_j are the *marginal distributions* of μ. If μ has a density f in R^k, then μ_j has over the line the density

(20.19) $f_j(x) =$

$$\int_{R^{k-1}} f(x_1, \ldots, x_{j-1}, x, x_{j+1}, \ldots, x_k) \, dx_1 \cdots dx_{j-1} dx_{j+1} \cdots dx_k,$$

since by Fubini's theorem the right side integrated over A comes to $\mu[(x_1, \ldots, x_k): x_j \in A]$.

Now suppose that $g: U \to V$ is one-to-one, where U and V are open subsets of R^k. Suppose that $T = g^{-1}$ is continuously differentiable, and let $J(x)$ be its Jacobian. If X has a density that vanishes outside U, then $P[g(X) \in A] = P[X \in TA] = \int_{TA} f(y) \, dy$, and so by (17.13),

(20.20) $$P[g(X) \in A] = \int_A f(Tx)|J(x)| \, dx$$

for $A \subset V$. Thus the density for $g(X)$ is $f(Tx)|J(x)|$ on V and 0 elsewhere.

Example 20.2. Suppose that (X_1, X_2) has density $f(x_1, x_2) = (2\pi)^{-1} \exp[-\frac{1}{2}(x_1^2 + x_2^2)]$, and let g be the transformation to polar coordinates. Then $U = R^2$, and V and T are as in Example 17.6. If R and Θ are the polar coordinates of (X_1, X_2), then $(R, \Theta) = g(X_1, X_2)$ has density $(2\pi)^{-1}re^{-r^2/2}$ in V. By (20.19), R has density $re^{-r^2/2}$ on $(0, \infty)$ and Θ is uniformly distributed over $(0, 2\pi)$. ∎

Independence

Random variables $X_1, \ldots, X_k$ are defined to be independent if the σ-fields $\sigma(X_1)$, $\ldots, \sigma(X_k)$ they generate are independent in the sense of Section 4. This concept for simple random variables was studied extensively in Chapter 1; the general case was touched on in Section 14. Since $\sigma(X_i)$ consists of the sets $[X_i \in H]$ for $H \in \mathcal{R}^1$, $X_1, \ldots, X_k$ are independent if and only if

(20.21) $P[X_1 \in H_1, \ldots, X_k \in H_k] = P[X_1 \in H_1] \cdots P[X_k \in H_k]$

for all linear Borel sets $H_1, \ldots, H_k$. Since the intervals $(-\infty, x]$ form a π-system generating $\mathcal{R}^1$, the sets $[X_i \leq x]$ form a π-system generating $\sigma(X_i)$. Therefore, by Theorem 4.2, $X_1, \ldots, X_k$ are independent if and only if

(20.22) $P[X_1 \leq x_1, \ldots, X_k \leq x_k] = P[X_1 \leq x_1] \cdots P[X_k \leq x_k]$

for all real $x_1, \ldots, x_k$. If, for example, the X_i are integer-valued, it is enough that $P[X_1 = n_1, \ldots, X_k = n_k] = P[X_1 = n_1] \cdots P[X_k = n_k]$ for integral $n_1, \ldots, n_k$.

Let $(X_1, \ldots, X_k)$ have distribution μ and distribution function F, and let the X_i have distributions μ_i and distribution functions F_i (the marginals). By (20.21), $X_1, \ldots, X_k$ are independent if and only if μ is product measure in the sense of Section 18:

$$(20.23) \qquad \mu = \mu_1 \times \cdots \times \mu_k.$$

By (20.22), $X_1, \ldots, X_k$ are independent if and only if

$$(20.24) \qquad F(x_1, \ldots, x_k) = F_1(x_1) \cdots F_k(x_k).$$

Suppose that each μ_i has density f_i; by Fubini's theorem, $f_1(y_1) \cdots f_k(y_k)$ integrated over $(-\infty, x_1] \times \cdots \times (-\infty, x_k]$ is just $F_1(x_1) \cdots F_k(x_k)$, so that μ has density

$$(20.25) \qquad f(x) = f_1(x_1) \cdots f_k(x_k)$$

in the case of independence.

If $\mathcal{G}_1, \ldots, \mathcal{G}_k$ are independent σ-fields and X_i is measurable $\mathcal{G}_i$, $i = 1, \ldots, k$, then certainly $X_1, \ldots, X_k$ are independent.

If X_i is a d_i-dimensional random vector, $i = 1, \ldots, k$, then $X_1, \ldots, X_k$ are by definition independent if the σ-fields $\sigma(X_1), \ldots, \sigma(X_k)$ are independent. The theory is just as for random variables: $X_1, \ldots, X_k$ are independent if and only if (20.21) holds for $H_1 \in \mathcal{R}^{d_1}, \ldots, H_k \in \mathcal{R}^{d_k}$. Now $(X_1, \ldots, X_k)$ can be regarded as a random vector of dimension $d = \Sigma_{i=1}^{k} d_i$; if μ is its distribution in $R^d = R^{d_1} \times \cdots \times R^{d_k}$ and μ_i is the distribution of X_i in R^{d_i}, then, just as before, $X_1, \ldots, X_k$ are independent if and only if $\mu = \mu_1 \times \cdots \times \mu_k$. In none of this need the d_i components of a single X_i be themselves independent random variables.

An infinite collection of random variables or random vectors is by definition independent if each finite subcollection is. The argument following (5.7) extends from collections of simple random variables to collections of random vectors:

Theorem 20.2. *Suppose that*

$$(20.26) \qquad \begin{array}{l} X_{11}, X_{12}, \ldots \\ X_{21}, X_{22}, \ldots \\ \cdots\cdots\cdots\cdots \end{array}$$

is an independent collection of random vectors. If $\mathcal{F}_i$ is the σ-field generated by the ith row, then $\mathcal{F}_1, \mathcal{F}_2, \ldots$ are independent.

PROOF.   Let $\mathcal{A}_i$ consist of the finite intersections of sets of the form $[X_{ij} \in H]$ with H a Borel set in a space of the appropriate dimension, and apply Theorem 4.2. The σ-fields $\mathcal{F}_i = \sigma(\mathcal{A}_i)$, $i = 1, \ldots, n$, are independent for each n, and the result follows.   ∎

Each row of (20.26) may be finite or infinite, and there may be finitely or infinitely many rows. As a matter of fact, rows may be uncountable and there may be uncountably many of them.

Suppose that X and Y are independent random vectors with distributions μ and ν in R^j and R^k. Then (X, Y) has distribution $\mu \times \nu$ in $R^j \times R^k = R^{j+k}$. Let x range over R^j and y over R^k. By Fubini's theorem,

$$(20.27) \quad (\mu \times \nu)(M) = \int_{R^j} \nu[y: (x, y) \in M]\mu(dx), \quad M \in \mathcal{R}^{j+k}.$$

Let $M = (A \times R^k) \cap B$, where $A \in \mathcal{R}^j$ and $B \in \mathcal{R}^{j+k}$. Then (20.27) reduces to

$$(20.28) \quad (\mu \times \nu)((A \times R^k) \cap B)$$

$$= \int_A \nu[y: (x, y) \in B]\mu(dx), \quad \begin{matrix} A \in \mathcal{R}^j, \\ B \in \mathcal{R}^{j+k}. \end{matrix}$$

Expressing those formulas in terms of the random vectors themselves gives this result:

Theorem 20.3. *If X and Y are independent random vectors with distributions μ and ν in R^j and R^k, then*

$$(20.29) \quad P[(X, Y) \in M] = \int_{R^j} P[(x, Y) \in M]\mu(dx), \quad M \in \mathcal{R}^{j+k}$$

and

$$(20.30) \quad P[X \in A, (X, Y) \in B]$$

$$= \int_A P[(x, Y) \in B]\mu(dx), \quad A \in \mathcal{R}^j, \quad B \in \mathcal{R}^{j+k}.$$

Of course, $P[(x, Y) \in M] = P[\omega: (x, Y(\omega)) \in M]$.

Example 20.3. Suppose that X and Y are independent exponentially distributed random variables. By (20.29), $P[Y/X \geq z] = \int_0^\infty P[Y/x \geq z]\alpha e^{-\alpha x} dx = \int_0^\infty e^{-\alpha xz}\alpha e^{-\alpha x} dx = (1 + z)^{-1}$. Thus Y/X has density $(1 + z)^{-2}$ for $z \geq 0$. Since $P[X \geq z_1, Y/X \geq z_2] = \int_{z_1}^\infty P[Y/x \geq z_2]\alpha e^{-\alpha x} dx$ by (20.30), the joint distribution of X and Y/X can be calculated as well. ∎

Formulas (20.29) and (20.30) are constantly applied as in this example. There is no virtue in making an issue of each case, however, and the appeal to Theorem 20.3 is usually silent.

Example 20.4. Here is a more complicated argument of the same sort. Let $X_1, \ldots,$ X_n be independent random variables, each uniformly distributed over $[0, t]$. Let Y_k be the kth largest among the X_i, so that $0 \le Y_1 \le \cdots \le Y_n \le t$. The X_i divide $[0, t]$ into $n + 1$ subintervals of lengths $Y_1, Y_2 - Y_1, \ldots, Y_n - Y_{n-1}, t - Y_n$; let M be the maximum of these lengths. Define $\psi_n(t, a) = P[M \le a]$. The problem is to show that

$$(20.31) \qquad \psi_n(t, a) = \sum_{k=0}^{n+1} (-1)^k \binom{n + 1}{k} \left(1 - k\frac{a}{t}\right)^n_+,$$

where $x_+ = (x + |x|)/2$ denotes positive part.

Separate consideration of the possibilities $0 \le a \le t/2, t/2 \le a \le t$, and $t \le a$ disposes of the case $n = 1$. Suppose it is shown that the probability $\psi_n(t, a)$ satisfies the recursion

$$(20.32) \qquad \psi_n(t, a) = n \int_0^a \psi_{n-1}(t - x, a) \left(\frac{t - x}{t}\right)^{n-1} \frac{dx}{t}.$$

Now (as follows by an integration together with Pascal's identity for binomial coefficients) the right side of (20.31) satisfies this same recursion, and so it will follow by induction that (20.31) holds for all n.

In intuitive form, the argument for (20.32) is this: If $[M \le a]$ is to hold, the smallest of the X_i must have some value x in $[0, a]$. If X_1 is the smallest of the X_i, then $X_2, \ldots,$ X_n must all lie in $[x, t]$ and divide it into subintervals of length at most a; the probability of this is $(1 - x/t)^{n-1}\psi_{n-1}(t - x, a)$ because $X_2, \ldots, X_n$ have probability $(1 - x/t)^{n-1}$ of all lying in $[x, t]$, and if they do, they are independent and uniformly distributed there. Now (20.32) results from integrating with respect to the density for X_1 and multiplying by n to account for the fact that any of $X_1, \ldots, X_n$ may be the smallest.

To make this argument rigorous, apply (20.30) for $j = 1$ and $k = n - 1$. Let A be the interval $[0, a]$, and let B consist of the points $(x_1, \ldots, x_n)$ for which $0 \le x_i \le t, x_1$ is the minimum of $x_1, \ldots, x_n$, and $x_2, \ldots, x_n$ divide $[x_1, t]$ into subintervals of length at most a. Then $P[X_1 = \min X_i, M \le a] = P[X_1 \in A, (X_1, \ldots, X_n) \in B]$. Take X_1 for X and $(X_2, \ldots, X_n)$ for Y in (20.30). Since X_1 has density $1/t$,

$$(20.33) \qquad P[X_1 = \min X_i, M \le a] = \int_0^a P[(x, X_2, \ldots, X_n) \in B] \frac{dx}{t}.$$

If C is the event that $x \le X_i \le t$ for $2 \le i \le n$, then $P(C) = (t - x/t)^{n-1}$. A simple calculation shows that $P[X_i - x \le s_i, 2 \le i \le n | C] = \Pi_{i=2}^n (s_i/(t - x))$; in other words, given $C, X_2 - x, \ldots, X_n - x$ are conditionally independent and uniformly distributed over $[0, t - x]$. Now $(X_2, \ldots, X_n)$ are random variables on some probability space $(\Omega, \mathcal{F}, P)$; replacing P by $P(\cdot|C)$ shows that the integrand in (20.33) is the same as that in (20.32). The same argument holds with the index 1 replaced by any k ($1 \le k \le n$), which gives (20.32). (The events $[X_k = \min X_i, Y \le a]$ are not disjoint, but any two intersect in a set of probability 0.) ∎

Sequences of Random Variables

Theorem 5.2 extends to general distributions μ_n.

Theorem 20.4. *If $\{\mu_n\}$ is a finite or infinite sequence of probability measures on $\mathcal{R}^1$, there exists on some probability space $(\Omega, \mathcal{F}, P)$ an independent sequence $\{X_n\}$ of random variables such that X_n has distribution μ_n.*

PROOF.   By Theorem 5.2 there exists on some probability space an independent sequence $Z_1, Z_2, \ldots$ of random variables assuming the values 0 and 1 with probabilities $P[Z_n = 0] = P[Z_n = 1] = \frac{1}{2}$. As a matter of fact, Theorem 5.2 is not needed: take the space to be the unit interval and the $Z_n(\omega)$ to be the digits of the dyadic expansion of ω—the functions $d_n(\omega)$ of Sections 1 and 4.

Relabel the countably many random variables Z_n so that they form a double array,

Put $U_n = \sum_{k=1}^{\infty} Z_{nk} 2^{-k}$. The series certainly converges, and U_n is a random variable by Theorem 13.4. Further, $U_1, U_2, \ldots$ is, by Theorem 20.2, an independent sequence.

Now $P[Z_{ni} = z_i, 1 \le i \le k] = 2^{-k}$ for each sequence $z_1, \ldots, z_k$ of 0's and 1's; hence the 2^k possible values $j 2^{-k}, 0 \le j < 2^k$, of $S_{nk} = \sum_{i=1}^{k} Z_{ni} 2^{-i}$ all have probability 2^{-k}. If $0 \le x < 1$, the number of the $j 2^{-k}$ that lie in $[0, x]$ is $[2^k x] + 1$, the brackets indicating integral part, and therefore $P[S_{nk} \le x] = ([2^k x] + 1)/2^k$. Since $S_{nk}(\omega) \uparrow U_n(\omega)$ as $k \uparrow \infty$, $[S_{nk} \le x] \downarrow [U_n \le x]$ as $k \uparrow \infty$, and so $P[U_n \le x] = \lim_k P[S_{nk} \le x] = \lim_k ([2^k x] + 1)/2^k = x$ for $0 \le x < 1$. Thus U_n is uniformly distributed over the unit interval.

The construction thus far establishes the existence of an independent sequence of random variables U_n each uniformly distributed over $[0, 1]$. Let F_n be the distribution function corresponding to μ_n, and put $\varphi_n(u) = \inf[x: u \le F_n(x)]$ for $0 < u < 1$. This is the inverse used in Section 14—see (14.5). Set $\varphi_n(u) = 0$, say, for u outside $(0, 1)$, and put $X_n(\omega) = \varphi_n(U_n(\omega))$. Since $\varphi_n(u) \le x$ if and only if $u \le F_n(x)$—see the argument following (14.5)—$P[X_n \le x] = P[U_n \le F_n(x)] = F_n(x)$. Thus X_n has distribution function F_n. And by Theorem 20.2, $X_1, X_2, \ldots$ are independent.   ∎

This theorem of course includes Theorem 5.2 as a special case, and its proof does not depend on the earlier result. It serves the same function as Theorem 5.2. The following sections contain many results on infinite sequences of independent random variables, and Theorem 20.4 ensures that they are not vacuous.

Convolution

Let X and Y be independent random variables with distributions μ and ν. Apply (20.27) and (20.29) to the planar set $M = [(x, y): x + y \in H]$ with $H \in \mathcal{R}^1$:

(20.34) $P[X + Y \in H]$

$$= \int_{-\infty}^{\infty} \nu(H - x)\mu(dx) = \int_{-\infty}^{\infty} P[Y \in H - x]\mu(dx).$$

The *convolution* of μ and ν is the measure $\mu * \nu$ defined by

(20.35) $(\mu * \nu)(H) = \int_{-\infty}^{\infty} \nu(H - x)\mu(dx),$ $H \in \mathcal{R}^1.$

If X and Y are independent and have distributions μ and ν, (20.34) shows that $X + Y$ has distribution $\mu * \nu$. Since addition of random variables is commutative and associative, the same is true of convolution: $\mu * \nu = \nu * \mu$ and $\mu * (\nu * \eta) = (\mu * \nu) * \eta$.

If F and G are the distribution functions corresponding to μ and ν, the distribution function corresponding to $\mu * \nu$ is denoted $F * G$. Taking $H = (-\infty, y]$ in (20.35) shows that

(20.36) $(F * G)(y) = \int_{-\infty}^{\infty} G(y - x)dF(x).$

(See (17.16) for the notation $dF(x)$.) If G has density g, then $G(y - x) = \int_{-\infty}^{y-x} g(s)\, ds = \int_{-\infty}^{y} g(t - x)\, dt$, and so the right side of (20.36) is $\int_{-\infty}^{y} [\int_{-\infty}^{\infty} g(t - x)dF(x)]\, dt$ by Fubini's theorem. Thus $F * G$ has density $F * g$, where

(20.37) $(F * g)(y) = \int_{-\infty}^{\infty} g(y - x)dF(x);$

this holds if G has density g. If, in addition, F has density f, (20.37) is denoted $f * g$ and reduces by (16.11) to

(20.38) $(f * g)(y) = \int_{-\infty}^{\infty} g(y - x)f(x)\, dx.$

This defines convolution for densities, and $\mu * \nu$ has density $f * g$ if μ and ν have densities f and g. The formula (20.38) can be used for many explicit calculations.

Example 20.5.   Let $X_1, \ldots, X_k$ be independent random variables, each with the exponential density (20.10). Define g_k by

(20.39) $g_k(x) = \alpha \dfrac{(\alpha x)^{k-1}}{(k - 1)!} e^{-\alpha x},$ $x \geq 0,$ $k = 1, 2, \ldots;$

put $g_k(x) = 0$ for $x \leq 0$. Now

$$(g_{k-1} * g_1)(y) = \int_0^y g_{k-1}(y - x)g_1(x)\, dx,$$

which reduces to $g_k(y)$. Thus $g_k = g_{k-1} * g_1$, and since g_1 coincides with (20.10), it follows by induction that the sum $X_1 + \cdots + X_k$ has density g_k. The corresponding distribution function is

$$(20.40) \quad G_k(x) = 1 - e^{-\alpha x} \sum_{i=0}^{k-1} \frac{(\alpha x)^i}{i!} = \sum_{i=k}^{\infty} e^{-\alpha x} \frac{(\alpha x)^i}{i!}, \qquad x \geq 0,$$

as follows by differentiation. ∎

Example 20.6. Suppose that X has the normal density (20.12) with $m = 0$ and that Y has the same density with τ in place of σ. If X and Y are independent, then $X + Y$ has density

$$\frac{1}{2\pi\sigma\tau} \int_{-\infty}^{\infty} \exp\left[-\frac{(y-x)^2}{2\sigma^2} - \frac{x^2}{2\tau^2} \right] dx.$$

A change of variable $u = x (\sigma^2 + \tau^2)^{1/2}/\sigma\tau$ reduces this to

$$\frac{1}{\sqrt{2\pi(\sigma^2 + \tau^2)}} \frac{1}{\sqrt{2\pi}} \int_{-\infty}^{\infty} \exp\left[-\frac{1}{2}\left(u - y\frac{\tau/\sigma}{\sqrt{\sigma^2 + \tau^2}} \right)^2 - \frac{y^2}{2(\sigma^2 + \tau^2)} \right] du$$

$$= \frac{1}{\sqrt{2\pi(\sigma^2 + \tau^2)}} e^{-y^2/2(\sigma^2 + \tau^2)}.$$

Thus $X + Y$ has the normal density with $m = 0$ and with $\sigma^2 + \tau^2$ in place of σ^2. ∎

If μ and ν are arbitrary finite measures on the line, their convolution is defined by (20.35) even if they are not probability measures.

Convergence in Probability

Random variables X_n converge in probability to X, written $X_n \to_P X$, if

$$(20.41) \quad \lim_n P[|X_n - X| \geq \epsilon] = 0$$

for each positive ϵ.* If $X_n \to X$ with probability 1, then $\lim \sup_n [|X_n - X| \geq \epsilon]$ has probability 0, and it follows by Theorem 4.1 that $X_n \to_P X$. But the converse does not hold: There exist sets A_n such that $P(A_n) \to 0$ and $P(\lim \sup_n A_n) = 1$. For example in the unit interval, let A_1, A_2 be the dyadic intervals of order one, $A_3, \ldots, A_6$, those of order two, and so on (see (4.27) or (1.31)). If $X_n = I_{A_n}$ and $X \equiv 0$, then $X_n \to_P X$ but $P[\lim_n X_n = X] = 0$.

Theorem 20.5. *A necessary and sufficient condition for $X_n \to_P X$ is that each*

* This is often expressed p $\lim_n X_n = X$.

subsequence $\{X_{n_k}\}$ contain a further subsequence $\{X_{n_{k(i)}}\}$ such that $X_{n_{k(i)}} \to X$ with probability 1 as $i \to \infty$.

PROOF. If $X_n \to_P X$, given $\{n_k\}$ choose a subsequence $\{n_{k(i)}\}$ so that $k \geq k(i)$ implies that $P[|X_{n_k} - X| \geq i^{-1}] < 2^{-i}$. By the first Borel-Cantelli lemma there is probability 1 that $|X_{n_{k(i)}} - X| < i^{-1}$ for all but finitely many i. Therefore, $\lim_i X_{n_{k(i)}} = X$ with probability 1.

If X_n does not converge to X in probability, there is some positive ϵ for which $P[|X_{n_k} - X| \geq \epsilon] > \epsilon$ along some sequence $\{n_k\}$. No subsequence of $\{X_{n_k}\}$ can converge in probability to X, and hence none can converge to X with probability 1. ∎

It follows from this theorem that if f is continuous and $X_n \to_P X$, then $f(X_n) \to_P f(X)$.

In nonprobabilistic contexts, convergence in probability becomes *convergence in measure:* If f_n and f are real measurable functions on a measure space $(\Omega, \mathcal{F}, \mu)$, and if $\mu[\omega: |f(\omega) - f_n(\omega)| \geq \epsilon] \to 0$ for each $\epsilon > 0$, then f_n converges in measure to f.

The Glivenko-Cantelli Theorem*

The *empirical distribution function* for random variables $X_1, \ldots, X_n$ is the distribution function $F_n(x, \omega)$ with a jump of n^{-1} at each $X_k(\omega)$:

$$(20.42) \qquad F_n(x, \omega) = \frac{1}{n} \sum_{k=1}^{n} I_{(-\infty, x]}(X_k(\omega)).$$

If the X_k have a common unknown distribution function $F(x)$, $F_n(x, \omega)$ is its natural estimate. The estimate has the right limiting behavior, according to the *Glivenko-Cantelli theorem*:

Theorem 20.6. *Suppose that $X_1, X_2, \ldots$ are independent and have a common distribution function F; put $D_n(\omega) = \sup_x |F_n(x, \omega) - F(x)|$. Then $D_n \to 0$ with probability 1.*

For each x, $F_n(x, \omega)$ as a function of ω is a random variable. By right continuity, the supremum above is unchanged if x is restricted to the rationals, and therefore D_n is a random variable.

The summands in (20.42) are independent, identically distributed simple random variables, and so by the strong law of large numbers (Theorem 6.1), for each x there is a set A_x of probability 0 such that

---

* This topic may be omitted.

(20.43) $$\lim_n F_n(x, \omega) = F(x)$$

for $\omega \notin A_x$. But Theorem 20.6 says more, namely that (20.43) holds for ω outside some set A of probability 0, where A does not depend on x—as there are uncountably many of the sets A_x, conceivably their union might necessarily have positive measure. Further, the convergence in (20.43) is uniform in x. Of course, the theorem implies that with probability 1 there is weak convergence $F_n(x, \omega)$ $\Rightarrow F(x)$ in the sense of Section 14.

PROOF OF THE THEOREM. As already observed, the set A_x where (20.43) fails has probability 0. Another application of the strong law of large numbers, with $I_{(-\infty,x)}$ in place of $I_{(-\infty,x]}$ in (20.42), shows that (see (20.5)) $\lim_n F_n(x-, \omega)$ $= F(x-)$ except on a set B_x of probability 0. Let $\varphi(u) = \inf [x: u \le F(x)]$ for $0 < u < 1$ (see (14.5)), and put $x_{m,k} = \varphi(k/m), m \ge 1, 1 \le k \le m$. It is not hard to see that $F(\varphi(u)-) \le u \le F(\varphi(u))$; hence $F(x_{m,k}-) - F(x_{m,k-1}) \le m^{-1}$, $F(x_{m,1}-) \le m^{-1}$, and $F(x_{m,m}) \ge 1 - m^{-1}$. Let $D_{m,n}(\omega)$ be the maximum of the quantities $|F_n(x_{m,k}, \omega) - F(x_{m,k})|$ and $|F_n(x_{m,k}-, \omega) - F(x_{m,k}-)|$ for $k = 1, \ldots, m$.

If $x_{m,k-1} \le x < x_{m,k}$, then $F_n(x, \omega) \le F_n(x_{m,k}-, \omega) \le F(x_{m,k}-) + D_{m,n}(\omega)$ $\le F(x) + m^{-1} + D_{m,n}(\omega)$ and $F_n(x, \omega) \ge F_n(x_{m,k-1}, \omega) \ge F(x_{m,k-1}) -$ $D_{m,n}(\omega) \ge F(x) - m^{-1} - D_{m,n}(\omega)$. Together with similar arguments for the cases $x < x_{m,1}$ and $x \ge x_{m,m}$, this shows that

(20.44) $$D_n(\omega) \le D_{m,n}(\omega) + m^{-1}.$$

If ω lies outside the union A of all the $A_{x_{mk}}$ and $B_{x_{mk}}$, then $\lim_n D_{m,n}(\omega) = 0$ and hence $\lim_n D_n(\omega) = 0$ by (20.44). But A has probability 0. ∎

PROBLEMS

20.1. 2.10↑ A necessary and sufficient condition for a σ-field $\mathscr{G}$ to be separable is that $\mathscr{G} = \sigma(X)$ for some random variable X.

20.2. If X is a positive random variable with density f, X^{-1} has density $f(1/x)/x^2$. Prove this by (20.16) and by a direct argument.

20.3. Suppose that a two-dimensional distribution function F has a continuous density f. Show that $f(x, y) = \partial^2 F(x, y)/\partial x \, \partial y$.

20.4. The construction in Theorem 20.4 requires only Lebesgue measure on the unit interval. Use the theorem to prove the existence of Lebesgue measure on the unit cube in R^k and then extend Lebesgue measure to all of R^k by translation and addition.

20.5. 19.8↑ Let Θ and Φ be the longitude and latitude of a random point on the surface of the unit sphere in R^3; probability is proportional to surface area. Show that Θ and Φ are independent, Θ is uniformly distributed over $[0, 2\pi)$, and Φ is distributed over $[-\pi/2, +\pi/2]$ with density $\frac{1}{2} \cos \phi$.

20.6. Suppose that A, B, and C are positive, independent random variables with distribution function F. Show that the quadratic $Az^2 + Bz + C$ has real zeros with probability $\int_0^\infty \int_0^\infty F(x^2/4y) \, dF(x) \, dF(y)$.

20.7. Show that $X_1, X_2, \ldots$ are independent if $\sigma(X_1, \ldots, X_{n-1})$ and $\sigma(X_n)$ are independent for each n.

20.8. Let $X_0, X_1, \ldots$ be a persistent, irreducible Markov chain and for a fixed state j let $T_1, T_2, \ldots$ be the times of the successive passages through j. Let $Z_1 = T_1$ and $Z_n = T_n - T_{n-1}$, $n \geq 2$. Show that $Z_1, Z_2, \ldots$ are independent and that $P[Z_n = k] = f_{jj}^{(k)}$ for $n \geq 2$.

20.9. Suppose that $F_1, F_2, \ldots$ are distribution functions and $p_1, p_2, \ldots$ are nonnegative and add to 1. Show that $F(x) = \sum_{n=1}^\infty p_n F_n(x)$ is a distribution function. Show that, if $F_n(x)$ has density $f_n(x)$ for each n, then $F(x)$ has density $\sum_{n=1}^\infty p_n f_n(x)$.

20.10. Let $X_1, \ldots, X_n$ be independent, identically distributed random variables and let π be a permutation of $1, 2, \ldots, n$. Show that $P[(X_1, \ldots, X_n) \in H] = P[(X_{\pi 1}, \ldots, X_{\pi n}) \in H]$ for $H \in \mathcal{R}^n$.

20.11. $\uparrow$ *Ranks and records.* Let $X_1, X_2, \ldots$ be independent random variables with a common continuous distribution function. Let B be the ω-set where $X_m(\omega) = X_n(\omega)$ for some pair m, n of distinct integers, and show that $P(B) = 0$. Remove B from the space Ω on which the X_n are defined. This leaves the joint distributions of the X_n unchanged and makes ties impossible.

Let $T^{(n)}(\omega) = (T_1^{(n)}(\omega), \ldots, T_n^{(n)}(\omega))$ be that permutation $(t_1, \ldots, t_n)$ of $(1, \ldots, n)$ for which $X_{t_1}(\omega) < X_{t_2}(\omega) < \cdots < X_{t_n}(\omega)$. Let Y_n be the rank of X_n among $X_1, \ldots, X_n$: $Y_n = r$ if and only if $X_i < X_n$ for exactly $r - 1$ values of i preceding n.

(a)   Show that $T^{(n)}$ is uniformly distributed over the $n!$ permutations.

(b)   Show that $P[Y_n = r] = 1/n$, $1 \leq r \leq n$.

(c)   Show that Y_k is measurable $\sigma(T^{(n)})$ for $k \leq n$.

(d)   Show that $Y_1, Y_2, \ldots$ are independent.

20.12. $\uparrow$ *Record values.* Let A_n be the event that a *record* occurs at time n: $\max_{k<n} X_k < X_n$.

(a)   Show that $A_1, A_2, \ldots$ are independent and $P(A_n) = 1/n$.

(b)   Show that no record stands forever.

(c)   Let N_n be the time of the first record after time n. Show that $P[N_n = n + k] = n(n + k - 1)^{-1}(n + k)^{-1}$.

20.13. Use Fubini's theorem to prove that convolution of finite measures is commutative and associative.

20.14. In Example 20.6, remove the assumption that m is 0 for each of the normal densities.

20.15. Suppose that X and Y are independent and have densities. Use (20.20) to find the joint density for $(X + Y, X)$ and then use (20.19) to find the density for $X + Y$. Check with (20.38).

20.16. If $F(x - \epsilon) < F(x + \epsilon)$ for all positive ϵ, x is a *point of increase* of F (see Problem 12.8). If $F(x -) < F(x)$, then x is an *atom* of F.

(a) Show that, if x and y are points of increase of F and G, then $x + y$ is a point of increase of $F * G$.

(b) Show that, if x and y are atoms of F and G, then $x + y$ is an atom of $F * G$.

20.17. Suppose that μ and ν consist of masses α_n and β_n at n, $n = 0, 1, 2, \ldots$ Show that $\mu * \nu$ consists of a mass of $\sum_{k=0}^{n} \alpha_k \beta_{n-k}$ at n, $n = 0, 1, 2, \ldots$

20.18. ↑ Show that two Poisson distributions (the parameters may differ) convolve to a Poisson distribution.

20.19. Show that, if $X_1, \ldots, X_n$ are independent and have the normal distribution (20.12) with $m = 0$, then the same is true of $(X_1 + \cdots + X_n)/\sqrt{n}$.

20.20. The *Cauchy* distribution has density

(20.45) $$c_u(x) = \frac{1}{\pi} \frac{u}{u^2 + x^2}, \quad -\infty < x < \infty,$$

for $u > 0$.

(a) Show that $c_u * c_v = c_{u+v}$.

Hint: Expand the convolution integrand in partial fractions.

(b) Show that, if $X_1, \ldots, X_n$ are independent and have density c_u, then $(X_1 + \cdots + X_n)/n$ has density c_u as well. Compare with the convolution law in Problem 20.19.

20.21. ↑ (a) Show that, if X and Y are independent and have the standard normal density, then X/Y has the Cauchy density with $u = 1$.

(b) Show that, if X has the uniform distribution over $(-\pi/2, \pi/2)$, then $\tan X$ has the Cauchy distribution with $u = 1$.

20.22. 18.17↑ Let $X_1, \ldots, X_n$ be independent, each having the standard normal distribution. Show that

$$\chi_n^2 = X_1^2 + \cdots + X_n^2$$

has density

(20.46) $$\frac{1}{2^{n/2}\Gamma(n/2)} x^{(n/2)-1} e^{-x/2}$$

over $(0, \infty)$. This is called the *chi-squared distribution with n degrees of freedom*.

20.23. ↑ The *gamma distribution* has density

(20.47) $$f(x; \alpha, u) = \frac{\alpha^u}{\Gamma(u)} x^{u-1} e^{-\alpha x}$$

over $(0, \infty)$ for positive parameters α and u. Check that (20.47) integrates to 1. Show that

(20.48) $$f(\cdot; \alpha, u) * f(\cdot; \alpha, v) = f(\cdot; \alpha, u + v).$$

Note that (20.46) is $f(x; \frac{1}{2}, n/2)$, and from (20.48) deduce again that (20.46) is the density of χ_n^2. Note that the exponential density (20.10) is $f(x; \alpha, 1)$, and from (20.48) deduce (20.39) once again.

20.24. ↑ The *beta function* is defined for positive u and v by

$$B(u, v) = \int_0^1 (1 - x)^{u-1} x^{v-1} \, dx.$$

From (20.48) deduce that

$$B(u, v) = \frac{\Gamma(u)\Gamma(v)}{\Gamma(u + v)}.$$

Show that

$$B^{-1}(u, v) = \binom{u + v - 2}{u - 1}$$

for integral u and v.

20.25. 20.23↑ Let $N, X_1, X_2, \ldots$ be independent, where $P[N = n] = q^{n-1}p, n \geq 1$, and each X_k has the exponential density $f(x; \alpha, 1)$. Show that $X_1 + \cdots + X_N$ has density $f(x; \alpha p, 1)$.

20.26. Let $A_{nm}(\epsilon) = P[|Z_k - Z| < \epsilon, n \leq k \leq m]$. Show that $Z_n \to Z$ with probability 1 if and only if $\lim_n \lim_m P(A_{nm}(\epsilon)) = 1$ for all positive ϵ, whereas $Z_n \to_P Z$ if and only if $\lim_n P(A_{nn}(\epsilon)) = 1$ for all positive ϵ.

20.27. **(a)** Suppose that $f: R^2 \to R^1$ is continuous. Show that $X_n \to_P X$ and $Y_n \to_P Y$ imply that $f(X_n, Y_n) \to_P f(X, Y)$.
(b) Show that addition and multiplication preserve convergence in probability.

20.28. Suppose that the sequence $\{X_n\}$ is *fundamental in probability* in the sense that for ϵ positive there exists an N_ϵ such that $P[|X_m - X_n| > \epsilon] < \epsilon$ for $m, n > N_\epsilon$.
(a) Prove there is a subsequence $\{X_{n_k}\}$ and a random variable X such that $\lim_k X_{n_k} = X$ with probability 1.

Hint: Choose increasing n_k such that $P[|X_m - X_n| > 2^{-k}] < 2^{-k}$ for $m, n \geq n_k$. Analyze $P[|X_{n_{k-1}} - X_{n_k}| > 2^{-k}]$.

(b) Show that $X_n \to_P X$.

20.29. **(a)** Suppose that $X_1 \leq X_2 \leq \cdots$ and that $X_n \to_P X$. Show that $X_n \to X$ with probability 1.
(b) Show by example that in an infinite measure space functions can converge almost everywhere without converging in measure.

20.30. If $X_n \to 0$ with probability 1, then $n^{-1}\sum_{k=1}^n X_k \to 0$ with probability 1 by the standard theorem on Cesàro means. Show by example that this is not so if convergence with probability 1 is replaced by convergence in probability.

20.31. 2.17↑ **(a)** Show that in a discrete probability space convergence in probability is equivalent to convergence with probability 1.

(b)   Show that discrete spaces are essentially the only ones where this equivalence holds: Suppose that P has a nonatomic part in the sense that there is a set A such that $P(A) > 0$ and $P(\cdot|A)$ is nonatomic. Construct random variables X_n such that $X_n \to_P 0$ but X_n does not converge to 0 with probability 1.

20.32.   20.28   20.31↑   Let $d(X, Y)$ be the infimum of those positive ϵ for which $P[|X - Y| \geq \epsilon] \leq \epsilon$.

(a)   Show that $d(X, Y) = 0$ if and only if $X = Y$ with probability 1. Identify random variables that are equal with probability 1 and show that d is a metric on the resulting space.

(b)   Show that $X_n \to_P X$ if and only if $d(X_n, X) \to 0$.

(c)   Show that the space is complete.

(d)   Show that in general there is no metric d_0 on this space such that $X_n \to X$ with probability 1 if and only if $d_0(X_n, X) \to 0$.

SECTION 21.   EXPECTED VALUES

Expected Value as Integral

The expected value of a random variable X on $(\Omega, \mathcal{F}, P)$ is the integral of X with respect to the measure P:

$$E[X] = \int X \, dP = \int_\Omega X(\omega) P(d\omega).$$

All the definitions, conventions, and theorems of Chapter 3 apply. For nonnegative X, $E[X]$ is always defined (it may be infinite); for the general X, $E[X]$ is defined, or X has an expected value, if at least one of $E[X^+]$ and $E[X^-]$ is finite, in which case $E[X] = E[X^+] - E[X^-]$; and X is integrable if and only if $E[|X|] < \infty$. The integral $\int_A X \, dP$ over a set A is defined, as before, as $E[I_A X]$. In the case of simple random variables, the definition reduces to that used in Sections 5 through 9.

Expected Values and Distributions

Suppose that X has distribution μ. If g is a real function of a real variable, then by the change-of-variable formula (16.16),

(21.1) $$E[g(X)] = \int_{-\infty}^{\infty} g(x)\mu(dx).$$

(In applying (16.16), replace $T: \Omega \to \Omega'$ by $X: \Omega \to R^1$, μ by P, μT^{-1} by μ, and f by g.) This formula holds in the sense explained in Theorem 16.12: It holds in the nonnegative case, so that

(21.2) $$E[|g(X)|] = \int_{-\infty}^{\infty} |g(x)|\mu(dx);$$

if one side is infinite, then so is the other. And if the two sides of (21.2) are finite, then (21.1) holds.

If μ is discrete and $\mu\{x_1, x_2, \ldots\} = 1$, then (21.1) becomes (use Theorem 16.9)

(21.3) $$E[g(X)] = \sum_r g(x_r)\mu\{x_r\}.$$

If X has density f, then (21.1) becomes (use Theorem 16.10)

(21.4) $$E[g(X)] = \int_{-\infty}^{\infty} g(x)f(x)\, dx.$$

If F is the distribution function of X and μ, (21.1) can be written $E[g(X)] = \int_{-\infty}^{\infty} g(x)dF(x)$ in the notation (17.16).

Moments

By (21.2), μ and F determine all the *absolute moments* of X:

(21.5) $$E[|X|^k] = \int_{-\infty}^{\infty} |x|^k \mu(dx) = \int_{-\infty}^{\infty} |x|^k\, dF(x), \qquad k = 1, 2, \ldots.$$

Since $j \leq k$ implies that $|x|^j \leq 1 + |x|^k$, if X has a finite absolute moment of order k, then it has finite absolute moments of orders $1, 2, \ldots, k - 1$ as well. For each k for which (21.5) is finite, X has kth *moment*

(21.6) $$E[X^k] = \int_{-\infty}^{\infty} x^k \mu(dx) = \int_{-\infty}^{\infty} x^k\, dF(x).$$

These quantities are also referred to as the moments of μ and of F. They can be computed via (21.3) and (21.4) in the appropriate circumstances.

Example 21.1.   Consider the normal density (20.12) with $m = 0$ and $\sigma = 1$. For each k, $x^k e^{-x^2/2}$ goes to 0 exponentially as $x \to \pm\infty$, and so finite moments of all orders exist. Integration by parts shows that

$$\frac{1}{\sqrt{2\pi}} \int_{-\infty}^{\infty} x^k e^{-x^2/2} dx = \frac{k - 1}{\sqrt{2\pi}} \int_{-\infty}^{\infty} x^{k-2} e^{-x^2/2}\, dx, \qquad k = 2, 3, \ldots.$$

(Apply (18.16) to $g(x) = x^{k-2}$ and $f(x) = xe^{-x^2/2}$, and let $a \to -\infty$, $b \to \infty$.) Of course, $E[X] = 0$ by symmetry and $E[X^0] = 1$. It follows by induction that

(21.7) $$E[X^{2k}] = 1 \cdot 3 \cdot 5 \cdots (2k - 1), \qquad k = 1, 2, \ldots,$$

and that the odd moments all vanish.   ∎

If the first two moments of X are finite and $E[X] = m$, then just as in Section 5, the *variance* is

(21.8) $$\text{Var } [X] = E[(X - m)^2] = \int_{-\infty}^{\infty} (x - m)^2 \mu(dx)$$

$$= E[X^2] - m^2.$$

From Example 21.1 and a change of variable, it follows that a random variable with the normal density (20.12) has mean m and variance σ^2.

Consider for nonnegative X the relation

(21.9) $$E[X] = \int_0^{\infty} P[X > t] \, dt = \int_0^{\infty} P[X \geq t] \, dt.$$

Since $P[X = t]$ can be positive for at most countably many values of t, the two integrands differ only on a set of Lebesgue measure 0 and hence the integrals are the same. For X simple and nonnegative, (21.9) was proved in Section 5; see (5.25). For the general nonnegative X, let X_n be simple random variables for which $0 \leq X_n \uparrow X$ (see (20.1)). By the monotone convergence theorem $E[X_n] \uparrow E[X]$; moreover, $P[X_n > t] \uparrow P[X > t]$, and so $\int_0^{\infty} P[X_n > t] \, dt \uparrow \int_0^{\infty} P[X > t] \, dt$, again by the monotone convergence theorem. Since (21.9) holds for each X_n, a passage to the limit establishes (21.9) for X itself. Note that both sides of (21.9) may be infinite. If the integral on the right is finite, then X is integrable.

Replacing X by $XI_{[X > \alpha]}$ leads from (21.9) to

(21.10) $$\int_{[X > \alpha]} X \, dP = \alpha P[X > \alpha] + \int_{\alpha}^{\infty} P[X > t] \, dt, \qquad \alpha \geq 0.$$

As long as $\alpha \geq 0$, this holds even if X is not nonnegative.

Inequalities

Since the final term in (21.10) is nonnegative, $\alpha P[X \geq \alpha] \leq \int_{[X \geq \alpha]} X \, dP \leq E[X]$. Thus

(21.11) $$P[X \geq \alpha] \leq \frac{1}{\alpha} \int_{[X \geq \alpha]} X \, dP \leq \frac{1}{\alpha} E[X], \qquad \alpha > 0$$

for nonnegative X. As in Section 5, there follows the inequality

(21.12) $$P[|X| \geq \alpha] \leq \frac{1}{\alpha^k} \int_{|X| \geq \alpha} |X|^k \, dP \leq \frac{1}{\alpha^k} E[|X|^k].$$

It is the inequality between the two extreme terms here that usually goes under the name of Markov; but the left-hand inequality is often useful, too. As a special case there is Chebyshev's inequality,

(21.13) $$P[|X - m| \geq \alpha] \leq \frac{1}{\alpha^2} \text{Var} [X]$$

$(m = E[X])$.

Jensen's inequality

(21.14) $$\varphi(E[X]) \leq E[\varphi(X)]$$

holds if φ is convex on an interval containing the range of X and if X and $\varphi(X)$ both have expected values. To prove it, let $l(x) = ax + b$ be a supporting line through $(E[X], \varphi(E[X]))$—a line lying entirely under the graph of φ. Then $aX(\omega) + b \leq \varphi(X(\omega))$, so that $aE[X] + b \leq E[\varphi(X)]$. But the left side of this inequality is $\varphi(E[X])$.

Hölder's inequality is

(21.15) $$E[|XY|] \leq E^{1/p}[|X|^p]E^{1/q}[|Y|^q], \qquad \frac{1}{p} + \frac{1}{q} = 1.$$

For discrete random variables, this was proved in Section 5; see (5.31). For the general case, choose simple random variables X_n and Y_n satisfying $0 \leq |X_n| \uparrow |X|$ and $0 \leq |Y_n| \uparrow |Y|$; see (20.2). Then (5.31) and the monotone convergence theorem gives (21.15). Notice that (21.15) implies that if $|X|^p$ and $|Y|^q$ are integrable, then so is XY. Schwarz's inequality is the case $p = q = 2$:

(21.16) $$E[|XY|] \leq E^{1/2}[X^2]E^{1/2}[Y^2].$$

If X and Y have second moments, then XY must have a first moment.

The same reasoning shows that (5.33) carries over from the discrete to the general case.

Joint Integrals[*]

The relation (21.1) extends to random vectors. Suppose that $(X_1, \ldots, X_k)$ has distribution μ in k-space and $g: R^k \to R^1$. By Theorem 16.12,

(21.17) $$E[g(X_1, \ldots, X_k)] = \int_{R^k} g(x)\mu(dx),$$

with the usual provisos about infinite values. For example, $E[X_iX_j] = \int_{R^k} x_ix_j\mu(dx)$. If $E[X_i] = m_i$, the *covariance* of X_i and X_j is

$$E[(X_i - m_i)(X_j - m_j)] = \int_{R^k} (x_i - m_i)(x_j - m_j)\mu(dx).$$

Independence and Expected Value

Suppose that X and Y are independent. If they are also simple, then $E[XY] = E[X]E[Y]$, as proved in Section 5—see (5.21). Define X_n by (20.2) and similarly define $Y_n = \psi_n(Y^+) - \psi_n(Y^-)$. Then X_n and Y_n are independent and simple, so that $E[|X_nY_n|] = E[|X_n|]E[|Y_n|]$, and $0 \le |X_n| \uparrow |X|$, $0 \le |Y_n| \uparrow |Y|$. If X and Y are integrable, then $E[|X_nY_n|] = E[|X_n|]E[|Y_n|] \le E[|X|] \cdot E[|Y|]$, and if follows by the monotone convergence theorem that $E[|XY|] < \infty$; since $X_nY_n \to XY$ and $|X_nY_n| \le |XY|$, it follows further by the dominated convergence theorem that $E[XY] = \lim_n E[X_nY_n] = \lim_n E[X_n]E[Y_n] = E[X]E[Y]$. Therefore, XY is integrable if X and Y are (which is by no means true for dependent random variables) and $E[XY] = E[X]E[Y]$.

This argument obviously extends inductively: If $X_1, \ldots, X_k$ are independent and integrable, then the product $X_1 \cdots X_k$ is also integrable and

(21.18) $$E[X_1 \cdots X_k] = E[X_1] \cdots E[X_k].$$

Suppose that $\mathcal{G}_1$ and $\mathcal{G}_2$ are independent σ-fields, A lies in $\mathcal{G}_1$, X_1 is measurable $\mathcal{G}_1$, and X_2 is measurable $\mathcal{G}_2$. Then $I_A X_1$ and X_2 are independent, so that (21.18) gives

(21.19) $$\int_A X_1 X_2 \, dP = \int_A X_1 \, dP \cdot E[X_2]$$

if the random variables are integrable. In particular,

(21.20) $$\int_A X_2 \, dP = P(A)E[X_2].$$

From (21.18) it follows just as for simple random variables (see (5.24)) that variances add for sums of independent random variables.

Moment Generating Functions

The *moment generating function* is defined as

(21.21) $$M(s) = E[e^{sX}] = \int_{-\infty}^{\infty} e^{sx}\mu(dx) = \int_{-\infty}^{\infty} e^{sx} \, dF(x)$$

for all s for which this is finite (note that the integrand is nonnegative). Section 9 shows in the case of simple random variables how useful the moment generating function is. This function is also called the *Laplace transform* of μ, especially in nonprobabilistic contexts.

Now $\int_0^\infty e^{sx}\mu(dx)$ is finite for $s \le 0$, and if it is finite for a positive s, then it is finite for all smaller s. Together with the corresponding result for the left half-line, this shows that $M(s)$ is defined on some interval containing 0. If X is nonnegative, this interval contains $(-\infty, 0]$ and perhaps part of $(0, \infty)$; if X is nonpositive, it contains $[0, \infty)$ and perhaps part of $(-\infty, 0)$. It is possible that the interval consists of 0 alone; this happens for example, if μ is concentrated on the integers and $\mu\{n\} = \mu\{-n\} = C/n^2$ for $n = 1, 2, \ldots$.

Suppose that $M(s)$ is defined throughout an interval $[-s_0, s_0], s_0 > 0$. Since $e^{|sx|} \le e^{sx} + e^{-sx}$ and the latter function is integrable for $|s| \le s_0$, $\int e^{|sx|} \mu(dx) < \infty$ on $[-s_0, s_0]$. As $x \to \pm\infty$, $|x|^k$ is of smaller order than $e^{s|x|}$ for fixed positive s, and so μ has finite moments of all orders. Moreover, the partial sums of the series $e^{sx} = \sum_{k=0}^{\infty} s^k x^k / k!$ are bounded by $\sum_{k=0}^{\infty} |sx|^k / k! = e^{|sx|}$, which is integrable with respect to μ for $|s| \le s_0$. By Theorem 16.7,

$$(21.22) \quad M(s) = \sum_{k=0}^{\infty} \frac{s^k}{k!} E[X^k] = \sum_{k=0}^{\infty} \frac{s^k}{k!} \int_{-\infty}^{\infty} x^k \mu(dx), \qquad |s| \le s_0.$$

Thus $M(s)$ has a Taylor expansion about 0 with positive radius of convergence if it is defined in some $[-s_0, s_0], s_0 > 0$. If $M(s)$ can somehow be calculated and expanded in a series $\sum_k a_k s^k$, and if the coefficients a_k can be identified, then, since a_k must coincide with $E[X^k]/k!$, the moments of X can be computed: $E[X^k] = a_k k!$. It also follows from the theory of Taylor expansions that $a_k k!$ is the kth derivative $M^{(k)}(s)$ evaluated at $s = 0$:

$$(21.23) \qquad M^{(k)}(0) = E[X^k] = \int_{-\infty}^{\infty} x^k \mu(dx).$$

This holds if $M(s)$ exists in some neighborhood of 0.

This relation also follows by differentiation under the integral. Suppose that the moment generating function exists in some neighborhood of s. Choose points such that $a < a_0 < s < b_0 < b$ and M exists in $[a, b]$. There exists a K such that $|x| \le K e^{(b-b_0)x}$ for $x \ge 0$ and $|x| \le K e^{(a-a_0)x}$ for $x \le 0$. For each t in (a_0, b_0),

$$\left| \frac{\partial}{\partial t} e^{tx} \right| = |x| e^{tx} \le \begin{cases} K e^{bx} & \text{if } x \ge 0, \\ K e^{ax} & \text{if } x \le 0. \end{cases}$$

By Theorem 16.8(ii), $M'(t) = \int_{-\infty}^{\infty} x e^{tx} \mu(dx)$ for t in (a_0, b_0). The argument can be repeated inductively, and so

$$(21.24) \qquad M^{(k)}(s) = \int_{-\infty}^{\infty} x^k e^{sx} \mu(dx).$$

This holds as long as the moment generating function exists in some neighborhood of s. If $s = 0$, this gives (21.23), and taking $k = 2$ shows that $M(s)$ is convex in its interval of definition.

Example 21.2. For the standard normal density,

$$M(s) = \frac{1}{\sqrt{2\pi}} \int_{-\infty}^{\infty} e^{sx} e^{-x^2/2} dx = \frac{1}{\sqrt{2\pi}} e^{s^2/2} \int_{-\infty}^{\infty} e^{-(x-s)^2/2} dx,$$

and a change of variable gives

$$(21.25) \qquad\qquad M(s) = e^{s^2/2}.$$

The moment generating function is in this case defined for all s. Since $e^{s^2/2}$ has the expansion

$$e^{s^2/2} = \sum_{k=0}^{\infty} \frac{1}{k!}\left(\frac{s^2}{2}\right)^k = \sum_{k=0}^{\infty} \frac{1 \cdot 3 \cdots (2k-1)}{(2k)!} s^{2k},$$

the moments can be read off from (21.22), which proves (21.7) once more. ∎

Example 21.3. In the exponential case (20.10) the moment generating function

$$(21.26) \qquad M(s) = \int_0^{\infty} e^{sx} \alpha e^{-\alpha x}\, dx = \frac{\alpha}{\alpha - s} = \sum_{k=0}^{\infty} \frac{s^k}{\alpha^k}$$

is defined for $s < \alpha$. By (21.22) the kth moment is $k!\alpha^{-k}$. The mean and variance are thus α^{-1} and α^{-2}. ∎

Example 21.4. For the Poisson distribution (20.7),

$$(21.27) \qquad M(s) = \sum_{r=0}^{\infty} e^{rs} e^{-\lambda} \frac{\lambda^r}{r!} = e^{\lambda(e^s - 1)}.$$

Since $M'(s) = \lambda e^s M(s)$ and $M''(s) = (\lambda^2 e^{2s} + \lambda e^s) M(s)$, the first two moments are $M'(0) = \lambda$ and $M''(0) = \lambda^2 + \lambda$; the mean and variance are both λ. ∎

Let $X_1, \ldots, X_k$ be independent random variables, and suppose that each X_i has a moment generating function $M_i(s) = E[e^{sX_i}]$ in $[-s_0, s_0]$. For $|s| \le s_0$, each $\exp(sX_i)$ is integrable and, since they are independent, their product $\exp(s\sum_{i=1}^k X_i)$ is also integrable (see (21.18)). The moment generating function of $X_1 + \cdots + X_k$ is therefore

$$(21.28) \qquad M(s) = M_1(s) \cdots M_k(s)$$

in $[s_0, s_0]$. This relation for the discrete case was essential to the arguments in Section 9.

In Section 9 it was shown in the discrete case that the moment generating function determines the distribution. This will later be proved for general random variables. See (22.8) for the case of nonnegative random variables, and see Section 30 for the case in which the moment generating function exists in a neighborhood of 0.

PROBLEMS

21.1. Prove

$$\frac{1}{\sqrt{2\pi}} \int_{-\infty}^{\infty} e^{-tx^2/2} dx = t^{-1/2},$$

differentiate k times with respect to t inside the integral (justify), and derive (21.7) again.

21.2. Show that, if X has the standard normal distribution, then $E[|X|^{2n+1}] = 2^n n! \sqrt{2/\pi}$.

21.3. 16.18↑ Suppose that $X_1, X_2, \ldots$ are identically distributed (not necessarily independent). Show that $E[\max_{k \le n} |X_k|] = o(n)$.

21.4. 20.12↑ *Records.* Consider the sequence of records in the sense of Problem 20.12. Show that the waiting time to the next record is infinite.

21.5. 20.20↑ Show that the Cauchy distribution has no mean.

21.6. Prove the first Borel-Cantelli lemma by applying Theorem 16.6 to indicator random variables. Why is Theorem 16.6 not enough for the second Borel-Cantelli lemma?

21.7. Derive (21.11) by imitating the proof of (5.26). Derive Jensen's inequality (21.14) for bounded X and φ by passing to the limit from the case of simple random variables (see (5.29)).

21.8. Prove (21.9) by Fubini's theorem.

21.9. Prove for integrable X that

$$E[X] = \int_0^\infty P[X > t]\,dt - \int_{-\infty}^0 P[X < t]\,dt.$$

21.10. **(a)** Suppose that X and Y have first moments and prove

$$E[Y] - E[X] = \int_{-\infty}^\infty (P[X < t \le Y] - P[Y < t \le X])\,dt.$$

(b) Let $(X, Y]$ be a nondegenerate random interval. Show that its expected length is the integral with respect to t of the probability that it covers t.

21.11. Suppose that X and Y are random variables with continuous distribution functions F and G. Show that $E[F(Y)] + E[G(X)] = 1$. In particular, $E[F(X)] = \frac{1}{2}$.

21.12. Let X have distribution function F.
(a) Show that $E[X^+] < \infty$ if and only if $\int_\alpha^\infty (-\log F(t))\,dt < \infty$ for some α.

(b) Show more precisely that, if α and $F(\alpha)$ are positive, then

$$\int_{X > \alpha} X\,dP \le \alpha(-\log F(\alpha)) + \int_\alpha^\infty (-\log F(t))\,dt \le \frac{1}{F(\alpha)} \int_{X > \alpha} X\,dP.$$

21.13. Suppose that X and Y are nonnegative random variables, $r > 1$, and $P[Y \ge t] \le t^{-1} \int_{Y \ge t} X\,dP$ for $t > 0$. Use (21.9), Fubini's theorem, and Hölder's inequality to prove $E[Y^r] \le (r/(r-1))^r E[X^r]$.

21.14. Let $X_1, X_2 \ldots$ be independent, identically distributed random variables with finite second moment. Prove successively that

   (a) $nP[|X_1| \geq \epsilon\sqrt{n}] \to 0$;

   (b) $(1 - P[|X_1| \geq \epsilon\sqrt{n}])^n \to 1$; and

   (c) $n^{-1/2} \max_{k \leq n} |X_k| \to_P 0$.

21.15. Random variables X and Y with second moments are *uncorrelated* if $E[XY]$ = $E[X]E[Y]$.

   (a) Show that independent random variables are uncorrelated but the converse is false.

   (b) Show that, if $X_1, \ldots, X_n$ are uncorrelated (in pairs), then Var $[X_1 + \cdots + X_n]$ = Var $[X_1] + \cdots +$ Var $[X_n]$.

21.16. ↑ Let X, Y, and Z be independent random variables such that X and Y assume the values 0, 1, 2 with probability $\frac{1}{3}$ each and Z assumes the values 0 and 1 with probabilities $\frac{1}{3}$ and $\frac{2}{3}$. Let $X' = X$ and $Y' = X + Z$ (mod 3).

   (a) Show that X', Y', and $X' + Y'$ have the same one-dimensional distributions as X, Y, and $X + Y$, respectively, even though (X', Y') and (X, Y) have different distributions.

   (b) Show that X' and Y' are dependent but uncorrelated.

   (c) Show that, despite dependence, the moment generating function of $X' + Y'$ is the product of the moment generating functions of X' and Y'.

21.17. Suppose that X and Y are independent, nonnegative random variables and that $E[X] = \infty$ and $E[Y] = 0$. What is the value common to $E[XY]$ and $E[X]E[Y]$? Use the conventions (15.2) for both the product of the random variables and the product of their expected values. What if $E[X] = \infty$ and $0 < E[Y] < \infty$?

21.18. Deduce (21.18) for $k = 2$ from (21.17), (20.23), and Fubini's theorem. Do not in the nonnegative case assume integrability.

21.19. Suppose that X and Y are independent and that $f(x, y)$ is nonnegative. Put $g(x)$ = $E[f(x, Y)]$ and show that $E[g(X)] = E[f(X, Y)]$. Show more generally that $\int_{X \in A} g(X)\, dP = \int_{X \in A} f(X, Y)\, dP$. Extend to f that may be negative.

21.20. ↑ The integrability of $X + Y$ does not imply that of X and Y separately. Show that it does if X and Y are independent.

21.21. Fatou's lemma and Lebesgue's dominated convergence theorem of course apply to random variables. Show that they hold if convergence with probability 1 is replaced by convergence in probability.

21.22. ↑ If $E[|X_n - X|] \to 0$, then $X_n \to_P 0$. The converse is false but holds under the added hypothesis that the X_n are uniformly bounded, or are uniformaly integrable. If $E[|X_n - X|] \to 0$, then X_n is said to converge to X *in the mean*.

21.23. ↑ Write $d_1(X, Y) = E[|X - Y|/(1 + |X - Y|)]$. Show that this is a metric equivalent to the one in Problem 20.32.

21.24. 5.10↑ Extend Minkowski's inequality (5.36) to arbitrary random variables.

21.25. 20.28 21.21 21.24↑ Suppose that $p \geq 1$. For a fixed probability space $(\Omega, \mathcal{F}, P)$, consider the set of random variables satisfying $E[|X|^p] < \infty$ and identify random variables that differ only on a set of probability 0. This is the space L^p = $L^p(\Omega, \mathcal{F}, P)$. Put $d_p(X, Y) = E^{1/p}[|X - Y|^p]$ for $X, Y \in L^p$.

(a)   Show that L^p is a metric space under d_p.

(b)   Show that L^p is complete.

21.26.   2.10 16.7 21.25↑ Show that $L^p(\Omega, \mathcal{F}, P)$ is separable if $\mathcal{F}$ is separable, or if $\mathcal{F}$ is the completion of a separable σ-field.

21.27.   21.25↑ For X and Y in L^2 put $\langle X, Y \rangle = E[XY]$. Show that $\langle \cdot , \cdot \rangle$ is an inner product for which $d_2(X, Y) = \langle X - Y, X - Y \rangle$. Thus L^2 is a Hilbert space.

21.28.   5.11↑ Extend Problem 5.11 concerning essential suprema to random variables that need not be simple.

21.29.   Consider on $(1, \infty)$ densities that are multiples of $e^{-\alpha x}$, e^{-x^2}, $x^{-2}e^{-\alpha x}$, x^{-2}; also consider reflections of these through 0 and linear combinations of them. Show that the convergence interval for a moment generating function can be any interval containing 0. There are 16 possibilities, depending on whether the endpoints are 0, finite and different from 0, or infinite, and whether they are contained in the interval or not.

21.30.   For the density $C \exp(-|x|^{1/2})$, $-\infty < x < \infty$, show that moments of all orders exist but that the moment generating function exists only at $s = 0$.

21.31.   16.8↑ Show that a moment generating function $M(s)$ defined in $(-s_0, s_0)$, $s_0 > 0$, can be extended to a function analytic in the strip $[z: -s_0 < \mathrm{Re}\, z < s_0]$. If $M(s)$ is defined in $[0, s_0)$, $s_0 > 0$, show that it can be extended to a function continuous in $[z: 0 \le \mathrm{Re}\, z < s_0]$ and analytic in $[z: 0 < \mathrm{Re}\, z < s_0]$.

21.32.   Use (21.28) to find the generating function of (20.39).

21.33.   The joint moment generating function of (X, Y) is defined as $M(s, t) = E[e^{sX+tY}]$. Assume that it exists for (s, t) in some neighborhood of the origin, expand it in a double series, and conclude that $E[X^m Y^n]$ is $\partial^{m+n}M/\partial s^m \partial t^n$ evaluated at $s = t = 0$.

21.34.   For independent random variables having moment generating functions, show by (21.28) that the variances add.

21.35.   20.23↑ Show that the gamma density (20.47) has moment generating function $(1 - s/\alpha)^{-u}$ for $s < \alpha$. Show that the kth moment is $u(u + 1) \ldots (u + k - 1)/\alpha^k$. Show that the chi-squared distribution with n degrees of freedom has mean n and variance $2n$.

21.36.   The cumulant generating function was defined by (9.5) for a simple random variable X. Even if X is not simple, the definition applies in the interval in which the moment generating function is defined. Extend (9.6) through (9.10).

21.37.   ↑ Suppose that Y is a random variable, not necessarily simple, satisfying (9.11). Suppose that its moment generating function $M(t)$ exists in a neighborhood of 0. Show that it exists in an interval $I = (a, b)$, $-\infty \le a < 0 < b \le \infty$, where $\lim_{t \to b} M(t) = \infty$. The analogue of (9.12) is $\rho = M(\tau) = \inf_{t \in I} M(t)$, $0 < \rho < 1$, $\tau > 0$, $\tau \in I$.

21.38. ↑ Take Y, ρ, τ as in Problem 21.37 and let Z be a random variable such that

$$P[Z \in A] = \int_A \rho^{-1} e^{\tau y} \mu(dy),$$

where μ is the distribution of Y. Show such a Z exists. Prove the analogues of Theorems 9.1 through 9.4.

SECTION 22. SUMS OF INDEPENDENT RANDOM VARIABLES

Let $X_1, X_2, \ldots$ be a sequence of independent random variables on some probability space. It is natural to ask whether the infinite series $\Sigma_{n=1}^{\infty} X_n$ converges with probability 1, or as in Section 6 whether $n^{-1} \Sigma_{k=1}^{n} X_k$ converges to some limit with probability 1. It is to questions of this sort that the present section is devoted.

Kolmogorov's Zero-One Law

Consider the set A of ω for which $n^{-1} \Sigma_{k=1}^{n} X_k(\omega) \to 0$ as $n \to \infty$. For each m, the values of $X_1(\omega), \ldots, X_{m-1}(\omega)$ are irrelevant to the question of whether or not ω lies in A, and so A ought to lie in the σ-field $\sigma(X_m, X_{m+1}, \ldots)$. In fact, $\lim_n n^{-1} \Sigma_{k=0}^{m-1} X_k(\omega) = 0$ for fixed m, and hence ω lies in A if and only if $\lim_n n^{-1} \Sigma_{k=m}^{n} X_k(\omega) = 0$. Therefore,

$$(22.1) \qquad A = \bigcap_{\epsilon} \bigcup_{N \geq m} \bigcap_{n \geq N} \left[\omega : \left| n^{-1} \sum_{k=m}^{n} X_k(\omega) \right| < \epsilon \right],$$

the first intersection extending over positive rational ϵ. The set on the inside lies in $\sigma(X_m, X_{m+1}, \ldots)$, and hence so does A. Similarly, the ω-set where the series $\Sigma_n X_n(\omega)$ converges lies in each $\sigma(X_m, X_{m+1}, \ldots)$.

The intersection $\mathcal{T} = \bigcap_{n=1}^{\infty} \sigma(X_n, X_{n+1}, \ldots)$ is the *tail* σ-field associated with the sequence $X_1, X_2, \ldots$; its elements are *tail events*. In the case $X_n = I_{A_n}$, this is the σ-field (4.26) studied in Section 4. The following general form of *Kolmogorov's zero-one law* extends Theorem 4.5.

Theorem 22.1. *If $\{X_n\}$ is independent, and if $A \in \mathcal{T} = \bigcap_{n=1}^{\infty} \sigma(X_n, X_{n+1}, \ldots)$, then either $P(A) = 0$ or $P(A) = 1$.*

PROOF. Let $\mathcal{F}_0 = \bigcup_{k=1}^{\infty} \sigma(X_1, \ldots, X_k)$. The first thing to establish is that $\mathcal{F}_0$ is a field generating the σ-field $\sigma(X_1, X_2, \ldots)$. If B and C lie in $\mathcal{F}_0$, then $B \in \sigma(X_1, \ldots, X_j)$ and $C \in \sigma(X_1, \ldots, X_k)$ for some j and k; if $m = \max\{j, k\}$, then B and C both lie in $\sigma(X_1, \ldots, X_m)$, so that $B \cup C \in \sigma(X_1, \ldots, X_m) \subset \mathcal{F}_0$. Thus $\mathcal{F}_0$ is closed under the formation of finite unions; since it is similarly closed under complementation, $\mathcal{F}_0$ is a field. If $\mathcal{G}$ is a σ-field satisfying $\mathcal{F}_0 \subset \mathcal{G}$, then each X_n is measurable $\mathcal{G}$ and hence $\sigma(X_1, X_2, \ldots) \subset \mathcal{G}$. Thus $\mathcal{F}_0$ generates $\sigma(X_1, X_2, \ldots)$ (which in general is much larger than $\mathcal{F}_0$).

Suppose that A lies in $\mathcal{T}$. Then A lies in $\sigma(X_{k+1}, X_{k+2}, \ldots)$ for each k. Therefore, if $B \in \sigma(X_1, \ldots, X_k)$, then A and B are independent by Theorem 20.2. Therefore, A is independent of $\mathcal{F}_0$ and hence by Theorem 4.2 is also independent of $\sigma(X_1, X_2, \ldots)$. But then A is independent of itself: $P(A \cap A) = P(A)P(A)$. Therefore, $P(A) = P^2(A)$, which implies that $P(A)$ is either 0 or 1. ∎

As noted above, the set where $\Sigma_n X_n(\omega)$ converges satisfies the hypothesis of Theorem 22.1, and so does the set where $n^{-1} \Sigma_{k=1}^n X_k(\omega) \to 0$. In many similar cases it is very easy to prove by this theorem that a set at hand must have probability either 0 or 1. As was seen in Chapter 1, however, much effort may be required to determine which of 0 and 1 is, in fact, the probability of the set.

In connection with the problem of verifying the hypotheses of Theorem 22.1, consider this example:

Example 22.1. Let $\Omega = R^1$, let A be a subset of R^1 not contained in $\mathcal{R}^1$ (see the end of Section 3), and let $\mathcal{F}$ be the σ-field generated by $\mathcal{R}^1 \cup \{A\}$. Take $X_1(\omega) = I_A(\omega)$ and $X_2(\omega) = X_3(\omega) = \cdots = \omega$. Then A lies in $\sigma(X_1, X_2, \ldots)$.

Fix m. If one knows $X_m(\omega), X_{m+1}(\omega), \ldots$, then he knows ω itself and hence knows whether or not ω lies in A. To put it another way, if $X_j(\omega') = X_j(\omega'')$ for all $j \ge m$, then $I_A(\omega') = I_A(\omega'')$; therefore, the values of $X_1(\omega), \ldots, X_{m-1}(\omega)$ are irrelevant to the question of whether or not $\omega \in A$. Does A lie in $\sigma(X_m, X_{m+1}, \ldots)$? For $m \ge 2$ the answer is no, because $\sigma(X_m, X_{m+1}, \ldots) = \mathcal{R}^1$ and $A \notin \mathcal{R}^1$. ∎

Although the issue is a technical one of measurability, this example shows that in checking the hypothesis of Theorem 22.1 it is not enough to show for each m that A does not depend on the values of $X_1, \ldots, X_{m-1}$. It must for each m be shown by some construction like (22.1) that A does depend in a measurable way on $X_m, X_{m+1}, \ldots$ alone.

Kolmogorov's Inequality

Essential to the study of random series is an inequality due to Kolmogorov.

Theorem 22.2. *Suppose that $X_1, \ldots, X_n$ are independent with mean 0 and finite variances, and let $S_k = X_1 + \cdots + X_k$. For $\alpha > 0$*

$$(22.2) \qquad P\left[\max_{1 \le k \le n} |S_k| \ge \alpha \right] \le \frac{1}{\alpha^2} \operatorname{Var}[S_n].$$

PROOF.   Let A_k be the set where $|S_k| \geq \alpha$ but $|S_j| < \alpha$ for $j < k$. Since the A_k are disjoint,

$$E[S_n^2] \geq \sum_{k=1}^{n} \int_{A_k} S_n^2 \, dP$$

$$= \sum_{k=1}^{n} \int_{A_k} [S_k^2 + 2S_k(S_n - S_k) + (S_n - S_k)^2] \, dP$$

$$\geq \sum_{k=1}^{n} \int_{A_k} [S_k^2 + 2S_k(S_n - S_k)] \, dP.$$

Since A_k and S_k are measurable $\sigma(X_1, \ldots, X_k)$ and $S_n - S_k$ is measurable $\sigma(X_{k+1}, \ldots, X_n)$, and since the means are all 0, it follows by (21.19) and independence that $\int_{A_k} S_k(S_n - S_k) \, dP = 0$. Therefore,

$$E[S_n^2] \geq \sum_{k=1}^{n} \int_{A_k} S_k^2 \, dP \geq \sum_{k=1}^{n} \alpha^2 P(A_k)$$

$$= \alpha^2 P\left[\max_{1 \leq k \leq n} |S_k| \geq \alpha \right]. \qquad \blacksquare$$

By Chebyshev's inequality, $P[|S_n| \geq \alpha] \leq \alpha^{-2} \operatorname{Var}[S_n]$. That this can be strengthened to (22.2) is an instance of a general phenomenon: For sums of independent variables, if $\max_{k \leq n} |S_k|$ is large, then $|S_n|$ is probably large as well. Another instance is Theorem 9.6.

Convergence of Random Series

Kolmogorov proved many notable results in this subject, among them the following one.

Theorem 22.3. *Suppose that $\{X_n\}$ is an independent sequence and $E[X_n] = 0$. If $\Sigma_n \operatorname{Var}[X_n] < \infty$, then $\Sigma_n X_n$ converges with probability 1.*

PROOF.   Let $S_n = X_1 + \cdots + X_n$. By (22.2),

$$P\left[\max_{1 \leq k \leq r} |S_{n+k} - S_n| > \epsilon \right] \leq \frac{1}{\epsilon^2} \sum_{k=1}^{r} \operatorname{Var}[X_{n+k}].$$

Since the sets on the left are nondecreasing in r, letting $r \to \infty$ gives

$$(22.3) \qquad P\left[\sup_{k \geq 1} |S_{n+k} - S_n| > \epsilon \right] \leq \frac{1}{\epsilon^2} \sum_{k=1}^{\infty} \operatorname{Var}[X_{n+k}].$$

If $M = \inf_{n \geq 1} \sup_{k \geq 1} |S_{n+k} - S_n|$, then $P[M > \epsilon]$ is for each n at most the right side of (22.3) and hence (let $n \to \infty$) $P[M > \epsilon] = 0$. This being true for each positive ϵ, $P[M > 0] = 0$. But $\Sigma_n X_n$ converges if $M = 0$. $\qquad \blacksquare$

Example 22.2.   Let $X_n(\omega) = r_n(\omega)a_n$, where the r_n are the Rademacher functions on the unit interval—see (1.12). Then X_n has variance a_n^2, and so Σ_n $a_n^2 < \infty$ implies that $\Sigma_n r_n(\omega)a_n$ converges with probability 1. An interesting special case is $a_n = n^{-1}$. If the signs in $\Sigma \pm n^{-1}$ are chosen on the toss of a coin, then the series converges with probability 1. The alternating harmonic series $1 - 2^{-1} + 3^{-1} - \cdots$ is thus typical in this respect.   ∎

The Strong Law of Large Numbers

Another consequence of Theorem 22.2 is *Kolmogorov's criterion* for the strong law of large numbers.

Theorem 22.4.   *Suppose that $\{X_n\}$ is an independent sequence and $E[X_n] = 0$. If Σ_n Var $[X_n]/n^2 < \infty$, then $n^{-1} \Sigma_{k=1}^n X_k \to 0$ with probability 1.*

PROOF.   Put $S_n = X_1 + \cdots + X_n$ and $\sigma_n^2 = E[X_n^2]$. It suffices to show that for each positive ϵ there is probability 0 that $|S_n| > n\epsilon$ infinitely often. Let B_k be the ω-set where $|S_n| > n\epsilon$ for some n in the range $2^k \le n < 2^{k+1}$. The theorem will follow by the first Borel-Cantelli lemma if $\Sigma_k P(B_k) < \infty$.

Now B_k implies that $|S_n| > \epsilon 2^k$ for some $n \le 2^{k+1}$, and so by Kolmogorov's inequality,

$$P(B_k) \le \sum_{n=1}^{2^{k+1}} \frac{1}{\epsilon^2 2^{2k}} \sigma_n^2.$$

Therefore,

$$\sum_{k=0}^{\infty} P(B_k) \le \frac{1}{\epsilon^2} \sum_{n=1}^{\infty} \sigma_n^2 \sum_{2^{k+1} \ge n} \frac{1}{2^{2k}}.$$

If k_n is the smallest integer k for which $2^{k+1} \ge n$, the final sum here is $2^{-2k_n}/(1 - \frac{1}{4}) \le 8/n^2$. Thus $\Sigma_k P(B_k) \le 8 \epsilon^{-2} \Sigma_n \sigma_n^2/n$.   ∎

The proof of Theorem 6.1 was for simple random variables only, but after the theory of the preceding two sections it obviously applies to X_n that are not necessarily discrete, provided that they are independent and identically distributed with a finite fourth moment. Since Σn^{-2} converges, Theorem 22.4 is stronger—it requires only second moments. But in the case of identically distributed random variables, there is a still stronger result: they only need have finite first moments. This is *Khinchine's* version of the strong law of large numbers.

Theorem 22.5.   *Suppose that X_1, X_2, ... are independent and identically distributed and $E[X_1] = m$. Then $n^{-1}\Sigma_{k=1}^n X_k \to m$ with probability 1.*

PROOF.   It is no restriction to assume $m = 0$. Consider the truncated random variables

$$Y_k = \begin{cases} X_k & \text{if } |X_k| \leq k, \\ 0 & \text{if } |X_k| > k. \end{cases}$$

Since the X_k all have the same distribution as X_1,

$$\sum_{k=1}^{\infty} P[X_k \neq Y_k] = \sum_{k=1}^{\infty} P[|X_1| > k] \leq \int_0^{\infty} P[|X_1| > t] \, dt.$$

Since $E[|X_1|] < \infty$, it follows by (21.9) that the integral here is finite. By the first Borel-Cantelli lemma there is thus probability 1 that $X_k = Y_k$ for all sufficiently large k. It is therefore enough to show that $n^{-1}\sum_{k=1}^{n} Y_k \to 0$ with probability 1.

Since the X_k are identically distributed,

$$\text{Var}\,[Y_k] \leq E[Y_k^2] = \int_{|X_1|\leq k} X_1^2 \, dP.$$

From the inequality $\sum_{k=i}^{\infty} k^{-2} \leq i^{-1} + \int_i^{\infty} x^{-2} \, dx = 2i^{-1}$, it follows that

$$\sum_{k=1}^{\infty} \frac{1}{k^2} \text{Var}\,[Y_k] \leq \sum_{k=1}^{\infty} \frac{1}{k^2} \sum_{i=1}^{k} \int_{i-1<|X_1|\leq i} X_1^2 \, dP$$

$$\leq \sum_{k=1}^{\infty} \frac{1}{k^2} \sum_{i=1}^{k} i \int_{i-1<|X_1|\leq i} |X_1| \, dP$$

$$\leq 2 \sum_{i=1}^{\infty} \int_{i-1<|X_1|\leq i} |X_1| \, dP = 2E[|X_1|].$$

By Kolmogorov's criterion, if $m_k = E[Y_k]$, then $n^{-1}\sum_{k=1}^{n}(Y_k - m_k) \to 0$ with probability 1. But as $k \to \infty$, $m_k = \int_{|X_1|\leq k} X_1 dP \to E[X_1] = 0$ by the dominated convergence theorem, and so $n^{-1}\sum_{k=1}^{n} m_k \to 0$ by the theorem on Cesàro means. ∎

Corollary. *Suppose that $X_1, X_2, \ldots$ are independent and identically distributed and $E[X_1^-] < \infty$, $E[X_1^+] = \infty$ (so that $E[X_1] = \infty$). Then $n^{-1}\sum_{k=1}^{n} X_k \to \infty$ with probability 1.*

PROOF. By the theorem, $n^{-1}\sum_{k=1}^{n} X_k^- \to E[X_1^-]$ with probability 1, and so it suffices to prove the corollary for the case $X_1 = X_1^+ \geq 0$. If

$$X_n^{(u)} = \begin{cases} X_n & \text{if } 0 \leq X_n \leq u, \\ 0 & \text{if } X_n > u, \end{cases}$$

then $n^{-1}\sum_{k=1}^{n} X_k \geq n^{-1}\sum_{k=1}^{n} X_k^{(u)} \to E[X_1^{(u)}]$ by the theorem. Let $u \to \infty$. ∎

Example 22.3. The sample mean $\overline{X} = n^{-1}\sum_{k=1}^{n} X_k$ converges with probability 1 to the population mean $E[X_1]$ if the latter is finite. The sample variance is $s^2 = (n-1)^{-1}\sum_{k=1}^{n}(X_k - \overline{X})^2$. Since

$$s^2 = \frac{n}{n-1}\left[\frac{1}{n}\sum_{k=1}^{n} X_k^2 - \overline{X}^2\right],$$

s^2 converges with probability 1 to the population variance in the case of finite second moments.   ∎

The Weak Law and Moment Generating Functions

The statement and proof of Theorem 6.2, the weak law of large numbers, carry over word for word to the case of general random variables with second moments—only Chebyshev's inequality is required. The idea can be used to prove in a very simple way that a distribution concentrated on $[0, \infty)$ is uniquely determined by its moment generating function or Laplace transform.

For each λ, let Y_λ be a random variable (on some probability space) having the Poisson distribution with parameter λ. Since Y_λ has mean and variance λ (Example 21.4), Chebyshev's inequality gives

$$P\left[\left|\frac{Y_\lambda - \lambda}{\lambda}\right| \geq \epsilon\right] \leq \frac{\lambda}{\lambda^2 \epsilon^2} \to 0, \qquad \lambda \to \infty.$$

Let G_λ be the distribution function of Y_λ/λ, so that

$$G_\lambda(t) = \sum_{k=0}^{[\lambda t]} e^{-\lambda}\frac{\lambda^k}{k!}.$$

The result above can be restated as

(22.4)
$$\lim_{\lambda \to \infty} G_\lambda(t) = \begin{cases} 1 & \text{if } t > 1, \\ 0 & \text{if } t < 1. \end{cases}$$

In the notation of Section 14, $G_\lambda(x) \Rightarrow \Delta(x - 1)$ as $\lambda \to \infty$.

Now consider a probability distribution μ concentrated on $[0, \infty)$. Let F be the corresponding distribution function. Define

(22.5)
$$M(s) = \int_0^\infty e^{-sx}\mu(dx), \qquad s \geq 0;$$

here 0 is included in the range of integration. This is the moment generating function (21.21), but the argument has been reflected through the origin. It is defined for all nonnegative s.

For positive s, (21.24) gives

(22.6)
$$M^{(k)}(s) = (-1)^k \int_0^\infty y^k e^{-sy}\mu(dy).$$

Therefore, for positive x and s,

$$(22.7) \quad \sum_{k=0}^{[sx]} \frac{(-1)^k}{k!} s^k M^{(k)}(s) = \int_0^\infty \sum_{k=0}^{[sx]} e^{-sy} \frac{(sy)^k}{k!} \mu(dy)$$

$$= \int_0^\infty G_{sy}\left(\frac{x}{y}\right) \mu(dy).$$

Fix $x > 0$. If* $0 \le y < x$, then $G_{sy}(x/y) \to 1$ as $s \to \infty$ by (22.4); if $y > x$, the limit is 0. If $\mu\{x\} = 0$, the integrand on the right in (22.7) thus converges as $s \to \infty$ to $I_{[0,x]}(y)$ except on a set of μ-measure 0. The bounded convergence theorem then gives

$$(22.8) \quad \lim_{s \to \infty} \sum_{k=0}^{[sx]} \frac{(-1)^k}{k!} s^k M^{(k)}(s) = \mu[0, x] = F(x).$$

Thus $M(s)$ determines the value of F at x if $x > 0$ and $\mu\{x\} = 0$, which covers all but countably many values of x in $[0, \infty)$. Since F is right-continuous, F itself and hence μ are determined through (22.8) by $M(s)$:

If $\mu[0, \infty) = 1$, then $M(s)$ determines μ.

Random Taylor Series[†]

Consider a power series $\Sigma_n \pm z^n$, where the signs are chosen on the toss of a coin. The radius of convergence being 1, the series represents an analytic function in the open unit disc $D_0 = [z: |z| < 1]$ in the complex plane. The question arises whether this function can be extended analytically beyond D_0. The answer is no: with probability 1 the unit circle is the natural boundary.

Theorem 22.6. *Let $\{X_n\}$ be an independent sequence such that*

$$(22.9) \quad P[X_n = 1] = P[X_n = -1] = \frac{1}{2}, \qquad n = 0, 1, \ldots.$$

There is probability 0 that

$$(22.10) \quad F(\omega, z) = \sum_{n=0}^\infty X_n(\omega) z^n$$

coincides in D_0 with a function analytic in an open set properly containing D_0.

It will be seen in the course of the proof that the ω-set in question lies in $\sigma(X_0,$

---

* If $y = 0$, the integrand in (22.6) is 1 for $k = 0$ and 0 for $k \ge 1$; hence for $y = 0$, the integrand in the middle term of (22.7) is 1.
† This topic, which requires complex variable theory, may be omitted.

$X_1, \ldots$) and hence has a probability. It is intuitively clear that if the set is measurable at all, it must depend only on the X_n for large n and hence must have probability either 0 or 1.

PROOF.   Since

$$(22.11) \qquad |X_n(\omega)| = 1, \qquad n = 0, 1, \ldots$$

with probability 1, the series in (22.10) has radius of convergence 1 outside a set of measure 0.

Consider an open disc $D = [z: |z - \zeta| < r]$, where $\zeta \in D_0$ and $r > 0$. Now (22.10) coincides in D_0 with a function analytic in $D_0 \cup D$ if and only if its expansion

$$F(\omega, z) = \sum_{m=0}^{\infty} \frac{1}{m!} F^{(m)}(\omega, \zeta)(z - \zeta)^m$$

about ζ converges at least for $|z - \zeta| < r$. Let A_D be the set of ω for which this holds. The coefficient

$$a_m(\omega) = \frac{1}{m!} F^{(m)}(\omega, \zeta) = \sum_{n=m}^{\infty} \binom{n}{m} X_n(\omega) \zeta^{n-m}$$

is a complex-valued random variable measurable $\sigma(X_m, X_{m+1}, \ldots)$. By the root test, $\omega \in A_D$ if and only if $\lim \sup_m \sqrt[m]{|a_m(\omega)|} \leq r^{-1}$. For each m_0, the condition for $\omega \in A_D$ can thus be expressed in terms of $a_{m_0}(\omega), a_{m_0+1}(\omega), \ldots$ alone, and so $A_D \in \sigma(X_{m_0}, X_{m_0+1}, \ldots)$. Thus A_D has a probability, and in fact $P(A_D)$ is 0 or 1 by the zero-one law.

Of course, $P(A_D) = 1$ if $D \subset D_0$. The central step in the proof is to show that $P(A_D) = 0$ if D contains points not in D_0. Assume on the contrary that $P(A_D) = 1$ for such a D. Consider that part of the circumference of the unit circle that lies in D, and let k be an integer large enough that this arc has length exceeding $2\pi/k$. Define

$$Y_n(\omega) = \begin{cases} X_n(\omega) & \text{if } n \not\equiv 0 \pmod{k}, \\ -X_n(\omega) & \text{if } n \equiv 0 \pmod{k}. \end{cases}$$

Let B_D be the ω-set where the function

$$(22.12) \qquad G(\omega, z) = \sum_{n=0}^{\infty} Y_n(\omega) z^n$$

coincides in D_0 with a function analytic in $D_0 \cup D$.

The sequence $\{Y_0, Y_1, \ldots\}$ has the same structure as the original sequence: the Y_n are independent and assume the values ± 1 with probability $\frac{1}{2}$ each. Since B_D is defined in terms of the Y_n in the same way as A_D is defined in terms of the X_n, it is intuitively clear that $P(B_D)$ and $P(A_D)$ must be the same. Assume for the moment the truth of this statement, which is somewhat more obvious than its proof.

If for a particular ω each of (22.10) and (22.12) coincides in D_0 with a function analytic in $D_0 \cup D$, the same must be true of

$$(22.13) \qquad F(\omega, z) - G(\omega, z) = 2 \sum_{m=0}^{\infty} X_{mk}(\omega) z^{mk}.$$

Let $D_l = [ze^{2\pi i l/k}: z \in D]$. Since replacing z by $ze^{2\pi i/k}$ leaves the function (22.13) unchanged, it can be extended analytically to each $D_0 \cup D_l$, $l = 1, 2,$ .... Because of the choice of k, it can therefore be extended analytically to $[z: |z| < 1 + \epsilon]$ for some positive ϵ; but this is impossible if (22.11) holds, since the radius of convergence must then be 1.

Therefore, $A_D \cap B_D$ cannot contain a point ω satisfying (22.11). Since (22.11) holds with probability 1, this rules out the possibility $P(A_D) = P(B_D) = 1$ and by the zero-one law leaves only the possibility $P(A_D) = P(B_D) = 0$. Let A be the ω-set where (22.10) extends to a function analytic in some open set larger than D_0. Then $\omega \in A$ if and only if (22.10) extends to $D_0 \cup D$ for some $D = [z: |z - \zeta| < r]$ for which $D - D_0 \neq 0$, r is rational, and ζ has rational real and imaginary parts; in other words, A is the countable union of A_D for such D. Therefore, A lies in $\sigma(X_0, X_1, \ldots)$ and has probability 0.

It remains only to show that $P(A_D) = P(B_D)$, and this is most easily done by comparing $\{X_n\}$ and $\{Y_n\}$ with a canonical sequence having the same structure. Put $X'_n(\omega) = (X_n(\omega) + 1)/2$, and let $T\omega$ be $\sum_{n=0}^{\infty} X'_n(\omega) 2^{-n-1}$ on the ω-set A^* where this sum lies in $(0, 1]$; on $\Omega - A^*$ let $T\omega$ be 1, say. Because of (22.9), $P(A^*) = 1$. Let $\mathcal{F} = \sigma(X_0, X_1, \ldots)$ and let $\mathcal{B}$ be the σ-field of Borel subsets of $(0, 1]$; then $T: \Omega \to (0, 1]$ is measurable $\mathcal{F}/\mathcal{B}$. Let $r_n(x)$ be the nth Rademacher function. If $M = [x: r_i(x) = u_i, i = 1, \ldots, n]$, where $u_i = \pm 1$ for each i, then $P(T^{-1}M) = P[\omega: X_i(\omega) = u_i, i = 0, 1, \ldots, n - 1] = 2^{-n}$, which is the Lebesgue measure $\lambda(M)$ of M. Since these sets form a π-system generating $\mathcal{B}$, $P(T^{-1}M) = \lambda(M)$ for all M in $\mathcal{B}$ (Theorem 3.3).

Let M_D be the set of x for which $\sum_{n=0}^{\infty} r_{n+1}(x) z^n$ extends analytically to $D_0 \cup D$. Then M_D lies in $\mathcal{B}$, this being a special case of the fact that A_D lies in $\mathcal{F}$. Moreover, if $\omega \in A^*$, then $\omega \in A_D$ if and only if $T\omega \in M_D$: $A^* \cap A_D = A^* \cap T^{-1}M_D$. Since $P(A^*) = 1$, it follows that $P(A_D) = \lambda(M_D)$.

This argument only uses (22.9), and therefore it applies to $\{Y_n\}$ and B_D as well. Therefore, $P(B_D) = \lambda(M_D) = P(A_D)$. ∎

PROBLEMS

22.1. Suppose that $X_1, X_2, \ldots$ is an independent sequence and Y is measurable $\sigma(X_n, X_{n+1}, \ldots)$ for each n. Show that there exists a constant a such that $P[Y = a] = 1$.

22.2. For the empirical distribution function $F_n(x, \omega)$ (see (20.42)) and a distribution function F, let A_F be the ω-set where there is weak convergence $F_n(x, \omega) \Rightarrow F(x)$ in the sense of Section 14. Show that A_F is a tail set.

22.3. 21.12↑ Let $X_1, X_2, \ldots$ be independent random variables with distribution functions $F_1, F_2, \ldots$.

(a) Show that $P[\sup_n X_n < \infty]$ is 0 or 1 and is in fact 1 if and only if $\Sigma_n (1 - F_n(x)) < \infty$ for some x.

(b) Suppose that $X = \sup_n X_n$ is finite with probability 1 and let F be its distribution function. Show that $F(x) = \Pi_{n=1}^\infty F_n(x)$.

(c) Show further that $E[X^+] < \infty$ if and only if $\Sigma_n \int_{X_n > \alpha} X_n \, dP < \infty$ for some α.

22.4. Let $X_1, X_2, \ldots$ be independent, and suppose that X_n takes the values $-n^4, 0, n^4$ with respective probabilities $n^{-2}, 1 - 2n^{-2}, n^{-2}$. Show that $\Sigma_n X_n$ (even $\Sigma_n |X_n|$) converges with probability 1 even though the hypothesis of Theorem 22.3 fails.

22.5. The series $\Sigma_n X_n$ in Theorem 22.3 is in general only conditionally convergent. Show that it converges absolutely with probability 1 if $\Sigma_n E[|X_n|] < \infty$ and hence if $\Sigma_n (\text{Var} [X_n])^{1/2} < \infty$ (assume that $E[X_n] = 0$). This last condition is of course stronger than $\Sigma_n \text{Var} [X_n] < \infty$; but observe that it implies that $\Sigma_n |X_n| < \infty$ with probability 1 even without the assumption of independence. The point of Theorem 22.3 is that even if $\Sigma_n |X_n|$ diverges, the independent X_n, because of their signs, tend to cancel each other in such a way that $\Sigma_n X_n$ converges conditionally. Compare Problem 22.10.

22.6. Show under the hypotheses of Theorem 22.3 that $\Sigma_n X_n$ has finite variance and extend Theorem 22.2 to infinite sums.

22.7. 22.4↑ Show that $n^{-1}\Sigma_{k=1}^n X_k$ may converge to 0 with probability 1 even though the hypothesis of Theorem 22.4 fails.

22.8. Suppose that $X_1, X_2, \ldots$ are independent random variables and that $\alpha > 0$, and let

$$Y_n^{(\alpha)} = \begin{cases} X_n & \text{if } |X_n| \le \alpha, \\ 0 & \text{if } |X_n| > \alpha. \end{cases}$$

The *three-series theorem*: In order that $\Sigma_n X_n$ converge with probability 1, it is necessary and sufficient that there exist an α such that the three series

$$(22.14) \qquad \sum_n P[|X_n| \ge \alpha], \qquad \sum_n E[Y_n^{(\alpha)}], \qquad \sum_n \text{Var} [Y_n^{(\alpha)}]$$

converge. More precisely, in order that $\Sigma_n X_n$ converge with probability 1 it is necessary that the three series converge for all positive α and sufficient that they converge for some positive α. See Problem 27.12 for the necessity. Prove sufficiency: Assume the three series in (22.14) converge for some positive α.

(a) Let $m_n^{(\alpha)} = E[Y_n^{(\alpha)}]$ and show that $\Sigma_n (Y_n^{(\alpha)} - m_n^{(\alpha)})$ converges with probability 1.

Hint: Use Theorem 22.3.

(b) Show that $\Sigma_n Y_n^{(\alpha)}$ converges with probability 1.

(c) Show that $\Sigma_n X_n$ converges with probability 1.

Hint: Show that $P[X_n \neq Y_n^{(\alpha)} \text{ i.o.}] = 0$.

22.9. ↑ Assume both necessity and sufficiency in the three-series theorem.

(a) Let $r_n(\omega)$ be the Rademacher functions. Show that $\Sigma_n a_n r_n(\omega)$ converges almost everywhere on the unit interval if and only if $\Sigma_n a_n^2 < \infty$.

(b) Show that $\Sigma_n \pm n^{-\theta}$ (signs chosen on the toss of a coin) converges with probability 1 if and only if $\theta > \frac{1}{2}$.

22.10. ↑ (a) Suppose that $X_1, X_2, \ldots$ are independent and that for all choices of the signs there is probability 1 that $\Sigma_n \pm X_n$ converges. Show that $\Sigma_n X_n^2 < \infty$ with probability 1.

(b) Change the order of quantification and suppose that with probability 1 the series $\Sigma_n \pm X_n$ converges for all choices of the signs. Show that $\Sigma_n |X_n| < \infty$ with probability 1.

22.11. 20.20↑ Suppose that $X_1, X_2, \ldots$ are independent, each with the Cauchy distribution (20.45) for a common value of u.

(a) Show that $n^{-1}\Sigma_{k=1}^n X_k$ does not converge with probability 1. Contrast with Theorem 22.5.

(b) Show that $P[n^{-1}\max_{k<n}X_k \leq x] \to e^{-u/\pi x}$. Relate to Theorem 14.3.

22.12. If $X_1, X_2, \ldots$ are independent and identically distributed, and if $P[X_1 \geq 0] = 1$ and $P[X_1 > 0] > 0$, then $\Sigma_n X_n = \infty$ with probability 1. Deduce this from Theorem 22.5 and its corollary and also directly: find a positive ϵ such that $X_n > \epsilon$ infinitely often with probability 1.

22.13. Suppose that $X_1, X_2, \ldots$ are independent and identically distributed and $E[|X_1|] = \infty$. Use (21.9) to show that $\Sigma_n P[|X_n| \geq an] = \infty$ for each a, and conclude that $\sup_n n^{-1}|X_n| = \infty$ with probability 1. Now show that $\sup_n n^{-1}|S_n| = \infty$ with probability 1. Compare with the corollary to Theorem 22.5.

22.14. *Wald's equation.* Let $X_1, X_2, \ldots$ be independent and identically distributed with finite mean, and put $S_n = X_1 + \cdots + X_n$. Suppose that τ is a stopping time: τ has positive integers as values and $[\tau = n] \in \sigma(X_1, \ldots, X_n)$; see Section 7 for examples. Suppose also that $E[\tau] < \infty$.

(a) Prove that

(22.15) $$E[S_\tau] = E[X_1]E[\tau].$$

(b) Suppose that X_n is ± 1 with probabilities p and q, $p \neq q$, let τ be the first n for which S_n is $-a$ or b (a and b positive integers), and calculate $E[\tau]$. This gives the expected duration of the game in the gambler's ruin problem for unequal p and q.

22.15. Extend to general random variables the results in Problems 6.6, 6.7, 6.8, and 6.11.

22.16. Deduce

$$\lim_{s \to \infty} \sum_{k=0}^{[sx]} s^k(1+s)^{-k-1} = 1 - e^{-x}$$

from (22.8). Prove it directly.

22.17. 20.12↑ Let Z_n be 1 or 0 according as at time n there is or is not a record in the sense of Problem 20.12. Let $R_n = Z_1 + \cdots + Z_n$ be the number of records up to time n. Show that $R_n/\log n \to_P 1$.

22.18. 22.1↑ (a) Show that for an independent sequence $\{X_n\}$ the radius of convergence of the random Taylor series $\Sigma_n X_n z^n$ is r with probability 1 for some nonrandom r.

(b) Suppose that the X_n have the same distribution and $P[X_1 \neq 0] > 0$. Show that r is 1 or 0 according as $\log^+|X_1|$ has finite mean or not.

22.19. Suppose that $X_0, X_1, \ldots$ are independent and each is uniformly distributed over $[0, 2\pi]$. Show that with probability 1 the series $\Sigma_n e^{iX_n} z^n$ has the unit circle and its natural boundary.

22.20. 21.38↑ Prove the law of the iterated logarithm for random variables satisfying the conditions of Problems 21.37 and 21.38.

22.21. 9.5 22.20↑ Extend the result in Problem 9.5 to the random variables of the preceding problem. This requires proving that $S_n a_n - S_{n-1} a_{n-1} \to 0$ with probability 1, where $a_n^{-2} = n \log \log n$. Represent the difference as $X_n a_n + S_{n-1}(a_n - a_{n-1})$ and treat the two terms separately.

SECTION 23. THE POISSON PROCESS

Characterization of the Exponential Distribution

Suppose that X has the exponential distribution with parameter α:

(23.1) $$P[X > x] = e^{-\alpha x}, \quad x \geq 0.$$

The definition (4.1) of conditional probability then gives

(23.2) $$P[X > x + y | X > x] = P[X > y], \quad x, y \geq 0.$$

Imagine X as the waiting time for the occurrence of some event such as the arrival of the next customer at a queue or telephone call at an exchange. As observed in Section 14 (see (14.6)), (23.2) attributes to the waiting-time mechanism a lack of memory or aftereffect. And as shown in Section 14, the condition (23.2) implies that X has the distribution (23.1) for some positive α. Thus if in the sense of (23.2) there is no aftereffect in the waiting-time mechanism, then the waiting time itself necessarily follows the exponential law.

The Poisson Process

Consider next a stream or sequence of events, say arrivals of calls at an exchange. Let X_1 be the waiting time to the first event, let X_2 be the waiting time between the first and second events, and so on. The formal model consists of an infinite sequence $X_1, X_2, \ldots$ of random variables on some probability space, and $S_n = X_1 + \cdots + X_n$ represents the time of occurrence of the nth event; it is convenient to write $S_0 = 0$. The stream of events itself remains intuitive and unformalized, and the mathematical definitions and arguments are framed exclusively in terms of the X_n.

If no two of the events are to occur simultaneously, the S_n must be strictly increasing, and if only finitely many of the events are to occur in each finite interval of time, S_n must go to infinity:

(23.3) $$0 = S_0 < S_1 < S_2 < \cdots, \qquad \sup_n S_n = \infty.$$

This condition is the same thing as

(23.4) $$X_1 > 0, X_2 > 0, \ldots, \qquad \sum_n X_n = \infty.$$

Throughout this section it will be assumed that these conditions do hold:

Condition $0°$.   *The event defined by* (23.3) *and* (23.4) *has probability* 1.

There are at the outset no further restrictions on the X_i; they are not assumed independent, for example, or identically distributed.

The number N_t of events that occur in the time interval $[0, t]$ is the largest integer n such that $S_n \leq t$:

(23.5) $$N_t = \max [n: S_n \leq t].$$

On the ω-set (23.3) define N_t by (23.5); elsewhere set $N_t = 0$ to be definite. Note that $N_t = 0$ if $t < S_1 = X_1$; in particular, $N_0 = 0$. The number of events in $(s, t]$ is the increment $N_t - N_s$.

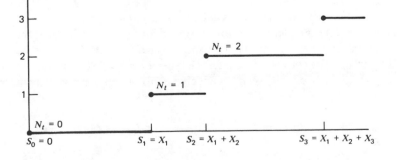

From (23.5) follows the basic relation connecting the N_t with the S_n:

(23.6) $[N_t \geq n] = [S_n \leq t]$.

From this follows

(23.7) $[N_t = n] = [S_n \leq t < S_{n+1}]$.

Each N_t is thus a random variable. The program is to study the joint distributions of the N_t under conditions on the waiting times X_n and vice versa. The most common model specifies the independence of the waiting times and the absence of aftereffect:

Condition 1°.   The X_n are independent and each is exponentially distributed with parameter α.

In this case $P[X_n > 0] = 1$ for each n and $n^{-1}S_n \to \alpha^{-1}$ by the strong law of large numbers (Theorem 22.5), and so Condition 0° does hold.

Under Condition 1°, S_n has the distribution function specified by (20.40), so that $P[N_t \geq n] = \sum_{i=n}^{\infty} e^{-\alpha t}(\alpha t)^i/i!$ by (23.6), and

(23.8) $P[N_t = n] = e^{-\alpha t}\dfrac{(\alpha t)^n}{n!}, \qquad n = 0, 1, \ldots$.

Thus N_t has the Poisson distribution with mean αt. More will be proved presently.

Condition 2°.   (i)   For $0 < t_1 < \cdots < t_k$ the increments $N_{t_1}, N_{t_2} - N_{t_1}, \ldots, N_{t_k} - N_{t_{k-1}}$ are independent.
(ii)   The individual increments have the Poisson distribution:

(23.9) $P[N_t - N_s = n] = e^{-\alpha(t-s)}\dfrac{(\alpha(t-s))^n}{n!}, \qquad n = 0, 1, \ldots, 0 \leq s < t$;

and $N_0 = 0$ with probability 1.

If $N_0 = 0$, (23.8) is a special case of (23.9). A collection $[N_t : t \geq 0]$ of random variables satisfying Condition 2° is called a *Poisson process,* and α is the *rate* of the process. As the increments are independent by (i), if $r < s < t$, then the distributions of $N_s - N_r$ and $N_t - N_s$ must convolve to that of $N_t - N_r$. But this requirement is consistent with (ii) because Poisson distributions with parameters u and v are easily seen to convolve to a Poisson distribution with parameter $u + v$.

Theorem 23.1.   *Conditions 1° and 2° are equivalent in the presence of Condition 0°.*

PROOF OF 1° → 2°.   Fix t and consider the events that happen after time t. By (23.5), $S_{N_t} \leq t < S_{N_t+1}$, and the waiting time from t to the first event following t is $S_{N_t+1} - t$; the waiting time between the first and second events following t is X_{N_t+2}; and so on. Thus

$$(23.10) \qquad X_1^{(t)} = S_{N_t+1} - t, \quad X_2^{(t)} = X_{N_t+2}, \quad X_3^{(t)} = X_{N_t+3}, \ldots$$

define the waiting times following t. By (23.6), $N_{t+s} - N_t \geq m$, or $N_{t+s} \geq N_t + m$, if and only if $S_{N_t+m} \leq t + s$, which is the same thing as $X_1^{(t)} + \ldots + X_m^{(t)} \leq s$. Thus

$$(23.11) \qquad\qquad N_{t+s} - N_t = \max [m: X_1^{(t)} + \cdots + X_m^{(t)} \leq s].$$

Hence $[N_{t+s} - N_t = m] = [X_1^{(t)} + \cdots + X_m^{(t)} \leq s < X_1^{(t)} + \cdots + X_{m+1}^{(t)}]$. A comparison of (23.11) and (23.5) shows that for fixed t the random variables $N_{t+s} - N_t$ for $s \geq 0$ are defined in terms of the sequence (23.10) in exactly the same way as the N_s are defined in terms of the original sequence of waiting times.

The idea now is to show that conditionally on the event $[N_t = n]$ the random variables (23.10) are independent and exponentially distributed. Because of the independence of the X_k and the basic property (23.2) of the exponential distribution, this seems intuitively clear. For a proof, apply (20.30). Suppose $y \geq 0$; if G_n is the distribution function of S_n, then since X_{n+1} has the exponential distribution,

$$P[S_n \leq t < S_{n+1}, S_{n+1} - t > y] = P[S_n \leq t, X_{n+1} > t + y - S_n]$$

$$= \int_{x \leq t} P[X_{n+1} > t + y - x] \, dG_n(x)$$

$$= e^{-\alpha y} \int_{x \leq t} P[X_{n+1} > t - x] \, dG_n(x)$$

$$= e^{-\alpha y} P[S_n \leq t, X_{n+1} > t - S_n].$$

By the assumed independence of the X_n,

$$P[S_{n+1} - t > y_1, X_{n+2} > y_2, \ldots, X_{n+j} > y_j, S_n \leq t < S_{n+1}]$$

$$= P[S_{n+1} - t > y_1, S_n \leq t < S_{n+1}] e^{-\alpha y_2} \cdots e^{-\alpha y_j}$$

$$= P[S_n \leq t < S_{n+1}] e^{-\alpha y_1} \cdots e^{-\alpha y_j}.$$

If $H = (y_1, \infty) \times \cdots \times (y_j, \infty)$, this is

$$(23.12) \quad P[N_t = n, (X_1^{(t)}, \ldots, X_j^{(t)}) \in H]$$
$$= P[N_t = n] P[(X_1, \ldots, X_j) \in H].$$

By Theorem 10.3 the equation extends from H of the special form above to all H in $\mathcal{R}^j$.

Now the event $[N_{s_i} = m_i, 1 \le i \le u]$ can be put in the form $[(X_1, \ldots, X_j)$ $\in H]$, where $j = m_u + 1$ and H is the set of x in R^j for which $x_1 + \cdots + x_{m_i} \le$ $s_i < x_1 + \cdots + x_{m_i+1}, 1 \le i \le u$. But then $[(X_1^{(t)}, \ldots, X_j^{(t)}) \in H]$ is by (23.11) the same as the event $[N_{t+s_i} - N_t = m_i, 1 \le i \le u]$. Thus (23.12) gives

$$P[N_t = n, N_{t+s_i} - N_t = m_i, 1 \le i \le u] = P[N_t = n]P[N_{s_i} = m_i, 1 \le i \le u].$$

From this it follows by induction on k that if $0 = t_0 < t_1 < \cdots < t_k$, then

$$(23.13) \quad P[N_{t_i} - N_{t_{i-1}} = n_i, 1 \le i \le k] = \prod_{i=1}^{k} P[N_{t_i - t_{i-1}} = n_i].$$

Thus Condition 1° implies (23.13) and, as already seen, (23.8). But from (23.13) and (23.8) follow the two parts of Condition 2°.

PROOF OF 2° → 1°. If 2° holds, then by (23.6), $P[X_1 > t] = P[N_t = 0] = e^{-\alpha t}$, so that X_1 is exponentially distributed. To find the joint distribution of X_1 and X_2, suppose that $0 \le s_1 < t_1 < s_2 < t_2$ and perform the calculation

$P[s_1 < S_1 \le t_1, s_2 < S_2 \le t_2]$

$\quad = P[N_{s_1} = 0, N_{t_1} - N_{s_1} = 1, N_{s_2} - N_{t_1} = 0, N_{t_2} - N_{s_2} \ge 1]$

$\quad = e^{-\alpha s_1} \times \alpha(t_1 - s_1)e^{-\alpha(t_1-s_1)} \times e^{-\alpha(s_2-t_1)} \times (1 - e^{-\alpha(t_2-s_2)})$

$\quad = \alpha(t_1 - s_1)(e^{-\alpha s_2} - e^{-\alpha t_2}) = \iint\limits_{\substack{s_1 < y_1 \le t_1 \\ s_2 < y_2 \le t_2}} \alpha^2 e^{-\alpha y_2} \, dy_1 \, dy_2.$

Thus for a rectangle A contained in the open set $G = [(y_1, y_2): 0 < y_1 < y_2]$,

$$P[(S_1, S_2) \in A] = \int_A \alpha^2 e^{-\alpha y_2} \, dy_1 \, dy_2.$$

By inclusion-exclusion, this holds for finite unions of such rectangles and hence by a passage to the limit for countable ones. Therefore, it holds for $A = G \cap G'$ if G' is open. Since the open sets form a π-system generating the Borel sets, (S_1, S_2) has density $\alpha^2 e^{-\alpha y_2}$ on G (of course, the density is 0 outside G).

By a similar argument in R^k (the notation only is more complicated), $(S_1, \ldots, S_k)$ has density $\alpha^k e^{-\alpha y_k}$ on $[y: 0 < y_1 < \cdots < y_k]$. If $g(y) = x$ is defined by $x_i = y_i - y_{i-1}$, then $(X_1, \ldots, X_k) = g(S_1, \ldots, S_k)$ has by (20.20) the density $\prod_{i=1}^{k} \alpha e^{-\alpha x_i}$ (the Jacobian is identically 1). This proves Condition 1°. ∎

Two Other Characterizations of the Poisson Process

The condition (23.2) is an interesting characterization of the exponential dis-

tribution because it is essentially qualitative. There are qualitative character-
izations of the Poisson process as well. (Condition $0°$ is still assumed.)

Condition 3°.   (i) *For $0 < t_1 < \cdots < t_k$ the increments $N_{t_1}, N_{t_2} - N_{t_1}, \ldots,$
$N_{t_k} - N_{t_{k-1}}$ are independent.*
   (ii) *The distribution of $N_t - N_s$ depends only on the difference $t - s$, and
$N_0 = 0$ with probability 1.*

Condition 4°.   *If $0 < t_1 < \cdots < t_k$, and if $n_1, \ldots, n_k$ are nonnegative integers, then*

$$(23.14) \qquad P[N_{t_k+h} - N_{t_k} = 1 \,|\, N_{t_j} = n_j, j \le k] = \alpha h + o(h)$$

and

$$(23.15) \qquad P[N_{t_k+h} - N_{t_k} \ge 2 \,|\, N_{t_j} = n_j, j \le k] = o(h)$$

as $h \downarrow 0$. Moreover,

$$(23.16) \qquad \lim_{s \to t} P[N_s \ne N_t] = 0,$$

and $N_0 = 0$ with probability 1.

The occurrences of $o(h)$ in (23.14) and (23.15) denote functions, say $\phi_1(h)$,
and $\phi_2(h)$, such that $h^{-1}\phi_i(h) \to 0$ as $h \downarrow 0$; the ϕ_i may a priori depend on k,
$t_1, \ldots, t_k$, and $n_1, \ldots, n_k$ as well as on h. It is assumed in (23.14) and (23.15)
that the conditioning events have positive probability, so that the conditional
probabilities are well defined. (If the $\phi_i(h)$ do not depend on the n_j, then (23.14)
and (23.15) together imply the weakened form of (23.16) in which s decreases
to t.)

Theorem 23.2.   *Conditions $1°$ through $4°$ are all equivalent in the presence
of Condition $0°$.*

Only $3°$ does not specify α; it implies that $1°, 2°$, and $4°$ hold for some positive
α.

Condition $3°$ in some cases makes the Poisson model quite plausible. The
increments will be essentially independent if the arrivals to time s cannot seri-
ously deplete the population of potential arrivals, so that N_s has for $t > s$ negligible effect on $N_t - N_s$. And the distribution of $N_{t+u} - N_t$ will essentially be
a function of u alone if conditions do not materially change with time. These
conditions obtain in the telephony example if the rate of calls arriving at the
exchange is small in comparison with the population of telephone subscribers
and if attention is confined, say, to the busy hours of the day (conditions not being
constant around the clock).

Condition 4° is "more qualitative" than 3°, since it involves only the approximate conditional distribution of $N_{t+h} - N_t$ for small values of h.

Because of Theorem 23.1, it suffices in proving Theorem 23.2 to establish the implications 2° → 3° → 4° → 2°. The first of these, 2° → 3°, is of course obvious.

PROOF OF 3° → 4°.   Because of the assumption of independence and the assumption that the distribution of $N_{t+h} - N_t$ is that of N_h, (23.14), (23.15), and (23.16) will follow if $P[N_h = 1] = \alpha h + o(h)$ and $P[N_h \geq 2] = o(h)$ (and the functions ϕ_i alluded to above will, in fact, not depend on the t_j and n_j). If $U(t) = P[N_t = 0]$, then $U(s + t) = P[N_s = 0, N_{s+t} - N_s = 0] = U(s)U(t)$. By the corollary to Theorem 14.4, U has the form $U(t) = e^{-\alpha t}$; since $P[N_t = 0] = P[X_1 > t] \to 0$ as $t \to \infty$, α must be positive.

If $\phi_2(t) = P[N_t \geq 2]$, and if A_n is the ω-set where $N_{k/n}(\omega) - N_{(k-1)/n}(\omega) \geq 2$ for some $k = 1, \ldots, n$, then by independence

$$(23.17) \qquad P(A_n) = 1 - (1 - \phi_2(n^{-1}))^n.$$

Now $\min_{i \leq N_1(\omega)} X_i(\omega)$ is positive on the set (23.3), and if n^{-1} is less than this quantity, then $\omega \notin A_n$. Thus for ω in the set (23.3), there exists an $n_0(\omega)$ such that $n > n_0(\omega)$ implies that $\omega \notin A_n$. Hence $P[A_n \text{ i.o.}] = 0$, and so $P(A_n) \to 0$ by Theorem 4.1. Since $\log(1 + x) \leq x$, (23.17) implies that $n\phi_2(1/n) \leq -\log(1 - P(A_n)) \to 0$; since $\phi_2(h)$ is nondecreasing, $\phi_2(h)/h \leq ([h^{-1}] + 1) \cdot \phi_2(1/[h^{-1}]) \to 0$. Therefore, $\phi_2(h) = o(h)$.

Since $P[N_h = 1] = 1 - P[N_h = 0] - P[N_h \geq 2] = 1 - e^{-\alpha h} + o(h) = \alpha h + o(h)$, the proof is complete.   ∎

PROOF OF 4° → 2°.   Fix k, the t_j, and the n_j, denote by A the event $[N_{t_j} = n_j, j \leq k]$, and for $t \geq 0$ put $p_n(t) = P[N_{t_k+t} - N_{t_k} = n | A]$. It suffices to prove that

$$(23.18) \qquad p_n(t) = e^{-\alpha t} \frac{(\alpha t)^n}{n!}, \qquad n = 0, 1, \ldots,$$

since 2° will then follow by induction.

For any events B and C, $|P(B) - P(C)| \leq P(B \triangle C)$. Therefore, $|p_n(t) - p_n(s)| \leq P^{-1}(A)P[N_{t_k+s} \neq N_{t_k+t}]$. By (23.16), $p_n(t)$ is thus continuous in t.

To simplify the notation, put $D_t = N_{t_k+t} - N_{t_k}$. If $D_{t+h} = n$, then $D_t = m$ for some $m \leq n$. By the rules for conditional probabilities,

$$p_n(t + h) = p_n(t)P[D_{t+h} - D_t = 0 | A \cap [D_t = n]]$$
$$+ p_{n-1}(t)P[D_{t+h} - D_t = 1 | A \cap [D_t = n - 1]]$$
$$+ \sum_{m=0}^{n-2} p_m(t)P[D_{t+h} - D_t = n - m | A \cap [D_t = m]].$$

For $n \leq 1$, the final sum is absent, and for $n = 0$, the middle term is absent as well. By (23.15), the final sum is $o(h)$ for each fixed n. Applying (23.14) and (23.15) once more leads to

$$p_n(t + h) = p_n(t)(1 - \alpha h) + p_{n-1}(t)\alpha h + o(h),$$

and letting $h \downarrow 0$ gives

(23.19) $p_n'(t) = -\alpha p_n(t) + \alpha p_{n-1}(t).$

This right-hand derivative is actually a two-sided derivative because the right side of the equation is continuous.* In the case $n = 0$, $p_{-1}(t)$ is to be taken identically 0.

Now (23.19) gives†

$$p_n(t) = e^{-\alpha t}[p_n(0) + \alpha \int_0^t p_{n-1}(s)e^{\alpha s}\, ds].$$

Since $p_n(0) = P[D_0 = n | A]$ is 1 or 0 according as $n = 0$ or $n \neq 0$, (23.18) follows by induction.   ∎

Stochastic Processes

The Poisson process $[N_t : t \geq 0]$ is an example of a *stochastic process*—that is, a collection of random variables (on some probability space $(\Omega, \mathcal{F}, P)$) indexed by a parameter regarded as representing time. In the Poisson case, time is *continuous*. In some cases the time is *discrete*: Section 7 concerns the sequence $\{F_n\}$ of a gambler's fortunes; there n represents time, but time that increases in jumps.

Part of the structure of a stochastic process is specified by its *finite-dimensional distributions*. For any finite sequence $t_1, \ldots, t_k$ of time points, the k-dimensional random vector $(N_{t_1}, \ldots, N_{t_k})$ has a distribution $\mu_{t_1 \ldots t_k}$ over R^k. These measures $\mu_{t_1 \ldots t_k}$ are the finite-dimensional distributions of the process. Condition 2° of this section in effect specifies them for the Poisson case:

$$(23.20) \quad P[N_{t_j} = n_j, j \leq k] = \prod_{j=1}^{k} e^{-\alpha(t_j - t_{j-1})} \frac{(\alpha(t_j - t_{j-1}))^{n_j - n_{j-1}}}{(n_j - n_{j-1})!}$$

if $0 \leq n_1 \leq \cdots \leq n_k$ and $0 \leq t_1 < \cdots < t_k$ (take $n_0 = t_0 = 0$).

---

* To prove that a continuous $f(t)$ with continuous right-hand derivative $f^+(t)$ for $t \geq 0$ necessarily has a two-sided derivative for $t > 0$, it suffices to prove that $F(t) = f(t) - f(0) - \int_0^t f^+(s)\, ds$ vanishes identically. If (say), $F(t_0) < 0$, then $G(t) = F(t) - tF(t_0)/t_0$ satisfies $G(0) = G(t_0) = 0$ and (since F^+ vanishes) $G^+(t) > 0$, so that G must over $[0, t_0]$ have a positive maximum at some interior point s_0; but $G^+(s_0) \leq 0$ at a maximum, a contradiction.

† The differential equation $f'(t) = Af(t) + g(t)$ for continuous g has the particular solution $f_0(t) = e^{At} \int_0^t g(s)e^{-As}\, ds$; for an arbitrary solution f, $(f(t) - f_0(t))e^{-At}$ has derivative 0 and hence equals $f(0)$ identically.

The finite-dimensional distributions do not, however, contain all the mathematically interesting information about the process in the case of continuous time. Because of Condition 0° and the definition (23.5) (and the remark following it), for each fixed ω, $N_t(\omega)$ is, as a function of t, nondecreasing, right-continuous, and integer-valued. Suppose that $f(t)$ is t or 0 according as t is rational or irrational. Let N_t be defined as before, and let

$$(23.21) \qquad M_t(\omega) = N_t(\omega) + f(t + X_1(\omega)).$$

If R is the set of rationals, then $P[\omega: f(t + X_1(\omega)) \neq 0] = P[\omega: X_1(\omega) \in \bar{R} - t] = 0$ for each t because $R - t$ is countable and X_1 has a density. Thus $P[M_t = N_t] = 1$ for each t, and so the stochastic process $[M_t: t \geq 0]$ has the same finite-dimensional distributions as $[N_t: t \geq 0]$. For ω fixed, however, $M_t(\omega)$ as a function of t is everywhere discontinuous and is neither monotone nor exclusively integer-valued.

The functions obtained by fixing ω and letting t vary are called the *path functions* or *sample paths* of the process. The example above shows that the finite-dimensional distributions do not suffice to determine the character of the path functions. In specifying a stochastic process as a model for some phenomenon, it is natural to place conditions on the character of the sample paths as well as on the finite-dimensional distributions. Condition 0° was imposed throughout this section to ensure that the sample paths are with probability 1 nondecreasing, right-continuous, integer-valued step functions, a natural condition if N_t is to represent the number of events in $[0, t]$. Stochastic processes in continuous time are studied further in Chapter 7.

PROBLEMS

23.1. Show that the minimum of independent exponential waiting times is again exponential and that the parameters add.

23.2. 20.23 ↑ Show that the time S_n of the nth event in a Poisson stream has the gamma density $f(x; \alpha, n)$ as defined by (20.47). This is sometimes called the *Erlang* density.

23.3. Let $A_t = t - S_{N_t}$ be the time back to the most recent event in the Poisson stream (or to 0) and let $B_t = S_{N_t+1} - t$ be the time forward to the next event. Show that A_t and B_t are independent, that B_t is distributed as X_1 (exponentially with parameter α), and that A_t is distributed as min $\{X_1, t\}$: $P[A_t \leq t]$ is 0, $1 - e^{-\alpha x}$, or 1 as $x < 0$, $0 \leq x < t$, or $x \geq t$.

23.4. ↑ Let $L_t = A_t + B_t = S_{N_t+1} - S_{N_t}$ be the length of the interarrival interval covering t.

(a) Show that L_t has density

$$d_t(x) = \begin{cases} \alpha^2 x e^{-\alpha x} & \text{if } 0 < x < t, \\ \alpha(1 + \alpha t)e^{-\alpha x} & \text{if } x \geq t. \end{cases}$$

(b) Show that $E[L_t]$ converges to $2E[X_1]$ as $t \to \infty$. This seems paradoxical because L_t is one of the X_n. Give an intuitive resolution of the apparent paradox.

23.5. *Merging Poisson streams.* Define a process $\{N_t\}$ by (23.5) for a sequence $\{X_n\}$ of random variables satisfying (23.4). Let $\{X_n'\}$ be a second sequence of random variables, on the same probability space, satisfying (23.4) and define $\{N_t'\}$ by $N_t' = \max [n: X_1' + \cdots + X_n' \le t]$. Define $\{N_t''\}$ by $N_t'' = N_t + N_t'$. Show that, if $\sigma(X_1, X_2, \ldots)$ and $\sigma(X_1', X_2', \ldots)$ are independent and $\{N_t\}$ and $\{N_t'\}$ are Poisson processes with respective rates α and β, then $\{N_t''\}$ is a Poisson process with rate $\alpha + \beta$.

23.6. ↑ The nth and $(n + 1)$st events in the process $\{N_t\}$ occur at times S_n and S_{n+1}.
(a) Find the distribution of the number $N_{S_{n+1}}' - N_{S_n}'$ of events in the other process during this time interval.
(b) Generalize to $N_{S_m}' - N_{S_n}'$.

23.7. Suppose that X and Y are independent and nonnegative. Suppose that X and $X + Y$ have Poisson distributions with means α and β. Show that either $\alpha = \beta$ and $Y = 0$ with probability 1 or else $\alpha < \beta$ and Y has the Poisson distribution with mean $\beta - \alpha$.

23.8. ↑ For a Poisson stream and a bounded Borel set A let $N(A)$ be the number of events that occur at times lying in the set A. Show that $N(A)$ is a random variable having the Poisson distribution with mean α times the Lebesgue measure of A.

23.9. Suppose that $X_1, X_2, \ldots$ are independent and exponentially distributed with parameter α, so that (23.5) defines a Poisson process $\{N_t\}$. Suppose that $Y_1, Y_2, \ldots$ are independent and identically distributed and that $\sigma(X_1, X_2, \ldots)$ and $\sigma(Y_1, Y_2, \ldots)$ are independent. Put $Z_t = \Sigma_{k \le N_t} Y_k$. This is the *compound Poisson process*. If, for example, the event at time S_n in the original process represents an insurance claim, and if Y_n represents the amount of the claim, then Z_t represents the total claims to time t.
(a) If $Y_k = 1$ with probability 1, then $\{Z_t\}$ is an ordinary Poisson process.
(b) Show that Z_t has independent increments and that $Z_{s+t} - Z_s$ has the same distribution as Z_t.
(c) Show that, if Y_k assumes the values 1 and 0 with probabilities p and $1 - p$, then $\{Z_t\}$ is a Poisson process with rate $p\alpha$.

23.10. The process $\{N_t\}$ defined by (23.5) is a *generalized* Poisson process if the increments are independent and have Poisson distributions. (Count the unit mass at 0 as a Poisson distribution with mean 0.)
(a) Suppose that $f(t)$ is nonnegative, nondecreasing, and continuous for $t \ge 0$, and suppose that $\{N_t\}$ is an ordinary Poisson process. Show that $\{N_{f(t)}\}$ is a generalized Poisson process.
(b) Show that every generalized Poisson process arises in this way.

23.11. If the waiting times X_n are independent and exponentially distributed with parameter α, then $S_n/n \to \alpha^{-1}$ with probability 1, by the strong law of large numbers. From $\lim_{t \to \infty} N_t = \infty$ and $S_{N_t} \le t < S_{N_t+1}$ deduce that $\lim_{t \to \infty} N_t/t = \alpha$ with probability 1.

23.12. ↑ (a)   Suppose that $X_1, X_2, \ldots$ are positive and assume directly that $S_n/n \to m$ with probability 1, as happens if the X_n are independent and identically distributed with mean m. Show that $\lim_t N_t/t = 1/m$ with probability 1.

(b)   Suppose now that $S_n/n \to \infty$ with probability 1, as happens if the X_n are independent and identically distributed and have infinite mean. Show that $\lim_t N_t/t = 0$ with probability 1.

The results in Problem 23.12 are theorems in *renewal theory*: A component of some mechanism is replaced each time it fails or wears out. The X_n are the lifetimes of the successive components, and N_t is the number of replacements, or renewals, to time t.

23.13.   20.8 23.12 ↑   Consider a persistent, irreducible Markov chain and for a fixed state j let N_n be the number of passages through j up to time n. Show that $N_n/n \to 1/m$ with probability 1, where $m = \Sigma_{k=1}^{\infty} k f_{jj}^{(k)}$ is the mean return time (replace $1/m$ by 0 if this mean is infinite).

SECTION 24.   QUEUES AND RANDOM WALK*

Results in queueing theory can be derived from facts about random walk, facts having an independent interest of their own.

The Single-Server Queue

Suppose that customers arrive at a server in sequence, their arrival times being $0, A_1, A_1 + A_2, \ldots$. The customers are numbered $0, 1, 2, \ldots$, and customer n arrives at time $A_1 + \cdots + A_n$, customer 0 arriving by convention at time 0. Assume that the sequence $\{A_n\}$ of *interarrival times* is independent and that the A_n all have the same distribution concentrated on $(0, \infty)$. The A_n may be exponentially distributed, in which case the arrivals form a Poisson process as in the preceding section. Another possibility is $P[A_n = a] = 1$ for a positive constant a; here the arrival stream is deterministic. At the outset, no special conditions are placed on the distribution common to the A_n.

Let $B_0, B_1, \ldots$ be the *service times* of the successive customers. The B_n are by assumption independent of each other and of the A_n, and they have a common distribution concentrated on $(0, \infty)$. Nothing further is assumed initially of this distribution: it may be concentrated at a single point for example, or it may be exponential. Another possibility is that the service-time distribution is a convolution of several exponential distributions; this represents a service consisting

* This section may be omitted.

of several stages, the service times of the constituent stages being independent and themselves exponentially distributed.

If a customer arrives to find the server busy, he joins the end of a queue and his service begins when all the preceding customers have been served. If he arrives to find the server free, his service begins immediately; the server is by convention free at time 0 when customer 0 arrives.

Airplanes coming into an airport form a queueing system. Here the A_n are the times between arrivals at the holding pattern, B_n is the time the plane takes to land and clear the runway, and the queue is the holding pattern itself.

Let $W_0, W_1, W_2, \ldots$ be the *waiting times* of the successive customers: W_n is the length of time customer n must wait until his service *begins*; thus $W_n + B_n$ is the total time spent at the service facility. Because of the convention that the server is free at time 0, $W_0 = 0$. Customer n arrives at time $A_1 + \cdots + A_n$; abbreviate this as t for the moment. His service begins at time $t + W_n$ and ends at time $t + W_n + B_n$. Moreover, customer $n + 1$ arrives at time $t + A_{n+1}$. If customer $n + 1$ arrives before $t + W_n + B_n$ (that is, if $t + A_{n+1} \leq t + W_n + B_n$, or $W_n + B_n - A_{n+1} \geq 0$), then his waiting time is $W_{n+1} = (t + W_n + B_n) - (t + A_{n+1}) = W_n + B_n - A_{n+1}$; if he arrives after $t + W_n + B_n$, he finds the server free and his waiting time is $W_{n+1} = 0$. Therefore, $W_{n+1} = \max [0, W_n + B_n - A_{n+1}]$. Define

$$(24.1) \qquad X_n = B_{n-1} - A_n, \qquad n = 1, 2, \ldots;$$

the recursion then becomes

$$(24.2) \qquad W_0 = 0, \quad W_{n+1} = \max [0, W_n + X_{n+1}], \qquad n = 0, 1, \ldots.$$

Note that W_n is measurable $\sigma(X_1, \ldots, X_n)$ and hence is measurable $\sigma(B_0, \ldots, B_{n-1}, A_1, \ldots, A_n)$.

The equations (24.1) and (24.2) also describe inventory systems. Suppose that on day n the inventory at a storage facility is augmented by the amount B_{n-1} and the day's demand is A_n. Since the facility cannot supply what it does not have, (24.1) and (24.2) describe the inventory W_n at the end of day n.

A problem of obvious interest in queuing theory is to compute or approximate the distribution of the waiting time W_n. Write

$$(24.3) \qquad S_0 = 0, \qquad S_n = X_1 + \cdots + X_n.$$

The following induction argument shows that

$$(24.4) \qquad W_n = \max_{0 \leq k \leq n} (S_n - S_k).$$

As this obvious for $n = 0$, assume that it holds for a particular n. If $W_n + X_{n+1}$ is nonnegative, it coincides with W_{n+1}, which is therefore the maximum of $S_n - S_k + X_{n+1} = S_{n+1} - S_k$ over $0 \leq k \leq n$; this maximum, being nonnegative, is unchanged if extended over $0 \leq k \leq n + 1$, whence (24.4) follows for $n + 1$.

On the other hand, if $W_n + X_{n+1} \leq 0$, then $S_{n+1} - S_k = S_n - S_k + X_{n+1} \leq 0$ for $0 \leq k \leq n$, so that $\max_{0 \leq k \leq n+1} (S_{n+1} - S_k)$ and W_{n+1} are both 0.

Thus (24.4) holds for all n. Since the components of $(X_1, X_2, \ldots, X_n)$ are independent and have a common distribution, this random vector has the same distribution over R^n as $(X_n, X_{n-1}, \ldots, X_1)$. Therefore, the partial sums $S_0, S_1, S_2, \ldots, S_n$ have the same joint distribution as $0, X_n, X_n + X_{n-1}, \ldots, X_n + \cdots + X_1$; that is, they have the same joint distribution as $S_n - S_n, S_n - S_{n-1}, S_n - S_{n-2}, \ldots, S_n - S_0$. Therefore, (24.4) has the same distribution as $\max_{0 \leq k \leq n} S_k$. The following theorem sums up these facts.

Theorem 24.1. *Let $\{X_n\}$ be an independent sequence of random variables with common distribution, and define W_n by (24.2) and S_n by (24.3). Then (24.4) holds and W_n has the same distribution as $\max_{0 \leq k \leq n} S_k$.*

In this and the other theorems of this section, the X_n are any independent random variables with a common distribution; it is only in the queueing-theory context that they are assumed defined by (24.1) in terms of positive, independent random variables A_n and B_n.

The problem is to investigate the distribution function

$$(24.5) \quad M_n(x) = P[W_n \leq x] = P[\max_{0 \leq k \leq n} S_k \leq x], \qquad x \geq 0.$$

(Since $W_n \geq 0$, $M(x) = 0$ for $x < 0$.) By the zero-one law (Theorem 22.1), the event $[\sup_k S_k = \infty]$ has probability 0 or 1. In the latter case $\max_{0 \leq k \leq n} S_k \to \infty$ with probability 1, so that

$$(24.6) \qquad \qquad \lim_{n \to \infty} M_n(x) = 0$$

for every x, however large. On the other hand, if $[\sup_k S_k = \infty]$ has probability 0, then

$$(24.7) \qquad \qquad M(x) = P[\sup_{k \geq 0} S_k \leq x]$$

is a well-defined distribution function. Moreover, $[\max_{0 \leq k \leq n} S_k \leq x] \downarrow [\sup_k S_k \leq x]$ for each x and hence

$$(24.8) \qquad \qquad \lim_{n \to \infty} M_n(x) = M(x).$$

Thus $M(x)$ provides an approximation for $M_n(x)$.

If the A_n and B_n of queuing theory have finite means, the ratio

$$(24.9) \qquad \qquad \rho = \frac{E[B_n]}{E[A_n]}$$

is called the *traffic intensity*. The larger ρ is, the more congested is the queueing system. If $\rho > 1$, then $E[X_n] > 0$ by (24.1), and since $n^{-1}S_n \to E[X_1]$ with probability 1 by the strong law of large numbers, $\sup_n S_n = \infty$ with probability 1. In this case (24.6) holds: the waiting time for the nth customer has a distribution that may be said to move out to $+\infty$; the service time is so large with respect to the arrival rate that the server cannot accommodate the customers.

It can be shown that $\sup_n S_n = \infty$ with probability 1 also if $\rho = 1$, in which case $E[X_n] = 0$. The analysis to follow covers only the case $\rho < 1$. Here the strong law gives $n^{-1}S_n \to E[X_1] < 0$, $S_n \to -\infty$, and $\sup_n S_n < \infty$ with probability 1, so that (24.8) applies.

Random Walk and Ladder Indices

To calculate (24.7) it helps to visualize the S_n as the successive positions in a random walk. The integer n is a *ladder index* for the random walk if S_n is the maximum position up to time n:

$$(24.10) \qquad \max_{0 \le k < n} S_k < S_n.$$

Note that the maximum is nonnegative because $S_0 = 0$.

Define a finite measure L by

$$(24.11) \quad L(A) = \sum_{n=1}^{\infty} P[\max_{0 \le k < n} S_k = 0 < S_n \in A], \qquad A \subset (0, \infty).$$

Thus L is a measure on the Borel sets of the line and has support $(0, \infty)$. The probability that there is at least one ladder index is then

$$(24.12) \qquad p = L(0, \infty) = P[\sup_n S_n > 0].$$

Let $T_1, T_2, \ldots$ be the times between the ladder indices: T_1 is the smallest n satisfying (24.10), $T_1 + T_2$ is the next smallest, and so on. To complete the definition, if there are only k ladder indices (k may be 0), set $T_{k+1} = T_{k+2} = \cdots = \infty$. Define the kth *ladder height* by $H_1 + \cdots + H_k = S_{T_1 + \cdots + T_k}$. Thus $H_1 + \cdots + H_k$ is the position in the random walk at the kth ladder index; if there are only k ladder indices, set $H_{k+1} = H_{k+2} = \cdots = 0$. As long as ladder indices occur, the ladder heights must increase and the H_k must be positive.

Whether there are infinitely many ladder indices or only finitely many, the definitions are so framed that

$$(24.13) \qquad \sup_k S_k = \sum_{n=1}^{\infty} H_n.$$

Therefore, (24.7) can be studied through the H_n. The distribution of H_1 consists of the measure L defined by (24.11) together with a mass of $1 - p$ at the point 0. If there is a ladder index and n is the first one ($T_1 = n$), then H_2 is the first positive term (if there is one) in the sequence $S_{n+1} - S_n, S_{n+2} - S_n, \ldots$; but this sequence has the same structure as the original random walk. Therefore, under the condition that there is a ladder index and hence a first one, the distribution of H_2 is the original distribution of H_1. An extension of this idea is the key to understanding the distribution of (24.13).

Let L^{n*} denote the n-fold convolution of L with itself: $L^{n*} = L^{(n-1)*} * L$, and L^{0*} is a unit mass at the point 0. Let ψ be the measure (see Example 16.5)

$$(24.14) \qquad \psi(A) = \sum_{n=0}^{\infty} L^{n*}(A), \qquad A \subset [0, \infty).$$

Note that ψ has a unit mass at 0 and the rest of the mass is confined to $(0, \infty)$.

Theorem 24.2.   (i)   *If $A_1, \ldots, A_n$ are subsets of $(0, \infty)$, then*

$$(24.15) \qquad P[H_i \in A_i, i \le n] = L(A_1) \cdots L(A_n).$$

(ii)   *If $p = 1$, then with probability 1 there are infinitely many ladder indices and $\sup_n S_n = \infty$.*

(iii)   *If $p < 1$, then with probability $p^n(1 - p)$ there are exactly n ladder indices; with probability 1 there are only finitely many ladder indices and $\sup_n S_n < \infty$; finally,*

$$(24.16) \qquad P[\sup_{n \ge 0} S_n \in A] = (1 - p)\,\psi(A), \qquad A \subset [0, \infty).$$

If $A \subset (0, \infty)$, then $P[H_n \in A] = p^{n-1}L(A)$, as will be seen in the course of the proof; thus (24.15) does not assert the independence of the H_n—obviously $H_n = 0$ implies that $H_{n+1} = 0$.

PROOF.   (i)   For $n = 1$, (24.15) is just the definition (24.11). Fix subsets $A_1, \ldots, A_n$ of $(0, \infty)$ and for $k \ge n - 1$ let B_k be the ω-set where $H_i \in A_i$ for $i \le n - 1$ and $T_1 + \ldots + T_{n-1} = k$. Then $B_k \in \sigma(X_1, \ldots, X_k)$, and so by independence and the assumption that the X_i all have the same distribution,

$$P(B_k \cap [\max_{0 \le j < m}(S_{k+j} - S_k) = 0 < S_{k+m} - S_k \in A_n])$$

$$= P(B_k)P[\max_{0 \le j < m} S_j = 0 < S_m \in A_n].$$

Summing over $m = 1, 2, \ldots$ and then over $k = n - 1, n, n + 1, \ldots$ gives $P[H_i \in A_i, i \le n] = P[H_i \in A_i, i \le n - 1]L(A_n)$. Thus (24.15) follows by induction.

Taking $A_1 = \cdots = A_n = (0, \infty)$ in (24.15) shows that the probability of at least n ladder indices is $P[H_i > 0, i \leq n] = p^n$.

(ii) If $p = 1$, then there must be infinitely many ladder indices with probability 1. In this case the H_n are independent random variables with distribution L concentrated on $(0, \infty)$; since $0 < E[H_n] \leq \infty$, it follows by the strong law of large numbers that $n^{-1} \sum_{k=1}^{n} H_k$ converges to a positive number or to ∞, so that (24.13) is infinite with probability 1.

(iii) Suppose that $p < 1$. The chance that there are exactly n ladder indices is $p^n - p^{n+1} = p^n(1 - p)$, and with probability 1 there are only finitely many. Moreover, the chance that there are exactly n ladder indices and that $H_i \in A_i$ for $i \leq n$ is $L(A_1) \cdots L(A_n) - L(A_1) \cdots L(A_n)L(0, \infty) = L(A_1) \cdots L(A_n) (1 - p)$. Conditionally on there being exactly n ladder indices, $H_1, \ldots, H_n$ are therefore independent, each with distribution $p^{-1}L$. Thus the probability that there are exactly n ladder indices and $H_1 + \cdots + H_n \in A$ is $(1 - p)L^{n*} (A)$ for $A \subset (0, \infty)$. This holds also for $n = 0$: $(1 - p)L^{0*}$ consists of a mass $1 - p = P[\sup_k S_k = 0]$ at the point 0. Adding over $n = 0, 1, 2, \ldots$ gives (24.16). Note that in this case there are with probability 1 only finitely many positive terms on the right in (24.13). ∎

Exponential Right Tail

Further information about the p and ψ in (24.16) can be obtained by a time-reversal argument like that leading to Theorem 24.1. As observed before, the two random vectors $(S_0, S_1, \ldots, S_n)$ and $(S_n - S_n, S_n - S_{n-1}, \ldots, S_n - S_0)$ have the same distribution over R^{n+1}. Therefore, the event $[\max_{0 \leq k < n} S_k < S_n \in A]$ has the same probability as the event $[\max_{0 \leq k < n} (S_n - S_{n-k}) < S_n - S_0 \in A]$:

$$(24.17) \qquad P[\max_{0 \leq k < n} S_k < S_n \in A] = P[\min_{1 \leq k \leq n} S_k > 0, S_n \in A]$$

for $n = 1, 2, \ldots$ and $A \subset (0, \infty)$.

Denote by $\psi_n(A)$ the common value of the two sides of (24.17); ψ_n is a measure supported by $(0, \infty)$. Identifying $\psi_n(A)$ with the left side of (24.17) gives for $A \subset (0, \infty)$,

$$\sum_{n=1}^{\infty} \psi_n(A) = \sum_{n=1}^{\infty} \sum_{j=1}^{\infty} P[T_1 + \cdots + T_j = n, H_1 + \cdots + H_j \in A]$$

$$= \sum_{j=1}^{\infty} L^{j*} (A).$$

Let ψ_0 be a unit mass at 0; it follows by (24.14) that

$$(24.18) \qquad \sum_{n=0}^{\infty} \psi_n(A) = \psi(A), \qquad A \subset [0, \infty).$$

Let F be the distribution function common to the X_n. If μ_n is the distribution in the plane of $(\min_{1 \le k \le n} S_k, S_n)$, then by (20.30), $P[\min_{1 \le k \le n} S_k > 0, S_{n+1} \le x] = \int_{y_1 > 0} F(x - y_2) \, d\mu_n(y_1, y_2)$. Identifying $\psi_n(A)$ with the right side of (24.17) gives*

$$(24.19) \qquad \int_{-\infty}^{\infty} F(x - y)\psi_n(dy) = P[\min_{1 \le k \le n} S_k > 0, S_{n+1} \le x].$$

If the minimum is suppressed, this holds for $n = 0$ as well as for $n \ge 1$. Since convolution commutes, the left side here is $\int_{-\infty}^{\infty} \psi_n(-\infty, x - y] \, dF(y) = \int_{y \le x} \psi_n[0, x - y] \, dF(y)$. Adding over $n = 0, 1, \ldots$ leads to this result:

Theorem 24.3. *The measure* (24.14) *satisfies*

$$(24.20) \qquad \int_{y \le x} \psi[0, x - y] \, dF(y) = \sum_{n=1}^{\infty} P[\min_{1 \le k < n} S_k > 0, S_n \le x].$$

The important case is $x \le 0$; here the sets on the right in (24.20) are disjoint and their union is the set where $S_n \le 0$ for some $n \ge 1$ and $S_n \le x$ for the first such n. This theorem makes it possible to calculate (24.16) in a case of interest for queueing theory.

Theorem 24.4. *Suppose that $E[X_1] < 0$ and that the right tail of F is exponential:*

$$(24.21) \qquad P[X_1 > x] = 1 - F(x) = \xi e^{-\lambda x}, \qquad x \ge 0,$$

where $0 < \xi < 1$ and $\lambda > 0$. Then $p < 1$ and

$$(24.22) \qquad P[\sup_{n \ge 0} S_n > x] = p e^{-\lambda(1-p)x}, \qquad x \ge 0.$$

Moreover, $\lambda(1 - p)$ is the unique root of the equation

$$(24.23) \qquad \int_{-\infty}^{\infty} e^{sx} \, dF(x) = 1$$

in the range $0 < s < \lambda$.

PROOF.   Because of the absence of memory (property (23.2) of the exponential distribution), it is intuitively clear that (even though only the right tail of F is exponential), if $S_{n-1} \le 0 < S_n$, then the amount by which S_n exceeds 0 should follow the exponential distribution. If so,

---

* That $\int_{y_1 > 0} f(y_2) \, d\mu_n(y_1, y_2) = \int_{-\infty}^{\infty} f(y) \, \psi_n(dy)$ follows by the usual argument starting with indicators f.

(24.24) $L(x, \infty) = pe^{-\lambda x}$, $x \geq 0$.

To prove this, apply (20.30):

$$P[\max_{k<n} S_k \leq 0, S_n > x] = P[\max_{k<n} S_k \leq 0, X_n > x - S_{n-1}]$$

$$= \int_{\substack{\max S_k \leq 0 \\ k<n}} \xi e^{-\lambda(x-S_{n-1})} dP$$

$$= e^{-\lambda x} P[\max_{k<n} S_k \leq 0 < S_n].$$

Adding over n gives (24.24).

Thus L is the exponential distribution multiplied by p. Therefore, L^{k*} can be computed explicitly: $L^{k*} [0, x]$ is p^k times the right side of (20.40) with $\lambda = \alpha$, so that

(24.25) $\psi[0, x] = \sum_{k=0}^{\infty} p^k \sum_{i=k}^{\infty} e^{-\lambda x} \frac{(\lambda x)^i}{i!}$.

Note that (24.24) and (24.25) follow from (24.21) alone—the condition $E[X_1] < 0$ has not been used.

From $E[X_1] < 0$ it follows that $S_n \to -\infty$ with probability 1, so that p must be less than 1. Summing (24.25) in the other order reduces it to

(24.26) $\psi[0, x] = \dfrac{1}{1-p} - \dfrac{p}{1-p} e^{-\lambda(1-p)x}$, $x \geq 0$,

whence (24.22) follows via (24.16).

It remains to analyze (24.23). Since $S_n \to -\infty$, the right side of (24.20) is 1 for $x = 0$, and so by (24.26)

(24.27) $\int_{y \leq 0} [1 - pe^{\lambda(1-p)y}] \, dF(y) = 1 - p$.

Let $f(s)$ denote the function on the left in (24.23). Because of (24.21), it is defined for $0 \leq s < \lambda$ and (see (21.26)) $f(s) = \xi\lambda(\lambda - s)^{-1} + \int_{x \leq 0} e^{sx} dF(x)$. From this, (24.27), and $F(0) = 1 - \xi$, it follows that $f(\lambda(1 - p)) = 1$.

Now f has value 1 at $s = 0$ also. Since f is continuous on $[0, \lambda)$ and $f''(s) = \int_{-\infty}^{\infty} x^2 e^{sx} dF(x)$ is positive on $(0, \lambda)$ because F does not concentrate at 0, (24.23) cannot have more than one root in $(0, \lambda)$.   ∎

Example 24.1. Suppose in the queuing context—see (24.1)—that the service times B_n are exponentially distributed with parameter β and that the A_n have the distribution function A over $(0, \infty)$. Assume that $\beta^{-1} < E[A_1]$, so that the traffic intensity (24.9) is less than 1 and $E[X_1] < 0$. Since $P[X_1 > x] = e^{-\beta x} \int_0^\infty e^{-\beta y} dA(y)$, Theorem 24.4 applies for $\lambda = \beta$. The B_n have moment generating function $\beta/(\beta - s)$ by (21.26); the moment generating function for F can by the product rule be expressed in terms of that for A. Therefore, (24.23) is in this case

(24.28)
$$\int_0^\infty e^{-sy}\, dA(y) = \frac{\beta - s}{\beta}.$$

This can be solved explicitly if the interarrival times A_n are also exponential, say with parameter $\alpha < \beta$. The left side of (24.28) is then $\alpha/(\alpha + s)$, the only positive root is $\beta - \alpha$, and so $p = \alpha/\beta$. In this case (see (24.8))

(24.29)
$$\lim_{n\to\infty} P[W_n \le x] = 1 - \frac{\alpha}{\beta} e^{-(\beta-\alpha)x}, \qquad x \ge 0.$$

This limit is a mixture of discrete and continuous distributions: there is a mass of $1 - \alpha/\beta$ at 0 and a density $\alpha\beta^{-1}(\beta - \alpha)\, e^{-(\beta-\alpha)x}$ over $(0, \infty)$. ■

Exponential Left Tail

If F is continuous, reflecting the relation (24.20) through 0 makes it possible to treat the case in which the left tail of F is exponential.

Theorem 24.5. *Suppose that $E[X_1] < 0$, that F is continuous, and that the left tail of F is exponential:*

(24.30)
$$P[X_1 \le x] = F(x) = \xi e^{\lambda x}, \qquad x \le 0,$$

where $0 < \xi < 1$ and $\lambda > 0$. Then

(24.31)
$$p = 1 + \lambda E[X_1]$$

and

(24.32)
$$L(0, x] = F(x) - F(0) + \lambda \int_0^x (1 - F(y))\, dy.$$

PROOF. First assume not (24.30) but (24.21), and assume that $E[X_1] > 0$. Then $S_n \to +\infty$, and so $p = 1$. As pointed out after the proof of (24.25), that relation follows from (24.21) alone; in the case $p = 1$ it reduces to (reverse the sum and use the Poisson mean) $\psi[0, x] = 1 + \lambda x, x \ge 0$. By Theorem 24.3,

(24.33)
$$\sum_{n=1}^\infty P[\min_{1\le k<n} S_k > 0, S_n \le x] = \int_{y\le x} (1 + \lambda(x - y))\, dF(y)$$

$$= (1 + \lambda x)P[X_1 \le x] - \lambda \int_{X_1 \le x} X_1\, dP$$

for $x \le 0$.

Now if (24.30) holds and if $E[X_1] < 0$, the analysis above applies to $\{-X_n\}$. If X_n is replaced by $-X_n$ and x by $-x$, then (24.33) becomes

$$\sum_{n=1}^{\infty} P[\max_{1 \le k < n} S_k < 0, S_n \ge x] = (1 - \lambda x)P[X_1 \ge x] + \lambda \int_{X_1 \ge x} X_1 \, dP,$$

valid for $x \ge 0$. Now F is assumed continuous, and it follows by the convolution formula that the same must be true of the distribution of each S_k. By (24.11) and (21.10), $L(x, \infty) = L[x, \infty) = 1 - F(x) + \lambda \int_x^\infty (1 - F(y)) \, dy$. By (24.30), $\int_{X_1 \le 0} X_1 \, dP = -\lambda^{-1}\xi = -\lambda^{-1}F(0)$, and so $\int_0^\infty (1 - F(y)) \, dy = \int_{X_1 \ge 0} X_1 \, dP = E[X_1] + \lambda^{-1}F(0)$. Since $p = L(0, \infty)$, this gives (24.31) and (24.32). ■

This theorem gives formulas for p and L and hence in principle determines the distribution of $\sup_{n \ge 0} S_n$ via (24.14) and (24.16). The relation becomes simpler if expressed in terms of the Laplace transforms

$$\mathcal{L}(s) = \int_0^\infty e^{-sx} L(dx)$$

and

$$\mathcal{M}(s) = E[\exp(-s \sup_{n \ge 0} S_n)].$$

These are moment generating functions with the argument reflected through 0 as in (22.5); they exist for $s \ge 0$. Let $\mathcal{F}_+(s)$ denote the transform of the restriction to $(0, \infty)$ of the measure corresponding to F:

$$\mathcal{F}_+(s) = \int_0^\infty e^{-sx} \, dF(x).$$

By (24.32), L is this last measure plus the measure with density $\lambda(1 - F(x))$ over $(0, \infty)$. An integration by parts shows that

$$\mathcal{L}(s) = \mathcal{F}_+(s) + \lambda \int_0^\infty e^{-sx}(1 - F(x)) \, dx$$

$$= \frac{\lambda}{s}(1 - F(0)) + \left(1 - \frac{\lambda}{s}\right)\mathcal{F}_+(s).$$

By (24.14) and (24.16), $\mathcal{M}(s) = (1 - p)\sum_{n=0}^\infty \mathcal{L}^n(s)$; the series converges because $\mathcal{L}(s) \le \mathcal{L}(0) = p < 1$. By (24.31),

$$(24.34) \quad \mathcal{M}(s) = \frac{\lambda E[X_1] s}{(s - \lambda)\,\mathcal{F}_+(s) - s + \lambda(1 - F(0))}, \qquad s > 0.$$

This, a form of the *Khinchine-Pollaczek formula,* is valid under the hypotheses of Theorem 24.5. Because of (22.8), it determines the distribution of $\sup_{n \ge 0} S_n$.

Example 24.2. Suppose that the queueing variables A_n are exponentially distributed with parameter α, so that the arrival stream is a Poisson process. Let $b = E[B_n]$, and assume that the traffic intensity $\rho = \alpha b$ is less than 1. If the B_n have distribution function $B(x)$ and transform $\mathcal{B}(s) = \int_0^\infty e^{-sx} dB(x)$, then $F(x)$ has density*

$$F'(x) = \begin{cases} \alpha e^{\alpha x} \mathcal{B}(\alpha), & x \le 0 \\ \\ \alpha e^{\alpha x} \displaystyle\int_{y \ge x} e^{-\alpha y} dB(y), & x \ge 0. \end{cases}$$

Note that $F(0) = \mathcal{B}(\alpha)$. Thus Theorem 24.5 applies for $\lambda = \alpha$, and reversing integrals shows that $\mathcal{F}_+(s) = \alpha(\alpha - s)^{-1}(\mathcal{B}(s) - F(0))$. Therefore, (24.34) becomes

$$(24.35) \qquad\qquad \mathcal{M}(s) = \frac{(1 - \rho)s}{s - \alpha + \alpha\,\mathcal{B}(s)},$$

the usual form of the Khinchine-Pollaczek formula. If the B_n are exponentially distributed with parameter β, this checks with the Laplace transform of the distribution function M derived in Example 24.1. ∎

Queue Size

Suppose that $\rho < 1$ and let Q_n be the size of the queue left behind by customer n when his service terminates. His service terminates at time $A_1 + \cdots + A_n + W_n + B_n$, and customer $n + k$ arrives at time $A_1 + \cdots + A_{n+k}$. Therefore, $Q_n < k$ if and only if $A_1 + \cdots + A_{n+k} \ge A_1 + \cdots + A_n + W_n + B_n$. (This requires the convention that a customer arriving exactly when service terminates is not counted as being left behind in the queue.) Thus

$$P[Q_n < k] = P[W_n + B_n - (A_{n+1} + \cdots + A_{n+k}) \le 0], \qquad k \ge 1.$$

Since W_n is measurable $\sigma(B_0, \ldots, B_{n-1}, A_1, \ldots A_n)$ (see (24.2)), the three random variables W_n, B_n, and $-(A_{n+1} + \cdots + A_{n+k})$ are independent. If M_n, B, and V_k are the distribution functions of these random variables, then by the convolution formula, $P[Q_n < k] = \int_{-\infty}^\infty M_n(-y)(B*V_k)(dy)$. If M is the distribution function of $\sup_n S_n$, then (see (24.8)) M_n converges to M and hence $P[Q_n < k]$ converges to $\int_{-\infty}^\infty M(-y)(B*V_k)(dy)$, which is the same as $\int_{-\infty}^\infty V_k(-y)(M*B)(dy)$ because convolution is commutative. Differencing for successive values of k now shows that

$$(24.36) \qquad\qquad \lim_{n \to \infty} P[Q_n = k] = q_k, \qquad k \ge 1,$$

---

* Just as (20.37) gives the density for $X + Y$, $\int_{-\infty}^\infty g(x - y) dF(x)$ gives the density for $X - Y$.

where

$$(24.37) \quad q_k = \int_0^\infty P[A_1 + \cdots + A_k < y \le A_1 + \cdots + A_{k+1}](M*B)(dy).$$

These relations hold under the sole assumption that $\rho < 1$. If the A_n are exponential with parameter α, the integrand in (24.37) can be written down explicitly:

$$(24.38) \qquad\qquad q_k = \int_0^\infty e^{-\alpha y} \frac{(\alpha y)^k}{k!} (M*B)(dy).$$

Example 24.3. Suppose that the A_n and B_n are exponential with parameters α and β, as in Example 24.1. Then (see (24.29)) M corresponds to a mass of $1 - \alpha/\beta = 1 - \rho$ at 0 together with a density $\rho(\beta - \alpha)e^{-(\beta-\alpha)x}$ over $(0, \infty)$. By (20.37), $M*B$ has density

$$\int_{[0,y]} \beta e^{-\beta(y-x)} \, dM(x) = (\beta - \alpha)e^{-(\beta-\alpha)y}.$$

By (24.38),

$$q_k = \rho^k(1 - \rho)\frac{1}{k!} \int_0^\infty x^k e^{-x} \, dx.$$

Use inductive integration by parts (or the gamma function) to reduce the last integral to $k!$. Therefore, $q_k = (1 - \rho)\rho^k$, a fundamental formula of queuing theory. See Examples 8.11 and 8.12. ■

CHAPTER 5

Convergence of Distributions

SECTION 25. WEAK CONVERGENCE

Many of the best known theorems in probability have to do with the asymptotic behavior of distributions. This chapter covers both general methods for deriving such theorems and specific applications. The present section concerns the general limit theory for distributions on the real line, and the methods of proof use in an essential way the order structure of the line. For the theory in R^k, see Section 29.

Definitions

Distribution functions F_n were in Section 14 defined to *converge weakly* to the distribution function F if

$$(25.1) \qquad\qquad \lim_n F_n(x) = F(x)$$

for every continuity point x of F; this is expressed by writing $F_n \Rightarrow F$. Examples 14.1, 14.2, and 14.3 illustrate this concept in connection with the asymptotic distribution of maxima. Example 14.4 shows the point of allowing (25.1) to fail if F is discontinuous at x.

If μ_n and μ are the probability measures on $(R^1, \mathscr{R}^1)$ corresponding to F_n and F, then $F_n \Rightarrow F$ if and only if

$$(25.2) \qquad\qquad \lim_n \mu_n(A) = \mu(A)$$

for every A of the form $A = (-\infty, x]$ for which $\mu\{x\} = 0$—see (20.5). In this case the distributions themselves are said to converge weakly, which is expressed by writing $\mu_n \Rightarrow \mu$. Thus $F_n \Rightarrow F$ and $\mu_n \Rightarrow \mu$ are only different expressions of the same fact. From weak convergence it follows that (25.2) holds for many sets A besides half-infinite intervals; see Theorem 25.8.

280

Example 25.1.   Let F_n be the distribution function corresponding to a unit mass at n: $F_n = I_{[n,\infty)}$. Then $\lim_n F_n(x) = 0$ for every x, so that (25.1) is satisfied if $F(x) \equiv 0$. But $F_n \Rightarrow F$ does not hold, because F is not a distribution function. Weak convergence is in this section defined only for functions F_n and F that rise from 0 at $-\infty$ to 1 at $+\infty$—that is, it is defined only for *probability* measures μ_n and μ.*

For another example of the same sort, take F_n as the distribution function of the waiting time W_n for the nth customer in a queue. As observed after (24.9), $F_n(x) \to 0$ for each x if the traffic intensity exceeds 1.   ∎

Example 25.2.   *Poisson approximation to the binomial.* Let μ_n be the binomial distribution (20.6) for $p = \lambda/n$ and let μ be the Poisson distribution (20.7). For nonnegative integers k,

$$\mu_n\{k\} = \frac{\lambda^k(1 - \lambda/n)^n}{k!} \times \frac{1}{(1 - \lambda/n)^k} \prod_{i=0}^{k-1} \left(1 - \frac{i}{n}\right)$$

if $n \geq k$. As $n \to \infty$ the second factor on the right goes to 1 for fixed k, and so $\mu_n\{k\} \to \mu\{k\}$. By the series form of Scheffé's theorem (the corollary to Theorem 16.11), (25.2) holds for every set A of nonnegative integers. Since the nonnegative integers support μ and the μ_n, (25.2) even holds for every linear Borel set A. Certainly μ_n converges weakly to μ in this case.   ∎

Example 25.3.   Let μ_n correspond to a mass of n^{-1} at each point k/n, $k = 0, 1, \ldots, n - 1$; let μ be Lebesgue measure confined to the unit interval. The corresponding distribution functions satisfy $F_n(x) = [nx]/n \to x = F(x)$ for $0 \leq x \leq 1$, and so $F_n \Rightarrow F$. In this case (25.1) holds for every x, but (25.2) does not as in the preceding example hold for every Borel set A: if A is the set of rationals, then $\mu_n(A) = 1$ does not converge to $\mu(A) = 0$. Despite this, μ_n does converge weakly to μ.   ∎

Example 25.4.   If μ_n is a unit mass at x_n and μ is a unit mass at x, then $\mu_n \Rightarrow \mu$ if and only if $x_n \to x$.   ∎

Uniform Distribution Modulo 1†

For a sequence $x_1, x_2, \ldots$ of real numbers, consider the corresponding sequence of their

---

* There is (see Section 28) a related notion of *vague* convergence in which μ may be *defective* in the sense that $\mu(R^1) < 1$. Weak convergence is in this context sometimes called *complete* convergence.

† This topic may be omitted.

fractional parts $\{x_n\} = x_n - [x_n]$. For each n, define a probability measure μ_n by

(25.3) $$\mu_n(A) = \frac{1}{n} \cdot \#[k: 1 \leq k \leq n, \{x_k\} \in A];$$

μ_n has mass n^{-1} at the points $\{x_1\}, \ldots, \{x_n\}$, and if several of these points coincide, the masses add. The problem is to find the weak limit of $\{\mu_n\}$ in number-theoretically interesting cases.

Suppose that $x_n = np/q$, where p/q is a rational fraction in lowest terms. Then $\{kp/q\}$ and $\{k'p/q\}$ coincide if and only if $(k - k')p/q$ is an integer, and since p and q have no factor in common, this holds if and only if q divides $k - k'$. Thus for any q successive values of k the fractional parts $\{kp/q\}$ consist of the q points

(25.4) $$\frac{0}{q}, \frac{1}{q}, \ldots, \frac{q-1}{q}$$

in some order. If $0 \leq x < 1$, then $\mu_n[0, x] = \mu_n[0, y]$, where y is the point i/q just left of x, namely $y = [qx]/q$. If $r = [n/q]$, then the number of $\{kp/q\}$ in $[0, y]$ for $1 \leq k \leq n$ is between $r([qx] + 1)$ and $(r + 1)([qx] + 1)$ and so differs from $nq^{-1}([qx] + 1)$ by at most $2q$. Therefore,

(25.5) $$\left| \mu_n[0, x] - \frac{[qx] + 1}{q} \right| \leq \frac{2q}{n}, \qquad 0 \leq x < 1.$$

If $\nu^{(q)}$ consists of a mass of q^{-1} at each of the points (25.4), then (25.5) implies that $\mu_n \Rightarrow \nu^{(q)}$.

Suppose next that $x_n = n\theta$, where θ is irrational. If p/q is a highly accurate rational approximation to θ (q large), then the measures (25.3) corresponding to $\{n\theta\}$ and to $\{np/q\}$ should be close to one another. If ν is Lebesgue measure restricted to the unit interval, then $\nu^{(q)}$ is for large q near ν (see Example 25.3), and so there is reason to believe that the measures (25.3) for $\{n\theta\}$ will converge weakly to ν.

If the μ_n defined by (25.3) converge weakly to Lebesgue measure restricted to the unit interval, the sequence $x_1, x_2, \ldots$ is said to be *uniformly distributed modulo* 1. In this case every subinterval has asymptotically its proportional share of the points $\{x_n\}$; by Theorem 25.8 below, the same is true of every subset whose boundary has Lebesgue measure 0.

Theorem 25.1. *For θ irrational, $\{n\theta\}$ is uniformly distributed modulo* 1.

PROOF. Define μ_n by (25.3) for $x_n = n\theta$. It suffices to show that

(25.6) $$\mu_n(x, y] \to y - x, \qquad 0 < x < y < 1.$$

Suppose that

(25.7) $$0 < \epsilon < \min \{x, 1 - y\}.$$

It follows by the elementary theory of continued fractions* that for n exceeding some $n_0 = n_0(\epsilon, \theta)$ there exists an irreducible fraction p/q (depending on ϵ, n, and θ) such that

$$\frac{1}{q} < \epsilon, \qquad \frac{q}{n} < \epsilon, \qquad n \left| \theta - \frac{p}{q} \right| < \epsilon.$$

---

* Let p_ν/q_ν be the convergents of θ and choose $\nu = \nu_n$ so that $q_\nu < n\epsilon \leq q_{\nu+1}$. If n is sufficiently large, this forces ν to be large enough that $q_\nu > \epsilon^{-2}$. By the theory of continued fractions, $|\theta - p_\nu/q_\nu| \leq 1/q_\nu q_{\nu+1} < 1/n\epsilon q_\nu < \epsilon/n$. Take $p/q = p_\nu/q_\nu$.

If μ'_n is the measure (25.3) for the sequence $\{np/q\}$, then by (25.5),

$$(25.8) \qquad |\mu'_n(x, y] - (y - x)| \le \frac{2}{q} + \left|\mu'_n(x, y] - \left(\frac{[qy] + 1}{q} - \frac{[qx] + 1}{q}\right)\right|$$

$$\le \frac{2}{q} + \frac{4q}{n} < 6\epsilon.$$

Now $|k\theta - kp/q| < \epsilon$ for $k = 1, \dots, n$. If $\{k\theta\}$ lies in $(x, y]$, then $k\theta$ and kp/q have the same integral part because of (25.7), so that $\{kp/q\}$ lies in $(x - \epsilon, y + \epsilon]$. Thus $\mu_n(x, y] \le \mu'_n(x - \epsilon, y + \epsilon]$. Similarly, $\mu'_n(x + \epsilon, y - \epsilon] \le \mu_n(x, y]$. Two applications of (25.8) give $|\mu_n(x, y] - (y - x)| < 8\epsilon$. Since this holds for $n > n_0$, (25.6) follows. ∎

Theorem 25.1 implies Kronecker's theorem, according to which $\{n\theta\}$ is dense in the unit interval if θ is irrational.

Convergence in Distribution

Let X_n and X be random variables with respective distribution functions F_n and F. If $F_n \Rightarrow F$, then X_n is said to *converge in distribution* or *in law* to X, written $X_n \Rightarrow X$. This dual use of the double arrow will cause no confusion. Because of the defining conditions (25.1) and (25.2), $X_n \Rightarrow X$ if and only if

$$(25.9) \qquad \lim_n P[X_n \le x] = P[X \le x]$$

for every x such that $P[X = x] = 0$.

Example 25.5. Let $X_1, X_2, \dots$ be independent random variables, each with the exponential distribution: $P[X_n \ge x] = e^{-\alpha x}$, $x \ge 0$. Put $M_n = \max\{X_1, \dots, X_n\}$ and $b_n = \alpha^{-1} \log n$. The relation (14.9) established in Example 14.1 can be restated as $P[M_n - b_n \le x] \to e^{-e^{-\alpha x}}$. If X is any random variable with distribution function $e^{-e^{-\alpha x}}$, this can be written $M_n - b_n \Rightarrow X$. ∎

One is usually interested in proving weak convergence of the distributions of some given sequence of random variables, such as the $M_n - b_n$ in this example, and the result is often most clearly expressed in terms of the random variables themselves rather than in terms of their distributions or distribution functions. Although the $M_n - b_n$ here arise naturally from the problem at hand, the random variable X is simply constructed to make it possible to express the asymptotic relation compactly by $M_n - b_n \Rightarrow X$. Recall that by Theorem 14.1 there does exist a random variable with any prescribed distribution.

Example 25.6. For each n, let Ω_n be the space of n-tuples of 0's and 1's, let $\mathcal{F}_n$ consist of all subsets of Ω_n, and let P_n assign probability $(\lambda/n)^k(1 - \lambda/n)^{n-k}$ to each ω consisting of k 1's and $n - k$ 0's. Let $X_n(\omega)$ be the number of 1's in ω; then X_n, a random variable on $(\Omega_n, \mathcal{F}_n, P_n)$, represents the number of successes in n Bernoulli trials having probability λ/n of success at each.

Let X be a random variable, on some $(\Omega, \mathcal{F}, P)$, having the Poisson distribution with parameter λ. According to Example 25.2, $X_n \Rightarrow X$. ∎

As this example shows, the random variables X_n may be defined on entirely different probability spaces. To allow for this possibility, the P on the left in (25.9) really should be written P_n. Suppressing the n causes no confusion if it is understood that P refers to whatever probability space it is that X_n is defined on; the underlying probability space enters into the definition only via the distribution μ_n it induces on the line. Any instance of $F_n \Rightarrow F$ or of $\mu_n \Rightarrow \mu$ can be rewritten in terms of convergence in distribution: There exist random variables X_n and X (on some probability spaces) with distribution functions F_n and F, and $F_n \Rightarrow F$ and $X_n \Rightarrow X$ express the same fact.

Convergence in Probability

Suppose that $X_1, X_2, \ldots$ are random variables all defined on the same probability space $(\Omega, \mathcal{F}, P)$. If $X_n \to X$ with probability 1, then $P[\,|X_n - X| \ge \epsilon \text{ i.o.}] = 0$ for $\epsilon > 0$, and hence

$$(25.10) \qquad \lim_n P[\,|X_n - X| > \epsilon] = 0$$

by Theorem 4.1. Thus there is *convergence in probability* $X_n \to_P X$ as discussed in Section 20; see (20.41). That convergence with probability 1 implies convergence in probability follows also from Theorem 20.5.

Suppose that (25.10) holds for each positive ϵ. Now $P[X \le x - \epsilon] - P[\,|X_n - X| \ge \epsilon] \le P[X_n \le x] \le P[X \le x + \epsilon] + P[\,|X_n - X| \ge \epsilon]$; letting n tend to ∞ and then letting ϵ tend to 0 shows that $P[X < x] \le \liminf_n P[X_n \le x] \le \limsup_n P[X_n \le x] \le P[X \le x]$. Thus $P[X_n \le x] \to P[X \le x]$ if $P[X = x] = 0$, and so $X_n \Rightarrow X$:

Theorem 25.2. *Suppose that X_n and X are random variables on the same probability space. If $X_n \to X$ with probability 1, then $X_n \to_P X$. If $X_n \to_P X$, then $X_n \Rightarrow X$.*

Of the two implications in this theorem, neither converse holds. As observed after (20.41), convergence in probability does not imply convergence with probability 1. Neither does convergence in distribution imply convergence in probability: if X and Y are independent and assume the values 0 and 1 with probability $\frac{1}{2}$ each, and if $X_n = Y$, then $X_n \Rightarrow X$, but $X_n \to_P X$ cannot hold because $P[\,|X - Y| = 1] = \frac{1}{2}$. What is more, (25.10) is impossible if X and the X_n are defined on different probability spaces, as may happen in the case of convergence in distribution.

Although (25.10) in general makes no sense unless X and the X_n are defined on the same probability space, suppose that X is replaced by a constant real number a—that is, suppose that $X(\omega) \equiv a$. Then (25.10) becomes

$$(25.11) \qquad \lim_n P[|X_n - a| \geq \epsilon] = 0,$$

and this condition makes sense even if the space for X_n does vary with n. Now a can be regarded as a random variable (on any probability space at all), and it is easy to show that (25.11) implies that $X_n \Rightarrow a$: Put $\epsilon = |x - a|$; if $x > a$, then $P[X_n \leq x] \geq P[|X_n - a| < \epsilon] \to 1$, and if $x < a$, then $P[X_n \leq x] \leq P[|X_n - a| \geq \epsilon] \to 0$. If a is regarded as a random variable, its distribution function is 0 for $x < a$ and 1 for $x \geq a$. Thus (25.11) implies that the distribution function of X_n converges weakly to that of a.

Suppose, on the other hand, that $X_n \Rightarrow a$. Then $P[|X_n - a| > \epsilon] \leq P[X_n \leq a - \epsilon] + 1 - P[X_n \leq a + \epsilon] \to 0$, so that (25.11) holds:

Theorem 25.3. *Condition* (25.11) *holds for all positive ϵ if and only if $X_n \Rightarrow a$.*

If (25.11) holds for all positive ϵ, X_n may be said to *converge to a in probability*. As this does not require that the X_n be defined on the same space, it is not really a special case of convergence in probability as defined by (25.10). Convergence in probability in this new sense will be denoted $X_n \Rightarrow a$, in accordance with the theorem just proved.

Example 14.4 restates the weak law of large numbers in terms of this concept. Indeed, if $X_1, X_2, \ldots$ are independent, identically distributed random variables with finite mean m, and if $S_n = X_1 + \cdots + X_n$, the weak law of large numbers is the assertion $n^{-1}S_n \Rightarrow m$. Example 6.4 provides another illustration: If S_n is the number of cycles in a random permutation on n letters, then $S_n/\log n \Rightarrow 1$.

Example 25.7. Suppose that $X_n \Rightarrow X$ and $\delta_n \to 0$. Given ϵ and η choose x so large that $P[|X| \geq x] < \eta$ and $P[X = \pm x] = 0$, and then choose n_0 so that $n \geq n_0$ implies that $|\delta_n| < \epsilon/x$ and $|P[X_n \leq y] - P[X \leq y]| < \eta$ for $y = \pm x$. Then $P[|\delta_n X_n| \geq \epsilon] < 3\eta$ for $n \geq n_0$. Thus $X_n \Rightarrow X$ and $\delta_n \to 0$ *imply that* $\delta_n X_n \Rightarrow 0$, a restatement of Lemma 2 of Section 14 (page 164). ∎

The asymptotic properties of a random variable should remain unaffected if it is altered by the addition of a random variable that goes to 0 in probability. Let (X_n, Y_n) to a two-dimensional random vector.

Theorem 25.4. *If $X_n \Rightarrow X$ and $X_n - Y_n \Rightarrow 0$, then $Y_n \Rightarrow X$.*

PROOF.   Suppose that $y' < x < y''$ and $P[X = y'] = P[X = y''] = 0$. If $\epsilon < x - y'$ and $\epsilon < y'' - x$, then

$$(25.12) \qquad P[X_n \leq y'] - P[|X_n - Y_n| \geq \epsilon] \leq P[Y_n \leq x]$$

$$\leq P[X_n \leq y''] + P[|X_n - Y_n| \geq \epsilon].$$

Since $X_n \Rightarrow X$, letting $n \to \infty$ gives

$$(25.13) \qquad P[X \leq y'] \leq \lim_n \inf P[Y_n \leq x]$$

$$\leq \lim_n \sup P[Y_n \leq x] \leq P[X \leq y''].$$

Since $P[X = y] = 0$ for all but countably many y, if $P[X = x] = 0$, then y' and y'' can further be chosen so that $P[X \leq y']$ and $P[X \leq y'']$ are arbitrarily near $P[X \leq x]$; hence $P[Y_n \leq x] \to P[X \leq x]$. ∎

Theorem 25.4 has an extension that is sometimes useful. Suppose that $(X_n^{(u)}, Y_n)$ is a two-dimensional random vector.

Theorem 25.5.   *If, for each u, $X_n^{(u)} \Rightarrow X^{(u)}$ as $n \to \infty$, if $X^{(u)} \Rightarrow X$ as $u \to \infty$, and if*

$$(25.14) \qquad \lim_u \lim_n \sup P[|X_n^{(u)} - Y_n| \geq \epsilon] = 0$$

for positive ϵ, then $Y_n \Rightarrow X$.

PROOF.   Replace X_n by $X_n^{(u)}$ in (25.12). If $P[X = y'] = 0 \equiv P[X^{(u)} = y']$ and $P[X = y''] = 0 \equiv P[X^{(u)} = y'']$, letting $n \to \infty$ and then $u \to \infty$ gives (25.13) once again. Since $P[X = y] = 0 \equiv P[X^{(u)} = y]$ for all but countably many y, the proof can be completed as before. ∎

Fundamental Theorems

Some of the fundamental properties of weak convergence were established in Section 14. It was shown there (page 163) that a sequence cannot have two distinct weak limits: *If $F_n \Rightarrow F$ and $F_n \Rightarrow G$, then $F = G$.* Another simple fact is this: *If $\lim_n F_n(d) = F(d)$ for d in a set D dense in R^1, then $F_n \Rightarrow F$.* Indeed, if F is continuous at x, there are in D points d' and d'' such that $d' < x < d''$ and $F(d'') - F(d') < \epsilon$, and it follows that the limits superior and inferior of $F_n(x)$ are within ϵ of $F(x)$.

For any probability measure on $(R^1, \mathcal{R}^1)$ there is on some probability space a random variable having that measure as its distribution. Therefore, for probability measures satisfying $\mu_n \Rightarrow \mu$, there exist random variables Y_n and Y having these measures as distributions and satisfying $Y_n \Rightarrow Y$. According to the following theorem, the Y_n and Y can be constructed on the same probability space, and even in such a way that $Y_n(\omega) \to Y(\omega)$ for every ω—a condition which is of course much stronger than $Y_n \Rightarrow Y$. This result, *Skorohod's theorem*, makes possible very simple and transparent proofs of many important facts.

Theorem 25.6. *Suppose that μ_n and μ are probability measures on $(R^1, \mathcal{R}^1)$ and $\mu_n \Rightarrow \mu$. There exist random variables Y_n and Y on a common probability space $(\Omega, \mathcal{F}, P)$ such that Y_n has distribution μ_n, Y has distribution μ, and $Y_n(\omega) \to Y(\omega)$ for each ω.*

PROOF. For the probability space $(\Omega, \mathcal{F}, P)$, take $\Omega = (0, 1)$, let $\mathcal{F}$ consist of the Borel subsets of $(0, 1)$, and for $P(A)$ take the Lebesgue measure of A.

The construction is related to that in the proofs of Theorems 14.1 and 20.4. Consider the distribution functions F_n and F corresponding to μ_n and μ. For $0 < \omega < 1$, put $Y_n(\omega) = \inf [x: \omega \leq F_n(x)]$ and $Y(\omega) = \inf [x: \omega \leq F(x)]$. Since $\omega \leq F_n(x)$ if and only if $Y_n(\omega) \leq x$ (see the argument following (14.5)), $P[\omega: Y_n(\omega) \leq x] = P[\omega: \omega \leq F_n(x)] = F_n(x)$. Thus Y_n has distribution function F_n; similarly, Y has distribution function F.

It remains to show that $Y_n(\omega) \to Y(\omega)$. The idea is that Y_n and Y are essentially inverse functions to F_n and F; if the direct functions converge, so must the inverses.

Suppose that $0 < \omega < 1$. Given ϵ, choose x so that $Y(\omega) - \epsilon < x < Y(\omega)$ and $\mu\{x\} = 0$. Then $F(x) < \omega$; $F_n(x) \to F(x)$ now implies that, for n large enough, $F_n(x) < \omega$ and hence $Y(\omega) - \epsilon < x < Y_n(\omega)$. Thus $\liminf_n Y_n(\omega) \geq Y(\omega)$. If $\omega < \omega'$ and ϵ is positive, choose a y for which $Y(\omega') < y < Y(\omega') + \epsilon$ and $\mu\{y\} = 0$. Now $\omega < \omega' \leq F(Y(\omega')) \leq F(y)$, and so, for n large enough, $\omega \leq F_n(y)$ and hence $Y_n(\omega) \leq y < Y(\omega') + \epsilon$. Thus $\limsup_n Y_n(\omega) \leq Y(\omega')$ if $\omega < \omega'$. Therefore, $Y_n(\omega) \to Y(\omega)$ if Y is continuous at ω.

Since Y is nondecreasing on $(0, 1)$, it has at most countably many discontinuities. At discontinuity points ω of Y, redefine $Y_n(\omega) = Y(\omega) = 0$. With this change, $Y_n(\omega) \to Y(\omega)$ for every ω. Since Y and the Y_n have been altered only on a set of Lebesgue measure 0, their distributions are still μ_n and μ. ∎

Note that this proof uses the order structure of the real line in an essential way. The proof of the corresponding result in R^k (Theorem 29.6) is more complicated.

The following *mapping theorem* is of very frequent use.

Theorem 25.7. *Suppose that $h: R^1 \to R^1$ is measurable and that the set D_h of its discontinuities is measurable.* If $\mu_n \Rightarrow \mu$ and $\mu(D_h) = 0$, then $\mu_n h^{-1} \Rightarrow \mu h^{-1}$.*

---

* That D_h lies in $\mathcal{R}^1$ is generally obvious in applications. In point of fact, it always holds (even if h is not measurable): Let $A(\epsilon, \delta)$ be the set of x for which there exist y and z such that $|x - y| < \delta$, $|x - z| < \delta$, and $|h(y) - h(z)| \geq \epsilon$. Then $A(\epsilon, \delta)$ is open and $D_h = \bigcup_\epsilon \bigcap_\delta A(\epsilon, \delta)$, where ϵ and δ range over the positive rationals.

Recall (see (13.7)) that μh^{-1} has value $\mu(h^{-1}A)$ at A.

PROOF.   Consider the random variables Y_n and Y of Theorem 25.6. Since $Y_n(\omega) \to Y(\omega)$, $Y(\omega) \notin D_h$ implies that $h(Y_n(\omega)) \to Y(\omega)$. Since $P[\omega: Y(\omega) \in D_h] = \mu(D_h) = 0$, $h(Y_n(\omega)) \to h(Y(\omega))$ holds with probability 1. Hence $h(Y_n) \Rightarrow h(Y)$ by Theorem 25.2. Since $P[h(Y) \in A] = P[Y \in h^{-1}A] = \mu(h^{-1}A)$, $h(Y)$ has distribution μh^{-1}; similarly, $h(Y_n)$ has distribution $\mu_n h^{-1}$. Thus $h(Y_n) \Rightarrow h(Y)$ is the same thing as $\mu_n h^{-1} \Rightarrow \mu h^{-1}$.   ∎

Because of the definition of convergence in distribution, this result has an equivalent statement in terms of random variables:

Corollary 1.   *If $X_n \Rightarrow X$ and $P[X \in D_h] = 0$, then $h(X_n) \Rightarrow h(X)$.*

Take $X \equiv a$:

Corollary 2.   *If $X_n \Rightarrow a$ and h is continuous at a, then $h(X_n) \Rightarrow h(a)$.*

Example 25.8.   From $X_n \Rightarrow X$ it follows directly by the theorem that $aX_n + b \Rightarrow aX + b$. Suppose also that $a_n \to a$ and $b_n \to b$. Then $(a_n - a) X_n \Rightarrow 0$ by Example 25.7, and so $(a_n X_n + b_n) - (aX_n + b) \Rightarrow 0$. And now $a_n X_n + b_n \Rightarrow aX + b$ follows by Theorem 25.4: *If $X_n \Rightarrow X$, $a_n \to a$, and $b_n \to b$, then $a_n X_n + b_n \Rightarrow aX + b$.* This fact was stated and proved differently in Section 14—see Lemma 1 on page 164.   ∎

By definition, $\mu_n \Rightarrow \mu$ means that the corresponding distribution functions converge weakly. The following theorem characterizes weak convergence without reference to distribution functions. The boundary ∂A of A consists of the points that are limits of sequences in A and are also limits of sequences in A^c; alternatively, ∂A is the closure of A minus its interior. A Borel set A is a *μ-continuity set* if $\mu(\partial A) = 0$.

Theorem 25.8.   *The following three conditions are equivalent.*
   (i) $\mu_n \Rightarrow \mu$;
   (ii) $\int f \, d\mu_n \to \int f \, d\mu$ *for every bounded, continuous real function f;*
   (iii) $\mu_n(A) \to \mu(A)$ *for every μ-continuity set A.*

PROOF.   Suppose that $\mu_n \Rightarrow \mu$ and consider the random variables Y_n and Y of Theorem 25.6. Suppose that f is a bounded function such that $\mu(D_f) = 0$. Since $P[Y \in D_f] = \mu(D_f) = 0$, $f(Y_n) \to f(Y)$ with probability 1, and so by change of variable (see (21.1)) and the bounded convergence theorem, $\int f \, d\mu_n = E[f(Y_n)] \to E[f(Y)] = \int f \, d\mu$. Thus $\mu_n \Rightarrow \mu$ and $\mu(D_f) = 0$ together imply that $\int f \, d\mu_n \to \int f \, d\mu$ if f is bounded. In particular, (i) implies (ii). Further, if $f = I_A$, then

$D_f = \partial A$, and from $\mu(\partial A) = 0$ and $\mu_n \Rightarrow \mu$ follows $\mu_n(A) = \int f \, d\mu_n \to \int f \, d\mu = \mu(A)$. Thus (i) also implies (iii).

Since $\partial(-\infty, x] = \{x\}$, obviously (iii) implies (i). It therefore remains only to deduce $\mu_n \Rightarrow \mu$ from (ii). Consider the corresponding distribution functions. Suppose that $x < y$, and let $f(t)$ be 1 for $t \leq x$, 0 for $t \geq y$, and interpolate linearly on $[x, y]$: $f(t) = (y - t)/(y - x)$ for $x \leq t \leq y$. Since $F_n(x) \leq \int f \, d\mu_n \leq F_n(y)$ and $F(x) \leq \int f \, d\mu \leq F(y)$, it follows from (ii) that $\limsup_n F_n(x) \leq F(y)$; letting $y \downarrow x$ shows that $\limsup_n F_n(x) \leq F(x)$. Similarly, $F(x) \leq \liminf_n F_n(y)$ and hence $F(y-) \leq \liminf_n F_n(y)$. This implies convergence at continuity points. ∎

The function f in this last part of the proof is uniformly continuous. Hence $\mu_n \Rightarrow \mu$ follows if $\int f \, d\mu_n \to \int f \, d\mu$ for every bounded and *uniformly* continuous f.

Example 25.9. The distributions of Example 25.3 satisfy $\mu_n \Rightarrow \mu$, but $\mu_n(A)$ does not converge to $\mu(A)$ if A is the set of rationals. Hence this A cannot be a μ-continuity set; in fact, of course, $\partial A = R^1$. ∎

The concept of weak convergence would be nearly useless if (25.2) were not allowed to fail when $\mu(\partial A) = 0$. Since $F(x) - F(x-) = \mu\{x\} = \mu(\partial(-\infty, x])$, it is therefore natural in the original definition to allow (25.1) to fail when x is not a continuity point of F.

Helley's Theorem

One of the most frequently used results in analysis is the *Helley selection theorem*:

Theorem 25.9. *For every sequence $\{F_n\}$ of distribution functions there exists a subsequence $\{F_{n_k}\}$ and a nondecreasing, right-continuous function F such that $\lim_k F_{n_k}(x) = F(x)$ at continuity points x of F.*

PROOF. An application of the diagonal method (Theorem 25.13 below) gives a sequence $\{n_k\}$ of integers along which the limit $G(r) = \lim_k F_{n_k}(r)$ exists for every rational r. Define $F(x) = \inf [G(r): x < r]$. Clearly F is nondecreasing.

To each x and ϵ there is an r for which $x < r$ and $G(r) < F(x) + \epsilon$. If $x \leq y < r$, then $F(y) \leq G(r) < F(x) + \epsilon$. Hence F is continuous from the right.

If F is continuous at x, choose $y < x$ so that $F(x) - \epsilon < F(y)$; now choose rational r and s so that $y < r < x < s$ and $G(s) < F(x) + \epsilon$. From $F(x) - \epsilon < G(r) \leq G(s) < F(x) + \epsilon$ and $F_n(r) \leq F_n(x) \leq F_n(s)$ it follows that as k goes to infinity $F_{n_k}(x)$ has limits superior and inferior within ϵ of $F(x)$. ∎

The F in this theorem necessarily satisfies $0 \leq F(x) \leq 1$. But F need not be a distribution function; if F_n has a unit jump at n, for example, $F(x) \equiv 0$ is the only possibility. It is important to have a condition which ensures that for some subsequence the limit F is a distribution function.

A sequence of probability measures μ_n on $(R^1, \mathcal{R}^1)$ is said to be *tight* if for each ϵ there exists a finite interval $(a, b]$ such that $\mu_n (a, b] > 1 - \epsilon$ for all n. In terms of the corresponding distribution functions F_n, the condition is that for each ϵ there exist x and y such that $F_n(x) < \epsilon$ and $F_n(y) > 1 - \epsilon$ for all n. If μ_n is a unit mass at n, $\{\mu_n\}$ is not tight in this sense—the mass of μ_n "escapes to infinity." Tightness is a condition preventing this escape of mass.

Theorem 25.10.   *Tightness is a necessary and sufficient condition that for every subsequence $\{\mu_{n_k}\}$ there exist a further subsequence $\{\mu_{n_{k(j)}}\}$ and a probability measure μ such that $\mu_{n_{k(j)}} \Rightarrow \mu$ as $j \to \infty$.*

PROOF: SUFFICIENCY.   Apply Helley's theorem to the subsequence $\{F_{n_k}\}$ of corresponding distribution functions. There exists a further subsequence $\{F_{n_{k(j)}}\}$ such that $\lim_j F_{n_{k(j)}} = F(x)$ at continuity points of F, where F is nondecreasing and right-continuous. There exists by Theorem 12.4 a measure μ on $(R^1, \mathcal{R}^1)$ such that $\mu(a, b] = F(b) - F(a)$. Given ϵ, choose a and b so that $\mu_n(a, b] > 1 - \epsilon$ for all n, which is possible by tightness. By decreasing a and increasing b, one can ensure that they are continuity points of F. But then $\mu(a, b] \geq 1 - \epsilon$. Therefore, μ is a *probability* measure, and of course $\mu_n \Rightarrow \mu$.

NECESSITY.   If $\{\mu_n\}$ is not tight, there exists a positive ϵ such that for each finite interval $(a, b]$, $\mu_n(a, b] \leq 1 - \epsilon$ for some n. Choose n_k so that $\mu_{n_k}(-k, k] \leq 1 - \epsilon$. Suppose that some subsequence $\{\mu_{n_{k(j)}}\}$ of $\{\mu_{n_k}\}$ were to converge weakly to some probability measure μ. Choose $(a, b]$ so that $\mu\{a\} = \mu\{b\} = 0$ and $\mu(a, b] > 1 - \epsilon$. For large enough j, $(a, b] \subset (-k(j), k(j)]$, and so $1 - \epsilon \geq \mu_{n_{k(j)}}(-k(j), k(j)] \geq \mu_{n_{k(j)}}(a, b] \to \mu(a, b]$. Thus $\mu(a, b] \leq 1 - \epsilon$, a contradiction.   ■

Corollary.   *If $\{\mu_n\}$ is a tight sequence of probability measures, and if each subsequence that converges weakly at all converges weakly to the probability measure μ, then $\mu_n \Rightarrow \mu$.*

PROOF.   By the theorem, each subsequence $\{\mu_{n_k}\}$ contains a further subsequence $\{\mu_{n_{k(j)}}\}$ converging weakly $(j \to \infty)$ to some limit, and that limit must by hypothesis be μ. Thus every subsequence $\{\mu_{n_k}\}$ contains a further subsequence $\{\mu_{n_{k(j)}}\}$ converging weakly to μ.

Suppose that $\mu_n \Rightarrow \mu$ is false. Then there exists some x such that $\mu\{x\} = 0$ but $\mu_n(-\infty, x]$ does not converge to $\mu(-\infty, x]$. But then there exists a positive ϵ such that $|\mu_{n_k}(-\infty, x] - \mu(-\infty, x]| \geq \epsilon$ for an infinite sequence $\{n_k\}$ of integers, and no subsequence of $\{\mu_{n_k}\}$ can converge weakly to μ. This contradiction shows that $\mu_n \Rightarrow \mu$.   ■

If μ_n is a unit mass at x_n, then $\{\mu_n\}$ is tight if and only if $\{x_n\}$ is bounded. The theorem above and its corollary reduce in this case to standard facts about the real line; see Example 25.4.

Example 25.10. Let μ_n be the normal distribution with mean m_n and variance σ_n^2. If m_n and σ_n^2 are bounded, then the second moment of μ_n is bounded, and it follows by Markov's inequality (21.12) that $\{\mu_n\}$ is tight. The conclusion of Theorem 25.10 can also be checked directly: If $\{n_{k(j)}\}$ is chosen so that $\lim_j m_{n_{k(j)}} = m$ and $\lim_j \sigma_{n_{k(j)}}^2 = \sigma^2$, then $\mu_{n_{k(j)}} \Rightarrow \mu$, where μ is normal with mean m and variance σ^2 (a unit mass at m if $\sigma^2 = 0$).

If $m_n > b$, then $\mu_n(b, \infty) \geq \frac{1}{2}$; if $m_n < a$, then $\mu_n(-\infty, a] \geq \frac{1}{2}$. Hence $\{\mu_n\}$ cannot be tight if m_n is unbounded. If m_n is bounded, say by K, then $\mu_n(-\infty, a] \geq \nu(-\infty, (a - K)\sigma_n^{-1}]$, where ν is the standard normal distribution. If σ_n is unbounded, then $\nu(-\infty, (a - K)\sigma_n^{-1}] \to \frac{1}{2}$ along some subsequence, and $\{\mu_n\}$ cannot be tight. Thus a sequence of normal distributions is tight if and only if the means and variances are bounded. ∎

Integration to the Limit

Theorem 25.11. *If $X_n \Rightarrow X$, then $E[|X|] \leq \lim\inf_n E[|X_n|]$.*

PROOF. Apply Skorohod's Theorem 25.6 to the distributions of X_n and X: There exist on a common probability space random variables Y_n and Y such that $Y = \lim_n Y_n$ with probability 1, Y_n has the distribution of X_n, and Y has the distribution of X. By Fatou's lemma, $E[|Y|] \leq \lim\inf_n E[|Y_n|]$. Since $|X|$ and $|Y|$ have the same distribution, they have the same expected value (see (21.6)), and similarly for $|X_n|$ and $|Y_n|$. ∎

The random variables X_n are said to be *uniformly integrable* if

$$(25.15) \qquad \lim_{\alpha \to \infty} \sup_n \int_{[|X_n| \geq \alpha]} |X_n|\, dP = 0;$$

see (16.20). Clearly, this implies that

$$(25.16) \qquad \sup_n E[|X_n|] < \infty.$$

Theorem 25.12. *If $X_n \Rightarrow X$ and the X_n are uniformly integrable, then X is integrable and*

$$(25.17) \qquad E[X_n] \to E[X].$$

PROOF.   Construct random variables Y_n and Y as in the preceding proof. Since $Y_n \to Y$ with probability 1 and the Y_n are uniformly integrable in the sense of (16.20), $E[X_n] = E[Y_n] \to E[Y] = E[Y]$ by Theorem 16.13.   ∎

If $\sup_n E[|X_n|^{1+\epsilon}] < \infty$ for some positive ϵ, then the X_n are uniformly integrable because

$$(25.18) \qquad \int_{[|X_n| \ge \alpha]} |X_n|\, dP \le \frac{1}{\alpha^\epsilon} E[|X_n|^{1+\epsilon}].$$

Since $X_n \Rightarrow X$ implies that $X_n^r \Rightarrow X^r$ by Theorem 25.7, there is the following consequence of the theorem.

Corollary.   *If $X_n \Rightarrow X$ and $\sup_n E[|X_n|^{r+\epsilon}] < \infty$, where $\epsilon > 0$, then $E[|X|^r] < \infty$ and $E[X_n^r] \to E[X^r]$.*

The X_n are also uniformly integrable if there is an integrable random variable Z such that $P[|X_n| \ge t] \le P[|Z| \ge t]$ for $t > 0$, because then (21.10) gives

$$\int_{[|X_n| \ge \alpha]} |X_n|\, dP \le \int_{[|Z| \ge \alpha]} |Z|\, dP.$$

From this the dominated convergence theorem follows again.

The Diagonal Method

If $x_1, x_2, \ldots$ is a bounded sequence of real numbers, then there exists an increasing sequence $n_1, n_2, \ldots$ of integers for which the limit $\lim_k x_{n_k}$ exists. That is, it is possible to select a convergent subsequence $x_{n_1}, x_{n_2}, \ldots$. This is one of the fundamental properties of the real line. From it there follows one of the basic principles of analysis.

Theorem 25.13.   *Suppose that each row of the array*

$$x_{1,1}, x_{1,2}, x_{1,3}, \cdots$$

$$(25.19) \qquad x_{2,1}, x_{2,2}, x_{2,3}, \cdots$$

$$\cdots \cdots \cdots \cdots \cdots$$

is a bounded sequence of real numbers. Then there exists an increasing sequence $n_1, n_2,$ ... of integers such that the limit $\lim_k x_{r,n_k}$ exists for $r = 1, 2, \ldots$.

PROOF.   From the first row, select a convergent subsequence

$$(25.20) \qquad x_{1,n_{1,1}}, x_{1,n_{1,2}}, x_{1,n_{1,3}}, \cdots;$$

here $\{n_{1,k}\}$ is an increasing sequence of integers and $\lim_k x_{1,n_{1,k}}$ exists. Look next at the second row of (25.19) along the sequence $n_{1,1}, n_{1,2}, \cdots$:

(25.21) $x_{2,n_{1,1}}, x_{2,n_{1,2}}, x_{2,n_{1,3}}, \ldots$

As a subsequence of the second row of (25.19), (25.21) is bounded. Select from it a convergent subsequence

$$x_{2,n_{2,1}}, x_{2,n_{2,2}}, x_{2,n_{2,3}}, \ldots;$$

here $\{n_{2,k}\}$ is an increasing sequence of integers, a subsequence of $\{n_{1,k}\}$, and $\lim_k x_{2,n_{2,k}}$ exists.

Continue inductively in the same way. This gives an array

$$n_{1,1}, n_{1,2}, n_{1,3}, \ldots$$

(25.22) $n_{2,1}, n_{2,2}, n_{2,3}, \ldots$

$$\ldots \ldots \vdots \ldots \ldots \ldots$$

with three properties. (i) Each row of (25.22) is an increasing sequence of integers. (ii) The rth row is a subsequence of the $(r - 1)$st. (iii) For each r, $\lim_k x_{r,n_{r,k}}$ exists. Thus

(25.23) $x_{r,n_{r,1}}, x_{r,n_{r,2}}, x_{r,n_{r,3}}, \ldots$

is a convergent subsequence of the rth row of (25.19).

Put $n_k = n_{k,k}$. Since each row of (25.22) is increasing and is contained in the preceding row, $n_1, n_2, n_3, \ldots$ is an increasing sequence of integers. Furthermore, $n_r, n_{r+1}, n_{r+2}, \ldots$ is a subsequence of the rth row of (25.22). Thus $x_{r,n_r}, x_{r,n_{r+1}}, x_{r,n_{r+2}}, \ldots$ is a subsequence of (25.23) and is therefore convergent. Thus $\lim_k n_{r,n_k}$ does exist.   ∎

Since $\{n_k\}$ is the diagonal of the array (25.22), application of this theorem is called the *diagonal method*.

PROBLEMS

25.1.   (a)   Show by example that distribution functions having densities can converge weakly even if the densities do not converge.

Hint: Consider $f_n(x) = 1 + \cos 2\pi nx$ on $[0, 1]$.

(b)   Show that distributions with densities can converge weakly to a limit that has no density (even to a unit mass).

(c)   Show that discrete distributions can converge weakly to a distribution that has a density.

(d)   Construct an example, like that of Example 25.3, in which $\mu_n(A) \to \mu(A)$ fails but in which all the measures come from continuous densities on $[0, 1]$.

25.2.   14.12 ↑   Give a simple proof of the Glivenko-Cantelli theorem (Theorem 20.6) under the extra hypothesis that F is continuous.

25.3.   *Initial digits.* (a)   Show that the first significant digit of a positive number x is d (in the scale of 10) if and only if $\{\log_{10} x\}$ lies between $\log_{10} d$ and $\log_{10} (d + 1)$, $d = 1, \ldots, 9$, where the brackets denote fractional part.

(b)   For positive numbers $x_1, x_2, \ldots,$ let $N_n(d)$ be the number among the first n that have initial digit d. Show that

$$(25.24) \qquad \lim_n \frac{1}{n} N_n(d) = \log_{10}(d+1) - \log_{10} d, \qquad d = 1, \ldots, 9,$$

if the sequence $\log_{10} x_n$, $n = 1, 2, \ldots$, is uniformly distributed modulo 1. This is true, for example, of $x_n = \vartheta^n$ if $\log_{10} \vartheta$ is irrational.

(c) Let D_n be the first significant digit of a positive random variable X_n. Show that

$$(25.25) \qquad \lim_n P[D_n = d] = \log_{10}(d+1) - \log_{10} d, \qquad d = 1, \ldots, 9,$$

if $\{\log_{10} X_n\} \Rightarrow U$, where U is uniformly distributed over the unit interval.

25.4. Show that for each probability measure μ there exist probability measures μ_n with finite support such that $\mu_n \Rightarrow \mu$. Show further that $\mu_n\{x\}$ can be taken rational and that each point in the support can be taken rational. Thus there exists a countable set of probability measures such that every μ is the weak limit of some sequence from the set. The space of distribution functions is thus separable in the Lévy metric (see Problem 14.8).

25.5. Show that (25.10) implies that $P([X \leq x] \triangle [X_n \leq x]) \to 0$ if $P[X = x] = 0$.

25.6. For arbitrary random variables X_n there exist positive constants a_n such that $a_n X_n \Rightarrow 0$.

25.7. (a) Generalize Example 25.8 by showing for three-dimensional random vectors (A_n, B_n, X_n) and constants a and b, $a > 0$, that, if $A_n \Rightarrow a$, $B_n \Rightarrow b$, and $X_n \Rightarrow X$, then $A_n X_n + B_n \Rightarrow aX + b$.

(b) Suppose that $X_n \Rightarrow X$ and $A_n \Rightarrow a$. Show that $X_n + A_n \Rightarrow X + a$, $X_n A_n \Rightarrow Xa$, and $X_n/A_n \Rightarrow X/a$ (if $a \neq 0$).

25.8. Suppose that $X_n \Rightarrow X$ and that h_n and h are Borel functions. Let E be the set of x for which $h_n x_n \to hx$ fails for some sequence $x_n \to x$. Suppose that $E \in \mathcal{R}^1$ and $P[X \in E] = 0$. Show that $h_n X_n \Rightarrow hX$.

25.9. Suppose that the distributions of random variables X_n and X have densities f_n and f. Show that if $f_n(x) \to f(x)$ for x outside a set of Lebesgue measure 0, then $X_n \Rightarrow X$.

25.10. ↑ Suppose that X_n assumes as values only integer multiples of a positive δ_n. Suppose that $\delta_n \to 0$ and that, if k_n is an integer varying with n in such a way that $k_n \delta_n \to x$, then $P[X_n = k_n \delta_n] \delta_n^{-1} \to f(x)$, where f is the density of a random variable X. Show that $X_n \Rightarrow X$.

25.11. ↑ Let S_n have the binomial distribution with parameters n and p. Assume as known that

$$(25.26) \qquad P[S_n = k_n](np(1-p))^{1/2} \to \frac{1}{\sqrt{2\pi}} e^{-x^2/2}$$

if $(k_n - np)(np(1-p))^{-1/2} \to x$. Deduce the DeMoivre-Laplace theorem: $(S_n - np)(np(1-p))^{-1/2} \Rightarrow N$, where N has the standard normal distribution. This is a special case of the central limit theorem; see Section 27.

25.12. Prove weak convergence in Example 25.3 by using Theorem 25.8 and the theory of the Riemann integral.

25.13. (a) Show that probability measures satisfy $\mu_n \Rightarrow \mu$ if $\mu_n(a, b] \to \mu(a, b]$ whenever $\mu\{a\} = \mu\{b\} = 0$.

(b) Show that, if $\int f \, d\mu_n \to \int f \, d\mu$ for all continuous f with bounded support, then $\mu_n \Rightarrow \mu$.

25.14. ↑ Show that, if $\mu_n \Rightarrow \mu$ and f is bounded and continuous except on a set of Lebesgue measure 0, then $\int f \, d\mu_n \to \int f \, d\mu$.

25.15. ↑ Let μ be Lebesgue measure confined to the unit interval; let μ_n correspond to a mass of $x_{n,i} - x_{n,i-1}$ at some point in $(x_{n,i-1}, x_{n,i}]$, where $0 = x_{n0} < x_{n1} < \cdots < x_{nn} = 1$. Show by considering the distribution functions that $\mu_n \Rightarrow \mu$ if $\max_{i < n} (x_{n,i} - x_{n,i-1}) \to 0$. Deduce that a bounded function continuous almost everywhere on the unit interval is Riemann integrable. See Problem 17.4.

25.16. 2.15 5.16 ↑ A function f of positive integers *has distribution function F* if F is the weak limit of the distribution function $P_n[m: f(m) \leq x]$ of f under the measure having probability $1/n$ at each of $1, \ldots, n$ (see (2.20)). In this case $D[m: f(m) \leq x] = F(x)$ (see (2.21)) for continuity points x of F. Show that $\varphi(m)/m$ (see (2.23)) has a distribution:

(a) Show by the mapping theorem that it suffices to prove that $f(m) = \log (\varphi(m)/m) = \Sigma_p \, \delta_p(m) \log (1 - 1/p)$ has a distribution.

(b) Let $f_u(m) = \Sigma_{p \leq u} \delta_p(m) \log (1 - 1/p)$, and show by (5.41) that f_u has distribution function $F_u(x) = P[\Sigma_{p \leq u} X_p \log (1 - 1/p) \leq x]$, where the X_p are independent random variables (one for each prime p) such that $P[X_p = 1] = 1/p$ and $P[X_p = 0] = 1 - 1/p$.

(c) Show that $\Sigma_p X_p \log (1 - 1/p)$ converges with probability 1.

Hint: Use Theorem 22.3.

(d) Show that $\lim_{u \to \infty} \sup_n E_n[|f - f_u|] = 0$ (see (5.42) for the notation).

(e) Conclude by Markov's inequality and Theorem 25.5 that f has the distribution of the sum in (c).

25.17. For $A \in \mathcal{R}^1$ and $T > 0$, put $\lambda_T(A) = \lambda([-T, T] \cap A]/2T$, where λ is Lebesgue measure. The *relative measure* of A is

(25.27) $$\rho(A) = \lim_{T \to \infty} \lambda_T(A),$$

provided that this limit exists. This is a continuous analogue of density (see (2.21)) for sets of integers. A Borel function f has a distribution under λ_T; if this converges weakly to F, then

(25.28) $$\rho [x: f(x) \leq u] = F(u)$$

for continuity points u of F, and F is called the distribution function of f. Show that all periodic functions have distributions.

25.18. Suppose that $\sup_n \int f \, d\mu_n < \infty$ for a nonnegative f such that $f(x) \to \infty$ as $x \to \pm\infty$. Show that $\{\mu_n\}$ is tight.

25.19. ↑ Let $f(x) = |x|^\alpha$, $\alpha > 0$; this is an f of the kind in Problem 25.18. Construct a tight sequence for which $\int |x|^\alpha \mu_n(dx) \to \infty$; construct one for which $\int |x|^\alpha \mu_n(dx) = \infty$.

25.20.   23.4 ↑   Show that the random variables A_t and L_t in Problems 23.3 and 23.4 converge in distribution. Show that the moments converge.

25.21.   In the applications of Theorem 9.2, only a weaker result is actually needed: For each K there exists a positive $\alpha = \alpha(K)$ such that if $E[X] = 0$, $E[X^2] = 1$, and $E[X^4] \leq K$, then $P[X \geq 0] \geq \alpha$. Prove this by using tightness and the corollary to Theorem 25.12.

SECTION 26.   CHARACTERISTIC FUNCTIONS

Definition

The *characteristic function* of a probability measure μ on the line is defined for real t by

$$\varphi(t) = \int_{-\infty}^{\infty} e^{itx}\mu(dx)$$

$$= \int_{-\infty}^{\infty} \cos tx\,\mu(dx) + i \int_{-\infty}^{\infty} \sin tx\,\mu(dx);$$

see the end of Section 16 for integrals of complex-valued functions.* A random variable X with distribution μ has characteristic function

$$\varphi(t) = E[e^{itX}] = \int_{-\infty}^{\infty} e^{itx}\mu(dx).$$

The characteristic function is thus defined as the moment generating function but with the real argument s replaced by it; it has the advantage that it always exists because e^{itx} is bounded. The characteristic function in nonprobabilistic contexts is called the *Fourier transform*.

The characteristic function has three fundamental properties to be established here:

   (i)   If μ_1 and μ_2 have respective characteristic functions $\varphi_1(t)$ and $\varphi_2(t)$, then $\mu_1 * \mu_2$ has characteristic function $\varphi_1(t)\varphi_2(t)$. Although convolution is essential to the study of sums of independent random variables, it is a complicated operation, and it is often simpler to study the products of the corresponding characteristic functions.

   (ii)   The characteristic function uniquely determines the distribution. This shows that in studying the products in (i), no information is lost.

   (iii)   From the pointwise convergence of characteristic functions follows the weak convergence of the corresponding distributions. This makes it possible, for example, to investigate the asymptotic distributions of sums of independent random variables by means of their characteristic functions.

---

* From complex variable theory only De Moivre's formula and the simplest properties of the exponential function are needed here.

Moments and Derivatives

It is convenient first to study the relation between a characteristic function and the moments of the distribution it comes from.

Of course, $\varphi(0) = 1$, and by (16.4) (which applies to complex-valued integrands—see the end of Section 16) $|\varphi(t)| \leq 1$ for all t. By Theorem 16.8(i), $\varphi(t)$ is continuous in t. In fact, $|\varphi(t+h) - \varphi(t)| \leq \int |e^{ihx} - 1| \mu(dx)$, and so it follows by the bounded convergence theorem that $\varphi(t)$ *is uniformly continuous.*

Integration by parts shows that

$$(26.1) \qquad \int_0^x (x-s)^n e^{is}\, ds = \frac{x^{n+1}}{n+1} + \frac{i}{n+1} \int_0^x (x-s)^{n+1} e^{is}\, ds,$$

and it follows by induction that

$$(26.2) \qquad e^{ix} = \sum_{k=0}^n \frac{(ix)^k}{k!} + \frac{i^{n+1}}{n!} \int_0^x (x-s)^n e^{is}\, ds$$

for $n \geq 0$. Replace n by $n-1$ in (26.1), solve for the integral on the right, and substitute this for the integral in (26.2); this gives

$$(26.3) \qquad e^{ix} = \sum_{k=0}^n \frac{(ix)^k}{k!} + \frac{i^n}{(n-1)!} \int_0^x (x-s)^{n-1}(e^{is} - 1)\, ds.$$

Estimating the integrals in (26.2) and (26.3) now leads to

$$(26.4) \qquad \left| e^{ix} - \sum_{k=0}^n \frac{(ix)^k}{k!} \right| \leq \min \left\{ \frac{|x|^{n+1}}{(n+1)!}, \frac{2|x|^n}{n!} \right\}$$

for $n \geq 0$. The first term on the right gives a sharp estimate for $|x|$ small, the second a sharp estimate for $|x|$ large.

If X has a moment of order n, it follows that

$$(26.5) \qquad \left| \varphi(t) - \sum_{k=0}^n \frac{(it)^k}{k!} E[X^k] \right| \leq E\left[\min \left\{ \frac{|tX|^{n+1}}{(n+1)!}, \frac{2|tX|^n}{n!} \right\} \right].$$

For any t satisfying

$$(26.6) \qquad \lim_n \frac{|t|^n E[|X|^n]}{n!} = 0,$$

$\varphi(t)$ must therefore have the expansion

$$(26.7) \qquad \varphi(t) = \sum_{k=0}^\infty \frac{(it)^k}{k!} E[X^k];$$

compare (21.22). If

$$\sum_{k=0}^\infty \frac{|t|^k}{k!} E[|X|^k] = E[e^{|tX|}] < \infty,$$

then (see (16.23)) (26.7) must hold. Thus (26.7) holds if X has a moment generating function over the whole line.

Example 26.1. Since $E[e^{|tX|}] < \infty$ if X has the standard normal distribution, by (26.7) its characteristic function is

$$(26.8) \quad \varphi(t) = \sum_{k=0}^{\infty} \frac{(it)^{2k}}{(2k)!} 1 \cdot 3 \cdots (2k-1)! = \sum_{k=0}^{\infty} \frac{1}{k!} \left(-\frac{t^2}{2}\right)^k = e^{-t^2/2}.$$

This and (21.25) formally coincide if $s = it$. ∎

If the power series expansion (26.7) holds, the moments of X can be read off from it:

$$(26.9) \qquad \varphi^{(k)}(0) = i^k E[X^k].$$

This is the analogue of (21.23). It holds, however, under the weakest possible assumption, namely that $E[|X^k|] < \infty$. Indeed,

$$\frac{\varphi(t+h) - \varphi(t)}{h} - E[iXe^{itX}] = E\left[e^{itX}\frac{e^{ihX} - 1 - ihX}{h}\right].$$

The integrand on the right is dominated by $2|X|$ (use (26.4)) and goes to 0 with h, and so the expected value goes to 0 by the dominated convergence theorem. Thus $\varphi'(t) = E[iXe^{itX}]$. Repeating this argument inductively gives

$$(26.10) \qquad \varphi^{(k)}(t) = E[(iX)^k e^{itX}]$$

if $E[|X^k|] < \infty$. Hence (26.9) holds if $E[|X^k|] < \infty$. The proof of uniform continuity for $\varphi(t)$ works for $\varphi^{(k)}(t)$ as well.

The right side of (26.5) is at most $2E[\min\{|t| \cdot |X|^{n+1}, |X|^n\}]|t|^n/n!$ Since the integrand here goes to 0 with t and is dominated by $|X|^n$,

$$(26.11) \qquad \varphi(t) = \sum_{k=0}^{n} \frac{(it)^k}{k!} E[X^k] + o(t^n), \qquad t \to 0,$$

if $E[|X|^n] < \infty$. If X has mean 0 and finite variance σ^2, for example, then $\varphi(t) = 1 - \frac{1}{2}\sigma^2 t^2 + \theta(t)$, where $|\theta(t)| \le t^2 E[\min\{|t| \cdot |X|^3, X^2\}]$; $\theta(t) = o(t^2)$ in any case, and since $|\theta(t)| \le |t|^3 E[|X|^3]$, $\theta(t) = O(t^3)$ if there is a third moment. Such estimates are useful in proving limit theorems.

The more moments μ has, the more derivatives φ has. This is one sense in which lightness of the tails of μ is reflected by smoothness of φ. There are results which connect the behavior of $\varphi(t)$ as $|t| \to \infty$ with smoothness properties of μ. The *Riemann-Lebesgue theorem* is the most important of these:

Theorem 26.1. *If μ has a density, then $\varphi(t) \to 0$ as $|t| \to \infty$.*

PROOF. The problem is to prove for integrable f that $\int f(x)e^{itx}\,dx \to 0$ as $|t|$

$\rightarrow \infty$. There exists by Theorem 17.1 a step function $g = \Sigma_k \alpha_k I_{A_k}$, a finite linear combination of indicators of intervals $A_k = (a_k, b_k]$, for which $\int |f - g|\, dx < \epsilon$. Now $\int f(x)e^{itx}\, dx$ differs by at most ϵ from $\int g(x)e^{itx}\, dx = \Sigma_k \alpha_k (e^{itb_k} - e^{ita_k})/it$, and this goes to 0 as $|t| \rightarrow \infty$. ■

Independence

The multiplicative property (21.28) of moment generating functions extends to characteristic functions. Suppose that X_1 and X_2 are independent random variables with characteristic functions φ_1 and φ_2. If $Y_j = \cos tX_j$ and $Z_j = \sin tX_j$, then (Y_1, Z_1) and (Y_2, Z_2) are independent; by the rules for integrating complex-valued functions,

$$\varphi_1(t)\varphi_2(t) = (E[Y_1] + iE[Z_1])(E[Y_2] + iE[Z_2])$$

$$= E[Y_1]E[Y_2] - E[Z_1]E[Z_2] + i(E[Y_1]E[Z_2] + E[Z_1]E[Y_2])$$

$$= E[Y_1Y_2 - Z_1Z_2 + i(Y_1Z_2 + Z_1Y_2)]$$

$$= E[e^{it(X_1+X_2)}].$$

This extends to sums of three or more: If $X_1, \ldots, X_n$ are independent, then

$$(26.12) \qquad E[e^{it \Sigma_{k=1}^n X_k}] = \prod_{k=1}^n E[e^{itX_k}].$$

If X has characteristic function $\varphi(t)$, then $aX + b$ has characteristic function

$$(26.13) \qquad E[e^{it(aX+b)}] = e^{itb}\varphi(at).$$

In particular, $-X$ has characteristic function $\varphi(-t)$, which is the complex conjugate of $\varphi(t)$.

Inversion and the Uniqueness Theorem

A characteristic function φ uniquely determines the measure μ it comes from. This fundamental fact will be derived by means of an inversion formula through which μ can in principle be recovered from φ.

Define

$$S(T) = \int_0^T \frac{\sin x}{x}\, dx, \qquad T \geq 0.$$

In Example 18.3 it is shown that

$$(26.14) \qquad \lim_{T \rightarrow \infty} S(T) = \frac{\pi}{2};$$

$S(T)$ is therefore bounded. If sgn θ is $+1, 0,$ or -1 as θ is positive, 0, or negative, then

$$(26.15) \qquad \int_0^T \frac{\sin t\theta}{t}\, dt = \text{sgn } \theta \cdot S(T|\theta|), \qquad T \geq 0.$$

Theorem 26.2 *If the probability measure μ has characteristic function φ, and if $\mu\{a\} = \mu\{b\} = 0$, then*

$$(26.16) \qquad \mu(a, b] = \lim_{T \to \infty} \frac{1}{2\pi} \int_{-T}^{T} \frac{e^{-ita} - e^{-itb}}{it}\, \varphi(t)\, dt.$$

Distinct measures cannot have the same characteristic function.

Note: As $t \to 0$, the integrand here converges to $b - a$, which is to be taken as its value for $t = 0$. For fixed a and b the integrand is thus continuous in t and bounded. If μ is a unit mass at 0, then $\varphi(t) \equiv 1$ and the integral in (26.16) cannot be extended over the whole line.

PROOF. The inversion formula will imply uniqueness: It will imply that if μ and ν have the same characteristic function, then $\mu(a, b] = \nu(a, b]$ if $\mu\{a\} = \nu\{a\} = \mu\{b\} = \nu\{b\} = 0$; but such intervals $(a, b]$ form a π-system generating $\mathcal{R}^1$.

Denote by I_T the quantity inside the limit in (26.16). By Fubini's theorem

$$(26.17) \qquad I_T = \frac{1}{2\pi} \int_{-\infty}^{\infty} \left[\int_{-T}^{T} \frac{e^{it(x-a)} - e^{it(x-b)}}{it}\, dt \right] \mu(dx).$$

This interchange is legitimate because the double integral extends over a set of finite product measure and by (26.4) for $n = 0$ the integrand is bounded by $|b - a|$. Rewrite the integrand by DeMoivre's formula. Since $\sin s$ and $\cos s$ are odd and even, respectively, (26.15) gives

$$I_T = \int_{-\infty}^{\infty} \left[\frac{\text{sgn }(x - a)}{\pi} S(T \cdot |x - a|) \right.$$

$$\left. - \frac{\text{sgn }(x - b)}{\pi} S(T \cdot |x - b|)\mu\, (dx). \right.$$

The integrand here is bounded and converges as $T \to \infty$ to the function

$$(26.18) \qquad \psi_{a,b}(x) = \begin{cases} 0 & \text{if } x < a, \\ 1/2 & \text{if } x = a, \\ 1 & \text{if } a < x < b, \\ 1/2 & \text{if } x = b, \\ 0 & \text{if } b < x. \end{cases}$$

Thus $I_T \to \int \psi_{a,b}\, d\mu$, which implies that (26.16) holds if $\mu\{a\} = \mu\{b\} = 0$. ∎

The inversion formula contains further information. Suppose that

(26.19) $$\int_{-\infty}^{\infty} |\varphi(t)| \, dt < \infty.$$

In this case the integral in (26.16) can be extended over R^1. By (26.4), then, $\mu(a, b] \le (b-a)\int_{-\infty}^{\infty} |\varphi(t)| \, dt$, and so there can be no point masses. By (26.16) the corresponding distribution function satisfies

$$\frac{F(x+h) - F(x)}{h} = \frac{1}{2\pi} \int_{-\infty}^{\infty} \frac{e^{-itx} - e^{-it(x+h)}}{ith} \varphi(t) \, dt$$

(whether h is positive or negative). The integrand is by (26.4) dominated by $|\varphi(t)|$ and goes to $e^{-itx}\varphi(t)$ as $h \to 0$. Therefore, F has derivative

(26.20) $$f(x) = \frac{1}{2\pi} \int_{-\infty}^{\infty} e^{-itx}\varphi(t) \, dt.$$

Since f is continuous for the same reason φ is, it integrates to F by the fundamental theorem of the calculus (see (17.6)). Thus (26.19) *implies that μ has the continuous density* (26.20). Moreover, this is the only continuous density. In this result, as in the Riemann-Lebesgue theorem, conditions on the size of $\varphi(t)$ for large $|t|$ are connected with smoothness properties of μ.

The inversion formula (26.20) has many applications. In the first place, it can be used for a new derivation of (26.14). As pointed out in Example 17.3, the existence of the limit in (26.14) is easy to prove. Denote this limit temporarily by $\pi_0/2$—without assuming that $\pi_0 = \pi$. Then (26.16) and (26.20) follow as before if π is replaced by π_0. Applying the latter to the standard normal density (see (26.8)) gives

(26.21) $$\frac{1}{\sqrt{2\pi}} e^{-x^2/2} = \frac{1}{2\pi_0} \int_{-\infty}^{\infty} e^{-itx} e^{-t^2/2} dt,$$

where the π on the left is that of analysis and geometry—it comes ultimately from the quadrature (18.10). An application of (26.8) with x and t interchanged reduces the right side of (26.21) to $(\sqrt{2\pi}/2\pi_0)e^{-x^2/2}$, and therefore π_0 does equal π.

Consider the densities in the table on page 302. The characteristic function for the *normal* distribution has already been calculated. For the *uniform* distribution over $(0, 1)$, the computation is of course straightforward; note that in this case the density cannot be recovered from (26.20) because $\varphi(t)$ is not integrable; this is reflected in the fact that the density has discontinuities at 0 and 1.

The characteristic function for the *exponential* distribution is easily calculated; compare Example 21.3. As for the *double exponential* or *Laplace* distribution, $e^{-|x|}e^{itx}$ integrates over $(0, \infty)$ to $(1 - it)^{-1}$ and over $(-\infty, 0)$ to $(1 + it)^{-1}$, which gives the result. By (26.20), then,

Distribution	Density	Interval	Characteristic Function
1. Normal	$\dfrac{1}{\sqrt{2\pi}}e^{-x^2/2}$	$-\infty < x < \infty$	$e^{-t^2/2}$
2. Uniform	1	$0 < x < 1$	$\dfrac{e^{it} - 1}{it}$
3. Exponential	e^{-x}	$0 < x < \infty$	$\dfrac{1}{1 - it}$
4. Double exponential or Laplace	$\dfrac{1}{2}e^{-|x|}$	$-\infty < x < \infty$	$\dfrac{1}{1 + t^2}$
5. Cauchy	$\dfrac{1}{\pi}\dfrac{1}{1 + x^2}$	$-\infty < x < \infty$	$e^{-|t|}$
6. Triangular	$1 - |x|$	$-1 < x < 1$	$2\dfrac{1 - \cos t}{t^2}$
7.	$\dfrac{1}{\pi}\dfrac{1 - \cos x}{x^2}$	$-\infty < x < \infty$	$(1 - |t|)I_{(-1,1)}(t)$

$$e^{-|x|} = \frac{1}{\pi}\int_{-\infty}^{\infty} e^{-itx}\frac{dt}{1 + t^2}.$$

For $x = 0$ this gives the standard integral $\int_{-\infty}^{\infty} dt/(1 + t^2) = \pi$; see Example 17.5. Thus the *Cauchy* density in the table integrates to 1 and has characteristic function $e^{-|t|}$. This distribution has no first moment and the characteristic function is not differentiable at the origin.

A straightforward integration shows that the *triangular* density has the characteristic function given in the table, and by (26.20),

$$(1 - |x|)I_{(-1,1)}(x) = \frac{1}{\pi}\int_{-\infty}^{\infty} e^{-itx}\frac{1 - \cos t}{t^2}\, dt;$$

For $x = 0$ this is $\int_{-\infty}^{\infty} (1 - \cos t)t^{-2}\, dt = \pi$; hence the last line of the table.

Each density and characteristic function in the table can be transformed by (26.13), which gives a family of distributions.

The Continuity Theorem

Because of (26.12), the characteristic function provides a powerful means of studying the distributions of sums of independent random variables. It is often easier to work with products of characteristic functions than with convolutions, and knowing the characteristic function of the sum is by Theorem 26.2 in principle the same thing as knowing the distribution itself. Because of the following *continuity theorem*, characteristic functions can be used to study limit distributions.

Theorem 26.3. *Let μ_n, μ be probability measures with characteristic functions φ_n, φ. A necessary and sufficient condition for $\mu_n \Rightarrow \mu$ is that $\varphi_n(t) \to \varphi(t)$ for each t.*

PROOF: NECESSITY.   For each t, e^{itx} has bounded modulus and is continuous in x. The necessity therefore follows by an application of Theqrem 25.8 (to the real and imaginary parts of e^{itx}).

SUFFICIENCY.   By Fubini's theorem,

$$\frac{1}{u} \int_{-u}^{u} (1 - \varphi_n(t)) \, dt = \int_{-\infty}^{\infty} \left[\frac{1}{u} \int_{-u}^{u} (1 - e^{itx}) \, dt \right] \mu_n(dx)$$

$$= 2 \int_{-\infty}^{\infty} \left(1 - \frac{\sin ux}{ux} \right) \mu_n(dx)$$

(26.22)

$$\geq 2 \int_{|x| \geq 2/u} \left(1 - \frac{1}{|ux|} \right) \mu_n(dx)$$

$$\geq \mu_n \left[x \colon |x| \geq \frac{2}{u} \right].$$

(Note that the first integral is real.) Since φ is continuous at the origin and $\varphi(0) = 1$, there is for positive ϵ a u for which $u^{-1} \int_{-u}^{u} (1 - \varphi(t)) \, dt < \epsilon$. Since φ_n converges to φ, the bounded convergence theorem implies that there exists an n_0 such that $u^{-1} \int_{-u}^{u} (1 - \varphi_n(t)) dt < 2\epsilon$ for $n \geq n_0$. If $a = 2/u$ in (26.22), then $\mu_n[x \colon |x| \geq a] < 2\epsilon$ for $n \geq n_0$. Increasing a if necessary will ensure that this inequality also holds for the finitely many n preceding n_0. Therefore, $\{\mu_n\}$ is tight.

By the corollary to Theorem 25.10, $\mu_n \Rightarrow \mu$ will follow if it is shown that each subsequence $\{\mu_{n_k}\}$ that converges weakly at all converges weakly to μ. But if $\mu_{n_k} \Rightarrow \nu$ as $k \to \infty$, then by the necessity half of the theorem, already proved, ν has characteristic function $\lim_k \varphi_{n_k}(t) = \varphi(t)$. By Theorem 26.2, ν and μ must coincide. ∎

Two corollaries, interesting in themselves, will make clearer the structure of the proof of sufficiency given above. In each, let μ_n be probability measures on the line with characteristic functions φ_n.

Corollary 1. *Suppose that $\lim_n \varphi_n(t) = g(t)$ for each t, where the limit function* *g is continuous at* 0. *Then there exists a* μ *such that* $\mu_n \Rightarrow \mu$, *and* μ *has characteristic function g.*

PROOF. The point of the corollary is that g is not assumed at the outset to be a characteristic function. But in the argument following (26.22), only $\varphi(0) = 1$ and the continuity of φ at 0 were used; hence $\{\mu_n\}$ is tight under the present hypothesis. If $\mu_{n_k} \Rightarrow \nu$ as $k \to \infty$, then ν must have characteristic function $\lim_k \varphi_{n_k}(t) = g(t)$. Thus g is, in fact, a characteristic function, and the proof goes through as before.   ∎

In this proof the continuity of g was used only to establish tightness. Hence if $\{\mu_n\}$ is assumed tight in the first place, the hypothesis of continuity can be suppressed:

Corollary 2. *Suppose that* $\lim_n \varphi_n(t) = g(t)$ *exists for each t and that* $\{\mu_n\}$ *is tight. Then there exists a* μ *such that* $\mu_n \Rightarrow \mu$, *and* μ *has characteristic function g.*

This second corollary applies, for example, if the μ_n have a common bounded support.

Example 26.2. If μ_n is the uniform distribution over $(-n, n)$, its characteristic function is $(nt)^{-1} \sin tn$ for $t \neq 0$, and hence it converges to $\delta_{\{0\}}(t)$. In this case $\{\mu_n\}$ is not tight, the limit function is not continuous at 0, and μ_n does not converge weakly.   ∎

PROBLEMS

26.1. A random variable has a *lattice distribution* if for some a and b, $b > 0$, the lattice $[a + nb: n = 0, \pm 1, \ldots]$ supports the distribution of X. Let X have characteristic function φ.

(a) Show that a necessary condition for X to have a lattice distribution is that $|\varphi(t)| = 1$ for some $t \neq 0$.

(b) Show that the condition is sufficient as well.

(c) Suppose that $|\varphi(t)| = |\varphi(t')| = 1$ for incommensurable t and t' (distinct from 0). Show that $P[X = c] = 1$ for some constant c.

26.2. If $\mu(-\infty, x] = \mu[-x, \infty)$ for all x (which implies that $\mu(A) = \mu(-A)$ for all $A \in \mathcal{R}^1$), then μ is *symmetric*. Show that this holds if and only if the characteristic function is real.

26.3. Consider functions φ that are real and nonnegative and satisfy $\varphi(-t) = \varphi(t)$ and $\varphi(0) = 1$.

(a) Suppose that $d_1, d_2, \ldots$ are positive and $\sum_{k=1}^{\infty} d_k = \infty$, that $s_1 \geq s_2 \geq \cdots$

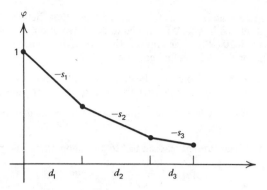

≥ 0 and $\lim_k s_k = 0$, and that $\Sigma_{k=1}^{\infty} s_k d_k = 1$. Let φ be the convex polygon whose successive sides have slopes $-s_1, -s_2, \ldots$ and lengths $d_1, d_2, \ldots$ when projected on the horizontal axis: φ has value $1 - \Sigma_{j=1}^{k} s_j d_j$ at $t_k = d_1 + \cdots + d_k$. If $s_n = 0$, there are in effect only n sides. Let $\varphi_0(t) = (1 - |t|)I_{(-1,1)}(t)$ be the characteristic function in the last line in the table on page 302 and show that $\varphi(t)$ is a convex combination of the characteristic functions $\varphi_0(t/t_k)$ and hence is itself a characteristic function.

(b) Construct a continuous density for which the characteristic function is not integrable. In such a case the density cannot be recovered from (26.20).

(c) *Polya's criterion.* Suppose that φ is continuous and convex on $[0, \infty)$. Show that φ is a characteristic function.

26.4. ↑ Let φ_1 and φ_2 be characteristic functions, and show that the set $A = [t: \varphi_1(t) = \varphi_2(t)]$ is closed, contains 0, and is symmetric about 0. Show that every set with these three properties can be such an A. What does this say about the uniqueness theorem?

26.5. Show by Theorem 26.1 and integration by parts that if μ has a density f with integrable derivative f', then $\varphi(t) = o(t^{-1})$ as $|t| \to \infty$. Extend to higher derivatives.

26.6. Show for independent random variables uniformly distributed over $(-1, +1)$ that $X_1 + \cdots + X_n$ has density $\pi^{-1}\int_0^{\infty} ((\sin t)/t)^n \cos tx \, dt$ for $n \geq 2$.

26.7. 21.31↑ *Uniqueness theorem for moment generating functions.* Suppose that F has a moment generating function in $(-s_0, s_0)$, $s_0 > 0$. From the fact that $\int_{-\infty}^{\infty} e^{zx} \, dF(x)$ is analytic in the strip $-s_0 < \operatorname{Re} z < s_0$, prove that the moment generating function determines F. Show that it is enough that the moment generating function exist in $[0, s_0)$, $s_0 > 0$.

26.8. 21.35 26.7↑ Show that the gamma density (20.47) has characteristic function

$$\frac{1}{(1 - it/\alpha)^u} = \exp\left[-u \log\left(1 - \frac{it}{\alpha}\right)\right],$$

where the logarithm is the principal part. Show that $\int_0^{\infty} e^{zx} f(x; \alpha, u) \, dx$ is analytic for $\operatorname{Re} z < \alpha$.

26.9.   Use characteristic functions for a simple proof that the family of Cauchy distributions defined by (20.45) is closed under convolution; compare the argument in Problem 20.20 (a). Do the same for the normal distribution (compare Example 20.6) and for the Poisson and gamma distributions.

26.10.   Suppose that $F_n \Rightarrow F$ and that the characteristic functions are uniformly integrable. Show that F has a density that is the limit of the densities of the F_n.

26.11.   Show for all a and b that the right side of (26.16) is $\mu(a, b) + \frac{1}{2}\mu\{a\} + \frac{1}{2}\mu\{b\}$.

26.12.   By the kind of argument leading to (26.16), show that

$$(26.23) \qquad \mu\{a\} = \lim_{T \to \infty} \frac{1}{2T} \int_{-T}^{T} e^{-ita}\, \varphi(t)\, dt.$$

26.13.   ↑   Let $x_1, x_2, \ldots$ be the points of positive μ measure. By the following steps prove that

$$(26.24) \qquad \lim_{T \to \infty} \frac{1}{2T} \int_{-T}^{T} |\varphi(t)|^2\, dt = \sum_k (\mu\{x_k\})^2.$$

Let X and Y be independent and have characteristic function φ.
(a)   Show by (26.23) that the left side of (26.24) is $P[X - Y = 0]$.
(b)   Show (Theorem 20.3) that $P[X - Y = 0] = \int_{-\infty}^{\infty} P[X = y]\, \mu(dy) = \sum_k (\mu\{x_k\})^2$.

26.14.   ↑   Show that μ has no point masses if $\varphi^2(t)$ is integrable.

26.15.   **(a)**   Show that $\mu_n \Rightarrow \mu$ implies that the characteristic functions $\varphi_n(t)$ are uniformly equicontinuous (for each ϵ there is a δ such that $|s - t| < \delta$ implies that $|\varphi_n(s) - \varphi_n(t)| < \epsilon$ for all n).
(b)   Show that $\mu_n \Rightarrow \mu$ implies that $\varphi_n(t) \to \varphi(t)$ uniformly on bounded sets.
(c)   Show that the convergence in part (b) need not be uniform over the entire line.

26.16.   14.8 26.15↑   For distribution functions F and G, define $d'(F, G) = \sup_t |\varphi(t) - \psi(t)|/(1 + |t|)$, where φ and ψ are the corresponding characteristic functions. Show that this is a metric and equivalent to the Lévy metric.

26.17.   25.17↑   A real function f has *mean value*

$$(26.25) \qquad M[f(x)] = \lim_{T \to \infty} \frac{1}{2T} \int_{-T}^{T} f(x)\, dx,$$

provided that f is integrable over each $[-T, T]$ and the limit exists.
(a)   Show that, if f is bounded and $e^{itf(x)}$ has a mean value for each t, then f has a distribution in the sense of (25.28).
(b)   Show that

$$(26.26) \qquad M[e^{itx}] = \begin{cases} 1 & \text{if } t = 0, \\ 0 & \text{if } t \neq 0. \end{cases}$$

Of course, $f(x) = x$ has no distribution.

26.18. Suppose that X is irrational with probability 1. Let μ_n be the distribution of the fractional part $\{nX\}$. Use the continuity theorem and Theorem 25.1 to show that $n^{-1}\Sigma_{k=1}^n \mu_k$ converges weakly to the uniform distribution on $[0, 1]$.

26.19. 25.13↑ The uniqueness theorem for characteristic functions can be derived from the Weierstrass approximation theorem. Fill in the details of the following argument. Let μ and ν be probability measures on the line. For continuous f with bounded support choose a so that $\mu(-a, a)$ and $\nu(-a, a)$ are nearly 1 and f vanishes outside $(-a, a)$. Let g be periodic and agree with f in $(-a, a)$, and by the Weierstrass theorem uniformly approximate $g(x)$ by a trigonometric sum $p(x) = \Sigma_{k=1}^N a_k e^{it_k x}$. If μ and ν have the same characteristic function, then $\int f\,d\mu \approx \int g\,d\mu \approx \int p\,d\mu = \int p\,d\nu \approx \int g\,d\nu \approx \int f\,d\nu$.

26.20. ↑ If g has period 2π, then g can be uniformly approximated by a trigonometric polynomial $\Sigma_{k=1}^N a_k e^{im_k x}$, where the m_k are integers. Modify the preceding argument to show that a probability measure μ supported by $(0, 2\pi)$ is determined by its Fourier coefficients $\int e^{imx}\,(dx)$, $m = 0, \pm 1, \pm 2, \ldots$—that is, by its characteristic function for integer arguments. Note that shifting mass from 2π to 0 does not affect the Fourier coefficients.

26.21. ↑ Suppose that μ_n and μ are probability measures supported by $(0, 2\pi)$. Use the result in Problem 26.20 to show that $\mu_n \Rightarrow \mu$ if the Fourier coefficients converge:

$$\lim_n \int e^{imx}\mu_n(dx) = \int e^{imx}\mu(dx), \qquad m = 0, \pm 1, \ldots.$$

26.22. ↑ Show that $n^{-1}\Sigma_{k=1}^n e^{im\theta k} \to 0$ if $m \neq 0$ and $\theta/2\pi$ is irrational. Use the result of Problem 26.21 (rescale from $(0, 2\pi)$ to $(0, 1)$) to give another proof of Theorem 25.1.

26.23. Use the continuity theorem to prove the result in Example 25.2 concerning the convergence of the binomial distribution to the Poisson.

26.24. According to Example 25.8, if $X_n \Rightarrow X$, $a_n \to a$, and $b_n \to b$, then $a_n X_n + b_n \Rightarrow aX + b$. Prove this by means of characteristic functions.

26.25. *Infinite convolutions.* Write $F = F_1 * F_2 * \cdots$ if $F_1 * \cdots * F_n \Rightarrow F$. Show that this holds if and only if the corresponding characteristic functions satisfy $\varphi(t) = \Pi_{n=1}^\infty \varphi_n(t)$ in the usual sense of infinite products.

26.26. 26.1 26.15↑ According to Theorem 14.2, if $X_n \Rightarrow X$ and $a_n X_n + b_n \Rightarrow Y$, where $a_n > 0$ and the distributions of X and Y are nondegenerate, then $a_n \to a > 0$, $b_n \to b$, and $aX + b$ and Y have the same distribution. Prove this by characteristic functions. Let φ_n, φ, ψ be the characteristic functions of X_n, X, Y.

(a) Show that $|\varphi_n(a_n t)| \to |\psi(t)|$ uniformly on bounded sets and hence that a_n cannot converge to 0 along a subsequence.

(b) Interchange the roles of φ and ψ and show that a_n cannot converge to infinity along a subsequence.

(c) Show that a_n converges to some $a > 0$.

(d) Show that $e^{itb_n} \to \psi(t)/\varphi(at)$ in a neighborhood of 0 and hence that $\int_0^t e^{isb_n}\,ds \to \int_0^t (\psi(s)/\varphi(as))\,ds$. Conclude that b_n converges.

26.27. Prove a continuity theorem for moment generating functions as defined by (22.5) for probability measures on $[0, \infty)$. For uniqueness, see (22.8); the analogue of (26.22) is

$$\frac{2}{u} \int_0^u (1 - M(s)) \, ds \geq \mu\left(\frac{2}{u}, \infty\right).$$

SECTION 27. THE CENTRAL LIMIT THEOREM

Identically Distributed Summands

The central limit theorem says roughly that the sum of many independent random variables will be approximately normally distributed if each summand has high probability of being small. Theorem 27.1, the *Lindeberg-Lévy theorem*, will give an idea of the techniques and hypotheses needed for the more general results that follow.

Throughout, N will denote a random variable with the standard normal distribution:

(27.1) $$P[N \in A] = \frac{1}{\sqrt{2\pi}} \int_A e^{-x^2/2} \, dx.$$

Theorem 27.1. *Suppose that $\{X_n\}$ is an independent sequence of random variables having the same distribution with mean c and finite positive variance σ^2. If $S_n = X_1 + \cdots + X_n$, then*

(27.2) $$\frac{S_n - nc}{\sigma\sqrt{n}} \Rightarrow N.$$

By the argument in Example 25.7, (27.2) implies that $n^{-1}S_n \Rightarrow c$. The central limit theorem and the strong law of large numbers thus refine the weak law of large numbers in different directions.

Replacing X_n by $X_n - c$ reduces the theorem to the case in which $E[X_n] = 0$. Hence assume that $c = 0$.

Let φ be the characteristic function common to the X_n. By (26.12) and (26.13), $S_n/\sigma\sqrt{n}$ has characteristic function $\varphi^n(t/\sigma\sqrt{n})$. By the continuity theorem, it suffices to prove that this converges to the characteristic function $e^{-t^2/2}$ of N. By (26.11) with $n = 2$, $\varphi(t) = 1 - \frac{1}{2}t^2\sigma^2 + \beta(t)$, where $\beta(t)/t^2 \to 0$ as $t \to 0$. The idea is to show that

$$\varphi^n\left(\frac{t}{\sigma\sqrt{n}}\right) = \left(1 - \frac{t^2}{2n} + \beta\left(\frac{t}{\sigma\sqrt{n}}\right)\right)^n = \left(1 - \frac{t^2}{2n}\right)^n + o(1) \to e^{-t^2/2}.$$

This can be proved by using logarithms for complex arguments,* but a simple device avoids them.

Lemma 1. *Let $z_1, \ldots, z_m$ and $z'_1, \ldots, z'_m$ be complex numbers of modulus at most 1; then*

$$(27.3) \qquad |z_1 \cdots z_m - z'_1 \cdots z'_m| \leq \sum_{k=1}^{m} |z_k - z'_k|.$$

PROOF. This follows by induction from $z_1 \cdots z_m - z'_1 \cdots z'_m = (z_1 - z'_1)(z_2 \cdots z_m) + z'_1(z_2 \cdots z_m - z'_2 \cdots z'_m)$. ■

Since the $\beta(t)$ above satisfies $\beta(t)/t^2 \to 0$,

$$(27.4) \qquad n \left| \varphi\left(\frac{t}{\sigma\sqrt{n}}\right) - \left(1 - \frac{t^2}{2n}\right) \right| \to 0$$

for each fixed t. In the lemma take $m = n$, $z_k = 1 - t^2/2n$, and $z'_k = \varphi(t/\sigma\sqrt{n})$; (27.3) implies that (for n large enough that $t^2/2n < 1$)

$$(27.5) \qquad \left| \varphi^n\left(\frac{t}{\sigma\sqrt{n}}\right) - \left(1 - \frac{t^2}{2n}\right)^n \right| \leq n \left| \varphi\left(\frac{t}{\sigma\sqrt{n}}\right) - \left(1 - \frac{t^2}{2n}\right) \right|.$$

This goes to 0 by (27.4), and Theorem 27.1 follows because $(1 - t^2/2n)^n \to e^{-t^2/2}$.

Example 27.1. In the classical De Moivre-Laplace theorem, X_n takes the values 1 and 0 with probabilities p and $q = 1 - p$, so that $c = p$ and $\sigma^2 = pq$. Here S_n is the number of successes in n Bernoulli trials, and $(S_n - np)/\sqrt{npq} \Rightarrow N$. ■

Example 27.2. Suppose that one wants to estimate the parameter α of an exponential distribution (20.10) on the basis of an independent sample $X_1, \ldots, X_n$. As $n \to \infty$ the sample mean $\overline{X}_n = n^{-1} \Sigma_{k=1}^n X_k$ converges in probability to the mean $1/\alpha$ of the distribution, and hence it is natural to use $1/\overline{X}_n$ to estimate α itself. How good is the estimate? The variance of the exponential distribution being $1/\alpha^2$ (Example 21.3), $\alpha\sqrt{n}(\overline{X}_n - 1/\alpha) \Rightarrow N$ by the Lindeberg-Lévy theorem. Thus $\overline{X}_n$ is approximately normally distributed with mean $1/\alpha$ and standard deviation $1/\alpha\sqrt{n}$.

By Skorohod's Theorem 25.6 there exist on a single probability space random variables $\overline{Y}_n$ and Y having the respective distributions of $\overline{X}_n$ and N and satisfying $\alpha\sqrt{n}(\overline{Y}_n(\omega) - 1/\alpha) \to Y(\omega)$ for each ω. Now $\overline{Y}_n(\omega) \to 1/\alpha$ and $\alpha^{-1}\sqrt{n}(\overline{Y}_n(\omega)^{-1} - \alpha) = \alpha\sqrt{n}(\alpha^{-1} - \overline{Y}_n(\omega))/\alpha\overline{Y}_n(\omega) \to -Y(\omega)$. Since $-Y$ has the distribution of N and $\overline{Y}_n$ has the distribution of $\overline{X}_n$, it follows that

$$\frac{\sqrt{n}}{\alpha}\left(\frac{1}{\overline{X}_n} - \alpha\right) \Rightarrow N;$$

---

* Fix t. Since $\gamma(t) = -t^2/2n + \beta(t/\sigma\sqrt{n})$ is $O(n^{-1})$, $\log \varphi^n(t/\sigma\sqrt{n}) = n \log(1 + \gamma(t)) = n(\gamma(t) + O(n^{-2})) \to -\frac{1}{2}t^2$.

thus $1/\overline{X}_n$ is approximately normally distributed with mean α and standard deviation $\alpha/\sqrt{n}$. In effect, $1/\overline{X}_n$ has been studied through the local linear approximation to the function $1/x$. This is called the *delta method*.   ∎

The Lindeberg and Lyapounov Theorems

Suppose that for each n

(27.6) $$X_{n1}, \ldots, X_{nr_n}$$

are independent; the probability space for the sequence may change with n. Such a collection is called a *triangular array* of random variables. Put $S_n = X_{n1} + \cdots + X_{nr_n}$. Theorem 27.1 covers the special case in which $r_n = n$ and $X_{nk} = X_k$. Theorem 6.2 is the weak law of large numbers for triangular arrays, and Example 6.4 on the number of cycles in a random permutation shows its usefulness. The central limit theorem for triangular arrays will be applied below to the same example.

To establish the asymptotic normality of S_n by means of the ideas in the preceding proof requires expanding the characteristic function of each X_{nk} to second-order terms and estimating the remainder. Suppose that

(27.7) $$E[X_{nk}] = 0, \qquad \sigma_{nk}^2 = E[X_{nk}^2], \qquad s_n^2 = \sum_{k=1}^{r_n} \sigma_{nk}^2.$$

The assumption that X_{nk} has mean 0 entails no loss of generality. A successful remainder estimate is possible under the assumption of the *Lindeberg condition*:

(27.8) $$\lim_{n \to \infty} \sum_{k=1}^{r_n} \frac{1}{s_n^2} \int_{|X_{nk}| \ge \epsilon s_n} X_{nk}^2 \, dP = 0$$

for $\epsilon > 0$.

Theorem 27.2. *Suppose that for each n the sequence $X_{n1}, \ldots, X_{nr_n}$ is independent and satisfies (27.7). If (27.8) holds for all positive ϵ, then $S_n/s_n \Rightarrow N$.*

This theorem contains the preceding one: Suppose that $X_{nk} = X_k$ and $r_n = n$, where the entire sequence $\{X_k\}$ is independent and the X_k all have the same distribution with mean 0 and variance σ^2. Then (27.8) reduces to

(27.9) $$\lim_{n \to \infty} \frac{1}{\sigma^2} \int_{|X_1| \ge \epsilon \sigma \sqrt{n}} X_1^2 \, dP = 0,$$

which holds because $[|X_1| \ge \epsilon \sigma \sqrt{n}] \downarrow 0$ as $n \uparrow \infty$.

PROOF OF THE THEOREM.   Replacing X_{nk} by X_{nk}/s_n shows that there is no loss of generality in assuming

(27.10) $$s_n^2 = \sum_{k=1}^{r_n} \sigma_{nk}^2 = 1.$$

By (26.4) for $n = 2$,

$$\left| e^{itx} - \left(1 + itx - \frac{1}{2} t^2 x^2 \right) \right| \le \min \{ |tx|^2, |tx|^3 \}.$$

Therefore, the characteristic function φ_{nk} of X_{nk} satisfies

(27.11) $$\left| \varphi_{nk}(t) - \left(1 - \frac{1}{2} t^2 \sigma_{nk}^2 \right) \right| \le E[\min \{ |tX_{nk}|^2, |tX_{nk}|^3 \}].$$

Note that the expected value is finite.

For positive ϵ the right side of (27.11) is at most

$$\int_{|X_{nk}| < \epsilon} |tX_{nk}|^3 \, dP + \int_{|X_{nk}| \ge \epsilon} |tX_{nk}|^2 \, dP$$

$$\le \epsilon |t|^3 \sigma_{nk}^2 + t^2 \int_{|X_{nk}| \ge \epsilon} X_{nk}^2 \, dP.$$

Since the σ_{nk}^2 add to 1 and ϵ is arbitrary, it follows by the Lindeberg condition that

(27.12) $$\sum_{k=1}^{r_n} \left| \varphi_{nk}(t) - \left(1 - \frac{1}{2} t^2 \sigma_{nk}^2 \right) \right| \to 0$$

for each fixed t. This is the analogue of (27.4), and the objective now is to show that

(27.13) $$\prod_{k=1}^{r_n} \varphi_{nk}(t) = \prod_{k=1}^{r_n} \left(1 - \frac{1}{2} t^2 \sigma_{nk}^2 \right) + o(1)$$

$$= \prod_{k=1}^{r_n} e^{-t^2 \sigma_{nk}^2 / 2} + o(1) = e^{-t^2/2} + o(1).$$

For ϵ positive

$$\sigma_{nk}^2 \le \epsilon^2 + \int_{|X_{nk}| \ge \epsilon} X_{nk}^2 \, dP,$$

and so it follows by the Lindeberg condition (recall that s_n is now 1) that

(27.14) $$\max_{1 \le k \le r_n} \sigma_{nk}^2 \to 0.$$

For large enough n, $1 - \frac{1}{2} t^2 \sigma_{nk}^2$ are all between 0 and 1, and by (27.3), $\prod_{k=1}^{r_n} \varphi_{nk}(t)$ and $\prod_{k=1}^{r_n} (1 - \frac{1}{2} t^2 \sigma_{nk}^2)$ differ by at most the sum in (27.12). This establishes the first of the asymptotic relations in (27.13).

Now (27.3) also implies that

$$\left| \prod_{k=1}^{r_n} e^{-t^2\sigma_{nk}^2/2} - \prod_{k=1}^{r_n} \left(1 - \frac{1}{2}t^2\sigma_{nk}^2\right) \right| \le \sum_{k=1}^{r_n} \left| e^{-t^2\sigma_{nk}^2/2} - 1 + \frac{1}{2}t^2\sigma_{nk}^2 \right|.$$

For real x,

$$(27.15) \qquad |e^x - 1 + x| \le \frac{1}{2}\sum_{v=2}^{\infty} |x|^v \le x^2 \qquad \text{if } |x| \le \tfrac{1}{2}.$$

Using this in the right member of the preceding inequality bounds it by $t^4 \Sigma_{k=1}^{r_n} \sigma_{nk}^4$; by (27.14) and (27.10) this sum goes to 0, from which the second equality in (27.13) follows. ■

Example 27.3. *Goncharov's Theorem.* Consider the sum $S_n = \Sigma_{k=1}^n X_{nk}$ in Example 6.4. Here S_n is the number of cycles in a random permutation, the X_{nk} are independent, and

$$P[X_{nk} = 1] = \frac{1}{n - k + 1} = 1 - P[X_{nk} = 0].$$

The mean m_n is $L_n = \Sigma_{k=1}^n k^{-1}$, and the variance s_n^2 is $L_n + O(1)$. Lindeberg's condition for $X_{nk} - (n - k + 1)^{-1}$ is easily verified because these random variables are bounded by 1.

The theorem gives $(S_n - L_n)/s_n \Rightarrow N$. Now, in fact, $L_n = \log n + O(1)$, and so (see Example 25.8) the sum can be renormalized: $(S_n - \log n)/\sqrt{\log n} \Rightarrow N$. ■

Suppose that the $|X_{nk}|^{2+\delta}$ are integrable for some positive δ and that *Lyapounov's condition*

$$(27.16) \qquad \lim_n \sum_{k=1}^{r_n} \frac{1}{s_n^{2+\delta}} E[|X_{nk}|^{2+\delta}] = 0$$

holds. Then Lindeberg's condition follows because the sum in (27.8) is bounded by

$$\sum_{k=1}^{r_n} \frac{1}{s_n^2} \int_{|X_{nk}| \ge \epsilon s_n} (|X_{nk}|^{2+\delta}/\epsilon^\delta s_n^\delta)\, dP \le \frac{1}{\epsilon^\delta}\sum_{k=1}^{r_n} \frac{1}{s_n^{2+\delta}} E[|X_{nk}|^{2+\delta}].$$

Hence Theorem 27.2 has this corollary:

Theorem 27.3. *Suppose that for each n the sequence $X_{n1}, \ldots, X_{nr_n}$ is independent and satisfies (27.7). If (27.16) holds for some positive δ, then $S_n/s_n \Rightarrow N$.*

Example 27.4. Elements are drawn from a population of size n, randomly and with replacement, until the number of distinct elements that have been sampled is r_n, where

$1 \leq r_n \leq n$. Let S_n be the drawing on which this first happens. A coupon collector requires S_n purchases to fill out a given portion of the complete set. Suppose that r_n varies with n in such a way that $r_n/n \to \rho$, $0 < \rho < 1$. What is the approximate distribution of S_n?

Let Y_p be the trial on which success first occurs in a Bernoulli sequence with probability p for success: $P[Y_p = k] = q^{k-1}p$, where $q = 1 - p$. Since the moment generating function is $pe^s/(1 - qe^s)$, $E[Y_p] = p^{-1}$ and $\text{Var}[Y_p] = qp^{-2}$. If $k - 1$ distinct items have thus far entered the sample, the waiting time until the next distinct one enters is distributed as Y_p for $p = (n - k + 1)/n$. Therefore, S_n can be represented as $\sum_{k=1}^{r_n} X_{nk}$ for independent summands, X_{nk} distributed as $Y_{(n-k+1)/n}$. Since $r_n \sim pn$, the mean and variance above give

$$m_n = E[S_n] = \sum_{k=1}^{r_n} \left(1 - \frac{k-1}{n}\right)^{-1} \sim n \int_0^\rho \frac{dx}{1-x}$$

and

$$s_n^2 = \sum_{k=1}^{r_n} \frac{k-1}{n} \left(1 - \frac{k-1}{n}\right)^{-2} \sim n \int_0^\rho \frac{x\,dx}{(1-x)^2}.$$

Lyapounov's theorem applies for $\delta = 2$, and to check (27.16) requires the inequality

$$(27.17) \qquad\qquad E[(Y_p - p^{-1})^4] \leq Kp^{-4}$$

for some K independent of p. (A calculation with the moment generating function shows that the left side is $qp^{-4}(1 + 7q + q^2)$, but the estimate will suffice.) Since $(Y_p - p^{-1})^4 \leq 8Y_p^4 + 8p^{-4}$ by Hölder's inequality, it is enough to estimate

$$(27.18) \quad E[Y_p^4] = \int_0^\infty P[Y_p \geq t^{1/4}]dt = 4p^{-4} \int_0^\infty P\left[Y_p \geq \frac{s}{p}\right]s^3\,ds.$$

If $s \geq 4p$, then $P[Y_p \geq s/p] = q^{\lceil s/p \rceil - 1} \leq q^{s/2p}$; since $p = 1 - q \leq -\log q$, $q^{s/2p} \leq e^{-s/2}$. This and (27.18) give (27.17).

It now follows that

$$\sum_{k=1}^{r_n} E\left[\left(X_{nk} - \frac{n}{n-k+1}\right)^4\right] \leq K \sum_{k=1}^{r_n} \left(1 - \frac{k-1}{n}\right)^{-4}$$

$$\sim Kn \int_0^\rho \frac{dx}{(1-x)^4}.$$

Since (27.16) follows, Theorem 27.3 applies: $(S_n - m_n)/s_n \Rightarrow N$. ∎

Feller's Theorem*

Consider a triangular array of normally distributed random variables; in this case of course $S_n/s_n \Rightarrow N$. Suppose further that $r_n = n$ and $X_{nk} = X_k$, and that $\sigma_{nk}^2 = \sigma_k^2$ is defined by $\sigma_1^2 = 1$ and the recursion $\sigma_n^2 = ns_{n-1}^2$, so that $s_n^2 \sim \sigma_n^2$. The sum in (27.8) is then at least

$$\frac{1}{s_n^2} \int_{|\sigma_n N| \geq \epsilon s_n} \sigma_n^2 N^2\,dP \to \int_{|N| \geq \epsilon} N^2\,dP > 0.$$

*This topic may be omitted.

Thus the Lindeberg condition is not necessary for the central limit theorem. (For another example, take $r_n \equiv 1$ and X_{n1} normal—but here the array does not arise from truncating a single infinite sequence.)

In this example, one of the summands dominates the others. If this sort of phenomenon is excluded, the Lindeberg condition becomes necessary as well as sufficient. Suppose that

$$(27.19) \qquad \max_{1 \leq k \leq r_n} \frac{\sigma_{nk}^2}{s_n^2} \to 0.$$

This is, in fact, a consequence of the Lindeberg condition—see (27.14), where s_n^2 is assumed 1. If $E[X_{nk}] = 0$, (27.19) and Chebyshev's inequality imply that

$$(27.20) \qquad \max_{1 \leq k \leq r_n} P\left[\left| \frac{X_{nk}}{s_n} \right| \geq \epsilon \right] \to 0$$

for positive ϵ. This condition ensures that no one summand contributes much to S_n; if it holds, the random variables X_{nk}/s_n form an *infinitesimal* array, or are *uniformly asymptotically negligible*. *Feller's theorem* is essentially the converse of Lindeberg's theorem:

Theorem 27.4. *Suppose that for each n the sequence $X_{n1}, \ldots, X_{nr_n}$ is independent and satisfies (27.7). If $S_n/s_n \Rightarrow N$ and (27.20) holds, then the Lindeberg condition (27.8) is satisfied.*

Since (27.8) implies (27.19), which in turn implies (27.20), Theorems 27.2 and 27.4 together show that (27.20) and the condition $S_n/s_n \Rightarrow N$ are jointly equivalent to the Lindeberg condition.

PROOF OF THE THEOREM.   As before, it is no restriction to assume that $s_n^2 = 1$.

By (2.64) for $n = 0$, $|\varphi_{nk}(t) - 1| \leq 2E[\min\{1, |tX_{nk}|\}]$. Splitting the expected value according as $|X_{nk}|$ exceeds ϵ or not shows that $|\varphi_{nk}(t) - 1| \leq 2P[|X_{nk}| \geq \epsilon] + 2\epsilon|t|$. By (27.20), $\max_{k \leq k_n} |\varphi_{nk}(t) - 1| \to 0$. By (26.4) for $n = 1$, $|\varphi_{nk}(t) - 1| \leq t^2 \sigma_{nk}^2$. Since the σ_{nk}^2 add to 1 and $\Sigma_k |\varphi_{nk}(t) - 1|^2 \leq \max_k |\varphi_{nk}(t) - 1| \cdot \Sigma_k |\varphi_{nk}(t) - 1|$, it follows that

$$(27.21) \qquad \sum_{k=1}^{r_n} |\varphi_{nk}(t) - 1|^2 \to 0$$

for each t.

If $|z| \leq 1$, then $|e^{z-1}| = e^{\mathrm{Re}z-1} \leq 1$. Hence (27.3) applies to $z_k = e^{\varphi_{nk}(t)-1}$ and $z_k' = \varphi_{nk}(t)$:

$$(27.22) \qquad \left| e^{\Sigma_k(\varphi_{nk}(t)-1)} - \prod_k \varphi_{nk}(t) \right| \leq \sum_k \left| e^{\varphi_{nk}(t)-1} - \varphi_{nk}(t) \right|.$$

If $|z| \leq \frac{1}{2}$, then $|e^z - 1 - z| \leq |z|^2$; for large n, $|\varphi_{nk}(t) - 1| \leq \frac{1}{2}$, and the right side of (27.22) is at most $\Sigma_k |\varphi_{nk}(t) - 1|^2$. Since $S_n \Rightarrow N$ by hypothesis, $\Pi_k \varphi_{nk}(t) \to e^{-t^2/2}$. By (27.21) and (27.22),

$$e^{\Sigma_k(\varphi_{nk}(t)-1)} \to e^{-t^2/2}.$$

Since $|e^z| = e^{\mathrm{Re}z}$, taking absolute values and then logarithms leads to

$$(27.23) \qquad \sum_{k=1}^{r_n} \mathrm{Re}(\varphi_{nk}(t) - 1) \to -\frac{1}{2}t^2.$$

The real part here is $E[\cos tX_{nk} - 1]$, and so (27.23) can be written

$$(27.24) \qquad \sum_{k=1}^{r_n} E\left[\cos tX_{nk} - 1 + \frac{1}{2}t^2 X_{nk}^2\right] \to 0.$$

Since $\cos x \geq 1 - \frac{1}{2}x^2$, the integrand is nonnegative and the sum not less than

$$\sum_{k=1}^{r_n} \int_{|X_{nk}| \geq \epsilon} \left(\frac{1}{2}t^2 X_{nk}^2 - 2\right) dP \geq \left(\frac{t^2}{2} - \frac{2}{\epsilon^2}\right) \sum_{k=1}^{r_n} \int_{|X_{nk}| \geq \epsilon} X_{nk}^2 \, dP.$$

Given ϵ, choose t so large that the first factor on the right is positive. By (27.24), the left side goes to 0, and hence so does the sum on the right. ∎

Dependent Variables*

The assumption of independence in the preceding theorems can be relaxed in various ways. Here a central limit theorem will be proved for sequences in which random variables far apart from one another are nearly independent in a sense to be defined.

For a sequence $X_1, X_2, \ldots$ of random variables, let α_n be a number such that

$$(27.25) \qquad |P(A \cap B) - P(A)P(B)| \leq \alpha_n$$

for $A \in \sigma(X_1, \ldots, X_k)$, $B \in \sigma(X_{k+n}, X_{k+n+1}, \ldots)$, and $k \geq 1$, $n \geq 1$. Suppose that $\alpha_n \to 0$, the idea being that X_k and X_{k+n} are then approximately independent for large n. In this case the sequence $\{X_n\}$ is said to be α-*mixing*. If the distribution of the random vector $(X_n, X_{n+1}, \ldots, X_{n+j})$ does not depend on n, the sequence is said to be *stationary*.

Example 27.5. Let $\{Y_n\}$ be a Markov chain with finite state space and positive transition probabilities p_{ij}, and suppose that $X_n = f(Y_n)$, where f is some real function on the state space. If the initial probabilities p_i are the stationary ones (see Theorem 8.7), then clearly $\{X_n\}$ is stationary. Moreover, by (8.37), $|p_{ij}^{(n)}|$

---

* This topic may be omitted.

$- p_j| \leq \rho^n$, where $\rho < 1$. By (8.11), $P[Y_1 = i_1, \ldots, Y_k = i_k, Y_{k+n} = j_0, \ldots, Y_{k+n+l} = j_l] = p_{i_1} p_{i_1 i_2} \cdots p_{i_{k-1} i_k} p_{i_k j_0}^{(n)} p_{j_0 j_1} \cdots p_{j_{l-1} j_l}$, which differs from $P[Y_1 = i_1, \ldots, Y_k = i_k] P[Y_{k+n} = j_0, \ldots, Y_{k+n+l} = j_l]$ by at most $p_{i_1} p_{i_1 i_2} \cdots p_{i_{k-1} i_k} \rho^n p_{j_0 j_1} \cdots p_{j_{l-1} j_l}$. It follows by addition that, if r is the number of states, then for sets of the form $A = [(Y_1, \ldots, Y_k) \in H]$ and $B = [(Y_{k+n}, \ldots, Y_{k+n+l}) \in H']$, (27.25) holds with $\alpha_n = r\rho^n$. These sets (for k and n fixed) form fields generating σ-fields which contain $\sigma(X_1, \ldots, X_k)$ and $\sigma(X_{k+n}, X_{k+n+1}, \ldots)$, respectively. For fixed A the set of B satisfying (27.25) is a monotone class, and similarly if A and B are interchanged. It follows by the monotone class theorem (Theorem 3.4) that $\{X_n\}$ is α-mixing with $\alpha_n = r\rho^n$. ■

The sequence is *m-dependent* if $(X_1, \ldots, X_k)$ and $(X_{k+n}, \ldots, X_{k+n+l})$ are independent whenever $n > m$. In this case the sequence is α-mixing with $\alpha_n = 0$ for $n > m$. In this terminology an independent sequence is 0-dependent.

Example 27.6. Let $Y_1, Y_2, \ldots$ be independent and identically distributed, and put $X_n = f(Y_n, \ldots, Y_{n+m})$ for a real function f on R^{m+1}. Then $\{X_n\}$ is stationary and m-dependent. ■

Theorem 27.5. *Suppose that $X_1, X_2, \ldots$ is stationary and α-mixing with $\alpha_n = O(n^{-5})$ and that $E[X_n] = 0$ and $E[X_n^{12}] < \infty$. If $S_n = X_1 + \cdots + X_n$, then*

$$(27.26) \qquad n^{-1} \operatorname{Var}[S_n] \to \sigma^2 = E[X_1^2] + 2 \sum_{k=1}^{\infty} E[X_1 X_{1+k}],$$

where the series converges absolutely. If $\sigma > 0$, then $S_n/\sigma\sqrt{n} \Rightarrow N$.

The conditions $\alpha_n = O(n^{-5})$ and $E[X_n^{12}] < \infty$ are stronger than necessary; they are imposed to avoid technical complications in the proof. The idea of the proof, which goes back to Markov, is this: Split the sum $X_1 + \cdots + X_n$ into alternate blocks of length b_n (the big blocks) and l_n (the little blocks). Namely, let

$$(27.27) \quad U_{ni} = X_{(i-1)(b_n+l_n)+1} + \cdots + X_{(i-1)(b_n+l_n)+b_n}, \qquad 1 \leq i \leq r_n,$$

where r_n is the largest integer i for which $(i-1)(b_n + l_n) + b_n < n$. Further, let

$$(27.28) \quad \begin{cases} V_{ni} = X_{(i-1)(b_n+l_n)+b_n+1} + \cdots + X_{i(b_n+l_n)}, & 1 \leq i < r_n, \\ V_{nr_n} = X_{(r_n-1)(b_n+l_n)+b_n+1} + \cdots + X_n. \end{cases}$$

Then $S_n = \sum_{i=1}^{r_n} U_{ni} + \sum_{i=1}^{r_n} V_{ni}$, and the technique will be to choose the l_n small enough that $\sum_i V_{ni}$ is small in comparison with $\sum_i U_{ni}$ but large enough that the

U_{ni} are nearly independent, so that Lyapounov's theorem can be adapted to prove $\Sigma_i U_{ni}$ asymptotically normal.

Lemma 2. *If Y is measurable $\sigma(X_1, \ldots, X_k)$ and bounded by C, and if Z is measurable $\sigma(X_{k+n}, X_{k+n+1}, \ldots)$ and bounded by D, then*

$$(27.29) \qquad |E[YZ] - E[Y]E[Z]| \le 4CD\alpha_n.$$

PROOF.   It is no restriction to take $C = D = 1$ and (by the usual approximation method) to take $Y = \Sigma_i y_i I_{A_i}$ and $Z = \Sigma_j z_j I_{B_j}$ simple ($|y_i|, |z_j| \le 1$). If $d_{ij} = P(A_i \cap B_j) - P(A_i)P(B_j)$, the left side of (27.29) is $|\Sigma_i y_i \Sigma_j z_j d_{ij}|$; if ξ_i is $+1$ or -1 as $\Sigma_j z_j d_{ij}$ is positive or not, this is at most $\Sigma_{ij} \xi_i z_j d_{ij}$. This last sum, in turn, is at most $\Sigma_{ij} \xi_i \eta_j d_{ij}$, where $\eta_j = \pm 1$. If $A^{(0)}$ is the union of the A_i for which $\xi_i = +1$ and $A^{(1)} = \Omega - A^{(0)}$, and if $B^{(0)}$ and $B^{(1)}$ are defined similarly in terms of the η_j, then $\Sigma_{ij} \xi_i \eta_j d_{ij} \le \Sigma_{uv} |P(A^{(u)} \cap B^{(v)}) - P(A^{(u)})P(B^{(v)})| \le 4\alpha_n.$ ∎

Lemma 3. *If Y is measurable $\sigma(X_1, \ldots, X_k)$ and $E[Y^4] \le C$, and if Z is measurable $\sigma(X_{k+n}, X_{k+n+1}, \ldots)$ and $E[Z^4] \le D$, then*

$$(27.30) \qquad |E[YZ] - E[Y]E[Z]| \le 8(1 + C + D)\alpha_n^{1/2}.$$

PROOF.   Let $Y_0 = YI_{[|Y| \le N]}$, $Y_1 = YI_{[|Y| > N]}$, $Z_0 = ZI_{[|Z| \le N]}$, $Z_1 = ZI_{[|Z| > N]}$. By Lemma 2, $|E[Y_0 Z_0] - E[Y_0]E[Z_0]| \le 4N^2 \alpha_n$. Further, $|E[Y_0 Z_1] - E[Y_0]E[Z_1]| \le E[|Y_0 - E[Y_0]| \cdot |Z_1 - E[Z_1]|] \le 2N \cdot 2E[|Z_1|] \le 4NE[|Z_1| \cdot |Z_1/N|^3] \le 4D/N^2$. Similarly, $|E[Y_1 Z_0] - E[Y_1]E[Z_0]| \le 4C/N^2$. Finally, $|E[Y_1 Z_1] - E[Y_1]E[Z_1]| \le \mathrm{Var}^{1/2}[Y_1]\mathrm{Var}^{1/2}[Z_1] \le E^{1/2}[Y_1^2]E^{1/2}[Z_1^2] \le E^{1/2}[Y_1^4/N^2]E^{1/2}[Z_1^4/N^2] \le C^{1/2}D^{1/2}/N^2$. Adding these inequalities gives $4N^2 \alpha_n + 4(C + D)N^{-2} + C^{1/2}D^{1/2}N^{-2}$ as a bound for the left side of (27.30). Take $N = \alpha_n^{-1/4}$ and observe that $4 + 4(C + D) + C^{1/2}D^{1/2} \le 4 + 4(C^{1/2} + D^{1/2})^2 \le 4 + 8(C + D)$. ∎

PROOF OF THEOREM 27.5.   By Lemma 3, $|E[X_1 X_{1+k}]| \le 8(1 + 2E[X_1^4])\alpha_n^{1/2} = O(n^{-5/2})$, and so the series in (27.26) converges absolutely. If $\rho_k = E[X_1 X_{1+k}]$, then by stationarity $E[S_n^2] = n\rho_0 + 2\Sigma_{k=1}^{n-1} (n - k)\rho_k$ and therefore $|\sigma^2 - n^{-1}E[S_n^2]| \le 2\Sigma_{k=n}^{\infty} |\rho_k| + 2n^{-1}\Sigma_{i=1}^{n-1} \Sigma_{k=i}^{\infty} |\rho_k|$; hence (27.26).

By stationarity again,

$$E[S_n^4] \le 4! n\Sigma |E[X_1 X_{1+i} X_{1+i+j} X_{1+i+j+k}]|,$$

where the indices in the sum are constrained by $i, j, k \ge 0$ and $i + j + k < n$. By Lemma 3 the summand is at most $8(1 + E[X_1^4] + E[X_{1+i}^4 X_{1+i+j}^4 X_{1+i+j+k}^4])\alpha_i^{1/2}$, which is at most* $8(1 + E[X_1^4] + E[X_1^{12}])\alpha_i^{1/2} = K_1 \alpha_i^{1/2}$. Similarly, $K_1 \alpha_k^{1/2}$ is a bound. Hence

* $E|XYZ| \le E^{1/3}|X|^3 \cdot E^{2/3}|YZ|^{3/2} \le E^{1/3}|X|^3 \cdot E^{1/3}|Y|^3 \cdot E^{1/3}|Z|^3$.

$$E[S_n^4] \le 4!n^2 \sum_{\substack{i,k \ge 0 \\ i+k<n}} K_1 \min\{\alpha_i^{1/2}, \alpha_k^{1/2}\}$$

$$\le K_2 n^2 \sum_{0 \le i \le k} \alpha_k^{1/2} = K_2 n^2 \sum_{k=0}^{\infty} (k+1)\alpha_k^{1/2}.$$

Since $\alpha_k = O(k^{-5})$, the series here converges, and therefore

(27.31) $E[S_n^4] \le Kn^2$

for some K independent of n.

Let $b_n = [n^{3/4}]$ and $l_n = [n^{1/4}]$. If r_n is the largest integer i such that $(i-1)(b_n + l_n) + b_n < n$, then

(27.32) $b_n \sim n^{3/4}, \qquad l_n \sim n^{1/4}, \qquad r_n \sim n^{1/4}.$

Consider the random variables (27.27) and (27.28). By (27.31), (27.32), and stationarity,

$$P\left[\left|\frac{1}{\sigma\sqrt{n}} \sum_{i=1}^{r_n-1} V_{ni}\right| \ge \epsilon\right] \le \sum_{i=1}^{r_n-1} P\left[|V_{ni}| \ge \frac{\epsilon\sigma\sqrt{n}}{r_n}\right]$$

$$\le \frac{r_n^4}{\epsilon^4\sigma^4 n^2} r_n K l_n^2 \sim \frac{K}{\epsilon^4\sigma^4 n^{1/4}} \to 0;$$

(27.31) and (27.32) also give

$$P\left[\frac{1}{\sigma\sqrt{n}} |V_{nr_n}| \ge \epsilon\right] \le \frac{K(b_n+l_n)^2}{\epsilon^4\sigma^4 n^2} \sim \frac{K}{\epsilon^4\sigma^4 n^{1/2}} \to 0.$$

Therefore, $\Sigma_{i=1}^{r_n} V_{ni}/\sigma\sqrt{n} \Rightarrow 0$, and by Theorem 25.4 it suffices to prove that $\Sigma_{i=1}^{r_n} U_{ni}/\sigma\sqrt{n} \Rightarrow N$.

Let U'_{ni}, $1 \le i \le r_n$, be independent random variables having the distribution common to the U_{ni}. By Lemma 2 extended inductively the characteristic functions of $\Sigma_{i=1}^{r_n} U_{ni}/\sigma\sqrt{n}$ and of $\Sigma_{i=1}^{r_n} U'_{ni}/\sigma\sqrt{n}$ differ by at most* $16 r_n \alpha_{l_n}$. Since $\alpha_n = O(n^{-5})$, this difference is $O(n^{-1})$ by (27.32).

The characteristic function of $\Sigma_{i=1}^{r_n} U_{ni}/\sigma\sqrt{n}$ will thus approach $e^{-t^2/2}$ if that of $\Sigma_{i=1}^{r_n} U'_{ni}/\sigma\sqrt{n}$ does. It therefore remains only to show that $\Sigma_{i=1}^{r_n} U'_{ni}/\sigma\sqrt{n} \Rightarrow N$. Now $E[|U'_{ni}|^2] = E[U_{n1}^2] \sim b_n\sigma^2$ by (27.26). Further, $E[|U'_{ni}|^4] \le Kb_n^2$ by (27.31). Lyapounov's condition (27.16) for $\delta = 2$ therefore follows because

$$\frac{r_n E[|U'_{n1}|^4]}{(r_n E[|U'_{n1}|^2])^2} \sim \frac{E[|U'_{n1}|^4]}{r_n b_n^2 \sigma^4} \le \frac{K}{r_n \sigma^4} \to 0. \qquad \blacksquare$$

Example 27.7.   Let $\{Y_n\}$ be the stationary Markov process of Example 27.5. Let f be a function on the state space, put $m = \Sigma_i p_i f(i)$, and define $X_n = f(Y_n) - m$. Then $\{X_n\}$ satisfies the conditions of Theorem 27.5. If $\beta_{ij} = \delta_{ij}p_i - p_i p_j +$

---

* The 4 in (27.29) has become 16 to allow for splitting into real and imaginary parts.

$2p_i \sum_{k=1}^{\infty} (p_{ij}^{(k)} - p_j)$, then the σ^2 in (27.26) is $\sum_{ij} \beta_{ij} (f(i) - m)(f(j) - m)$, and $\sum_{k=1}^{n} f(Y_k)$ is approximately normally distributed with mean nm and standard deviation $\sigma \sqrt{n}$.

If $f(i) = \delta_{i_0 i}$, then $\sum_{k=1}^{n} f(Y_k)$ is the number of passages through the state i_0 in the first n steps of the process. In this case $m = p_{i_0}$ and $\sigma^2 = p_{i_0}(1 - p_{i_0}) + 2p_{i_0} \sum_{k=1}^{\infty} (p_{i_0 i_0}^{(k)} - p_{i_0})$. ∎

Example 27.8. If the X_n are stationary and m-dependent and have mean 0, Theorem 27.5 applies and $\sigma^2 = E[X_1^2] + 2\sum_{k=1}^{m} E[X_1 X_{1+k}]$. Example 27.6 is a case in point. Taking $m = 1$ and $f(x, y) = x - y$ in that example gives an instance where $\sigma^2 = 0$. ∎

PROBLEMS

27.1. Suppose that $X_{n1}, \ldots, X_{nr_n}$ are independent and take on only the values 0 and 1, and put $p_{nk} = P[X_{nk} = 1]$. If $\sum_{k=1}^{r_n} p_{nk} \to \lambda$ and $\max_{k \le r_n} p_{nk} \to 0$, then S_n converges in distribution to the Poisson distribution with parameter λ.

Hint: Use (27.3) to compare the characteristic function of S_n with $\exp[\sum_k p_{nk}(e^{it} - 1)]$.

27.2. If $\{X_n\}$ is independent and the X_n all have the same distribution with finite first moment, then $n^{-1}S_n \to E[X_1]$ with probability 1 (Theorem 22.5), so that $n^{-1}S_n \Rightarrow E[X_1]$. Prove the latter fact by characteristic functions.

Hint: Use (27.3).

27.3. For a Poisson variable Y_λ with mean λ, show that $(Y_\lambda - \lambda)/\sqrt{\lambda} \Rightarrow N$. Show that (22.4) fails for $t = 1$.

27.4. Suppose that $|X_{nk}| \le M_n$ with probability 1 and $M_n/s_n \to 0$. Verify Lyapounov's condition and then Lindeberg's condition.

27.5. Suppose that the random variables in any single row of the triangular array are identically distributed. To what do Lindeberg's and Lyapounov's conditions reduce?

27.6. Suppose that $Z_1, Z_2, \ldots$ are independent and identically distributed with mean 0 and variance 1 and suppose that $X_{nk} = \sigma_{nk} Z_k$. Write down the Lindeberg condition and show that it holds if $\max_{k \le r_n} \sigma_{nk}^2 = o(\sum_{k=1}^{r_n} \sigma_{nk}^2)$.

27.7. Construct an example where Lindeberg's condition holds but Lyapounov's does not.

27.8. 22.17↑ Prove a central limit theorem for the number R_n of records up to time n.

27.9. 6.3↑ Let S_n be the number of inversions in a random permutation on n letters. Prove a central limit theorem for S_n.

27.10. *The δ-method.* Suppose that Theorem 27.1 applies to $\{X_n\}$, so that $\sqrt{n}\,\sigma^{-1}(\overline{X}_n - c) \Rightarrow N$, where $\overline{X}_n = n^{-1}\Sigma_{k=1}^n X_k$. Use Theorem 25.6 as in Example 27.2 to show that, if $f(x)$ has a nonzero derivative at c, then $\sqrt{n}(f(\overline{X}_n) - f(c))/\sigma f'(c) \Rightarrow N$. Example 27.2 is the case $f(x) = 1/x$.

27.11. ↑ The mean volume of grains of sand can be estimated by measuring the amount of water a sample of them displaces. Use Problem 27.10 to find a method for estimating the mean diameter of the grains and analyze the variance of the estimator.

27.12. 22.8↑ *The three-series theorem.* Suppose that $X_1, X_2, \ldots$ are independent and $\Sigma_n X_n$ converges with probability 1. Put $Y_n^{(\alpha)} = X_n I_{[|X_n| \le \alpha]}$, and show by the following steps that for each positive α the three series

$$\sum_n P[|X_n| > \alpha], \qquad \sum_n E[Y_n^{(\alpha)}], \qquad \sum_n \text{Var}\,[Y_n^{(\alpha)}]$$

converge. Put $Z_n^{(\alpha)} = X_n - Y_n^{(\alpha)}$.

(a) From the convergence of $\Sigma_n X_n$ conclude that $\Sigma_n P[|X_n| > \alpha] < \infty$.

Hint: Use the second Borel-Cantelli lemma.

(b) Show that $\Sigma_n Z_n^{(\alpha)}$ converges with probability 1.

Hint: By part (a), $\Sigma_n P[Z_n^{(\alpha)} \ne 0] = \Sigma_n P[|X_n| > \alpha] < \infty$.

(c) Show that $\Sigma_n Y_n^{(\alpha)}$ converges with probability 1.

(d) Let $S_n^{(\alpha)} = \Sigma_{k=1}^n Y_k^{(\alpha)}$, $M_n^{(\alpha)} = E[S_n^{(\alpha)}]$, and $s_n^{(\alpha)} = \text{Var}^{1/2}[S_n^{(\alpha)}]$. Suppose that $s_n^{(\alpha)} \to \infty$ and draw the contradictory conclusions that $(S_n^{(\alpha)} - M_n^{(\alpha)})/s_n^{(\alpha)} \Rightarrow N$ and that $S_n^{(\alpha)}/s_n^{(\alpha)} \to 0$ with probability 1.

Hint: For the first, use the central limit theorem (Problem 27.4); for the second use the fact that $S_n^{(\alpha)}$ converges with probability 1.

(e) Let $m_n^{(\alpha)} = E[Y_n^{(\alpha)}]$. Show that $\Sigma_n (Y_n^{(\alpha)} - m_n^{(\alpha)})$ converges with probability 1 and hence that $\Sigma_n m_n^{(\alpha)}$ converges.

Hint: Use Theorem 22.3 and part (c).

27.13. Let $d_n(\omega)$ be the dyadic digits of a point ω drawn at random from the unit interval. For a k-tuple $(u_1, \ldots, u_k)$ of 0's and 1's, let $N_n(u_1, \ldots, u_k; \omega)$ be the number of $m \le n$ for which $(d_m(\omega), \ldots, d_{m+k-1}(\omega)) = (u_1, \ldots, u_k)$. Prove a central limit theorem for $N_n(u_1, \ldots, u_k; \omega)$. (See Problem 6.12.)

27.14. *The central limit theorem for a random number of summands.* Let $X_1, X_2, \ldots$ be independent, identically distributed random variables with mean 0 and variance σ^2, and let $S_n = X_1 + \cdots + X_n$. For each positive t, let ν_t be a random variable assuming positive integers as values; it need not be independent of the X_n. Suppose that there exist positive constants a_t and θ such that

$$a_t \to \infty, \qquad \frac{\nu_t}{a_t} \Rightarrow \theta$$

as $t \to \infty$. Show by the following steps that

(27.33) $$\frac{S_{\nu_t}}{\sigma\sqrt{\nu_t}} \Rightarrow N, \qquad \frac{S_{\nu_t}}{\sigma\sqrt{\theta a_t}} \Rightarrow N.$$

(a) Show that it may be assumed that $\theta = 1$ and the a_t are integers.

(b) Show that it suffices to prove the second relation in (27.33).

(c) Show that it suffices to prove $(S_{\nu_t} - S_{a_t})/\sqrt{a_t} \Rightarrow 0$.

(d) Show that

$$P[|S_{\nu_t} - S_{a_t}| \geq \epsilon\sqrt{a_t}] \leq P[|\nu_t - a_t| \geq \epsilon^3 a_t]$$

$$+ P[\max_{|k - a_t| \leq \epsilon^3 a_t} |S_k - S_{a_t}| \geq \epsilon\sqrt{a_t}].$$

and conclude from Kolmogorov's inequality that the last probability is at most $2\epsilon\sigma^2$.

27.15. 21.14 23.12 27.14↑ *A central limit theorem in renewal theory.* Let $X_1, X_2,$ ... be independent, identically distributed postive random varaiables with mean m and variance σ^2, and as in Problem 23.12 let N_t be the maximum n for which $S_n \leq t$. Prove by the following steps that

$$\frac{N_t - tm^{-1}}{\sigma t^{1/2} m^{-3/2}} \Rightarrow N.$$

(a) Show by the results in Problems 21.14 and 23.12 that $(S_{N_t} - t)/\sqrt{t} \Rightarrow 0$.

(b) Show that it suffices to prove that

$$\frac{N_t - S_{N_t} m^{-1}}{\sigma t^{1/2} m^{-3/2}} = \frac{-(S_{N_t} - mN_t)}{\sigma t^{1/2} m^{-1/2}} \Rightarrow N.$$

(c) Show (Problem 23.12) that $N_t/t \Rightarrow m^{-1}$ and apply the theorem in Problem 27.14.

27.16. Show by partial integration that

(27.34) $$\frac{1}{\sqrt{2\pi}} \int_x^\infty e^{-u^2/2}\, du \sim \frac{1}{\sqrt{2\pi}} \frac{1}{x} e^{-x^2/2}$$

as $x \to \infty$.

27.17. ↑ Suppose that, $X_1, X_2, \ldots$ are independent and identically distributed with mean 0 and variance 1, and suppose that $a_n \to \infty$. A *formal* combination of the central limit theorem and (27.34) gives

(27.35) $$P[S_n \geq a_n\sqrt{n}] \sim \frac{1}{\sqrt{2\pi}} \frac{1}{a_n} e^{-a_n^2/2} = e^{-a_n^2(1 + \zeta_n)/2},$$

where $\zeta_n \to 0$ if $a_n \to \infty$. Compare Theorem 9.4 and Problem 9.3.

27.18. 21.2↑ *Stirling's formula.* Let $S_n = X_1 + \cdots + X_n$, where the X_n are independent and each has the Poisson distribution with parameter 1. Prove successively:

(a) $$E\left[\left(\frac{S_n - n}{\sqrt{n}}\right)^-\right] = e^{-n} \sum_{k=0}^n \left(\frac{n - k}{\sqrt{n}}\right) \frac{n^k}{k!} = \frac{n^{n+(1/2)} e^{-n}}{n!}$$

(b) $\left(\dfrac{S_n - n}{\sqrt{n}}\right)^- \Rightarrow N^-$

(c) $E\left[\left(\dfrac{S_n - n}{\sqrt{n}}\right)^-\right] \to E[N^-] = \dfrac{1}{\sqrt{2\pi}}$

(d) $n! \sim \sqrt{2\pi}\, n^{n+(1/2)} e^{-n}$

27.19. Let $l_n(\omega)$ be the length of the run of 0's starting at the nth place in the dyadic expansion of a point ω drawn at random from the unit interval; see Example 4.1.

  (a) Show that $l_1, l_2, \ldots$ is an α-mixing sequence, where $\alpha_n = 4/2^n$.

  (b) Show that $\sum_{k=1}^{n} l_k(\omega)$ is approximately normally distributed with mean n and variance $6n$.

SECTION 28. INFINITELY DIVISIBLE DISTRIBUTIONS*

Suppose that Z_λ has the Poisson distribution with parameter λ and that X_{n1}, $\ldots$, X_{nn} are independent and $P[X_{nk} = 1] = \lambda/n$, $P[X_{nk} = 0] = 1 - \lambda/n$. According to Example 25.2, $X_{n1} + \cdots + X_{nn} \Rightarrow Z_\lambda$. This contrasts with the central limit theorem, in which the limit law is normal. What is the class of all possible limit laws for independent triangular arrays? A suitably restricted form of this question will be answered here.

Vague Convergence

The theory requires two preliminary facts about convergence of measures. Let μ_n and μ be finite measures on $(R^1, \mathcal{R}^1)$. If $\mu_n(a, b] \to \mu(a, b]$ for every finite interval for which $\mu\{a\} = \mu\{b\} = 0$, then μ_n *converges vaguely* to μ, written $\mu_n \to_v \mu$. If μ_n and μ are probability measures, it is not hard to see that this is equivalent to weak convergence: $\mu_n \Rightarrow \mu$. On the other hand, if μ_n is a unit mass at n and $\mu(R^1) = 0$, then $\mu_n \to_v \mu$, but $\mu_n \Rightarrow \mu$ makes no sense because μ is not a probability measure.

  The first fact needed is this: *Suppose that $\mu_n \to_v \mu$ and*

(28.1) $\sup_n \mu_n(R^1) < \infty$;

then

(28.2) $$\int f\, d\mu_n \to \int f\, d\mu$$

for every continuous real f that vanishes at $\pm\infty$ in the sense that $\lim_{|x| \to \infty} f(x) = 0$. Indeed, choose M so that $\mu(R^1) < M$ and $\mu_n(R^1) < M$ for all n. Given ϵ, choose a and b so that $\mu\{a\} = \mu\{b\} = 0$ and $|f(x)| < \epsilon/M$ if $x \notin A = (a, b]$. Then $|\int_{A^c} f\, d\mu_n| < \epsilon$ and $|\int_{A^c} f\, d\mu| < \epsilon$. If $\mu(A) > 0$, define $\nu(B) = \mu(B \cap A)/\mu(A)$ and $\nu_n(B) = \mu_n(B \cap A)/\mu_n(A)$. It is easy to see that $\nu_n \Rightarrow \nu$, so that $\int f\, d\nu_n \to \int f\, d\nu$. But then $|\int_A f\, d\mu_n - \int_A f\, d\mu| < \epsilon$ for large n, and hence $|\int f\, d\mu_n - \int f\, d\mu| < 3\epsilon$ for large n. If $\mu(A) = 0$, then $\int_A f\, d\mu_n \to 0$, and the argument is even simpler.

* This section may be omitted.

The other fact needed below is this: *If* (28.1) *holds, then there is a subsequence* $\{\mu_{n_k}\}$ *and a finite measure* μ *such that* $\mu_{n_k} \to_v \mu$ *as* $k \to \infty$. Indeed, let $F_n(x) = \mu_n(-\infty, x]$. Since the F_n are uniformly bounded because of (28.1), the proof of Helley's theorem shows there exists a subsequence $\{F_{n_k}\}$ and a bounded, nondecreasing, right-continuous function F such that $\lim_k F_{n_k}(x) = F(x)$ at continuity points x of F. If μ is the measure for which $\mu(a, b] = F(b) - F(a)$ (Theorem 12.4), then clearly $\mu_{n_k} \to_v \mu$.

The Possible Limits

Let $X_{n1}, \ldots, X_{nr_n}$, $n = 1, 2, \ldots$, be a triangular array as in the preceding section. The random variables in each row are independent, and

$$(28.3) \qquad E[X_{nk}] = 0, \qquad \sigma_{nk}^2 = E[X_{nk}^2], \qquad s_n^2 = \sum_{k=1}^{r_n} \sigma_{nk}^2.$$

Put $S_n = X_{n1} + \cdots + X_{nr_n}$. Here it will be assumed that the total variance is bounded:

$$(28.4) \qquad \qquad \sup_n s_n^2 < \infty.$$

In order that the X_{nk} be small compared with S_n, assume that

$$(28.5) \qquad \qquad \lim_n \max_{k \leq r_n} \sigma_{nk}^2 = 0.$$

The arrays in the preceding section were normalized by replacing X_{nk} by X_{nk}/s_n. This has the effect of replacing s_n by 1, in which case of course (28.4) holds, and (28.5) is the same thing as (27.19).

A distribution function F is *infinitely divisible* if for each n there is a distribution function F_n such that F is the n-fold convolution $F_n * \cdots * F_n$ (n copies) of F_n. The class of possible limit laws will turn out to consist of the infinitely divisible distributions with mean 0 and finite variance.*

Theorem 28.1. *Suppose that*

$$(28.6) \qquad \varphi(t) = \exp \int_{R^1} (e^{itx} - 1 - itx) \frac{1}{x^2} \mu(dx),$$

where μ *is a finite measure. Then* φ *is the characteristic function of an infinitely divisible distribution with mean 0 and variance* $\mu(R^1)$.

By (26.4) for $n = 2$, the integrand in (28.6) converges to $-t^2/2$ as $x \to 0$; take this as its value at $x = 0$. By (26.4) for $n = 1$, the integrand is at most $t^2/2$ in modulus and so is integrable.

The formula (28.6) is the *canonical representation* of φ, and μ is the *canonical measure*.

---

* There do exist infinitely divisible distributions without moments (see Problems 28.3 and 28.4), but they do not figure in the theory of this section.

Before proceeding to the proof, consider three examples.

Example 28.1. If μ consists of a mass of σ^2 at the origin, (28.6) is $e^{-\sigma^2 t^2/2}$, the characteristic function of a centered normal distribution F. It is certainly infinitely divisible—take F_n normal wtih variance σ^2/n.   ∎

Example 28.2. The Poisson distribution is infinitely divisible: divide the parameter λ by n. If Z_λ has this distribution, its characteristic function is exp $\lambda(e^{it} - 1)$ and hence $E[\exp(itx(Z_\lambda - \lambda))] = \exp \lambda(e^{itx} - 1 - itx)$. If μ consists of a mass of $x^2 \lambda$ at $x \neq 0$, (28.6) has this form and hence is the characteristic function of $x(Z_\lambda - \lambda)$. It is therefore infinitely divisible.   ∎

Example 28.3. If $\varphi_j(t)$ is given by (28.6) with μ_j for the measure, and if $\mu = \Sigma_{j=1}^k \mu_j$, then (28.6) is $\varphi_1(t) \cdots \varphi_k(t)$. It follows by the preceding two examples that (28.6) is a characteristic function if μ consists of finitely many point masses. It is easy to check in the preceding two examples that the distribution corresponding to $\varphi(t)$ has mean 0 and variance $\mu(R^1)$, and since the means and variances add, the same must be true in the present example.   ∎

PROOF OF THEOREM 28.1. Let μ_k have mass $\mu(j2^{-k}, (j + 1)2^{-k}]$ at $j2^{-k}$ for $j = 0, \pm 1, \ldots, \pm 2^{2k}$. Then $\mu_k \to_v \mu$. As observed in Example 28.3, if $\varphi_k(t)$ is (28.6) with μ_k in place of μ, then φ_k is a characteristic function. For each t the integrand in (28.6) vanishes at $\pm\infty$; since $\sup_k \mu_k(R^1) < \infty$, $\varphi_k(t) \to \varphi(t)$ follows (see (28.2)). Therefore, $\varphi(t)$, as a limit of characteristic functions, is itself a characteristic function. Further, the distribution corresponding to $\varphi_k(t)$ has second moment $\mu_k(R^1)$, and since this is bounded, it follows (Theorem 25.11) that the distribution corresponding to $\varphi(t)$ has a finite second moment. Differentiation (use Theorem 16.8) shows that the mean is $\varphi'(0) = 0$ and the variance is $- \varphi''(0) = \mu(R^1)$. Thus (28.6) is always the characteristic function of a distribution with mean 0 and variance $\mu(R^1)$.

If $\varphi_n(t)$ is (28.6) with μ/n in place of μ, then $\varphi(t) = \varphi_n^n(t)$, so that the distribution corresponding to $\varphi(t)$ is indeed infinitely divisible.   ∎

The representation (28.6) shows that the normal and Poisson distributions are special cases in a very large class of infinitely divisible laws.

Theorem 28.2. *Every infinitely divisible distribution with mean 0 and finite variance is the limit law of S_n for some independent triangular array satisfying* (28.3), (28.4), *and* (28.5).

The proof requires this preliminary result:

Lemma. *If X and Y are independent and $X + Y$ has a second moment, then X and Y have second moments as well.*

PROOF. Since $X^2 + Y^2 \leq (X + Y)^2 + 2|XY|$, it suffices to prove $|XY|$ integrable, and by Fubini's theorem applied to the joint distribution of X and Y it suffices to prove $|X|$ and $|Y|$ individually integrable. Since $|Y| \leq |x| + |x + Y|$, $E[|Y|] = \infty$ would imply $E[|x + Y|] = \infty$ for each x; by Fubini's theorem again $E[|Y|] = \infty$ would therefore imply $E[|X + Y|] = \infty$, which is impossible. Hence $E[|Y|] < \infty$, and similarly $E[|X|] < \infty$. ∎

PROOF OF THEOREM 28.2. Let F be infinitely divisible with mean 0 and variance σ^2. If F is the n-fold convolution of F_n, then by the lemma extended inductively F_n has finite mean and variance and these must be 0 and σ^2/n. Take $r_n = n$ and take $X_{n1}, \ldots, X_{nn}$ independent, each with distribution function F_n. ∎

Theorem 28.3. *If F is the limit law of S_n for an independent triangular array satisfying (28.3), (28.4), and (28.5), then F has characteristic function of the form (28.6) for some finite measure μ.*

PROOF. The proof will yield information making it possible to identify the limit. Let $\varphi_{nk}(t)$ be the characteristic function of X_{nk}. The first step is to prove that

$$(28.7) \qquad \prod_{k=1}^{r_n} \varphi_{nk}(t) - \exp \sum_{k=1}^{n} (\varphi_{nk}(t) - 1) \to 0$$

for each t. Since $|z| \leq 1$ implies that $|e^{z-1}| = e^{\operatorname{Re} z - 1} \leq 1$, it follows by (27.3) that the difference $\delta_n(t)$ in (28.7) satisfies $|\delta_n(t)| \leq \sum_{k=1}^{r_n} |\varphi_{nk}(t) - \exp(\varphi_{nk}(t) - 1)|$. Fix t. If $\varphi_{nk}(t) - 1 = \theta_{nk}$, then $|\theta_{nk}| \leq \sigma_{nk}^2/2$, and it follows by (28.4) and (28.5) that $\max_k |\theta_{nk}| \to 0$ and $\sum_k |\theta_{nk}| = O(1)$. Therefore, for sufficiently large n, $|\delta_n(t)| \leq \sum_k |1 + \theta_{nk} - e^{\theta_{nk}}| \leq \sum_k |\theta_{nk}|^2 \leq \max_k |\theta_{nk}| \cdot \sum_k |\theta_{nk}|$. Hence (28.7).

If F_{nk} is the distribution function of X_{nk}, then

$$\sum_{k=1}^{r_n} (\varphi_{nk}(t) - 1) = \sum_{k=1}^{r_n} \int_{R^1} (e^{itx} - 1) \, dF_{nk}(x)$$

$$= \sum_{k=1}^{r_n} \int_{R^1} (e^{itx} - 1 - itx) \, dF_{nk}(x).$$

Let μ_n be the finite measure satisfying

$$(28.8) \qquad \mu_n(-\infty, x] = \sum_{k=1}^{r_n} \int_{y \leq x} y^2 \, dF_{nk}(y),$$

and put

(28.9) $\varphi_n(t) = \exp \int_{R^1} (e^{itx} - 1 - itx) \frac{1}{x^2} \mu_n(dx).$

Then (28.7) can be written

(28.10) $\prod_{k=1}^{r_n} \varphi_{nk}(t) - \varphi_n(t) \to 0.$

By (28.8), $\mu_n(R^1) = s_n^2$, and this is bounded by assumption. Thus (28.1) holds, and some subsequence $\{\mu_{n_k}\}$ converges vaguely to a finite measure μ. Since the integrand in (28.9) vanishes at $\pm\infty$, $\varphi_{n_k}(t)$ converges to (28.6). But, of course, $\lim_n \varphi_n(t)$ must coincide with the characteristic function of the limit law F, which exists by hypothesis. Thus F must have characteristic function of the form (28.6). ■

Theorems 28.1, 28.2, and 28.3 together show that the possible limit laws are exactly the infinitely divisible distributions with mean 0 and finite variance, and they give explicitly the form the characteristic functions of such laws must have.

Characterizing the Limit

Theorem 28.4. *Suppose that F has characteristic function* (28.6) *and that an independent triangular array satisfies* (28.3), (28.4), *and* (28.5). *Then S_n has limit law F if and only if $\mu_n \to_v \mu$, where μ_n is defined by* (28.8).

PROOF.   Since (28.7) holds as before, S_n has limit law F if and only if $\varphi_n(t)$ (defined by (28.9)) converges for each t to $\varphi(t)$ (defined by (28.6)). If $\mu_n \to_v \mu$, then $\varphi_n(t) \to \varphi(t)$ follows because the integrand in (28.9) and (28.6) vanishes at $\pm\infty$ and because (28.1) follows from (28.4).

Now suppose that $\varphi_n(t) \to \varphi(t)$. Since $\mu_n(R^1) = s_n^2$ is bounded, each subsequence $\{\mu_{n_k}\}$ contains a further subsequence $\{\mu_{n_{k(j)}}\}$ converging vaguely to some ν. If it can be shown that ν necessarily coincides with μ, it will follow by the usual argument that $\mu \to_v \mu$. But by the definition (28.9) of $\varphi_n(t)$, it follows that $\varphi(t)$ must coincide with $\psi(t) = \exp \int_{R^1} (e^{itx} - 1 - itx)x^{-2} \nu(dx)$. Now $\varphi'(t) = i\varphi(t) \int_{R^1} (e^{itx} - 1)x^{-1} \lambda(dx)$, and similarly for $\psi'(t)$. Hence $\varphi(t) = \psi(t)$ implies that $\int_{R^1}(e^{itx} - 1)x^{-1}\nu(dx) = \int_{R^1}(e^{itx} - 1)x^{-1}\mu(dx)$. A further differentiation gives $\int_{R^1}e^{itx}\mu(dx) = \int_{R^1} e^{itx}\nu(dx)$. This implies that $\mu(R^1) = \nu(R^1)$, and so $\mu = \nu$ by the uniqueness theorem for characteristic functions. ■

Example 28.4. *The Normal Case.* According to the theorem, $S_n \Rightarrow N$ if and only if μ_n converges vaguely to a unit mass at 0. If $s_n^2 = 1$, this holds if and only if $\Sigma_{k=1}^{r_n} \int_{|x| \geq \epsilon} x^2 \, dF_{nk}(x) \to 0$, which is exactly Lindeberg's condition. Thus Theorem 28.4 contains Theorems 27.2 and 27.4.   ∎

Example 28.5.   *The Poisson Case.* Let $Y_{n1}, \ldots, Y_{nr_n}$ be an independent triangular array, and suppose that $X_{nk} = Y_{nk} - m_{nk}$ satisfies the conditions of the theorem, where $m_{nk} = E[Y_{nk}]$. If Z_λ has the Poisson distribution with parameter λ, then $\Sigma_k X_{nk} \Rightarrow Z_\lambda - \lambda$ if and only if μ_n converges vaguely to a mass of λ at 1 (see Example 28.2). If $s_n^2 \to \lambda$, the requirement is $\mu_n[1 - \epsilon, 1 + \epsilon] \to \lambda$ or

$$(28.11) \qquad \sum_k \int_{|Y_{nk} - m_{nk} - 1| \geq \epsilon} (Y_{nk} - m_{nk})^2 \, dP \to 0$$

for positive ϵ. If s_n^2 and $m_n = \Sigma_k m_{nk}$ both converge to λ, (28.11) is a necessary and sufficient condition that $\Sigma_n Y_{nk} \Rightarrow Z_\lambda$. The conditions are easily checked if $r_n = n$ and Y_{nk} assumes the values 1 and 0 with probabilities λ/n and $1 - \lambda/n$.   ∎

PROBLEMS

28.1.   Show that $\mu_n \to_v \mu$ implies that $\mu(R^1) \leq \liminf_n \mu_n(R^1)$. Thus in vague convergence mass can "escape to infinity" but mass cannot "enter from infinity."

28.2.   (a)   Show by example that, if (28.1) fails, then $\mu_n \to_v \mu$ does not imply (28.2) for all continuous f vanishing at $\pm\infty$.

(b)   Show that $\mu_n \to_v \mu$ if and only if (28.2) holds for every continuous f with bounded support.

28.3.   23.9↑   Suppose that $N, Y_1, Y_2, \ldots$ are independent, the Y_n have a common distribution function F, and N has the Poisson distribution with mean α. Then $S = Y_1 + \cdots + Y_N$ has the *compound Poisson distribution*.

(a)   Show that the distribution of S is infinitely divisible. Note that S may not have a mean.

(b)   The distribution function of S is $\Sigma_{n=0}^{\infty} e^{-\alpha} \alpha^n F^{n*}(x)/n!$, where F^{n*} is the n-fold convolution of F (a unit jump at 0 for $n = 0$). The characteristic function of S is $\exp \alpha \int_{-\infty}^{\infty} (e^{itx} - 1) \, dF(x)$.

(c)   Show that, if F has mean 0 and finite variance, then the canonical measure μ in (28.6) is specified by $\mu(A) = \int_A x^2 \, dF(x)$.

28.4.   (a)   Let ν be a finite measure and define

$$(28.12) \qquad \varphi(t) = \exp \left[i\gamma t + \int_{-\infty}^{\infty} \left(e^{itx} - 1 - \frac{itx}{1 + x^2} \right) \frac{1 + x^2}{x^2} \nu(dx) \right],$$

where the integrand is $-t^2/2$ at the origin. Show that this is the characteristic function of an infinitely divisible distribution.

(b)   Show that the Cauchy distribution (see the table on page 302) is the case where $\gamma = 0$ and ν has density $\pi^{-1}(1 + x^2)^{-1}$ with respect to Lebesgue measure.

28.5. Show that the Cauchy, exponential, and gamma (see (20.47)) distributions are infinitely divisible.

28.6. Find the canonical representation (28.6) of the exponential distribution with mean 1:

(a) The characteristic function is $\int_0^\infty e^{itx}e^{-x}\,dx = (1 - it)^{-1} = \varphi(t)$.

(b) Show that (use the principal branch of the logarithm or else operate formally for the moment) $d(\log \varphi(t))/dt = i\varphi(t) = i\int_0^\infty e^{itx}e^{-x}\,dx$. Integrate with respect to t to obtain

$$(28.13) \qquad \frac{1}{1 - it} = \exp \int_0^\infty (e^{itx} - 1)\frac{e^{-x}}{x}\,dx.$$

Verify (28.13) after the fact by showing that the ratio of the two sides has derivative 0.

(c) Multiply (28.13) by e^{-it} to center the exponential distribution at its mean: the canonical measure μ has density xe^{-x} over $(0, \infty)$.

28.7. ↑ If X and Y are independent and each has the exponential density e^{-x}, $X - Y$ has the double exponential density $\frac{1}{2}e^{-|x|}$ (see the table on page 302). Show that its characteristic function is

$$\frac{1}{1 + t^2} = \exp \int_{-\infty}^\infty (e^{itx} - 1 - itx)\frac{1}{x^2}|x|e^{-|x|}\,dx.$$

28.8. ↑ Suppose $X_1, X_2, \ldots$ are independent and each has the double exponential density. Show that $\sum_{n=1}^\infty X_n/n$ converges with probability 1. Show that the distribution of the sum is infinitely divisible and that its canonical measure has density $|x|e^{-|x|}/(1 - e^{-|x|}) = \sum_{n=1}^\infty |x|e^{-|nx|}$.

28.9. 26.8↑ Show that for the gamma density $e^{-x}x^{u-1}/\Gamma(u)$ the canonical measure has density uxe^{-x} over $(0, \infty)$.

The remaining problems require the notion of a *stable law*. A distribution function F is stable if for each n there exist constants a_n and b_n, $a_n > 0$, such that, if $X_1, \ldots, X_n$ are independent and have distribution function F, then $a_n^{-1}(X_1 + \cdots + X_n) + b_n$ also has distribution function F.

28.10. Suppose that for all a, a', b, b' there exist a'', b'' (here a, a', a'' are all positive) such that $F(ax + b)*F(a'x + b') = F(a''x + b'')$. Show that F is stable.

28.11. Show that a stable law is infinitely divisible.

28.12. Show that the Poisson law, although infinitely divisible, is not stable.

28.13. Show that the normal and Cauchy laws are stable.

28.14. 28.10↑ Suppose that F has mean 0 and variance 1 and that the dependence of a'', b'' on a, a', b, b' is such that

$$F\left(\frac{x}{\sigma_1}\right) * F\left(\frac{x}{\sigma_2}\right) = F\left(\frac{x}{\sqrt{\sigma_1^2 + \sigma_2^2}}\right).$$

Show that F is the standard normal distribution.

28.15. **(a)**   Let Y_{nk} be independent random variables having the Poisson distribution with mean $cn^\alpha/|k|^{1+\alpha}$, where $c > 0$ and $0 < \alpha < 2$. Let $Z_n = n^{-1}\Sigma_{k=-n^2}^{n^2} kY_{nk}$ (omit $k = 0$ in the sum) and show that if c is properly chosen then the characteristic function of Z_n converges to $e^{-|t|^\alpha}$.

(b)   Show for $0 < \alpha \leq 2$ that $e^{-|t|^\alpha}$ is the characteristic function of a symmetric stable distribution; it is called the *symmetric stable law of exponent* α. The case $\alpha = 2$ is the normal law, and $\alpha = 1$ is the Cauchy law.

SECTION 29.   LIMIT THEOREMS IN R^k

If F_n and F are distribution functions on R^k, F_n *converges weakly* to F, written $F_n \Rightarrow F$, if $\lim_n F_n(x) = F(x)$ for all continuity points x of F. The corresponding distributions μ_n and μ are in this case also said to converge weakly: $\mu_n \Rightarrow \mu$. If X_n and X are k-dimensional random vectors (possibly on different probability spaces), X_n *converges in distribution* to X, written $X_n \Rightarrow X$, if the corresponding distribution functions converge weakly. The definitions are thus exactly as for the line.

The Basic Theorems

The closure A^- of a set in R^k is the set of limits of sequences in A; the interior is $A^\circ = R^k - (R^k - A)^-$; and the boundary is $\partial A = A^- - A^\circ$. A Borel set A is a μ-*continuity set* if $\mu(\partial A) = 0$. The first theorem is the k-dimensional version of Theorem 25.8.

Theorem 29.1.   *For probability measures μ_n and μ on $(R^k, \mathcal{R}^k)$, each of the following conditions is equivalent to the weak convergence of μ_n to μ.*

(i) $\lim_n \int f \, d\mu_n = \int f \, d\mu$ *for bounded continuous f;*

(ii) $\lim \sup_n \mu_n(C) \leq \mu(C)$ *for closed C;*

(iii) $\lim \inf_n \mu_n(G) \geq \mu(G)$ *for open G;*

(iv) $\lim_n \mu_n(A) = \mu(A)$ *for μ-continuity sets A.*

PROOF.   It will first be shown that (i) through (iv) are all equivalent.

(i) *implies* (ii): Consider the distance dist $(x, C) = \inf[|x - y| : y \in C]$ from x to C. It is continuous in x. Let

$$\varphi_j(t) = \begin{cases} 1 & \text{if } t \leq 0, \\ 1 - jt & \text{if } 0 \leq t \leq j^{-1}, \\ 0 & \text{if } j^{-1} \leq t. \end{cases}$$

Then $f_j(x) = \varphi_j (\text{dist } (x, C))$ is continuous and bounded by 1, and $f_j(x) \downarrow I_C(x)$

as $j \uparrow \infty$ because C is closed. If (i) holds, then $\lim \sup_n \mu_n(C) \leq \lim_n \int f_j \, d\mu_n$ $= \int f_j \, d\mu$. As $j \uparrow \infty$, $\int f_j d\mu \downarrow \int I_C d\mu = \mu(C)$.

(ii) *is equivalent to* (iii): Take $C = R^k - G$.

(ii) *and* (iii) *imply* (iv): From (ii) and (iii) follows

$$\mu(A^\circ) \leq \lim \inf_n \mu_n(A^\circ) \leq \lim \inf_n \mu_n(A)$$
$$\leq \lim \sup_n \mu_n(A) \leq \lim \sup_n \mu_n(A^-) \leq \mu(A^-).$$

Clearly (iv) follows from this.

(iv) *implies* (i): Suppose that f is continuous and $|f(x)|$ is bounded by K. Given ϵ choose reals $\alpha_0 < \alpha_1 < \cdots < \alpha_l$ so that $\alpha_0 < -K < K < \alpha_l$, $\alpha_i - \alpha_{i-1} < \epsilon$, and $\mu[x: f(x) = \alpha_i] = 0$. The last condition can be achieved because the sets $[x: f(x) = \alpha]$ are disjoint for different α. Put $A_i = [x: \alpha_{i-1} < f(x) \leq \alpha_i]$. Since f is continuous, $A_i^- \subset [x: \alpha_{i-1} \leq f(x) \leq \alpha_i]$ and $A_i^\circ \supset [x: \alpha_{i-1} < f(x) < \alpha_i]$. Therefore, $\partial A_i \subset [x: f(x) = \alpha_{i-1}] \cup [x: f(x) = \alpha_i]$, and therefore $\mu(\partial A_i) = 0$. Now $|\int f \, d\mu_n - \sum_{i=1}^l \alpha_i \mu_n(A_i)| \leq \epsilon$ and similarly for μ, and $\sum_{i=1}^l \alpha_i \mu_n(A_i) \to \sum_{i=1}^l \alpha_i \mu(A_i)$ because of (iv). Since ϵ was arbitrary, (i) follows.

It remains to prove these four conditions equivalent to weak convergence.

(iv) *implies* $\mu_n \Rightarrow \mu$: Consider the corresponding distribution functions. Since F is everywhere continuous from above, and since it is nondecreasing in each variable, it is continuous at x if and only if $\lim_m F(x_1 - m^{-1}, \ldots, x_k - m^{-1}) = F(x_1, \ldots, x_k)$. Of course, $F(x_1, \ldots, x_k)$ is the μ-measure of the closed set $A = [y: y_i \leq x_i, i = 1, \ldots, k]$, and the limit above is the μ-measure of $A^\circ = [y: y_i < x_i, i = 1, \ldots, k]$. Thus F is continuous at x if and only if $\mu(A) = \mu(A^\circ)$, or $\mu(\partial A) = 0$. Therefore, (iv) implies that $\lim_n F_n(x) = \lim_n \mu_n(A) = \mu(A) = F(x)$ if F is continuous at x.

$\mu_n \Rightarrow \mu$ *implies* (iii): Since only countably many parallel hyperplanes can have positive μ-measure, there is a dense set D of reals such that $\mu[x: x_i = d] = 0$ for $d \in D$ and $i = 1, \ldots, k$. Let $\mathcal{A}$ be the class of rectangles $A = [x: a_i < x_i \leq b_i, i = 1, \ldots, k]$ for which the a_i and the b_i all lie in D. All 2^k vertices of such a rectangle are continuity points of F, and so $F_n \Rightarrow F$ implies (see (12.12)) that $\mu_n(A) = \Delta_A F_n \to \Delta_A F = \mu(A)$. It follows by the inclusion-exclusion formula that $\mu_n(B) \to \mu(B)$ for finite unions B of elements of $\mathcal{A}$. Since D is dense on the line, an open set G in R^k is a countable union of sets A_m in $\mathcal{A}$. But $\mu(\bigcup_{m \leq M} A_m) = \lim_n \mu_n(\bigcup_{m \leq M} A_m) \leq \lim \inf_n \mu_n(G)$. Letting $M \to \infty$ gives (iii). ∎

Theorem 29.2. *Suppose that $h: R^k \to R^j$ is measurable and that the set D_h of its discontinuities is measurable.[*] If $\mu_n \Rightarrow \mu$ in R^k and $\mu(D_h) = 0$, then $\mu_n h^{-1} \Rightarrow \mu h^{-1}$ in R^j.*

PROOF. Let C be a closed set in R^j. The closure $(h^{-1}C)^-$ in R^k satisfies

---

[*] The argument in the footnote on page 287 shows that in fact $D_h \in \mathcal{R}^k$ always holds.

$(h^{-1}C)^- \subset D_h \cup h^{-1}C$. If $\mu_n \Rightarrow \mu$, then part (ii) of Theorem 29.1 gives lim sup$_n$ $\mu_n h^{-1}(C) \leq$ lim sup$_n \mu_n((h^{-1}C)^-) \leq \mu((h^{-1}C)^-) \leq \mu(D_h) + \mu(h^{-1}C)$. Using (ii) again gives $\mu_n h^{-1} \Rightarrow \mu h^{-1}$ if $\mu(D_h) = 0$. ∎

Theorem 29.2 is the k-dimensional version of the *mapping theorem*—Theorem 25.7. The two proofs just given provide in the case $k = 1$ a second approach to the theory of Section 25, which there was based on Skorohod's theorem (Theorem 25.6).* Skorohod's theorem does extend to R^k (see Theorem 29.6), but the proof is harder.

Theorems 29.1 and 29.2 can of course be stated in terms of random vectors. For example, $X_n \Rightarrow X$ if and only if $P[X \in G] \leq$ lim inf$_n P[X_n \in G]$ for all open sets G.

A sequence $\{\mu_n\}$ of probability measures on $(R^k, \mathcal{R}^k)$ is *tight* if for every ϵ there is a bounded rectangle A such that $\mu_n(A) > 1 - \epsilon$ for all n.

Theorem 29.3. *If $\{\mu_n\}$ is a tight sequence of probability measures, there is a subsequence $\{\mu_{n_i}\}$ and a probability measure μ such that $\mu_{n_i} \Rightarrow \mu$ as $i \to \infty$.*

FIRST PROOF. The proof of Helley's theorem (Theorem 25.9) carries over to R^k; it is only necessary to let the points r there be vectors whose coordinates are all rational and for vectors x and y to interpret $x \leq y$ as $x_i \leq y_i, i = 1, \ldots,$ k, and $x < y$ as $x_i < y_i, i = 1, \ldots, k$. In the proof of sufficiency in Theorem 25.10 replace the application of Theorem 12.4 by an application of its multidimensional version, Theorem 12.5. ∎

There is a proof that avoids an appeal to Theorem 12.5.[†]

SECOND PROOF. Let $\mathcal{A}$ be the class of bounded, *closed* rectangles whose vertices have rational coordinates, and let $\mathcal{H}$ consist of the empty set and the finite unions of $\mathcal{A}$-sets. Then $\mathcal{H}$ is countable and closed under the formation of finite unions, and each set in $\mathcal{H}$ is compact.

Select by the diagonal procedure a subsequence $\{\mu_{n_i}\}$ along which the limits

(29.1) $\qquad\qquad\qquad \alpha(H) = \lim_i \mu_{n_i}(H)$

exist for all H in $\mathcal{H}$. Suppose it can be shown that there exists a probability measure μ such that

(29.2) $\qquad\qquad\qquad \mu(G) = \sup [\alpha(H): H \subset G, H \in \mathcal{H}]$

for all open sets G. Then $H \subset G$ will imply that $\alpha(H) = \lim_i \mu_{n_i}(H) \leq$ lim inf$_i \mu_{n_i}(G)$, so that $\mu(G) \leq$ lim inf$_i \mu_{n_i}(G)$ for open G by (29.2). Therefore (condition (iii) in Theorem 29.1), it suffices to produce a probability measure satisfying (29.2).

---

* The approach of this section carries over to general metric spaces; for this theory and its applications, see BILLINGSLEY'.

[†] It extends to the general metric space; see LOÈVE.

Clearly, $\alpha(H)$, defined by (29.1) for all H in $\mathcal{H}$, has these properties:

(29.3) $\alpha(H_1) \leq \alpha(H_2)$ if $H_1 \subset H_2$;

(29.4) $\alpha(H_1 \cup H_2) = \alpha(H_1) + \alpha(H_2)$ if $H_1 \cap H_2 = 0$;

(29.5) $\alpha(H_1 \cup H_2) \leq \alpha(H_1) + \alpha(H_2)$.

Define

(29.6) $\beta(G) = \sup\,[\alpha(H) \colon H \subset G, H \in \mathcal{H}]$

for open sets G, and then for arbitrary A define

$$\gamma(A) = \inf\,[\beta(G) \colon A \subset G],$$

where G ranges over open sets. Clearly, $\gamma(G) = \beta(G)$ for open G.

Suppose it is shown that γ is an outer measure and that each closed set is γ-measurable. In the notation of Theorem 11.1, it will then follow that $\mathcal{R}^k \subset \mathcal{M}(\gamma)$, and the restriction μ of γ to $\mathcal{R}^k$ will be a measure satisfying $\mu(G) = \gamma(G) = \beta(G)$, and so (29.2) will hold for open G, as required. By tightness, there is for given ϵ an $\mathcal{H}$-set H such that $\mu_n(H) > 1 - \epsilon$ for all n. But then $1 \geq \alpha(H) \geq 1 - \epsilon$ by (29.1), and $\mu(R^k) = \beta(R^k) = 1$ by (29.2).

The first step in proving that γ is an outer measure is to show that β *is finitely subadditive* (on open sets): If $H \subset G_1 \cup G_2$ and $H \in \mathcal{H}$, define $C_1 = [x \in H \colon \mathrm{dist}\,(x, G_1^c) \geq \mathrm{dist}\,(x, G_2^c)]$ and $C_2 = [x \in H \colon \mathrm{dist}\,(x, G_1^c) \leq \mathrm{dist}\,(x, G_2^c)]$. If $x \in C_1$ and $x \notin G_1$, then $x \in G_2$, so that, since G_2^c is closed, $\mathrm{dist}\,(x, G_2^c) > 0 = \mathrm{dist}\,(x, G_1^c)$, a contradiction. Thus $C_1 \subset G_1$; similarly, $C_2 \subset G_2$.

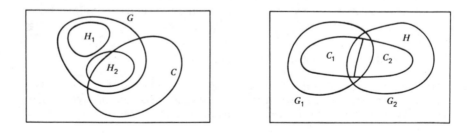

If $x \in C_1$, there is an $\mathcal{A}$-set A_x such that $x \in A_x^{\circ} \subset G_1$. These A_x° cover C_1, which is compact (a closed subset of the compact set H), and so finitely many of them cover C_1; the union H_1 of the corresponding sets A_x is in $\mathcal{H}$ and satisfies $C_1 \subset H_1 \subset G_1$. Similarly, there is and $\mathcal{H}$-set H_2 for which $C_2 \subset H_2 \subset G_2$. But then $\alpha(H) \leq \alpha(H_1 \cup H_2) \leq \alpha(H_1) + \alpha(H_2) \leq \beta(G_1) + \beta(G_2)$ by (29.3), (29.5), and (29.6). Since H can be any $\mathcal{H}$-set in $G_1 \cup G_2$, another application of (29.6) gives $\beta(G_1 \cup G_2) \leq \beta(G_1) + \beta(G_2)$.

Next, β *is countably subadditive* (on open sets): If $H \subset \bigcup_n G_n$, then by compactness $H \subset \bigcup_{n \leq n_0} G_n$ for some n_0, and therefore by finite subadditivity $\alpha(H) \leq \beta(\bigcup_{n \leq n_0} G_n) \leq \sum_{n \leq n_0} \beta(G_n) \leq \sum_n \beta(G_n)$. Letting H vary over the $\mathcal{H}$-sets in $\bigcup_n G_n$ gives $\beta(\bigcup_n G_n) \leq \sum_n \beta(G_n)$.

And now γ *is an outer measure*: Clearly, γ is monotone. To prove it subadditive, suppose that A_n are arbitrary sets and $\epsilon > 0$. Choose open sets G_n such that $A_n \subset G_n$ and

$\beta(G_n) < \gamma(A_n) + \epsilon/2^n$. By the countable subadditivity of β, $\gamma(\bigcup_n A_n) \leq \beta(\bigcup_n G_n) \leq \Sigma_n \beta(G_n) < \Sigma_n \gamma(A_n) + \epsilon$; ϵ being arbitrary, $\gamma(\bigcup_n A_n) \leq \Sigma_n \gamma(A_n)$ follows.

It remains only to prove that *each closed set is γ-measurable*:

$$(29.7) \qquad \gamma(A) \geq \gamma(A \cap C) + \gamma(A \cap C^c)$$

for C closed and A arbitrary (see (11.3)). To prove (29.7) it suffices to prove

$$(29.8) \qquad \beta(G) \geq \gamma(G \cap C) + \gamma(G \cap C^c)$$

for open G, because then $G \supset A$ implies that $\beta(G) \geq \gamma(A \cap C) + \gamma(A \cap C^c)$ and taking the infimum over G will give (29.7).

To prove (29.8), for given ϵ choose an $\mathcal{H}$-set H_1 for which $H_1 \subset G \cap C^c$ and $\alpha(H_1) > \beta(G \cap C^c) - \epsilon$. Now choose an $\mathcal{H}$-set H_2 for which $H_2 \subset G \cap H_1^c$ and $\alpha(H_2) > \beta(G \cap H_1^c) - \epsilon$. Since H_1 and H_2 are disjoint and are contained in G, it follows by (29.4) that $\beta(G) \geq \alpha(H_1 \cup H_2) = \alpha(H_1) + \alpha(H_2) > \beta(G \cap C^c) + \beta(G \cap H_1^c) - 2\epsilon \geq \gamma(G \cap C^c) + \gamma(G \cap C) - 2\epsilon$. Hence (29.8). ∎

Obviously Theorem 29.3 implies that tightness is a sufficient condition that each subsequence of $\{\mu_n\}$ contain a further subsequence converging weakly to some probability measure. (An easy modification of the proof of Theorem 25.10 shows that tightness is necessary for this as well.) And clearly the corollary to Theorem 25.10 now goes through:

Corollary. *If $\{\mu_n\}$ is a tight sequence of probability measures, and if each subsequence that converges weakly at all converges weakly to the probability measure μ, then $\mu_n \Rightarrow \mu$.*

Characteristic Functions

Consider a random vector $X = (X_1, \ldots, X_k)$ and its distribution μ in R^k. Let $t \cdot x = \Sigma_{u=1}^k t_u x_u$ denote inner product. The characteristic function of X and of μ is defined over R^k by

$$(29.9) \qquad \varphi(t) = \int_{R^k} e^{it \cdot x} \mu(dx) = E[e^{it \cdot X}].$$

To a great extent its properties parallel those of the one-dimensional characteristic function and can be deduced by parallel arguments.

The inversion formula (26.16) takes this form: For a bounded rectangle $A = [x : a_u < x_u \leq b_u, u \leq k]$ such that $\mu(\partial A) = 0$,

$$(29.10) \qquad \mu(A) = \lim_{T \to \infty} \frac{1}{(2\pi)^k} \int_{B_T} \prod_{u=1}^k \frac{e^{-it_u a_u} - e^{-it_u b_u}}{it_u} \varphi(t) \, dt,$$

where $B_T = [t \in R^k : |t_u| \leq T, u \leq k]$ and dt is short for $dt_1 \cdots dt_k$. To prove it, replace $\varphi(t)$ by the middle term in (29.9) and reverse the integrals as in

(26.17): The integral in (29.10) is

$$I_T = \frac{1}{(2\pi)^k} \int_{R^k} \left[\int_{B_T} \prod_{u=1}^{k} \frac{e^{-it_u a_u} - e^{-it_u b_u}}{it_u} e^{it_u x_u} \, dt \right] \mu(dx).$$

The inner integral may be evaluated by Fubini's theorem in R^k, which gives

$$I_T = \int_{R^k} \prod_{u=1}^{k} \left[\frac{\text{sgn}\,(x_u - a_u)}{\pi} S(T \cdot |x_u - a_u|) \right.$$

$$\left. - \frac{\text{sgn}\,(x_u - b_u)}{\pi} S(T \cdot |x_u - b_u|) \right] \mu(dx).$$

Since the integrand converges to $\prod_{u=1}^{k} \psi_{a_u, b_u}(x_u)$ (see (26.18)), (29.10) follows as in the case $k = 1$.

The proof that weak convergence implies (iii) in Theorem 29.1 shows that for probability measures μ and ν on R^k there exists a dense set D of reals such that $\mu(\partial A) = \nu(\partial A) = 0$ for all rectangles A whose vertices have coordinates in D. If $\mu(A) = \nu(A)$ for such rectangles, then μ and ν are identical by Theorem 3.3.

Thus the characteristic function φ uniquely determines the probability measure μ. Further properties of the characteristic function can be derived from the one-dimensional case by means of the following device of Cramér and Wold. For $t \in R^k$, define $h_t : R^k \to R^1$ by $h_t(x) = t \cdot x$. For real α, $[x: t \cdot x \le \alpha]$ is a half-space, and its μ-measure is

$$(29.11) \qquad \mu[x: t \cdot x \le \alpha] = \mu h_t^{-1}(-\infty, \alpha].$$

By change of variable, the characteristic function of μh_t^{-1} is

$$(29.12) \qquad \int_{R^1} e^{isy} \mu h_t^{-1}(dy) = \int_{R^k} e^{is(t \cdot x)} \mu(dx)$$

$$= \varphi(st_1, \ldots, st_k), \qquad s \in R^1.$$

To know the μ-measure of every half-space is (by (29.11)) to know each μh_t^{-1} and hence is (by (29.12) for $s = 1$) to know $\varphi(t)$ for every t; and to know the characteristic function φ of μ is to know μ. *Thus μ is uniquely determined by the values it gives to the half-spaces.* This result, very simple in its statement, seems to require Fourier methods—no elementary proof is known.

If $\mu_n \Rightarrow \mu$ for probability measures on R^k, then $\varphi_n(t) \to \varphi(t)$ for the corresponding characteristic functions by Theorem 29.1. But suppose that the characteristic functions converge pointwise. It follows by (29.12) that for each t the characteristic function of $\mu_n h_t^{-1}$ converges pointwise on the line to the characteristic function of μh_t^{-1}; by the continuity theorem for characteristic functions on the line then, $\mu_n h_t^{-1} \Rightarrow \mu h_t^{-1}$. Take the uth component of t to be 1 and the others 0; then the $\mu_n h_t^{-1}$ are the marginals for the uth coordinate. Since

$\{\mu_n h_t^{-1}\}$ is weakly convergent, there is a bounded interval $(a_u, b_u]$ such that $\mu_n[x \in R^k: a_u < x_u \le b_u] = \mu_n h_t^{-1}(a_u, b_u] > 1 - \epsilon/k$ for all n. But then $\mu_n(A) > 1 - \epsilon$ for the bounded rectangle $A = [x: a_u < x_u \le b_u, u = 1, \ldots, k]$. The sequence $\{\mu_n\}$ is therefore tight. If a subsequence $\{\mu_{n_k}\}$ converges weakly to ν, then $\varphi_{n_k}(t)$ converges to the characteristic function of ν, which is therefore $\varphi(t)$. By uniqueness, $\nu = \mu$, so that $\mu_n \Rightarrow \mu$. By the corollary to Theorem 29.3, $\mu_n \Rightarrow \mu$. This proves the continuity theorem for k-dimensional characteristic functions: $\mu_n \Rightarrow \mu$ *if and only if* $\varphi_n(t) \to \varphi(t)$ *for all* t.

The Cramér-Wold idea leads also to the following result, by means of which certain limit theorems can in a routine way be reduced to the one-dimensional case.

Theorem 29.4. *For random vectors* $X_n = (X_{n1}, \ldots, X_{nk})$ *and* $Y = (Y_1, \ldots, Y_k)$, *a necessary and sufficient condition for* $X_n \Rightarrow Y$ *is that* $\sum_{u=1}^k t_u X_{nu} \Rightarrow \sum_{u=1}^k t_u Y_u$ *for each* $(t_1, \ldots, t_k)$ *in* R^k.

PROOF. The necessity follows from a consideration of the continuous mapping h_t above—use Theorem 29.2. As for sufficiency, the condition implies by the continuity theorem for one-dimensional characteristic functions that for each $(t_1, \ldots, t_k)$

$$E[e^{is\sum_{u=1}^k t_u X_{nu}}] \to E[e^{is\sum_{u=1}^k t_u Y_u}]$$

for all real s. Taking $s = 1$ shows that the characteristic function of X_n converges pointwise to that of Y. ∎

Normal Distributions in R^k

By Theorem 20.4 there is (on some probability space) a random vector $X = (X_1, \ldots, X_k)$ with independent components each having the standard normal distribution. Since each X_u has density $e^{-x^2/2}/\sqrt{2\pi}$, X has density (see (20.25))

$$(29.13) \qquad f(x) = \frac{1}{(2\pi)^{k/2}} e^{-|x|^2/2},$$

where $|x|^2 = \sum_{u=1}^k x_u^2$ denotes Euclidean norm. This distribution plays the role of the standard normal distribution in R^k. Its characteristic function is

$$(29.14) \qquad E\left[\prod_{u=1}^k e^{it_u X_u}\right] = \prod_{u=1}^k e^{-t_u^2/2} = e^{-|t|^2/2}.$$

Let $A = [a_{uv}]$ be a $k \times k$ matrix and put $Y = AX$, where X is viewed as a column vector. Since $E[X_\alpha X_\beta] = \delta_{\alpha\beta}$, the matrix $\Sigma = [\sigma_{uv}]$ of the covariances of Y has entries $\sigma_{uv} = E[Y_u Y_v] = \sum_{\alpha=1}^k a_{u\alpha} a_{v\alpha}$. Thus $\Sigma = AA'$, where the prime denotes transpose. The matrix Σ is symmetric and nonnegative definite:

$\Sigma_{uv}\sigma_{uv}x_u x_v = |A'x|^2 \geq 0$. View t also as a column vector with transpose t', and note that $t \cdot x = t'x$. The characteristic function of AX is thus

$$E[e^{it'(AX)}] = E[e^{i(A't)'X}] = e^{-|A't|^2/2} = e^{-t'\Sigma t/2}.$$

Define a centered *normal distribution* as any probability measure whose characteristic function has this form for some symmetric, nonnegative definite Σ.

If Σ is symmetric and nonnegative definite, then for an appropriate orthogonal matrix U, $U'\Sigma U = D$ is a diagonal matrix whose diagonal elements are the eigenvalues of Σ and hence are nonnegative. If D_0 is the diagonal matrix whose elements are the square roots of those of D, and if $A = UD_0$, then $\Sigma = AA'$. Thus for every nonnegative definite Σ there exists a unique centered normal distribution (namely the distribution of AX) with covariance matrix Σ and characteristic function $\exp(-\frac{1}{2} t'\Sigma t)$.

If Σ is nonsingular, so is the A just constructed. Since X has density (29.13), $Y = AX$ has by the Jacobian transformation formula (20.20) density $f(A^{-1}x)$ $|\det A^{-1}|$. Since $\Sigma = AA'$, $|\det A^{-1}| = (\det \Sigma)^{-1/2}$. Moreover, $\Sigma^{-1} = (A')^{-1}$ A^{-1}, so that $|A^{-1}x|^2 = x' \Sigma^{-1} x$. Thus the normal distribution has density $(2\pi)^{-k/2}(\det \Sigma)^{-1/2} \cdot \exp(-\frac{1}{2} x' \Sigma^{-1} x)$ if Σ is nonsingular. If Σ is singular, the A constructed above must be singular as well, so that AX is confined to some hyperplane of dimension $k - 1$ and the distribution can have no density.

If M is a $j \times k$ matrix and Y has in R^k the centered normal distribution with covariance matrix Σ, MY has in R^j the characteristic function $\exp(-\frac{1}{2}(M't)'$ $\Sigma(M't)) = \exp(-\frac{1}{2} t'(M\Sigma M')t)$ $(t \in R^j)$. Hence MY has the centered normal distribution in R^j with covariance matrix $M\Sigma M'$. Thus *a linear transformation of a normal distribution is itself normal.*

These normal distributions are special in that all the first moments vanish. The general normal distribution is a translation of one of these centered distributions.

The Central Limit Theorem

Let $X_n = (X_{n1}, \ldots, X_{nk})$ be independent random vectors all having the same distribution. Suppose that $E[X_{nu}^2] < \infty$; let the vector of means be $c = (c_1, \ldots, c_k)$, where $c_u = E[X_{nu}]$, and let the covariance matrix be $\Sigma = [\sigma_{uv}]$, where $\sigma_{uv} = E[(X_{nu} - c_u)(X_{nv} - c_v)]$. Put $S_n = X_1 + \cdots + X_n$.

Theorem 29.5. *Under these assumptions, the distribution of the random vector $(S_n - nc)/\sqrt{n}$ converges weakly to the centered normal distribution with covariance matrix Σ.*

PROOF. Let $Y = (Y_1, \ldots, Y_k)$ be a normally distributed random vector with

0 means and covariance matrix Σ. For given $t = (t_1, \ldots, t_k)$, let $Z_n = \Sigma_{u=1}^{k}$ $t_u(X_{nu} - c_u)$ and $Z = \Sigma_{u=1}^{k} t_u Y_u$. By Theorem 29.4, it suffices to prove that $n^{-1/2}$ $\Sigma_{j=1}^{n} Z_j \Rightarrow Z$ (for arbitrary t). But this is an instant consequence of the Linde-berg-Lévy theorem (Theorem 27.1). ■

Skorohod's Theorem in R^{k*}

Suppose that Y_n and Y are k-dimensional random vectors and $Y_n \to Y$ with probability 1. If G is open, then $[Y \in G] \subset \liminf_n [Y_n \in G]$, and Theorem 4.1 implies that $P[Y \in G] \le \liminf_n P[Y_n \in G]$, so that $Y_n \Rightarrow Y$ by Theorem 29.1. Thus convergence with probability 1 implies convergence in distribution.

Theorem 29.6. *Suppose that μ_n and μ are probability measures on $(R^k, \mathcal{R}^k)$ and $\mu_n \Rightarrow \mu$. Then there exist random vectors Y_n and Y on a common probability space $(\Omega, \mathcal{F}, P)$ such that Y_n has distribution μ_n, Y has distribution μ, and $Y_n(\omega)$ $\to Y(\omega)$ for each ω.*

PROOF. As in the proof of Theorem 25.6, the probability space $(\Omega, \mathcal{F}, P)$ will be taken to be the unit interval with Lebesgue measure. It will be convenient to take $\Omega = [0, 1)$—closed on the left and open on the right.

Let B_0 be the set of real s such that $\mu[x: x_i = s] > 0$ for some $i = 1, \ldots, k$, and let B consist of all rational multiples of elements of B_0. Then B is countable, and so there exists some positive t outside B. By the construction the set

(29.15) $D = [n2^{-u}t: u = 0, 1, 2, \ldots; n = 0, \pm1, \pm2, \ldots]$

is disjoint from B_0; of course, D is dense on the line.

Let $[C_{i_1}^{(1)}: i_1 = 1, 2, \ldots]$ be the countable decomposition of R^k into cubes of the form $[x: n_i t \le x_i < (n_i + 1)t, i = 1, \ldots, k]$ for integers $n_1, \ldots, n_k$. For each i_1 let $[C_{i_1 i_2}^{(2)}: i_2 = 1, \ldots, 2^k]$ be the decomposition of $C_{i_1}^{(1)}$ into 2^k cubes of side $t/2$. Now decompose each $C_{i_1 i_2}^{(2)}$ into 2^k cubes of side $t/2^2$. Continue inductively in this way to obtain for $u = 1, 2, \ldots$ decompositions $[C_{i_1}^{(u)} \ldots {}_{i_u}]$ of R^k into cubes of side $t/2^{u-1}$ for which

(29.16) $$C_{i_1 \ldots i_{u-1}}^{(u-1)} = \bigcup_{i=1}^{2^k} C_{i_1 \ldots i_{u-1}i}^{(u)}.$$

In all that follows, the first index, i_1, ranges over the positive integers, and the other indices range over $1, 2, \ldots, 2^k$. Since (29.15) is disjoint from B_0,

(29.17) $$\mu(\partial C_{i_1 \ldots i_u}^{(u)}) = 0.$$

* This topic may be omitted.

Construction of Y.   Next, decompose* $[0, 1)$ into subintervals $I_{i_1}^{(1)}$, $i_1 = 1$, $2, \ldots$, in such a way that $P(I_{i_1}^{(1)}) = \mu(C_{i_1}^{(1)})$, where μ is the given limit measure on R^k. Now decompose* $I_{i_1}^{(1)}$ into subintervals $I_{i_1 i_2}^{(2)}$, $i_2 = 1, \ldots, 2^k$, in such a way that $P(I_{i_1 i_2}^{(2)}) = \mu(C_{i_1 i_2}^{(2)})$. Continue inductively in this way to obtain for $u = 1$, $2, \ldots$ decompositions $[I_{i_1 \ldots i_u}^{(u)}]$ of $[0,1)$ into subintervals for which

$$(29.18) \qquad I_{i_1 \ldots i_{u-1}}^{(u-1)} = \bigcup_{i=1}^{2^k} I_{i_1 \ldots i_{u-1} i}^{(u)}$$

and

$$(29.19) \qquad P(I_{i_1 \ldots i_u}^{(u)}) = \mu(C_{i_1 \ldots i_u}^{(u)}).$$

In each $C_{i_1 \ldots i_u}^{(u)}$ choose an arbitrary point $x_{i_1 \ldots i_u}^{(u)}$. Define random vectors $Y^{(u)}$ on $[0, 1)$ by

$$(29.20) \qquad Y^{(u)}(\omega) = x_{i_1 \ldots i_u}^{(u)} \qquad \text{if } \omega \in I_{i_1 \ldots i_u}^{(u)}.$$

Since $C_{i_1 \ldots i_{u+1}}^{(u+1)} \subset C_{i_1 \ldots i_u}^{(u)}$, $Y^{(u+1)}(\omega)$ and $Y^{(u)}(\omega)$ lie in the same element of the uth decomposition of R^k. Therefore, $|Y^{(u+1)}(\omega) - Y^{(u)}(\omega)| \le kt/2^{u-1}$. The sequence $\{Y^{(u)}(\omega)\}$ is thus fundamental in the Euclidean metric for each ω, and so the limit

$$(29.21) \qquad Y(\omega) = \lim_u Y^{(u)}(\omega)$$

exists. Note that

$$(29.22) \qquad |Y(\omega) - Y^{(u)}(\omega)| \le \frac{kt}{2^{u-1}}.$$

The $Y^{(u)}$ and Y are measurable mappings from the unit interval into R^k—are random vectors.

Applying (29.16), (29.20), (29.19), (29.16) in succession gives

$$P[\omega: Y^{(u+v)}(\omega) \in C_{i_1 \ldots i_u}^{(u)}] = \sum_{j_1 \ldots j_v} P[\omega: Y^{(u+v)}(\omega) \in C_{i_1 \ldots i_u j_1 \ldots j_v}^{(u+v)}]$$

$$= \sum_{j_1 \ldots j_v} P(I_{i_1 \ldots i_u j_1 \ldots j_v}^{(u+v)}) = \sum_{j_1 \ldots j_v} \mu(C_{i_1 \ldots i_u j_1 \ldots j_v}^{(u+v)}) = \mu(C_{i_1 \ldots i_u}^{(u)}).$$

Therefore, $\lim_v P[\omega: Y^{(v)}(\omega) \in C_{i_1 \ldots i_u}^{(u)}] = \mu(C_{i_1 \ldots i_u}^{(u)})$. If $A = [y: y_i \le x_i, i = 1, \ldots, k]$ and the x_i all lie in the set (29.15), then A is for some u the union of countably many sets $C_{i_1 \ldots i_u}^{(u)}$, so that by Scheffé's theorem $\lim_v P[Y^{(v)} \in A]$

---

* If $b - a = \delta_1 + \delta_2 + \cdots$ (finite sum or infinite series) and $\delta_i \ge 0$, then $I_i = [a + \Sigma_{j<i}\delta_j, a + \Sigma_{j \le i} \delta_j)$ decomposes $[a, b)$ into subintervals of lengths $P(I_i) = \delta_i$.

$= \mu(A)$. Since (29.15) is dense, the distribution of $Y^{(v)}$ converges weakly to μ.* Since $Y^{(v)} \Rightarrow Y$ by (29.21), μ and the distribution of Y must coincide.

Construction of the Y_n. The construction is as for Y. For n, $u \geq 1$, let $[I_{i_1 \ldots i_u}^{(n,u)}]$ be a decomposition of $[0, 1)$ into subintervals for which

(29.23)
$$I_{i_1 \ldots i_{u-1}}^{(n,u-1)} = \bigcup_{i=1}^{2^k} I_{i_1 \ldots i_{u-1}i}^{(n,u)}$$

and

(29.24)
$$P(I_{i_1 \ldots i_u}^{(n,u)}) = \mu_n(C_{i_1 \ldots i_u}^{(u)}).$$

This time arrange inductively† that $i < j$ if and only if $I_{i_1 \ldots i_{u-1}i}^{(n,u)}$ lies to the left of $I_{i_1 \ldots i_{u-1}j}^{(n,u)}$. Order the $(i_1, \ldots, i_u)$ lexicographically: $(i_1, \ldots, i_u) < (j_1, \ldots, j_u)$ means that there exists an r, $1 \leq r < u$ such that $i_s = j_s$ for $1 \leq s < r$ and $i_r < j_r$. It follows by induction that $(i_1, \ldots, i_u) < (j_1, \ldots, j_u)$ if and only if $I_{i_1 \ldots i_u}^{(n,u)}$ lies to the left $I_{j_1 \ldots j_u}^{(n,u)}$.

For the same $x_{i_1 \ldots i_u}^{(u)}$ as in (29.20), define $Y_n^{(u)}(\omega)$ as $x_{i_1 \ldots i_u}^{(u)}$ for ω in $I_{i_1 \ldots i_u}^{(n,u)}$. As before, $|Y_n^{(u+1)}(\omega) - Y_n^{(u)}(\omega)| \leq kt/2^{u-1}$, so that the limit $Y_n(\omega) = \lim_u Y_n^{(u)}(\omega)$ exists,

(29.25)
$$|Y_n(\omega) - Y_n^{(u)}(\omega)| \leq \frac{kt}{2^{u-1}},$$

and Y_n has distribution μ_n.

Convergence. It remains to consider the convergence of $Y_n(\omega)$ to $Y(\omega)$. Let

$$I_{i_1 \ldots i_u}^{(n,u)} = [a_{i_1 \ldots i_u}^{(n,u)}, b_{i_1 \ldots i_u}^{(n,u)}), \qquad I_{i_1 \ldots i_u}^{(u)} = [a_{i_1 \ldots i_u}^{(u)}, b_{i_1 \ldots i_u}^{(u)}).$$

If Σ' denotes summation over those $(j_1, \ldots, j_u)$ for which $(j_1, \ldots, j_u) < (i_1, \ldots, i_u)$—here u, $i_1, \ldots, i_u$ are fixed—then by (29.24),

$$a_{i_1 \ldots i_u}^{(n,u)} = \Sigma' P(I_{j_1 \ldots j_u}^{(n,u)}) = \Sigma' \mu_n(C_{j_1 \ldots j_u}^{(u)}).$$

Since $\mu_n \Rightarrow \mu$ by hypothesis, it follows by (29.17) that this finite sum converges to

$$a_{i_1 \ldots i_u}^{(u)} = \Sigma' P(I_{j_1 \ldots j_u}^{(u)}) = \Sigma' \mu(C_{j_1 \ldots j_u}^{(u)}).$$

Similarly, $\lim_n b_{i_1 \ldots i_u}^{(n,u)} = b_{i_1 \ldots i_u}^{(u)}$ for each u and $i_1, \ldots, i_u$.

If ω is interior to $I_{i_1 \ldots i_u}^{(u)}$, it is therefore also interior to $I_{i_1 \ldots i_u}^{(n,u)}$ for large n, so that $Y^{(u)}(\omega) = Y_n^{(u)}(\omega)$. It follows by (29.22) and (29.25) that $|Y(\omega) - Y_n(\omega)| \leq kt/2^{u-2}$ for large n. Since u was arbitrary, $\lim_n Y_n(\omega) = Y(\omega)$ if ω is, for each u, interior to the $I_{i_1 \ldots i_u}^{(u)}$ that contains it.

---

* See the proof on page 330 that $\mu_n \Rightarrow \mu$ implies (iii).
† The construction in the footnote on page 338 has the property that $i < j$ if and only if I_i lies to the left of I_j.

The endpoints of all the $I_{i_1 \ldots i_u}^{(u)}$ make up a countable set; redefining $Y_n(\omega)$ $= Y(\omega) = 0$ on this set of Lebesgue measure 0 ensures that $Y_n(\omega) \to Y(\omega)$ for every ω and does not alter the distributions of Y_n and Y. This completes the proof of Skorohod's theorem for general k. ∎

Example 29.1. *The δ-method.* Suppose that

$$(29.26) \qquad v_n(X_n - \alpha, Y_n - \beta) \Rightarrow (X, Y),$$

where $v_n \to \infty$, and suppose that $f(x, y)$ is differentiable at (α, β) with partial derivatives f_1 and f_2. The problem is to prove that

$$(29.27) \qquad v_n[f(X_n, Y_n) - f(\alpha, \beta)] \Rightarrow f_1(\alpha, \beta)X + f_2(\alpha, \beta)Y.$$

By Skorohod's theorem in R^2 it is legitimate to proceed as though all these random variables were defined on the same probability space and (29.26) were ordinary convergence in R^2 for each sample point ω. Since f is differentiable, if ξ_n and η_n go to 0, then $f(\alpha + \xi_n, \beta + \eta_n) = f(\alpha, \beta) + f_1(\alpha, \beta)\xi_n + f_2(\alpha, \beta)\eta_n + \theta_n(\xi_n^2 + \eta_n^2)^{1/2}$, where $\theta_n \to 0$. Since $v_n \to \infty$, $(X_n(\omega), Y_n(\omega)) \to (\alpha, \beta)$ for each ω and hence

$$v_n[f(X_n(\omega), Y_n(\omega)) - f(\alpha, \beta)]$$

$$= f_1(\alpha, \beta)v_n(X_n(\omega) - \alpha) + f_2(\alpha, \beta)v_n(Y_n(\omega) - \beta)$$
$$+ \theta_n(\omega)[v_n^2(X_n(\omega) - \alpha)^2 + v_n^2(Y_n(\omega) - \beta)^2]^{1/2},$$

where $\theta_n(\omega) \to 0$ for each ω. The expression converges to $f_1(\alpha, \beta)X(\omega) + f_2(\alpha, \beta)Y(\omega)$ for each ω, and so (29.97) follows.

Therefore, if $v_n \to \infty$, (29.26) implies (29.27). The proof assumed of the random variables more than (29.26), but the result is true under the assumption of $(v_n \to \infty$ and) (29.26) alone. This is another case of the delta method; see Example 27.2. ∎

PROBLEMS

29.1. A real function f on R^k is everywhere upper semicontinuous (see Problem 13.11) if for each x and ϵ there is a δ such that $|x - y| < \delta$ implies that $f(y) < f(x) + \epsilon$; f is lower semicontinuous if $-f$ is upper semicontinuous.

(a) Use condition (iii) of Theorem 29.1, Fatou's lemma, and (21.9) to show that, if $\mu_n \Rightarrow \mu$ and f is bounded and lower semicontinuous, then

$$(29.28) \qquad \liminf_n \int f \, d\mu_n \geq \int f \, d\mu.$$

(b) Show that, if (29.28) holds for all bounded, lower semicontinuous functions f, then $\mu_n \Rightarrow \mu$.

(c) Prove the analogous results for upper semicontinuous functions.

29.2. (a) Show for probability measures on the line that $\mu_n \times \nu_n \Rightarrow \mu \times \nu$ if and only if $\mu_n \Rightarrow \mu$ and $\nu_n \Rightarrow \nu$.

(b) Suppose that X_n and Y_n are independent and that X and Y are independent. Show that, if $X_n \Rightarrow X$ and $Y_n \Rightarrow Y$, then $(X_n, Y_n) \Rightarrow (X, Y)$ and hence that $X_n + Y_n \Rightarrow X + Y$.

(c) Show that part (b) fails without independence.

(d)   If $F_n \Rightarrow F$ and $G_n \Rightarrow G$, then $F_n * G_n \Rightarrow F * G$. Prove this by part (b) and also by characteristic functions.

29.3.   **(a)**   Show that $\{\mu_n\}$ is tight if and only if for each ϵ there is a compact set K such that $\mu_n(K) > 1 - \epsilon$ for all n.

(b)   Show that $\{\mu_n\}$ is tight if and only if each of the k sequences of marginal distributions is tight on the line.

29.4.   Assume of (X_n, Y_n) that $X_n \Rightarrow X$ and $Y_n \Rightarrow c$. Show that $(X_n, Y_n) \Rightarrow (X, c)$. This is an example of Problem 29.2(b) where X_n and Y_n need not be assumed independent.

29.5.   Let μ and ν be a pair of centered normal distributions on R^2. Suppose that all the variances are 1 but that the two covariances differ. Show that the mixture $\frac{1}{2}\mu + \frac{1}{2}\nu$ is not normal even though its marginal distributions are.

29.6.   Prove analogues for R^k of the corollaries to Theorem 26.3.

29.7.   Devise a version of Lindeberg's theorem for R^k.

29.8.   Suppose that estimates U and V of α and β are approximately normally distributed with means α and β and small variances σ_u^2 and σ_v^2, respectively. Suppose that U and V are independent and $\beta > 0$. Use Example 29.1 for $f(x, y) = x/y$ to show that U/V is approximately normally distributed with mean α/β and variance

$$\frac{\alpha^2}{\beta^2} \left(\frac{\sigma_u^2}{\alpha^2} + \frac{\sigma_v^2}{\beta^2} \right).$$

A physical example is this: U estimates the charge $\alpha = e$ on the electron, V estimates the charge-to-mass ratio $\beta = e/m$, and U/V estimates the mass $\alpha/\beta = m$.

29.9.   To obtain a uniform distribution over the surface of the unit sphere in R^k, fill in the details in the following argument. Let X have the centered normal distribution with the identity as covariance matrix. For U orthogonal, UX has the same distribution. Let $\phi(x) = x/|x|$ on R^k (take $\phi(0) = 0$, say). Then $Y = \phi(X)$ lies on the unit sphere and, as $UY = \phi(UX)$ for U orthogonal, the distribution of Y is invariant under rotations.

29.10.   Assume that

$$\Sigma = \begin{bmatrix} \sigma_{11} & \sigma_{12} \\ \sigma_{12} & \sigma_{22} \end{bmatrix}$$

is positive definite, invert it explicitly, and show that the corresponding two-dimensional normal density is

$$(29.29) \quad f(x_1, x_2) = \frac{1}{2\pi D^{1/2}} \exp \left[-\frac{1}{2D} (\sigma_{22}x_1^2 - 2\sigma_{12}x_1x_2 + \sigma_{11}x_2^2) \right],$$

where $D = \sigma_{11}\sigma_{22} - \sigma_{12}^2$.

29.11.   21.15 29.10↑   Although uncorrelated random variables are not in general independent, they are if they are jointly normally distributed.

29.12.   ↑   Suppose that $f(X)$ and $g(Y)$ are uncorrelated for all bounded continuous f and g. Show that X and Y are independent.

Hint: Use characteristic functions.

29.13. 20.22↑ Suppose that the random vector X has a centered k-dimensional normal distribution whose covariance matrix has 1 as an eigenvalue of multiplicity r and 0 as an eigenvalue of multiplicity $k - r$. Show that $|X|^2$ has the chi-squared distribution with r degrees of freedom.

29.14. ↑ *Multinomial sampling.* Let $p_1, \ldots, p_k$ be positive and add to 1, and let Z_1, $Z_2, \ldots$ be independent k-dimensional random vectors such that Z_n has with probability p_i a 1 in the ith component and 0's elsewhere. Then $f_n = (f_{n1}, \ldots, f_{nk}) = \Sigma_{m=1}^n Z_m$ is the frequency count for a sample of size n from a multinomial population with cell probabilities p_i. Put $X_{ni} = (f_{ni} - np_i)/\sqrt{np_i}$ and $X_n = (X_{n1}, \ldots, X_{nk})$.

(a) Show that X_n has mean values 0 and covariances $\sigma_{ij} = (\delta_{ij}p_j - p_ip_j)/\sqrt{p_ip_j}$.

(b) Show that the chi-squared statistic $\Sigma_{i=1}^k (f_{ni} - np_i)^2/np_i$ has asymptotically the chi-squared distribution with $k - 1$ degrees of freedom.

29.15. 19.12↑ *A theorem of Poincaré.* (a) Suppose that $X_n = (X_{n1}, \ldots, X_{nn})$ is uniformly distributed over the surface of the sphere of radius $\sqrt{n}$ in R^n. Fix t and show that $X_{n1}, \ldots, X_{nt}$ are in the limit independent, each with the standard normal distribution.

(b) Suppose that the distribution of $X_n = (X_{n1}, \ldots, X_{nn})$ is spherically symmetric in the sense that $X_n/|X_n|$ is uniformly distributed over the unit sphere. Assume that $|X_n|^2/n \Rightarrow 1$ and show that $X_{n1}, \ldots, X_{nt}$ are asymptotically independent and normal.

SECTION 30.   THE METHOD OF MOMENTS*

The Moment Problem

For some distributions the characteristic function is intractable but moments can nonetheless be calculated. In these cases it is sometimes possible to prove weak convergence of the distributions by establishing that the moments converge. This approach requires conditions under which a distribution is uniquely determined by its moments, and this is for the same reason that the continuity theorem for characteristic functions requires for its proof the uniqueness theorem.

Theorem 30.1. *Let μ be a probability measure on the line having finite moments $\alpha_k = \int_{-\infty}^{\infty} x^k \mu(dx)$ of all orders. If the power series $\Sigma_k \alpha_k r^k/k!$ has a positive radius of convergence, then μ is the only probability measure with the moments $\alpha_1, \alpha_2, \ldots$.*

* This section may be omitted.

PROOF.   Let $\beta_k = \int_{-\infty}^{\infty} |x|^k \mu(dx)$ be the absolute moments. The first step is to show that

(30.1)
$$\frac{\beta_k r^k}{k!} \to 0, \qquad k \to \infty,$$

for some positive r. By hypothesis there exists an s, $0 < s < 1$, such that $\alpha_k s^k/k! \to 0$. Choose $0 < r < s$; then $2kr^{2k-1} < s^{2k}$ for large k. Since $|x|^{2k-1} \le 1 + |x|^{2k}$,

$$\frac{\beta_{2k-1} r^{2k-1}}{(2k-1)!} \le \frac{r^{2k-1}}{(2k-1)!} + \frac{\beta_{2k} s^{2k}}{(2k)!} ,$$

for large k. Hence (30.1) holds as k goes to infinity through odd values; since $\beta_k = \alpha_k$ for k even, (30.1) follows.

By (26.4),

$$\left| e^{itx} \left(e^{ihx} - \sum_{k=0}^{n} \frac{(ihx)^k}{k!} \right) \right| \le \frac{|hx|^{n+1}}{(n+1)!} ,$$

and therefore the characteristic function φ of μ satisfies

$$\left| \varphi(t+h) - \sum_{k=0}^{n} \frac{h^k}{k!} \int_{-\infty}^{\infty} (ix)^k e^{itx} \mu(dx) \right| \le \frac{|h|^{n+1} \beta_{n+1}}{(n+1)!} .$$

By (26.10), the integral here is $\varphi^{(k)}(t)$. By (30.1),

(30.2)
$$\varphi(t+h) = \sum_{k=0}^{\infty} \frac{\varphi^{(k)}(t)}{k!} h^k, \qquad |h| \le r.$$

If ν is another probability measure with moments α_k and characteristic function $\psi(t)$, the same argument gives

(30.3)
$$\psi(t+h) = \sum_{k=0}^{\infty} \frac{\psi^{(k)}(t)}{k!} h^k, \qquad |h| \le r.$$

Take $t = 0$; since $\varphi^{(k)}(0) = i^k \alpha_k = \psi^{(k)}(0)$ (see (26.9)), φ and ψ agree in $(-r, r)$ and hence have identical derivatives there. Taking $t = r - \epsilon$ and $t = -r + \epsilon$ in (30.2) and (30.3) shows that φ and ψ also agree in $(-2r + \epsilon, 2r - \epsilon)$ and hence in $(-2r, 2r)$. But then they must be the same argument agree in $(-3r, 3r)$ as well, and so on.* Thus φ and ψ coincide, and by the uniqueness theorem for characteristic functions, so do μ and ν.   ∎

A probability measure satisfying the conclusion of the theorem is said to be *determined by its moments.*

* This process is a version of analytic continuation.

Example 30.1.   For the standard normal distribution, $|\alpha_k| \leq k!$, and so the theorem implies that it is determined by its moments.   ∎

But not all measures are determined by their moments:

Example 30.2.   If N has the standard normal density, then e^N has the log-normal density

$$f(x) = \begin{cases} \dfrac{1}{\sqrt{2\pi}}\dfrac{1}{x}\,e^{-(\log x)^2/2} & \text{if } x > 0, \\ 0 & \text{if } x \leq 0. \end{cases}$$

Put $g(x) = f(x)\,(1 + \sin(2\pi \log x))$. If

$$\int_0^\infty x^k f(x) \sin\,(2\pi \log x)\,dx = 0, \qquad k = 0, 1, 2, \ldots,$$

then g, which is nonnegative, will be a probability density and will have the same moments as f. But a change of variable $\log x = s + k$ reduces the integral above to

$$\frac{1}{\sqrt{2\pi}}\,e^{k^2/2} \int_{-\infty}^\infty e^{-s^2/2} \sin 2\pi s\,ds,$$

which vanishes because the integrand is odd.   ∎

Theorem 30.2.   *Suppose that the distribution of X is determined by its moments, that the X_n have moments of all orders, and that $\lim_n E[X_n^r] = E[X^r]$ for $r = 1, 2, \ldots$. Then $X_n \Rightarrow X$.*

PROOF.   Let μ_n and μ be the distributions of X_n and X. Since $E[X_n^2]$ converges, it is bounded, say by K. By Markov's inequality, $P[|X_n| \geq x] \leq K/x^2$, which implies that the sequence $\{\mu_n\}$ is tight.

Suppose that $\mu_{n_k} \Rightarrow \nu$ and let Y be a random variable with distribution ν. If u is an even integer exceeding r, the convergence and hence boundedness of $E[X_n^u]$ implies that $E[X_{n_k}^r] \to E[Y^r]$ by the corollary to Theorem 25.12. By the hypothesis, then, $E[Y^r] = E[X^r]$—or ν and μ have the same moments. Since μ is by hypothesis determined by its moments, ν must be the same as μ, and so $\mu_{n_k} \Rightarrow \mu$. The conclusion now follows by the corollary to Theorem 25.10.   ∎

Convergence to the log-normal distribution cannot be proved by establishing convergence of moments (take X to have density f and the X_n to have density g in Example 30.2). Because of Example 30.1, however, this approach will work for a normal limit.

Moment Generating Functions

Suppose that μ has a moment generating function $M(s)$ for $s \in [-s_0, s_0]$, $s_0 > 0$. By (21.22), the hypothesis of Theorem 30.1 is satisfied, and so μ is determined by its moments, which are in turn determined by $M(s)$ via (21.23). Thus μ *is determined by $M(s)$ if it exists in a neighborhood of* 0.* The version of this for one-sided transforms was proved in Section 22—see (22.8).

Suppose that μ_n and μ have moment generating functions in a common interval $[-s_0, s_0]$, $s_0 > 0$, and suppose that $M_n(s) \to M(s)$ in this interval. Since $\mu_n[(-a, a)^c] \leq e^{-s_0 a}(M_n(-s_0) + M_n(s_0))$, it follows easily that $\{\mu_n\}$ is tight. Since $M(s)$ determines μ, the usual argument now gives $\mu_n \Rightarrow \mu$.

Central Limit Theorem by Moments

To understand the application of the method of moments, consider once again a sum $S_n = X_{n1} + \cdots + X_{nk_n}$, where $X_{n1}, \ldots, X_{nk_n}$ are independent and

$$(30.4) \qquad E[X_{nk}] = 0, \qquad E[X_{nk}^2] = \sigma_{nk}^2, \qquad s_n^2 = \sum_{k=1}^{k_n} \sigma_{nk}^2.$$

Suppose further that for each n there is an M_n such that $|X_{nk}| \leq M_n$, $k = 1, \ldots, k_n$, with probability 1. Finally, suppose that

$$(30.5) \qquad \frac{M_n}{s_n} \to 0.$$

All moments exist, and[†]

$$(30.6) \qquad S_n^r = \sum_{u=1}^{r} \Sigma' \frac{r!}{r_1! \ldots r_u! \, u!} \Sigma'' X_{ni_1}^{r_1} \cdots X_{ni_u}^{r_u},$$

where Σ' extends over the u-tuples of positive integers satisfying $r_1 + \cdots + r_u = r$ and Σ'' extends over the u-tuples $(i_1, \ldots, i_u)$ of distinct integers in the range $1 \leq i_\alpha \leq k_n$.

By independence, then,

$$(30.7) \qquad E\left[\left(\frac{S_n}{s_n}\right)^r\right] = \sum_{u=1}^{r} \Sigma' \frac{r!}{r_1! \ldots r_u! \, u!} A_n(r_1, \ldots, r_u),$$

---

* For another proof, see Problem 26.7. The present proof does not require the idea of analyticity.
† To deduce this from the multinomial formula, restrict the inner sum to u-tuples satisfying $1 \leq i_1 < \cdots < i_u \leq k_n$ and compensate by striking out the $1/u!$.

where

$$(30.8) \qquad A_n(r_1, \ldots, r_u) = \Sigma'' \frac{1}{s_n^r} E[X_{ni_1}^{r_1}] \cdots E[X_{ni_u}^{r_u}],$$

and Σ' and Σ'' have the same ranges as before. To prove that (30.7) converges to the rth moment of the standard normal distribution, it suffices to show that

$$(30.9) \qquad \lim_n A_n(r_1, \ldots, r_u) = \begin{cases} 1 & \text{if } r_1 = \cdots = r_u = 2, \\ 0 & \text{otherwise.} \end{cases}$$

Indeed, if r is even, all terms in (30.7) will then go to 0 except the one for which $u = r/2$ and $r_\alpha \equiv 2$, which will go to $r!/(r_1! \cdots r_u! u!) = 1 \cdot 3 \cdot 5 \cdots (r-1)$. And if r is odd, the terms will go to 0 without exception.

If $r_\alpha = 1$ for some α, then (30.9) holds because by (30.4) each summand in (30.8) vanishes. Suppose that $r_\alpha \geq 2$ for each α and $r_\alpha > 2$ for some α. Then $r > 2u$, and since $|E[X_{ni}^{r_\alpha}]| \leq M_n^{(r_\alpha - 2)} \sigma_{ni}^2$, it follows that $A_n(r_1, \ldots, r_u) \leq (M_n/s_n)^{r-2u} A_n(2, \ldots, 2)$. But this goes to 0 because (30.5) holds and because $A_n(2, \ldots, 2)$ is bounded by 1 (it increases to 1 if the sum in (30.8) is enlarged to include all the u-tuples $(i_1, \ldots, i_u)$).

It remains only to check (30.9) for $r_1 = \cdots = r_u = 2$. As just noted, $A_n(2, \ldots, 2)$ is at most 1, and it differs from 1 by $\Sigma s_n^{-2u} \sigma_{ni_1}^2 \cdots \sigma_{ni_u}^2$, the sum extending over the $(i_1, \ldots, i_u)$ with at least one repeated index. Since $\sigma_{ni}^2 \leq M_n^2$, the terms for example with $i_u = i_{u-1}$ sum to at most $M_n^2 s_n^{-2u} \Sigma \sigma_{ni_1}^2 \cdots \sigma_{ni_{u-1}}^2 \leq M_n^2 s_n^{-2}$. Thus $1 - A_n(2, \ldots, 2) \leq u^2 M_n^2 s_n^{-2} \to 0$.

This proves that the moments (30.7) converge to those of the normal distribution and hence that $S_n/s_n \Rightarrow N$.

Application to Sampling Theory

Suppose that n numbers

$$x_{n1}, x_{n2}, \ldots, x_{nn},$$

not necessarily distinct, are associated with the elements of a population of size n. Suppose that these numbers are normalized by the requirement

$$(30.10) \qquad \sum_{h=1}^n x_{nh} = 0, \qquad \sum_{h=1}^n x_{nh}^2 = 1, \qquad M_n = \max_{h \leq n} |x_{nh}|.$$

An ordered sample $X_{n1}, \ldots, X_{nk_n}$ is taken, where the sampling is without replacement. By (30.10), $E[X_{nk}] = 0$ and $E[X_{nk}^2] = 1/n$. Let $s_n^2 = k_n/n$ be the fraction of the population sampled. If the X_{nk} were independent, which they are not, $S_n = X_{n1} + \cdots + X_{nk_n}$ would have variance s_n^2. If k_n is small in comparison with n, the effects of dependence should be small. It will be shown that $S_n/s_n \Rightarrow N$ if

(30.11) $\qquad s_n^2 = \dfrac{k_n}{n} \to 0, \qquad \dfrac{M_n}{s_n} \to 0, \qquad k_n \to \infty.$

Since $M_n^2 \geq n^{-1}$ by (30.10), the second condition here in fact implies the third.

The moments again have the form (30.7), but this time $E[X_{ni_1}^{r_1} \cdots X_{ni_u}^{r_u}]$ cannot be factored as in (30.8). On the other hand, this expected value is by symmetry the same for each of the $(k_n)_u = k_n(k_n - 1) \cdots (k_n - u + 1)$ choices of the indices i_α in the sum Σ''. Thus

$$A_n(r_1, \ldots, r_u) = \frac{(k_n)_u}{s_n^r} E[X_{n1}^{r_1} \cdots X_{nu}^{r_u}].$$

The problem again is to prove (30.9).

The proof goes by induction on u. Now $A_n(r) = k_n s_n^{-r} n^{-1} \Sigma_{h=1}^n x_{nh}^r$, so that $A_n(1) = 0$ and $A_n(2) = 1$. If $r \geq 3$, then $|x_{nh}^r| \leq M_n^{r-2} x_{nh}^2$, and so $|A_n(r)| \leq (M_n/s_n)^{r-2} \to 0$ by (30.11).

Next suppose as induction hypothesis that (30.9) holds with $u - 1$ in place of u. Since the sampling is without replacement, $E[X_{n1}^{r_1} \cdots X_{nu}^{r_u}] = \Sigma x_{nh_1}^{r_1} \cdots x_{nh_u}^{r_u}/(n)_u$, where the summation extends over the u-tuples $(h_1, \ldots, h_u)$ of distinct integers in the range $1 \leq h_\alpha \leq n$. In this last sum enlarge the range by requiring of $(h_1, h_2, \ldots, h_u)$ only that $h_2, \ldots, h_u$ be distinct, and then compensate by subtracting away the terms where $h_1 = h_2$, where $h_1 = h_3$, and so on. The result is

$$E[X_{n1}^{r_1} \cdots X_{nu}^{r_u}] = \frac{n(n)_{u-1}}{(n)_u} E[X_{n1}^{r_1}] E[X_{n2}^{r_2} \cdots X_{nu}^{r_u}]$$

$$- \sum_{\alpha=2}^u \frac{(n)_{u-1}}{(n)_u} E[X_{n2}^{r_2} \cdots X_{n\alpha}^{r_1+r_\alpha} \cdots X_{nu}^{r_u}].$$

This takes the place of the factorization made possible in (30.8) by the assumed independence there. It gives

$$A_n(r_1, \ldots, r_u) = \frac{n}{n - u + 1} \frac{k_n - u + 1}{k_n} A_n(r_1) A_n(r_2, \ldots, r_u)$$

$$- \frac{k_n - u + 1}{n - u + 1} \sum_{\alpha=2}^u A_n(r_2, \ldots, r_1 + r_\alpha, \ldots, r_u).$$

By the induction hypothesis the last sum is bounded, and the factor in front goes to 0 by (30.11). As for the first term on the right, the factor in front goes to 1. If $r_1 \neq 2$, $A_n(r_1) \to 0$ and $A_n(r_2, \ldots, r_u)$ is bounded, and so $A_n(r_1, \ldots, r_u) \to 0$. The same holds by symmetry if $r_\alpha \neq 2$ for some α other than 1. If $r_1 = \cdots = r_u = 2$, then $A_n(r_1) = 1$, and $A_n(r_2, \ldots, r_u) \to 1$ by the induction hypothesis.

Thus (30.9) holds in all cases, and $S_n/s_n \Rightarrow N$ follows by the method of moments.

Application to Number Theory

Let $g(m)$ be the number of distinct prime factors of the integer m; for example $g(3^4 \cdot 5^2) = 2$. Since there are infinitely many primes, $g(m)$ is unbounded above; for the same reason, it drops back to 1 for infinitely many m (for the primes and their powers). Since g fluctuates in an irregular way, it is natural to inquire into its average behavior.

On the space Ω of positive integers, let P_n be the probability measure that places mass $1/n$ at each of $1, 2, \ldots, n$, so that among the first n positive integers the proportion that are contained in a given set A is just $P_n(A)$. The problem is to study $P_n[m: g(m) \le x]$ for large n.

If $\delta_p(m)$ is 1 or 0 according as the prime p divides m or not, then

$$(30.12) \qquad g(m) = \sum_p \delta_p(m).$$

Probability theory can be used to investigate this sum because under P_n the $\delta_p(m)$ behave somewhat like independent random variables.* If $p_1, \ldots, p_u$ are distinct primes, then by the fundamental theorem of arithmetic, $\delta_{p_1}(m) = \cdots = \delta_{p_u}(m) = 1$—that is, each p_i divides m—if and only if the product $p_1 \cdots p_u$ divides m. The probability under P_n of this is just n^{-1} times the number of m in the range $1 \le m \le n$ that are multiples of $p_1 \cdots p_u$, and this number is the integer part of $n/p_1 \cdots p_u$. Thus

$$(30.13) \qquad P_n[m: \delta_{p_i}(m) = 1, i = 1, \ldots, u] = \frac{1}{n} \left\lfloor \frac{n}{p_1 \cdots p_u} \right\rfloor$$

for distinct p_i.

Now let X_p be independent random variables (on some probability space, one variable for each prime p) satisfying

$$P[X_p = 1] = \frac{1}{p}, \qquad P[X_p = 0] = 1 - \frac{1}{p}.$$

If $p_1, \ldots, p_u$ are distinct, then

$$(30.14) \qquad P[X_{p_i} = 1, i = 1, \ldots, u] = \frac{1}{p_1 \cdots p_u}.$$

For fixed $p_1, \ldots, p_u$, (30.13) converges to (30.14) as $n \to \infty$. Thus the behavior of the X_p can serve as a guide to that of the $\delta_p(m)$. If $m \le n$, (30.12) is

---

* See also Problems 2.15, 5.16, and 6.17.

$\Sigma_{p \le n} \delta_p(m)$ because no prime exceeding m can divide it. The idea is to compare this sum with the corresponding sum $\Sigma_{p \le n} X_p$.

This will require from number theory the elementary estimate*

$$(30.15) \qquad \sum_{p \le x} \frac{1}{p} = \log \log x + O(1).$$

The mean and variance of $\Sigma_{p \le n} X_p$ are $\Sigma_{p \le n} p^{-1}$ and $\Sigma_{p \le n} p^{-1}(1 - p^{-1})$; since $\Sigma_p \, p^{-2}$ converges, each of these two sums is asymptotically $\log \log n$. Comparing $\Sigma_{p \le n} \, \delta_p(m)$ with $\Sigma_{p \le n} X_p$ then leads one to conjecture the *Erdös-Kac central limit theorem for the prime divisor function*:

Theorem 30.3.   *For all x,*

$$(30.16) \quad P_n \left[m: \frac{g(m) - \log \log n}{\sqrt{\log \log n}} \le x \right] \to \frac{1}{\sqrt{2\pi}} \int_{-\infty}^{x} e^{-u^2/2} \, du.$$

PROOF.   The argument uses the method of moments. The first step is show that (30.16) is unaffected if the range of p in (30.12) is further restricted. Let $\{\alpha_n\}$ be a sequence going to infinity slowly enough that

$$(30.17) \qquad \frac{\log \alpha_n}{\log n} \to 0$$

but fast enough that

$$(30.18) \qquad \sum_{\alpha_n < p \le n} \frac{1}{p} = o(\log \log n)^{1/2}.$$

Because of (30.15), these two requirements are met if, for example, $\log \alpha_n = \log n / \log \log n$.

Now define

$$(30.19) \qquad g_n(m) = \sum_{p \le \alpha_n} \delta_p(m).$$

For a function f of positive integers, let

$$E_n[f] = n^{-1} \sum_{m=1}^{n} f(m)$$

denote its expected value computed with respect to P_n. By (30.13) for $u = 1$,

$$E_n \left[\sum_{p > \alpha_n} \delta_p \right] = \sum_{\alpha_n < p \le n} P_n[m: \delta_p = 1] \le \sum_{\alpha_n < p \le n} \frac{1}{p}.$$

By (30.18) and Markov's inequality

---

* See, for example, HARDY & WRIGHT, Chapter XXII, or Problem 18.14.

$$P_n[m: |g(m) - g_n(m)| \geq \epsilon(\log \log n)^{1/2}] \to 0.$$

Therefore (Theorem 25.4), (30.16) is unaffected if $g_n(m)$ is substituted for $g(m)$.

Now compare (30.19) with the corresponding sum $S_n = \Sigma_{p \leq \alpha_n} X_p$. The mean and variance of S_n are

$$c_n = \sum_{p \leq \alpha_n} \frac{1}{p}, \qquad s_n^2 = \sum_{p \leq \alpha_n} \frac{1}{p}\left(1 - \frac{1}{p}\right),$$

and each is $\log \log n + o(\log \log n)^{1/2}$ by (30.18). Thus (see Example 25.8), (30.16) with $g(m)$ replaced as above is equivalent to

$$(30.20) \qquad P_n\left[m: \frac{g_n(m) - c_n}{s_n} \leq x\right] \to \frac{1}{\sqrt{2\pi}} \int_{-\infty}^{x} e^{-u^2/2} \, du.$$

It therefore suffices to prove (30.20).

Since the X_p are bounded, the analysis of the moments (30.7) applies here. The only difference is that the summands in S_n are indexed not by the integers k in the range $k \leq k_n$ but by the primes p in the range $p \leq \alpha_n$; also, X_p must be replaced by $X_p - p^{-1}$ to center it. Thus the rth moment of $(S_n - c_n)/s_n$ converges to that of the normal distribution, and so (30.20) and (30.16) will follow by the method of moments if it is shown that as $n \to \infty$,

$$(30.21) \qquad E\left[\left(\frac{S_n - c_n}{s_n}\right)^r\right] - E_n\left[\left(\frac{g_n - c_n}{s_n}\right)^r\right] \to 0$$

for each r.

Now $E[S_n^r]$ is the sum

$$(30.22) \qquad \sum_{u=1}^{r} \Sigma' \frac{r!}{r_1! \cdots r_u! \, u!} \Sigma'' E[X_{p_1}^{r_1} \cdots X_{p_u}^{r_u}],$$

where the range of Σ' is as in (30.6) and (30.7) and Σ'' extends over the u-tuples $(p_1, \ldots, p_u)$ of distinct primes not exceeding α_n. Since X_p assumes only the values 0 and 1, from the independence of the X_p and the fact that the p_i are distinct, it follows that the summand in (30.22) is

$$(30.23) \qquad E[X_{p_1} \cdots X_{p_u}] = \frac{1}{p_1 \cdots p_u}.$$

By the definition (30.19), $E_n[g_n^r]$ is just (30.22) with the summand replaced by $E_n[\delta_{p_1}^{r_1} \cdots \delta_{p_u}^{r_u}]$. Since $\delta_p(m)$ assumes only the values 0 and 1, from (30.13) and the fact that the p_i are distinct, it follows that this summand is

$$(30.24) \qquad E_n[\delta_{p_1} \cdots \delta_{p_u}] = \frac{1}{n}\left[\frac{n}{p_1 \cdots p_u}\right].$$

But (30.23) and (30.24) differ by at most $1/n$, and hence $E[S_n^r]$ and $E_n[g_n^r]$ differ by at most the sum (30.22) with the summand replaced by $1/n$. Therefore,

$$(30.25) \qquad |E[S_n^r] - E_n[g_n^r]| \le \frac{1}{n}\left(\sum_{p \le \alpha_n} 1\right)^r \le \frac{\alpha_n^r}{n}.$$

Now

$$E[(S_n - c_n)^r] = \sum_{k=0}^r \binom{r}{k} E[S_n^k](-c_n)^{r-k},$$

and $E_n[(g_n - c_n)^r]$ has the analogous expansion. Comparing the two expansions term for term and applying (30.25) shows that

$$(30.26) \quad |E[(S_n - c_n)^r] - E_n[(g_n - c_n)^r]|$$

$$\le \sum_{k=0}^r \binom{r}{k}\frac{\alpha_n^k}{n} c_n^{r-k} = \frac{1}{n}(\alpha_n + c_n)^r.$$

Since $c_n \le \alpha_n$, and since $\alpha_n^r/n \to 0$ by (30.17), (30.21) follows as required. ∎

The method of proof requires passing from (30.12) to (30.19). Without this, the α_n on the right in (30.26) would instead be n, and the difference on the left would not go to 0; hence the truncation (30.19) for an α_n small enough to satisfy (30.17). On the other hand, α_n must be large enough to satisfy (30.18) in order that the truncation leave (30.16) unaffected.

PROBLEMS

30.1. From the central limit theorem under the assumption (30.5) get the full Lindeberg theorem by a truncation argument.

30.2. For a sample of size k_n with replacement from a population of size n, the probability of no duplicates is $\prod_{j=0}^{k_n-1} (1 - j/n)$. Under the assumption $k_n/\sqrt{n} \to 0$ in addition to (30.10), deduce the asymptotic normality of S_n by a reduction to the independent case.

30.3. Let μ be a probability measure on R^k for which $\int_{R^k} |x_i|^r \mu(dx) < \infty$ for $i = 1, \ldots, k$ and $r = 1, 2, \ldots$. Consider the cross moments

$$\alpha(r_1, \ldots, r_k) = \int_{R^k} x_1^{r_1} \cdots x_k^{r_k} \mu(dx)$$

for nonnegative integers r_i.

(a) Suppose for each i that

$$\sum_r \frac{\theta^r}{r!} \int_{R^k} |x_i|^r \mu(dx)$$

has a positive radius of convergence as a power series in θ. Show that μ is determined by its moments in the sense that, if a probability measure ν satisfies $\alpha(r_1, \ldots, r_k) = \int x_1^{r_1} \cdots x_k^{r_k} \nu(dx)$ for all $r_1, \ldots, r_k$, then ν coincides with μ.

(b)   Show that a k-dimensional normal distribution is determined by its moments.

30.4.   ↑   Suppose that X and Y are bounded random variables and that X^m and Y^n are uncorrelated for $m, n = 1, 2, \ldots$. Show that X and Y are independent.

30.5.   30.3↑   Let μ_n and μ be probability measures on R^k. Suppose that μ is determined by its moments and that

$$\int_{R^k} x_1^{r_1} \cdots x_k^{r_k} \mu_n(dx) \to \int_{R^k} x_1^{r_1} \cdots x_k^{r_k} \mu(dx)$$

for all nonnegative integers $r_1, \ldots, r_k$. Show that $\mu_n \Rightarrow \mu$.

30.6.   26.17 30.5↑   (a)   In the notation (26.25), show for $\lambda \neq 0$ that

$$(30.27) \qquad\qquad M[(\cos \lambda x)^r] = \binom{r}{r/2} \frac{1}{2^r}$$

for even r and that the mean is 0 for odd r. It follows by the method of moments that $\cos \lambda x$ has a distribution in the sense of (25.27), and in fact of course the relative measure is

$$(30.28) \qquad\qquad \rho[x: \cos \lambda x \leq u] = 1 - \frac{1}{\pi} \arccos u, \qquad -1 < u < 1.$$

(b)   Suppose that $\lambda_1, \lambda_2, \ldots$ are linearly independent over the field of rationals in the sense that, if $n_1\lambda_1 + \cdots + n_m\lambda_m = 0$ for integers n_ν, then $n_1 = \cdots = n_m = 0$. Show that

$$(30.29) \qquad\qquad M\left[\prod_{\nu=1}^{k} (\cos \lambda_\nu x)^{r_\nu}\right] = \prod_{\nu=1}^{k} M[(\cos \lambda_\nu x)^{r_\nu}]$$

for nonnegative integers $r_1, \ldots, r_k$.

(c)   Let $X_1, X_2, \ldots$ be independent and have the distribution on the right in (30.28). Show that

$$(30.30) \qquad\qquad \rho\left[x: \sum_{j=1}^{k} \cos \lambda_j x \leq u\right] = P[X_1 + \cdots + X_k \leq u].$$

(d)   Show that

$$(30.31) \qquad \lim_{k \to \infty} \rho\left[x: u_1 < \sqrt{\frac{2}{k}} \sum_{j=1}^{k} \cos \lambda_j x \leq u_2\right] = \frac{1}{\sqrt{2\pi}} \int_{u_1}^{u_2} e^{-v^2/2}\, dv.$$

For a signal that is the sum of a large number of pure cosine signals with incommensurable frequencies, (30.31) describes the relative amount of time the signal is between u_1 and u_2.

30.7.   6.17↑   From (30.16), deduce once more the Hardy-Ramanujan theorem (see (6.14)).

30.8.   ↑   (a)   Prove that (if P_n puts probability $1/n$ at $1, \ldots, n$)

(30.32) $\lim_n P_n \left[m: \left| \dfrac{\log \log m - \log \log n}{\sqrt{\log \log n}} \right| \geq \epsilon \right] = 0.$

(b)   From (30.16) deduce that (see (2.21) for the notation)

(30.33) $D \left[m: \dfrac{g(m) - \log \log m}{\sqrt{\log \log m}} \leq x \right] = \dfrac{1}{\sqrt{2\pi}} \int_{-\infty}^{x} e^{-u^2/2} \, du.$

30.9. ↑ Let $G(m)$ be the number of prime factors in m with multiplicity counted. In the notation of Problem 5.16, $G(m) = \Sigma_p \, \alpha_p(m)$.

(a)   Show for $k \geq 1$ that $P_n[m: \alpha_p(m) - \delta_p(m) \geq k] \leq 1/p^{k+1}$; hence $E_n[\alpha_p - \delta_p] \leq 2/p^2$.

(b)   Show that $E_n[G - g]$ is bounded.

(c)   Deduce from (30.16) that

$$P_n \left[m: \dfrac{G(m) - \log \log n}{\sqrt{\log \log n}} \leq x \right] \to \dfrac{1}{\sqrt{2\pi}} \int_{-\infty}^{x} e^{-u^2/2} \, du.$$

(d)   Prove for G the analogue of (30.33).

30.10. ↑   Prove the Hardy-Ramanujan theorem in the form

$$D \left[m: \left| \dfrac{g(m)}{\log \log m} - 1 \right| \geq \epsilon \right] = 0.$$

Prove this with G in place of g.

CHAPTER 6

Derivatives and
Conditional Probability

SECTION 31. DERIVATIVES ON THE LINE*

This section on Lebesgue's theory of derivatives for real functions of a real variable serves to introduce the general theory of Radon-Nikodym derivatives, which underlies the modern theory of conditional probability. The results here are interesting in themselves and will be referred to later for purposes of illustration and comparison, but they will not be required in subsequent proofs.

The Fundamental Theorem of Calculus

To what extent are the operations of integration and differentiation inverse to one another? A function F is by definition an *indefinite integral* of another function f on $[a, b]$ if

$$(31.1) \qquad F(x) - F(a) = \int_a^x f(t)\, dt$$

for $a \leq x \leq b$; F is by definition a *primitive* of f if it has derivative f:

$$(31.2) \qquad F'(x) = f(x)$$

for $a \leq x \leq b$. According to the *fundamental theorem of calculus* (see (17.5)), these concepts coincide in the case of continuous f:

Theorem 31.1. *Suppose that f is continuous on $[a, b]$.*

(i) *An indefinite integral of f is a primitive of f: if* (31.1) *holds for all x in $[a, b]$, then so does* (31.2).

(ii) *A primitive of f is an indefinite integral of f: if* (31.2) *holds for all x in $[a, b]$, then so does* (31.1).

---

* This section may be omitted.

354

A basic problem is to investigate the extent to which this theorem holds if f is not assumed continuous. First consider part (i). Suppose f is integrable, so that the right side of (31.1) makes sense. If f is 0 for $x < m$ and 1 for $x \geq m$ ($a < m < b$), then an F satisfying (31.1) has no derivative at m. It is thus too much to ask that (31.2) hold for all x. On the other hand, according to a famous theorem of Lebesgue, if (31.1) holds for all x, then (31.2) holds almost everywhere—that is, except for x in a set of Lebesgue measure 0. In this section *almost everywhere* will refer to Lebesgue measure only. This result, the most one could hope for, will be proved below (Theorem 31.3).

Now consider part (ii) of Theorem 31.1. Suppose that (31.2) holds almost everywhere, as in Lebesgue's theorem, just stated. Does (31.1) follow? The answer is no: If f is identically 0, and if $F(x)$ is 0 for $x < m$ and 1 for $x \geq m$ ($a < m < b$), then (31.2) holds almost everywhere, but (31.1) fails for $x \geq m$. The question was wrongly posed, and the trouble is not far to seek: If f is integrable and (31.1) holds, then

$$(31.3) \qquad F(x + h) - F(x) = \int_a^b I_{(x,x+h)}(t)f(t)\, dt \to 0$$

as $h \downarrow 0$ by the dominated convergence theorem. Together with a similar argument for $h \uparrow 0$ this shows that F must be continuous. Hence the question becomes this: if F is continuous and f is integrable, and if (31.2) holds almost everywhere, does (31.1) follow? The answer strangely enough is still no: In Example 31.1 there is constructed a continuous, strictly increasing F for which $F'(x) = 0$ except on a set of Lebesgue measure 0, and (31.1) is of course impossible if f vanishes almost everywhere and F is strictly increasing. This leads to the problem of characterizing those F for which (31.1) does follow if (31.2) holds outside a set of Lebesgue measure 0 and f is integrable. In other words, which functions are the integrals of their (almost everywhere) derivatives? Theorem 31.7 gives the characterization.

It is possible to extend part (ii) of Theorem 31.1 in a different direction. Suppose that (31.2) holds for every x, not just almost everywhere. In Example 17.4 there was given a function F, everywhere differentiable, whose derivative f is not integrable, and in this case the right side of (31.1) has no meaning. If, however, (31.2) holds for every x, and if f is integrable, then (31.1) does hold for all x. For most purposes of probability theory, it is natural to impose conditions only almost everywhere, and so this theorem will not be proved here.*

The program then is first to show that (31.1) with f integrable implies that (31.2) holds almost everywhere, and second to characterize those F for which the reverse implication is valid. It will be convenient first to consider only non-

---

* For a proof, see RUDIN', p. 179.

negative f and nondecreasing F. This is the case of greatest interest for probability theory; F can be regarded as a distribution function and f as a density.

In Chapters 4 and 5 many distribution functions F were either shown to have a density f with respect to Lebesgue measure or were assumed to have one, but such F's were never intrinsically characterized, as they will be in this section.

Derivatives of Integrals

The first step is to show that a nondecreasing function has a derivative almost everywhere. This requires two preliminary results. Let λ denote Lebesgue measure.

Lemma 1. *Let A be a bounded linear Borel set, and let $\mathcal{J}$ be a collection of open intervals covering A. Then $\mathcal{J}$ has a finite, disjoint subcollection $I_1, \ldots, I_k$ for which $\sum_{i=1}^{k} \lambda(I_i) \geq \lambda(A)/6$.*

PROOF. By regularity (Theorem 12.3) A contains a compact subset K satisfying $\lambda(K) \geq \lambda(A)/2$. Choose in $\mathcal{J}$ a finite subcollection $\mathcal{J}_0$ covering K. Let I_1 be an interval in $\mathcal{J}_0$ of maximal length; discard from $\mathcal{J}_0$ the interval I_1 and all the others that intersect I_1. Among the intervals remaining in $\mathcal{J}_0$, let I_2 be one of maximal length; discard I_2 and all intervals that intersect it. Continue in this way until $\mathcal{J}_0$ is exhausted. The I_i are disjoint. Let J_i be the interval with the same midpoint as I_i and three times the length. If I is an interval in $\mathcal{J}_0$ that is cast out because it meets I_i, then $I \subset J_i$. Thus each discarded interval is contained in one of the J_i, and so the J_i cover K. Hence $\sum \lambda(I_i) = \sum \lambda(J_i)/3 \geq \lambda(K)/3 \geq \lambda(A)/6$.■

If

(31.4) $$\Delta: a = a_0 < a_1 < \cdots < a_k = b$$

is a partition of an interval $[a, b]$ and F is a function over $[a, b]$, let

(31.5) $$\|F\|_\Delta = \sum_{i=1}^{k} |F(a_i) - F(a_{i-1})|.$$

Lemma 2. *Consider a partition* (31.4) *and a positive θ. If*

(31.6) $$F(a) \leq F(b),$$

and if

(31.7) $$\frac{F(a_i) - F(a_{i-1})}{a_i - a_{i-1}} \leq -\theta$$

for a set of intervals $[a_{i-1}, a_i]$ *of total length d, then*

$$\|F\|_\Delta \geq |F(b) - F(a)| + 2\theta d.$$

This also holds if the inequalities in (31.6) and (31.7) are reversed and $-\theta$ is replaced by θ in the latter.

PROOF. The figure shows the case where $k = 2$ and the left-hand interval satisfies (31.7). Here F falls at least θd over $[a, a + d]$, rises the same amount over $[a + d, u]$, and then rises $F(b) - F(a)$ over $[u, b]$.

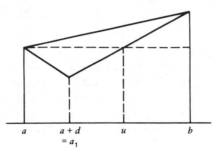

For the general case, let Σ' denote summation over those i satisfying (31.7) and let Σ'' denote summation over the remaining i ($1 \leq i \leq k$). Then

$$\|F\|_\Delta = \Sigma'(F(a_{i-1}) - F(a_i)) + \Sigma''|F(a_i) - F(a_{i-1})|$$

$$\geq \Sigma'(F(a_{i-1}) - F(a_i)) + |\Sigma''(F(a_i) - F(a_{i-1}))|$$

$$= \Sigma'(F(a_{i-1}) - F(a_i)) + |(F(b) - F(a)) + \Sigma'(F(a_{i-1}) - F(a_i))|.$$

As all the differences in this last expression are nonnegative, the absolute value bars can be suppressed; therefore,

$$\|F\|_\Delta \geq F(b) - F(a) + 2\Sigma'(F(a_{i-1}) - F(a_i))$$

$$\geq F(b) - F(a) + 2\theta\Sigma'(a_i - a_{i-1}). \qquad \blacksquare$$

A function F has at each x four *derivates*, the upper and lower right derivates

$$D^F(x) = \lim_{h \downarrow 0} \sup \frac{F(x + h) - F(x)}{h}, \quad D_F(x) = \lim_{h \downarrow 0} \inf \frac{F(x + h) - F(x)}{h},$$

and the upper and lower left derivates

$$^FD(x) = \lim_{h \downarrow 0} \sup \frac{F(x) - F(x - h)}{h}, \quad _FD(x) = \lim_{h \downarrow 0} \inf \frac{F(x) - F(x - h)}{h}.$$

There is a derivative at x if and only if these four quantities have a common value. Suppose that F has derivative $F'(x)$ at x. If $u \le x \le v$, then

$$\left| \frac{F(v) - F(u)}{v - u} - F'(x) \right|$$

$$\le \frac{v - x}{v - u} \left| \frac{F(v) - F(x)}{v - x} - F'(x) \right| + \frac{x - u}{v - u} \left| \frac{F(x) - F(u)}{x - u} - F'(x) \right|.$$

Therefore,

(31.8)
$$\frac{F(v) - F(u)}{v - u} \to F'(x)$$

if u and v approach x in such a way that $u \le x \le v$ and $u < v$.

Suppose that F is measurable and that it is continuous except possibly at countably many points. This will be true if F is nondecreasing or is the difference of two nondecreasing functions. Let M be a countable, dense set containing all the discontinuity points of F; let $r_n(x)$ be the smallest number of the form k/n exceeding x. Then

$$D^F(x) = \lim_{n \to \infty} \sup_{\substack{x < y < r_n(x) \\ y \in M}} \frac{F(y) - F(x)}{y - x};$$

the function inside the limit is measurable because the x-set where it exceeds α is

$$\bigcup_{y \in M} [x : x < y < r_n(x), F(y) - F(x) > \alpha (y - x)].$$

Thus $D^F(x)$ is measurable, as are the other three derivates. This does not exclude infinite values.

The following theorems concern real functions defined over the whole line. They could also be formulated for functions over an interval.

Theorem 31.2. *A nondecreasing function F is differentiable except on a set of Lebesgue measure 0. The derivative F' is measurable* and nonnegative, and*

(31.9)
$$\int_a^b F'(t) \, dt \le F(b) - F(a)$$

for all a and b.

---

* Outside some Borel set A such that $\lambda(A) = 0$, F has finite derivative $F' = D^F$; since D^F is measurable, F' is measurable if it is set equal to 0 (say) on A.

PROOF. If it can be shown that

$$(31.10) \qquad D^F(x) \le {}_FD(x)$$

except on a set of Lebesgue measure 0, then by the same result applied to $G(x)$ $= -F(-x)$ it will follow that ${}^FD(x) = D^G(-x) \le {}_GD(-x) = D_F(x)$ almost everywhere. This will imply that $D_F(x) \le D^F(x) \le {}_FD(x) \le {}^FD(x) \le D_F(x)$ almost everywhere, since first and third of these inequalities are obvious, and so outside a set of Lebesgue measure 0, F will have a derivative, possibly infinite. Since F is nondecreasing, F' must be nonnegative, and once (31.9) is proved, it will follow that F' is finite almost everywhere.

If (31.10) is violated for a particular x, then for some pair α, β of rationals satisfying $\alpha < \beta$, x will lie in the set $A_{\alpha\beta} = [x: {}_FD(x) < \alpha < \beta < D^F(x)]$. Since there are only countably many of these sets, (31.10) will hold outside a set of Lebesgue measure 0 if $\lambda(A_{\alpha\beta}) = 0$ for all α and β.

Put $G(x) = F(x) - \frac{1}{2}(\alpha + \beta)x$ and $\theta = \frac{1}{2}(\beta - \alpha)$. Since differentiation is linear, $A_{\alpha\beta} = B_\theta = [x: {}_GD(x) < -\theta < 0 < D^G(x)]$. Since F and G have only countably many discontinuities, it suffices to prove that $\lambda(C_\theta) = 0$, where C_θ is the set of points in B_θ that are continuity points of G. Consider an interval (a, b), and suppose for the moment that $G(a) \le G(b)$. For each x in C_θ satisfying $a < x < b$, from ${}_GD(x) < -\theta$ it follows that there exists an open interval (a_x, b_x) for which $x \in (a_x, b_x) \subset (a, b)$ and

$$(31.11) \qquad \frac{G(b_x) - G(a_x)}{b_x - a_x} < -\theta.$$

There exists by Lemma 1 a finite, disjoint collection (a_{x_i}, b_{x_i}) of these intervals of total length $\Sigma\, (b_{x_i} - a_{x_i}) \ge \lambda((a, b) \cap C_\theta)/6$. Let Δ be the partition (31.4) of $[a, b]$ with the points a_{x_i} and b_{x_i} in the role of the $a_1, \ldots, a_{k-1}$. By Lemma 2,

$$(31.12) \qquad \|G\|_\Delta \ge |G(b) - G(a)| + \frac{1}{3}\theta\lambda((a, b) \cap C_\theta).$$

If instead of $G(a) \le G(b)$ the reverse inequality holds, choose a_x and b_x so that the ratio in (31.11) exceeds θ, which is possible because $D^G(x) > \theta$ for $x \in C_\theta$. Again (31.12) follows.

In each interval $[a, b]$ there is thus a partition (31.4) satisfying (31.12). Apply this to each interval $[a_{i-1}, a_i]$ in the partition. This gives a partition Δ_1 that refines Δ, and adding the corresponding inequalities (31.12) leads to

$$\|G\|_{\Delta_1} \ge \|G\|_\Delta + \frac{1}{3}\theta\lambda((a, b) \cap C_\theta).$$

Continuing leads to a sequence of finer partitions Δ_n such that

$$(31.13) \qquad \|G\|_{\Delta_n} \ge n\frac{\theta}{3}\,\lambda((a, b) \cap C_\theta).$$

Now $\|G\|_\Delta$ is bounded by $|F(b) - F(a)| + \frac{1}{2}|\alpha + \beta|(b - a)$ because F is monotonic. Thus (31.13) is impossible unless $\lambda((a, b) \cap C_\theta) = 0$. Since (a, b) can be any interval, $\lambda(C_\theta) = 0$. This proves (31.10) and establishes the differentiability of F almost everywhere.

It remains to prove (31.9). Let

$$(31.14) \qquad\qquad f_n(x) = \frac{F(x + n^{-1}) - F(x)}{n^{-1}}.$$

Now f_n is nonnegative, and by what has been shown, $f_n(x) \to F'(x)$ except on a set of Lebesgue measure 0. By Fatou's lemma and the fact that F is nondecreasing,

$$\int_a^b F'(x)\, dx \leq \liminf_n \int_a^b f_n(x)\, dx$$

$$= \liminf_n \left[n \int_b^{b+n^{-1}} F(x)\, dx - n \int_a^{a+n^{-1}} F(x)\, dx \right]$$

$$\leq \liminf_n [F(b + n^{-1}) - F(a)] = F(b+) - F(a).$$

Replacing b by $b - \epsilon$ and letting $\epsilon \to 0$ gives (31.9). $\blacksquare$

Theorem 31.3. *If f is nonnegative and integrable, and if $F(x) = \int_{-\infty}^x f(t)\, dt$, then $F'(x) = f(x)$ except on a set of Lebesgue measure 0.*

Since f is nonnegative, F is nondecreasing and hence by Theorem 31.2 is differentiable almost everywhere. The problem is to show that the derivative F' coincides with f almost everywhere.

PROOF FOR BOUNDED f. Suppose first that f is bounded by M. Define f_n by (31.14). Then $f_n(x) = n \int_x^{x+n^{-1}} f(t)\, dt$ is bounded by M and converges almost everywhere to $F'(x)$, so that the bounded convergence theorem gives

$$\int_a^b F'(x)\, dx = \lim_n \int_a^b f_n(x)\, dx$$

$$= \lim_n \left[n \int_b^{b+n^{-1}} F(x)\, dx - n \int_a^{a+n^{-1}} F(x)\, dx \right].$$

Since F is continuous (see (31.3)), this last limit is $F(b) - F(a) = \int_a^b f(x)\, dx$.

Thus $\int_A F'(x)\, dx = \int_A f(x)\, dx$ for bounded intervals $A = (a, b]$. Since these form a π-system, it follows (see Example 16.9) that $F' = f$ almost everywhere.

PROOF FOR INTEGRABLE f. Apply the result for bounded functions to f truncated at n: If $h_n(x)$ is $f(x)$ or n as $f(x) \leq n$ or $f(x) > n$, then $H_n(x) = \int_{-\infty}^{x} h_n(t)\, dt$ differentiates almost everywhere to $h_n(x)$ by the case already treated. Now $F(x) = H_n(x) + \int_{-\infty}^{x} (f(t) - h_n(t))\, dt$; the integral here is nondecreasing because the integrand is nonnegative, and it follows by Theorem 31.2 that it has almost everywhere a nonnegative derivative. Since differentiation is linear, $F'(x) \geq H_n'(x) = h_n(x)$ almost everywhere. As n was arbitrary, $F'(x) \geq f(x)$ almost everywhere, and so $\int_a^b F'(x)\, dx \geq \int_a^b f(x)\, dx = F(b) - F(a)$. But the reverse inequality is a consequence of (31.9). Therefore, $\int_a^b (F'(x) - f(x))\, dx = 0$, and as before $F = f'$ except on a set of Lebesgue measure 0. ∎

Singular Functions

If $f(x)$ is nonnegative and integrable, differentiating its indefinite integral $\int_{-\infty}^{x} f(t)\, dt$ leads back to $f(x)$ except perhaps on a set of Lebesgue measure 0. That is the content of Theorem 31.3. The converse question is this: If $F(x)$ is nondecreasing and hence has almost everywhere a derivative $F'(x)$, does integrating $F'(x)$ lead back to $F(x)$? As stated before, the answer turns out to be no even if $F(x)$ is assumed continuous:

Example 31.1. Let $X_1, X_2, \ldots$ be independent, identically distributed random variables such that $P[X_n = 0] = p_0$ and $P[X_n = 1] = p_1 = 1 - p_0$, and let $X = \sum_{n=1}^{\infty} X_n 2^{-n}$. Let $F(x) = P[X \leq x]$ be the distribution function of X. For an arbitrary sequence $u_1, u_2, \ldots$ of 0's and 1's, $P[X_n = u_n, n = 1, 2, \ldots] = \lim_n p_{u_1} \cdots p_{u_n} = 0$; since x can have at most two dyadic expansions $x = \sum_n u_n 2^{-n}$, $P[X = x] = 0$. Thus F is everywhere continuous. Of course, $F(0) = 0$ and $F(1) = 1$. For $0 \leq k < 2^n$, $k 2^{-n}$ has the form $\sum_{i=1}^{n} u_i 2^{-i}$ for some n-tuple $(u_1, \ldots, u_n)$ of 0's and 1's. Since F is continuous,

$$(31.15) \qquad F\left(\frac{k+1}{2^n}\right) - F\left(\frac{k}{2^n}\right) = P\left[\frac{k}{2^n} < X < \frac{k+1}{2^n}\right]$$

$$= P[X_i = u_i, i \leq n] = p_{u_1} \cdots p_{u_n}.$$

This shows that F is strictly increasing over the unit interval.

If $p_0 = p_1 = \frac{1}{2}$, the right side of (31.15) is 2^{-n}, and a passage to the limit shows that $F(x) = x$ for $0 \leq x \leq 1$. Assume, however, that $p_0 \neq p_1$. It will be shown that $F'(x) = 0$ except on a set of Lebesgue measure 0 in this case. Obviously the derivative is 0 outside the unit interval, and by Theorem 31.2 it exists almost everywhere inside it. Suppose then that $0 < x < 1$ and that F has a derivative $F'(x)$ at x. It will be shown that $F'(x) = 0$.

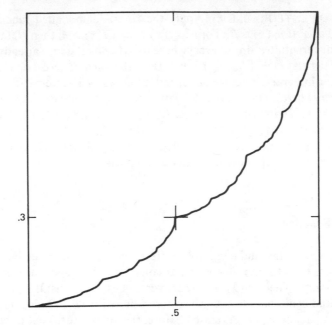

Graph of $F(x)$ for $p_0 = 3$, $p_1 = .7$. Because of the recursion (31.17), the part of the graph over [0, .5] and the part over [.5, 1] are identical, apart from changes in scale, with the whole graph. Each segment of the curve therefore contains scaled copies of the whole; the extreme irregularity which this implies is obscured by the fact that the accuracy is only to within the width of the printed line.

For each n choose k_n so that x lies in the interval $I_n = (k_n 2^{-n}, (k_n + 1)2^{-n}]$; I_n is that dyadic interval of rank n that contains x. By (31.8),

$$\frac{P[X \in I_n]}{2^{-n}} = \frac{F((k_n + 1)2^{-n}) - F(k_n 2^{-n})}{2^{-n}} \to F'(x).$$

If $F'(x)$ is distinct from 0, the ratio of two successive terms here must go to 1, so that

$$(31.16) \qquad\qquad \frac{P[X \in I_{n+1}]}{P[X \in I_n]} \to \frac{1}{2}.$$

If I_n consists of the reals with nonterminating base-2 expansions beginning with the digits $u_1, \ldots, u_n$, then $P[X \in I_n] = p_{u_1} \cdots p_{u_n}$ by (31.15). But I_{n+1} must for some u_{n+1} consist of the reals beginning $u_1, \ldots, u_n, u_{n+1}$ (u_{n+1} is 1 or 0 according as x lies to the right of the midpoint of I_n or not). Thus $P[X \in I_{n+1}]/P[X \in I_n] = p_{u_{n+1}}$ is either p_0 or p_1, and (31.16) is possible only if $p_0 = p_1$, which was excluded by hypothesis.

Thus F is continuous and strictly increasing over $[0, 1]$ but $F'(x) = 0$ except on a set of Lebesgue measure 0. For $0 \leq x \leq \frac{1}{2}$ independence gives $F(x) = P[X_1 = 0, \sum_{n=2}^{\infty} X_n 2^{-n+1} \leq 2x] = p_0 F(2x)$. Similarly, $F(x) - p_0 = p_1 F(2x - 1)$ for $\frac{1}{2} \leq x \leq 1$. Thus

$$(31.17) \qquad F(x) = \begin{cases} p_0 F(2x) & \text{if } 0 \leq x \leq \dfrac{1}{2}, \\ p_0 + p_1 F(2x - 1) & \text{if } \dfrac{1}{2} \leq x \leq 1. \end{cases}$$

In Section 7, $F(x)$ (there denoted $Q(x)$) entered as the probability of success at bold play; see (7.32) and (7.35).   ∎

A function is *singular* if it has derivative 0 except on a set of Lebesgue measure 0. Of course, a step function constant over intervals is singular. What is remarkable (indeed, singular) about the function in the preceding example is that it is continuous and strictly increasing but nonetheless has derivative 0 except on a set of Lebesgue measure 0. Note that there is strict inequality in (31.9) for this F.

Further properties of nondecreasing functions can be discovered through a study of the measures they generate. Assume from now on that F is nondecreasing, that F is continuous from the right (this is only a normalization), and that $0 = \lim_{x \to -\infty} F(x) \leq \lim_{x \to +\infty} F(x) = m < \infty$. Call such an F a *distribution function*, even though m need not be 1. By Theorem 12.4 there exists a unique measure μ on the Borel sets of the line for which

$$(31.18) \qquad \mu(a, b] = F(b) - F(a).$$

Of course, $\mu(R^1) = m$ is finite.

The larger F' is, the larger μ is:

Theorem 31.4. *Suppose that F and μ are related by* (31.18) *and that $F'(x)$ exists throughout A.*

   (i)   *If $F'(x) \leq \alpha$ for $x \in A$, then $\mu(A) \leq \alpha\lambda(A)$.*

   (ii) *If $F'(x) \geq \alpha$ for $x \in A$, then $\mu(A) \geq \alpha\lambda(A)$.*

PROOF. It is no restriction to assume A bounded. Fix ϵ for the moment and let A_n consist of those x in A such that

$$(31.19) \qquad \mu(I) \leq (\alpha + \epsilon)\lambda(I)$$

holds for every interval $I = (u, v]$ for which $x \in I$, $v - u \leq n^{-1}$, and u, v are rational. Then A_n is a Borel set and (see (31.8)) $A_n \uparrow A$ under the hypothesis of (i). By Theorem 11.4 there exist disjoint intervals I_{nk} such that $A_n \subset \bigcup_k I_{nk}$ and

(31.20) $$\sum_k \lambda(I_{nk}) < \lambda(A_n) + \epsilon.$$

It is no restriction to assume that each I_{nk} has length less than n^{-1}, has rational endpoints, and meets A_n. Then (31.19) applies to each I_{nk} and hence

$$\mu(A_n) \leq \sum_k \mu(I_{nk}) \leq (\alpha + \epsilon) \sum_k \lambda(I_{nk}) \leq (\alpha + \epsilon)(\lambda(A_n) + \epsilon).$$

In the extreme terms here let $n \to \infty$ and then $\epsilon \to 0$; (i) follows.

To prove (ii), use the same argument with $\mu(I) \geq (\alpha - \epsilon)\lambda(I)$ in place of (31.19) and $\sum_k \mu(I_{nk}) < \mu(A_n) + \epsilon$ in place of (31.20). In the definition of A_n let u and v range over some countable, dense set that includes all the discontinuity points of F; then the I_{nk} can be taken to have endpoints in this set. It follows that

$$\mu(A_n) + \epsilon > \sum_k \mu(I_{nk}) \geq (\alpha - \epsilon) \sum_k \lambda(I_{nk}) \geq (\alpha - \epsilon)\lambda(A_n).$$

Again let $n \to \infty$ and then $\epsilon \to 0$.   ■

The measures μ and λ have *disjoint supports* if there exist Borel sets S_μ and S_λ such that

(31.21) $$\begin{cases} \mu(R^1 - S_\mu) = 0, & \lambda(R^1 - S_\lambda) = 0, \\ S_\mu \cap S_\lambda = 0. \end{cases}$$

Theorem 31.5. *A necessary and sufficient condition that μ and λ have disjoint supports is that $F'(x) = 0$ except on a set of Lebesgue measure 0.*

PROOF. By Theorem 31.4, $\mu[x: |x| \leq a, F'(x) \leq \epsilon] \leq 2a\epsilon$, and so (let $\epsilon \to 0$ and then $a \to \infty$) $\mu[x: F'(x) = 0] = 0$. If $F'(x) = 0$ outside a set of Lebesgue measure 0, then $S_\lambda = [x: F'(x) = 0]$ and $S_\mu = R^1 - S_\lambda$ satisfy (31.21).

Suppose that there exist S_μ and S_λ satisfying (31.21). By the other half of Theorem 31.4, $\epsilon\lambda[x: F'(x) \geq \epsilon] = \epsilon\lambda[x: x \in S_\lambda, F'(x) \geq \epsilon] \leq \mu(S_\lambda) = 0$, and so (let $\epsilon \to 0$) $F'(x) = 0$ except on a set of Lebesgue measure 0.   ■

Example 31.2. Suppose that μ is discrete, consisting of a mass m_k at each of countably many points x_k. Then $F(x) = \sum m_k$, the sum extending over the k for which $x_k \leq x$. Certainly, μ and λ have disjoint supports, and so F' must vanish except on a set of Lebesgue measure 0. This is directly obvious if the x_k have no limit points but not, for example, if they are dense.   ■

Example 31.3. Consider again the distribution function F in Example 31.1. Here $\mu(A) = P[X \in A]$. Since F is singular, μ and λ have disjoint supports. This fact has an interesting direct probabilistic proof.

For x in the unit interval, let $d_1(x), d_2(x), \ldots$ be the digits in its nonterminating dyadic expansion, as in Section 1. If $(k2^{-n}, (k+1)2^{-n}]$ is the dyadic interval of rank n consisting of the reals whose expansions begin with the digits $u_1, \ldots, u_n$, then, by (31.15),

$$(31.22) \quad \mu\left(\frac{k}{2^n}, \frac{k+1}{2^n}\right] = \mu[x: d_i(x) = u_i, i \le n] = p_{u_1} \cdots p_{u_n}.$$

If the unit interval is regarded as a probability space under the measure μ, then the $d_i(x)$ become random variables and (31.22) says that these random variables are independent and identically distributed and $\mu[x: d_i(x) = 0] = p_0$, $\mu[x: d_i(x) = 1] = p_1$.

Since these random variables have expected value p_1, the strong law of large numbers implies that their averages go to p_1 with probability 1:

$$(31.23) \quad \mu\left[x \in (0, 1]: \lim_n \frac{1}{n} \sum_{i=1}^{n} d_i(x) = p_1\right] = 1.$$

On the other hand, by the normal number theorem

$$(31.24) \quad \lambda\left[x \in (0, 1]: \lim_n \frac{1}{n} \sum_{i=1}^{n} d_i(x) = \frac{1}{2}\right] = 1.$$

(Of course, (31.24) is just (31.23) for the special case $p_0 = p_1 = \frac{1}{2}$; in this case μ and λ coincide in the unit interval.) If $p_1 \ne \frac{1}{2}$, the sets in (31.23) and (31.24) are disjoint, so that μ and λ do have disjoint supports.

It was shown in Example 31.1 that if $F'(x)$ exists at all $(0 < x < 1)$, then it is 0. By part (i) of Theorem 31.4 the set where $F'(x)$ fails to exist therefore has μ-measure 1; in particular, this set is uncountable. ∎

Example 31.4. Restrict the F of Examples 31.1 and 31.3 to $(0, 1)$, and let g be the inverse. Thus F and g are continuous, strictly increasing mappings of $(0, 1)$ onto itself. If $A = [x \in (0, 1): F'(x) = 0]$, then $\lambda(A) = 1$, as shown in the examples, while $\mu(A) = 0$. Let H be a set in $(0, 1)$ that is not a Lebesgue set. Since $H - A$ is contained in a set of Lebesgue measure 0, it is a Lebesgue set; hence $H_0 = H \cap A$ is not a Lebesgue set, since otherwise $H = H_0 \cup (H - A)$ would be a Lebesgue set. If $B = (0, x]$, then $\lambda g^{-1}(B) = \lambda(0, F(x)] = F(x) = \mu(B)$, and it follows that $\lambda g^{-1}(B) = \mu(B)$ for all Borel sets B. Since $g^{-1}H_0$ is a subset of $g^{-1}A$ and $\lambda(g^{-1}A) = \mu(A) = 0$, $g^{-1}H_0$ is a Lebesgue set. On the other hand, if $g^{-1}H_0$ were a Borel set, $H_0 = F^{-1}(g^{-1}H_0)$ would also be a Borel set. Thus $g^{-1}H_0$ provides an example of a Lebesgue set that is not a Borel set.* ∎

Integrals of Derivatives

Return now to the problem of extending part (ii) of Theorem 31.1, to the problem

---

* For a different argument, see Problem 3.14.

of characterizing those distribution functions F for which F' integrates back to F:

$$(31.25) \qquad\qquad F(x) = \int_{-\infty}^{x} F'(t) \, dt.$$

The first step is easy: If (31.25) holds, then F has the form

$$(31.26) \qquad\qquad F(x) = \int_{-\infty}^{x} f(t) \, dt$$

for a nonnegative, integrable f (a density), namely $f = F'$. On the other hand, (31.26) implies by Theorem 31.3 that $F' = f$ outside a set of Lebesgue measure 0, whence (31.25) follows. Thus (31.25) holds if and only if F has the form (31.26) for some f, and the problem is to characterize functions of this form. The function of Example 31.1 is not among them.

As observed earlier (see (31.3)), an F of the form (31.26) with f integrable is continuous. It has a still stronger property: For each ϵ there exists a δ such that

$$(31.27) \qquad\qquad \int_{A} f(x) \, dx < \epsilon \qquad \text{if } \lambda(A) < \delta.$$

Indeed, if $A_n = [x : f(x) > n]$, then $A_n \downarrow 0$, and since f is integrable, the dominated convergence theorem implies that $\int_{A_n} f(x) dx < \epsilon/2$ for large n. Fix such an n and take $\delta = \epsilon/2n$. If $\lambda(A) < \delta$, then $\int_{A} f(x) \, dx = \int_{A-A_n} f(x) \, dx + \int_{A_n} f(x) \, dx \le n\lambda(A) + \epsilon/2 < \epsilon$.

If F is given by (31.26), then $F(b) - F(a) = \int_{a}^{b} f(x) \, dx$, and (31.27) has this consequence: For every ϵ there exists a δ such that for each finite collection $[a_i, b_i]$, $i = 1, \ldots, k$, of nonoverlapping* intervals

$$(31.28) \qquad \sum_{i=1}^{k} |F(b_i) - F(a_i)| < \epsilon \qquad \text{if } \sum_{i=1}^{k} (b_i - a_i) < \delta.$$

A function F with this property is said to be *absolutely continuous*.[†] A function of the form (31.26) (f integrable) is thus absolutely continuous.

A continuous distribution function is uniformly continuous, and so for every ϵ there is a δ such that the implication in (31.28) holds provided that $k = 1$. The definition of absolute continuity requires this to hold whatever k may be, which puts severe restrictions on F. Absolute continuity of F can be characterized in terms of the measure μ related to it by (31.18):

---

* Intervals are nonoverlapping if their interiors are disjoint. In this definition it is immaterial whether the intervals are regarded as closed or open or half-open, as this has no effect on (31.28).

† The definition applies to all functions, not just to distribution functions. If F is a distribution function as in the present discussion, the absolute-value bars in (31.28) are unnecessary.

Theorem 31.6. *The distribution function F is absolutely continuous in the sense of* (31.28) *if and only if $\mu(A) = 0$ for every A for which $\lambda(A) = 0$.*

PROOF. Suppose that F is absolutely continuous and that $\lambda(A) = 0$. Given ϵ, choose δ so that (31.28) holds. By Theorem 11.4 there exists a countable disjoint union $B = \bigcup_k I_k$ of intervals such that $A \subset B$ and $\lambda(B) < \delta$. By (31.28) it follows that $\mu(\bigcup_{k=1}^n I_k) < \epsilon$ for each n and hence that $\mu(A) \le \mu(B) \le \epsilon$. Since ϵ was arbitrary, $\mu(A) = 0$.

If F is not absolutely continuous, then there exists an ϵ such that for every δ some finite disjoint union A of intervals satisfies $\lambda(A) < \delta$ and $\mu(A) \ge \epsilon$. Choose A_n so that $\lambda(A_n) < n^{-2}$ and $\mu(A_n) \ge \epsilon$. Then $\lambda(\limsup_n A_n) = 0$ by the first Borel-Cantelli lemma (Theorem 4.3, the proof of which does not require P to be a probability measure or even finite). On the other hand, $\mu(\limsup_n A_n) \ge \epsilon > 0$ by Theorem 4.1 (the proof of which applies because μ is finite). ∎

This result leads to a characterization of indefinite integrals.

Theorem 31.7. *A distribution function $F(x)$ has the form $\int_{-\infty}^x f(t)\, dt$ for an integrable f if and only if it is absolutely continuous in the sense of* (31.28).

PROOF. That an F of the form (31.26) is absolutely continuous was proved in the argument leading to the definition (31.28). For another proof, apply Theorem 31.6: if F has this form, then $\lambda(A) = 0$ implies that $\mu(A) = \int_A f(t)\, dt = 0$.

To go the other way, define for any distribution function F

(31.29)
$$F_{ac}(x) = \int_{-\infty}^x F'(t)\, dt$$

and

(31.30)
$$F_s(x) = F(x) - F_{ac}(x).$$

Then F_s is right-continuous, and by (31.9) it is both nonnegative and nondecreasing. Since F_{ac} comes from a density, it is absolutely continuous. By Theorem 31.3, $F'_{ac} = F'$ and hence $F'_s = 0$ except on a set of Lebesgue measure 0. Thus F has a decomposition

(31.31)
$$F(x) = F_{ac}(x) + F_s(x),$$

where F_{ac} has a density and hence is absolutely continuous and F_s is singular. This is called the *Lebesgue decomposition*.

Suppose that F is absolutely continuous. Then the F_s of (31.30) must, as the difference of absolutely continuous functions, be absolutely continuous itself. If it can be shown that F_s is identically 0, it will follow that $F = F_{ac}$ has the required form. It thus suffices to show that a distribution function that is both absolutely continuous and singular must vanish.

If a distribution function F is singular, then by Theorem 31.5 there are disjoint supports S_μ and S_λ. But if F is also absolutely continuous, then from $\lambda(S_\mu) = 0$ it follows by Theorem 31.6 that $\mu(S_\mu) = 0$. But then $\mu(R^1) = 0$, and so $F(x) \equiv 0$. ∎

This theorem identifies the distribution functions that are integrals of their derivatives as the absolutely continuous functions. Theorem 31.6, on the other hand, characterizes absolute continuity in a way that extends to spaces Ω without the geometric structure of the line necessary to a treatment involving distribution functions and ordinary derivatives.* The extension is studied in the next section.

Functions of Bounded Variation

The remainder of the section briefly sketches the extension of the preceding theory to functions that are not monotone. The results are for simplicity given only for a finite interval $[a, b]$ and for functions F on $[a, b]$ satisfying $F(a) = 0$.

If $F(x) = \int_a^x f(t)\, dt$ is an indefinite integral, where f is integrable but not necessarily nonnegative, then $F(x) = \int_a^x f^+(t)\, dt - \int_a^x f^-(t)\, dt$ exhibits F as the difference of two nondecreasing functions. The problem of characterizing indefinite integrals thus leads to the preliminary problem of characterizing functions representable as a difference of nondecreasing functions.

Now F is said to be of *bounded variation* over $[a, b]$ if $\sup_\Delta \|F\|_\Delta$ is finite, where $\|F\|_\Delta$ is defined by (31.5) and Δ ranges over all partitions (31.4) of $[a, b]$. Clearly, a difference of nondecreasing functions is of bounded variation. But the converse holds as well: For every finite collection Γ of nonoverlapping intervals $[x_i, y_i]$ in $[a, b]$, put

$$P_\Gamma = \Sigma(F(y_i) - F(x_i))^+, \qquad N_\Gamma = \Sigma(F(y_i) - F(x_i))^-.$$

Now define

$$P(x) = \sup_\Gamma P_\Gamma, \qquad N(x) = \sup_\Gamma N_\Gamma,$$

where the suprema extend over partitions Γ of $[a, x]$. If F is of bounded variation, then $P(x)$ and $N(x)$ are finite. For each such Γ, $P_\Gamma = N_\Gamma + F(x)$. This gives the inequalities

$$P_\Gamma \leq N(x) + F(x), \qquad P(x) \geq N_\Gamma + F(x)$$

which in turn lead to the inequalities

$$P(x) \leq N(x) + F(x), \qquad P(x) \geq N(x) + F(x)$$

Thus

(31.32) $$F(x) = P(x) - N(x)$$

gives the required representation: *A function is the difference of two nondecreasing functions if and only if it is of bounded variation.*

---

* Theorems 31.3 and 31.7 do have geometric analogues in R^k; see RUDIN', Chapter 8.

If $T_\Gamma = P_\Gamma + N_\Gamma$, then $T_\Gamma = \Sigma |F(y_i) - F(x_i)|$. According to the definition (31.28), F is absolutely continuous if for every ϵ there exists a δ such that $T_\Gamma < \epsilon$ whenever the intervals in the collection Γ have total length less than δ. If F is absolutely continuous, take the δ corresponding to an ϵ of 1 and decompose $[a, b]$ into a finite number, say n, of subintervals $[u_{j-1}, u_j]$ of lengths less than δ. Any partition Δ of $[a, b]$ can by the insertion of the u_j be split into n sets of intervals each of total length less than δ, and it follows* that $\|F\|_\Delta \le n$. Therefore, an absolutely continuous function is necessarily of bounded variaion.

An absolutely continuous F thus has a representation (31.32). It follows by the definitions that $P(y) - P(x)$ is at most $\sup_\Gamma T_\Gamma$, where Γ ranges over the partitions of $[x, y]$. If $[x_i, y_i]$ are nonoverlapping intervals, then $\Sigma (P(y_i) - P(x_i))$ is at most $\sup_\Gamma T_\Gamma$, where now Γ ranges over the collections of intervals that partition *each* of the $[x_i, y_i]$. If F is absolutely continuous, there thus exists for each ϵ a δ such that $\Sigma (y_i - x_i) < \delta$ implies that $\Sigma (P(y_i) - P(x_i)) < \epsilon$. In other words, P is absolutely continuous. Similarly, N is absolutely continuous.

Thus an absolutely continuous F is the difference of two nondecreasing absolutely continuous functions. By Theorem 31.7, each of these is an indefinite integral, which implies that F is an indefinite integral as well: *For an F on $[a, b]$ satisfying $F(a) = 0$, absolute continuity is a necessary and sufficient condition that F be an indefinite integral—have the form $F(x) = \int_a^x f(t)\, dt$ for an integrable f.*

PROBLEMS

31.1. Extend Examples 31.1 and 31.3: Let $p_0, \ldots, p_{r-1}$ be nonnegative numbers adding to 1, where $r \ge 2$, let $X_1, X_2, \ldots$ be independent, identically distributed random variables such that $P[X_n = i] = p_i$, $0 \le i < r$, and put $X = \Sigma_{n=1}^\infty X_n r^{-n}$. Let F be the distribution function of X. Show that F is continuous. Show that F is strictly increasing over the unit interval if and only if all the p_i are strictly positive. Show that $F(x) \equiv x$ for $0 \le x \le 1$ if $p_i \equiv r^{-1}$ and that otherwise F is singular; prove singularity by extending the arguments both of Example 31.1 and of Example 31.3. What is the analogue of (31.17)?

31.2. ↑ In Problem 31.1 take $r = 3$ and $p_0 = p_2 = \frac{1}{2}, p_1 = 0$. The corresponding F is called the *Cantor* function. The complement in $[0, 1]$ of the Cantor set (see Problems 1.6 and 3.16) consists of the middle third $(\frac{1}{3}, \frac{2}{3})$, the middle thirds $(\frac{1}{9}, \frac{2}{9})$ and $(\frac{7}{9}, \frac{8}{9})$, and so on. Show that F is $\frac{1}{2}$ on the first of these intervals, $\frac{1}{4}$ on the second, $\frac{3}{4}$ on the third, and so on. Show by direct argument that $F' = 0$ except on a set of Lebesgue measure 0.

31.3. A real function f of a real variable is a *Lebesgue function* if $[x: f(x) \le \alpha]$ is a Lebesgue set for each α.

(a) Show that, if f_1 is a Borel function and f_2 is a Lebesgue function, then the composition $f_1 f_2$ is a Lebesgue function.

(b) Show that there exists a Lebesgue function f_1 and a Lebesgue (even Borel, even continuous) function f_2 such that $f_1 f_2$ is not a Lebesgue function.

Hint: Use Example 31.4.

---

* This uses the fact that $\|F\|_\Delta$ cannot decrease under passage to a finer partition.

31.4.   Let $r_1, r_2, \ldots$ be an enumeration of the rationals in $(0, 1)$ and put $F(x) = \Sigma_{k:r_k \leq x}$ 2^{-k}. Define φ by (14.5) and prove that it is continuous and singular.

31.5.   If the distribution function corresponding to μ is singular, then $\mu(A) > 0$ for some set A of Lebesgue measure 0. It is an interesting fact, however, that almost all translates of A must have μ-measure 0. From Fubini's theorem and the fact that λ is invariant under translation and reflection through 0 show that, if $\lambda(A)$ $= 0$ and μ is σ-finite, then $\mu(A + x) = 0$ for x outside a set of Lebesgue measure 0.

31.6.   From Theorem 31.4 deduce that, if $\mu[x: F'(x) = \infty] = \mu(R^1)$, then μ and λ have disjoint supports; they can be taken as $S_\mu = [x: F'(x) = \infty]$ and $S_\lambda = [x: F'(x) = 0]$.

31.7.   *Lebesgue's density theorem.* A point x is a *density point* of a Borel set A if $\lambda((u, v] \cap A)/(v - u) \to 1$ as $u \uparrow x$ and $v \downarrow x$. From Theorems 31.2 and 31.4 deduce that almost all points of A are density points—that is, that the points of A that are not density points form a set of Lebesgue measure 0. Similarly, $\lambda((u, v] \cap A)/(v - u) \to 0$ almost everywhere on A^c.

31.8.   19.3↑ Let $f: [a, b] \to R^k$ be an arc; $f(t) = (f_1(t), \ldots, f_k(t))$. Show that the arc is rectifiable if and only if each f_i is of bounded variation over $[a, b]$.

31.9.   ↑   Suppose that F is continuous and nondecreasing and that $F(0) = 0, F(1)$ $= 1$. Then $f(x) = (x, F(x))$ defines a curve $f: [0, 1] \to R^2$. It is easy to see by monotonicity that the curve is rectifiable and that, in fact, its length satisfies $L(f) \leq 2$. It is also easy, given ϵ, to produce functions F for which $L(f) > 2 - \epsilon$. Show by the arguments in the proof of Theorem 31.4 that $L(f) = 2$ if F is singular.

31.10.   Suppose that the characteristic function of F satisfies $\limsup_{t \to \infty} |\varphi(t)| = 1$. Show that F is singular. Compare the lattice case (Problem 26.1).

Hint: Use the Lebesgue decomposition and the Riemann-Lebesgue theorem.

31.11.   26.25↑   Suppose that $X_1, X_2, \ldots$ are independent and assume the values ± 1 with probability $\frac{1}{2}$ each, and let $X = \Sigma_{n=1}^\infty X_n/2^n$. Show that X is uniformly distributed over $[-1, +1]$. Calculate the characteristic functions of X and X_n and deduce (1.36). Conversely, establish (1.36) by trigonometry and conclude that X is uniformly distributed over $[-1, +1]$.

31.12.   (a)   Suppose that $X_1, X_2, \ldots$ are independent and assume the values 0 and 1 with probability $\frac{1}{2}$ each. Let F and G be the distribution functions of $\Sigma_{n=1}^\infty$ $X_{2n-1}/2^{2n-1}$ and $\Sigma_{n=1}^\infty X_{2n}/2^{2n}$. Show that F and G are singular but that $F *$ G is absolutely continuous.

(b)   Show that the convolution of an absolutely continuous distribution function with an arbitrary distribution function is absolutely continuous.

31.13.   31.2↑   Show that the Cantor function is the distribution function of $\Sigma_{n=1}^\infty X_n/3^n$, where the X_n are independent and assume the values 0 and 2 with probability $\frac{1}{2}$ each. Express its characteristic function as an infinite product.

In the next three problems the notation is that of Examples 31.1 and 31.3; x ranges only over the unit interval, $d_n(x)$ is the nth digit in the nonterminating binary expansion of x, and $s_n(x) = d_1(x) + \cdots + d_n(x)$.

31.14. Show that $_FD(1) = \infty$ and $D^F(0) = 0$ if $p_0 < \frac{1}{2}$. From (31.17) deduce that $_FD(x) = \infty$ and $D^F(x) = 0$ for all dyadic rationals x. Analyze the case $p_0 > \frac{1}{2}$ and sketch the graph.

31.15. **(a)** Show that

$$F(x) = \sum_{n=1}^{\infty} d_n(x)p_0^{n-s_{n-1}(x)}p_1^{s_{n-1}(x)}.$$

Show that the sum is unchanged if the terminating expansion of x is used instead of the nonterminating one.

(b) Now suppose that p_0 and p_1 are complex numbers satisfying $p_0 + p_1 = 1$ and $\max\{|p_0|, |p_1|\} < 1$. Show that the series above converges uniformly and absolutely and that the F thus defined is continuous. Specifically, let $p_0 = (1 + i)/2 = 1 - p_1$. Show that $|F((k + 1)2^{-n}) - F(k2^{-n})| = 2^{-n/2}$ and deduce that F is nowhere differentiable.

Let $R(x)$ be the real part of $F(x)$ and suppose that $(k2^{-n}, (k + 1)2^{-n}]$ is the dyadic interval of order n containing x; show that $R(k + 1)2^{-n}) - R(k2^{-n}) = 2^{-n/2} \cos((n - 2s_n(x))\pi/4)$. Show that $n - 2s_n(x)$ must be incongruent to $\pm 2 \mod 8$ for infinitely many n because its successive values differ by ± 1; conclude that the continuous real function $R(x)$ is nowhere differentiable.

31.16. 6.14↑ Return to the case in which p_0 and p_1 are real and let μ be the probability measure on the unit interval corresponding to F. If $I_n(x)$ is the dyadic interval of order n containing x, then

$$(31.33) \quad -\frac{1}{n}\log \mu(I_n(x)) = -\left(1 - \frac{s_n(x)}{n}\right)\log p_0 - \frac{s_n(x)}{n}\log p_1.$$

(a) Show that (31.33) converges on a set of μ-measure 1 to the entropy $h = -p_0 \log p_0 - p_1 \log p_1$. From the fact that this entropy is less than $\log 2$ if $p_0 \neq \frac{1}{2}$, deduce that on a set of μ-measure 1, F does not have a finite derivative.

(b) Show that (31.33) converges to $-\frac{1}{2}\log p_0 - \frac{1}{2}\log p_1$ on a set of Lebesgue measure 1. If $p_0 \neq \frac{1}{2}$ this limit exceeds $\log 2$ (arithmetic versus geometric means), and so $\mu(I_n(x))/2^{-n} \to 0$ except on a set of Lebesgue measure 0. This does not prove that $F'(x)$ exists almost everywhere, but it does show that, except for x in a set of Lebesgue measure 0, if $F'(x)$ does exist, then it is 0.

(c) Show that, if (31.33) converges to l, then

$$(31.34) \quad \lim_n \frac{\mu(I_n(x))}{(2^{-n})^\alpha} = \begin{cases} \infty & \text{if } \alpha > \dfrac{l}{\log 2}, \\ 0 & \text{if } \alpha < \dfrac{l}{\log 2}. \end{cases}$$

If (31.34) holds, then (roughly) F satisfies a Lipschitz condition of (exact) order $l/\log 2$. Thus F satisfies a Lipschitz condition of order $h/\log 2$ on a set of μ-measure 1 and a Lipschitz condition of order $(-\frac{1}{2}\log p_0 - \frac{1}{2}\log p_1)/\log 2$ on a set of Lebesgue measure 1.

31.17. Show (see (31.31)) that (apart from addition of constants) a function can have only one representation $F_1 + F_2$ with F_1 absolutely continuous and F_2 singular.

31.18. Show that the F_s in the Lebesgue decomposition can be further split into $F_d + F_{cs}$, where F_{cs} is continuous and singular and F_d increases only in jumps in the sense that the corresponding measure is discrete. The complete decomposition is then $F = F_{ac} + F_{cs} + F_d$.

31.19. (a) Suppose that $x_1 < x_2 < \cdots$ and $\Sigma_n |F(x_n)| = \infty$. Show that, if F assumes the value 0 in each interval (x_n, x_{n+1}), then it is of unbounded variation.

(b) Define F over $[0, 1]$ by $F(0) = 0$ and $F(x) = x^\alpha \sin x^{-1}$ for $x > 0$. For which values α is F of bounded variation?

31.20. 14.4↑ If f is nonnegative and Lebesgue integrable, then by Theorem 31.3 and (31.8), except for x in a set of Lebesgue measure 0,

$$(31.35) \qquad \frac{1}{v - u} \int_u^v f(t) \, dt \to f(x)$$

if $u \le x \le v, u < v, u, v \to x$. There is an analogue in which Lebesgue measure is replaced by a general probability measure μ: If f is nonnegative and integrable with respect to μ, then as $h \downarrow 0$,

$$(31.36) \qquad \frac{1}{\mu(x - h, x + h]} \int_{(x-h, x+h]} f(t)\mu(dt) \to f(x)$$

on a set of μ-measure 1. Let F be the distribution function corresponding to μ and put $\varphi(u) = \inf [x : u \le F(x)]$ for $0 < u < 1$ (see (14.5)). Deduce (31.36) from (31.35) by change of variable and Problem 14.4(d).

SECTION 32.   THE RADON-NIKODYM THEOREM

If f is a nonnegative function on a measure space $(\Omega, \mathcal{F}, \mu)$, then $\nu(A) = \int_A f \, d\mu$ defines another measure on $\mathcal{F}$. In the terminology of Section 16, ν has *density f* with respect to μ; see (16.10). For each A in $\mathcal{F}$, $\mu(A) = 0$ implies that $\nu(A) = 0$. The purpose of this section is to show conversely that if this last condition holds and ν and μ are σ-finite on $\mathcal{F}$, then ν has a density with respect to μ. This was proved for the case $(R^1, \mathcal{R}^1, \lambda)$ in Theorems 31.6 and 31.7. The theory of the preceding section, although illuminating, is not required here.

Additive Set Functions

Throughout this section $(\Omega, \mathcal{F})$ is a measurable space. All sets involved are assumed as usual to lie in $\mathcal{F}$.

An *additive set function* is a function φ from $\mathcal{F}$ to the reals for which

$$(32.1) \qquad \varphi(\cup_n A_n) = \sum_n \varphi(A_n)$$

if $A_1, A_2, \ldots$ is a finite or infinite sequence of disjoint sets. A set function differs from a measure in that the values $\varphi(A)$ may be negative but must be finite—the special values $+\infty$ and $-\infty$ are prohibited. It will turn out that the series on the right in (32.1) must in fact converge absolutely, but this need not be assumed. Note that $\varphi(0) = 0$.

The proof of the main theorem of this section (Theorem 32.2) requires certain facts about additive set functions, even though the statement of the theorem involves only measures.

The Hahn Decomposition

Theorem 32.1. *For any additive set function φ, there exist disjoint sets A^+ and A^- such that $A^+ \cup A^- = \Omega$, $\varphi(E) \geq 0$ for all E in A^+ and $\varphi(E) \leq 0$ for all E in A^-.*

A set A is *positive* if $\varphi(E) \geq 0$ for $E \subset A$ and *negative* if $\varphi(E) \leq 0$ for $E \subset A$. The A^+ and A^- in the theorem decompose Ω into a positive and a negative set. This is the *Hahn decomposition*.

PROOF.   Let $\alpha = \sup [\varphi(A) \colon A \in \mathcal{F}]$. Suppose that there exists a set A^+ with $\varphi(A^+) = \alpha$ (which implies that α is finite). Let $A^- = \Omega - A^+$. If $A \subset A^+$ and $\varphi(A) < 0$, then $\varphi(A^+ - A) > \alpha$, an impossibility; hence A^+ is a positive set. If $A \subset A^-$ and $\varphi(A) > 0$, then $\varphi(A^+ \cup A) > \alpha$, an impossibility; hence A^- is a negative set.

It is therefore only necessary to construct a set A^+ for which $\varphi(A^+) = \alpha$. Choose sets A_n such that $\varphi(A_n) \to \alpha$ and let $A = \bigcup_n A_n$. For each n consider the 2^n sets B_{ni} (some perhaps empty) that are intersections of the form $\bigcap_{k=1}^{n} A'_k$, where each A'_k is either A_k or else $A - A_k$. The collection $\mathcal{B}_n = [B_{ni} \colon 1 \leq i \leq 2^n]$ of these sets partitions A. Clearly, $\mathcal{B}_n$ refines $\mathcal{B}_{n-1}$: each B_{nj} is contained in exactly one of the $B_{n-1,i}$.

Let C_n be the union of those B_{ni} in $\mathcal{B}_n$ for which $\varphi(B_{ni}) > 0$. Since A_n is the union of certain of the B_{ni}, $\varphi(A_n) \leq \varphi(C_n)$. Since the partitions $\mathcal{B}_1, \mathcal{B}_2, \ldots$ are successively finer, $m < n$ implies that $(C_m \cup \cdots \cup C_{n-1} \cup C_n) - (C_m \cup \cdots \cup C_{n-1})$ is the union (perhaps empty) of certain of the sets B_{ni}; the B_{ni} in this union must satisfy $\varphi(B_{ni}) > 0$ because they are contained in C_n. Therefore, $\varphi(C_m \cup \cdots \cup C_{n-1}) \leq \varphi(C_m \cup \cdots \cup C_n)$, so that by induction $\varphi(A_m) \leq \varphi(C_m) \leq \varphi(C_m \cup \cdots \cup C_n)$. If $D_m = \bigcup_{n=m}^{\infty} C_n$, then* $\varphi(A_m) \leq \varphi(D_m)$. Let $A^+ = \bigcap_{m=1}^{\infty} D_m$, so that $D_m \downarrow A^+$. Then† $\alpha = \lim_m \varphi(A_m) \leq \lim_m (D_m) = \varphi(A^+)$. Thus A^+ does have maximal φ-value.   ∎

---

* If $E_u \uparrow E$, then $\varphi(E) = \varphi(E_1 \cup \bigcup_{u=1}^{\infty}(E_{u+1} - E_u)) = \varphi(E_1) + \Sigma_{u=1}^{\infty} \varphi(E_{u+1} - E_u) = \lim_v [\varphi(E_1) + \Sigma_{u=1}^{v-1} \varphi(E_{u+1} - E_u)] = \lim_v \varphi(E_v)$ by (32.1). Take $E_v = C_m \cup \cdots \cup C_{m+v}$.

† If $E_u \downarrow E$, then $\varphi(E_v) = \varphi(E) + \Sigma_{u=v+1}^{\infty} \varphi(E_{u-1} - E_u) \to \varphi(E)$. These limits need not be monotone unless φ happens to be a measure.

If $\varphi^+(A) = \varphi(A \cap A^+)$ and $\varphi^-(A) = -\varphi(A \cap A^-)$, then φ^+ and φ^- are finite measures. Thus

$$(32.2) \qquad \varphi(A) = \varphi^+(A) - \varphi^-(A)$$

represents the set function φ as the difference of two finite measures having disjoint supports. If $E \subset A$, then $\varphi(E) \leq \varphi^+(E) \leq \varphi^+(A)$, and there is equality if $E = A \cap A^+$. Therefore, $\varphi^+(A) = \sup_{E \subset A} \varphi(E)$. Similarly, $\varphi^-(A) = -\inf_{E \subset A} \varphi(E)$. The measures φ^+ and φ^- are called the *upper* and *lower variations* of φ, and the measure $|\varphi|$ with value $\varphi^+(A) + \varphi^-(A)$ at A is called the *total variation*. The representation (32.2) is the *Jordan decomposition*.

Absolute Continuity and Singularity

Measures μ and ν on $(\Omega, \mathcal{F})$ are by definition *mutually singular* if they have disjoint supports—that is, if these exist sets S_μ and S_ν such that

$$(32.3) \qquad \begin{cases} \mu(\Omega - S_\mu) = 0, & \nu(\Omega - S_\nu) = 0, \\ S_\mu \cap S_\nu = 0. \end{cases}$$

In this case μ is also said to be *singular with respect to* ν and ν singular with respect to μ. Singularity is sometimes indicated by $\mu \perp \nu$. Note that measures are automatically singular if one of them is identically 0.

According to Theorem 31.5 a finite measure on R^1 with distribution function F is in the sense of (32.3) singular with respect to Lebesgue measure if and only if $F'(x) = 0$ except on a set of Lebesgue measure 0. In Section 31 the latter condition was taken as the definition of singularity, but of course it is the requirement of disjoint supports that can be generalized from R^1 to an arbitrary Ω.

The measure ν is *absolutely continuous* with respect to μ if for each A in $\mathcal{F}$, $\mu(A) = 0$ implies $\nu(A) = 0$. In this case ν is also said to be *dominated* by μ, and the relation is indicated by $\nu \ll \mu$. If $\nu \ll \mu$ and $\mu \ll \nu$, the measures are *equivalent*, indicated by $\nu \equiv \mu$.

A finite measure on the line is by Theorem 31.6 absolutely continuous in this sense with respect to Lebesgue measure if and only if the corresponding distribution function F satisfies the condition (31.28). The latter condition, taken in Section 31 as the definition of singularity, is again not the one that generalizes from R^1 to Ω.

There is an ϵ-δ idea related to the definition of absolute continuity above. Suppose that for every ϵ there exists a δ such that

$$(32.4) \qquad \nu(A) < \epsilon \qquad \text{if } \mu(A) < \delta.$$

If this condition holds, $\mu(A) = 0$ implies that $\nu(A) < \epsilon$ for all ϵ, and so $\nu \ll \mu$. Suppose, on the other hand, that this condition fails and that ν is finite. Then for some ϵ there exist sets A_n such that $\mu(A_n) < n^{-2}$ and $\nu(A_n) \geq \epsilon$. If $A = \lim \sup_n A_n$, then $\mu(A) = 0$ by the first Borel-Cantelli lemma (which applies to arbitrary measures) but $\nu(A) \geq \epsilon > 0$ by the right-hand inequality in (4.10) (which applies because ν is finite). Hence $\nu \ll \mu$ fails, and so (32.4) follows if ν is finite and $\nu \ll \mu$. If ν is finite, in order that $\nu \ll \mu$ it is therefore necessary and sufficient that for every ϵ there is a δ satisfying (32.4). This condition is not suitable as a definition because it need not follow from $\nu \ll \mu$ if ν is infinite.*

The Main Theorem

If $\nu(A) = \int_A f \, d\mu$, then certainly $\nu \ll \mu$. The *Radon-Nikodym* theorem goes in the opposite direction:

Theorem 32.2.   *If μ and ν are σ-finite measures such that $\nu \ll \mu$, then there exists a nonnegative f, a density, such that $\nu(A) = \int_A f \, d\mu$ for all $A \in \mathcal{F}$. For two such densities f and g, $\mu[f \neq g] = 0$.*

The uniqueness of the density up to sets of μ-measure 0 was settled in Section 16 (see the remarks following (16.10)). It is only the existence that must be proved.

The density f is integrable μ if and only if ν is finite. By Theorem 16.10 integrals with respect to ν can be calculated by the formula

$$(32.5) \qquad \int_A h \, d\nu = \int_A hf \, d\mu.$$

The density whose existence is to be proved is called the *Radon-Nikodym derivative* of ν with respect to μ and is often denoted $d\nu/d\mu$. The term *derivative* is appropriate because of Theorems 31.3 and 31.7: For an absolutely continuous distribution function F on the line, the corresponding measure μ has with respect to Lebesgue measure the Radon-Nikodym derivative F'. Note that (32.5) can be written

$$(32.6) \qquad \int_A h \, d\nu = \int_A h \frac{d\nu}{d\mu} \, d\mu.$$

Suppose that Theorem 32.2 holds for finite μ and ν. It is not hard to see in the σ-finite case that there exists a countable decomposition of Ω into $\mathcal{F}$-sets A_n for which $\mu(A_n)$ and $\nu(A_n)$ are both finite. If $\mu_n(A) = \mu(A \cap A_n)$ and $\nu_n(A) = \nu(A \cap A_n)$, then $\nu_n \ll \mu_n$, and so $\nu_n(A) = \int_A f_n \, d\mu_n$ for some density f_n. Then $f = \Sigma_n f_n$ is the density sought.

It is therefore enough to treat finite μ and ν. This requires a preliminary result.

* See Problem 32.3.

Lemma. *If μ and ν are finite measures and are not mutually singular, then there exists a set A and a positive ϵ such that $\mu(A) > 0$ and $\epsilon\mu(E) \leq \nu(E)$ for all $E \subset A$.*

PROOF. Let $A_n^+ \cup A_n^-$ be a Hahn decomposition for the set function $\nu - n^{-1}\mu$; put $M = \bigcup_n A_n^+$, so that $M^c = \bigcap_n A_n^-$. Since M^c is in the negative set A_n^- for $\nu - n^{-1}\mu$, $\nu(M^c) \leq n^{-1}\mu(M^c)$; since this holds for all n, $\nu(M^c) = 0$. Thus M supports ν, and from the fact that μ and ν are not mutually singular it follows that M^c cannot support μ—that is, that $\mu(M)$ must be positive. Therefore, $\mu(A_n^+) > 0$ for some n. Take $A = A_n^+$ and $\epsilon = n^{-1}.$ ∎

PROOF OF THEOREM 32.2. Suppose that μ and ν are finite measures satisfying $\nu \ll \mu$. Let $\mathcal{G}$ be the class of nonnegative functions g such that $\int_E g \, d\mu \leq \nu(E)$ for all E. If g and g' lie in $\mathcal{G}$, then $\max(g, g')$ also lies in $\mathcal{G}$ because

$$\int_E \max(g, g') \, d\mu = \int_{E \cap [g \geq g']} g \, d\mu + \int_{E \cap [g' > g]} g' \, d\mu$$

$$\leq \nu(E \cap [g \geq g']) + \nu(E \cap [g' > g]) = \nu(E).$$

Thus $\mathcal{G}$ is closed under the formation of finite maxima. Suppose that functions g_n lie in $\mathcal{G}$ and $g_n \uparrow g$. Then $\int_E g \, d\mu = \lim_n \int_E g_n \, d\mu \leq \nu(E)$ by the monotone convergence theorem, so that g lies in $\mathcal{G}$. Thus $\mathcal{G}$ is closed under nondecreasing passages to the limit.

Let $\alpha = \sup \int g \, d\mu$ for g ranging over $\mathcal{G}$ ($\alpha \leq \nu(\Omega)$). Choose g_n in $\mathcal{G}$ so that $\int g_n \, d\mu > \alpha - n^{-1}$. If $f_n = \max(g_1, \ldots, g_n)$ and $f = \lim f_n$, then f lies in $\mathcal{G}$ and $\int f \, d\mu = \lim_n \int f_n \, d\mu \geq \lim_n \int g_n \, d\mu = \alpha$. Thus f is an element of $\mathcal{G}$ for which $\int f \, d\mu$ is maximal.

Define ν_{ac} by $\nu_{ac}(E) = \int_E f \, d\mu$ and ν_s by $\nu_s(E) = \nu(E) - \nu_{ac}(E)$. Thus

$$(32.7) \qquad \nu(E) = \nu_{ac}(E) + \nu_s(E) = \int_E f \, d\mu + \nu_s(E).$$

Since f is in $\mathcal{G}$, ν_s as well as ν_{ac} is a finite measure. Of course, ν_{ac} is absolutely continuous with respect to μ.

Suppose that ν_s fails to be singular with respect to μ. It then follows from the lemma that there is a set A and a positive ϵ such that $\mu(A) > 0$ and $\epsilon\mu(E) \leq \nu_s(E)$ for all $E \subset A$. Then for every E

$$\int_E (f + \epsilon I_A) \, d\mu = \int_E f \, d\mu + \epsilon\mu(E \cap A) \leq \int_E f \, d\mu + \nu_s(E \cap A)$$

$$= \int_{E \cap A} f \, d\mu + \nu_s(E \cap A) + \int_{E - A} f \, d\mu$$

$$= \nu(E \cap A) + \int_{E - A} f \, d\mu \leq \nu(E \cap A) + \nu(E - A) = \nu(E).$$

In other words, $f + \epsilon I_A$ lies in $\mathcal{G}$; since $\int (f + \epsilon I_A) \, d\mu = \alpha + \epsilon\mu(A) > \alpha$, this contradicts the maximality of f.

Therefore, μ and ν_s are mutually singular, and there exists an S such that $\nu_s(S) = \mu(S^c) = 0$. But since $\nu \ll \mu$, $\nu_s(S^c) \leq \nu(S^c) = 0$, and so $\nu_s(\Omega) = 0$. The rightmost term in (32.7) thus drops out. ∎

Absolute continuity was not used until the last step of the proof, and what the argument shows is that ν always has a decomposition (32.7) into an *absolutely continuous part* and a *singular part* with respect to μ. This is the *Lebesgue decomposition*, and it generalizes the one in the preceding section (see (31.31)).

PROBLEMS

32.1. There are two ways to show that the convergence in (32.1) must be absolute: Use the Jordan decomposition. Use the fact that a series converges absolutely if it has the same sum no matter what order the terms are taken in.

32.2. If $A^+ \cup A^-$ is a Hahn decomposition for φ, there may be other ones $A_1^+ \cup A_1^-$. Construct an example of this. Show that there is uniqueness to the extent that $\varphi(A^+ \,\Delta\, A_1^+) = \varphi(A^- \,\Delta\, A_1^-) = 0$. Prove the Jordan decomposition unique.

32.3. Show that absolute continuity does not imply the ϵ-δ condition (32.4) if ν is infinite.

Hint: Let $\mathcal{F}$ consist of all subsets of the space of integers, let ν be counting measure, and let μ have mass n^{-2} at n.

32.4. Show that the Radon-Nikodym theorem fails if μ is not σ-finite, even if ν is finite.

Hint: Let $\mathcal{F}$ consist of the countable and the co-countable sets in an uncountable Ω, let μ be counting measure, and let $\nu(A)$ be 0 or 1 as A is countable or co-countable. Or consider one-dimensional and two-dimensional Hausdorff measure in the plane.

32.5. Let μ be the restriction of planar Lebesgue measure λ_2 to the σ-field $\mathcal{F} = \{A \times R^1 : A \in \mathcal{R}^1\}$ of vertical strips. Define ν on $\mathcal{F}$ by $\nu(A \times R^1) = \lambda_2(A \times (0, 1))$. Show that ν is absolutely continuous with respect to μ but has no density. Why does this not contradict the Radon-Nikodym theorem?

32.6. Let μ, ν, and ρ be σ-finite measures on $(\Omega, \mathcal{F})$. Show that $\nu \ll \mu$ and $\mu \ll \rho$ imply that $\nu \ll \rho$ and

$$\frac{d\nu}{d\rho} = \frac{d\nu}{d\mu} \cdot \frac{d\mu}{d\rho}.$$

Show that $d\mu/d\mu = 1$ and that $\nu \equiv \mu$ implies that $d\nu/d\mu = (d\mu/d\nu)^{-1}$.

32.7. Show that there is a Lebesgue decomposition (32.7) in the σ-finite as well as the finite case. Prove that it is unique.

32.8. The Lebesgue decomposition (32.7) can be described differently. Show that there exists a set S (the singular set for ν with respect to μ) and a nonnegative function f (the density) such that $\mu(S) = 0$ and $\nu(A) = \int_A f \, d\mu$ for $A \subset \Omega - S$. Show that for any such S and f, $\nu_{ac}(A) = \int_A f \, d\mu$, and $\nu_s(A) = \nu(A \cap S)$.

32.9. Let μ and ν be finite measures on $(\Omega, \mathcal{F})$ and suppose that $\mathcal{F}^\circ$ is a σ-field contained in $\mathcal{F}$. Then the restrictions μ° and ν° of μ and ν to $\mathcal{F}^\circ$ are measures on $(\Omega, \mathcal{F}^\circ)$. Let ν_{ac}, ν_s, ν_{ac}°, ν_s° be, respectively, the absolutely continuous and singular parts of ν and ν° with respect to μ and μ°. Show that $\nu_{ac}^\circ(E) \geq \nu_{ac}(E)$ and $\nu_s^\circ(E) \leq \nu_s(E)$ for $E \in \mathcal{F}^\circ$.

32.10. Suppose that μ, ν, ν_n are finite measures on $(\Omega, \mathcal{F})$ and that $\nu(A) = \sum_n \nu_n(A)$ for all A. Let $\nu_n(A) = \int_A f_n \, d\mu + \nu_n'(A)$ and $\nu(A) = \int_A f \, d\mu + \nu'(A)$ be the decompositions (32.7); here ν' and ν_n' are singular with respect to μ. Show that $f = \sum_n f_n$ except on a set of μ-measure 0 and that $\nu'(A) = \sum_n \nu_n'(A)$ for all A. Show that $\nu \ll \mu$ if and only if $\nu_n \ll \mu$ for all n.

32.11. 32.2↑ Absolute continuity of a set function φ with respect to a measure μ is defined just as if φ were itself a measure: $\mu(A) = 0$ must imply that $\varphi(A) = 0$. Show that, if this holds and μ is σ-finite, then $\varphi(A) = \int_A f \, d\mu$ for some integrable f. Show that $A^+ = [\omega : f(\omega) \geq 0]$ and $A^- = [\omega : f(\omega) < 0]$ give a Hahn decomposition for φ. Show that the three variations satisfy $\varphi^+(A) = \int_A f^+ \, d\mu$, $\varphi^-(A) = \int_A f^- \, d\mu$, and $|\varphi|(A) = \int_A |f| \, d\mu$.

Hint: To construct f, start with (32.2).

32.12. ↑ A *signed measure* φ is a set function which satisfies (32.1) if $A_1, A_2, \ldots$ are disjoint and which may assume one of the values $+\infty$ and $-\infty$ but not both. Extend Theorems 32.1 and 32.2 to signed measures.

32.13. 16.11↑ Show that, if ν is not σ-finite, then the Radon-Nikodym derivative need not be unique even if it exists.

32.14. 31.20↑ Suppose that μ and ν are a probability measure and a σ-finite measure on the line and that $\nu \ll \mu$. Show that the Radon-Nikodym derivative f satisfies

$$\lim_{h \to 0} \frac{\nu(x - h, x + h]}{\mu(x - h, x + h]} = f(x)$$

on a set of μ-measure 1.

SECTION 33. CONDITIONAL PROBABILITY

The concepts of conditional probability and expected value with respect to a σ-field underlie much of modern probability theory. The difficulty in understanding these ideas has to do not with mathematical detail so much as with probabilistic meaning, and the way to get at this meaning is through calculations and examples, of which there are many in this section and the next.

The Discrete Case

Consider first the conditional probability of a set A with respect to another set B. It is defined of course by $P(A|B) = P(A \cap B)/P(B)$, unless $P(B)$ vanishes, in which case it is not defined at all.

It is helpful to consider conditional probability in terms of an observer in possession of partial information.* A probability space $(\Omega, \mathcal{F}, P)$ describes the working of a mechanism, governed by chance, which produces a result ω distributed according to P; $P(A)$ is for the observer the probability that the point ω produced lies in A. Suppose now that ω lies in B and that the observer learns this fact and no more. From the point of view of the observer, now in possession of this partial information about ω, the probability that ω also lies in A is $P(A|B)$ rather than $P(A)$. This is the idea lying back of the definition.

If, on the other hand, ω happens to lie in B^c and the observer learns of this, his probability instead becomes $P(A|B^c)$. These two conditional probabilities can be linked together by the simple function

$$(33.1) \qquad f(\omega) = \begin{cases} P(A|B) & \text{if } \omega \in B \\ P(A|B^c) & \text{if } \omega \in B^c. \end{cases}$$

The observer learns whether ω lies in B or in B^c; his new probability for the event $\omega \in A$ is then just $f(\omega)$. Although the observer does not in general know the argument ω of f, he can calculate the value $f(\omega)$ because he knows which of B and B^c contains ω. (Note conversely that from the value $f(\omega)$ it is possible to determine whether ω lies in B or in B^c, unless $P(A|B) = P(A|B^c)$—that is, unless A and B are independent, in which case the conditional probability coincides with the unconditional one anyway.)

The sets B and B^c partition Ω, and these ideas carry over to the general partition. Let $B_1, B_2, \ldots$ be a finite or countable partition of Ω into $\mathcal{F}$-sets, and let $\mathcal{G}$ consist of all the unions of the B_i. Then $\mathcal{G}$ is the σ-field generated by the B_i. For A in $\mathcal{F}$, consider the function with values

$$(33.2) \quad f(\omega) = P(A|B_i) = \frac{P(A \cap B_i)}{P(B_i)} \qquad \text{if } \omega \in B_i, \quad i = 1, 2, \ldots.$$

If the observer learns which element B_i of the partition it is that contains ω, then his new probability for the event $\omega \in A$ is $f(\omega)$. The partition $\{B_i\}$, or equivalently

---

* As always, *observer, information, know,* and so on are informal, nonmathematical terms; see the related discussion in Section 4 (page 45).

the σ-field $\mathscr{G}$, can be regarded as an experiment, and to learn which B_i it is that contains ω is to learn the outcome of the experiment. For this reason the function or random variable f defined by (33.2) is called the *conditional probability of A given $\mathscr{G}$* and is denoted $P[A\|\mathscr{G}]$. This is written $P[A\|\mathscr{G}]_\omega$ whenever the argument ω needs to be explicitly shown.

Thus $P[A\|\mathscr{G}]$ is the function whose value on B_i is the ordinary conditional probability $P(A|B_i)$. This definition needs to be completed, because $P(A|B_i)$ is not defined if $P(B_i) = 0$. In this case $P[A\|\mathscr{G}]$ will be taken to have any constant value on B_i; the value is arbitrary but must be the same over all of the set B_i. If there are nonempty sets B_i for which $P(B_i) = 0$, $P[A\|\mathscr{G}]$ therefore stands for any one of a family of functions on Ω. A specific such function is for emphasis often called a *version* of the conditional probability. Note that any two versions are equal except on a set of probability 0.

Example 33.1.   Consider the Poisson process. Suppose that $0 \le s \le t$, and let $A = [N_s = 0]$ and $B_i = [N_t = i]$, $i = 0, 1, \ldots$. Since the increments are independent (Section 23), $P(A|B_i) = P[N_s = 0]P[N_t - N_s = i]/P[N_t = i]$, and since they have Poisson distributions (see (23.9)), a simple calculation reduces this to

$$(33.3) \quad P[N_s = 0\|\mathscr{G}]_\omega = \left(1 - \frac{s}{t}\right)^i \qquad \text{if } \omega \in B_i, \quad i = 0, 1, 2, \ldots .$$

Since $i = N_t(\omega)$ on B_i, this can be written

$$(33.4) \qquad\qquad P[N_s = 0\|\mathscr{G}]_\omega = \left(1 - \frac{s}{t}\right)^{N_t(\omega)} .$$

Here the experiment or observation corresponding to $\{B_i\}$ or $\mathscr{G}$ determines the number of events—telephone calls, say—occurring in the time interval $[0, t]$. For an observer who knows this number but not the locations of the calls within $[0, t]$, (33.4) gives his probability for the event that none of them occurred before time s. Although this observer does not know ω, he knows $N_t(\omega)$, which is all he needs to calculate the right side of (33.4).   ∎

Example 33.2.   Suppose that $X_0, X_1, \ldots$ is a Markov chain with state space S as in Section 8. The events

$$(33.5) \qquad\qquad [X_0 = i_0, \ldots, X_n = i_n]$$

form a finite or countable partition of Ω as $i_0, \ldots, i_n$ range over S. If $\mathscr{G}_n$ is the σ-field generated by this partition, then by the defining condition (8.2) for Markov chains, $P[X_{n+1} = j\|\mathscr{G}_n]_\omega = p_{i_n j}$ holds for ω in (33.5). The sets

$$(33.6) \qquad\qquad [X_n = i]$$

for $i \in S$ also partition Ω, and they generate a σ-field $\mathcal{G}_n^0$ smaller than $\mathcal{G}_n$. Now (8.2) also stipulates $P[X_{n+1} = j \| \mathcal{G}_n^0] = p_{ij}$ for ω in (33.6), and the essence of the Markov property is that

$$(33.7) \qquad P[X_{n+1} = j \| \mathcal{G}_n] = P[X_{n+1} = j \| \mathcal{G}_n^0]. \qquad \blacksquare$$

The General Case

If $\mathcal{G}$ is the σ-field generated by a partition $B_1, B_2, \ldots$, then the general element of $\mathcal{G}$ is a disjoint union $B_{i_1} \cup B_{i_2} \cup \ldots$, finite or countable, of certain of the B_i. To know which set B_i it is that contains ω is the same thing as to know which sets in $\mathcal{G}$ contain ω and which do not. This second way of looking at the matter carries over to the general σ-field $\mathcal{G}$ contained in $\mathcal{F}$. (As always, the probability space is $(\Omega, \mathcal{F}, P)$.) The σ-field $\mathcal{G}$ will not in general come from a partition as above.

One can imagine an observer who knows for each G in $\mathcal{G}$ whether $\omega \in G$ or $\omega \in G^c$. Thus the σ-field $\mathcal{G}$ can in principle be identified with an experiment or observation. This is the point of view adopted in Section 4; see page 45. It is natural to try and define conditional probabilities $P[A \| \mathcal{G}]$ with respect to the experiment $\mathcal{G}$. To do this, fix an A in $\mathcal{F}$ and define a finite measure ν on $\mathcal{G}$ by

$$\nu(G) = P(A \cap G), \qquad G \in \mathcal{G}.$$

Then $P(G) = 0$ implies that $\nu(G) = 0$. The Radon-Nikodym theorem can be applied to the measures ν and P on the measureable space $(\Omega, \mathcal{G})$ because the first one is absolutely continuous with respect to the second.* It follows that there exists a function or random variable f, measurable $\mathcal{G}$ and integrable with respect to P, such that* $P(A \cap G) = \nu(G) = \int_G f \, dP$ for all G in $\mathcal{G}$.

Denote this function f by $P[A \| \mathcal{G}]$. It is a random variable with two properties:

(i) $P[A \| \mathcal{G}]$ *is measurable $\mathcal{G}$ and integrable.*

(ii) $P[A \| \mathcal{G}]$ *satisfies the functional equation*

$$(33.8) \qquad \int_G P[A \| \mathcal{G}] \, dP = P(A \cap G), \qquad G \in \mathcal{G}.$$

There will in general be many such random variables $P[A \| \mathcal{G}]$, but any two of them are equal with probability 1. A specific such random variable is called a *version* of the conditional probability.

If $\mathcal{G}$ is generated by a partition $B_1, B_2, \ldots$, the function f defined by (33.2) is measurable $\mathcal{G}$ because $[\omega: f(\omega) \in H]$ is the union of those B_i over which the

---

* Let P_0 be the restriction of P to $\mathcal{G}$ (Example 10.4) and find on $(\Omega, \mathcal{G})$ a density f for ν with respect to P_0. Then for $G \in \mathcal{G}$, $\nu(G) = \int_G f \, dP_0 = \int_G f \, dP$ (Example 16.4). If g is another such density, then $P[f \neq g] = P_0[f \neq g] = 0$.

constant value of f lies in H. Any G in $\mathscr{G}$ is a disjoint union $G = \bigcup_k B_{i_k}$, and $P(A \cap G) = \Sigma_k P(A | B_{i_k}) P(B_{i_k})$, so that (33.2) satisfies (33.8) as well. Thus the general definition is an extension of the one for the discrete case.

Condition (i) in the definition above in effect requires that the values of $P[A \| \mathscr{G}]$ depend only on the sets in $\mathscr{G}$. An observer who knows the outcome of $\mathscr{G}$ viewed as an experiment knows for each G in $\mathscr{G}$ whether it contains ω or not; for each x he knows this in particular for the set $[\omega': P[A \| \mathscr{G}]_{\omega'} = x]$, and hence he in principle knows the functional value $P[A \| \mathscr{G}]_{\omega}$ even if he does not know ω itself. In Example 33.1 a knowledge of $N_t(\omega)$ suffices to determine the value of (33.4)—ω itself is not needed.

Condition (ii) in the definition has a gambling interpretation. Suppose that the observer, after he has learned the outcome of $\mathscr{G}$, is offered the opportunity to bet on the event A (unless A lies in $\mathscr{G}$, he does not yet know whether or not it occurred). He is required to pay an entry fee of $P[A \| \mathscr{G}]$ units and will win 1 unit if A occurs and nothing otherwise. If the observer decides to bet and pays his fee, he gains $1 - P[A \| \mathscr{G}]$ if A occurs and $-P[A \| \mathscr{G}]$ otherwise, so that his gain is

$$(1 - P[A \| \mathscr{G}])I_A + (-P[A \| \mathscr{G}])I_{A^c} = I_A - P[A \| \mathscr{G}].$$

If he declines to bet, his gain is of course 0. Suppose that he adopts the strategy of betting if G occurs but not otherwise, where G is some set in $\mathscr{G}$. He can actually carry out this strategy, since after learning the outcome of the experiment $\mathscr{G}$ he knows whether or not G occurred. His expected gain with this strategy is his gain integrated over G:

$$\int_G (I_A - P[A \| \mathscr{G}]) \, dP.$$

But (33.8) is exactly the requirement that this vanish for each G in $\mathscr{G}$. Condition (ii) requires then that each strategy be fair in the sense that the observer stands neither to win nor to lose on the average. Thus $P[A \| \mathscr{G}]$ is the just entry fee, as intuition requires.

Example 33.3. Suppose that $A \in \mathscr{G}$, which will always hold if $\mathscr{G}$ coincides with the whole σ-field $\mathscr{F}$. Then I_A satisfies conditions (i) and (ii), so that $P[A \| \mathscr{G}] = I_A$ with probability 1. If $A \in \mathscr{G}$, then to know the outcome of $\mathscr{G}$ viewed as an experiment is in particular to know whether or not A has occurred.   ■

Example 33.4. If $\mathscr{G}$ is $\{0, \Omega\}$, the smallest possible σ-field, every function measurable $\mathscr{G}$ must be constant. Therefore, $P[A \| \mathscr{G}]_{\omega} = P(A)$ for all ω in this case. The observer learns nothing from the experiment $\mathscr{G}$.   ■

According to these two examples, $P[A \| \{0, \Omega\}]$ is identically $P(A)$, whereas I_A is a version of $P[A \| \mathscr{F}]$. For any $\mathscr{G}$, the function identically equal to $P(A)$

satisfies condition (i) in the definition of conditional probability, whereas I_A satisfies condition (ii). Condition (i) becomes more stringent as $\mathscr{G}$ decreases, and condition (ii) becomes more stringent as $\mathscr{G}$ increases. The two conditions work in opposite directions and between them delimit the class of versions of $P[A\|\mathscr{G}]$.

Example 33.5. Let Ω be the plane R^2 and let $\mathscr{F}$ be the class $\mathscr{R}^2$ of planar Borel sets. A point of Ω is a pair (x, y) of reals. Let $\mathscr{G}$ be the σ-field consisting of the vertical strips, the product sets $E \times R^1 = [(x, y): x \in E]$, where E is a linear Borel set. If the observer knows for each strip $E \times R^1$ whether or not it contains (x, y), then, as he knows this for each one-point set E, he knows the value of x. Thus the experiment $\mathscr{G}$ consists in the determination of the first coordinate of the sample point. Suppose now that P is a probability measure on $\mathscr{R}^2$ having a density $f(x, y)$ with respect to planar Lebesgue measure: $P(A) = \iint_A f(x, y)\, dx\, dy$. Let A be a horizontal strip $R^1 \times F = [(x, y): y \in F]$, F being a linear Borel set. The conditional probability $P[A\|\mathscr{G}]$ can be calculated explicitly.

Put

$$(33.9) \qquad \varphi(x, y) = \frac{\int_F f(x, t)\, dt}{\int_{R^1} f(x, t)\, dt}.$$

Set $\varphi(x, y) = 0$, say, at points where the denominator here vanishes; these points form a set of P-measure 0. Since $\varphi(x, y)$ is a function of x alone, it is measurable $\mathscr{G}$. The general element of $\mathscr{G}$ being $E \times R^1$, it will follow that $\varphi(x, y)$ is a version of $P[A\|\mathscr{G}]_{(x,y)}$ if it is shown that

$$(33.10) \qquad \int_{E \times R^1} \varphi(x, y)\, dP(x, y) = P(A \cap (E \times R^1)).$$

Since $A = R^1 \times F$, the right side here is $P(E \times F)$. Since P has density f, Theorem 16.10 and Fubini's theorem reduce the left side to

$$\int_E \left\{ \int_{R^1} \varphi(x, y) f(x, y)\, dy \right\} dx = \int_E \left\{ \int_F f(x, t)\, dt \right\} dx$$

$$= \iint_{E \times F} f(x, y)\, dx\, dy = P(E \times F).$$

Thus (33.9) does give a version of $P[R^1 \times F\|\mathscr{G}]_{(x,y)}$. ∎

The right side of (33.9) is the classical formula for the conditional probability of the event $R^1 \times F$ (the event that $y \in F$) given the event $\{x\} \times R^1$ (given the value of x). Since the event $\{x\} \times R^1$ has probability 0, the formula $P(A|B) = P(A \cap B)/P(B)$ does not work here. The whole point of this section is the systematic development of a notion of conditional probability that covers conditioning with respect to events of probability 0. This is accomplished by conditioning with respect to *collections* of events—that is, with respect to σ-fields $\mathcal{G}$.

Example 33.6.   The set A is by definition independent of the σ-field $\mathcal{G}$ if it is independent of each G in $\mathcal{G}$: $P(A \cap G) = P(A)P(G)$. This being the same thing as $P(A \cap G) = \int_G P(A)\, dP$, A is independent of $\mathcal{G}$ if and only if $P[A\|\mathcal{G}] = P(A)$ with probability 1.   ∎

The σ-field $\sigma(X)$ generated by a random variable X consists of the sets $[\omega: X(\omega) \in H]$ for $H \in \mathcal{R}^1$; see Theorem 20.1. The conditional probability of A given X is defined as $P[A\|\sigma(X)]$ and is denoted $P[A\|X]$. Thus $P[A\|X] = P[A\|\sigma(X)]$ by definition. From the experiment corresponding to the σ-field $\sigma(X)$, one learns which of the sets $[\omega': X(\omega') = x]$ contains ω and hence learns the value $X(\omega)$. Example 33.5 is a case of this: take $X(x, y) = x$ for (x, y) in the sample space $\Omega = R^2$ there.

This definition applies without change to a random vector, or, equivalently, to a finite set of random variables. It can be adapted to arbitrary sets of random variables as well. For any such set $[X_t, t \in T]$, the σ-field $\sigma[X_t, t \in T]$ it generates is the smallest σ-field with respect to which each X_t is measurable. It is generated by the collection of sets of the form $[\omega: X_t(\omega) \in H]$ for t in T and H in $\mathcal{R}^1$. The *conditional probability* $P[A\|X_t, t \in T]$ *of A with respect to this set of random variables* is by definition the conditional probability $P[A\|\sigma[X_t, t \in T]]$ of A with respect to the σ-field $\sigma[X_t, t \in T]$ it generates.

In this notation the property (33.7) of Markov chains becomes

$$(33.11)\qquad P[X_{n+1} = j\|X_0, \ldots, X_n] = P[X_{n+1} = j\|X_n].$$

The conditional probability of $[X_{n+1} = j]$ is the same for someone who knows the present state X_n as for someone who knows the present state X_n and the past states $X_0, \ldots, X_{n-1}$ as well.

Example 33.7.   Let X and Y be random vectors of dimensions j and k, let μ be the distribution of X over R^j, and suppose that X and Y are independent. According to (20.30),

$$P[X \in H, (X, Y) \in J] = \int_H P[(x, Y) \in J]\mu(dx)$$

for $H \in R^j$ and $J \in R^{j+k}$. This is a consequence of Fubini's theorem; it has a conditional probability interpretation. For each x in R^j put

(33.12) $f(x) = P[(x, Y) \in J] = P[\omega': (x, Y(\omega')) \in J].$

If X and Y are independent, then

(33.13) $f(X(\omega)) = P[(X, Y) \in J\|X]_\omega$

with probability 1. By the theory of Section 18, $f(x(\omega))$ is measurable $\sigma(X)$, and since μ is the distribution of X, a change of variable gives

$$\int_{[X \in H]} f(X(\omega))P(d\omega) = \int_H f(x)\mu(dx) = P([(X, Y) \in J] \cap [X \in H]).$$

Since $[X \in H]$ is the general element of $\sigma(X)$, (33.13) follows.

The fact just proved can be written

$$P[(X, Y) \in J\|X]_\omega = P[(X(\omega), Y) \in J]$$

$$= P[\omega': (X(\omega), Y(\omega')) \in J].$$

Replacing ω' by ω on the right here causes a notational collision analogous to the one replacing y by x causes in $\int_a^b f(x, y)\, dy$.   ∎

Suppose that X and Y are independent random variables and that Y has distribution function F. For $J = [(u, v): \max\{u, v\} \le m]$, (33.12) is 0 for $m < x$ and $F(m)$ for $m \ge x$; if $M = \max\{X, Y\}$, then (33.13) gives

(33.14) $P[M \le m\|X]_\omega = I_{[X \le m]}(\omega)F(m)$

with probability 1. All equations involving conditional probabilities must in this way be qualified by the phrase *with probability 1* because the conditional probability is unique only to within a set of probability 0.

The following theorem is helpful in calculating conditional probabilities.

Theorem 33.1.   *Let $\mathcal{P}$ be a π-system generating the σ-field $\mathcal{G}$. An integrable function f is a version of $P[A\|\mathcal{G}]$ if it is measurable $\mathcal{G}$ and if*

(33.15) $\int_G f\, dP = P(A \cap G)$

holds for all G in $\mathcal{P}$.

PROOF.   Each side of (33.15) is, as a function of G, a finite measure on $\mathcal{G}$. If they agree on the π-system $\mathcal{P}$, then they must agree on $\sigma(\mathcal{P}) = \mathcal{G}$.   ∎

Example 33.8.   Suppose that X and Y are independent random variables with a common distribution function F that is positive and continuous. What is the conditional probability of $[X \le x]$ given the random variable $M = \max\{X, Y\}$? As it should clearly be 1 if $M \le x$, suppose that $M > x$. Since $X \le x$

requires $M = Y$, the chance of which is $\frac{1}{2}$ by symmetry, the conditional probability of $[X \leq x]$ should by independence be $\frac{1}{2}F(x)/F(m) = \frac{1}{2}P[X \leq x | X \leq m]$ with the random variable M substituted for m. Intuition thus gives

$$(33.16) \quad P[X \leq x \| M]_\omega = I_{[M \leq x]}(\omega) + \frac{1}{2} I_{[M > x]}(\omega) \frac{F(x)}{F(M(\omega))}.$$

It suffices to check (33.15) for sets $G = [M \leq m]$, because these form a π-system generating $\sigma(M)$. The functional equation reduces to

$$(33.17) \quad P[M \leq \min\{x, m\}] + \frac{1}{2} \int_{x < M \leq m} \frac{F(x)}{F(M)} \, dP$$

$$= P[M \leq m, X \leq x].$$

Since the other case is easy, suppose that $x < m$. Since the distribution of (X, Y) is product measure, it follows by Fubini's theorem and the assumed continuity of F that

$$\int_{x < M \leq m} \frac{1}{F(M)} \, dP = \iint_{\substack{u \leq v \\ x < v \leq m}} \frac{1}{F(v)} \, dF(u) \, dF(v)$$

$$+ \iint_{\substack{v < u \\ x < u \leq m}} \frac{1}{F(u)} \, dF(u) \, dF(v) = 2(F(m) - F(x)),$$

which gives (33.17).   ∎

Example 33.9.   A collection $[X_t : t \geq 0]$ of random variables is a *Markov process in continuous time* if for $k \geq 1$, $0 \leq t_1 \leq \cdots \leq t_k \leq u$, and $H \in \mathcal{R}^1$,

$$(33.18) \quad P[X_u \in H \| X_{t_1}, \ldots, X_{t_k}] = P[X_u \in H \| X_{t_k}]$$

with probability 1. The analogue for discrete time is (33.11). (The X_n there have countable range as well, and the transition probabilities are constant in time, conditions that are not imposed here.)

Suppose that $t \leq u$. Looking on the right side of (33.18) as a version of the conditional probability on the left shows that

$$(33.19) \quad \int_G P[X_u \in H \| X_t] \, dP = P([X_u \in H] \cap G)$$

if $0 \leq t_1 \leq \cdots \leq t_k = t \leq u$ and $G \in \sigma(X_{t_1}, \ldots, X_{t_k})$. Fix t, u, and H and let k and $t_1, \ldots, t_k$ vary. Consider the class $\mathcal{P} = \bigcup \sigma(X_{t_1}, \ldots, X_{t_k})$, the union extending over all $k \geq 1$ and all k-tuples satisfying $0 \leq t_1 \leq \cdots \leq t_k = t$. If $A \in \sigma(X_{t_1}, \ldots, X_{t_k})$ and $B \in \sigma(X_{s_1}, \ldots, X_{s_i})$, then $A \cap B \in \sigma(X_{r_1}, \ldots, X_{r_j})$, where the r_α are the s_β and the t_γ merged together. Thus $\mathcal{P}$ is a π-system. Since $\mathcal{P}$ generates $\sigma[X_s : s \leq t]$ and $P[X_u \in H \| X_t]$ is measurable with respect to this

σ-field, it follows by (33.19) and Theorem 33.1 that $P[X_u \in H \| X_t]$ is a version of $P[X_u \in H \| X_s, s \leq t]$:

(33.20) $P[X_u \in H \| X_s, s \leq t] = P[X_u \in H \| X_t]$, $t \leq u$,

with probability 1.

This says that for calculating conditional probabilities about the future, the present $\sigma(X_t)$ is equivalent to the present and the entire past $\sigma[X_s : s \leq t]$. This follows from the much weaker condition (33.18). ∎

Example 33.10.   The Poisson process $[N_t : t \geq 0]$ has independent increments (Section 23). Suppose that $0 \leq t_1 \leq \cdots \leq t_k \leq u$. The random vector $(N_{t_1}, N_{t_2} - N_{t_1}, \ldots, N_{t_k} - N_{t_{k-1}})$ is independent of $N_u - N_{t_k}$, and so (Theorem 20.2) $(N_{t_1}, N_{t_2}, \ldots, N_{t_k})$ is independent of $N_u - N_{t_k}$. If J is the set of points $(x_1, \ldots, x_k, y)$ in R^{k+1} such that $x_k + y \in H$, where $H \in \mathcal{R}^1$, and if μ is the distribution of $N_u - N_{t_k}$, then (33.12) is $P[(x_1, \ldots, x_k, N_u - N_{t_k}) \in J] = P[x_k + N_u - N_{t_k} \in H] = \mu(H - x_k)$. Therefore, (33.13) gives $P[N_u \in H \| N_{t_1}, \ldots, N_{t_k}] = \mu(H - N_{t_k})$. This holds also if $k = 1$, and hence $P[N_u \in H \| N_{t_1}, \ldots, N_{t_k}] = P[N_u \in H \| N_{t_k}]$. The Poisson process thus has the Markov property (33.18); this is a consequence solely of the independence of the increments. The extended Markov property (33.20) follows. ∎

Properties of Conditional Probability

Theorem 33.2.   *For each A*

(33.21) $0 \leq P[A \| \mathcal{G}] \leq 1$

with probability 1. If $A_1, A_2, \ldots$ is a finite or countable sequence of disjoint sets, then

(33.22) $P[\bigcup_n A_n \| \mathcal{G}] = \sum_n P[A_n \| \mathcal{G}]$

with probability 1.

PROOF.   For each version of the conditional probability, $\int_G P[A \| \mathcal{G}] \, dP = P(A \cap G) \geq 0$ for each G in $\mathcal{G}$; since $P[A \| \mathcal{G}]$ is measurable $\mathcal{G}$, it must be nonnegative except on a set of P-measure 0. The other inequality in (33.21) is proved the same way.

If the A_n are disjoint and if G lies in $\mathcal{G}$, it follows (Theorem 16.6) that

$$\int_G \left(\sum_n P[A_n \| \mathcal{G}] \right) dP = \sum_n \int_G P[A_n \| \mathcal{G}] \, dP = \sum_n P(A_n \cap G)$$

$$= P((\bigcup_n A_n) \cap G).$$

Thus $\Sigma_n P[A_n \| \mathcal{G}]$, which is certainly measurable $\mathcal{G}$, satisfies the functional equation for $P[\bigcup_n A_n \| \mathcal{G}]$, and so must coincide with it except perhaps on a set of P-measure 0. Hence (33.22). ∎

Additional useful facts can be established by similar arguments. If $A \subset B$, then

(33.23) $P[B - A \| \mathcal{G}] = P[B \| \mathcal{G}] - P[A \| \mathcal{G}], \qquad P[A \| \mathcal{G}] \le P[B \| \mathcal{G}].$

The inclusion-exclusion formula

$$(33.24) \qquad P\left[\bigcup_{i=1}^{n} A_i \| \mathcal{G}\right] = \sum_i P[A_i \| \mathcal{G}] - \sum_{i<j} P[A_i \cap A_j \| \mathcal{G}] + \cdots$$

holds. If $A_n \uparrow A$, then

$$(33.25) \qquad P[A_n \| \mathcal{G}] \uparrow P[A \| \mathcal{G}],$$

and if $A_n \downarrow A$, then

$$(33.26) \qquad P[A_n \| \mathcal{G}] \downarrow P[A \| \mathcal{G}].$$

Further, $P(A) = 1$ implies that

$$(33.27) \qquad P[A \| \mathcal{G}] = 1$$

and $P(A) = 0$ implies that

$$(33.28) \qquad P[A \| \mathcal{G}] = 0.$$

Of course, (33.23) through (33.28) hold with probability 1 only.

Difficulties and Curiosities

This section has been devoted almost entirely to examples connecting the abstract definition (33.8) with the probabilistic idea lying back of it. There are pathological examples showing that the interpretation of conditional probability in terms of an observer with partial information breaks down in certain cases.

Example 33.11.   Let $(\Omega, \mathcal{F}, P)$ be the unit interval Ω with Lebesgue measure P on the σ-field $\mathcal{F}$ of Borel subsets of Ω. Take $\mathcal{G}$ to be the σ-field of sets that are either countable or co-countable. Then the function identically equal to $P(A)$ is a version of $P[A \| \mathcal{G}]$ because $P(G)$ is either 0 or 1 for every G in $\mathcal{G}$; therefore,

$$(33.29) \qquad P[A \| \mathcal{G}]_\omega = P(A)$$

with probability 1. But since $\mathcal{G}$ contains all one-point sets, to know which elements of $\mathcal{G}$ contain ω is to know ω itself. Thus $\mathcal{G}$ viewed as an experiment should be completely informative—the observer given the information in $\mathcal{G}$ should know ω exactly—and so it ought to be true that

$$(33.30) \qquad P[A \| \mathcal{G}]_\omega = \begin{cases} 1 & \text{if } \omega \in A, \\ 0 & \text{if } \omega \notin A. \end{cases}$$

This is Example 4.6 in a new form. ∎

The mathematical definition gives (33.29); the heuristic considerations lead to (33.30). Of course, (33.29) is right and (33.30) is wrong. The heuristic view breaks down in certain cases but is nonetheless illuminating and cannot, since it does not intervene in proofs, lead to any difficulties.

The point of view in this section has been "global." To each fixed A in $\mathcal{F}$ has been attached a function (actually a family of functions) $P[A\|\mathcal{G}]_\omega$ defined over all of Ω. What happens if the point of view is reversed—if ω is fixed and A varies over $\mathcal{F}$? Will this result in a probability measure on $\mathcal{F}$? Intuition says it should, and if it does, then (33.21) through (33.28) all reduce to standard facts about measures.

Suppose that $B_1, \ldots, B_r$ is a partition of Ω into $\mathcal{F}$-sets, and let $\mathcal{G} = \sigma(B_1, \ldots, B_r)$. If $P(B_1) = 0$ and $P(B_i) > 0$ for the other i, then one version of $P[A\|\mathcal{G}]$ is

$$P[A\|\mathcal{G}]_\omega = \begin{cases} 0 & \text{if } \omega \in B_1, \\ \dfrac{P(A \cap B_i)}{P(B_i)} & \text{if } \omega \in B_i, \quad i = 2, \ldots, r. \end{cases}$$

With this choice of version for each A, $P[A\|\mathcal{G}]_\omega$ is, as a function of A, a probability measure on $\mathcal{F}$ if $\omega \in B_2 \cup \cdots \cup B_r$ but not if $\omega \in B_1$. The "wrong" versions have been chosen. If, for example,

$$P[A\|\mathcal{G}]_\omega = \begin{cases} P(A) & \text{if } \omega \in B_1, \\ \dfrac{P(A \cap B_i)}{P(B_i)} & \text{if } \omega \in B_i, \quad i = 2, \ldots, r, \end{cases}$$

then $P[A\|\mathcal{G}]_\omega$ is a probability measure in A for each ω. Clearly, versions such as this one exist if $\mathcal{G}$ is finite.

It might be thought that for an arbitrary σ-field $\mathcal{G}$ in $\mathcal{F}$ versions of the various $P[A\|\mathcal{G}]$ can be so chosen that $P[A\|\mathcal{G}]_\omega$ is for each fixed ω a probability measure as A varies over $\mathcal{F}$. It is possible to construct a counterexample showing that this is not so.* The example is possible because the exceptional ω-set of probability 0 where (33.22) fails depends on the sequence $A_1, A_2, \ldots$; if there are uncountably many such sequences, it can happen that the union of these exceptional sets has positive probability whatever versions $P[A\|\mathcal{G}]$ are chosen.

The existence of such pathological examples turns out not to matter. Example 33.9 illustrates the reason why. From the assumption (33.18) the notably

---

* The argument is outlined in Problem 32.13. It depends on the construction of certain nonmeasurable sets.

stronger conclusion (33.20) was reached. Since the set $[X_u \in H]$ is fixed throughout the argument, it does not matter that conditional probabilities may not, in fact, be measures. What does matter for the theory is Theorem 33.2 and its extensions.

Consider a point ω_0 with the property that $P(G) > 0$ for every G in $\mathcal{G}$ that contains ω_0. This will be true if the one-point set $\{\omega_0\}$ lies in $\mathcal{F}$ and has positive probability. Fix any versions of the $P[A \| \mathcal{G}]$. For each A the set $[\omega: P[A \| \mathcal{G}]_\omega < 0]$ lies in $\mathcal{G}$ and has probability 0; it therefore cannot contain ω_0. Thus $P[A \| \mathcal{G}]_{\omega_0} \geq 0$. Similarly, $P[\Omega \| \mathcal{G}]_{\omega_0} = 1$, and, if the A_n are disjoint, $P[\bigcup_n A_n \| \mathcal{G}]_{\omega_0} = \Sigma_n P[A \| \mathcal{G}]_{\omega_0}$. Therefore, $P[A \| \mathcal{G}]_{\omega_0}$ is a probability measure as A ranges over $\mathcal{F}$.

Thus conditional probabilities behave like probabilities at points of positive probability. That they may not do so at points of probability 0 causes no problem because individual such points have no effect on the probabilities of sets. Of course, *sets* of points individually having probability 0 do have an effect, but here the global point of view reenters.

Conditional Probability Distributions

Let X be a random variable on $(\Omega, \mathcal{F}, P)$ and let $\mathcal{G}$ be a σ-field in $\mathcal{F}$.

Theorem 33.3. *There exists a function $\mu(H, \omega)$, defined for H in $\mathcal{R}^1$ and ω in Ω, with these two properties:*

(i)   *For each ω in Ω, $\mu(H, \omega)$ is, as a function of H, a probability measure on $\mathcal{R}^1$.*

(ii)   *For each H in $\mathcal{R}^1$, $\mu(H, \omega)$ is, as a function of ω, a version of $P[X \in H \| \mathcal{G}]_\omega$.*

The probability measure $\mu(\cdot, \omega)$ is a *conditional distribution* of X given $\mathcal{G}$. If $\mathcal{G} = \sigma(Z)$, it is a conditional distribution of X given Z.

PROOF.   For each rational r, let $F(r, \omega)$ be a version of $P[X \leq r \| \mathcal{G}]_\omega$. If $r \leq s$, then by (33.23),

$$(33.31) \qquad\qquad F(r, \omega) \leq F(s, \omega)$$

for ω outside a $\mathcal{G}$-set A_{rs} of probability 0. By (33.26),

$$(33.32) \qquad\qquad F(r, \omega) = \lim_n F(r + n^{-1}, \omega)$$

for ω outside a $\mathcal{G}$-set B_r of probability 0. Finally, by (33.25) and (33.26),

$$(33.33) \qquad \lim_{r \to -\infty} F(r, \omega) = 0, \qquad \lim_{n \to \infty} F(r, \omega) = 1$$

outside a $\mathcal{G}$-set C of probability 0. As there are only countably many of these exceptional sets, their union E lies in $\mathcal{G}$ and has probability 0.

For $\omega \notin E$ extend $F(\cdot, \omega)$ to all of R^1 by setting $F(x, \omega) = \inf [F(r, \omega): x < r]$. For $\omega \in E$ take $F(x, \omega) = F(x)$, where F is some arbitrary but fixed distribution function. Suppose that $\omega \notin E$. By (33.31) and (33.32), $F(x, \omega)$ agrees with the first definition on the rationals and is nondecreasing; it is right-continuous; and by (33.33) it is a probability distribution function. Therefore, there exists a probability measure $\mu(\cdot, \omega)$ on $(R^1, \mathcal{R}^1)$ with distribution function $F(\cdot, \omega)$. For $\omega \in E$, let $\mu(\cdot, \omega)$ be the probability measure corresponding to $F(x)$. Then condition (i) is satisfied.

The class of H for which $\mu(H, \omega)$ is measurable $\mathcal{G}$ is a λ-system containing the sets $H = (-\infty, r]$ for rational r; therefore $\mu(H, \omega)$ is measurable $\mathcal{G}$ for H in $\mathcal{R}^1$.

By construction, $\mu((-\infty, r], \omega) = P[X \le r \| \mathcal{G}]_\omega$ with probability 1 for rational r. For $H = (-\infty, r]$, then,

$$\int_G \mu(H, \omega) P(d\omega) = P([X \in H] \cap G)$$

for all G in $\mathcal{G}$. Fix G. Each side of this equation is a measure as a function of H, and so the equation must hold for all H in $\mathcal{R}^1$. ∎

Example 33.12. Let X and Y be random variables whose joint distribution ν in R^2 has density $f(x, y)$ with respect to Lebesgue measure: $P[(X, Y) \in A] = \nu(A) = \iint_A f(x, y)\, dx\, dy$. Let $g(x, y) = f(x, y)/\int_{R^1} f(x, t)\, dt$, and let $\mu(H, x) = \int_H g(x, y)\, dy$ have probability density $g(x, \cdot)$; if $\int_{R^1} f(x, t)\, dt = 0$, let $\mu(\cdot, x)$ be an arbitrary probability measure on the line. Then $\mu(H, X(\omega))$ will serve as the conditional distribution of Y given X. Indeed, (33.10) is the same thing as $\int_{E \times R^1} \mu(F, x)\, d\nu(x, y) = \nu(E \times F)$, and a change of variable gives $\int_{[X \in E]} \mu(F, X(\omega)) P(d\omega) = P[X \in E, Y \in F]$. Thus $\mu(F, X(\omega))$ is a version of $P[Y \in F \| X]_\omega$. This is Example 33.5 in a new guise. ∎

PROBLEMS

33.1. 20.5↑ *Borel's paradox.* Suppose that a random point on the sphere is specified by longitude Θ and latitude Φ, but restrict Θ by $0 \le \Theta < \pi$, so that Θ specifies only the complete meridian circle (not semicircle) containing the point, and compensate by letting Φ range over $(-\pi, \pi]$.

(a) Show that for given Θ the conditional distribution of Φ has density $\frac{1}{4} | \cos \phi |$ over $(-\pi, +\pi]$. If the point lies on, say, the meridian circle through Greenwich, it is thus not uniformly distributed over that great circle.

(b) Show that for given Φ the conditional distribution of Θ is uniform over $(0, \pi)$. If the point lies on the equator (Φ is 0 or π), it is thus uniformly distributed over that great circle.

Since the point is uniformly distributed over the spherical surface and great circles are indistinguishable, (a) and (b) stand in apparent contradiction. This shows again the inadmissibility of conditioning with respect to an isolated event of probability 0. The relevant σ-field must not be lost sight of.

33.2.   20.22↑   Let X and Y be independent, each having the standard normal distribution, and let (R, Θ) be the polar coordinates for (X, Y).

(a)   Show that $X + Y$ and $X - Y$ are independent and that $R^2 = [(X + Y)^2 + (X - Y)^2]/2$, and conclude that the conditional distribution of R^2 given $X - Y$ is the chi-squared distribution with one degree of freedom translated by $(X - Y)^2/2$.

(b)   Show that the conditional distribution of R^2 given Θ is chi-squared with two degrees of freedom.

(c)   If $X - Y = 0$, the conditional distribution of R^2 is chi-squared with one degree of freedom. If $\Theta = \pi/4$ or $\Theta = 5\pi/4$, the conditional distribution of R^2 is chi-squared with two degrees of freedom. But the events $[X - Y = 0]$ and $[\Theta = \pi/4] \cup [\Theta = 5\pi/4]$ are the same. Resolve the apparent contradiction.

33.3.   ↑   Paradoxes of a somewhat similar kind arise in very simple cases.

(a)   Of three prisoners, call them 1, 2, and 3, two have been chosen by lot for execution. Prisoner 3 says to the guard, "Which of 1 and 2 is to be executed? One of them will be, and you give me no information about myself in telling me which it is." The guard finds this reasonable and says, "Prisoner 1 is to be executed." And now 3 reasons, "I know that 1 is to be executed; the other will be either 2 or me, and so my chance of being executed is now only $\frac{1}{2}$, instead of the $\frac{2}{3}$ it was before." Apparently, the guard *has* given him information.

If one looks for a σ-field, it must be the one describing the guard's answer, and it then becomes clear that the sample space is incompletely specified. Suppose that, if 1 and 2 are to be executed, the guard's response is "1" with probability p and "2" with probability $1 - p$; and, of course, suppose that, if 3 is to be executed, the guard names the other victim. Calculate the conditional probabilities.

(b)   Assume that among families with two children the four sex distributions are equally likely. You have been introduced to one of the two children in such a family, and he is a boy. What is the conditional probability that the other is a boy as well?

33.4.   Extend Examples 33.4 and 33.11: If $P(G)$ is 0 or 1 for every G in $\mathcal{G}$, then $P[A\|\mathcal{G}] = P(A)$ with probability 1.

33.5.   There is a slightly different approach to conditional probability. Let $(\Omega, \mathcal{F}, P)$ be a probability space, $(\Omega', \mathcal{F}')$ a measurable space, and $T: \Omega \to \Omega'$ a mapping measurable $\mathcal{F}/\mathcal{F}'$. Define a measure ν on $\mathcal{F}'$ by $\nu(A') = P(A \cap T^{-1}A')$ for $A' \in \mathcal{F}'$. Prove that there exists a function $p(A|\omega')$ on Ω', measurable $\mathcal{F}'$ and integrable PT^{-1}, such that $\int_{A'} p(A|\omega')PT^{-1}(d\omega') = P(A \cap T^{-1}A')$ for all A' in $\mathcal{F}'$. Intuitively, $p(A|\omega')$ is the conditional probability that $\omega \in A$ for someone who knows that $T\omega = \omega'$. Let $\mathcal{G} = [T^{-1}A': A' \in \mathcal{F}']$; show that $\mathcal{G}$ is a σ-field and that $p(A|T\omega)$ is a version of $P[A\|\mathcal{G}]_\omega$.

33.6.   For the Poisson process (see Example 33.1) show that for $0 < s < t$,

$$P[N_s = k \| N_t] = \begin{cases} \binom{N_t}{k}\left(\dfrac{s}{t}\right)^k \left(1 - \dfrac{s}{t}\right)^{N_t - k}, & k \leq N_t, \\ 0, & k > N_t. \end{cases}$$

Thus the conditional distribution (in the sense of Theorem 33.3) of N_s given N_t is binomial with parameters N_t and s/t.

33.7.   29.10↑   Suppose that (X_1, X_2) has the centered normal distribution—has in the plane the distribution with density (29.29). Express the quadratic form in the exponential as

$$\frac{1}{\sigma_{11}} x_1^2 + \frac{\sigma_{11}}{D}\left(x_2 - \frac{\sigma_{12}}{\sigma_{11}} x_1\right)^2;$$

integrate out the x_2 and show that

$$\frac{f(x_1, x_2)}{\displaystyle\int_{-\infty}^{\infty} f(x_1, t)\, dt} = \frac{1}{\sqrt{2\pi\tau}} \exp\left[-\frac{1}{2\tau}\left(x_2 - \frac{\sigma_{12}}{\sigma_{11}} x_1\right)^2\right],$$

where $\tau = \sigma_{22} - \sigma_{12}^2 \sigma_{11}^{-1}$. Describe the conditional distribution of X_2 given X_1.

33.8.   Let $X_1, X_2, \ldots$ be independent random variables and let $\mathcal{G}$ be the σ-field $\bigcap_n \sigma(X_n, X_{n+1}, \ldots)$. Use Theorem 33.1 and Example 33.6 to prove Kolmogorov's zero-one law.

33.9.   Prove this form of Bayes' theorem:

$$P(G|A) = \frac{\displaystyle\int_G P[A \| \mathcal{G}]\, dP}{\displaystyle\int_\Omega P[A \| \mathcal{G}]\, dP}$$

for $G \in \mathcal{G}$.

33.10.   Suppose that $\mu(H, \omega)$ has property (i) in Theorem 33.3, and suppose that $\mu(H, \cdot)$ is a version of $P[X \in H \| \mathcal{G}]$ for H in a π-system generating $\mathcal{R}^1$. Show that $\mu(\cdot, \omega)$ is a conditional distribution of X given $\mathcal{G}$.

33.11.   ↑   Deduce from (33.16) that the conditional distribution of X given M is

$$\frac{1}{2} I_{[M \in H]}(\omega) + \frac{1}{2}\frac{\mu(H \cap (-\infty, M(\omega)])}{\mu(-\infty, M(\omega)])},$$

where μ is the distribution corresponding to F (positive and continuous).

Hint: First check $H = (-\infty, x]$.

33.12.   It is shown in Example 33.9 that (33.18) implies (33.20). Prove the reverse implication.

33.13.   4.13 12.4 ↑   The following construction shows that conditional probabilities may not give measures. Complete the details.

In Problem 4.13 it is shown that there exist a probability space $(\Omega, \mathcal{F}, P)$, a σ-field $\mathcal{G}$ in $\mathcal{F}$, and a set H in $\mathcal{F}$, such that $P(H) = \frac{1}{2}$, H and $\mathcal{G}$ are independent, $\mathcal{G}$ contains all the singletons, and $\mathcal{G}$ is generated by a countable subclass. The countable subclass generating $\mathcal{G}$ can be taken to be a π-system $\mathcal{P} = \{B_1, B_2, \ldots\}$ (pass to the finite intersections of the sets in the original class).

Assume that is is possible to choose versions $P[A \| \mathcal{G}]$ so that $P[A \| \mathcal{G}]_\omega$ is for each ω a probability measure as A varies over $\mathcal{F}$. Let C_n be the ω-set where $P[B_n \| \mathcal{G}]_\omega = I_{B_n}(\omega)$; show (Example 33.3) that $C = \bigcap_n C_n$ has probability 1. Show that $\omega \in C$ implies that $P[G \| \mathcal{G}]_\omega = I_G(\omega)$ for all G in $\mathcal{G}$ and hence that $P[\{\omega\} \| \mathcal{G}]_\omega = 1$.

Now $\omega \in H \cap C$ implies that $P[H \| \mathcal{G}]_\omega \geq P[\{\omega\} \| \mathcal{G}]_\omega = 1$ and $\omega \in H^c \cap C$ implies that $P[H \| \mathcal{G}]_\omega \leq P[\Omega - \{\omega\} \| \mathcal{G}]_\omega = 0$. Thus $\omega \in C$ implies that $P[H \| \mathcal{G}]_\omega = I_H(\omega)$. But since H and $\mathcal{G}$ are independent, $P[H \| \mathcal{G}]_\omega = P(H) = \frac{1}{2}$ with probability 1, a contradiction.

This example is related to Example 4.6 but concerns mathematical fact instead of heuristic interpretation.

33.14. Use Theorem 12.5 to extend Theorem 33.3 to the case in which X is a random vector.

33.15. Let α and β be σ-finite measures on the line and let $f(x, y)$ be a probability density with respect to $\alpha \times \beta$. Show that

$$(33.34) \qquad g_x(y) = \frac{f(x, y)}{\displaystyle\int_{R^1} f(x, t)\beta(dt)}$$

is a probability density with respect to β except for x in a set of α-measure 0. Show that, if (X, Y) has density f with respect to $\alpha \times \beta$, then the conditional distribution of Y given X has density $g_X(y)$ with respect to α. This generalizes Examples 33.5 and 33.12, where α and β are Lebesgue measure.

33.16. 18.19↑　Suppose that μ and ν_x (one for each real x) are probability measures on the line, and suppose that $\nu_x(B)$ is a Borel function in x for each $B \in \mathcal{R}^1$. Then (see Problem 18.19)

$$(33.35) \qquad \pi(E) = \int_{R^1} \nu_x[y \colon (x, y) \in E]\mu(dx)$$

defines a probability measure on $(R^2, \mathcal{R}^2)$.

Suppose that (X, Y) has distribution π and show that ν_X is a version of the conditional distribution of Y given X.

33.17. ↑　Let α and β be σ-finite measures on the line. Specialize the setup of Problem 33.16 by supposing that μ has density $f(x)$ with respect to α and ν_x has density $g_x(y)$ with respect to β. Assume that $g_x(y)$ is measurable $\mathcal{R}^2$ in the pair (x, y), so that $\nu_x(B)$ is automatically measurable in x. Show that (33.35) has density $f(x)g_x(y)$ with respect to $\alpha \times \beta$: $\pi(E) = \iint_E f(x)g_x(y)\alpha(dx)\beta(dy)$. Show that (33.34) is consistent with $f(x, y) = f(x)g_x(y)$. Put

$$p_y(x) = \frac{f(x)g_x(y)}{\displaystyle\int_{R^1} f(s)g_s(y)\alpha(ds)}.$$

Suppose that (X, Y) has density $f(x)g_x(y)$ with respect to $\alpha \times \beta$ and show that $p_Y(x)$ is a density with respect to α for the conditional distribution of X given Y.

In the language of Bayes, $f(x)$ is the prior density of a parameter x, $g_x(y)$ is the conditional density of the observation y given the parameter, and $p_y(x)$ is the posterior density of the parameter given the observation.

33.18. ↑ Now suppose that α and β are Lebesgue measure, that $f(x)$ is positive, continuous, and bounded, and that $g_x(y) = e^{-(y-x)^2 n/2}/\sqrt{2\pi/n}$. Thus the observation is distributed as the average of n independent normal variables with mean x and variance 1. Show that

$$\frac{1}{\sqrt{n}} p_y \left(y + \frac{x}{\sqrt{n}} \right) \rightarrow \frac{1}{\sqrt{2\pi}} e^{-x^2/2}$$

for fixed x and y. Thus the posterior density is approximately that of a normal distribution with mean y and variance $1/n$.

33.19. 32.14↑ Suppose that X has distribution μ. Now $P[A\|X]_\omega = f(X(\omega))$ for some Borel function f. Show that $\lim_{h \to 0} P[A|x - h < X \le x + h] = f(x)$ for x in a set of μ-measure 1. Roughly speaking, $P[A|x - h < X \le x + h] \rightarrow P[A|X = x]$.

Hint: Take $\nu(B) = P(A \cap [X \in B])$ in Problem 32.14.

SECTION 34. CONDITIONAL EXPECTATION

Definition

Suppose that X is an integrable random variable on $(\Omega, \mathcal{F}, P)$ and that $\mathcal{G}$ is a σ-field in $\mathcal{F}$. There exists a random variable $E[X\|\mathcal{G}]$, called the *conditional expected value* of X given $\mathcal{G}$, which has these two properties:

(i) $E[X\|\mathcal{G}]$ *is measurable $\mathcal{G}$ and integrable.*

(ii) $E[X\|\mathcal{G}]$ *satisfies the functional equation*

(34.1) $$\int_G E[X\|\mathcal{G}] \, dP = \int_G X \, dP, \qquad G \in \mathcal{G}.$$

To prove the existence of such a random variable, consider first the case of nonnegative X. Define a measure ν on $\mathcal{G}$ by $\nu(G) = \int_G X \, dP$. This measure is finite because X is integrable, and it is absolutely continuous with respect to P. By the Radon-Nikodym theorem there is a function f, measurable $\mathcal{G}$, such that $\nu(G) = \int_G f \, dP$.* This f has properties (i) and (ii).[†] If X is not necessarily nonnegative, $E[X^+\|\mathcal{G}] - E[X^-\|\mathcal{G}]$ clearly has the required properties.

---

* As in the case of conditional probabilities, the integral is the same on $(\Omega, \mathcal{F}, P)$ as on $(\Omega, \mathcal{G})$ with P restricted to $\mathcal{G}$ (Example 16.4).

[†] See Problem 34.19 for another construction.

There will in general be many such random variables $E[X\|\mathcal{G}]$; any one of them is called a *version* of the conditional expected value. Any two versions are equal with probability 1 (by Example 16.9 with P restricted to $\mathcal{G}$).

Arguments such as those in Examples 33.3 and 33.4 show that $E[X\|\{0, \Omega\}]$ $= E[X]$ and that $E[X\|\mathcal{F}] = X$ with probability 1. As $\mathcal{G}$ increases, condition (i) becomes weaker and condition (ii) becomes stronger.

The value $E[X\|\mathcal{G}]_\omega$ at ω is to be interpreted as the expected value of X for someone who knows for each G in $\mathcal{G}$ whether or not it contains the point ω, which itself in general remains unknown. Condition (i) ensures that $E[X\|\mathcal{G}]$ can in principle be calculated from this partial information alone. Condition (ii) can be restated as $\int_G (X - E[X\|\mathcal{G}]) \, dP = 0$; if the observer, in possession of the partial information contained in $\mathcal{G}$, is offered the opportunity to bet, paying an entry fee of $E[X\|\mathcal{G}]$ and being returned the amount X, and if he adopts the strategy of betting if G occurs, this equation says that the game is fair.

Example 34.1. Suppose that $B_1, B_2, \ldots$ is a finite or countable partition of Ω generating the σ-field $\mathcal{G}$. Then $E[X\|\mathcal{G}]$ must, since it is measurable $\mathcal{G}$, have some constant value over B_i, say α_i. Then (34.1) for $G = B_i$ gives $\alpha_i P(B_i) = \int_{B_i} X \, dP$. Thus

$$E[X\|\mathcal{G}]_\omega = \frac{1}{P(B_i)} \int_{B_i} X \, dP, \qquad \omega \in B_i, \quad P(B_i) > 0.$$

If $P(B_i) = 0$, the value of $E[X\|\mathcal{G}]$ over B_i is constant but arbitrary. ■

Example 34.2. For an indicator I_A the defining properties of $E[I_A\|\mathcal{G}]$ and $P[A\|\mathcal{G}]$ coincide; therefore, $E[I_A\|\mathcal{G}] = P[A\|\mathcal{G}]$ with probability 1. It is easily checked that, more generally, $E[X\|\mathcal{G}] = \Sigma_i \, \alpha_i P[A_i\|\mathcal{G}]$ with probability 1 for a simple function $X = \Sigma_i \, \alpha_i I_{A_i}$. ◢

Properties of Conditional Expectation

The first result has the same proof as Theorem 33.1.

Theorem 34.1. *Let $\mathcal{P}$ be a π-system generating the σ-field $\mathcal{G}$. An integrable function f is a version of $E[X\|\mathcal{G}]$ if it is measurable $\mathcal{G}$ and if*

$$\int_G f \, dP = \int_G X \, dP$$

holds for all G in $\mathcal{P}$.

All the equalities and inequalities in the following theorem hold with probability 1.

Theorem 34.2. *Suppose that X, Y, X_n are integrable.*

(i)   *If $X = a$ with probability 1, then $E[X\|\mathcal{G}] = a$.*

(ii) *For constants a and b, $E[aX + bY\|\mathcal{G}] = aE[X\|\mathcal{G}] + bE[Y\|\mathcal{G}]$.*

(iii) *If $X \le Y$ with probability 1, then $E[X\|\mathcal{G}] \le E[Y\|\mathcal{G}]$.*

(iv) *$|E[X\|\mathcal{G}]| \le E[|X|\|\mathcal{G}]$.*

(v)   *If $\lim_n X_n = X$ with probability 1, $|X_n| \le Y$, and Y is integrable, then $\lim_n E[X_n\|\mathcal{G}] = E[X\|\mathcal{G}]$ with probability 1.*

PROOF. If $X = a$ with probability 1, the function identically equal to a satisfies conditions (i) and (ii) in the definition of $E[X\|\mathcal{G}]$, and so (i) above follows by uniqueness.

As for (ii), $aE[X\|\mathcal{G}] + bE[Y\|\mathcal{G}]$ is integrable and measurable $\mathcal{G}$, and

$$\int_G (aE[X\|\mathcal{G}] + bE[Y\|\mathcal{G}])\, dP$$

$$= a \int_G E[X\|\mathcal{G}]\, dP + b \int_G E[Y\|\mathcal{G}]\, dP$$

$$= a \int_G X\, dP + b \int_G Y\, dP = \int_G (aX + bY)\, dP$$

for all G in $\mathcal{G}$, so that this function satisfies the functional equation.

If $X \le Y$ with probability 1, then $\int_G (E[Y\|\mathcal{G}] - E[X\|\mathcal{G}])\, dP = \int_G (Y - X)\, dP \ge 0$ for all G in $\mathcal{G}$. Since $E[Y\|\mathcal{G}] - E[X\|\mathcal{G}]$ is measurable $\mathcal{G}$, it must be nonnegative with probability 1 (see Example 16.9). This proves (iii), which clearly implies (iv) as well as the fact that $E[X\|\mathcal{G}] = E[Y\|\mathcal{G}]$ if $X = Y$ with probability 1.

To prove (v), consider $Z_n = \sup_{k \ge n} |X_k - X|$. Now $Z_n \downarrow 0$ with probability 1, and by (ii) and (iv), $|E[X_n\|\mathcal{G}] - E[X\|\mathcal{G}]| \le E[Z_n\|\mathcal{G}]$. It suffices, therefore, to show that $E[Z_n\|\mathcal{G}] \downarrow 0$ with probability 1. By (iii) the sequence $E[Z_n\|\mathcal{G}]$ is nondecreasing and hence has a limit Z; the problem is to prove that $Z = 0$ with probability 1, or, Z being nonnegative, that $E[Z] = 0$. But $|Z_n| \le 2Y$, and so the dominated convergence theorem gives $E[Z] = \int E[Z\|\mathcal{G}]\, dP \le \int E[Z_n\|\mathcal{G}]\, dP = \int Z_n\, dP \to 0$. ∎

The properties (33.21) through (33.28) can be derived anew from Theorem 34.2. Part (ii) shows once again that $E[\sum_i \alpha_i I_{A_i}\|\mathcal{G}] = \sum_i \alpha_i P[A_i\|\mathcal{G}]$ for simple functions.

If X is measurable $\mathcal{G}$, then clearly $E[X\|\mathcal{G}] = X$ with probability 1. The following generalization of this is used constantly.

Theorem 34.3. *If X is measurable $\mathcal{G}$, and if Y and XY are integrable, then*

(34.2) $$E[XY\|\mathcal{G}] = XE[Y\|\mathcal{G}]$$

with probability 1.

PROOF. It will be shown first that the right side of (34.2) is a version of the left side if $X = I_{G_0}$ and $G_0 \in \mathcal{G}$. Since $I_{G_0}E[Y\|\mathcal{G}]$ is certainly measurable $\mathcal{G}$, it suffices to show that it satisfies the functional equation $\int_G I_{G_0}E[Y\|\mathcal{G}]\,dP = \int_G I_{G_0}Y\,dP$, $G \in \mathcal{G}$. But this reduces to $\int_{G \cap G_0}E[Y\|\mathcal{G}]\,dP = \int_{G \cap G_0}Y\,dP$, which holds by the definition of $E[Y\|\mathcal{G}]$. Thus (34.2) holds if X is the indicator of an element of $\mathcal{G}$.

It follows by Theorem 34.2(ii) that (34.2) holds if X is a simple function measurable $\mathcal{G}$. For the general X that is measurable $\mathcal{G}$, there exist simple functions X_n, measurable $\mathcal{G}$, such that $|X_n| \leq |X|$ and $\lim_n X_n = X$ (Theorem 13.5). Since $|X_nY| \leq |XY|$ and $|XY|$ is integrable, Theorem 34.2(v) implies that $\lim_n E[X_nY\|\mathcal{G}] = E[XY\|\mathcal{G}]$ with probability 1. But $E[X_nY\|\mathcal{G}] = X_nE[Y\|\mathcal{G}]$ by the case already treated, and of course $\lim_n X_nE[Y\|\mathcal{G}] = XE[Y\|\mathcal{G}]$. (Note that $|X_nE[Y\|\mathcal{G}]| = |E[X_nY\|\mathcal{G}]| \leq E[|X_nY|\|\mathcal{G}] \leq E[|XY|\|\mathcal{G}]$, so that the limit $XE[Y\|\mathcal{G}]$ is integrable.) Thus (34.2) holds in general. Notice that X has not been assumed integrable. ∎

Taking a conditional expected value can be thought of as an averaging or smoothing operation. This leads one to expect that averaging X with respect to $\mathcal{G}_2$, and then averaging the result with respect to a coarser (smaller) σ-field $\mathcal{G}_1$ should lead to the same result as would averaging with respect to $\mathcal{G}_1$ in the first place:

Theorem 34.4. *If X is integrable and the σ-fields $\mathcal{G}_1$ and $\mathcal{G}_2$ satisfy $\mathcal{G}_1 \subset \mathcal{G}_2$, then*

$$(34.3) \qquad E[E[X\|\mathcal{G}_2]\|\mathcal{G}_1] = E[X\|\mathcal{G}_1]$$

with probability 1.

PROOF. The left side of (34.3) is measurable $\mathcal{G}_1$, and so to prove that it is a version of $E[X\|\mathcal{G}_1]$, it is enough to verify $\int_G E[E[X\|\mathcal{G}_2]\|\mathcal{G}_1]\,dP = \int_G X\,dP$ for $G \in \mathcal{G}_1$. But if $G \in \mathcal{G}_1$, then $G \in \mathcal{G}_2$, and the left side here is $\int_G E[X\|\mathcal{G}_2]\,dP = \int_G X\,dP$. ∎

If $\mathcal{G}_2 = \mathcal{F}$, then $E[X\|\mathcal{G}_2] = X$, so that (34.3) is trivial. If $\mathcal{G}_1 = \{0, \Omega\}$ and $\mathcal{G}_2 = \mathcal{G}$, then (34.3) becomes

$$(34.4) \qquad E[E[X\|\mathcal{G}]] = E[X],$$

the special case of (34.1) for $G = \Omega$.

If $\mathcal{G}_1 \subset \mathcal{G}_2$, then $E[X\|\mathcal{G}_1]$, being measurable $\mathcal{G}_1$, is also measurable $\mathcal{G}_2$, so that taking an expected value with respect to $\mathcal{G}_2$ does not alter it: $E[E[X\|\mathcal{G}_1]\|\mathcal{G}_2] = E[X\|\mathcal{G}_1]$. Therefore, if $\mathcal{G}_1 \subset \mathcal{G}_2$, taking iterated expected values in either order gives $E[X\|\mathcal{G}_1]$.

The remaining result of a general sort needed here is *Jensen's inequality* for conditional expected values: If φ is a convex function on the line and X and $\varphi(X)$ are both integrable, then

$$(34.5) \qquad \varphi(E[X\|\mathcal{G}]) \leq E[\varphi(X)\|\mathcal{G}]$$

with probability 1. For each x_0 take a support line through $(x_0, \varphi(x_0))$: $\varphi(x_0) + A(x_0)(x - x_0) \leq \varphi(x)$. The slope $A(x_0)$ can be taken as the right-hand derivative of φ, so that it is nondecreasing in x_0. Now

$$\varphi(E[X\|\mathcal{G}]) + A(E[X\|\mathcal{G}])(X - E[X\|\mathcal{G}]) \leq \varphi(X).$$

Suppose that $E[X\|\mathcal{G}]$ is bounded. Then all three terms here are integrable, and taking expected values with respect to $\mathcal{G}$ and using (34.2) on·the middle term gives (34.5).

To prove (34.5) in general, let $G_n = [|E[X\|\mathcal{G}]| \leq n]$. Then $E[I_{G_n}X\|\mathcal{G}] = I_{G_n}E[X\|\mathcal{G}]$ is bounded, and so (34.5) holds for $I_{G_n}X$: $\varphi(I_{G_n}E[X\|\mathcal{G}]) \leq E[\varphi(I_{G_n}X)\|\mathcal{G}]$. Now $E[\varphi(I_{G_n}X)\|\mathcal{G}] = E[I_{G_n}\varphi(X) + I_{G_n^c}\varphi(0)\|\mathcal{G}] = I_{G_n} \cdot E[\varphi(X)\|\mathcal{G}] + I_{G_n^c}\varphi(0) \to E[\varphi(X)\|\mathcal{G}]$. Since $\varphi(I_{G_n}E[X\|\mathcal{G}])$ converges to $\varphi(E[X\|\mathcal{G}])$ by the continuity of φ, (34.5) follows. If $\varphi(x) = |x|$, (34.5) gives part (iv) of Theorem 34.2 again.

Conditional Distributions and Expectations

Theorem 34.5. *Let $\mu(\cdot, \omega)$ be a conditional distribution with respect to $\mathcal{G}$ of a random variable X, in the sense of Theorem 33.3. If $\varphi: R^1 \to R^1$ is a Borel function for which $\varphi(X)$ is integrable, then $\int_{R^1} \varphi(x)\mu(dx, \omega)$ is a version of $E[\varphi(X)\|\mathcal{G}]_\omega$.*

PROOF. If $\varphi = I_H$ and $H \in \mathcal{R}^1$, this is an immediate consequence of the definition of conditional distribution, and by Theorem 34.2(ii) it follows for φ a simple function over R^1. For the general nonnegative φ, choose simple φ_n such that $0 \leq \varphi_n(x) \uparrow \varphi(x)$ for each x in R^1. By the case already treated, $\int_{R^1} \varphi_n(x)\mu(dx, \omega)$ is a version of $E[\varphi_n(X)\|\mathcal{G}]_\omega$. The integral converges by the monotone convergence theorem in $(R^1, \mathcal{R}^1, \mu(\cdot, \omega))$ to $\int_{R^1} \varphi(x)\mu(dx, \omega)$ for each ω, the value $+\infty$ not excluded, and $E[\varphi_n(X)\|\mathcal{G}]_\omega$ converges to $E[\varphi(X)\|\mathcal{G}]_\omega$ with probability 1 by Theorem 34.2(v). Thus the result holds for nonnegative φ, and the general case follows from splitting into positive and negative parts. ∎

It is a consequence of the proof above that $\int_{R^k} \varphi(x)\mu(dx, \omega)$ is measurable $\mathcal{G}$ and finite with probability 1. If X is itself integrable, it follows by the theorem for the case $\varphi(x) = x$ that

$$E[X\|\mathcal{G}]_\omega = \int_{-\infty}^{\infty} x\mu(dx, \omega)$$

with probability 1. If $\varphi(X)$ is integrable as well, then

$$(34.6) \qquad E[\varphi(X)\|\mathcal{G}]_\omega = \int_{-\infty}^{\infty} \varphi(x)\mu(dx, \omega)$$

with probability 1. By Jensen's inequality (21.14) for unconditional expected values, the right side of (34.6) is at least $\varphi(\int_{-\infty}^{\infty} x\mu(dx, \omega))$ if φ is convex. This gives another proof of (34.5).

Sufficient Subfields*

Suppose that for each θ in an index set Θ, P_θ is a probability measure on $(\Omega, \mathcal{F})$. In statistics the problem is to draw inferences about the unknown parameter θ from an observation ω.

Denote by $P_\theta[A\|\mathcal{G}]$ and $E_\theta[X\|\mathcal{G}]$ conditional probabilities and expected values calculated with respect to the probability measure P_θ on $(\Omega, \mathcal{F})$. A σ-field $\mathcal{G}$ in $\mathcal{F}$ is *sufficient* for the family $[P_\theta: \theta \in \Theta]$ if versions $P_\theta[A\|\mathcal{G}]$ can be chosen that are independent of θ—that is, if there exists a function $p(A, \omega)$ of $A \in \mathcal{F}$ and $\omega \in \Omega$ such that, for each $A \in \mathcal{F}$ and $\theta \in \Theta$, $p(A, \cdot)$ is a version of $P_\theta[A\|\mathcal{G}]$. There is no requirement that $p(\cdot, \omega)$ be a measure for ω fixed. The idea is that although there may be information in $\mathcal{F}$ not already contained in $\mathcal{G}$, this information is irrelevant to the drawing of inferences about θ. A *sufficient statistic* is a random variable or random vector T such that $\sigma(T)$ is a sufficient subfield.

A family $\mathcal{M}$ of measures *dominates* another family $\mathcal{N}$ if, for each A, from $\mu(A) = 0$ for all μ in $\mathcal{M}$, it follows that $\nu(A) = 0$ for all ν in $\mathcal{N}$. If each of $\mathcal{M}$ and $\mathcal{N}$ dominates the other, they are *equivalent*. For sets consisting of a single measure these are the concepts introduced in Section 32.

Theorem 34.6. *Suppose that $[P_\theta: \theta \in \Theta]$ is dominated by the σ-finite measure μ. A necessary and sufficient condition that $\mathcal{G}$ be sufficient is that the density f_θ of P_θ with respect to μ can be put in the form $f_\theta = g_\theta h$ for a g_θ measurable $\mathcal{G}$.*

It is assumed throughout that g_θ and h are nonnegative and of course that h is measurable $\mathcal{F}$. Theorem 34.6 is called the *factorization theorem,* the condition being that the density f_θ splits into a factor depending on ω only through $\mathcal{G}$ and a factor independent of θ. Although g_θ and h are not assumed integrable μ, their product f_θ, as the density of a finite measure, must be. Before proceeding to the proof, consider an application.

Example 34.3. Let $(\Omega, \mathcal{F}) = (R^k, \mathcal{R}^k)$ and for $\theta > 0$ let P_θ be the measure having with respect to k-dimensional Lebesgue measure the density

$$f_\theta(x) = f_\theta(x_1, \ldots, x_k) = \begin{cases} \theta^{-k} & \text{if } 0 \leq x_i \leq \theta, i = 1, \ldots, k, \\ 0 & \text{otherwise.} \end{cases}$$

If X_i is the function on R^k defined by $X_i(x) = x_i$, then under P_θ, $X_1, \ldots, X_k$ are independent random variables, each uniformly distributed over $[0, \theta]$. Let $T(x) = \max_{i \leq k} X_i(x)$. If $g_\theta(t)$ is θ^{-k} for $0 \leq t \leq \theta$ and 0 otherwise, and if $h(x)$ is 1 for all x, then $f_\theta(x) = g_\theta(T(x))h(x)$. The factorization criterion is thus satisfied, and T is a sufficient statistic.

---

* This topic may be omitted.

Sufficiency is clear on intuitive grounds as well: θ is not involved in the conditional distribution of $X_1, \ldots, X_k$ given T because, roughly speaking, a random one of them equals T and the others are independent and uniform over $[0, T]$. If this is true, the distribution of X_i given T ought to have a mass of k^{-1} at T and a uniform distribution of mass $1 - k^{-1}$ over $[0, T]$, so that

$$(34.7) \qquad E_\theta[X_i \| T] = \frac{1}{k} T + \frac{k-1}{k} \frac{T}{2} = \frac{k+1}{2k} T.$$

For a proof of this fact, needed later, note that by (21.9)

$$(34.8) \qquad \int_{T \le t} X_i \, dP_\theta = \int_0^\infty P_\theta[T \le t, X_i \ge u] \, du$$

$$= \int_0^t \frac{t-u}{\theta} \left(\frac{t}{\theta}\right)^{k-1} du = \frac{t^{k+1}}{2\theta^k}$$

if $0 \le t \le \theta$. On the other hand, $P_\theta[T \le t] = (t/\theta)^k$, so that under P_θ the distribution of T has density kt^{k-1}/θ^k over $[0, \theta]$. Thus

$$(34.9) \qquad \int_{T \le t} \frac{k+1}{2k} T \, dP_\theta = \frac{k+1}{2k} \int_0^t uk \frac{u^{k-1}}{\theta^k} \, du = \frac{t^{k+1}}{2\theta^k}.$$

Since (34.8) and (34.9) agree, (34.7) follows by Theorem 34.1. ∎

The essential ideas in the proof of Theorem 34.6 are most easily understood through a preliminary consideration of special cases. For the sufficiency of the condition, suppose first that h is integrable μ.

Lemma 1. *Suppose that $[P_\theta: \theta \in \Theta]$ is dominated by a measure μ and that each P_θ has with respect to μ a density $f_\theta = g_\theta h$, where g_θ is measurable $\mathcal{G}$ and h is integrable μ. Then $\mathcal{G}$ is sufficient.*

PROOF. Put $\alpha = \int h \, d\mu$ and replace g_θ by $g_\theta \alpha$ and h by h/α. This shows that it is no restriction to assume that $\int h \, d\mu = 1$. Define P on $\mathcal{F}$ by $P(A) = \int_A h \, d\mu$, so that $P_\theta(A) = \int_A g_\theta \, dP$. For G in $\mathcal{G}$, (34.2) gives

$$\int_G P[A \| \mathcal{G}] \, dP_\theta = \int_G E[I_A \| \mathcal{G}] g_\theta \, dP = \int_G E[I_A g_\theta \| \mathcal{G}] \, dP$$

$$= \int_G I_A g_\theta \, dP = \int_{A \cap G} g_\theta \, dP = P_\theta(A \cap G).$$

Therefore, $P[A \| \mathcal{G}]$—the conditional probability calculated with respect to P—serves as a version of $P_\theta[A \| \mathcal{G}]$ for each θ in Θ. Thus $\mathcal{G}$ is sufficient for the family $[P_\theta: \theta \in \Theta]$—even for this family augmented by P (which might happen to lie in the family to start with). ∎

For the necessity, suppose first that the family is dominated by one of its members.

Lemma 2. *Suppose that $[P_\theta: \theta \in \Theta]$ is dominated not only by μ but by P_{θ_0} for some $\theta_0 \in \Theta$. If $\mathcal{G}$ is sufficient, then each P_θ has with respect to μ a density $f_\theta = g_\theta h$, where g_θ is measurable $\mathcal{G}$ and h is integrable μ.*

PROOF. Let d_θ be any density of P_θ with respect to P_{θ_0}. By a number of applications of (34.2),

$$\int_A E_{\theta_0}[d_\theta \| \mathcal{G}] \, dP_{\theta_0} = \int I_A E_{\theta_0}[d_\theta \| \mathcal{G}] \, dP_{\theta_0}$$

$$= \int E_{\theta_0}\{I_A E_{\theta_0}[d_\theta \| \mathcal{G}] \| \mathcal{G}\} \, dP_{\theta_0} = \int E_{\theta_0}\{I_A \| \mathcal{G}\} E_{\theta_0}[d_\theta \| \mathcal{G}] \, dP_{\theta_0}$$

$$= \int E_{\theta_0}[E_{\theta_0}\{I_A \| \mathcal{G}\} d_\theta \| \mathcal{G}] \, dP_{\theta_0} = \int E_{\theta_0}\{I_A \| \mathcal{G}\} d_\theta \, dP_{\theta_0}$$

$$= \int P_{\theta_0}[A \| \mathcal{G}] \, dP_\theta = \int P_\theta[A \| \mathcal{G}] \, dP_\theta = P_\theta(A),$$

the next-to-last equality by sufficiency. Thus $g_\theta = E_{\theta_0}[d_\theta \| \mathcal{G}]$, which is measurable $\mathcal{G}$, can serve as a density for P_θ with respect to P_{θ_0}. Take for h a density for P_{θ_0} with respect to μ. ∎

To complete the proof of Theorem 34.6 requires one more lemma of a technical sort.

Lemma 3. *If $[P_\theta: \theta \in \Theta]$ is dominated by a σ-finite measure, then it is equivalent to some countable subfamily.*

PROOF. If μ is σ-finite, there is a countable partition of Ω into $\mathcal{F}$-sets A_n such that $0 < \mu(A_n) < \infty$. The measure with value $\Sigma_n \, 2^{-n}\mu(A \cap A_n)/\mu(A_n)$ at A is finite and equivalent to μ. In proving the lemma it is therefore no restriction to assume the family is dominated by a *finite* measure μ.

Each P_θ is dominated by μ and hence has a density f_θ with respect to it. Let $S_\theta = [\omega: f_\theta(\omega) > 0]$. Then $P_\theta(A) = P_\theta(A \cap S_\theta)$ for all A, and $P_\theta(A) = 0$ if and only if $\mu(A \cap S_\theta) = 0$. In particular, S_θ supports P_θ.

Call a set B in $\mathcal{F}$ a *kernel* if $B \subset S_\theta$ for some θ, and call a finite or countable union of kernels a *chain*. Let α be the supremum of $\mu(C)$ over chains C. Since μ is finite and a countable union of chains is a chain, α is finite and $\mu(C) = \alpha$ for some chain C. Suppose that $C = \bigcup_n B_n$, where each B_n is a kernel, and suppose that $B_n \subset S_{\theta_n}$.

The problem is to show that $[P_\theta: \theta \in \Theta]$ is dominated by $[P_{\theta_n}: n = 1, 2, \ldots]$ and hence equivalent to it. Suppose that $P_{\theta_n}(A) = 0$ for all n. Then $\mu(A \cap S_{\theta_n}) = 0$, as observed above. Since $C \subset \bigcup_n S_{\theta_n}$, $\mu(A \cap C) = 0$, and it follows that $P_\theta(A \cap C) = 0$ whatever θ may be. But suppose that $P_\theta(A - C) > 0$. Then $P_\theta((A - C) \cap S_\theta) = P_\theta(A - C)$ is positive, and so $(A - C) \cap S_\theta$ is a kernel, disjoint from C, of positive μ-measure; this is impossible because of the maximality of C. Thus $P_\theta(A - C)$ is 0 along with $P_\theta(A \cap C)$, and so $P_\theta(A) = 0$. ∎

Suppose that $[P_\theta: \theta \in \Theta]$ is dominated by a σ-finite μ, as in Theorem 34.6, and let $[P_{\theta_n}: n = 1, 2, \ldots]$ be the sequence whose existence is guaranteed by Lemma 3. Define a probability measure P on $\mathcal{F}$ by

(34.10) $$P(A) = \sum_n 2^{-n} P_{\theta_n}(A).$$

Clearly, P is equivalent to $[P_{\theta_n}: n = 1, 2, \ldots]$ and hence to $[P_\theta: \theta \in \Theta]$, and μ dominates P.

PROOF OF SUFFICIENCY IN THEOREM 34.6. If each P_θ has density $g_\theta h$ with respect to μ, then by the construction (34.10), P has density fh with respect to μ, where $f = \Sigma_n \, 2^{-n} g_{\theta_n}$. Put $r_\theta = g_\theta/f$ if $f > 0$ and $r_\theta = 0$ (say) if $f = 0$. If each g_θ is measurable $\mathcal{G}$, the same is true of f and hence of the r_θ. Since $P[f = 0] = 0$ and P is equivalent to the entire family, $P_\theta[f = 0] = 0$ for all θ. Therefore,

$$\int_A r_\theta \, dP = \int_A r_\theta \, fh \, d\mu = \int_{A \cap [f > 0]} r_\theta \, fh \, d\mu = \int_{A \cap [f > 0]} g_\theta h \, d\mu$$

$$= P_\theta(A \cap [f > 0]) = P_\theta(A).$$

Each P_θ thus has with respect to the probability measure P a density measurable $\mathcal{G}$, and it follows by Lemma 1 for $\mu = P$ and $h \equiv 1$ that $\mathcal{G}$ is sufficient. ∎

PROOF OF NECESSITY IN THEOREM 34.6. Let $p(A, \omega)$ be a function such that, for each A and θ, $p(A, \cdot)$ is a version of $P_\theta[A \| \mathcal{G}]$, as required by the definition of sufficiency. For P as in (34.10) and $G \in \mathcal{G}$,

$$\int_G p(A, \omega) P(d\omega) = \sum_n 2^{-n} \int_G p(A, \omega) P_{\theta_n}(d\omega)$$

$$= \sum_n 2^{-n} \int_G P_{\theta_n}[A \| \mathcal{G}] \, dP_{\theta_n} = \sum_n 2^{-n} P_{\theta_n}(A \cap G) = P(A \cap G).$$

Thus $p(A, \cdot)$ serves as a version of $P[A \| \mathcal{G}]$ as well, and $\mathcal{G}$ is still sufficient if P is added to the family. Since P dominates the augmented family, Lemma 2 gives the required factorization. ∎

Minimum-Variance Estimation*

To illustrate sufficiency, suppose that Θ is a subset of the line. An *estimate* of θ is a random variable Z, and the estimate is *unbiased* if $E_\theta[Z] = \theta$ for all θ. A measure of the accuracy of the estimate Z is $E_\theta[(Z - \theta)^2]$.

If $\mathcal{G}$ is sufficient, it follows by linearity (Theorem 34.2(ii)) that $E_\theta[X \| \mathcal{G}]$ has for X simple a version that is independent of θ. Since there are simple X_n such that $|X_n| \leq |X|$ and $X_n \to X$, the same is true of any X that is integrable with respect to each P_θ (use Theorem 34.2(v)). Suppose that $\mathcal{G}$ is, in fact, sufficient, and denote by $E[Z \| \mathcal{G}]$ a version of $E_\theta[X \| \mathcal{G}]$ that is independent of θ.

Theorem 34.7. *Suppose that $E_\theta[(Z - \theta)^2] < \infty$ for all θ and that $\mathcal{G}$ is sufficient. Then*

$$(34.11) \qquad E_\theta[(E[Z \| \mathcal{G}] - \theta)^2] \leq E_\theta[(Z - \theta)^2]$$

for all θ. If Z is unbiased, then so is $E[Z \| \mathcal{G}]$.

PROOF. By Jensen's inequality (34.5) for $\varphi(x) = (x - \theta)^2$, $(E[Z \| \mathcal{G}] - \theta)^2 \leq E_\theta[(Z - \theta)^2 \| \mathcal{G}]$. Applying E_θ to each side gives (34.11). The second statement follows from the fact that $E_\theta[E[Z \| \mathcal{G}]] = E_\theta[Z]$. ∎

This, the *Rao-Blackwell theorem,* says that $E[Z \| \mathcal{G}]$ is at least as good an estimate as Z if $\mathcal{G}$ is sufficient.

Example 34.4. Returning to Example 34.3, note that each X_i has mean $\theta/2$ under P_θ, so that if $\overline{X} = k^{-1} \sum_{i=1}^k X_i$ is the sample mean, then $2\overline{X}$ is an unbiased estimate of θ. But there is a better one. By (34.7), $E_\theta[2\overline{X} \| T] = (k + 1)T/k = T'$, and by the Rao-Blackwell theorem, T' is an unbiased estimate with variance at most that of $2\overline{X}$.

---

* This topic may be omitted.

In fact, for an arbitrary unbiased estimate Z, $E_\theta[(T' - \theta)^2] \leq E_\theta[(Z - \theta)^2]$. To prove this, let $\delta = T' - E[Z\|T]$. By Theorem 20.1(ii), $\delta = f(T)$ for some Borel function f, and $E_\theta[f(T)] = 0$ for all θ. Taking account of the density for T leads to $\int_0^\theta f(x)x^{k-1}\,dx = 0$, so that $f(x)x^{k-1}$ integrates to 0 over all intervals. Therefore, $f(x)$ along with $f(x)x^{k-1}$ vanishes for $x > 0$, except on a set of Lebesgue measure 0, and hence $P_\theta[f(T) = 0] = 1$ and $P_\theta[T' = E[Z\|T]] = 1$ for all θ. Therefore, $E_\theta[(T' - \theta)^2] = E_\theta[(E[Z\|T] - \theta)^2] \leq E_\theta[(Z - \theta)^2]$ for Z unbiased, and T' has minimum variance among all unbiased estimates of θ. ∎

PROBLEMS

34.1. Work out for conditional expected values the analogue of Problem 33.5.

34.2. In the context of Examples 33.5 and 33.12, show that the conditional expected value of Y (if it is integrable) given X is $g(X)$, where

$$g(x) = \frac{\displaystyle\int_{-\infty}^{\infty} f(x, y)y \, dy}{\displaystyle\int_{-\infty}^{\infty} f(x, y) \, dy}.$$

34.3. Show that the independence of X and Y implies that $E[Y\|X] = E[Y]$, which in turn implies that $E[XY] = E[X]E[Y]$. Show by examples on an Ω of three points that the reverse implications are both false.

34.4. Let B be an event with $P(B) > 0$. Define a probability measure P_1 by $P_1(A) = P(A|B)$. Show that $P_1[A\|\mathcal{G}] = P[A \cap B\|\mathcal{G}]/P[B\|\mathcal{G}]$ except on a set of P_1-measure 0.

34.5. The equation (34.3) was proved by showing that the left side is a version of the right side. Prove it by showing that the right side is a version of the left side.

34.6. 21.19↑ Let X and Y be independent random variables and let φ be a function on R^2. Show under appropriate conditions that, if $f(x) = E[\varphi(x, Y)]$, then $f(X)$ is a version of $E[\varphi(X, Y)\|X]$.

34.7. 33.14↑ Generalize Theorem 34.5 by replacing X with a random vector.

34.8. Assume that X is nonnegative but not necessarily integrable. Show that it is still possible to define a nonnegative random variable $E[X\|\mathcal{G}]$, measurable $\mathcal{G}$, such that (34.1) holds. Prove versions of the monotone convergence theorem and Fatou's lemma.

34.9. (a) Show for nonnegative X that $E[X\|\mathcal{G}] = \int_0^\infty P[X > t\|\mathcal{G}]\,dt$ with probability 1.

(b) Generalize Markov's inequality: $P[|X| \geq \alpha\|\mathcal{G}] \leq \alpha^{-k}E[|X|^k\|\mathcal{G}]$ with probability 1.

(c) Similarly generalize Chebyshev's and Hölder's inequalities.

34.10. Show that, if $\mathcal{G}_1 \subset \mathcal{G}_2$ and $E[X^2] < \infty$, then $E[(X - E[X\|\mathcal{G}_2])^2] \leq E[(X - E[X\|\mathcal{G}_1])^2]$. The dispersion of X about its conditional mean becomes smaller as the σ-field grows.

34.11. Let $\mathcal{G}_1, \mathcal{G}_2, \mathcal{G}_3$ be σ-fields in $\mathcal{F}$, let $\mathcal{G}_{ij}$ be the σ-field generated by $\mathcal{G}_i \cup \mathcal{G}_j$, and let A_i be the generic set in $\mathcal{G}_i$. Consider three conditions:

   (i) $P[A_3 \| \mathcal{G}_{12}] = P[A_3 \| \mathcal{G}_2]$ *for all A_3.*
   (ii) $P[A_1 \cap A_3 \| \mathcal{G}_2] = P[A_1 \| \mathcal{G}_2] P[A_3 \| \mathcal{G}_2]$ *for all A_1 and A_3.*
   (iii) $P[A_1 \| \mathcal{G}_{23}] = P[A_1 \| \mathcal{G}_2]$ *for all A_1.*

If $\mathcal{G}_1, \mathcal{G}_2$, and $\mathcal{G}_3$ are interpreted as descriptions of the past, present, and future, respectively, (i) is a general version of the Markov property: the conditional probability of a future event A_3 given the past and present $\mathcal{G}_{12}$ is the same as the conditional probability given the present $\mathcal{G}_2$ alone. Condition (iii) is the same with time reversed. And (ii) says that past and future events A_1 and A_3 are conditionally independent given the present $\mathcal{G}_2$. Prove the three conditions equivalent.

34.12. 33.6 34.11↑ Use Example 33.10 to calculate $P[N_s = k \| N_u, u \geq t]$ $(s \leq t)$ for the Poisson process.

34.13. Let L^2 be the Hilbert space of square-integrable random variables on $(\Omega, \mathcal{F}, P)$. For $\mathcal{G}$ a σ-field in $\mathcal{F}$, let $M_\mathcal{G}$ be the subspace of elements of L^2 that are measurable $\mathcal{G}$. Show that the operator $P_\mathcal{G}$ defined for $X \in L^2$ by $P_\mathcal{G} X = E[X \| \mathcal{G}]$ is the perpendicular projection on $M_\mathcal{G}$.

34.14. ↑ Suppose in Problem 34.13 that $\mathcal{G} = \sigma(Z)$ for a random variable Z in L^2. Let S_Z be the one-dimensional subspace spanned by Z. Show that S_Z may be much smaller than $M_{\sigma(Z)}$, so that $E[X \| Z]$ (for $X \in L^2$) is by no means the projection of X on Z.

Hint: Take Z the identity function on the unit interval with Lebesgue measure.

34.15. ↑ Problem 34.13 can be turned around to give an alternative approach to conditional probability and expected value. For a σ-field $\mathcal{G}$ in $\mathcal{F}$, let $P_\mathcal{G}$ be the perpendicular projection on the subspace $M_\mathcal{G}$. Show that $P_\mathcal{G} X$ has for $X \in L^2$ the two properties required of $E[X \| \mathcal{G}]$. Use this to *define* $E[X \| \mathcal{G}]$ for $X \in L^2$ and then extend it to all integrable X via approximation by random variables in L^2. Now define conditional probability.

34.16. *Mixing sequences.* A sequence of $\mathcal{F}$-sets in a probability space $(\Omega, \mathcal{F}, P)$ is *mixing* with constant α if

$$(34.12) \qquad \lim_n P(A_n \cap E) = \alpha P(E)$$

for every E in $\mathcal{F}$.

(a) Show that $\{A_n\}$ is mixing with constant α if and only if

$$(34.13) \qquad \lim_n \int_{A_n} X \, dP = \alpha \int X \, dP$$

for each integrable X (measurable $\mathcal{F}$).

(b) Suppose that (34.12) holds for $E \in \mathcal{P}$, where $\mathcal{P}$ is a π-system, $\Omega \in \mathcal{P}$, and $A_n \in \sigma(\mathcal{P})$ for all n. Show that $\{A_n\}$ is mixing.

Hint: First check (34.13) for X measurable $\sigma(\mathcal{P})$ and then use conditional expected values with respect to $\sigma(\mathcal{P})$.

(c) Show that, if P_0 is a probability measure on $(\Omega, \mathcal{F})$ and $P_0 \ll P$, then mixing is preserved if P is replaced by P_0.

34.17. ↑ *Application of mixing to the central limit theorem.* Let $X_1, X_2, \ldots$ be random variables on $(\Omega, \mathcal{F}, P)$, independent and identically distributed with mean 0 and variance σ^2, and put $S_n = X_1 + \ldots + X_n$. Then $S_n/\sigma\sqrt{n} \Rightarrow N$ by the Lindeberg-Lévy theorem. Show by the steps below that this still holds if P is replaced by any probability measure P_0 on $(\Omega, \mathcal{F})$ that P dominates. For example, the central limit theorem applies to the sums $\Sigma_{k=1}^n r_k(\omega)$ of Rademacher functions if ω is chosen according to the uniform density over the unit interval, and this result shows that the same is true if ω is chosen according to an arbitrary density.

Let $Y_n = S_n/\sigma\sqrt{n}$ and $Z_n = (S_n - S_{[\log n]})/\sigma\sqrt{n}$, and take $\mathcal{P}$ to consist of the sets of the form $[(X_1, \ldots, X_k) \in H]$, $k \geq 1$, $H \in \mathcal{R}^k$. Prove successively:

(a) $P[Y_n \leq x] \to P[N \leq x]$.

(b) $P[|Y_n - Z_n| \geq \epsilon] \to 0$.

(c) $P[Z_n \leq x] \to P[N \leq x]$.

(d) $P(E \cap [Z_n \leq x]) \to P(E)P[N \leq x]$ for $E \in \mathcal{P}$.

(e) $P(E \cap [Z_n \leq x]) \to P(E)P[N \leq x]$ for $E \in \mathcal{F}$.

(f) $P_0[Z_n \leq x] \to P[N \leq x]$.

(g) $P_0[|Y_n - Z_n| \geq \epsilon] \to 0$.

(h) $P_0[Y_n \leq x] \to P[N \leq x]$.

34.18. Let $(X, \mathcal{X}, \mu)$ and $(Y, \mathcal{Y}, \nu)$ be probability measure spaces and let $(\Omega, \mathcal{F}, P)$ be their product. Let $\mathcal{G} = [A \times Y: A \in \mathcal{X}]$ be the σ-field of vertical strips. For a random variable $f = f(x, y)$, put $g(x, y) = \int_Y f(x, z) \nu(dz)$ and show by Fubini's theorem that g is a version of $E[f \| \mathcal{G}]$. This is a general version of Problem 34.6.

34.19. Here is an alternative construction of conditional expectation. It uses the Hahn decomposition and in effect incorporates one of the proofs of the Radon-Nikodym theorem.

Suppose that X is nonnegative and integrable. For each nonnegative dyadic rational t, find by the Hahn result (Theorem 32.1) a $\mathcal{G}$-set A_t such that

$$\begin{cases} \int_G X\, dP \geq tP(G) & \text{if } G \subset A_t, \quad G \in \mathcal{G}, \\ \int_G X\, dP \leq tP(G) & \text{if } G \subset A_t^c, \quad G \in \mathcal{G}. \end{cases}$$

Show that $P(A_v - A_u) = 0$ if $v > u$ and replace A_t by $A_t - \bigcup_{v>u} (A_v - A_u)$. Show that it is possible to take $A_0 = \Omega$. Show by the integrability of X that $P(\bigcap_{s>0} A_s) = 0$ and replace A_t by $A_t - \bigcap_{s>0} A_s$ for $t > 0$. Thus it can be assumed that the A_t are nonincreasing (over dyadic rational t), that $A_0 = \Omega$, and that $\bigcap_{t>0} A_t = 0$.

Put $G_{nk} = A_{k/2^n} - A_{(k+1)/2^n}$, and consider the successively finer decompositions $\Delta_n = \{G_{n0}, G_{n1}, \ldots\}$ of Ω into $\mathcal{G}$-sets. To simplify details, suppose that the G_{nk} all have positive probability, and define $f_n(\omega) = \int_{G_{nk}} X\, dP/P(G_{nk})$ for $\omega \in G_{nk}$. (In fact, of course, f_n is a version of the conditional expected value of X with respect to the σ-field generated by Δ_n.) If $G \in \mathcal{G}$, then

(34.14) $\dfrac{k}{2^n} \le \dfrac{1}{P(G \cap G_{nk})} \displaystyle\int_{G \cap G_{nk}} X\, dP \le \dfrac{k+1}{2^n}.$

If $\omega \in G_{mj} \subset G_{nk}$ $(n \le m)$, then $(j2^{-m}, (j+1)2^{-m}] \subset (k2^{-n}, (k+1)2^{-n}]$ and (take $G = G_{nk}$ and then $G = G_{mj}$ in (34.14)) $f_n(\omega), f_m(\omega) \in [k2^{-n}, (k+1)2^{-n}]$. For all ω, $|f_m(\omega) - f_n(\omega)| \le 2^{-n}$ for $m \ge n$, and so $\lim_n f_n(\omega) = f(\omega)$, where $|f(\omega) - f_n(\omega)| \le 2^{-n}$. Clearly, $\int_G f_n\, dP \to \int_G f\, dP$. If $G \in \mathcal{G}$, then

$$\left| \int_G (f_n - X)\, dP \right| \le \sum_k P(G \cap G_{nk}) \left| \frac{1}{P(G_{nk})} \int_{G_{nk}} X\, dP \right.$$

$$\left. - \frac{1}{P(G \cap G_{nk})} \int_{G \cap G_{nk}} X\, dP \right| \le \sum_k P(G \cap G_{nk}) 2^{-n} \le 2^{-n}$$

by (34.14). Hence $\int_G f_n\, dP \to \int_G X\, dP$, and the $\mathcal{G}$-measurable function f satisfies $\int_G f\, dP = \int_G X\, dP$ for $G \in \mathcal{G}$.

SECTION 35. MARTINGALES

Definition

Let $X_1, X_2, \ldots$ be a sequence of random variables on a probability space $(\Omega, \mathcal{F}, P)$, and let $\mathcal{F}_1, \mathcal{F}_2, \ldots$ be a sequence of σ-fields in $\mathcal{F}$. The sequence $\{(X_n, \mathcal{F}_n): n = 1, 2, \ldots\}$ is a *martingale* if these four conditions hold:

(i) $\mathcal{F}_n \subset \mathcal{F}_{n+1}$;
(ii) X_n is measurable $\mathcal{F}_n$;
(iii) $E[|X_n|] < \infty$;
(iv) *with probability* 1,

(35.1) $$E[X_{n+1} \| \mathcal{F}_n] = X_n.$$

Alternatively, the sequence $X_1, X_2, \ldots$ is said to be a *martingale relative to the* σ-*fields* $\mathcal{F}_1, \mathcal{F}_2, \ldots$. Condition (ii) is expressed by saying the X_n are *adapted* to the $\mathcal{F}_n$.

If X_n represents the fortune of a gambler after the nth play and $\mathcal{F}_n$ represents his information about the game at that time, (35.1) says that his expected fortune after the next play is the same as his present fortune. Thus a martingale represents a fair game, and sums of independent random variables with mean 0 give one example. As will be seen below, martingales arise in very diverse connections.

The sequence $X_1, X_2, \ldots$ is defined to be a martingale if it is a martingale relative to *some* sequence $\mathcal{F}_1, \mathcal{F}_2, \ldots$. In this case, the σ-fields $\mathcal{G}_n = \sigma(X_1, \ldots, X_n)$ always work: Obviously, $\mathcal{G}_n \subset \mathcal{G}_{n+1}$ and X_n is measurable $\mathcal{G}_n$, and if (35.1) holds, then $E[X_{n+1} \| \mathcal{G}_n] = E[E[X_{n+1} \| \mathcal{F}_n] \| \mathcal{G}_n] = E[X_n \| \mathcal{G}_n] = X_n$ by (34.3). For these special σ-fields $\mathcal{G}_n$, (35.1) reduces to

(35.2) $$E[X_{n+1} \| X_1, \ldots, X_n] = X_n.$$

Since $\sigma(X_1, \ldots, X_n) \subset \mathscr{F}_n$ if and only if X_n is measurable $\mathscr{F}_n$ for each n, the $\sigma(X_1, \ldots, X_n)$ are the *smallest* σ-fields with respect to which the X_n are a martingale.

The essential condition is embodied in (35.1) and in its specialization (35.2). Condition (iii) is of course needed to ensure that $E[X_{n+1}\|\mathscr{F}_n]$ exists. Condition (iv) says that X_n is a version of $E[X_{n+1}\|\mathscr{F}_n]$; since X_n is measurable $\mathscr{F}_n$, the requirement reduces to

$$(35.3) \qquad \int_A X_{n+1}\, dP = \int_A X_n\, dP, \qquad A \in \mathscr{F}_n.$$

Since the $\mathscr{F}_n$ are nested, $A \in \mathscr{F}_n$ implies that $\int_A X_n\, dP = \int_A X_{n+1}\, dP = \cdots = \int_A X_{n+k}\, dP$. Therefore, X_n, being measurable $\mathscr{F}_n$, is a version of $E[X_{n+k}\|\mathscr{F}_n]$:

$$(35.4) \qquad E[X_{n+k}\|\mathscr{F}_n] = X_n$$

with probability 1 for $k \geq 1$. Note that for $A = \Omega$, (35.3) gives

$$(35.5) \qquad E[X_1] = E[X_2] = \cdots.$$

The defining conditions for a martingale can also be given in terms of the differences

$$(35.6) \qquad \Delta_n = X_n - X_{n-1}$$

$(\Delta_1 = X_1)$. By linearity (35.1) is the same thing as

$$(35.7) \qquad E[\Delta_{n+1}\|\mathscr{F}_n] = 0.$$

Note that, since $X_k = \Delta_1 + \cdots + \Delta_k$ and $\Delta_k = X_k - X_{k-1}$, the sets $X_1, \ldots, X_n$ and $\Delta_1, \ldots, \Delta_n$ generate the same σ-field:

$$(35.8) \qquad \sigma(X_1, \ldots, X_n) = \sigma(\Delta_1, \ldots, \Delta_n).$$

Example 35.1. Let $\Delta_1, \Delta_2, \ldots$ be independent, integrable random variables such that $E[\Delta_n] = 0$ for $n \geq 2$. If $\mathscr{F}_n$ is the σ-field (35.8), then by independence $E[\Delta_{n+1}\|\mathscr{F}_n] = E[\Delta_{n+1}] = 0$. If Δ is another random variable, independent of the Δ_n, and if $\mathscr{F}_n$ is replaced by $\sigma(\Delta, \Delta_1, \ldots, \Delta_n)$, then the $X_n = \Delta_1 + \cdots + \Delta_n$ are still a martingale relative to the $\mathscr{F}_n$. It is natural and convenient in the theory to allow σ-fields $\mathscr{F}_n$ larger than the minimal ones (35.8). ∎

Example 35.2. Let $(\Omega, \mathscr{F}, P)$ be a probability space, let ν be a finite measure on $\mathscr{F}$, and let $\mathscr{F}_1, \mathscr{F}_2, \ldots$ be a nondecreasing sequence of σ-fields in $\mathscr{F}$. Suppose that P dominates ν when both are restricted to $\mathscr{F}_n$—that is, suppose that $A \in \mathscr{F}_n$ and $P(A) = 0$ together imply that $\nu(A) = 0$. There is then a density or Radon-Nikodym derivative X_n of ν with respect to P when both are restricted to $\mathscr{F}_n$; X_n is a function that is measurable $\mathscr{F}_n$ and integrable with respect to P, and it satisfies

(35.9) $$\int_A X_n \, dP = \nu(A), \qquad A \in \mathscr{F}_n.$$

If $A \in \mathscr{F}_n$, then $A \in \mathscr{F}_{n+1}$ as well, so that $\int_A X_{n+1} \, dP = \nu(A)$; this and (35.9) give (35.3). Thus the X_n are a martingale with respect to the $\mathscr{F}_n$. ∎

Example 35.3. For a specialization of the preceding example, let P be Lebesgue measure on the σ-field $\mathscr{F}$ of Borel subsets of $\Omega = (0, 1]$ and let $\mathscr{F}_n$ be the finite σ-field generated by the partition of Ω into dyadic intervals $(k2^{-n}, (k + 1)2^{-n}]$, $0 \le k < 2^n$. If $A \in \mathscr{F}_n$ and $P(A) = 0$, then A is empty. Hence P dominates *every* finite measure ν on $\mathscr{F}_n$. The Radon-Nikodym derivative is

(35.10) $$X_n(\omega) = \frac{\nu(k2^{-n}, (k + 1)2^{-n}]}{2^{-n}} \qquad \text{if } \omega \in (k2^{-n}, (k + 1)2^{-n}].$$

There is no need here to assume that P dominates ν when they are viewed as measures on all of $\mathscr{F}$. Suppose that ν is the distribution of $\Sigma_{k=1}^{\infty} Z_k 2^{-k}$ for independent Z_k assuming values 1 and 0 with probabilities p and $1 - p$. This is the measure in Examples 31.1 and 31.3 (there denoted μ), and for $p \ne 1/2$, ν is singular with respect to Lebesgue measure P. It is nonetheless absolutely continuous with respect to P when both are restricted to $\mathscr{F}_n$. ∎

Example 35.4. For another specialization of Example 35.2, suppose that ν is a probability measure Q on $\mathscr{F}$ and that $\mathscr{F}_n = \sigma(Y_1, \ldots, Y_n)$ for random variables $Y_1, Y_2, \ldots$ on $(\Omega, \mathscr{F})$. Suppose that under the measure P the distribution of the random vector $(Y_1, \ldots, Y_n)$ has density $p_n(y_1, \ldots, y_n)$ with respect to n-dimensional Lebesgue measure and that under Q it has density $q_n(y_1, \ldots, y_n)$. To avoid technicalities, assume that p_n is everywhere positive. Then the Radon-Nikodym derivative for Q with respect to P on $\mathscr{F}_n$ is

(35.11) $$X_n = \frac{q_n(Y_1, \ldots, Y_n)}{p_n(Y_1, \ldots, Y_n)}.$$

To see this, note that the general element of $\mathscr{F}_n$ is $[(Y_1, \ldots, Y_n) \in H]$, $H \in \mathscr{R}^n$; by the change-of-variable formula,

$$\int_{[(Y_1, \ldots, Y_n) \in H]} X_n \, dP = \int_H \frac{q_n(y_1, \ldots, y_n)}{p_n(y_1, \ldots, y_n)} p_n(y_1, \ldots, y_n) \, dy_1 \cdots dy_n$$

$$= Q[(Y_1, \ldots, Y_n) \in H].$$

In statistical terms, (35.11) is a likelihood ratio: p_n and q_n are rival densities, and the larger X_n is, the more strongly one prefers q_n as an explanation of the observation $(Y_1, \ldots, Y_n)$. The analysis is carried out under the assumption that

P is the measure actually governing the Y_n; that is, X_n is a martingale under P and not in general under Q.

In the most common case the Y_n are independent and identically distributed under both P and Q: $p_n(y_1, \ldots, y_n) = p(y_1) \cdots p(y_n)$ and $q_n(y_1, \ldots, y_n) = q(y_1) \cdots q(y_n)$ for densities p and q on the line, where p is assumed everywhere positive for simplicity. Suppose that the measures corresponding to the densities p and q are not identical, so that $P[Y_n \in H] \neq Q[Y_n \in H]$ for some $H \in \mathcal{R}^1$. If $Z_n = I_{[Y_n \in H]}$, then by the strong law of large numbers, $n^{-1} \sum_{k=1}^n Z_k$ converges to $P[Y_1 \in H]$ on a set (in $\mathcal{F}$) of P-measure 1 and to $Q[Y_1 \in H]$ on a (disjoint) set of Q-measure 1. Thus P and Q are mutually singular on $\mathcal{F}$ even though P dominates Q on $\mathcal{F}_n$. ∎

Example 35.5. Suppose that Z is an integrable random variable on $(\Omega, \mathcal{F}, P)$ and $\mathcal{F}_n$ are nondecreasing σ-fields in $\mathcal{F}$. If

$$(35.12) \qquad\qquad X_n = E[Z \| \mathcal{F}_n],$$

then the first three conditions in the martingale definition are satisfied, and by (34.3), $E[X_{n+1} \| \mathcal{F}_n] = E[E[Z \| \mathcal{F}_{n+1}] \| \mathcal{F}_n] = E[Z \| \mathcal{F}_n] = X_n$. Thus X_n is a martingale relative to $\mathcal{F}_n$. ∎

Example 35.6. Let $N_{nk}, n, k = 1, 2, \ldots$ be an independent array of identically distributed random variables assuming the values $0, 1, 2, \ldots$. Define $Z_0, Z_1, Z_2, \ldots$ inductively by $Z_0(\omega) = 1$ and $Z_n(\omega) = N_{n,1}(\omega) + \cdots + N_{n,Z_{n-1}(\omega)}(\omega)$ $-Z_n(\omega) = 0$ if $Z_{n-1}(\omega) = 0$. If N_{nk} is thought of as the number of progeny of an organism, and if Z_{n-1} represents the size at time $n - 1$ of a population of these organisms, then Z_n represents the size at time n. If the expected number of progeny is $E[N_{nk}] = m$, then $E[Z_n \| Z_{n-1}] = Z_{n-1}m$, so that $X_n = Z_n/m^n$, $n = 0, 1, 2, \ldots$, is a martingale. The sequence $Z_0', Z_1, \ldots$ is a *branching process*. ∎

In the preceding definition and examples, n ranges over the positive integers. The definition makes sense if n ranges over $1, 2, \ldots, N$; here conditions (ii) and (iii) are required for $1 \leq n \leq N$ and conditions (i) and (iv) only for $1 \leq n < N$. It is, in fact, clear that the definition makes sense if the indices range over an arbitrary ordered set. Although martingale theory with an interval of the line as the index set is of interest and importance, here the index set will be discrete.

Submartingales

Random variables X_n are a *submartingale* relative to σ-fields $\mathcal{F}_n$ if (i), (ii), and (iii) of the definition above hold and if this condition holds in place of (iv):

    (iv′) *with probability 1,*

(35.13)
$$E[X_{n+1}\|\mathscr{F}_n] \geq X_n.$$

As before, the X_n are a submartingale if they are a submartingale with respect to some sequence $\mathscr{F}_n$, and the special sequence $\mathscr{F}_n = \sigma(X_1, \ldots, X_n)$ works if any does. The requirement (35.13) is the same thing as

(35.14)
$$\int_A X_{n+1} \, dP \geq \int_A X_n \, dP, \qquad A \in \mathscr{F}_n.$$

This extends inductively (see the argument for (35.4)), and so

(35.15)
$$E[X_{n+k}\|\mathscr{F}_n] \geq X_n$$

for $k \geq 1$. Taking expected values in (35.15) gives

(35.16)
$$E[X_1] \leq E[X_2] \leq \cdots.$$

Example 35.7. Suppose that the Δ_n are independent and integrable, as in Example 35.1, but assume that $E[\Delta_n]$ is for $n \geq 2$ nonnegative rather than 0. Then the partial sums $\Delta_1 + \cdots + \Delta_n$ form a submartingale. ∎

Example 35.8. Suppose that the X_n are a martingale relative to the $\mathscr{F}_n$. Then $|X_n|$ is measurable $\mathscr{F}_n$ and integrable, and by Theorem 34.2(iv), $E[|X_{n+1}|\|\mathscr{F}_n] \geq |E[X_{n+1}\|\mathscr{F}_n]| = |X_n|$. Thus the $|X_n|$ are a submartingale relative to the $\mathscr{F}_n$. Note that even if $X_1, \ldots, X_n$ generate $\mathscr{F}_n$, $|X_1|, \ldots, |X_n|$ will in general generate a σ-field smaller than $\mathscr{F}_n$. ∎

Reversing the inequality in (35.13) gives the definition of a *supermartingale*. The inequalities in (35.14), (35.15), and (35.16) become reversed as well. The theory for supermartingales is of course symmetric to that of submartingales.

Gambling

Consider again the gambler whose fortune after the nth play is X_n and whose information about the game at that time is represented by the σ-field $\mathscr{F}_n$. If $\mathscr{F}_n = \sigma(X_1, \ldots, X_n)$, he knows the sequence of his fortunes and nothing else, but $\mathscr{F}_n$ could be larger. The martingale condition (35.1) stipulates that his expected or average fortune after the next play equals his present fortune, and so the martingale is the model for a *fair game*. Since the condition (35.13) for a submartingale stipulates that he stands to gain (or at least not lose) on the average, a submartingale represents a game *favorable* to the gambler. Similarly, a supermartingale represents a game *unfavorable* to the gambler.*

---

* There is a reversal of terminology here: a subfair game (Section 7) is against the gambler, while a submartingale favors him.

Examples of such games were studied in Section 7, and some of the results there have immediate generalizations. Start the martingale at $n = 0$, X_0 representing the gambler's initial fortune. The difference $\Delta_n = X_n - X_{n-1}$ represents the amount the gambler wins on the nth play,[†] a negative win being of course a loss. Suppose instead that Δ_n represents the amount he wins if he puts up unit stakes. If instead of unit stakes he wagers the amount W_n on the nth play, $W_n \Delta_n$ represents his gain on that play. Suppose that $W_n \geq 0$, and that W_n is measurable $\mathcal{F}_{n-1}$ to exclude prevision: Before the nth play the information available to the gambler is that in $\mathcal{F}_{n-1}$, and his choice of stake W_n must be based on this alone. For simplicity take W_n bounded. Then $W_n \Delta_n$ is integrable and measurable $\mathcal{F}_n$ if Δ_n is, and if X_n is a martingale, then $E[W_n \Delta_n \| \mathcal{F}_{n-1}] = W_n E[\Delta_n \| \mathcal{F}_{n-1}] = 0$ by (34.2). Thus

$$(35.17) \qquad X_0 + W_1 \Delta_1 + \cdots + W_n \Delta_n$$

is a martingale relative to the $\mathcal{F}_n$. The sequence $W_1, W_2, \ldots$ represents a betting system, and transforming a fair game by a betting system preserves fairness; that is, transforming X_n into (35.17) preserves the martingale property.

The various betting systems discussed in Section 7 give rise to various martingales, and these martingales are not in general sums of independent random variables—are not in general the special martingales of Example 35.1. If W_n assumes only the values 0 and 1, the betting system is a selection system; see Section 7.

If the game is unfavorable to the gambler—that is, if X_n is a supermartingale—and if W_n is nonnegative, bounded, and measurable $\mathcal{F}_{n-1}$, then the same argument shows that (35.17) is again a supermartingale, is again unfavorable. Betting systems are thus of no avail in unfavorable games.

The stopping-time arguments of Section 7 also extend. Suppose that $\{X_n\}$ is a martingale relative to $\{\mathcal{F}_n\}$; it may have come from another martingale via transformation by a betting system. Let τ be a random variable taking on nonnegative integers as values, and suppose that

$$(35.18) \qquad [\tau = n] \in \mathcal{F}_n.$$

If τ is the time the gambler stops, $[\tau = n]$ is the event he stops just after the nth play, and (35.18) requires that his decision is to depend only on the information $\mathcal{F}_n$ available to him at that time. His fortune at time n for this stopping rule is

$$(35.19) \qquad X_n^* = \begin{cases} X_n & \text{if } n \leq \tau, \\ X_\tau & \text{if } n \geq \tau. \end{cases}$$

Here X_τ (which has value $X_{\tau(\omega)}(\omega)$ at ω) is the gambler's ultimate fortune, and it is his fortune for all times subsequent to τ.

The problem is to show that $X_0^*, X_1^*, \ldots$ is a martingale relative to $\mathcal{F}_0, \mathcal{F}_1, \ldots$. First,

[†] The notation has, of course, changed. The F_n and X_n of Section 7 have become X_n and Δ_n.

$$E[|X_n^*|] = \sum_{k=1}^{n-1} \int_{[\tau=k]} |X_k|\, dP + \int_{[\tau\geq n]} |X_n|\, dP \leq \sum_{k=1}^{n} E[|X_k|] < \infty.$$

Since $[\tau > n] = \Omega - [\tau \leq n] \in \mathcal{F}_n$,

$$[X_n^* \in H] = \bigcup_{k=0}^{n} [\tau = k, X_k \in H] \cup [\tau > n, X_n \in H] \in \mathcal{F}_n.$$

Moreover,

$$\int_A X_n^*\, dP = \int_{A\cap[\tau>n]} X_n\, dP + \int_{A\cap[\tau\leq n]} X_\tau\, dP$$

and

$$\int_A X_{n+1}^*\, dP = \int_{A\cap[\tau>n]} X_{n+1}\, dP + \int_{A\cap[\tau\leq n]} X_\tau\, dP.$$

Because of (35.3), the right sides here coincide if $A \in \mathcal{F}_n$; this establishes (35.3) for the sequence $X_1^*, X_2^*, \ldots$, which is thus a martingale. The same kind of argument works for supermartingales.

Since $X_n^* = X_\tau$ for $n \geq \tau$, $X_n^* \to X_\tau$. As pointed out in Section 7, it is not always possible to integrate to the limit here. Let $X_n = a + \Delta_1 + \cdots + \Delta_n$, where the Δ_n are independent and assume the values ± 1 with probability $\frac{1}{2}$ ($X_0 = a$), and let τ be the smallest n for which $\Delta_1 + \cdots + \Delta_n = 1$. Then $E[X_n^*] = a$ and $X_\tau = a + 1$. On the other hand, if the X_n are uniformly bounded or uniformly integrable, it is possible to integrate to the limit: $E[X_\tau] = E[X_1]$.

Functions of Martingales

Convex functions of martingales are submartingales:

Theorem 35.1. (i) *If $X_1, X_2, \ldots$ is a martingale relative to $\mathcal{F}_1, \mathcal{F}_2, \ldots$, if φ is convex, and if the $\varphi(X_n)$ are integrable, then $\varphi(X_1), \varphi(X_2), \ldots$ is a submartingale relative to $\mathcal{F}_1, \mathcal{F}_2, \ldots$.*

(ii) *If $X_1, X_2, \ldots$ is a submartingale relative to $\mathcal{F}_1, \mathcal{F}_2, \ldots$, if φ is nondecreasing and convex, and if the $\varphi(X_n)$ are integrable, then $\varphi(X_1), \varphi(X_2), \ldots$ is a submartingale relative to $\mathcal{F}_1, \mathcal{F}_2, \ldots$.*

PROOF. In the submartingale case, $X_n \leq E[X_{n+1}\|\mathcal{F}_n]$, and if φ is nondecreasing, then $\varphi(X_n) \leq \varphi(E[X_{n+1}\|\mathcal{F}_n])$. In the martingale case, $X_n = E[X_{n+1}\|\mathcal{F}_n]$, and so $\varphi(X_n) = \varphi(E[X_{n+1}\|\mathcal{F}_n])$. If φ is convex, then by Jensen's inequality (34.5) for conditional expectations $\varphi(E[X_{n+1}\|\mathcal{F}_n]) \leq E[\varphi(X_{n+1}) \|\mathcal{F}_n]$. ∎

Example 35.8 is the case of part (i) for $\varphi(x) = |x|$.

Inequalities

There are two inequalities that are fundamental to the theory of martingales.

Theorem 35.2. *If $X_1, \ldots, X_n$ is a submartingale, then for $\alpha > 0$,*

$$(35.20) \qquad P\left[\max_{i \leq n} X_i \geq \alpha\right] \leq \frac{1}{\alpha} E[|X_n|].$$

This extends Kolmogorov's inequality: If $S_1, S_2, \ldots$ are partial sums of independent random variables with mean 0, they form a martingale; if the variances are finite, then $S_1^2, S_2^2, \ldots$ is a submartingale by part (i) of the preceding theorem, and (35.20) for this submartingale is exactly Kolmogorov's inequality (22.2).

PROOF OF THE THEOREM.   For arbitrary α, let $A_1 = [X_1 \geq \alpha]$, $A_k = [\max_{i<k} X_i < \alpha \leq X_k]$, and $A = \bigcup_{k=1}^n A_k = [\max_{i \leq n} X_i \geq \alpha]$. Since $A_k \in \mathcal{F}_k = \sigma(X_1, \ldots, X_k)$, it follows (see (35.15)) that

$$\int_A X_n \, dP = \sum_{k=1}^n \int_{A_k} X_n \, dP = \sum_{k=1}^n \int_{A_k} E[X_n \| \mathcal{F}_k] \, dP$$

$$\geq \sum_{k=1}^n \int_{A_k} X_k \, dP \geq \alpha \sum_{k=1}^n P(A_k) = \alpha P(A).$$

Thus

$$\alpha P\left[\max_{i \leq n} X_i \geq \alpha\right] \leq \int_{[\max_{i \leq n} X_i \geq \alpha]} X_n \, dP \leq E[X_n^+] \leq E[|X_n|]. \qquad \blacksquare$$

If $X_1, \ldots, X_n$ is a martingale, $|X_1|, \ldots, |X_n|$ is a submartingale, and so (35.20) gives $P[\max_{i \leq n} |X_i| \geq \alpha] \leq \alpha^{-1} E[|X_n|]$.

The second fundamental inequality requires the notion of an *upcrossing*. Let $[\alpha, \beta]$ be an interval ($\alpha < \beta$) and let $X_1, \ldots, X_n$ be random variables. The number of upcrossings of $[\alpha, \beta]$ by $X_1(\omega), \ldots, X_n(\omega)$ is the number of times the sequence passes from below α to above β. In the diagram, $n = 20$ and there are three upcrossings, corresponding to the three strings of consecutive 1's under the graph.

To make all this precise, set $Y_1 = 0$ and, for $2 \leq k \leq n + 1$, set

$$Y_k = \begin{cases} 1 & \text{if } Y_{k-1} = 0 \text{ and } X_{k-1} \leq \alpha, \\ 0 & \text{if } Y_{k-1} = 0 \text{ and } X_{k-1} > \alpha, \\ 0 & \text{if } Y_{k-1} = 1 \text{ and } X_{k-1} \geq \beta, \\ 1 & \text{if } Y_{k-1} = 1 \text{ and } X_{k-1} < \beta. \end{cases}$$

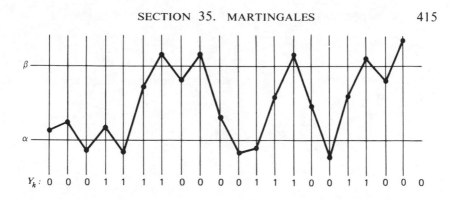

Y_k: 0 0 0 1 1 1 1 0 0 0 0 1 1 1 0 0 1 1 0 0 0

An upcrossing corresponds in $Y_2, \ldots, Y_n$ to an unbroken string of 1's with a 0 on either side. Put $Z_k = 1$ if $Y_k = 1$ and $Y_{k+1} = 0$, and put $Z_k = 0$ otherwise. Then the number of upcrossings is defined as $U = \Sigma_{k=2}^n Z_k$.

If $\mathcal{F}_k = \sigma(X_1, \ldots, X_k)$ and $\mathcal{F}_0 = \{0, \Omega\}$, then, as follows by induction from the definition, Y_k is measurable $\mathcal{F}_{k-1}$ for $k = 1, \ldots, n + 1$. Moreover, U is measurable $\mathcal{F}$.

Theorem 35.3. *For a submartingale $X_1, \ldots, X_n$, the number U of upcrossings of $[\alpha, \beta]$ satisfies*

(35.21) $$E[U] \leq \frac{E[|X_n|] + |\alpha|}{\beta - \alpha}.$$

PROOF. Let $\varphi(x) = \max\{\alpha, x\}$ and $X'_k = \varphi(X_k)$; since φ is nondecreasing and convex, the X'_k are a submartingale relative to the σ-fields $\mathcal{F}_k = \sigma(X_1, \ldots, X_k)$. Now $X'_k \leq \alpha$ if and only if $X_k \leq \alpha$, and $X'_k \geq \beta$ if and only if $X_k \geq \beta$. Therefore, replacing X_k by X'_k in the definition of the Y_k in no way changes these random variables; in particular, U is the number of upcrossings of $[\alpha, \beta]$ by the X'_k as well as the number of upcrossings by the X_k. (In the diagram, the points below the lower line are moved up to it.) It will be shown that

(35.22) $$(\beta - \alpha)E[U] \leq E[X'_n - X'_1].$$

This will give

$$(\beta - \alpha)E[U] \leq E[X'_n - \alpha] - E[X'_1 - \alpha]$$

$$\leq E[X'_n - \alpha] = \int_{[X_n > \alpha]} (X_n - \alpha)\, dP,$$

from which (35.21) will follow.

First, if $\Delta'_k = X'_k - X'_{k-1}$,

$$X'_n - X'_1 = \sum_{k=2}^{n} \Delta'_k = \sum_{k=2}^{n} Y_k \Delta'_k + \sum_{k=2}^{n} (1 - Y_k)\Delta'_k.$$

Since Y_k is measurable $\mathscr{F}_{k-1}$, as noted above, the submartingale property for the X'_k gives

$$E\left[\sum_{k=2}^{n} (1 - Y_k)\Delta'_k \right] = \sum_{k=2}^{n} \int_{[Y_k=0]} \Delta'_k \, dP$$

$$= \sum_{k=2}^{n} \int_{[Y_k=0]} E[\Delta'_k \| \mathscr{F}_{k-1}] \, dP \geq 0.$$

Therefore, (35.22) will follow if it is shown that

(35.23) $$\sum_{k=2}^{n} Y_k \Delta'_k \geq (\beta - \alpha)U.$$

Since Y_k is 1 "during" an upcrossing, (35.23) is intuitively clear. Consider a maximal stretch of 1's among $Y_2, \ldots, Y_n$. There are two possibilities: One is that $2 \leq u \leq v \leq n$ and $Y_{u-1} = 0$, $Y_u = \cdots = Y_v = 1$, $Y_{v+1} = 0$ (a string of 1's with 0's on either side). In this case $X'_{u-1} = \alpha$ and $X'_v \geq \beta$, so that $\Sigma_{k=u}^{v} Y_k \Delta'_k = X'_v - X'_{u-1} \geq \beta - \alpha$; moreover, $Z_v = 1$ (an upcrossing is completed). The only other possibility is that $2 \leq u \leq n$ and $Y_{u-1} = 0$, $Y_u = \cdots = Y_n = Y_{n+1} = 1$ (a string of 1's terminates $Y_2, \ldots, Y_n$ and Y_{n+1} is 1 as well). In this case $X'_{u-1} = \alpha$ and of course $X'_n \geq \alpha$, so that $\Sigma_{k=u}^{n} Y_k \Delta'_k = X'_n - X'_{u-1} \geq 0$; moreover, $Z_n = 0$ (no upcrossing). Thus summing $Y_k \Delta'_k$ over a string of 1's gives at least 0 in any case and gives at least $\beta - \alpha$ if it corresponds to an upcrossing. Hence (35.23). ∎

Convergence Theorems

The principal martingale convergence theorem is this:

Theorem 35.4. *Let $X_1, X_2, \ldots$ be a submartingale. If $K = \sup_n E[|X_n|] < \infty$, then $X_n \to X$ with probability 1, where X is a random variable satisfying $E[|X|] \leq K$.*

PROOF. Fix α and β for the moment and let U_n be the number of upcrossings of $[\alpha, \beta]$ by $X_1, \ldots, X_n$. By the upcrossing theorem $E[U_n] \leq (E[|X_n|] + |\alpha|)/(\beta - \alpha) \leq (K + |\alpha|)/(\beta - \alpha)$. Since U_n is nondecreasing and $E[U_n]$ is bounded, it follows by the monotone convergence theorem that $\sup_n U_n$ is integrable and hence finite-valued almost everywhere.

Let X^* and X_* be the limits superior and inferior of the sequence $X_1, X_2, \ldots$; they may be infinite. If $X_* < \alpha < \beta < X^*$, then U_n must go to infinity. Since U_n is bounded with probability 1, $P[X_* < \alpha < \beta < X^*] = 0$. Now

(35.24) $$[X_* < X^*] = \bigcup [X_* < \alpha < \beta < X^*],$$

where the union extends over all pairs of rationals α and β. The set on the left therefore has probability 0.

Thus X^* and X_* are equal with probability 1, and X_n converges to their common value X, which may be infinite. By Fatou's lemma $E[|X|] \le \liminf_n E[|X_n|] \le K$. Since it is integrable, X is finite with probability 1. ∎

If the X_n form a martingale, then by (35.16) applied to the submartingale $|X_1|, |X_2|, \ldots$ the $E[|X_n|]$ do not decrease, so that $K = \lim_n E[|X_n|]$. The hypothesis in the theorem that K be finite is essential: If $X_n = \Delta_1 + \cdots + \Delta_n$, where the Δ_n are independent and assume values ± 1 with probability $\frac{1}{2}$, then X_n does not converge; of course, $E[|X_n|]$ must go to infinity in this case.

If the X_n form a *nonnegative* martingale, then $E[|X_n|] = E[X_n] = E[X_1]$ by (35.5), and K is necessarily finite.

Example 35.9. The X_n in Example 35.6 are nonnegative, and so $X_n = Z_n/m^n \to X$, where X is nonnegative and integrable. Suppose that $m < 1$. If $Z_n \ge 1$, then $X_n \ge m^{-n}$, and so from the boundedness of X_n, it follows that $Z_n = 0$ for large n: The population eventually dies out if $m < 1$. Thus $X = 0$ with probability 1. Since $E[X_n] = E[X_0] = 1$, this shows that $E[X_n] \to E[X]$ may fail in Theorem 35.4. ∎

Theorem 35.4 has an important application to the martingale of Example 35.5, and this requires a lemma.

Lemma. *If Z is integrable and $\mathcal{F}_n$ are arbitrary σ-fields, then the random variables $E[Z\|\mathcal{F}_n]$ are uniformly integrable.*

For the definition of uniform integrability, see (16.20). The $\mathcal{F}_n$ must, of course, lie in the σ-field $\mathcal{F}$, but they need not, for example, be nondecreasing.

PROOF OF THE LEMMA. Since $|E[Z\|\mathcal{F}_n]| \le E[|Z|\,\|\mathcal{F}_n]$, Z may be assumed nonnegative. Let $A_{\alpha n} = [E[Z\|\mathcal{F}_n] \ge \alpha]$. Since $E[Z\|\mathcal{F}_n]$ is measurable $\mathcal{F}_n$,

$$\int_{A_{\alpha n}} E[Z\|\mathcal{F}_n]\, dP = \int_{A_{\alpha n}} Z\, dP.$$

It is therefore enough to find for given ϵ an α such that this last integral is less than ϵ for all n. Now $\int_A Z\, dP$ is, as a function of A, a finite measure dominated by P; by the ϵ-δ version of absolute continuity (see (32.4)) there is a δ such that $P(A) < \delta$ implies that $\int_A Z\, dP < \epsilon$. But $P[E[Z\|\mathcal{F}_n] \ge \alpha] \le \alpha^{-1} E[E[Z\|\mathcal{F}_n]]$ $= \alpha^{-1} E[Z] < \delta$ for large enough α. ∎

Suppose that $\mathcal{F}_n$ are σ-fields satisfying $\mathcal{F}_1 \subset \mathcal{F}_2 \subset \cdots$. If the union $\bigcup_{n=1}^{\infty} \mathcal{F}_n$

generates the σ-field $\mathcal{F}_\infty$, this is expressed by $\mathcal{F}_n \uparrow \mathcal{F}_\infty$. The requirement is not that $\mathcal{F}_\infty$ coincide with the union but that it be generated by it.

Theorem 35.5. *If $\mathcal{F}_n \uparrow \mathcal{F}_\infty$ and Z is integrable, then*

$$(35.25) \qquad\qquad E[Z\|\mathcal{F}_n] \to E[Z\|\mathcal{F}_\infty]$$

with probability 1.

PROOF. According to Example 35.5, the random variables $X_n = E[Z\|\mathcal{F}_n]$ form a martingale relative to the $\mathcal{F}_n$. By the lemma, the X_n are uniformly integrable. Since $E[|X_n|] \leq E[|Z|]$, by Theorem 35.4 the X_n converge to an integrable X. The problem is to identify X with $E[Z\|\mathcal{F}_\infty]$.

Because of the uniform integrability, it is possible (Theorem 16.13) to integrate to the limit: $\int_A X\, dP = \lim_n \int_A X_n\, dP$. If $A \in \mathcal{F}_k$ and $n \geq k$, then $\int_A X_n\, dP = \int_A E[Z\|\mathcal{F}_n]\, dP = \int_A Z\, dP$. Therefore, $\int_A X\, dP = \int_A Z\, dP$ for all A in the π-system $\bigcup_{k=1}^\infty \mathcal{F}_k$; since X is measurable $\mathcal{F}_\infty$, it follows by Theorem 34.1 that X is a version of $E[Z\|\mathcal{F}_\infty]$. ∎

Reversed Martingales

A left-infinite sequence $\ldots, X_{-2}, X_{-1}$ is a martingale relative to σ-fields $\ldots, \mathcal{F}_{-2}, \mathcal{F}_{-1}$ if conditions (ii) and (iii) in the definition of martingale are satisfied for $n \leq -1$ and conditions (i) and (iv) are satisfied for $n < -1$. Such a sequence is a reversed or backward martingale.

Theorem 35.6. *For a reversed martingale, $\lim_{n\to\infty} X_{-n} = X$ exists and is integrable, and $E[X] = E[X_{-n}]$ for all n.*

PROOF. The proof is almost the same as that for Theorem 35.4. Let X^* and X_* be the limits superior and inferior of the sequence $X_{-1}, X_{-2}, \ldots$. Again (35.24) holds. Let U_n be the number of upcrossings of $[\alpha, \beta]$ by $X_{-n}, \ldots, X_{-1}$. By the upcrossing theorem $E[U_n] \leq (E[|X_{-1}|] + |\alpha|)/(\beta - \alpha)$. Again $E[U_n]$ is bounded, and so $\sup_n U_n$ is finite with probability 1 and the sets in (35.24) have probability 0.

Therefore, $\lim_{n\to\infty} X_{-n} = X$ exists with probability 1. By the property (35.4) for martingales, $X_{-n} = E[X_{-1}\|\mathcal{F}_n]$ for $n = 1, 2, \ldots$. The lemma now implies that the X_n are uniformly integrable. Therefore, X is integrable and $E[X]$ is the limit of the $E[X_{-n}]$; these all have the same value by (35.5). ∎

If $\mathcal{F}_n$ are σ-fields satisfying $\mathcal{F}_1 \supset \mathcal{F}_2 \cdots$, then the intersection $\bigcap_{n=1}^\infty \mathcal{F}_n = \mathcal{F}_0$ is also a σ-field, and the relation is expressed $\mathcal{F}_n \downarrow \mathcal{F}_0$.

Theorem 35.7. *If $\mathcal{F}_n \downarrow \mathcal{F}_0$ and Z is integrable, then*

$$(35.26) \qquad\qquad E[Z \| \mathcal{F}_n] \to E[Z \| \mathcal{F}_0]$$

with probability 1.

PROOF. If $X_{-n} = E[Z \| \mathcal{F}_n]$, then $\ldots, X_{-2}, X_{-1}$ is a martingale relative to $\ldots,$ $\mathcal{F}_2, \mathcal{F}_1$. By the preceding theorem, $E[Z \| \mathcal{F}_n]$ converges as $n \to \infty$ to an integrable X, and by the lemma, the $E[Z \| \mathcal{F}_n]$ are uniformly integrable. As the limit of the $E[Z \| \mathcal{F}_n]$ for $n \geq k$, X is measurable $\mathcal{F}_k$; k being arbitrary, X is measurable $\mathcal{F}_0$.

By uniform integrability, $A \in \mathcal{F}_0$ implies that $\int_A X \, dP = \lim_n \int_A E[Z \| \mathcal{F}_n]$ $dP = \lim_n \int_A E[E[Z \| \mathcal{F}_n] \| \mathcal{F}_0] \, dP = \lim_n \int_A E[Z \| \mathcal{F}_0] \, dP = \int_A E[Z \| \mathcal{F}_0]$ dP. Thus X is a version of $E[Z \| \mathcal{F}_0]$. ∎

Theorems 35.5 and 35.7 are parallel. There is an essential difference between Theorems 35.4 and 35.6, however. In the latter, the martingale has a last random variable, namely X_{-1}, and so it is unnecessary in proving convergence to assume the $E[|X_n|]$ bounded. On the other hand, the proof in Theorem 35.6 that X is integrable would not work for a submartingale.

In the applications of Theorem 35.7, $\mathcal{F}_n$ is often $\sigma(Y_n, Y_{n+1}, \ldots)$ for random variables $Y_1, Y_2, \ldots$. In Theorem 35.5, $\mathcal{F}_n$ is often $\sigma(Y_1, \ldots, Y_n)$; in this case $\mathcal{F}_\infty = \sigma(Y_1, Y_2, \ldots)$ is in general strictly larger than $\bigcup_{n=1}^\infty \mathcal{F}_n$.

Applications: Derivatives

Theorem 35.8. *Suppose that $(\Omega, \mathcal{F}, P)$ is a probability space, ν is a finite measure on $\mathcal{F}$, and $\mathcal{F}_n \uparrow \mathcal{F}_\infty \subset \mathcal{F}$. Suppose that P dominates ν on $\mathcal{F}_n$, and let X_n be the corresponding Radon-Nikodym derivative. Then $X_n \to X$ with probability* 1, *where X is integrable.*

(i) If P dominates ν on $\mathcal{F}_\infty$, then X is the corresponding Radon-Nikodym derivative.

(ii) If P and ν are mutually singular on $\mathcal{F}_\infty$, then $X = 0$ with probability 1.

PROOF. The situation is that of Example 35.2. The density X_n is measurable $\mathcal{F}_n$ and satisfies (35.9). Since X_n is nonnegative, $E[|X_n|] = E[X_n] = \nu(\Omega)$, and it follows by Theorem 35.4 that X_n converges to an integrable X. The limit X is measurable $\mathcal{F}_\infty$.

Suppose that P dominates ν on $\mathcal{F}_\infty$ and let Z be the Radon-Nikodym derivative: Z is measurable $\mathcal{F}_\infty$, and $\int_A Z \, dP = \nu(A)$ for $A \in \mathcal{F}_\infty$. It follows that $\int_A Z \, dP = \int_A X_n \, dP$ for A in $\mathcal{F}_n$, and so $X_n = E[Z \| \mathcal{F}_n]$. Now Theorem 35.5 implies that $X_n \to E[Z \| \mathcal{F}_\infty] = Z$.

Suppose, on the other hand, that P and ν are mutually singular on $\mathcal{F}_\infty$, so that there exists a set S in $\mathcal{F}_\infty$ such that $\nu(S) = 0$ and $P(S) = 1$. By Fatou's lemma $\int_A X \, dP \leq \lim\inf_n \int_A X_n \, dP$. If $A \in \mathcal{F}_k$, then $\int_A X_n \, dP = \nu(A)$ for $n \geq k$, and so $\int_A X \, dP \leq \nu(A)$ for A in the field $\bigcup_{k=1}^\infty \mathcal{F}_k$. It follows by the monotone class theorem that this holds for all A in $\mathcal{F}_\infty$. Therefore, $\int X \, dP = \int_S X \, dP \leq \nu(S) = 0$, and X vanishes with probability 1.   ∎

Example 35.10. As in Example 35.3, let ν be a finite measure on the unit interval with Lebesgue measure $(\Omega, \mathcal{F}, P)$. For $\mathcal{F}_n$ the σ-field generated by the dyadic intervals of rank n, (35.10) gives X_n. In this case $\mathcal{F}_n \uparrow \mathcal{F}_\infty = \mathcal{F}$. For each ω and n choose the dyadic rationals $a_n(\omega) = k2^{-n}$ and $b_n(\omega) = (k+1)2^{-n}$ for which $a_n(\omega) < \omega \leq b_n(\omega)$. By Theorem 35.8, if F is the distribution function for ν, then

$$(35.27) \qquad \frac{F(b_n(\omega)) - F(a_n(\omega))}{b_n(\omega) - a_n(\omega)} \to X(\omega)$$

except on a set of Lebesgue measure 0.

According to Theorem 31.2, F has a derivative F' except on a set of Lebesgue measure 0, and since the intervals $(a_n(\omega), b_n(\omega)]$ contract to ω, the difference ratio in (35.27) converges almost everywhere to $F'(\omega)$ (see (31.8)). This identifies X. Since (35.27) involves intervals $(a_n(\omega), b_n(\omega)]$ of a special kind, it does not quite imply Theorem 31.2.

By Theorem 35.8, $X = F'$ is the density for ν in the absolutely continuous case, and $X = F' = 0$ (except on a set of Lebesgue measure 0) in the singular case, facts proved in a different way in Section 31. The singular case gives another example where $E[X_n] \to E[X]$ fails in Theorem 35.4.   ∎

Likelihood Ratios

Return to Example 35.4: $\nu = Q$ is a probability measure, $\mathcal{F}_n = \sigma(Y_1, \ldots, Y_n)$ for random variables Y_n, and the Radon-Nikodym derivative or likelihood ratio X_n has the form (35.11) for densities p_n and q_n on R^n. By Theorem 35.8 the X_n converge to some X which is integrable and measurable $\mathcal{F}_\infty = \sigma(Y_1, Y_2, \ldots)$.

If the Y_n are independent under P and under Q, and if the densities are different, then P and Q are mutually singular on $\sigma(Y_1, Y_2, \ldots)$, as shown in Example 35.4. In this case $X = 0$ and the likelihood ratio converges to 0 on a set of P-measure 1. The statistical relevance of this is that the smaller X_n is, the more strongly one prefers P over Q as an explanation of the observation $(Y_1, \ldots, Y_n)$, and X_n goes to 0 with probability 1 if P is the measure governing the Y_n.

It might be thought that a disingenuous experimenter could bias his results by stopping at an X_n he likes—a large value if his prejudices favor Q, a small value if they favor P. This is not so, as the following analysis shows. For this argument P must dominate Q on each $\mathcal{F}_n = \sigma(Y_1, \ldots, Y_n)$, but the likelihood ratio X_n need have no special form.

Let τ be a positive-integer-valued random variable representing the experimenter's stopping time. Assume that $[\tau = n] \in \mathcal{F}_n$, which, like the analogous assumption (35.18) for the gambler's stopping time, excludes prevision on the part of the experimenter. Let $\mathcal{F}_\tau$ be the class of sets G for which

(35.28) $$G \cap [\tau = n] \in \mathcal{F}_n, \qquad n \geq 1.$$

Now $\mathcal{F}_\tau$ is a σ-field. It is to be interpreted as representing the information the experimenter has when he stops: to know $(Y_1(\omega), \ldots, Y_{\tau(\omega)}(\omega))$ is the same thing as to know for each G satisfying (35.28) whether or not it contains ω.

What is to be shown is that X_τ is the likelihood ratio (Radon-Nikodym derivative) for Q with respect to P on $\mathcal{F}_\tau$. If $H \in \mathcal{R}^1$, then $[X_\tau \in H] = \bigcup_{n=1}^\infty ([\tau = n] \cap [X_n \in H])$ lies in $\mathcal{F}_\tau$, so that X_τ is measurable $\mathcal{F}_\tau$. Moreover, if G satisfies (35.28), then

$$\int_G X_\tau \, dP = \sum_{n=1}^\infty \int_{G \cap [\tau = n]} X_n \, dP = \sum_{n=1}^\infty Q(G \cap [\tau = n]) = Q(G),$$

as required.

Bayes Estimation

Let $\theta, Y_1, Y_2, \ldots$ be random variables such that $0 < \theta < 1$ and, conditionally on θ, the Y_n are independent and assume the values 1 and 0 with probabilities θ and $1 - \theta$: For $u_1, \ldots, u_k$ a sequence of 0's and 1's,

(35.29) $$P[Y_k = u_k, k \leq n \| \theta] = \theta^s (1 - \theta)^{n-s},$$

where $s = u_1 + \cdots + u_k$.

To see that such sequences exist, let $\theta, Z_1, Z_2, \ldots$ be independent random variables, where θ has an arbitrarily prescribed distribution supported by $(0, 1)$ and the Z_n are uniformly distributed over $(0, 1)$. Put $Y_n = I_{[Z_n \leq \theta]}$. If, for instance, $f(x) = x(1 - x) = P[Z_1 \leq x, Z_2 > x]$, then $P[Y_1 = 1, Y_2 = 0 \| \theta]_\omega = f(\theta(\omega))$ by (33.13). The obvious extension establishes (35.29).

From the Bayes point of view in statistics, θ is a parameter governed by some a priori distribution known to the statistician. For given $Y_1, \ldots, Y_n$ the Bayes estimate of θ is $E[\theta \| Y_1, \ldots, Y_n]$. The problem is to show that this estimate is consistent in the sense that

(35.30) $$E[\theta \| Y_1, \ldots, Y_n] \to \theta.$$

By Theorem 35.5, $E[\theta \| Y_1, \ldots, Y_n] \to E[\theta \| \mathcal{F}_\infty]$, where $\mathcal{F}_\infty = \sigma(Y_1, Y_2, \ldots)$, and so what must be shown is that $E[\theta \| \mathcal{F}_\infty] = \theta$ with probability 1.

By an elementary argument that parallels the unconditional case, it follows from (35.29) that for $S_n = Y_1 + \cdots + Y_n$, $E[S_n \| \theta] = n\theta$ and $E[(S_n - n\theta)^2 \| \theta]$ $= n\theta(1 - \theta)$. Hence $E[(n^{-1}S_n - \theta)^2] \leq n^{-1}$, and by Chebyshev's inequality $n^{-1}S_n$ converges in probability to θ. Therefore (Theorem 20.5), $\lim_k n_k^{-1}S_{n_k}$ $= \theta$ with probability 1 for some subsequence. Thus $\theta' = \theta$ with probability 1 for a θ' measurable $\mathcal{F}_\infty$, and $E[\theta \| \mathcal{F}_\infty] = E[\theta' \| \mathcal{F}_\infty] = \theta' = \theta$ with probability 1.

The Zero-One Law

For a sequence $X_1, X_2, \ldots$, let

(35.31) $\mathcal{T}_n = \sigma(X_n, X_{n+1}, \ldots)$, $\mathcal{T} = \bigcap_{n=1}^{\infty} \mathcal{T}_n$.

An event in $\mathcal{T}$, a tail event, can for each n be described in terms of the X_n, X_{n+1}, ... alone. According to the zero-one law (Theorem 22.1), *if $X_1, X_2, \ldots$ are independent and A lies in $\mathcal{T}$, then $P(A)$ is either 0 or 1.* For a proof by martingale theory, suppose that $X = I_A$ and $A \in \mathcal{T}$, and note that since X is independent of $\sigma(X_1, \ldots, X_n)$, $E[X] = E[X \| X_1, \ldots, X_n]$. By Theorem 35.5, $E[X \| X_1, \ldots, X_n]$ converges to $E[X \| X_1, X_2, \ldots]$, which coincides with X because X is measurable $\sigma(X_1, X_2, \ldots)$. Therefore, $E[X] = X$ with probability 1, which is possible only if A has probability either 0 or 1.

The argument actually shows that if X is measurable $\mathcal{T}$ and integrable, so that the conditional expectations exist, then $X = c$ with probability 1 for some constant c. But the integrability is irrelevant: Let $f_k(x)$ be x if $|x| \leq k$ and 0 otherwise; if X is measurable $\mathcal{T}$, then $f_k(X)$ is constant with probability 1, and this is impossible for all k unless X itself is constant with probability 1.

The Strong Law of Large Numbers*

There is an interesting martingale proof of the law of large numbers:

(35.32) $\dfrac{1}{n} S_n = \dfrac{X_1 + \cdots + X_n}{n} \to E[X_1]$

with probability 1 if $X_1, X_2, \ldots$ are independent and identically distributed random variables with finite mean. Let

(35.33) $\mathcal{T}_n' = \sigma(S_n, S_{n+1}, \ldots) = \sigma(S_n, X_{n+1}, X_{n+2}, \ldots)$, $\mathcal{T}' = \bigcap_{n=1}^{\infty} \mathcal{T}_n'$.

The two forms for $\mathcal{T}_n'$ are clearly identical. The second form shows that $\mathcal{T}_{n+1}$ $\subset \mathcal{T}_n'$, so that $\mathcal{T} \subset \mathcal{T}'$. The first form shows that $\mathcal{T}_n' \supset \mathcal{T}_{n+1}'$, and so by Theorem 35.7, $E[X_1 \| \mathcal{T}_n'] \to E[X_1 \| \mathcal{T}']$. From the second form for $\mathcal{T}_n'$, it seems intuitively clear because of symmetry that

---

* This topic may be omitted.

(35.34) $E[X_k \| \mathcal{T}_n'] = E[X_1 \| \mathcal{T}_n'], \quad k \le n.$

If so, then $n^{-1} S_n = E[n^{-1} S_n \| \mathcal{T}_n'] = n^{-1} \Sigma_{k=1}^n E[X_k \| \mathcal{T}_n'] = E[X_1 \| \mathcal{T}_n']$, and hence $n^{-1} S_n \to E[X_1 \| \mathcal{T}']$. Now $\lim_n n^{-1} S_n = \lim_n n^{-1} \Sigma_{k=m}^n X_k$ for each m, and so the limit is measurable $\mathcal{T}$. By the zero-one law, $E[X_1 \| \mathcal{T}'] = c$ with probability 1 for some c, and c must be $E[X_1]$. Hence (35.32) will follow if (35.34) is proved.

Suppose that

(35.35) $E[f(X_1, \ldots, X_n) \| \mathcal{T}_n'] = E[f(X_{\pi 1}, \ldots, X_{\pi n}) \| \mathcal{T}_n']$

for permutations π of $(1, 2, \ldots, n)$ and integrable $f(X_1, \ldots, X_n)$. If π is the permutation that interchanges 1 and k and if $f(x_1, \ldots, x_n) = x_1$, this will give (35.34). The sets

(35.36) $A = [X_1 + \cdots + X_n \in H] \cap [X_{n+1} \in H_1] \cap \cdots \cap [X_{n+j} \in H_j]$

for $j \ge 1$ and $H, H_1, \ldots, H_j$ in $\mathcal{R}^1$ form a π-system generating $\mathcal{T}_n'$, and so to prove (35.35) it suffices to check

(35.37) $\int_A f(X_1, \ldots, X_n)\, dP = \int_A f(X_{\pi 1}, \ldots, X_{\pi n})\, dP$

for such A. If g is the indicator of

$$[x \in R^{n+j} : x_1 + \cdots + x_n \in H, x_{n+1} \in H_1, \ldots, x_{n+j} \in H_j],$$

then

(35.38) $g(x_1, \ldots, x_n, x_{n+1}, \ldots, x_{n+j}) = g(x_{\pi 1}, \ldots, x_{\pi n}, x_{n+1}, \ldots, x_{n+j}).$

Since the X_k are independent and identically distributed, $(X_1, \ldots, X_n, X_{n+1}, \ldots, X_{n+j})$ and $(X_{\pi 1}, \ldots, X_{\pi n}, X_{n+1}, \ldots, X_{n+j})$ have the same distribution over R^n and therefore

$$\begin{aligned}
E[f(X_1, &\ldots, X_n)g(X_1, \ldots, X_n, X_{n+1}, \ldots, X_{n+j})] \\
&= E[f(X_{\pi 1}, \ldots, X_{\pi n})g(X_{\pi 1}, \ldots, X_{\pi n}, X_{n+1}, \ldots, X_{n+j})] \\
&= E[f(X_{\pi 1}, \ldots, X_{\pi n})g(X_1, \ldots, X_n, X_{n+1}, \ldots, X_{n+j})].
\end{aligned}$$

But this gives (35.37) and hence (35.35), as required.

This approach to the strong law of large numbers leads to an extension. Let $\mathcal{S}_n$ be the σ-field of sets depending symmetrically on $X_1, \ldots, X_n$ and arbitrarily on $X_{n+1}, X_{n+2}, \ldots$. To be precise, consider the sets of the form .

(35.39) $A = [(X_1, \ldots, X_n) \in M] \cap [X_{n+1} \in H_1] \cap \cdots \cap [X_{n+j} \in H_j]$

where $H_1, \ldots, H_j$ are in $\mathcal{R}^1$ and M is an $\mathcal{R}^n$-set symmetric in the sense that $(x_1, \ldots, x_n) \in M$ if and only if $(x_{\pi 1}, \ldots, x_{\pi n}) \in M$ for all π. Let $\mathcal{S}_n$ be the σ-field generated by these sets (j arbitrary). The information in $\mathscr{S}_n$ gives the values of $X_{n+1}, X_{n+2}, \ldots$ exactly and gives the n values of $X_1, \ldots, X_n$ to within a permutation. Since (35.36) is a special case of (35.39), $\mathcal{T}'_n \subset \mathscr{S}_n$. If g is the indicator function of

$$[x \in R^{n+j}: (x_1, \ldots, x_n) \in M, x_{n+1} \in H_1, \ldots, x_{n+j} \in H_j]$$

and M is symmetric as above, then (35.38) holds, and so the argument following it goes through as before. Therefore, (35.37) holds for A in the π-system of sets (35.39), so that

(35.40) $$E[f(X_1, \ldots, X_n)\|\mathscr{S}_n] = E[f(X_{\pi 1}, \ldots, X_{\pi n})\|\mathscr{S}_n]$$

for integrable $f(X_1, \ldots, X_n)$.

Now $\mathscr{S}_n$ can also be characterized as the σ-field genererated by the random variables $g(X_1, \ldots, X_{n+j})$, where g is a Borel function on R^{n+j} (j arbitrary) satisfying (35.38) for all π. Therefore, $\mathscr{S}_n \supset \mathscr{S}_{n+1}$, and so

(35.41) $$\mathscr{S}_n \downarrow \mathscr{S} = \bigcap_{n=1}^{\infty} \mathscr{S}_n.$$

For a function φ of k variables, put

(35.42) $$A_n(\varphi) = \frac{1}{(n)_k} \sum \varphi(X_{i_1}, \ldots, X_{i_k}),$$

where the sum extends over all $(n)_k = n(n-1)\cdots(n-k+1)$ k-tuples of distinct integers not exceeding n. If $X_1, X_2, \ldots$ *are independent and identically distributed and* $\varphi(X_1, \ldots, X_k)$ *is integrable, then*

(35.43) $$A_n(\varphi) \to E[\varphi(X_1, \ldots, X_k)]$$

with probability 1. For $k = 1$ this is the usual strong law. To prove it generally, assume as induction hypothesis that it holds for $k - 1$.

Since $A_n(\varphi)$ is symmetric in $X_1, \ldots, X_n$ and hence measurable $\mathscr{S}_n$,

(35.44) $A_n(\varphi) = E[A_n(\varphi)\|\mathscr{S}_n]$

$$= \frac{1}{(n)_k} \sum E[\varphi(X_{i_1}, \ldots, X_{i_k})\|\mathscr{S}_n] = E[\varphi(X_1, \ldots, X_k)\|\mathscr{S}_n]$$

by (35.40). Now (35.41) gives

(35.45) $$E[\varphi(X_1, \ldots, X_k)\|\mathscr{S}_n] \to E[\varphi(X_1, \ldots, X_k)\|\mathscr{S}].$$

The terms in (35.42) involving the index 1 contribute

$$\frac{1}{n}\frac{1}{(n-1)_{k-1}} \sum E[\varphi(X_1, X_{j_1}, \ldots, X_{j_{k-1}}) + \cdots + \varphi(X_{j_1}, \ldots, X_{j_{k-1}}, X_1)],$$

where the sum extends over the $(k-1)$-tuples of distinct integers in the range $2 \le i \le n$. Since this goes to 0 by the induction hypothesis,[*] the limit $E[\varphi(X_1, \ldots, X_k)\|\mathscr{S}]$ of (35.44) is measurable $\mathcal{T}_2$. Similarly, it is measurable $\mathcal{T}_3, \mathcal{T}_4, \ldots$ and so is measurable $\mathcal{T}$. By the zero-one law, it is constant with probability 1:

(35.46) $$E[\varphi(X_1, \ldots, X_k)\|\mathscr{S}] = E[\varphi(X_1, \ldots, X_k)].$$

Now (35.43) follows from (35.44), (35.45), (35.46).

---

[*] For bounded φ, this is directly obvious.

The Hewitt-Savage Theorem*

The limiting σ-fields in (35.31), (35.33), and (35.41) stand in the relation $\mathcal{T} \subset \mathcal{T}' \subset \mathcal{S}$. According to a theorem of Hewitt and Savage, *if $X_1, X_2, \ldots$ are independent and identically distributed and A lies in $\mathcal{S}$, then $P(A)$ is either 0 or 1.* Since $\mathcal{T} \subset \mathcal{S}$, the conclusion here is stronger than that of the zero-one law; but the hypothesis is stronger also, because the X_n must be identically distributed as well as independent. In general, $\mathcal{S}$ is strictly larger than $\mathcal{T}$: $[X_1 + \cdots + X_n = 0$ i.o.$]$ lies in $\mathcal{S}$ but in general not in $\mathcal{T}$.

To prove the Hewitt-Savage theorem, use (35.46), which is valid if the X_n are independent and identically distributed. Since this holds for all indicators φ of k-dimensional Borel sets, $\sigma(X_1, \ldots, X_k)$ is independent of $\mathcal{S}$. Therefore, (Theorem 4.2) $\sigma(X_1, X_2, \ldots)$ is independent of $\mathcal{S}$. But since $\mathcal{S}$ is contained in $\sigma(X_1, X_2, \ldots)$, it is independent of itself: $A \in \mathcal{S}$ implies that $P(A) = P(A \cap A) = P(A)P(A)$ and hence that $P(A)$ is 0 or 1.

Exchangeable Random Variables†

A sequence $X_1, X_2, \ldots$ is *exchangeable* if for each n the distribution of $X_1, \ldots, X_n$ is symmetric in the sense that $(X_1, \ldots, X_n)$ and $(X_{\pi 1}, \ldots, X_{\pi n})$ have the same distribution for every permutation π. The random variables Y_k in (35.29) are conditionally independent given θ, and taking expected values shows that $P[Y_k = u_k, k \leq n]$ depends only on $u_1 + \cdots + u_n$. Therefore, the Y_k are exchangeable although not independent.

Define $\mathcal{S}$ as in (35.41). According to a theorem of de Finetti, *if $X_1, X_2, \ldots$ are exchangeable, then they are conditionally independent given $\mathcal{S}$ in the sense that*

$$(35.47) \qquad P[X_i \in H_i, i \leq k \| \mathcal{S}] = \prod_{i \leq k} P[X_i \in H_i \| \mathcal{S}].$$

To prove this, notice first that the proof of (35.40) goes through as before under the sole assumption that $X_1, X_2, \ldots$ is exchangeable. The calculation (35.44) thus goes through as before, and of course (35.45) still holds. Therefore,

$$(35.48) \qquad A_n(\varphi) \to E[\varphi(X_1, \ldots, X_k) \| \mathcal{S}].$$

This is the analogue of (35.43); the limit here cannot be replaced by $E[\varphi(X_1, \ldots, X_k)]$ because random variables measurable $\mathcal{S}$ or even $\mathcal{T}$ need not be constant if the X_n are not independent.

Suppose that f and g are indicators of sets in R^{k-1} and R^1. If $\varphi(x_1, \ldots, x_k) = f(x_1, \ldots, x_{k-1}) g(x_k)$ and $\varphi_\nu(x_1, \ldots, x_{k-1}) = f(x_1, \ldots, x_{k-1})g(x_\nu)$, then

$$A_n(\varphi) = \frac{n}{n-k+1} A_n(f)A_n(g) - \frac{1}{n-k+1} \sum_{\nu=1}^{k-1} A_n(\varphi_\nu).$$

By (35.48) for $k - 1$ and for $k = 1$

$$A_n(\varphi) \to E[f(X_1, \ldots, X_{k-1}) \| \mathcal{S}] E[g(X_k) \| \mathcal{S}].$$

By this, (35.40), and (35.48) itself,

$$E[f(X_1, \ldots, X_{k-1})g(X_k) \| \mathcal{S}] = E[f(X_1, \ldots, X_{k-1}) \| \mathcal{S}] E[g(X_k) \| \mathcal{S}].$$

But (35.47) follows easily by induction from this.

---

* This topic may be omitted.
† This topic may be omitted.

PROBLEMS

35.1. Suppose that $\Delta_1, \Delta_2, \ldots$ are independent random variables with mean 0. Let $X_1 = \Delta_1$ and $X_{n+1} = X_n + \Delta_{n+1} f_n(X_1, \ldots, X_n)$ and suppose that the X_n are integrable. Show that $\{X_n\}$ is a martingale. The martingales of gambling have this form.

35.2. Let $Y_1, Y_2, \ldots$ be independent random variables with mean 0 and variance σ^2. Let $X_n = (\Sigma_{k=1}^n Y_k)^2 - n\sigma^2$ and show that $\{X_n\}$ is a martingale.

35.3. Suppose that $\{Y_n\}$ is a finite-state Markov chain with transition matrix $[p_{ij}]$. Suppose that $\Sigma_j p_{ij} x(j) = \lambda x(i)$ for all i (the $x(i)$ are the components of a right eigenvector of the transition matrix). Put $X_n = \lambda^{-n} x(Y_n)$ and show that $\{X_n\}$ is a martingale.

35.4. Suppose that $Y_1, Y_2, \ldots$ are independent, positive random variables and that $E[Y_n] = 1$. Put $X_n = Y_1 \cdots Y_n$.

(a) Show that $\{X_n\}$ is a martingale and converges with probability 1 to an integrable X.

(b) Suppose specifically that Y_n assumes the values $\frac{1}{2}$ and $\frac{3}{2}$ with probability $\frac{1}{2}$ each. Show that $X = 0$ with probability 1. This gives an example where $E[\prod_{n=1}^\infty Y_n] \neq \prod_{n=1}^\infty E[Y_n]$ for independent, integrable, positive random variables. Show, however, that $E[\prod_{n=1}^\infty Y_n] \le \prod_{n=1}^\infty E[Y_n]$ always holds.

35.5. Suppose that $X_1, X_2, \ldots$ is a martingale satisfying $E[X_1] = 0$ and $E[X_n^2] < \infty$. Show that $E[(X_{n+r} - X_n)^2] = \Sigma_{k=1}^r E[(X_{n+k} - X_{n+k-1})^2]$ (the variance of the sum is the sum of the variances). Assume that $\Sigma_n E[(X_n - X_{n-1})^2] < \infty$ and prove that X_n converges with probability 1. Do this first by Theorem 35.4 and then (see Theorem 22.3) by Theorem 35.2.

35.6. Show that a submartingale X_n can be represented as $X_n = Y_n + Z_n$, where Y_n is a martingale and $0 \le Z_1 \le Z_2 \le \cdots$.

Hint: Take $X_0 = 0$ and $\Delta_n = X_n - X_{n-1}$, and define $Z_n = \Sigma_{k=1}^n E[\Delta_k \| \mathcal{F}_{k-1}]$ ($\mathcal{F}_0 = \{0, \Omega\}$).

35.7. If $X_1, X_2, \ldots$ is a martingale and bounded either above or below, then $\sup_n E[|X_n|] < \infty$.

35.8. ↑ Let $X_n = \Delta_1 + \cdots + \Delta_n$, where the Δ_n are independent and assume the values ± 1 with probability $\frac{1}{2}$ each. Let τ be the smallest n such that $X_n = 1$ and define X_n^* by (35.19). Show that the hypotheses of Theorem 35.4 are satisfied by $\{X_n^*\}$ but that is is impossible to integrate to the limit.

Hint: Use (7.8) and Problem 35.7.

35.9. Let $X_1, X_2, \ldots$ be a martingale and assume that $|X_1(\omega)|$ and $|X_n(\omega) - X_{n-1}(\omega)|$ are bounded by a constant independent of ω and n. Let τ be a stopping time (see (35.18)) with finite mean. Show that X_τ is integrable and that $E[X_\tau] = E[X_1]$.

35.10. 35.8 35.9↑ Use the preceding result to show that the τ in Problem 35.8 has infinite mean. Thus the waiting time until a symmetric random walk moves one step up from the starting point has infinite expected value.

35.11. Let $X_1, X_2, \ldots$ be a Markov chain with countable state space S and transition probabilities p_{ij}. A function φ on S is excessive or superharmonic if $\varphi(i) \geq \Sigma_j$ $p_{ij}\varphi(j)$. Show by martingale theory that $\varphi(X_n)$ converges with probability 1 if φ is bounded and excessive. Deduce from this that if the chain is irreducible and persistent, then φ must be constant. Compare Problem 8.25.

35.12. ↑ A function φ on the integer lattice in R^k is superharmonic if for each lattice point x, $\varphi(x) \geq (2k)^{-1} \Sigma \varphi(y)$, the sum extending over the $2k$ nearest neighbors y. Show for $k = 1$ and $k = 2$ that a bounded superharmonic function is constant. Show for $k \geq 3$ that there exist nonconstant bounded harmonic functions.

35.13. 32.9↑ Let $(\Omega, \mathcal{F}, P)$ be a probability space, let ν be a finite measure on $\mathcal{F}$, and suppose that $\mathcal{F}_n \uparrow \mathcal{F}_\infty \subset \mathcal{F}$. For $n \leq \infty$, let X_n be the Radon-Nikodym derivative with respect to P of the absolutely continuous part of ν when P and ν are both restricted to $\mathcal{F}_n$. Use the result of Problem 32.9 to show that $X_1, X_2, \ldots$ is a nonnegative supermartingale. Extend Theorem 35.8 by showing that $X_n \to X_\infty$ with probability 1.

35.14. (a) Show that $\{X_n\}$ is a martingale with respect to $\{\mathcal{F}_n\}$ if and only if for all n and all stopping times τ such that $\tau \leq n$, $E[X_n \| \mathcal{F}_\tau] = X_\tau$, where $\mathcal{F}_\tau$ is the class of G such that $G \cap [\tau = k] \in \mathcal{F}_k$ for all k (see (35.28)).

(b) Show that, if $\{X_n\}$ is a martingale and τ is a bounded stopping time, then $E[X_\tau] = E[X_1]$.

(c) Show that, if $\{X_n\}$ is a submartingale and τ is a stopping time bounded by n, then $E[X_\tau] \leq E[X_n]$.

35.15. 31.7↑ Suppose that $\mathcal{F}_n \uparrow \mathcal{F}_\infty$ and $A \in \mathcal{F}_\infty$, and prove that $P[A \| \mathcal{F}_n] \to I_A$ with probability 1. Compare Lebesgue's density theorem.

35.16. 21.13 ↑ Suppose that $X_1, \ldots, X_n$ is a martingale and that the $|X_k|^r$ are integrable, $r > 1$. If $M = \max_{k \leq n} |X_k|$, then $P[M \geq t] \leq t^{-1} \int_{M \geq t} |X_n| \, dP$, as follows by the proof of Theorem 35.2. Show that $E[M^r] \leq (r/(r-1))^r \cdot E[|X_n|^r]$.

35.17. (a) Extend Theorem 35.5: Replace the condition $\mathcal{F}_n \uparrow \mathcal{F}_\infty$ by the condition that each set in $\mathcal{F}_\infty$ differs by a set of measure 0 from some set in $\sigma(\cup_n \mathcal{F}_n)$ and conversely.

(b) Extend Theorem 35.7 analogously.

35.18. Theorems 35.5 and 35.7 have analogues in Hilbert space. For $n \leq \infty$, let P_n be the perpendicular projection on a subspace M_n. Then $P_n x \to Px$ for all x if either (a) $M_1 \subset M_2 \subset \cdots$ and M_∞ is the closure of $\cup_{n=1}^\infty M_n$ or (b) $M_1 \supset M_2 \supset \cdots$ and $M = \cap_{n=1}^\infty M_n$.

35.19. Suppose that θ has an arbitrary distribution, and suppose that, conditionally on θ, $Y_1, Y_2, \ldots$ are independent and normally distributed with mean θ and variance σ^2. Construct such a sequence $\{\theta, Y_1, Y_2, \ldots\}$. Prove (35.30).

35.20. It is shown on page 421 that optional stopping has no effect on likelihood ratios. This is not true of tests of significance. Suppose that $X_1, X_2, \ldots$ are independent and identically distributed and assume the values 1 and 0 with probabilities p and $1 - p$. Consider the null hypothesis that $p = \frac{1}{2}$ and the alternative that $p > \frac{1}{2}$. The usual .05-level test of significance is to reject the null hypothesis if

$$(35.49) \qquad \frac{2}{\sqrt{n}} \left(X_1 + \cdots + X_n - \frac{1}{2} n \right) > 1.645.$$

For this test the chance of falsely rejecting the null hypothesis is approximately $P[N > 1.645] \approx .05$ if n is large and fixed. Suppose that n is not fixed in advance of sampling and show by the law of the iterated logarithm that, even if p is, in fact, $\frac{1}{2}$, there are with probability 1 infinitely many n for which (35.49) holds.

35.21. Extend the sum in (35.42) to all n^k k-tuples from $\{1, \ldots, n\}$. Show that (35.43) still holds.

Amarts. In the remaining problems of this section $X_1, X_2, \ldots$ are integrable random variables, X_n is measurable $\mathcal{F}_n$, $\mathcal{F}_1 \subset \mathcal{F}_2 \subset \cdots$, and $\mathcal{F}_\infty = \sigma(\bigcup_{n=1}^\infty \mathcal{F}_n)$. Let T be the class of bounded stopping times: $\tau \in T$ if τ is a random variable whose range is a finite set of positive integers and if $[\tau = n] \in \mathcal{F}_n$ for $n \geq 1$. If τ is bounded by m, then $E[|X_\tau|] \leq \Sigma_{k=1}^m E[|X_k|]$, and so X_τ is always integrable. Let T_n consist of the τ in T for which $\tau(\omega) \geq n$ for all ω. The sequence $\{X_n\}$ is of course uniformly bounded if $\sup_{n,\omega} |X_n(\omega)| < \infty$. It is called L_1-*bounded* if $\sup_n E[|X_n|] < \infty$. Maxima and minima will be denoted by $\vee$ and $\wedge$.

The sequence is an *amart* (asymptotic martingale) if for every sequence $\{\tau_n\}$ of stopping times such that $\tau_n \in T_n$, $\lim_n E[X_{\tau_n}]$ exists and is finite. Since $E[X_\tau] = E[X_1]$ for $\tau \in T$ in the case of a martingale (see Problem 35.14(b)), every martingale is an amart. The following arguments, which do not depend on martingale theory, show that an L_1-bounded amart converges with probability 1.

35.22. (a) If $\{X_n\}$ is an amart, the limits $\lim_n E[X_{\tau_n}]$ ($\tau_n \in T_n$) all have a common value c, $\lim_n \sup_{\tau \in T_n} |E[X_\tau] - c| = 0$, and $\sup_{\tau \in T} |E[X_\tau]| < \infty$.

(b) Show that $\{X_n\}$ is an amart if $\lim_n E[X_{\tau_n}]$ exists and is finite for each sequence $\{\tau_n\}$ such that $\tau_n \in T$ and $\tau_n \leq \tau_{n-1}$.

35.23. ↑ Define Y to be a *cluster* random variable if Y is measurable $\mathcal{F}_\infty$ and if for each ω, $Y(\omega)$ is a limit point (the limit of a convergent subsequence) of $X_1(\omega), X_2(\omega), \ldots$. By the following steps show that, if Y is a cluster random variable, then there are stopping times τ_n such that $\tau_n \in T_n$ and $X_{\tau_n} \to Y$ with probability 1. (This does not require the amart property.)

(a) Choose for each k an integer $n_k \geq k$ and an $\mathcal{F}_{n_k}$-measurable random variable Y_k such that $P[|Y - Y_k| \geq 1/k] \leq 1/k^2$.

(b) Choose an integer $q_k \geq n_k$ such that $P[|X_n - Y| \geq 1/k, n_k \leq n \leq q_k] \leq 1/k^2$.

(c) Let τ_k be the smallest n in the range $n_k \leq n \leq q_k$ for which $|X_n - Y_k| < 2/k$ ($\tau_k = q_k$ if there are none) and show that $X_{\tau_k} \to X$ with probability 1.

35.24. ↑ Prove that a uniformly bounded amart converges with probability 1.

Hint: Apply the result in Problem 35.23 to $\lim \sup_n X_n$ and $\lim \inf_n X_n$.

The remaining problems show how to extend this last result from the uniformly bounded to the L_1-bounded case.

35.25. ↑ Suppose that $\{X_n\}$ is an L_1-bounded amart and show that

(a) $\{X_n^+\}$ is an L_1-bounded amart; and

(b) $\{X_n \vee \alpha\}$ and $\{X_n \wedge \alpha\}$ are L_1-bounded amarts.

35.26. ↑ Prove the maximal inequality $P[\sup_n |X_n| > \alpha] \leq \alpha^{-1} \sup_{\tau \in T} E[|X_\tau|]$, $\alpha > 0$. (This does not require the amart property.)

35.27. ↑ Let $\{X_n\}$ be an L_1-bounded amart, and define

$$X_n^{(\alpha)} = \begin{cases} -\alpha & \text{if } X_n \leq -\alpha, \\ X_n & \text{if } -\alpha \leq X_n \leq \alpha, \\ \alpha & \text{if } \alpha \leq X_n. \end{cases}$$

(a) Show that $\{X_n^{(\alpha)}\}$ is a uniformly bounded amart.

(b) Show that $\lim_{\alpha \to \infty} P[\sup_n |X_n| > \alpha] = 0$.

(c) Show that $\{X_n\}$ converges with probability 1.

35.28. ↑ As noted above, every martingale is an amart.

(a) Show that a submartingale is an amart if $E[X_n]$ is bounded.

(b) Find an amart that is not a martingale or even a submartingale.

CHAPTER 7

Stochastic Processes

SECTION 36. KOLMOGOROV'S EXISTENCE THEOREM

Stochastic Processes

A *stochastic process* is a collection $[X_t: t \in T]$ of random variables on a probability space $(\Omega, \mathcal{F}, P)$. The sequence of gambler's fortunes in Section 7, the sequences of independent random variables in Section 22, the queueing process of Section 24, the martingales of Section 35—all these are stochastic processes for which $T = \{1, 2, \ldots\}$. For the Poisson process $[N_t: t \geq 0]$ of Section 23, $T = [0, \infty)$. For all these processes the points of T are thought of as representing *time*. In most cases, T is the set of integers and time is *discrete*, or else T is an interval of the line and time is *continuous*. For the general theory of this section, however, T can be quite arbitrary.

Finite-Dimensional Distributions

A process is usually described in terms of distributions it induces in Euclidean spaces. For each k-tuple $(t_1, \ldots, t_k)$ of distinct elements of T, the random vector $(X_{t_1}, \ldots, X_{t_k})$ has over R^k some distribution $\mu_{t_1 \cdots t_k}$:

$$(36.1) \qquad \mu_{t_1 \cdots t_k}(H) = P[(X_{t_1}, \ldots, X_{t_k}) \in H], \qquad H \in \mathcal{R}^k.$$

These probability measures $\mu_{t_1 \cdots t_k}$ are the *finite-dimensional distributions* of the stochastic process $[X_t: t \in T]$. The system of finite-dimensional distributions does not completely determine the properties of the process. For example, the Poisson process $[N_t: t \geq 0]$ as defined by (23.5) has sample paths (functions $N_t(\omega)$ with ω fixed and t varying) that are step functions. But (23.21) defines a process which has the same finite-dimensional distributions and has sample paths that are *not* step functions. Nevertheless, the first step in a general theory is to construct processes for given systems of finite-dimensional distributions.

430

Now (36.1) implies two consistency properties of the system $\mu_{t_1 \ldots t_k}$. Suppose that π is a permutation of $(1, 2, \ldots, k)$, and define $\varphi_\pi: R^k \to R^k$ by $\varphi_\pi(x_{\pi 1}, \ldots, x_{\pi k}) = (x_1, \ldots, x_k)$; φ_π applies the inverse permutation to the coordinates. The random vector $(X_{t_1}, \ldots, X_{t_k}) = \varphi_\pi(X_{t_{\pi 1}}, \ldots, X_{t_{\pi k}})$ must have distribution $\mu_{t_1 \ldots t_k}$ on the one hand and distribution $\mu_{t_{\pi 1} \ldots t_{\pi k}} \varphi_\pi^{-1}$ on the other. Therefore,

$$(36.2) \qquad \mu_{t_1 \cdots t_k} = \mu_{t_{\pi 1} \cdots t_{\pi k}} \varphi_\pi^{-1}.$$

For example, if $\mu_{s,t} = \nu \times \nu'$, then necessarily $\mu_{t,s} = \nu' \times \nu$.

The second consistency condition is

$$(36.3) \qquad \mu_{t_1 \ldots t_k}(H) = \mu_{t_1 \ldots t_k t_{k+1}}(H \times R^1), \qquad H \in \mathcal{R}^k.$$

This is clear because $P[(X_{t_1}, \ldots, X_{t_k}) \in H] = P[(X_{t_1}, \ldots, X_{t_k}', X_{t_{k+1}}) \in H \times R^1]$. It can be expressed as $\mu_{t_1 \ldots t_k} = \mu_{t_1 \ldots t_k t_{k+1}} \varphi^{-1}$ if $\varphi(x_1, \ldots, x_k, x_{k+1}) = (x_1, \ldots, x_k)$.

The conditions (36.2) and (36.3) have a common extension. Suppose that $(u_1, \ldots, u_m)$ is an m-tuple of distinct elements of T and that each element of $(t_1, \ldots, t_k)$ is also an element of $(u_1, \ldots, u_m)$. Then $(t_1, \ldots, t_k)$ must be the initial segment of some permutation of $(u_1, \ldots, u_m)$; that is, $k \le m$ and there is a permutation π of $(1, 2, \ldots, m)$ such that $(u_{\pi 1}, \ldots, u_{\pi m}) = (t_1, \ldots, t_k, t_{k+1}, \ldots, t_m)$, where $t_{k+1}, \ldots, t_m$ are the elements of $(u_1, \ldots, u_m)$ that do not appear in $(t_1, \ldots, t_k)$. Define $\psi: R^m \to R^k$ by

$$(36.4) \qquad \psi(x_1, \ldots, x_m) = (x_{\pi 1}, \ldots, x_{\pi k});$$

ψ applies π to the coordinates and projects onto the first k of them. Since $(X_{t_1}, \ldots, X_{t_k}) = (X_{u_{\pi 1}}, \ldots, X_{u_{\pi k}}) = \psi(X_{u_1}, \ldots, X_{u_m})$,

$$(36.5) \qquad \mu_{t_1 \ldots t_k} = \mu_{u_1 \ldots u_m} \psi^{-1}.$$

This contains (36.2) and (36.3) as special cases, but as ψ is a coordinate permutation followed by a sequence of projections of the form $(x_1, \ldots, x_l) \to (x_1, \ldots, x_{l-1})$, it is also a consequence of these special cases.

Measures $\mu_{t_1 \ldots t_k}$ coming from a process via (36.1) necessarily satisfy (36.2) and (36.3). The problem is to show conversely that, if (36.2) and (36.3) hold for a given system of measures, then there exists a process having these finite-dimensional distributions. Proving this theorem is the main objective of the section.

Product Spaces

The standard construction of the general process involves product spaces. Let T be an arbitrary index set and let R^T be the collection of all real functions on T—all maps from T into the real line. If $T = \{1, 2, \ldots, k\}$, a real function on T can be identified with a k-tuple $(x_1, \ldots, x_k)$ of real numbers, and so R^T can be identified with k-dimensional Euclidean space R^k. If $T = \{1, 2, \ldots\}$, a real function on T is a sequence $\{x_1, x_2, \ldots\}$ of real numbers. If T is an interval, R^T consists of all real functions, however irregular, on the interval.

Whatever the set T may be, an element of R^T will be denoted x. The value of x at t will be denoted $x(t)$ or x_t, depending on whether x is viewed as a function of t with domain T or as a vector with components indexed by the elements t of T. Just as R^k can be regarded as the cartesian product of k copies of the real line, R^T can be regarded as a *product space*—a product of copies of the real line, one copy for each t in T.

For each t define a mapping $Z_t: R^T \to R^1$ by

$$(36.6) \qquad\qquad Z_t(x) = x(t) = x_t.$$

The Z_t are called the *coordinate functions* or *projections*. When later on a probability measure has been defined on R^T, the Z_t will be random variables, the *coordinate variables*. Frequently, the value $Z_t(x)$ is instead denoted $Z(t, x)$. If x is fixed, $Z(\cdot, x)$ is a real function on T and is, in fact, nothing other than $x(\cdot)$—that is, x itself. If t is fixed, $Z(t, \cdot)$ is a real function on R^T and is identical with the function Z_t defined by (36.6).

There is a natural generalization to R^T of the idea of the σ-field of k-dimensional Borel sets. Let $\mathcal{R}^T$ be the σ-field generated by all the coordinate functions Z_t, $t \in T$: $\mathcal{R}^T = \sigma[Z_t: t \in T]$. It is generated by the sets of the form

$$[x \in R^T: Z_t(x) \in H] = [x \in R^T: x_t \in H]$$

for $t \in T$ and $H \in \mathcal{R}^1$. If $T = \{1, 2, \ldots, k\}$, then $\mathcal{R}^T$ coincides with $\mathcal{R}^k$.

Consider the class $\mathcal{R}_0^T$ consisting of the sets of the form

$$(36.7) \qquad A = [x \in R^T: (Z_{t_1}(x), \ldots, Z_{t_k}(x)) \in H]$$

$$= [x \in R^T: (x_{t_1}, \ldots, x_{t_k}) \in H],$$

where k is an integer, $(t_1, \ldots, t_k)$ is a k-tuple of distinct points of T, and $H \in \mathcal{R}^k$. Sets of this form, elements of $\mathcal{R}_0^T$, are called *finite-dimensional sets,* or *cylinders.* Of course, $\mathcal{R}_0^T$ generates $\mathcal{R}^T$. Now $\mathcal{R}_0^T$ is not a σ-field, does not coincide with $\mathcal{R}^T$ (unless T is finite), but the following argument shows that it is a field.

The complement of (36.7) is $R^T - A = [x \in R^T: (x_{t_1}, \ldots, x_{t_k}) \in R^k - H]$, and so $\mathcal{R}_0^T$ is closed under complementation. Suppose that A is given by (36.7) and B is given by

$$(36.8) \qquad B = [x \in R^T: (x_{s_1}, \ldots, x_{s_j}) \in I],$$

where $I \in R^j$. Let $(u_1, \ldots, u_m)$ be an m-tuple containing all the t_α and all the s_β. Now $(t_1, \ldots, t_k)$ must be the initial segment of some permutation π of $(u_1, \ldots, u_m)$, and if ψ is as in (36.4) and $H' = \psi^{-1}H$, then $H' \in \mathcal{R}^m$ and A is given

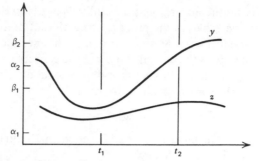

If T is an interval, the cylinder $[x \in R^T: \alpha_1 < x(t_1) \le \beta_1, \alpha_2 < x(t_2) \le \beta_2]$ consists of the functions that go through the two gates shown; y lies in the cylinder and z does not (they need not be continuous functions, of course).

by

(36.9) $$A = [x \in R^T: (x_{u_1}, \ldots, x_{u_m}) \in H']$$

as well as by (36.7). Similarly, B can be put in the form

(36.10) $$B = [x \in R^T: (x_{u_1}, \ldots, x_{u_m}) \in I'],$$

where $I' \in \mathcal{R}^m$. But then

(36.11) $$A \cup B = [x \in R^T: (x_{u_1}, \ldots, x_{u_m}) \in H' \cup I'].$$

Since $H' \cup I' \in \mathcal{R}^m$, $A \cup B$ is a cylinder. This proves that $\mathcal{R}_0^T$ is *a field such that* $\mathcal{R}^T = \sigma(\mathcal{R}_0^T)$.

The Z_t are measurable functions on the measurable space $(R^T, \mathcal{R}^T)$. If P is a probability measure on $\mathcal{R}^T$, then $[Z_t: t \in T]$ is a stochastic process on $(R^T, \mathcal{R}^T, P)$, the *coordinate-variable process*.

Kolmogorov's Existence Theorem

The existence theorem can be stated two ways:

Theorem 36.1. *If* $\mu_{t_1 \ldots t_k}$ *are a system of distributions satisfying the consistency conditions* (36.2) *and* (36.3), *then there is a probability measure P on* $\mathcal{R}^T$ *such that the coordinate-variable process* $[Z_t: t \in T]$ *on* $(R^T, \mathcal{R}^T, P)$ *has the* $\mu_{t_1 \ldots t_k}$ *as its finite-dimensional distributions.*

Theorem 36.2. *If* $\mu_{t_1 \ldots t_k}$ *are a system of distributions satisfying the consistency conditions* (36.2) *and* (36.3), *then there exists on some probability space* $(\Omega, \mathcal{F}, P)$ *a stochastic process* $[X_t: t \in T]$ *having the* $\mu_{t_1 \ldots t_k}$ *as its finite-dimensional distributions.*

For many purposes the underlying probability space is irrelevant, the joint distributions of the variables in the process being all that matters, so that the two theorems are equally useful. As a matter of fact, they are equivalent anyway. Obviously, the first implies the second. To prove the converse, suppose that the process $[X_t: t \in T]$ on $(\Omega, \mathcal{F}, P)$ has finite-dimensional distributions $\mu_{t_1 \ldots t_k}$, and define a map $\xi: \Omega \to R^T$ by the requirement

$$(36.12) \qquad\qquad Z_t(\xi(\omega)) = X_t(\omega), \qquad t \in T.$$

For each ω, $\xi(\omega)$ is an element of R^T, a real function on T, and the requirement is that $X_t(\omega)$ be its value at t. Clearly,

$$\xi^{-1}[x \in R^T: (Z_{t_1}(x), \ldots, Z_{t_k}(x)) \in H]$$

$$(36.13) \qquad = [\omega \in \Omega: (Z_{t_1}(\xi(\omega)), \ldots, Z_{t_k}(\xi(\omega))) \in H]$$

$$= [\omega \in \Omega: (X_{t_1}(\omega), \ldots, X_{t_k}(\omega)) \in H];$$

since the X_t are random variables, measurable $\mathcal{F}$, this set lies in $\mathcal{F}$ if $H \in \mathcal{R}^k$. Thus $\xi^{-1}A \in \mathcal{F}$ for $A \in \mathcal{R}_0^T$, and so (Theorem 13.1) ξ is measurable $\mathcal{F}/\mathcal{R}^T$. By (36.13) and the fact that $[X_t: t \in T]$ has finite-dimensional distributions $\mu_{t_1 \ldots t_k}$, $P\xi^{-1}$ (see (13.7)) satisfies

$$(36.14) \qquad P\xi^{-1}[x \in R^T: (Z_{t_1}(x), \ldots, Z_{t_k}(x)) \in H]$$

$$= P[\omega \in \Omega: (X_{t_1}(\omega), \ldots, X_{t_k}(\omega)) \in H] = \mu_{t_1 \ldots t_k}(H).$$

Thus the coordinate-variable process $[Z_t: t \in T]$ on $(R^T, \mathcal{R}^T, P\xi^{-1})$ also has finite-dimensional distributions $\mu_{t_1 \ldots t_k}$.

Thus to prove either of the two versions of Kolmogorov's existence theorem above is to prove the other one as well.

Example 36.1. Suppose that T is finite, say $T = \{1, 2, \ldots, k\}$. Then $(R^T, \mathcal{R}^T)$ is $(R^k, \mathcal{R}^k)$, and taking $P = \mu_{1,2,\ldots,k}$ satisfies the requirements of Theorem 36.1. ∎

Example 36.2. Suppose that $T = \{1, 2, \ldots\}$ and

$$(36.15) \qquad\qquad \mu_{t_1 \ldots t_k} = \mu_{t_1} \times \cdots \times \mu_{t_k},$$

where $\mu_1, \mu_2, \ldots$ are distributions on the line. The consistency conditions are easily checked. By Theorem 20.4 there exists on some $(\Omega, \mathcal{F}, P)$ an independent sequence $X_1, X_2, \ldots$ of random variables with respective distributions $\mu_1, \mu_2, \ldots$. But then (36.15) is the distribution of $(X_{t_1}, \ldots, X_{t_k})$. For the special case (36.15), Theorem 36.2 was thus proved in Section 20. The existence of independent sequences with prescribed distributions was, in fact, the measure-theoretic basis of all the probabilistic developments of Chapters 4, 5, and 6—even dependent processes like the Poisson and queueing processes were constructed from independent sequences. The existence of independent sequences can also be made the basis of a proof of Theorem 36.2 in its full generality; see the second proof below.   ∎

Two proofs of Kolmogorov's existence theorem will be given. The first is based on the extension theorem of Section 3.

FIRST PROOF OF KOLMOGOROV'S THEOREM.   Consider the first formulation, Theorem 36.1. If A is the cylinder (36.7), define

$$(36.16) \qquad\qquad P(A) = \mu_{t_1\ldots t_k}(H).$$

This gives rise to the question of consistency because A will have other representations as a cylinder. Suppose, in fact, that A coincides with the cylinder B defined by (36.8). As observed before, if $(u_1, \ldots, u_m)$ contains all the t_α and s_β, A is also given by (36.9), where $H' = \psi^{-1}H$ and ψ is defined in (36.4). Since the consistency conditions (36.2) and (36.3) imply the more general one (36.5), $P(A) = \mu_{t_1\ldots t_k}(H) = \mu_{u_1\ldots u_m}(H')$. Similarly, (36.8) has the form (36.10), and $P(B) = \mu_{s_1\ldots s_j}(I) = \mu_{u_1\ldots u_m}(I')$. Since the u_γ are distinct, for any real numbers $z_1, \ldots, z_m$ there are points x of R^T for which $(x_{u_1}, \ldots, x_{u_m}) = (z_1 \ldots, z_m)$. From this it follows that if the cylinders (36.9) and (36.10) coincide, then $H' = I'$. Hence $A = B$ implies that $P(A) = \mu_{u_1\ldots u_m}(H') = \mu_{u_1\ldots u_m}(I') = P(B)$, and the definition (36.16) is indeed consistent.

Now consider disjoint cylinders A and B. As usual, the index sets may be taken identical. Assume then that A is given by (36.9) and B by (36.10), so that (36.11) holds. If $H' \cap I'$ were nonempty, then $A \cap B$ would be nonempty as well. Therefore, $H' \cap I' = 0$, and

$$P(A \cup B) = \mu_{u_1\ldots u_m}(H' \cup I')$$
$$= \mu_{u_1\ldots u_m}(H') + \mu_{u_1\ldots u_m}(I') = P(A) + P(B).$$

Therefore, P is finitely additive on $\mathcal{R}_0^T$. Clearly, $P(R^T) = 1$.

Suppose that P can be shown to be countably additive on $\mathcal{R}_0^T$. By Theorem 3.1, P will then extend to a probability measure on $\mathcal{R}^T$. By the way P was defined on $\mathcal{R}_0^T$,

$$P[x \in R^T: (Z_{t_1}(x), \ldots, Z_{t_k}(x)) \in H] = \mu_{t_1\ldots t_k}(H),$$

and therefore the coordinate process $[Z_t: t \in T]$ will have the required finite-dimensional distributions.

It suffices, then, to prove P countably additive on $\mathcal{R}_0^T$, and this will follow if $A_n \in \mathcal{R}_0^T$ and $A_n \downarrow 0$ together imply $P(A_n) \downarrow 0$ (see Example 2.10). Suppose that $A_1 \supset A_2 \supset \cdots$ and that $P(A_n) \geq \epsilon > 0$ for all n. The problem is to show that $\cap_n A_n$ must be nonempty. Since $A_n \in \mathcal{R}_0^T$, and since the index set involved in the specification of a cylinder can always be permuted and expanded, there exists a sequence $t_1, t_2, \ldots$ of points in T for which

$$A_n = [x \in R^T: (x_{t_1}, \ldots, x_{t_n}) \in H_n],$$

where* $H_n \in \mathcal{R}^n$.

Of course, $P(A_n) = \mu_{t_1 \ldots t_n}(H_n)$. By Theorem 12.3 (regularity), there exists inside H_n a compact set K_n such that $\mu_{t_1 \ldots t_n}(H_n - K_n) < \epsilon/2^{n+1}$. If $B_n = [x \in R^T: (x_{t_1}, \ldots, x_{t_n}) \in K_n]$, then $P(A_n - B_n) < \epsilon/2^{n+1}$. Put $C_n = \cap_{k=1}^n B_k$. Then $C_n \subset B_n \subset A_n$ and $P(A_n - C_n) < \epsilon/2$, so that $P(C_n) > \epsilon/2 > 0$. Therefore, C_n is nonempty.

Choose a point $x^{(n)}$ of R^T in C_n. If $n \geq k$, then $x^{(n)} \in C_n \subset C_k \subset B_k$ and hence $(x_{t_1}^{(n)}, \ldots, x_{t_k}^{(n)}) \in K_k$. Since K_k is bounded, the sequence $\{x_{t_k}^{(1)}, x_{t_k}^{(2)}, \ldots\}$ is bounded for each k. By the diagonal method (Theorem 25.13) select an increasing sequence $n_1, n_2, \ldots$ of integers such that $\lim_i x_{t_k}^{(n_i)}$ exists for each k. There is in R^T some point x whose t_kth coordinate is this limit for each k. But then, for each k, $(x_{t_1}, \ldots x_{t_k})$ is the limit as $i \to \infty$ of $(x_{t_1}^{(n_i)}, \ldots, x_{t_k}^{(n_i)})$ and hence lies in K_k. But that means that x itself lies in B_k and hence in A_k. Thus $x \in \cap_{k=1}^\infty A_k$, which completes the proof.[†] ∎

The second proof[‡] of Kolmogorov's theorem goes in two stages, first for countable T, then for general T.

SECOND PROOF FOR COUNTABLE T. The result for countable T will be proved in its second formulation, Theorem 36.2. It is no restriction to enumerate T as $\{t_1, t_2, \ldots\}$ and then to identify t_n with n; in other words, it is no restriction to assume that $T = \{1, 2, \ldots\}$. Write μ_n in place of $\mu_{1,2,\ldots,n}$.

By Theorem 20.4 there exists on a probability space $(\Omega, \mathcal{F}, P)$ (which can be taken to be the unit interval) an independent sequence $U_1, U_2, \ldots$ of random variables each uniformly distributed over $(0, 1)$. Let F_1 be the distribution function corresponding to μ_1. If the "inverse" g_1 of F_1 is defined over $(0, 1)$ by $g_1(s) = \inf[x: s \leq F_1(x)]$, then $X_1 = g_1(U_1)$ has distribution μ_1 by the usual argument: $P[g_1(U_1) \leq x] = P[U_1 \leq F_1(x)] = F_1(x)$.

The problem is to construct $X_2, X_3, \ldots$ inductively in such a way that

$$(36.17) \qquad X_k = h_k(U_1, \ldots, U_k)$$

for a Borel function h_k and $(X_1, \ldots, X_n)$ has the distribution μ_n. Assume that $X_1, \ldots, X_{n-1}$ have been defined ($n \geq 2$): they have joint distribution μ_{n-1} and (36.17) holds for $k \leq n - 1$. The idea now is to construct an appropriate conditional distribution function $F_n(x|x_1, \ldots, x_n)$; here $F_n(x|X_1(\omega), \ldots, X_{n-1}(\omega))$ will have the value $P[X_n \leq x \| X_1, \ldots, X_{n-1}]_\omega$ would have if X_n were already defined. If $g_n(\cdot|x_1, \ldots, x_{n-1})$ is the "inverse" function, then $X_n(\omega) = g_n(U_n(\omega)|X_1(\omega), \ldots, X_{n-1}(\omega))$ will by the usual argument have

---

* In general, A_n will involve indices $t_1, \ldots, t_{a_n}$ satisfying $a_1 < a_2 < \cdots$. For notational simplicity a_n is taken as n; as a matter of fact, this can be arranged: start off by repeating Ω $a_1 - 1$ times and then for $n \geq 1$ repeat A_n $a_{n+1} - a_n$ times.

† The last part of the argument is, in effect, the proof that a countable product of compact sets is compact.

‡ This second proof, which may be omitted, uses the conditional probability theory of Section 33.

the right conditional distribution given $X_1, \ldots, X_{n-1}$, so that $(X_1, \ldots, X_{n-1}, X_n)$ will have the right distribution over R^n.

To construct the conditional distribution function, apply Theorem 33.3 in $(R^n, \mathcal{R}^n, \mu_n)$ to get a conditional distribution of the last coordinate of $(x_1, \ldots, x_n)$ given the first $n-1$ of them. This will have (Theorem 20.1) the form $\nu(H; x_1, \ldots, x_{n-1})$; it is a probability measure as H varies over $\mathcal{R}^1$, and

$$\int_{(x_1, \ldots, x_{n-1}) \in M} \nu(H; x_1, \ldots, x_{n-1}) \, d\mu_n(x_1, \ldots, x_n)$$

$$= \mu_n[x \in R^n: (x_1, \ldots, x_{n-1}) \in M, x_n \in H].$$

Since the integrand involves only $x_1, \ldots, x_{n-1}$ and μ_n by consistency projects to μ_{n-1} under the map $(x_1, \ldots, x_n) \to (x_1, \ldots, x_{n-1})$, a change of variable gives

$$\int_M \nu(H; x_1, \ldots, x_{n-1}) \, d\mu_{n-1}(x_1, \ldots, x_{n-1})$$

$$= \mu_n[x \in R^n: (x_1, \ldots, x_{n-1}) \in M, x_n \in H].$$

Define $F_n(x | x_1, \ldots, x_{n-1}) = \nu((-\infty, x]; x_1, \ldots, x_{n-1})$. Then $F_n(\cdot | x_1, \ldots, x_{n-1})$ is a probability distribution function over the line, $F_n(x | \cdot)$ is a Borel function over R^{n-1}, and

$$\int_M F_n(x | x_1, \ldots, x_{n-1}) \, d\mu_{n-1}(x_1, \ldots, x_{n-1})$$

$$= \mu_n[x \in R^n: (x_1, \ldots, x_{n-1}) \in M, x_n \leq x].$$

Put $g_n(u | x_1, \ldots, x_{n-1}) = \inf[x: u \leq F_n(x | x_1, \ldots, x_{n-1})]$ for $0 < u < 1$. Since $F_n(x | x_1, \ldots, x_{n-1})$ is nondecreasing and right-continuous in x, $g_n(u | x_1, \ldots, x_{n-1}) \leq x$ if and only if $u \leq F_n(x | x_1, \ldots, x_{n-1})$. Set $X_n = g_n(U_n | X_1, \ldots, X_{n-1})$. Since $(X_1, \ldots, X_{n-1})$ has distribution μ_{n-1} and by (36.17) is independent of U_n, an application of (20.30) gives

$$P[(X_1, \ldots, X_{n-1}) \in M, X_n \leq x]$$

$$= P[(X_1, \ldots, X_{n-1}) \in M, U_n \leq F_n(x | X_1, \ldots, X_n)]$$

$$= \int_M P[U_n \leq F_n(x | x_1, \ldots, x_{n-1})] d\mu_{n-1}(x_1, \ldots, x_{n-1})$$

$$= \int_M F_n(x | x_1, \ldots, x_{n-1}) d\mu_{n-1}(x_1, \ldots, x_{n-1})$$

$$= \mu_n[x \in R^n: (x_1, \ldots, x_{n-1}) \in M, x_n \leq x].$$

Thus $(X_1, \ldots, X_n)$ has distribution μ_n. Note that X_n, as a function of $X_1, \ldots, X_{n-1}$ and U_n, is a function of $U_1, \ldots, U_n$ because (36.17) was assumed to hold for $k < n$. Hence (36.17) holds for $k = n$ as well. ∎

SECOND PROOF FOR GENERAL T. Consider $(R^T, \mathcal{R}^T)$ once again. If $S \subset T$, let $\mathcal{F}_S = \sigma[Z_t: t \in S]$. Then $\mathcal{F}_S \subset \mathcal{F}_T = \mathcal{R}^T$.

Suppose that S is countable. By the case just treated, there exists a process $[X_t: t \in S]$ on some $(\Omega, \mathcal{F}, P)$—the process depends on S—such that $(X_{t_1}, \ldots, X_{t_k})$ has distribution $\mu_{t_1 \ldots t_k}$ for every k-tuple $(t_1, \ldots, t_k)$ from S. Define a map $\xi: \Omega \to R^T$ by requiring that

$$Z_t(\xi(\omega)) = \begin{cases} X_t(\omega) & \text{if } t \in S, \\ 0 & \text{if } t \notin S. \end{cases}$$

Now (36.13) holds as before if $t_1, \ldots, t_k$ all lie in S, and so ξ is measurable $\mathcal{F}/\mathcal{F}_S$. Further, (36.14) holds for $t_1, \ldots, t_k$ in S. Put $P_S = P\xi^{-1}$ on $\mathcal{F}_S$. Then P_S is a probability measure on $(R^T, \mathcal{F}_S)$, and

$$(36.18) \qquad P_S[x \in R^T: (Z_{t_1}(x), \ldots, Z_{t_k}(x)) \in H] = \mu_{t_1 \ldots t_k}(H)$$

if $H \in \mathcal{R}^k$ and $t_1, \ldots, t_k$ all lie in S. (The various spaces $(\Omega, \mathcal{F}, P)$ and processes $[X_t: t \in S]$ now become irrelevant.)

If $S_0 \subset S$, and if A is a cylinder (36.7) for which the $t_1, \ldots, t_k$ lie in S_0, then $P_{S_0}(A)$ and $P_S(A)$ coincide, their common value being $\mu_{t_1 \ldots t_k}(H)$. Since these cylinders generate $\mathcal{F}_{S_0}$, $P_{S_0}(A) = P_S(A)$ for all A in $\mathcal{F}_{S_0}$. If A lies both in $\mathcal{F}_{S_1}$ and $\mathcal{F}_{S_2}$, then $P_{S_1}(A) = P_{S_1 \cup S_2}(A) = P_{S_2}(A)$. Thus $P(A) = P_S(A)$ consistently defines a set function on the class $\bigcup_S \mathcal{F}_S$, the union extending over the countable subsets S of T. If A_n lies in this union and $A_n \in \mathcal{F}_{S_n}$ (S_n countable), then $S = \bigcup_n S_n$ is countable and $\bigcup_n A_n$ lies in $\mathcal{F}_S$. Thus $\bigcup_S \mathcal{F}_S$ is a σ-field and so must coincide with $\mathcal{R}^T$. Therefore, P is a probability measure on $\mathcal{R}^T$, and by (36.18) the coordinate process has under P the required finite-dimensional distributions. ∎

The Inadequacy of $\mathcal{R}^T$

Theorem 36.3. *Let $[X_t: t \in T]$ be a family of real functions on Ω.*

(i) *If $A \in \sigma[X_t: t \in T]$ and $\omega \in A$, and if $X_t(\omega) = X_t(\omega')$ for all $t \in T$, then $\omega' \in A$.*

(ii) *If $A \in \sigma[X_t: t \in T]$, then $A \in \sigma[X_t: t \in S]$ for some countable subset S of T.*

PROOF. Define $\xi: \Omega \to R^T$ by $Z_t(\xi(\omega)) = X_t(\omega)$. Let $\mathcal{F} = \sigma[X_t: t \in T]$. By (36.13), ξ is measurable $\mathcal{F}/\mathcal{R}^T$ and hence $\mathcal{F}$ contains the class $[\xi^{-1}M: M \in \mathcal{R}^T]$. The latter class is a σ-field, however, and by (36.13) it contains the sets $[\omega \in \Omega: (X_{t_1}(\omega), \ldots, X_{t_k}(\omega)) \in H]$, $H \in \mathcal{R}^k$, and hence contains the σ-field $\mathcal{F}$ they generate. Therefore

$$(36.19) \qquad \sigma[X_t: t \in T] = [\xi^{-1}M: M \in \mathcal{R}^T].$$

This is an infinite-dimensional analogue of Theorem 20.1(i).

As for (i), the hypotheses imply that $\omega \in A = \xi^{-1}M$ and $\xi(\omega) = \xi(\omega')$, so that $\omega' \in A$ certainly follows.

For $S \subset T$, let $\mathcal{F}_S = [X_t: t \in S]$; (ii) says that $\mathcal{F} = \mathcal{F}_T$ coincides with $\mathcal{G} = \bigcup_S \mathcal{F}_S$, the union extending over the countable subsets S of T. If $A_1, A_2, \ldots$ lie in $\mathcal{G}$, A_n lies in $\mathcal{F}_{S_n}$ for some countable S_n, and so $\bigcup_n A_n$ lies in $\mathcal{G}$ because it lies in $\mathcal{F}_S$ for $S = \bigcup_n S_n$. Thus $\mathcal{G}$ is a σ-field, and since it contains the sets $[X_t \in H]$, it contains the σ-field $\mathcal{F}$ they generate. (This part of the argument was used in the second proof of the existence theorem.) ∎

From this theorem it follows that various important sets lie outside the class $\mathcal{R}^T$. Suppose that $T = [0, \infty)$. Of obvious interest is the subset C of R^T consisting of the functions continuous over $[0, \infty)$. But C is not in $\mathcal{R}^T$. For suppose it were.

By part (ii) of the theorem (let $\Omega = R^T$ and put $[Z_t: t \in T]$ in the role of $[X_t: t \in T)]$, C would lie in $\sigma[Z_t: t \in S]$ for some countable $S \subset [0, \infty)$. But then by part (i) of the theorem (let $\Omega = R^T$ and put $[Z_t: t \in S]$ in the role of $[X_t: t \in T)]$, if $x \in C$ and $Z_t(x) = Z_t(y)$ for all $t \in S$, then $y \in C$. From the assumption that C lies in $\mathcal{R}^T$ thus follows the existence of a countable set S such that, if $x \in C$ and $x(t) = y(t)$ for all t in S, then $y \in C$. But whatever countable set S may be, for every continuous x there obviously exist functions y which have discontinuities but which agree with x on S. Therefore, C cannot lie in $\mathcal{R}^T$.

What the argument shows is this: A set A in R^T cannot lie in $\mathcal{R}^T$ unless there exists a countable subset S of T with the property that, if $x \in A$ and $x(t) = y(t)$ for all t in S, then $y \in A$. Thus A cannot lie in $\mathcal{R}^T$ if it effectively involves all the points t in the sense that, for each x in A and each t in T, it is possible to move x out of A by changing its value at t alone. And C is such a set. For another, consider the set of functions x over $T = [0, \infty)$ that are nondecreasing and assume as values $x(t)$ only nonnegative integers:

(36.20) $\quad [x \in R^{[0,\infty)}: x(s) \leq x(t),\ s \leq t;\ x(t) \in \{0, 1, \ldots\},\ t \geq 0]$.

This, too, lies outside $\mathcal{R}^T$.

In Section 23 the Poisson process was defined as follows: Let $X_1, X_2, \ldots$ be independent and identically distributed with the exponential distribution (the probability space Ω on which they are defined may by Theorem 20.4 be taken to be the unit interval with Lebesgue measure). Put $S_0 = 0$ and $S_n = X_1 + \cdots + X_n$. If $S_n(\omega) < S_{n+1}(\omega)$ for $n \geq 0$ and $S_n(\omega) \to \infty$, put $N(t, \omega) = N_t(\omega) = \max[n: S_n(\omega) \leq t]$ for $t \geq 0$; otherwise, put $N(t, \omega) = N_t(\omega) = 0$ for $t \geq 0$. Then the stochastic process $[N_t: t \geq 0]$ has the finite-dimensional distributions described by the equations (23.20). The function $N(\cdot, \omega)$ is the *path function* or *sample function**** corresponding to ω, and by the construction every path function lies in the set (36.20). This is a good thing if the process is to be a model for, say, calls arriving at a telephone exchange: The sample path represents the history of the calls, its value at t being the number of arrivals up to time t, and so it ought to be nondecreasing and integer-valued.

According to Theorem 36.1, there exists a measure P on R^T for $T = [0, \infty)$ such that the coordinate process $[Z_t: t \geq 0]$ on $(R^T, \mathcal{R}^T, P)$ has the finite-dimensional distributions of the Poisson process. This time does the path function $Z(\cdot, x)$ lie in the set (36.20) with probability 1? Since $Z(\cdot, x)$ is just x itself, the question is whether the set (36.20) has P-measure 1. But this set does not lie in $\mathcal{R}^T$, and so it has no measure at all.

An application of Theorem 36.1 will always yield a stochastic process with prescribed finite-dimensional distributions, but the process may lack certain path-function properties which it is reasonable to require of it as a model for some natural phenomenon. The special construction of Section 23 gets around this

---

* Other terms are *realization* of the process and *trajectory*.

difficulty for the Poisson process, and in the next section a special construction will yield a model for Brownian motion with continuous paths. Section 38 treats a general method for producing stochastic processes that have prescribed finite-dimensional distributions and at the same time have path functions with desirable regularity properties.

PROBLEMS

36.1. 13.6↑   Generalize Theorem 20.1(ii), replacing $(X_1, \ldots, X_k)$ by $[X_t : t \in T]$ for an arbitrary T.

36.2. A process $(\ldots, X_{-1}, X_0, X_1, \ldots)$ (here T is the set of all integers) is *stationary* if the distribution of $(X_n, X_{n+1}, \ldots, X_{n+k-1})$ over R^k is the same for all $n = 0, \pm 1, \pm 2, \ldots$. Define $\tau \colon R^T \to R^T$ by $Z_n(\tau x) = Z_{n+1}(x)$; thus τ moves a doubly infinite sequence (that is, an element of R^T) one place left: $\tau(\ldots, x_{-1}, x_0, x_1, \ldots) = (\ldots, x_0, x_1, x_2, \ldots)$. Show that τ is measurable $\mathcal{R}^T / \mathcal{R}^T$ and show that the coordinate process $(\ldots, Z_{-1}, Z_0, Z_1, \ldots)$ on $(R^T, \mathcal{R}^T, P)$ is stationary if and only if τ preserves the measure P in the sense that $P\tau^{-1} = P$.

36.3. Show that, if X is measurable $\sigma[X_t : t \in T]$, then X is measurable $\sigma[X_t : t \in S]$ for some countable subset S of T.

36.4. Suppose that $[X_t : t \in T]$ is a stochastic process on $(\Omega, \mathcal{F}, P)$ and $A \in \mathcal{F}$. Show that there is a countable subset S of T for which $P[A \| X_t, t \in T] = P[A \| X_t, t \in S]$ with probability 1. Replace A by a random variable and prove a similar result.

36.5. Let T be arbitrary and let $K(s, t)$ be a real function over $T \times T$. Suppose that K is symmetric in the sense that $K(s, t) = K(t, s)$ and nonnegative definite in the sense that $\sum_{i,j=1}^k K(t_i, t_j) x_i x_j \geq 0$ for $k \geq 1, t_1, \ldots, t_k$ in $T, x_1, \ldots, x_k$ real. Show that there exists a process $[X_t : t \in T]$ for which $(X_{t_1}, \ldots, X_{t_k})$ has the centered normal distribution with covariances $K(t_i, t_j), i, j = 1, \ldots, k$.

36.6. 8.4↑   Suppose that $p_n(u_1, \ldots, u_n)$ is a nonnegative real for each n and each n-long sequence $u_1, \ldots, u_n$ of 0's and 1's. Suppose that $p_1(0) + p_1(1) = 1$ and $p_{n+1}(u_1, \ldots, u_n, 0) + p_{n+1}(u_1, \ldots, u_n, 1) = p_n(u_1, \ldots, u_n)$. Prove that on the σ-field $\mathcal{C}$ generated by the cylinders in sequence space there exists a probability measure P such that $P[\omega : a_i(\omega) = u_i, i \leq n] = p_n(u_1, \ldots, u_n)$. Problems 2.16, 4.18, and 8.4 cover the coin-tossing, independent, and Markov cases, respectively.

36.7. ↑   Let L be a Borel set on the line, and let L^T consist of all maps from T into L. Define the appropriate notion of cylinder, and let $\mathcal{L}^T$ be the σ-field generated by the cylinders. State a version of Theorem 36.1 for $(L^T, \mathcal{L}^T)$. Assume T countable and prove this theorem not by imitating the previous proof but by observing that L^T is a subset of R^T and lies in $\mathcal{R}^T$. If L consists of 0 and 1, and if T is the set of positive integers, then L^T is the space considered in Problem 36.6.

36.8.   Suppose that the random variables $X_1, X_2, \ldots$ assume the values 0 and 1 and $P[X_n = 1 \text{ i.o.}] = 1$. Let μ be the distribution over $(0, 1]$ of $\Sigma_{n=1}^{\infty} X_n/2^n$. Show that on the unit interval with the measure μ the digits of the nonterminating dyadic expansion form a stochastic process with the same finite-dimensional distributions as $X_1, X_2, \ldots$.

36.9.   36.7↑   There is an infinite-dimensional version of Fubini's theorem. Let $L = I = (0, 1)$ and $T = \{1, 2, \ldots\}$ in the construction in Problem 36.7, and suppose that each k-dimensional distribution is the k-fold product of Lebesgue measure over the unit interval. Then I^T is a countable product of copies of $(0, 1)$, its elements are sequences $x = (x_1, x_2, \ldots)$ of points of $(0, 1)$, and Kolmogorov's theorem ensures the existence on $(I^T, \mathcal{I}^T)$ of a *product* probability measure $\pi \colon \pi[x \colon x_i \le \alpha_i, i \le n] = \alpha_1 \cdots \alpha_n$ for $0 \le \alpha_i \le 1$. Let I^n denote the n-dimensional unit cube.

(a)   Define $\psi \colon I^n \times I^T \to I^T$ by $\psi((x_1, \ldots, x_n), (y_1, y_2, \ldots)) = (x_1, \ldots, x_n, y_1, y_2, \ldots)$. Show that ψ is measurable $\mathcal{I}^n \times \mathcal{I}^T / \mathcal{I}^T$ and ψ^{-1} is measurable $\mathcal{I}^T / \mathcal{I}^n \times \mathcal{I}^T$. Show that $\psi^{-1}(\lambda_n \times \pi) = \pi$, where λ_n is n-dimensional Lebesgue measure restricted to I^n.

(b)   Let f be a function measurable $\mathcal{I}^T$ and, for simplicity, bounded. Define

$$f_n(x_{n+1}, x_{n+2}, \ldots) = \int_0^1 \cdots \int_0^1 f(y_1, \ldots, y_n, x_{n+1}, \ldots) \, dy_1 \cdots dy_n;$$

in other words, integrate out the coordinates one by one. Show by Problem 34.18, martingale theory, and the zero-one law that

$$(36.21) \qquad f_n(x_{n+1}, x_{n+2}, \ldots) \to \int_{I^T} f(y) \pi(dy)$$

except for x in a set of π-measure 0.

(c)   Adopting the point of view of part (a), let $g_n(x_1, \ldots, x_n)$ be the result of integrating $(y_{n+1}, y_{n+2}, \ldots)$ out (with respect to π) from $f(x_1, \ldots, x_n, y_{n+1}, \ldots)$. This may suggestively be written as

$$g_n(x_1, \ldots, x_n)$$
$$= \int_0^1 \int_0^1 \cdots f(x_1, \ldots, x_n, y_{n+1}, y_{n+2}, \ldots) \, dy_{n+1} \, dy_{n+2} \cdots.$$

Show that $g_n(x_1, \ldots, x_n) \to f(x_1, x_2, \ldots)$ except for x in a set of π-measure 0.

36.10.   **(a)**   Let T be an interval of the line. Show that $\mathcal{R}^T$ fails to contain the sets of: linear functions, polynomials, constants, nondecreasing functions, functions of bounded variation, differentiable functions, analytic functions, functions continuous at a fixed t_0, Borel measurable functions. Show that it fails to contain the set of functions that: vanish somewhere in T, satisfy $x(s) < x(t)$ for some pair with $s < t$, have a local maximum anywhere, fail to have a local maximum.

(b)   Let C be the set of continuous functions on $T = [0, \infty)$. Show that $A \in \mathcal{R}^T$ and $A \subset C$ imply that $A = 0$. Show, on the other hand, that $A \in \mathcal{R}^T$ and $C \subset A$ do not imply that $A = R^T$.

36.11. Not all systems of finite-dimensional distributions can be realized by stochastic processes for which Ω is the unit interval. Show that there is on the unit interval with Lebesgue measure no process $[X_t: t \geq 0]$ for which the X_t are independent and assume the values 0 and 1 with probability $\frac{1}{2}$ each. Compare Problem 1.1.

36.12. By Kolmogorov's existence theorem there is a process $[N_t: t \geq 0]$ having the finite-dimensional distributions (23.20) specified for the Poisson process. As pointed out at the end of this section and in Section 23, the sample paths may be very irregular. Let D be the set of dyadic rationals, and let A be the ω-set where the sample path when restricted to D (i) satisfies $N_0(\omega) = 0$, (ii) is non-decreasing, (iii) assumes only nonnegative integers as values, and (iv) has for each t in D the property that $\inf_{t \leq s \in D} N_s(\omega) \leq N_t(\omega) + 1$. Show that A is measurable and $P(A) = 1$. For $\omega \in A$ define $N_t'(\omega) = \inf_{t \leq s \in D} N_s(\omega)$; for $\omega \notin A$ define $N_t'(\omega) = 0$. Show that $[N_t': t \geq 0]$ is a Poisson process each of whose paths is integer-valued and nondecreasing and has no jump exceeding 1.

This kind of argument is used in the next section to construct a Brownian motion with continuous paths.

36.13. Finite-dimensional distributions can be specified by conditional distributions. Suppose that ν is a probability measure on the line, and suppose that for $n \geq 2$, $\nu_n(H; x_1, \ldots, x_{n-1})$ is a probability measure as H varies over $\mathcal{R}^1$ and is a Borel function as $(x_1, \ldots, x_{n-1})$ varies over R^{n-1}. Show that there exists a stochastic process $\{X_1, X_2, \ldots\}$ such that X_1 has distribution ν and $P[X_n \in H \| X_1, \ldots, X_{n-1}]_\omega = \nu_n(H; X_1(\omega), \ldots, X_{n-1}(\omega))$. Show that $E[f(X_n)\|X_1, \ldots, X_{n-1}]_\omega = \int_{-\infty}^{\infty} f(x)\nu_n(dx; X_1(\omega), \ldots, X_{n-1}(\omega))$ if $f(X_n)$ is integrable.

SECTION 37.   BROWNIAN MOTION

Definition

A *Brownian motion* or *Wiener process* is a stochastic process $[W_t: t \geq 0]$, on some $(\Omega, \mathcal{F}, P)$, with these three properties:

(i)   *The process starts at 0:*

(37.1) $P[W_0 = 0] = 1.$

(ii)   *The increments are independent: If*

(37.2) $0 \leq t_0 \leq t_1 \leq \cdots \leq t_k,$

then

(37.3) $P[W_{t_i} - W_{t_{i-1}} \in H_i, i \leq k] = \prod_{i \leq k} P[W_{t_i} - W_{t_{i-1}} \in H_i].$

(iii)   *For $0 \leq s < t$ the increment $W_t - W_s$ is normally distributed with mean 0 and variance $t - s$:*

$$(37.4) \qquad P[W_t - W_s \in H] = \frac{1}{\sqrt{2\pi(t-s)}} \int_H e^{-x^2/2(t-s)} \, dx.$$

The existence of such processes will be proved presently.

Imagine suspended in a fluid a particle bombarded by molecules in thermal motion. The particle will perform a seemingly random movement first described by the nineteenth-century botanist Robert Brown. Consider a single component of this motion—imagine it projected on a vertical axis—and by W_t denote the height at time t of the particle above a fixed horizontal plane. Condition (i) is merely a convention: the particle starts at 0. Condition (ii) reflects a kind of lack of memory. The displacements $W_{t_1} - W_{t_0}, \ldots, W_{t_{k-1}} - W_{t_{k-2}}$ the particle undergoes during the intervals $[t_0, t_1], \ldots, [t_{k-2}, t_{k-1}]$ in no way influence the displacement $W_{t_k} - W_{t_{k-1}}$ it undergoes during $[t_{k-1}, t_k]$. Although the future behavior of the particle depends on its present position, it does not depend on how the particle got there. As for (iii), that $W_t - W_s$ has mean 0 reflects the fact that the particle is as likely to go up as to go down—there is no drift. The variance grows as the length of the interval $[s, t]$; the particle tends to wander away from its position at time s, and having done so suffers no force tending to restore it to that position. To Norbert Wiener are due the mathematical foundations of the theory of such motion.

The increments of the Brownian motion process are *stationary* in the sense that the distribution of $W_t - W_s$ depends only on the difference $t - s$. Since $W_0 = 0$, the distribution of these increments is described by saying that W_t is normally distributed with mean 0 and variance t. This implies (37.1). If $0 \le s \le t$, then by the independence of the increments $E[W_s W_t] = E[(W_s(W_t - W_s)] + E[W_s^2] = E[W_s]E[W_t - W_s] + E[W_s^2] = s$. This specifies all the means, variances, and covariances:

$$(37.5) \qquad E[W_t] = 0, \qquad E[W_t^2] = t, \qquad E[W_s W_t] = \min\{s, t\}.$$

If $0 < t_1 < \cdots < t_k$, the joint density of $(W_{t_1}, W_{t_2} - W_{t_1}, \ldots, W_{t_k} - W_{t_{k-1}})$ is by (20.25) the product of the corresponding normal densities. By the Jacobian formula (20.20), $(W_{t_1}, \ldots, W_{t_k})$ has density

$$(37.6) \quad f_{t_1 \cdots t_k}(x_1, \ldots, x_k) = \prod_{i=1}^{k} \frac{1}{\sqrt{2\pi(t_i - t_{i-1})}} \exp\left[-\frac{(x_i - x_{i-1})^2}{2(t_i - t_{i-1})} \right],$$

where $t_0 = x_0 = 0$.

Sometimes W_t will be denoted $W(t)$, and its value at ω will be $W(t, \omega)$. The nature of the path functions $W(\cdot, \omega)$ will be of great importance.

The existence of the Brownian motion process follows from Kolmogorov's theorem. For $0 < t_1 < \cdots < t_k$ let $\mu_{t_1 \cdots t_k}$ be the distribution in R^k with density (37.6). To put it another way, let $\mu_{t_1 \cdots t_k}$ be the distribution of $(S_1, \ldots, S_k)$, where $S_i = X_1 + \cdots + X_i$ and where $X_1, \ldots, X_k$ are independent, normally distributed random variables with mean 0 and variances $t_1, t_2 - t_1, \ldots, t_k - t_{k-1}$. If $g(x_1, \ldots, x_k) = (x_1, \ldots, x_{i-1}, x_{i+1}, \ldots, x_k)$, then $g(S_1, \ldots, S_k) = (S_1, \ldots, S_{i-1}, S_{i+1}, \ldots, S_k)$ has the distribution prescribed for $\mu_{t_1 \ldots t_{i-1} t_{i+1} \ldots t_k}$. This is because $X_i + X_{i+1}$ is normally distributed with mean 0 and variance $t_{i+1} - t_{i-1}$; see Example 20.6. Therefore,

$$(37.7) \qquad \mu_{t_1 \ldots t_{i-1} t_{i+1} \ldots t_k} = \mu_{t_1 \ldots t_k} g^{-1}.$$

If $t_1, \ldots, t_k$ are not in increasing order, take $\mu_{t_1 \ldots t_k} = \mu_{t_{\pi 1} \ldots t_{\pi k}} \varphi_\pi^{-1}$, where π is the permutation that puts them in increasing order and $\varphi_\pi(x_{\pi 1}, \ldots, x_{\pi k}) = (x_1, \ldots, x_k)$. This of course guarantees (36.2), the first of Kolmogorov's consistency conditions. And (36.3), the second condition, is an immediate consequence of (37.7).*

By Kolmogorov's theorem there does exist a process $[W_t: t > 0]$ corresponding to the $\mu_{t_1 \ldots t_k}$. Taking $W_t = 0$ for $t = 0$ shows that there exists on some $(\Omega, \mathcal{F}, P)$ a process $[W_t: \geq 0]$ with the finite-dimensional distributions specified by the conditions (i), (ii), and (iii).

Continuity of Paths

If the Brownian motion process is to represent the motion of a particle, it is natural to require that the path functions $W(\cdot, \omega)$ be continuous. But Kolmogorov's theorem does not guarantee continuity. Indeed, for $T = [0, \infty)$ the space $(\Omega, \mathcal{F})$ in the proof of Kolmogorov's theorem is $(R^T, \mathcal{R}^T)$, and as shown in the last section, the set of continuous functions does not lie in $\mathcal{R}^T$.

A special construction gets around this difficulty. The idea is to use for dyadic rational t the random variables W_t as already defined and then to redefine the other W_t in such a way as to ensure continuity. To carry this through requires proving that with probability 1 the sample path is uniformly continuous for dyadic rational arguments in bounded intervals.

Fix a space $(\Omega, \mathcal{F}, P)$ and on it a process $[W_t: t \geq 0]$ having the finite-dimensional distributions prescribed for Brownian motion. Let D be the set of nonnegative dyadic rationals, let $I_{nk} = [k2^{-n}, (k + 1)2^{-n}]$, and put

$$(37.8) \quad B_n = \left[\omega: \max_{0 \leq k < n2^n} \sup_{r \in I_{nk} \cap D} |W(r, \omega) - W(k2^{-n}, \omega)| > \frac{1}{n} \right].$$

Suppose it is shown that $\Sigma P(B_n)$ converges. The first Borel-Cantelli lemma will them imply that $B = \lim \sup_n B_n$ has probability 0. But suppose that ω lies outside

---

* Alternatively, define $\mu_{t_1 \ldots t_k}$ as the centered normal distribution (Section 29) in R^k with covariances $\min\{t_i, t_j\}$. Consistency is then obvious.

B. Then for every t and every ϵ there exists an n such that $t < n$, $3n^{-1} < \epsilon$, and $\omega \in B_n^c$. Fix such an n and suppose that r and r' are dyadic rationals in $[0, t]$ and $|r - r'| < 2^{-n}$. Then r and r' must lie in the same or in adjacent dyadic intervals I_{nk}. If $r, r' \in I_{nk}$, then $|W(r, \omega) - W(r', \omega)| \leq |W(r, \omega) - W(k2^{-n}, \omega)| + |W(k2^{-n}, \omega) - W(r', \omega)| \leq 2n^{-1} < \epsilon$. If $r \in I_{nk}$, $r' \in I_{n,k+1}$, then $|W(r, \omega) - W(r', \omega)| \leq |W(r, \omega) - W(k2^{-n}, \omega)| + |W(k2^{-n}, \omega) - W((k + 1)2^{-n}, \omega)| + |W((k + 1)2^{-n}, \omega) - W(r', \omega)| \leq 3n^{-1} < \epsilon$. Therefore, $\omega \notin B$ implies that $W(r, \omega)$ is for every t uniformly continuous as r ranges over the dyadic rationals in $[0, t]$, and hence $W(\cdot, \omega)$ will have a continuous extension to $[0, \infty)$.

To prove that $\Sigma P(B_n)$ converges requires an inequality similar to those in Theorems 9.6, 22.2 (Kolmogorov's inequality), and 35.2.

Lemma 1. *Suppose that $X_1, \ldots, X_n$ are independent, normally distributed random variables with mean 0, and put $S_k = X_1 + \cdots + X_k$. Then for positive α and ϵ,*

$$(37.9) \quad 2P[S_n \geq \alpha + 2\epsilon] - 2 \sum_{k=1}^{n} P[X_k \geq \epsilon]$$

$$\leq P\left[\max_{k \leq n} S_k \geq \alpha\right] \leq 2P[S_n \geq \alpha].$$

PROOF.   Clearly,

$$(37.10) \qquad P\left[\max_{k \leq n} S_k \geq \alpha, S_n \geq \alpha\right] = P[S_n \geq \alpha].$$

Let $A_k = [\max_{i < k} S_i < \alpha \leq S_k]$. Since $S_n - S_k$ is independent of A_k and has distribution symmetric about the origin,

$$P\left[\max_{k \leq n} S_k \geq \alpha, S_n < \alpha\right] = \sum_{k=1}^{n-1} P(A_k \cap [S_n < \alpha])$$

$$\leq \sum_{k=1}^{n-1} P(A_k \cap [S_n - S_k < 0]) = \sum_{k=1}^{n-1} P(A_k \cap [S_n - S_k > 0])$$

$$\leq \sum_{k=1}^{n-1} P(A_k \cap [S_n > \alpha]) \leq P[S_n \geq \alpha].$$

Adding this to (37.10) gives the right-hand inequality in (37.9).

The other inequality is needed only for the calculation (37.16) and its consequence (37.51). Now $S_{k-1} < \alpha$, $X_k < \epsilon$, $S_n - S_k < -\epsilon$ imply that $S_n < \alpha$, and $S_{k-1} < \alpha$, $X_k < \epsilon$, $S_n \geq \alpha + 2\epsilon$ imply that $S_n - S_k > \epsilon$. Therefore,

$$\sum_{k=1}^{n-1} P(A_k \cap [S_n < \alpha])$$

$$\geq \sum_{k=1}^{n-1} \{P(A_k \cap [S_n - S_k < -\epsilon]) - P[X_k \geq \epsilon]\}$$

$$= \sum_{k=1}^{n-1} \{P(A_k \cap [S_n - S_k > \epsilon]) - P[X_k \geq \epsilon]\}$$

$$\geq \sum_{k=1}^{n} \{P(A_k \cap [S_n \geq \alpha + 2\epsilon]) - 2P[X_k \geq \epsilon]\}$$

$$\geq P[S_n \geq \alpha + 2\epsilon] - 2\sum_{k=1}^{n} P[X_k \geq \epsilon].$$

Adding this to (37.10) gives the other inequality in (37.9). ∎

The argument goes through if the X_k are independent and $S_n - S_k$ is symmetrically distributed about 0.

Since $S_1, \ldots, S_n$ have the same joint distribution as $-S_1, \ldots, -S_n$, the right-hand inequality in (37.9) implies that

(37.11) $$P\left[\max_{k \leq n} |S_k| \geq \alpha\right] \leq 2P[|S_n| \geq \alpha].$$

To analyze the probability of the event (37.8), fix δ and t for the moment. Since the increments of Brownian motion are independent and normally distributed with mean 0, (37.11) implies that

$$P\left[\max_{i \leq 2^m} |W(t + \delta i 2^{-m}) - W(t)| \geq \alpha\right] \leq 2P[|W(t + \delta) - W(t)| \geq \alpha]$$

$$\leq \frac{2}{\alpha^4} E[(W(t + \delta) - W(t))^4] = \frac{6\delta^2}{\alpha^4}$$

(see (21.7) for the moments of the normal distribution). The sets on the left here increase with m, and letting $m \to \infty$ leads to

(37.12) $$P\left[\sup_{0 \leq r \leq 1,\, r \in D} |W(t + r\delta) - W(t)| > \alpha\right] \leq \frac{6\epsilon^2}{\alpha^4}.$$

Therefore,

(37.13) $$P(B_n) \leq n 2^n \frac{6 \cdot (2^{-n})^2}{(n^{-1})^4} = \frac{6n^5}{2^n},$$

and $\Sigma P(B_n)$ does converge.

Therefore, there exists a measurable set B such that $P(B) = 0$ and such that for ω outside B, $W(r, \omega)$ is uniformly continuous as r ranges over the dyadic rationals in any bounded interval. If $\omega \notin B$ and r decreases to t through dyadic rational values, then $W(r, \omega)$ has the Cauchy property and hence converges. Put

$$W'_t(\omega) = W'(t, \omega) = \begin{cases} \lim_{r \downarrow t} W(r, \omega) & \text{if } \omega \notin B, \\ 0 & \text{if } \omega \in B. \end{cases}$$

By construction $W'(t, \omega)$ is continuous in t for each ω in Ω. If $\omega \notin B$, $W(r, \omega) = W'(r, \omega)$ for dyadic rationals, and $W'(\cdot, \omega)$ is the continuous extension to all of $[0, \infty)$.

The next thing is to show that the W'_t have the same joint distributions as the W_t. Since (37.6) is continuous in the t_i, the same is by Scheffé's theorem true of $P[(W_{t_1}, \ldots, W_{t_k}) \in H]$ for fixed H. If $0 < t_1 < \cdots < t_k$, choose dyadic rationals $r_i(n)$ approaching the t_i. Let F_n, F, and F' be the respective distribution functions of $(W_{r_1(n)}, \ldots, W_{r_k(n)})$, $(W_{t_1}, \ldots, W_{t_k})$, and $(W'_{t_1}, \ldots, W'_{t_k})$. Then $\lim_n F_n(x_1, \ldots, x_k) = F(x_1, \ldots, x_k)$. Moreover, since $W_{r_i(n)} \to W'_{t_i}$ with probability 1, Theorem 4.1 gives $\lim \sup_n F_n(x_1, \ldots, x_k) \le F'(x_1, \ldots, x_k)$ and $F'(x_1, \ldots, x_k) \le \lim \inf_n F_n(x_1 + \epsilon, \ldots, x_k + \epsilon)$. Thus

$$(37.14) \qquad F(x_1, \ldots, x_k) \le F'(x_1, \ldots, x_k) \le F(x_1 + \epsilon, \ldots, x_k + \epsilon),$$

so that F and F' coincide, as required.*

Thus $[W'_t: t \ge 0]$ is a stochastic process, on the same probability space as $[W_t, t \ge 0]$, which has the finite-dimensional distributions required for Brownian motion and moreover has a continuous sample path $W'(\cdot, \omega)$ for every ω. By enlarging the set B in the definition of $W'_t(\omega)$ to include all the ω for which $W(0, \omega) \ne 0$, one can also ensure that $W'(0, \omega) = 0$. Now discard the original random variables W_t and relabel W'_t as W_t. The new $[W_t: t \ge 0]$ is a stochastic process satisfying conditions (i), (ii), and (iii) for Brownian motion and this one as well:

(iv)   *For each ω, $W(t, \omega)$ is continuous in t and $W(0, \omega) = 0$.*

From now on, by a Brownian motion will be meant a process satisfying (iv) as well as (i), (ii), and (iii). What has been proved is this:

Theorem 37.1.   *There exist processes $[W_t: t \ge 0]$ satisfying conditions* (i), (ii), (iii), *and* (iv)—*Brownian motion processes.*

In the construction above, W_r for dyadic r was used to define W_t in general. For that reason it suffices to apply Kolmogorov's theorem for a countable index set. By the second proof of that theorem the space $(\Omega, \mathcal{F}, P)$ can be taken as the unit interval with Lebesgue measure.

The next section treats a general scheme for dealing with path-function questions by in effect replacing an uncountable time set by a countable one.

---

* This follows more directly from the weak-convergence theory of Section 29.

Since the paths are continuous, $\sup_{s \le t} |W_{t_0+s} - W_{t_0}|$ is a random variable because it is unchanged if s is restricted to dyadic rationals. By (37.11), $P[\max_{k \le 2^m} |W(t_0 + k2^{-m}t) - W(t_0)| \ge \alpha - \epsilon] \le 2P[|W(t_0 + t) - W(t_0)| \ge \alpha - \epsilon]$. Letting $m \to \infty$ and then $\epsilon \to 0$ gives

$$(37.15) \quad P\left[\sup_{s \le t} |W_{t_0+s} - W_{t_0}| \ge \alpha\right] \le 2P[|W_{t_0+t} - W_{t_0}| \ge \alpha].$$

From (37.9) it follows in the same way that $P[\sup_{s \le t} W_s \ge \alpha] \le 2P[W_t \ge \alpha]$ for $\alpha \ge 0$. On the other hand,

$$P\left[\sup_{s \le t} W_s \ge \alpha\right] \ge P\left[\max_{k \le 2^m} W(k2^{-m}t) \ge \alpha\right]$$

$$\ge 2P[W_t \ge \alpha + 2m^{-1}] - 2 \cdot 2^m P[W(2^{-m}t) \ge m^{-1}].$$

Since the last term here is by Markov's inequality at most $12m^4 2^{-m} t^2$, letting $m \to \infty$ leads to

$$(37.16) \qquad P\left[\sup_{s \le t} W_s \ge \alpha\right] = 2P[W_t \ge \alpha], \qquad \alpha \ge 0.$$

Measurable Processes

Let T be a Borel set on the line, let $[X_t : t \in T]$ be a stochastic process on an $(\Omega, \mathcal{F}, P)$, and consider the mapping

$$(37.17) \qquad\qquad (t, \omega) \to X_t(\omega) = X(t, \omega)$$

carrying $T \times \Omega$ into R^1. Let $\mathcal{T}$ be the σ-field of Borel subsets of T. The process is said to be *measurable* if the mapping (37.17) is measurable $\mathcal{T} \times \mathcal{F}/\mathcal{R}^1$.

In the presence of measurability, each sample path $X(\cdot, \omega)$ is measurable $\mathcal{T}$ by Theorem 18.1. Then, for example, $\int_a^b \varphi(X(t, \omega)) \, dt$ makes sense if $(a, b) \subset T$ and φ is a Borel function, and by Fubini's theorem $E[\int_a^b \varphi(X(t, \cdot)) \, dt] = \int_a^b E[\varphi(X_t)] \, dt$ if $\int_a^b E[|\varphi(X_t)|] \, dt < \infty$. Hence the usefulness of this result:

Theorem 37.2. *Brownian motion is measurable.*

PROOF. If

$$W^{(n)}(t, \omega) = W(k2^{-n}, \omega) \quad \text{for } k2^{-n} \le t < (k+1)2^{-n}, \quad k = 0, 1, 2, \ldots,$$

then the mapping $(t, \omega) \to W^{(n)}(t, \omega)$ is measurable $\mathcal{T} \times \mathcal{F}$. But by the continuity of the sample paths, this mapping converges to the mapping (37.17) pointwise (for every (t, ω)), and so by Theorem 13.4(ii) the latter mapping is also measurable $\mathcal{T} \times \mathcal{F}/\mathcal{R}^1$. ∎

Irregularity of Brownian Motion Paths

Starting with a Brownian motion $[W_t: t \geq 0]$ define

(37.18) $$W_t'(\omega) = c^{-1}W_{c^2t}(\omega),$$

where $c > 0$. Since $t \to c^2t$ is an increasing function, it is easy to see that the process $[W_t': t \geq 0]$ has independent increments. Moreover, $W_t' - W_s' = c^{-1}(W_{c^2t} - W_{c^2s})$, and for $s \leq t$ this is normally distributed with mean 0 and variance $c^{-2}(c^2t - c^2s) = t - s$. Since the paths $W'(\cdot, \omega)$ all start from 0 and are continuous, $[W_t': t \geq 0]$ is another Brownian motion. In (37.18) the time scale is contracted by the factor c^2 but the other scale only by the factor c.

That the transformation (37.18) preserves the properties of Brownian motion implies that the paths, although continuous, must be highly irregular. It seems intuitively clear that for c large enough the path $W(\cdot, \omega)$ must with probability nearly 1 have somewhere over the time interval $[0, c]$ a chord with slope exceeding, say, 1. But then $W'(\cdot, \omega)$ has over $[0, c^{-1}]$ a chord with slope exceeding c. Since the W_t' are distributed as the W_t, this makes it plausible that $W(\cdot, \omega)$ must in arbitrarily small intervals $[0, \delta]$ have chords with arbitrarily great slopes, which in turn makes it plausible that $W(\cdot, \omega)$ cannot be differentiable at 0. More generally, mild irregularities in the path will become ever more extreme under the transformation (37.18) with ever larger values of c. It is shown below that, in fact, the paths are with probability 1 nowhere differentiable.

Also interesting in this connection is the transformation

(37.19) $$W_t''(\omega) = \begin{cases} tW_{1/t}(\omega) & \text{if } t > 0, \\ 0 & \text{if } t = 0. \end{cases}$$

Again it is easily checked that the increments are independent and normally distributed with the means and variances appropriate to Brownian motion. Moreover, the path $W''(\cdot, \omega)$ is continuous except possibly at $t = 0$. But (37.12) holds with W_s'' in place of W_s because it depends only on the finite-dimensional distributions, and by the continuity of $W''(\cdot, \omega)$ over $(0, \infty)$ the supremum is the same if not restricted to dyadic rationals. Therefore, $P[\sup_{s \leq n^{-3}} |W_s''| > n^{-1}] \leq 6/n^2$, and it follows by the first Borel-Cantelli lemma that $W''(\cdot, \omega)$ is continuous also at 0 for ω outside a set M of probability 0. For $\omega \in M$, redefine $W''(t, \omega) \equiv 0$; then $[W_t'': t \geq 0]$ is a Brownian motion and (37.19) holds with probability 1. The behavior of $W(\cdot, \omega)$ near 0 can be studied through the behavior of $W''(\cdot, \omega)$ near ∞ and vice versa.

Since $W_t''/t = W_{1/t}$, $W''(\cdot, \omega)$ cannot have a derivative at 0 if $W(\cdot, \omega)$ has no limit at ∞. Now, in fact,

(37.20) $$\inf_n W_n = -\infty, \qquad \sup_n W_n = +\infty$$

with probability 1. To prove this, note that $W_n = X_1 + \cdots + X_n$, where the $X_k = W_k - W_{k-1}$ are independent. Put

$$(37.21) \qquad \left[\sup_n W_n < \infty\right] = \bigcup_{u=1}^{\infty} \bigcap_{m=1}^{\infty} \left[\max_{i \leq m} W_i \leq u\right];$$

this is a tail set and hence by the zero-one law has probability 0 or 1. Now $-X_1$, $-X_2, \ldots$ have the same joint distributions as $X_1, X_2, \ldots$, and so (37.21) has the same probability as

$$(37.22) \qquad \left[\inf_n W_n > -\infty\right] = \bigcup_{u=1}^{\infty} \bigcap_{m=1}^{\infty} \left[\max_{i \leq m} (-W_i) \leq u\right].$$

If these two sets have probability 1, so has $[\sup_n |W_n| < \infty]$, so that $P[\sup_n |W_n| < x] > 0$ for some x. But $P[|W_n| < x] = P[|W_1| < x/n^{1/2}] \to 0$. This proves (37.20).

Since (37.20) holds with probability 1, $W''(\cdot, \omega)$ has with probability 1 upper and lower right derivatives of $+\infty$ and $-\infty$ at $t = 0$. The same must be true of every Brownian motion. A similar argument shows that, for each fixed t, $W(\cdot, \omega)$ is nondifferentiable at t with probability 1. In fact, $W(\cdot, \omega)$ is nowhere differentiable:

Theorem 37.3. *For ω outside a set of probability 0, $W(\cdot, \omega)$ is nowhere differentiable.*

PROOF. The proof is direct—it makes no use of the transformations (37.18) and (37.19). Let

$$(37.23) \quad X_{nk} = \max\left\{\left|W\left(\frac{k}{2^n}\right) - W\left(\frac{k-1}{2^n}\right)\right|,\right.$$
$$\left.\left|W\left(\frac{k+1}{2^n}\right) - W\left(\frac{k}{2^n}\right)\right|, \left|W\left(\frac{k+2}{2^n}\right) - W\left(\frac{k+1}{2^n}\right)\right|\right\}.$$

By independence and the fact that the differences here have the distribution of $2^{-n/2}W_1$, $P[X_{nk} \leq \epsilon] = P^3[|W_1| \leq 2^{n/2}\epsilon]$; since the standard normal density is bounded by 1, $P[X_{nk} \leq \epsilon] \leq (2 \cdot 2^{n/2}\epsilon)^3$. If $Y_n = \min_{k \leq n2^n} X_{nk}$, then

$$(37.24) \qquad P[Y_n \leq \epsilon] \leq n2^n(2 \cdot 2^{n/2} \epsilon)^3.$$

Now define A as the ω-set where $W(\cdot, \omega)$ has a derivative somewhere. Suppose that ω lies in A; suppose specifically that $W(\cdot, \omega)$ is differentiable at the point t and that the derivative there is D. There exists a positive $\delta = \delta(\omega, t)$ such that $|s - t| \leq \delta$ implies that $|W(s, \omega) - W(t, \omega)| \leq (|D| + 1)|s - t|$. If n exceeds some $n_0 = n_0(\omega, t)$, then $2^{-n} < \delta/2$, $n > 2(|D| + 1)$, and $n > t$. For such an n, choose k so that $k2^{-n} \leq t < (k + 1)2^{-n}$. Then $|t - i2^{-n}| < \delta$ for $i = k - 1$, $k, k + 1, k + 2$, and therefore $X_{nk} \leq 2(|D| + 1)2^{-n} \leq n2^{-n}$. Since $n > t$, $Y_n \leq n2^{-n}$.

What has been shown is that if ω lies in A, then ω lies in $A_n = [Y_n \le n2^{-n}]$ for all sufficiently large n: $A \subset \lim \inf_n A_n$. By (37.24),

$$P(A_n) \le n2^n(2 \cdot 2^{n/2} \cdot n2^{-n})^3 \to 0.$$

By Theorem 4.1, $\lim \inf_n A_n$ has probability 0, and outside this set $W(\cdot, \omega)$ is nowhere differentiable.   ∎

If A is the set of ω for which $W(\cdot, \omega)$ has a derivative somewhere, what has been shown is that $A \subset B$ for a measurable B such that $P(B) = 0$; $P(A) = 0$ if A is measurable, but this has not been proved. To avoid such problems in the study of continuous-time processes, it is convenient to work in a *complete* probability space. The space $(\Omega, \mathcal{F}, P)$ is complete (see page 35) if $A \subset B, B \in \mathcal{F}$, and $P(B) = 0$ together imply that $A \in \mathcal{F}$ (and then, of course, $P(A) = 0$). If the space is not already complete, it is possible to enlarge $\mathcal{F}$ to a new σ-field and extend P to it in such a way that the new space is complete. The following assumption therefore entails no loss of generality: *For the rest of this section the space $(\Omega, \mathcal{F}, P)$ on which the Brownian motion is defined is assumed complete.* Theorem 37.3 now becomes: $W(\cdot, \omega)$ is with probability 1 nowhere differentiable.

A nowhere-differentiable path represents the motion of a particle that at no time has a velocity. Since a function of bounded variation is differentiable almost everywhere (Section 31), $W(\cdot, \omega)$ is of unbounded variation with probability 1. Such a path represents the motion of a particle that in its wanderings back and forth travels an infinite distance. The Brownian motion model thus does not in its fine structure represent physical reality. The irregularity of the Brownian motion paths is of considerable mathematical interest, however. A continuous, nowhere-differentiable function is regarded as pathological, or used to be, and constructing one is somewhat complicated,* but from the Brownian-motion point of view such functions are the rule not the exception.

The set of zeros of the Brownian motion is also interesting. By property (iv), $t = 0$ is a zero of $W(\cdot, \omega)$ for each ω. Now $[W_t'': t \ge 0]$ as defined by (37.19) is another Brownian motion, and so by (37.20) the sequence $\{W_n'': n = 1, 2, \ldots\}$ $= \{nW_{1/n}: n = 1, 2, \ldots\}$ has supremum $+\infty$ and infimum $-\infty$ for ω outside a set of probability 0; for such an ω, $W(\cdot, \omega)$ changes sign infinitely often near 0 and hence by continuity has zeros arbitrarily near 0. Let $Z(\omega)$ denote the set of zeros of $W(\cdot, \omega)$. What has just been shown is that $0 \in Z(\omega)$ for each ω and that 0 is with probability 1 a limit of positive points in $Z(\omega)$. More is true:

Theorem 37.4.   *The set $Z(\omega)$ is with probability 1 perfect,[†] nowhere dense, and of Lebesgue measure 0.*

---

* See RIESZ & SZ.-NAGY, page 3, for example, or Problem 31.15.
[†] A set Z on the line is *perfect* if it is nonempty and closed and if for $z \in Z$ and $\epsilon > 0$ there is in Z an x such that $0 < |z - x| < \epsilon$. A perfect set is uncountable; see HAUSDORFF, p. 156.

PROOF. Since $W(\cdot, \omega)$ is continuous, $Z(\omega)$ is closed for every ω. Let λ denote Lebesgue measure. Since Brownian motion is measurable (Theorem 37.2), Fubini's theorem applies:

$$\int_\Omega \lambda(Z(\omega))P(d\omega) = (\lambda \times P)[(t, \omega): W(t, \omega) = 0]$$

$$= \int_0^\infty P[\omega: W(t, \omega) = 0]\, dt = 0.$$

Thus $\lambda(Z(\omega)) = 0$ with probability 1.

If $W(\cdot, \omega)$ is nowhere differentiable, it cannot vanish throughout an interval I and hence must by continuity be nonzero throughout some subinterval of I. By Theorem 37.3, then, $Z(\omega)$ is with probability 1 nowhere dense.

It remains to show that each point of $Z(\omega)$ is a limit of other points of $Z(\omega)$. As observed above, this is true of the point 0 of $Z(\omega)$. For the general point of $Z(\omega)$, a stopping-time argument is required. Fix $r \geq 0$ and let $\tau(\omega) = \inf[t: t \geq r, W(t, \omega) = 0]$; note that this set is with probability 1 nonempty by (37.20). Thus $\tau(\omega)$ is the first zero following r. Now

$$[\omega: \tau(\omega) \leq t] = \left[\omega: \inf_{r \leq s \leq t} |W(s, \omega)| = 0\right],$$

and by continuity the infimum here is unchanged if s is restricted to rationals. This shows that τ is a random variable and that

$$[\omega: \tau(\omega) \leq t] \in \sigma[W_u: u \leq t].$$

A nonnegative random variable with this property is called a *stopping time*.

To know the value of τ is to know at most the values of W_u for $u \leq \tau$. Since the increments are independent, it therefore seems intuitively clear that the process

$$(37.25) \quad W_t^*(\omega) = W_{\tau(\omega)+t}(\omega) - W_{\tau(\omega)}(\omega) = W_{\tau(\omega)+t}(\omega), \qquad t \geq 0,$$

ought itself to be a Brownian motion. This is, in fact, true by the next result, Theorem 37.5 below. What is proved there is that the finite-dimensional distributions of $[W_t^*: t \geq 0]$ are the right ones for Brownian motion. The other properties are obvious: $W^*(\cdot, \omega)$ is continuous and vanishes at 0 by construction, and the space on which $[W_t^*; t \geq 0]$ is defined is complete because it is the original space $(\Omega, \mathcal{F}, P)$, assumed complete.

If $[W_t^*: t \geq 0]$ is indeed a Brownian motion, then, as observed above, for ω outside a set B_r of probability 0 there is a positive sequence $\{t_n\}$ such that $t_n \to 0$ and $W^*(t_n, \omega) = 0$. But then $W(\tau(\omega) + t_n, \omega) = 0$, so that $\tau(\omega)$, a zero of $W(\cdot, \omega)$, is the limit of other larger zeros of $W(\cdot, \omega)$. Now $\tau(\omega)$ was the first zero following r. (There is a different stopping time τ for each r, but the notation does not show this.) If B is the union of the B_r for rational r, then $P(B) = 0$. Moreover, if ω lies outside B, then for every rational r, the first point of $Z(\omega)$ following r is a limit of other larger points of $Z(\omega)$. Suppose that $\omega \notin B$ and $t \in Z(\omega)$, where $t > 0$; it is to be shown that t is a limit of other points of $Z(\omega)$. If t is the limit of smaller points of $Z(\omega)$, there is of course nothing to prove. Otherwise, there is a rational r such that $r < t$ and $W(\cdot, \omega)$ does not vanish in $[r, t)$; but then, since $\omega \notin B_r$, t is a limit of larger points s that lie in $Z(\omega)$. This completes the proof of Theorem 37.4 under the provisional assumption that (37.25) is a Brownian motion.   ∎

The Strong Markov Property

Fix $t_0 \geq 0$ and put

(37.26) $$W'_t = W_{t_0 + t} - W_{t_0}, \qquad t \geq 0.$$

It is easily checked that $[W'_t: t \geq 0]$ has the finite-dimensional distributions appropriate to Brownian motion. As the other properties are obvious, it is in fact a Brownian motion.

Let

(37.27) $$\mathcal{F}_t = \sigma[W_s: s \leq t].$$

The random variables (37.26) are independent of $\mathcal{F}_{t_0}$. To see this, suppose that $0 \leq s_1 \leq \cdots \leq s_j \leq t_0$ and $0 \leq t_1 \leq \cdots \leq t_k$. Put $u_i = t_0 + t_i$. Since the increments are independent, $(W'_{t_1}, W'_{t_2} - W'_{t_1}, \ldots, W'_{t_k} - W'_{t_{k-1}}) = (W_{u_1} - W_{t_0}, W_{u_2} - W_{u_1}, \ldots, W_{u_k} - W_{u_{k-1}})$ is independent of $(W_{s_1}, W_{s_2} - W_{s_1}, \ldots, W_{s_j} - W_{s_{j-1}})$. But then $(W'_{t_1}, W'_{t_2}, \ldots, W'_{t_k})$ is independent of $(W_{s_1}, W_{s_2}, \ldots, W_{s_j})$. By Theorem 4.2, $(W'_{t_1}, \ldots, W'_{t_k})$ is independent of $\mathcal{F}_{t_0}$. Thus

(37.28) $$P([W'_{t_1}, \ldots, W'_{t_k}] \in H] \cap A) = P[(W'_{t_1}, \ldots, W'_{t_k}) \in H]P(A)$$
$$= P[(W_{t_1}, \ldots, W_{t_k}) \in H]P(A), \qquad A \in \mathcal{F}_{t_0},$$

where the second equality follows because (37.26) is a Brownian motion. This holds for all H in $\mathcal{R}^k$.

The problem now is to prove all this when t_0 is replaced by a *stopping time* τ—a nonnegative random variable for which

(37.29) $$[\omega: \tau(\omega) \leq t] \in \mathcal{F}_t, \qquad t \geq 0.$$

It will be assumed that τ is finite, at least with probability 1. Since $[\tau = t] = [\tau \leq t] - \cup_n[\tau \leq t - n^{-1}]$, (37.29) implies that

(37.30) $$[\omega: \tau(\omega) = t] \in \mathcal{F}_t, \qquad t \geq 0.$$

The conditions (37.29) and (37.30) are analogous to the conditions (7.20) and (35.18), which prevent prevision on the part of the gambler.

Now $\mathcal{F}_{t_0}$ contains the information on the past of the Brownian motion up to time t_0, and the analogue for τ is needed. Let $\mathcal{F}_\tau$ consist of all measurable sets M for which

$$(37.31) \qquad M \cap [\omega: \tau(\omega) \le t] \in \mathcal{F}_t$$

for all t. (See (35.28) for an analogue in discrete time.) Note that $\mathcal{F}_\tau$ is a σ-field and τ is measurable $\mathcal{F}_\tau$. Since $M \cap [\tau = t] = M \cap [\tau = t] \cap [\tau \le t]$,

$$(37.32) \qquad M \cap [\omega: \tau(\omega) = t] \in \mathcal{F}_t$$

for M in $\mathcal{F}_\tau$. For example, $\tau = \inf[t: W_t = 1]$ is a stopping time and $[\inf_{s \le \tau} W_s > -1]$ is in $\mathcal{F}_\tau$.

Theorem 37.5. *Let τ be a stopping time and put*

$$(37.33) \qquad W_t^*(\omega) = W_{\tau(\omega)+t}(\omega) - W_{\tau(\omega)}(\omega), \qquad t \ge 0.$$

Then $[W_t^*: t \ge 0]$ is a Brownian motion, and it is independent of $\mathcal{F}_\tau$—that is, $\sigma[W_t^*: t \ge 0]$ is independent of $\mathcal{F}_\tau$:

$$(37.34) \quad P([(W_{t_1}^*, \ldots, W_{t_k}^*) \in H] \cap M) = P[(W_{t_1}^*, \ldots, W_{t_k}^*) \in H]P(M)$$

$$= P[(W_{t_1}, \ldots, W_{t_k}) \in H]P(M)$$

for H in $\mathcal{R}^k$ and M in $\mathcal{F}_\tau$.

That the transformation (37.33) preserves Brownian motion is the *strong Markov property*.[†] Part of the conclusion is that the W_t^* are random variables.

PROOF. Suppose first that τ has countable range V and let t_0 be the general point of V. Since

$$[\omega: W_t^*(\omega) \in H] = \bigcup_{t_0 \in V} [\omega: W_{t_0+t}(\omega) - W_{t_0}(\omega) \in H, \tau(\omega) = t_0],$$

W_t^* is a random variable. Also,

$$P([(W_{t_1}^*, \ldots, W_{t_k}^*) \in H] \cap M)$$

$$= \sum_{t_0 \in V} P([(W_{t_1}^*, \ldots, W_{t_k}^*) \in H] \cap M \cap [\tau = t_0]).$$

If $M \in \mathcal{F}_\tau$, then $M \cap [\tau = t_0] \in \mathcal{F}_{t_0}$ by (37.32). Further, if $\tau = t_0$, then W_t^* coincides with W_t' as defined by (37.26). Therefore, (37.28) reduces this last sum to

---

[†] Since the Brownian motion has independent increments, it is a Markov process (see Example 33.9); hence the terminology.

$$\sum_{t_0 \in V} P[(W_{t_1}, \ldots, W_{t_k}) \in H] P(M \cap [\tau = t_0])$$

$$= P[(W_{t_1}, \ldots, W_{t_k}) \in H] P(M).$$

This proves the first and third terms in (37.34) equal; to prove equality with the middle term, simply consider the case $M = \Omega$.

Thus the theorem holds if τ has countable range. For the general τ, put

$$(37.35) \quad \tau_n = \begin{cases} k2^{-n} & \text{if } (k-1)2^{-n} < \tau \le k2^{-n}, \quad k = 1, 2, \ldots \\ 0 & \text{if } \tau = 0. \end{cases}$$

If $k2^{-n} \le t < (k+1)2^{-n}$, then $[\tau_n \le t] = [\tau \le k2^{-n}] \in \mathcal{F}_{k2^{-n}} \subset \mathcal{F}_t$. Thus each τ_n is a stopping time. Suppose that $M \in \mathcal{F}_\tau$ and $k2^{-n} \le t < (k+1)2^{-n}$. Then $M \cap [\tau_n \le t] = M \cap [\tau \le k2^{-n}] \in \mathcal{F}_{k2^{-n}} \subset \mathcal{F}_t$. Thus $\mathcal{F}_\tau \subset \mathcal{F}_{\tau_n}$. Let $W_t^{(n)}(\omega) = W_{\tau_n(\omega)+t}(\omega) - W_{\tau_n(\omega)}(\omega)$—that is, let $W_t^{(n)}$ be the W_t^* corresponding to the stopping time τ_n. If $M \in \mathcal{F}_\tau$ then $M \in \mathcal{F}_{\tau_n}$, and by an application of (37.34) to the discrete case already treated,

$$(37.36) \quad P([(W_{t_1}^{(n)}, \ldots, W_{t_k}^{(n)}) \in H] \cap M)$$

$$= P[(W_{t_1}, \ldots, W_{t_k}) \in H] P(M).$$

But $\tau_n(\omega) \to \tau(\omega)$ for each ω, and by continuity of the sample paths $W_t^{(n)}(\omega) \to W_t^*(\omega)$ for each ω. By a simple modification of the argument leading to (37.14) (condition on M), (37.34) follows from (37.36). ∎

The τ in the proof of Theorem 37.4 is a stopping time, and so (37.25) is a Brownian motion, as required in that proof. Further applications will be given below.

If $\mathcal{F}^* = \sigma[W_t^*: t \ge 0]$, then according to (37.34) (and Theorem 4.2) the σ-fields $\mathcal{F}_\tau$ and $\mathcal{F}^*$ are independent:

$$(37.37) \quad P(A \cap B) = P(A)P(B), \quad A \in \mathcal{F}_\tau, \quad B \in \mathcal{F}^*.$$

For fixed t define τ_n by (37.35) but with $t2^{-n}$ in place of 2^{-n} at each occurrence. Then $[W_\tau < x] \cap [\tau \le t]$ is the limit superior of the sets $[W_{\tau_n} < x] \cap [\tau \le t]$, each of which lies in $\mathcal{F}_t$. This proves that $[W_\tau < x]$ lies in $\mathcal{F}_\tau$ and hence that W_τ is measurable $\mathcal{F}_\tau$. Since τ is measurable $\mathcal{F}_\tau$,

$$(37.38) \quad [(\tau, W_\tau) \in H] \in \mathcal{F}_\tau$$

for planar Borel sets H.

Skorohod Embedding[†]

Suppose that $X_1, X_2, \ldots$ are independent and identically distributed random variables with mean 0 and variance σ^2. A powerful method, due to Skorohod,

---

[†] The rest of this section, which requires martingale theory, may be omitted.

of studying the partial sums $S_n = X_1 + \cdots + X_n$ is to construct an increasing sequence $\tau_0 = 0, \tau_1, \tau_2, \ldots$ of stopping times such that $W(\tau_n)$ has the same distribution as S_n. The differences $\tau_k - \tau_{k-1}$ will turn out to be independent and identically distributed with mean σ^2, so that by the law of large numbers $n^{-1}\tau_n = n^{-1}\Sigma_{k=1}^n (\tau_k - \tau_{k-1})$ is likely to be near σ^2. But if τ_n is near $n\sigma^2$, then by the continuity of Brownian motion paths $W(\tau_n)$ will be near $W(n\sigma^2)$, and so the distribution of $S_n/\sigma\sqrt{n}$, which coincides with the distribution of $W(\tau_n)/\sigma\sqrt{n}$, will be near the distribution of $W(n\sigma^2)/\sigma\sqrt{n}$—that is, will be near the standard normal distribution. The method will thus yield another proof of the central limit theorem, one independent of the characteristic-function arguments of Section 27.

But it will also give more. For example, the distribution of $\max_{k \leq n} S_k/\sigma\sqrt{n}$ is exactly the distribution of $\max_{k \leq n} W(\tau_k)/\sigma\sqrt{n}$, and this in turn is near the distribution of $\sup_{t \leq n\sigma^2} W(t)/\sigma\sqrt{n}$, which can be written down explicitly because of (37.16). It will thus be possible to derive the limiting distribution of $\max_{k \leq n} S_k$. The joint behavior of the partial sums is closely related to the behavior of Brownian motion paths.

The Skorohod construction involves the class T of stopping times for which

(37.39) $$E[W_\tau] = 0,$$

(37.40) $$E[\tau] = E[W_\tau^2],$$

and

(37.41) $$E[\tau^2] \leq 4E[W_\tau^4].$$

Lemma 2. *All bounded stopping times are members of T.*

PROOF. Define $Y_{\theta,t} = \exp(\theta W_t - \frac{1}{2}\theta^2 t)$ for all θ and for $t \geq 0$. Suppose that $s \leq t$ and $A \in \mathcal{F}_s$ (see (37.27)). Since Brownian motion has independent increments,

$$\int_A Y_{\theta,t}\, dP = \int_A e^{\theta W_s - \theta^2 s/2}\, dP \cdot E[e^{\theta(W_t - W_s) - \theta^2(t-s)/2}],$$

and a calculation with moment generating functions (see Example 21.2) shows that

(37.42) $$\int_A Y_{\theta,s}\, dP = \int_A Y_{\theta,t}\, dP, \qquad s \leq t, \quad A \in \mathcal{F}_s.$$

This says that for θ fixed, $[Y_{\theta,t}: t \geq 0]$ is a continuous-time martingale adapted to the σ-fields $\mathcal{F}_t$. It is the *moment generating function martingale* associated with the Brownian motion.

Let $f(\theta, t)$ denote the right side of (37.42). By Theorem 16.8,

$$\frac{\partial}{\partial \theta} f(\theta, t) = \int_A Y_{\theta, t}(W_t - \theta t) \, dP,$$

$$\frac{\partial^2}{\partial \theta^2} f(\theta, t) = \int_A Y_{\theta, t}[(W_t - \theta t)^2 - t] \, dP,$$

$$\frac{\partial^4}{\partial \theta^4} f(\theta, t) = \int_A Y_{\theta, t}[(W_t - \theta t)^4 - 6(W_t - \theta t)^2 t + 3t^2] \, dP.$$

Differentiate the other side of the equation (37.42) the same way and set $\theta = 0$. The result is

$$\int_A W_s \, dP = \int_A W_t \, dP, \qquad\qquad\qquad s \le t, \quad A \in \mathcal{F}_s,$$

$$\int_A (W_s^2 - s) \, dP = \int_A (W_t^2 - t) \, dP, \qquad\qquad s \le t, \quad A \in \mathcal{F}_s,$$

$$\int_A (W_s^4 - 6W_s^2 s + 3s^2) \, dP = \int_A (W_t^4 - 6W_t^2 + 3t^2) \, dP, \quad s \le t, \quad A \in \mathcal{F}_s.$$

This gives three more martingales: If Z_t is any of the three random variables

(37.43) $\qquad\qquad W_t, \qquad W_t^2 - t, \qquad W_t^4 - 6W_t^2 t + 3t^2,$

then $Z_0 = 0$, Z_t is integrable and measurable $\mathcal{F}_t$, and

(37.44) $\qquad\qquad \int_A Z_s \, dP = \int_A Z_t \, dP, \qquad s \le t, \quad A \in \mathcal{F}_s.$

In particular, $E[Z_t] = E[Z_0] = 0$.

If τ is a stopping time with finite range $\{t_1, \ldots, t_m\}$ bounded by t, then (37.44) implies that

$$E[Z_\tau] = \sum_i \int_{[\tau = t_i]} Z_{t_i} \, dP = \sum_i \int_{[\tau = t_i]} Z_t \, dP = E[Z_t] = 0.$$

Suppose that τ is bounded by t but does not necessarily have finite range. Put $\tau_n = k2^{-n}t$ if $(k - 1)2^{-n}t < \tau \le k2^{-n}t$, $1 \le k \le 2^n$, and put $\tau_n = 0$ if $\tau = 0$. Then τ_n is a stopping time and $E[Z_{\tau_n}] = 0$. For each of the three possibilities (37.43) for Z_t, $\sup_{s \le t} |Z_s|$ is integrable because of (37.15). It therefore follows by the dominated convergence theorem that $E[Z_\tau] = \lim_n E[Z_{\tau_n}] = 0$.

Thus $E[Z_\tau] = 0$ for every bounded stopping time τ. The three cases (37.43) give

$$E[W_\tau] = E[W_\tau^2 - \tau] = E[W_\tau^4 - 6W_\tau^2 \tau + 3\tau^2] = 0.$$

This implies (37.39), (37.40), and

$$0 = E[W_\tau^4] - 6E[W_\tau^2\tau] + 3E[\tau^2] \geq E[W_\tau^4] - 6E^{1/2}[W_\tau^4]E^{1/2}[\tau^2] + 3E[\tau^2].$$

If $C = E^{1/2}[W_\tau^4]$ and $x = E^{1/2}[\tau^2]$, the inequality is $0 \geq q(x) = 3x^2 - 6Cx + C^2$. Each zero of q is at most $2C$, and q is negative only between these two zeros. Therefore, $x \leq 2C$, which implies (37.41). ∎

Lemma 3. *Suppose that τ and τ_n are stopping times, that each τ_n is a member of $\mathcal{T}$, and that $\tau_n \to \tau$ with probability 1. If* (i) $E[W_{\tau_n}^4] \leq E[W_\tau^4]$ *for all n, or if* (ii) *the $W_{\tau_n}^4$ are uniformly integrable, then τ is a member of $\mathcal{T}$.*

PROOF. Since Brownian motion paths are continuous, $W_{\tau_n} \to W_\tau$ with probability 1. Each of the two hypotheses (i) and (ii) implies that $E[W_{\tau_n}^4]$ is bounded and hence that $E[\tau_n^2]$ is bounded, and it follows (see (16.21)) that the sequences $\{\tau_n\}$, $\{W_{\tau_n}\}$, and $\{W_{\tau_n}^2\}$ are uniformly integrable. Hence (37.39) and (37.40) for τ follow by Theorem 16.13 from the same relations for the τ_n. The first hypothesis implies that $\liminf_n E[W_{\tau_n}^4] \leq E[W_\tau^4]$, and the second implies that $\lim_n E[W_{\tau_n}^4] = E[W_\tau^4]$. In either case it follows by Fatou's lemma that $E[\tau^2] \leq \liminf_n E[\tau_n^2] \leq 4\liminf_n E[W_{\tau_n}^4] \leq 4E[W_\tau^4]$. ∎

Suppose that $a, b \geq 0$ and $a + b > 0$, and let $\tau(a, b)$ be the *hitting time* for the set $\{-a, b\}$: $\tau(a, b) = \inf\{t: W_t \in \{-a, b\}\}$. By (37.20), $\tau(a, b)$ is finite with probability 1, and it is a stopping time because $\tau(a, b) \leq t$ if and only if for every m there is a rational $r \leq t$ for which W_r is within m^{-1} of $-a$ or of b. From $|W(\min\{\tau(a, b), n\})| \leq \max\{a, b\}$ it follows by Lemma 3 that $\tau(a, b)$ is a member of $\mathcal{T}$. Since $W_{\tau(a,b)}$ assumes only the values $-a$ and b, $E[W_{\tau(a,b)}] = 0$ implies that

$$(37.45) \quad P[W_{\tau(a,b)} = -a] = \frac{b}{a+b}, \qquad P[W_{\tau(a,b)} = b] = \frac{a}{a+b}.$$

This is obvious on grounds of symmetry in the case $a = b$.

Let μ be a probability measure on the line with mean 0. The program is to construct a stopping time τ for which W_τ has distribution μ. Assume that $\mu\{0\} < 1$, since otherwise $\tau \equiv 0$ obviously works. If μ consists of two point masses, they must for some a and b be a mass of $b/(a + b)$ at $-a$ and a mass of $a/(a + b)$ at b; in this case $\tau_{(a,b)}$ is by (37.45) the required stopping time. The general case will be treated by adding together stopping times of this sort.

Consider a random variable X having distribution μ. (The probability space for X has nothing to do with the space the given Brownian motion is defined on.) The technique will be to represent X as the limit of a martingale $X_1, X_2, \ldots$ of a simple form and then to duplicate the martingale by $W_{\tau_1}, W_{\tau_2}, \ldots$ for stopping times τ_n; the τ_n will have a limit τ such that W_τ has the same distribution as X.

The first step is to construct sets

$$\Delta_n: a_0^{(n)} < a_1^{(n)} < \cdots < a_{r_n}^{(n)}$$

and corresponding partitions

$$\mathcal{P}_n: \begin{cases} I_0^n = (-\infty, a_0^{(n)}] \\ I_k^n = (a_{k-1}^{(n)}, a_k^{(n)}], & 1 \le k \le r_n, \\ I_{r_n+1}^n = (a_{r_n}^{(n)}, \infty). \end{cases}$$

Let $M(H)$ be the conditional mean:

$$M(H) = \frac{1}{\mu(H)} \int_H x\, \mu(dx) \qquad \text{if } \mu(H) > 0.$$

Let Δ_1 consist of the single point $M(R^1) = E[X] = 0$, so that $\mathcal{P}_1$ consists of $I_0^1 = (-\infty, 0]$ and $I_1^1 = (0, \infty)$. Suppose that Δ_n and $\mathcal{P}_n$ are given. If $\mu((I_k^n)^\circ) > 0$, split I_k^n by adding to Δ_n the point $M(I_k^n)$, which lies in $(I_k^n)^\circ$; if $\mu((I_k^n)^\circ) = 0$, I_k^n appears again in $\mathcal{P}_{n+1}$. From $\mu\{0\} < 1$ it follows by induction that $\mu(I_k^n) > 0$ for all n and k.

Let $\mathcal{G}_n$ be the σ-field generated by the sets $[X \in I_k^n]$ and put $X_n = E[X\|\mathcal{G}_n]$. Then $X_1, X_2, \ldots$ is a martingale and $X_n = M(I_k^n)$ on $[X \in I_k^n]$. The X_n have finite range and their joint distributions can be written out explicitly. In fact, $[X_1 = M(I_{k_1}^1), \ldots, X_n = M(I_{k_n}^n)] = [X \in I_{k_1}^1, \ldots, X \in I_{k_n}^n]$, and this set is empty unless $I_{k_1}^1 \supset \cdots \supset I_{k_n}^n$, in which case it is $[X_n = M(I_{k_n}^n)] = [X \in I_{k_n}^n]$. Therefore, if $k_{n-1} = j$ and $I_j^{n-1} = I_{k-1}^n \cup I_k^n$,

$$P[X_n = M(I_{k-1}^n)\|X_1 = M(I_{k_1}^1), \ldots, X_{n-1} = M(I_{k_{n-1}}^{n-1})] = \frac{\mu(I_{k-1}^n)}{\mu(I_j^{n-1})}$$

and

$$P[X_n = M(I_k^n) \| X_1 = M(I_{k_1}^1), \ldots, X_{n-1} = M(I_{k_{n-1}}^{n-1})] = \frac{\mu(I_k^n)}{\mu(I_j^{n-1})},$$

provided the conditioning event has postive probability. Thus the martingale $\{X_n\}$ has the Markov property, and if $x = M(I_j^{n-1})$, $u = M(I_{k-1}^n)$, and $v = M(I_k^n)$, then the conditional distribution of X_n given $X_{n-1} = x$ is concentrated at the two points u and v and has mean x. The structure of $\{X_n\}$ is determined by these conditional probabilities together with the distribution

$$P[X_1 = M(I_0^1)] = \mu(I_0^1), \qquad P[X_1 = M(I_1^1)] = \mu(I_1^1).$$

of X_1.

If $\mathcal{G}_\infty = \sigma(\bigcup_n \mathcal{G}_n)$, then $X_n \to E[X\|\mathcal{G}]$ with probability 1 by the martingale theorem (Theorem 35.5). But, in fact, $X_n \to X$ with probability 1, as the following argument shows. Let B be the union of all open sets of μ-measure 0. Then B is a countable disjoint union of open intervals; enlarge B by adding to it any endpoints of μ-measure 0 these intervals may have. Then $\mu(B) = 0$ and $x \notin B$ implies that $\mu(x - \epsilon, x] > 0$ and $\mu[x, x + \epsilon) > 0$ for all positive ϵ. Suppose that $x = X(\omega) \notin B$ and let $x_n = X_n(\omega)$. Let $I^n_{k_n}$ be the element of $\mathcal{P}_n$ containing x; then $x_{n+1} = M(I^n_{k_n})$ and $I^n_{k_n} \downarrow I$ for some interval I. Suppose that $x_{n+1} < x - \epsilon$ for n in an infinite sequence N of integers. Then x_{n+1} is the left endpoint of $I^{n+1}_{k_{n+1}}$ for n in N and converges along N to the left endpoint, say a, of I, and $(x - \epsilon, x] \subset I$. Further, $x_{n+1} = M(I^n_{k_n}) \to M(I)$ along along N, so that $M(I) = a$. But this is impossible because $\mu(x - \epsilon, x] > 0$. Therefore, $x_n \geq x - \epsilon$ for large n. Similarly, $x_n \leq x + \epsilon$ for large n, and so $x_n \to x$. Thus $X_n(\omega) \to X(\omega)$ if $X(\omega) \notin B$, the probability of which is 1.

Now $X_1 = E[X \| \mathcal{G}_1]$ has mean 0 and its distribution consists of point masses at $-a = M(I^1_0)$ and $b = M(I^1_1)$. If $\tau_1 = \tau(a, b)$ is the hitting time to $\{-a, b\}$, then (see (37.45)) τ_1 is a stopping time, a member of $\mathcal{T}$, and W_{τ_1} has the same distribution as X_1.

Let τ_2 be the infimum of those t for which $t \geq \tau_1$ and W_t is one of the points $M(I^2_k)$, $0 \leq k \leq r_2 + 1$. By (37.20), τ_2 is finite with probability 1; it is a stopping time because $\tau_2 \leq t$ if and only if for every m there are rationals r and s such that $r \leq s + m^{-1}$, $r \leq t$, $s \leq t$, W_r is within m^{-1} of one of the points $M(I^1_j)$, and W_s is within m^{-1} of one of the points $M(I^2_k)$. Since $|W(\min\{\tau_2, n\})|$ is at most the maximum of the values $|M(I^2_k)|$, it follows by Lemma 3 that τ_2 is a member of $\mathcal{T}$.

Define W_t^* by (37.33) with τ_1 for τ. If $x = M(I^1_j)$, then x is an endpoint common to two adjacent intervals I^2_{k-1} and I^2_k; put $u = M(I^2_{k-1})$ and $v = M(I^2_k)$. If $W_{\tau_1} = x$, then u and v are the only possible values of W_{τ_2}. It τ^* is the first time the Brownian motion $[W_t^*: t \geq 0]$ hits $u - x$ or $v - x$, then by (37.45),

$$P[W^*_{\tau^*} = u - x] = \frac{v - x}{v - u}, \qquad P[W^*_{\tau^*} = v - x] = \frac{x - u}{v - u}.$$

On the set $[W_{\tau_1} = x]$, τ_2 coincides with $\tau_1 + \tau^*$, and it follows by (37.37) that

$$P[W_{\tau_1} = x, \quad W_{\tau_2} = v] = P[W_{\tau_1} = x, \quad x + W^*_{\tau^*} = v]$$

$$= P[W_{\tau_1} = x]P[W^*_{\tau^*} = v - x] = P[W_{\tau_1}] \frac{x - u}{v - u}.$$

This, together with the same computation with u in place of v, shows that for $W_{\tau_1} = x$ the conditional distribution of W_{τ_2} is concentrated at the two points u and v and has mean x. Thus the conditional distribution of W_{τ_2} given W_{τ_1} coincides with the conditional distribution of X_2 given X_1. Since W_{τ_1} and X_1 have the same distribution, the random vectors (W_{τ_1}, W_{τ_2}) and (X_1, X_2) also have the same distribution.

An inductive extension of this argument proves the existence of a sequence of stopping times τ_n such that $\tau_1 \leq \tau_2 \leq \cdots$, each τ_n is a member of $\mathcal{T}$, and for each n, $W_{\tau_1}, \ldots, W_{\tau_n}$ have the same joint distribution as $X_1, \ldots, X_n$. Now suppose that X has finite variance. Since τ_n is a member of T, $E[\tau_n] = E[W_{\tau_n}^2]$ $= E[X_n^2] = E[E^2[X\|\mathcal{G}_n]] \le E[X^2]$ by Jensen's inequality (34.5). Thus $\tau = \lim_n \tau_n$ is finite with probability 1. Obviously it is a stopping time, and by path continuity, $W_{\tau_n} \to W_\tau$ with probability 1. Since $X_n \to X$ with probability 1, it follows by the argument leading to (37.14) that W_τ has the distribution of X. Since $X_n^2 \le E[X^2\|\mathcal{G}_n]$, the X_n are uniformly integrable by the lemma preceding Theorem 35.5. By the monotone convergence theorem and Theorem 16.13, $E[\tau]$ $= \lim_n E[\tau_n] = \lim_n E[W_{\tau_n}^2] = \lim_n E[X_n^2] = E[X^2] = E[W_\tau^2]$. If $E[X^4] < \infty$, then $E[W_{\tau_n}^4] = E[X_n^4] \le E[X^4] = E[W_\tau^4]$ (Jensen's inequality again), and so τ is a member of T. Hence $E[\tau^2] \le 4E[W_\tau^4]$.

This construction establishes the first of Skorohod's embedding theorems:

Theorem 37.6. *Suppose that X is a random variable with mean 0 and finite variance. There is a stopping time τ such that W_τ has the same distribution as X, $E[\tau] = E[X^2]$, and $E[\tau^2] \le 4E[X^4]$.*

Of course, the last inequality is trivial unless $E[X^4]$ is finite. The theorem could be stated in terms not of X but of its distribution, the point being that the probability space X is defined on is irrelevant. Skorohod's second embedding theorem is this:

Theorem 37.7. *Suppose that $X_1, X_2, \ldots$ are independent and identically distributed random variables with mean 0 and finite variance, and put $S_n = X_1 + \cdots + X_n$. There is a nondecreasing sequence $\tau_1, \tau_2, \ldots$ of stopping times such that the W_{τ_n} have the same joint distributions as the S_n and $\tau_1, \tau_2 - \tau_1, \tau_3 - \tau_2, \ldots$ are independent and identically distributed random variables satisfying $E[\tau_n - \tau_{n-1}] = E[X_1^2]$ and $E[(\tau_n - \tau_{n-1})^2] \le 4E[X_1^4]$.*

PROOF.   The method is to repeat the construction above inductively. For notational clarity write $W_t = W_t^{(1)}$ and put $\mathcal{F}_t^{(1)} = \sigma[W_s^{(1)}: 0 \le s \le t]$ and $\mathcal{F}^{(1)}$ $= \sigma[W_t^{(1)}: t \ge 0]$. Let δ_1 be the stopping time of Theorem 37.6, so that $W_{\delta_1}^{(1)}$ and X_1 have the same distribution. Let $\mathcal{F}_{\delta_1}^{(1)}$ be the class of M such that $M \cap [\delta_1 \le t] \in \mathcal{F}_t^{(1)}$ for all t.

Now put $W_t^{(2)} = W_{\delta_1+t}^{(1)} - W_{\delta_1}^{(1)}$, $\mathcal{F}_t^{(2)} = \sigma[W_s^{(2)}: 0 \le s \le t]$, and $\mathcal{F}^{(2)} = \sigma[W_t^{(2)}: t \ge 0]$. By another application of Theorem 37.6, construct a stopping time δ_2 for the Brownian motion $[W_t^{(2)}: t \ge 0]$ in such a way that $W_{\delta_2}^{(2)}$ has the same distribution as X_1. In fact, use for δ_2 the very same martingale construction as for δ_1, so that $(\delta_1, W_{\delta_1}^{(1)})$ and $(\delta_2, W_{\delta_2}^{(2)})$ have the same distribution. Since $\mathcal{F}_{\delta_1}^{(1)}$ and $\mathcal{F}^{(2)}$ are independent (see (37.37)), it follows (see (37.38)) that $(\delta_1, W_{\delta_1}^{(1)})$ and $(\delta_2, W_{\delta_2}^{(2)})$ are independent.

Let $\mathcal{F}_{\delta_2}^{(2)}$ be the class of M such that $M \cap [\delta_2 \le t] \in \mathcal{F}_t^{(2)}$ for all t. If $W_t^{(3)} =$

$W_{\delta 2+t}^{(2)} - W_{\delta 2}^{(2)}$ and $\mathcal{F}^{(3)}$ is the σ-field generated by these random variables, then again $\mathcal{F}_{\delta 2}^{(2)}$ and $\mathcal{F}^{(3)}$ are independent. These two σ-fields are contained in $\mathcal{F}^{(2)}$, which is independent of $\mathcal{F}_{\delta 1}^{(1)}$. Therefore, the three σ-fields $\mathcal{F}_{\delta 1}^{(1)}$, $\mathcal{F}_{\delta 2}^{(2)}$, $\mathcal{F}^{(3)}$ are independent. The procedure therefore extends inductively to give independent, identically distributed random vectors $(\delta_n, W_{\delta n}^{(n)})$. If $\tau_n = \delta_1 + \cdots + \delta_n$, then $W_{\tau_n}^{(1)} = W_{\delta 1}^{(1)} + \cdots + W_{\delta n}^{(n)}$ has the distribution of $X_1 + \cdots + X_n$. ∎

Invariance*

If $E[X_1^2] = \sigma^2$, then, since the random variables $\tau_n - \tau_{n-1}$ of Theorem 37.7 are independent and identically distributed, the strong law of large numbers (Theorem 22.5) applies and hence so does the weak one:

$$(37.46) \qquad P[|n^{-1}\tau_n - \sigma^2| \geq \epsilon] \to 0.$$

(If $E[X_1^4] < \infty$, so that the $\tau_n - \tau_{n-1}$ have second moments, this follows immediately by Chebyshev's inequality.) Now S_n has the distribution of $W(\tau_n)$, and τ_n is near $n\sigma^2$ by (37.46); hence S_n should have nearly the distribution of $W(n\sigma^2)$, namely the normal distribution with mean 0 and variance $n\sigma^2$.

To prove this, choose an increasing sequence of integers N_k such that $P[|n^{-1}\tau_n - \sigma^2| \geq k^{-1}] < k^{-1}$ for $n \geq N_k$ and put $\epsilon_n = k^{-1}$ for $N_k \leq n < N_{k+1}$. Then $\epsilon_n \to 0$ and $P[|n^{-1}\tau_n - \sigma^2| \geq \epsilon_n] < \epsilon_n$. By two applications of (37.15),

$$\delta_n(\epsilon) = P[|W(n\sigma^2) - W(\tau_n)|/\sigma\sqrt{n} \geq \epsilon]$$

$$\leq P[|n^{-1}\tau_n - \sigma^2| \geq \epsilon_n] + P\left[\sup_{|t - n\sigma^2| \leq \epsilon_n n} |W(t) - W(n\sigma^2)| \geq \epsilon\sigma\sqrt{n}\right]$$

$$\leq \epsilon_n + 4P[|W(\epsilon_n n)| \geq \epsilon\,\sigma\sqrt{n}],$$

and it follows by Chebyshev's inequality that $\lim_n \delta_n(\epsilon) = 0$. Since S_n is distributed as $W(\tau_n)$,

$$P\left[\frac{W(n\sigma^2)}{\sigma\sqrt{n}} \leq x - \epsilon\right] - \delta_n(\epsilon) \leq P\left[\frac{S_n}{\sigma\sqrt{n}} \leq x\right]$$

$$\leq P\left[\frac{W(n\sigma^2)}{\sigma\sqrt{n}} \leq x + \epsilon\right] + \delta_n(\epsilon).$$

Here $W(n\sigma^2)/\sigma\sqrt{n}$ can be replaced by a random variable N with the standard normal distribution, and letting $n \to \infty$ and then $\epsilon \to 0$ shows that

$$\lim_n P\left[\frac{S_n}{\sigma\sqrt{n}} \leq x\right] = P[N \leq x].$$

* This topic may be omitted.

This gives a new proof of the central limit theorem for independent, identically distributed random variables with second moments (the Lindeberg-Lévy theorem—Theorem 27.1). Observe that none of the convergence theory of Chapter 5 has been used.

This proof of the central limit theorem is an application of the *invariance principle*: S_n has nearly the distribution of $W(n\sigma^2)$, and the distribution of the latter does not depend on (vary with) the distribution common to the X_n. More can be said if the X_n have fourth moments.

For each n, define a stochastic process $[Y_n(t): 0 \le t \le 1]$ by $Y_n(0, \omega) = 0$ and

$$(37.47) \quad Y_n(t, \omega) = \frac{1}{\sigma\sqrt{n}} S_k(\omega) \quad \text{if} \quad \frac{k-1}{n} < t \le \frac{k}{n}, \quad k = 1, \ldots, n.$$

If $k/n = t > 0$ and n is large, then k is large, too, and $Y_n(t) = t^{1/2}S_k/\sigma\sqrt{k}$ is by the central limit theorem approximately normally distributed with mean 0 and variance t. Since the X_n are independent, the increments of (37.47) should be approximately independent, and so the process should behave approximately as a Brownian motion does.

Let τ_n be the stopping times of Theorem 37.7, and in analogy with (37.47) put $Z_n(0) = 0$ and

$$(37.48) \quad Z_n(t) = \frac{1}{\sigma\sqrt{n}} W(\tau_k) \quad \text{if} \quad \frac{k-1}{n} < t \le \frac{k}{n}, \quad k = 1, \ldots, n.$$

By construction, the finite-dimensional distributions of $[Y_n(t): 0 \le t \le 1]$ coincide with those of $[Z_n(t): 0 \le t \le 1]$. It will be shown that the latter process nearly coincides with $[W(tn\sigma^2)/\sigma\sqrt{n}: 0 \le t \le 1]$, which is itself a Brownian motion over the time interval $[0,1]$—see (37.18). Put $W_n(t) = W(tn\sigma^2)/\sigma\sqrt{n}$.

Let $B_n(\delta)$ be the event that $|\tau_k - k\sigma^2| \ge \delta n\sigma^2$ for some $k \le n$. By Kolmogorov's inequality (22.2),

$$(37.49) \qquad P(B_n(\delta)) \le \frac{\text{Var} [\tau_n]}{\delta^2 n^2 \sigma^4} \le \frac{4E[X_1^4]}{\delta^2 n\sigma^4} \to 0.$$

If $(k-1)n^{-1} < t \le kn^{-1}$ and $n > \delta^{-1}$, then

$$\left| \frac{\tau_k}{n\sigma^2} - t \right| \le \left| \frac{\tau_k}{n\sigma^2} - \frac{k}{n} \right| + \frac{1}{n} \le 2\delta$$

on the event $(B_n(\delta))^c$, and so

$$|Z_n(t) - W_n(t)| = \left| W_n\left(\tau\frac{k}{n\sigma^2}\right) - W_n(t) \right| \le \sup_{|s-t| \le 2\delta} |W_n(s) - W_n(t)|$$

on $(B_n(\delta))^c$. Since the distribution of this last random variable is unchanged if the $W_n(t)$ are replaced by $W(t)$,

$$P\left[\sup_{t \leq 1} |Z_n(t) - W_n(t)| \geq \epsilon\right]$$

$$\leq P(B_n(\delta)) + P\left[\sup_{t \leq 1} \sup_{|s-t| \leq 2\delta} |W(s) - W(t)| \geq \epsilon\right].$$

Let $n \to \infty$ and then $\delta \to 0$; it follows by (37.49) and the continuity of Brownian motion paths that

$$(37.50) \qquad \lim_n P\left[\sup_{t \leq 1} |Z_n(t) - W_n(t)| \geq \epsilon\right] = 0$$

for positive ϵ. Since the processes (37.47) and (37.48) have the same finite-dimensional distributions, this proves the following general invariance principle or *functional central limit theorem*.

Theorem 37.8. *Suppose that $X_1, X_2, \ldots$ are independent, identically distributed random variables with mean 0, variance σ^2, and finite fourth moments, and define $Y_n(t)$ by (37.47). There exist (on another probability space) for each n processes $[Z_n(t): 0 \leq t \leq 1]$ and $[W_n(t): 0 \leq t \leq 1]$ such that the first has the same finite-dimensional distributions as $[Y_n(t): 0 \leq t \leq 1]$, the second is a Brownian motion, and $P[\sup_{t \leq 1} |Z_n(t) - W_n(t)| \geq \epsilon] \to 0$ for positive ϵ.*

As an application, consider the maximum $M_n = \max_{k \leq n} S_k$. Now $M_n/\sigma\sqrt{n} = \sup_t Y_n(t)$ has the same distribution as $\sup_t Z_n(t)$, and it follows by (37.50) that

$$P\left[\left|\sup_{t \leq 1} Z_n(t) - \sup_{t \leq 1} W_n(t)\right| \geq \epsilon\right] \to 0.$$

But $P[\sup_{t \leq 1} W_n(t) \geq x] = P[\sup_{t \leq 1} W(t) \geq x] = 2P[N \geq x]$ for $x \geq 0$ by (37.16). Therefore,

$$(37.51) \qquad P\left[\frac{M_n}{\sigma\sqrt{n}} \leq x\right] \to 2P[N \leq x], \qquad x \geq 0.$$

PROBLEMS

37.1. Let $X(t)$ be independent, standard normal variables, one for each dyadic rational t (Theorem 20.4; the unit interval can be used as the probability space). Let $W(0) = 0$ and $W(1) = X(1)$. Suppose that $W(t)$ is already defined for dyadic rationals in $[0, 1]$ of rank n, and put

$$W\left(\frac{2k+1}{2^{n+1}}\right) = \frac{1}{2} W\left(\frac{k}{2^n}\right) + \frac{1}{2} W\left(\frac{k+1}{2^n}\right) + \frac{1}{2^{1+1/2n}} X\left(\frac{2k+1}{2^{n+1}}\right).$$

Prove by induction that the $W(t)$ for dyadic t have the finite-dimensional distributions prescribed for Brownian motion. Now construct a Brownian motion for $0 \leq t \leq 1$ with continuous paths by the method of Theorem 37.1. This avoids an appeal to Kolmogorov's existence theorem.

37.2. 36.10↑ Let $T = [0, \infty)$ and let P be a probability measure on $(R^T, \mathscr{R}^T)$ having the finite-dimensional distributions prescribed for Brownian motion. Let C consist of the continuous elements of R^T.

(a) Show that $P_*(C) = 0$, or $P^*(R^T - C) = 1$ (see (3.9) and (3.10)). Thus completing $(R^T, \mathscr{R}^T, P)$ will not give C probability 1.

(b) Show that $P^*(C) = 1$.

37.3. Suppose that $[W_t: t \geq 0]$ is some stochastic process having independent, stationary increments satisfying $E[W_t] = 0$ and $E[W_t^2] = t$. Show that if the finite-dimensional distributions are preserved by the transformation (37.18), then they must be those of Brownian motion.

37.4. Show that $\bigcap_{t>0}\sigma[W_s: s \geq t]$ contains only sets of probability 0 and 1. Do the same for $\bigcap_{\epsilon>0}[W_t: 0 < t < \epsilon]$; give examples of sets in this σ-field.

37.5. Show by a direct argument that $W(\cdot, \omega)$ is with probability 1 of unbounded variation on $[0, 1]$: Let $Y_n = \sum_{i=1}^{2^n}|W(i2^{-n}) - W((i-1)2^{-n})|$. Show that Y_n has mean $2^{n/2}E[|W_1|]$ and variance $\mathrm{Var}[|W_1|]$. Conclude that $\sum P[Y_n < n] < \infty$.

37.6. Show that the Poisson process as defined by (23.5) is measurable.

37.7. Show that for $T = [0, \infty)$ the coordinate variable process $[Z_t: t \in T]$ on $(R^T, \mathscr{R}^T)$ is not measurable.

37.8. Extend Theorem 37.4 to the set $[t: W(t, \omega) = \alpha]$.

37.9. Let τ_α be the first time the Brownian motion hits $\alpha > 0$: $\tau_\alpha = \inf[t: W_t \geq \alpha]$. Show by (37.16) that

(37.52) $$P[\tau_\alpha \leq t] = \frac{2}{\sqrt{2\pi}} \int_{\alpha/\sqrt{t}}^{\infty} e^{-u^2/2}\, du$$

and hence that the distribution of τ_α has over $(0, \infty)$ the density

(37.53) $$h_\alpha(t) = \frac{\alpha}{\sqrt{2\pi}}\frac{1}{t^{3/2}} e^{-\alpha^2/2t}.$$

Show that $E[\tau_\alpha] = \infty$.

37.10. ↑ (a) Show by the strong Markov property that τ_α and $\tau_{\alpha+\beta} - \tau_\alpha$ are independent and that the latter has the same distribution as τ_β. Conclude that $h_\alpha * h_\beta = h_{\alpha+\beta}$. Show that $\beta\tau_\alpha$ has the same distribution as $\tau_{\alpha\sqrt{\beta}}$.

(b) Show that each h_α is stable—see Problem 28.10.

37.11. ↑ Suppose that $X_1, X_2, \ldots$ are independent and each has the distribution (37.53).

(a) Show that $(X_1 + \cdots + X_n)/n^2$ also has the distribution (37.53). Constrast this with the law of large numbers.

(b) Show that $P[n^{-2}\max_{k \leq n}X_k \leq x] \to \exp(-\alpha\sqrt{2/\pi x})$. Relate this to Theorem 14.3.

37.12. 37.9↑ Let $\rho(s, t)$ be the probability that a Brownian path has at least one zero in (s, t). From (37.52), (37.53), and the Markov property, deduce

(37.54) $$\rho(s, t) = \frac{2}{\pi} \arccos \sqrt{\frac{s}{t}}.$$

Hint: Condition with respect to W_s.

37.13. ↑ **(a)** Show that the probability of no zero in $(t, 1)$ is $(2/\pi)$ arcsin $\sqrt{t}$ and hence that the position of the last zero preceding 1 is distributed over $(0, 1)$ with density $\pi^{-1}(t(1 - t))^{-1/2}$.

(b) Similarly calculate the distribution of the position of the first zero following time 1.

(c) Calculate the joint distribution of the two zeros in (a) and (b).

37.14. ↑ **(a)** Show by Theorem 37.8 that $\inf_{s \leq u \leq t} Y_n(u)$ and $\inf_{s \leq u \leq t} Z_n(u)$ both converge in distribution to $\inf_{s \leq u \leq t} W(u)$ for $0 \leq s \leq t \leq 1$. Prove a similar result for the supremum.

(b) Let $A_n(s, t)$ be the event that S_k is 0 for at least one k in the range $sn \leq k \leq tn$, and show that $P(A_n(s, t)) \to (2/\pi)$ arccos $\sqrt{s/t}$.

(c) Let T_n be the maximum k such that $k \leq n$ and $S_k = 0$. Show that T_n/n has asymptotically the distribution with density $\pi^{-1}(t(1 - t))^{-1/2}$ over $(0, 1)$. As this density is larger at the ends of the interval than in the middle, the last time during a night's play a gambler was even is more likely to be either early or late than to be around midnight.

37.15. ↑ Show that $\rho(s, t) = \rho(t^{-1}, s^{-1}) = \rho(cs, ct)$. Check this by (37.54) and also by the fact that the transformations (37.18) and (37.19) preserve the properties of Brownian motion.

37.16. Show by means of the transformation (37.19) that for positive a and b the probability is 1 that the process is within the boundary $-at < W_t < bt$ for all sufficiently large t. Show that $a/(a + b)$ is the probability that it last touches above rather than below.

37.17. The martingale calculation used for (37.45) also works for slanting boundaries. For positive a, b, r, let τ be the smallest t such that either $W_t = -a + rt$ or $W_t = b + rt$, and let $p(a, b, r)$ be the probability that the exit is through the upper barrier—that $W_\tau = b + r\tau$.

(a) For the martingale $Y_{\theta,t}$ in the proof of Lemma 2, show that $E[Y_{\theta,t}] = 1$. Operating formally at first, conclude that

$$(37.55) \qquad\qquad E[e^{\theta W_\tau - \theta^2 \tau/2}] = 1.$$

Take $\theta = 2r$ and note that $\theta W_\tau - \frac{1}{2}\theta^2 \tau$ is then $2rb$ if the exit is above (probability $p(a, b, r)$) and $-2ra$ if the exit is below (probability $1 - p(a, b, r)$). Deduce

$$p(a, b, r) = \frac{1 - e^{2ra}}{e^{2rb} - e^{-2ra}}.$$

(b) Show that $p(a, b, r) \to a/(a + b)$ as $r \to 0$, in agreement with (37.45).

(c) It remains to justify (37.55) for $\theta = 2r$. From $E[Y_{\theta,t}] = 1$ deduce

$$(37.56) \qquad\qquad E[e^{2r(W_\sigma - r^2\sigma)}] = 1$$

for nonrandom σ. By the arguments in the proofs of Lemmas 2 and 3, show that (37.56) holds for simple stopping times σ, for bounded ones, for $\sigma = \tau \wedge n$, for $\sigma = \tau$.

SECTION 38. SEPARABILITY*

Introduction

As observed a number of times above, the finite-dimensional distributions do not suffice to determine the character of the sample paths of a process. To obtain paths with natural regularity properties, the Poisson and Brownian motion processes were constructed by ad hoc methods. It is always possible to ensure that the paths have a certain very general regularity property called *separability,* and from this property will follow in appropriate circumstances various other desirable regularity properties.

Section 4 dealt with "denumerable" probabilities; questions about path functions involve all the time points and hence concern "nondenumerable" probabilities.

Example 38.1. For a mathematically simple illustration of the fact that path properties are not entirely determined by the finite-dimensional distributions, consider a probability space $(\Omega, \mathcal{F}, P)$ on which is defined a positive random variable V with continuous distribution: $P[V = x] = 0$ for each x. For $t \geq 0$, put $X(t, \omega) = 0$ for all ω, and put

$$(38.1) \qquad Y(t, \omega) = \begin{cases} 1 & \text{if } V(\omega) = t, \\ 0 & \text{if } V(\omega) \neq t. \end{cases}$$

Since V has continuous distribution, $P[X_t = Y_t] = 1$ for each t, and so $[X_t: t \geq 0]$ and $[Y_t: t \geq 0]$ are stochastic processes with identical finite-dimensional distributions; for each $t_1, \ldots, t_k$, the distribution $\mu_{t_1 \ldots t_k}$ common to $(X_{t_1}, \ldots, X_{t_k})$ and $(Y_{t_1}, \ldots, Y_{t_k})$ concentrates all its mass at the origin of R^k. But what about the sample paths? Of course, $X(\cdot, \omega)$ is identically 0, but $Y(\cdot, \omega)$ has a discontinuity—it is 1 at $t = V(\omega)$ and 0 elsewhere. It is because the position of this discontinuity has a continuous distribution that the two processes have the same finite-dimensional distributions. ∎

Definitions

The idea of separability is to make a countable set of time points serve to determine the properties of the process. In all that follows, the time set T will for definiteness be taken as $[0, \infty)$. Most of the results hold with an arbitrary subset of the line in the role of T.

As in Section 36, let R^T be the set of all real functions over $T = [0, \infty)$. Let D be a countable, dense subset of T. A function x— an element of R^T— is

* This section may be omitted.

separable D, or *separable with respect to D*, if for each t in T there exists a sequence $t_1, t_2, \ldots$ of points such that

$$(38.2) \qquad t_n \in D, \qquad t_n \to t, \qquad x(t_n) \to x(t).$$

(Because of the middle condition here, it was redundant to require D dense at the outset.) For t in D, (38.2) imposes no condition on x, since t_n may be taken as t. An x separable with respect to D is determined by its values at the points of D. Note, however, that separability requires that (38.2) hold for every t—an uncountable set of conditions. It is not hard to show that the set of functions separable with respect to D lies outside $\mathcal{R}^T$.

Example 38.2. If x is everywhere continuous or right-continuous, then it is separable with respect to every countable, dense D.

Suppose that $x(t)$ is 0 for $t \neq v$ and 1 for $t = v$, where $v > 0$. Then x is not separable with respect to D unless v lies in D. The paths $Y(\cdot, \omega)$ in Example 38.1 are of this form. ∎

The condition for separability can be stated another way: x is separable D if and only if for every t and every open interval I containing t, $x(t)$ lies in the closure of $[x(s): s \in I \cap D]$.

Suppose that x is separable D and that I is an open interval in T. If $\epsilon > 0$, then $x(t_0) + \epsilon > \sup_{t \in I} x(t) = u$ for some t_0 in I. By separability $|x(s_0) - x(t_0)| < \epsilon$ for some s_0 in $I \cap D$. But then $x(s_0) + 2\epsilon > u$, so that

$$(38.3) \qquad \sup_{t \in I} x(t) = \sup_{t \in I \cap D} x(t).$$

Similarly,

$$(38.4) \qquad \inf_{t \in I} x(t) = \inf_{t \in I \cap D} x(t)$$

and

$$(38.5) \qquad \sup_{t_0 \leq t < t_0 + \delta} |x(t) - x(t_0)| = \sup_{\substack{t_0 \leq t < t_0 + \delta \\ t \in D}} |x(t) - x(t_0)|.$$

A *stochastic process* $[X_t: t \geq 0]$ on $(\Omega, \mathcal{F}, P)$ is *separable D* if D is a countable, dense subset of $T = [0, \infty)$ and there is an $\mathcal{F}$-set N such that $P(N) = 0$ and such that the sample path $X(\cdot, \omega)$ is separable with respect to D for ω outside N. Finally, the process is separable if it is separable with respect to some D; this D is sometimes called a *separant*. In these definitions it is assumed for the moment that $X(t, \omega)$ is a *finite* real number for each t and ω.

Example 38.3. If the sample path $X(\cdot, \omega)$ is continuous for each ω, then the process is separable with respect to each countable, dense D. This covers Brownian motion as constructed in the preceding section. ∎

Example 38.4.   Suppose that $[W_t: t \ge 0]$ has the finite-dimensional distributions of Brownian motion, but do not assume as in the preceding section that the paths are necessarily continuous. Assume, however, that $[W_t: t \ge 0]$ is separable with respect to D. Fix t_0 and δ. Choose sets $D_m = \{t_{m1}, \ldots, t_{mm}\}$ of D-points such that $t_0 < t_{m1} < \cdots < t_{mm} < t_0 + \delta$ and $D_m \uparrow D \cap (t_0, t_0 + \delta)$. By (37.11) and Markov's inequality,

$$P\left[\max_{i \le m} |W(t_{mi}) - W(t_0)| \ge \alpha \right]$$

$$\le 2P[|W(t_{mm}) - W(t_0)| \ge \alpha] \le \frac{6\delta^2}{\alpha^4}.$$

Letting $m \to \infty$ gives

(38.6) $$P\left[\sup_{\substack{t_0 \le t < t_0 + \delta \\ t \in D}} |W_t - W_{t_0}| > \alpha \right] \le \frac{6\delta^2}{\alpha^4}.$$

For sample points outside the N in the definition of separability, the supremum in (38.6) is because of (38.5) unaltered if the restriction $t \in D$ is dropped. Since $P(N) = 0$,

$$P\left[\sup_{t_0 < t < t_0 + \delta} |W_t - W_{t_0}| > \alpha \right] \le \frac{6\delta^2}{\alpha^4}.$$

Define B_n by (37.8) but with r ranging over all the reals (not just over the dyadic rationals) in $[k2^{-n}, (k+1)2^{-n}]$. Then $P(B_n) \le 6n^5/2^n$ follows just as (37.13) does. But for ω outside $B = \limsup_n B_n$, $W(\cdot, \omega)$ is continuous. Since $P(B) = 0$, $W(\cdot, \omega)$ is continuous for ω outside an $\mathscr{F}$-set of probability 0. If $(\Omega, \mathscr{F}, P)$ is complete, then the set of ω for which $W(\cdot, \omega)$ is continuous is an $\mathscr{F}$-set of probability 1. Thus paths are continuous with probability 1 for any separable process having the finite-dimensional distributions of Brownian motion—provided that the underlying space is complete, which can of course always be arranged. ∎

As it will be shown below that there exists a separable process with any consistently prescribed set of finite-dimensional distributions, Example 38.4 provides another approach to the construction of continuous Brownian motion. The value of the method lies in its generality. It must not, however, be imagined that separability automatically ensures smooth sample paths:

Example 38.5.   Suppose that the random variables X_t, $t \ge 0$, are independent, each having the standard normal distribution. Let D be any countable set dense in $T = [0, \infty)$. Suppose that I and J are open intervals with rational endpoints. Since the random variables X_t with $t \in D \cap I$ are independent, and since the value common to the $P[X_t \in J]$ is positive, the second Borel-Cantelli lemma implies that with probability 1, $X_t \in J$ for some t in $D \cap I$. Since there are only countably many pairs I and J with rational endpoints, there is an $\mathcal{F}$-set N such that $P(N) = 0$ and such that for ω outside N the set $[X(t, \omega): t \in D \cap I]$ is everywhere dense on the line for every open interval I in T. This implies that $[X_t: t \geq 0]$ is separable with respect to D. But also of course it implies that the paths are highly irregular. This irregularity is not a shortcoming of the concept of separability—it is a necessary consequence of the properties of the finite-dimensional distributions specified in this example. ∎

Example 38.6. The process $[Y_t: t \geq 0]$ in Example 38.1 is not separable: The path $Y(\cdot, \omega)$ is not separable D unless D contains the point $V(\omega)$. The set of ω for which $Y(\cdot, \omega)$ is separable D is thus contained in $[\omega: V(\omega) \in D]$, a set of probability 0 since D is countable and V has a continuous distribution. ∎

Existence Theorems

It will be proved in stages that for every consistent system of finite-dimensional distributions there exists a separable process having those distributions. Define x to be separable D *at the point* t if there exist points t_n in D such that $t_n \to t$ and $x(t_n) \to x(t)$. Note that this is no restriction on x if t lies in D, and note that separability is the same thing as separability at every t.

Lemma 1. *Let $[X_t: t \geq 0]$ be a stochastic process on $(\Omega, \mathcal{F}, P)$. There exists a countable, dense set D in $[0, \infty)$, and there exists for each t an $\mathcal{F}$-set $N(t)$, such that $P(N(t)) = 0$ and such that for ω outside $N(t)$ the path function $X(\cdot, \omega)$ is separable D at t.*

PROOF. Fix open intervals I and J and consider the probability

$$p(U) = P\left(\bigcap_{s \in U} [X_s \notin J]\right)$$

for countable subsets U of $I \cap T$. As U increases, the intersection here decreases and so does $p(U)$. Choose U_n so that $p(U_n) \to \inf_U p(U)$. If $U(I, J) = \bigcup_n U_n$, then $U(I, J)$ is a countable subset of $I \cap T$ making $p(U)$ minimal:

$$(38.7) \qquad P\left(\bigcap_{s \in U(I,J)} [X_s \notin J]\right) \leq P\left(\bigcap_{s \in U} [X_s \notin J]\right)$$

for every countable subset U of $I \cap T$. If $t \in I \cap T$, then

$$(38.8) \qquad P\left([X_t \in J] \cap \bigcap_{s \in U(I,J)} [X_s \notin J]\right) = 0,$$

because otherwise (38.7) would fail for $U = U(I, J) \cup \{t\}$.

Let $D = \bigcup U(I, J)$, where the union extends over all open intervals I and J with rational endpoints. Then D is a countable, dense subset of T. For each t let

$$(38.9) \qquad N(t) = \bigcup \left\{ [X_t \in J] \cap \bigcap_{s \in U(I,J)} [X_s \notin J] \right\},$$

where the union extends over all open intervals J that have rational endpoints and over all open intervals I that have rational endpoints and contain t. Then $N(t)$ is by (38.8) an $\mathcal{F}$-set such that $P(N(t)) = 0$.

Fix t and $\omega \notin N(t)$. The problem is to show that $X(\cdot, \omega)$ is separable with respect to D at t. Given n, choose open intervals I and J that have rational endpoints and lengths less than n^{-1} and satisfy $t \in I$ and $X(t, \omega) \in J$. Since ω lies outside (38.9), there must be an s_n in $U(I, J)$ such that $X(s_n, \omega) \in J$. But then $s_n \in D$, $|s_n - t| < n^{-1}$, and $|X(s_n, \omega) - X(t, \omega)| < n^{-1}$. Thus $s_n \to t$ and $X(s_n, \omega) \to X(t, \omega)$ for a sequence $s_1, s_2, \ldots$ in D. ∎

For any countable D, the set of ω for which $X(\cdot, \omega)$ is separable with respect to D at t is

$$(38.10) \qquad \bigcap_{n=1}^{\infty} \bigcup_{\substack{|s-t| < n^{-1} \\ s \in D}} [\omega : |X(t, \omega) - X(s, \omega)| < n^{-1}].$$

This set lies in $\mathcal{F}$ for each t, and the point of the lemma is that it is possible to choose D in such a way that each of these sets has probability 1.

Lemma 2. *Let $[X_t : t \geq 0]$ be a stochastic process on $(\Omega, \mathcal{F}, P)$. Suppose that for all t and ω*

$$(38.11) \qquad a < X(t, \omega) < b.$$

Then there exists on $(\Omega, \mathcal{F}, P)$ a process $[X_t' : t \geq 0]$ having these three properties:

(i) *$P[X_t' = X_t] = 1$ for each t.*

(ii) *For some countable, dense subset D of $[0, \infty)$, $X'(\cdot, \omega)$ is separable D for every ω in Ω.*

(iii) *For all t and ω,*

$$(38.12) \qquad a \leq X'(t, \omega) \leq b.$$

PROOF. Choose a countable, dense set D and $\mathcal{F}$-sets $N(t)$ of probability 0 as in Lemma 1. If $t \in D$ or if $\omega \notin N(t)$, define $X'(t, \omega) = X(t, \omega)$. If $t \notin D$, fix some sequence $\{s_n^{(t)}\}$ in D for which $\lim_n s_n^{(t)} = t$ and define $X'(t, \omega) = \limsup_n X(s_n^{(t)}, \omega)$ for $\omega \in N(t)$. To sum up,

$$(38.13) \quad X'(t, \omega) = \begin{cases} X(t, \omega) & \text{if } t \in D \text{ or } \omega \notin N(t) \\ \limsup_n X(s_n^{(t)}, \omega) & \text{if } t \notin D \text{ and } \omega \in N(t). \end{cases}$$

Since $N(t) \in \mathcal{F}$, X_t' is measurable $\mathcal{F}$ for each t. Since $P(N(t)) = 0$, $P[X_t = X_t']$ $= 1$ for each t.

Fix t and ω. If $t \in D$, then certainly $X'(\cdot, \omega)$ is separable D at t, and so assume $t \notin D$. If $\omega \notin N(t)$, then by the construction of $N(t)$, $X(\cdot, \omega)$ is separable with respect to D at t, so that there exist points s_n in D such that $s_n \to t$ and $X(s_n, \omega)$ $\to X(t, \omega)$. But $X(s_n, \omega) = X'(s_n, \omega)$ because $s_n \in D$, and $X(t, \omega) = X'(t, \omega)$ because $\omega \notin N(t)$. Hence $X'(s_n, \omega) \to X'(t, \omega)$, and so $X'(\cdot, \omega)$ is separable with respect to D at t. Finally, suppose that $t \notin D$ and $\omega \in N(t)$. Then $X'(t, \omega) = \lim_k X(s_{n_k}^{(t)}, \omega)$ for some sequence $\{n_k\}$ of integers. As $k \to \infty$, $s_{n_k}^{(t)} \to t$ and $X'(s_{n_k}^{(t)}, \omega) = X(s_{n_k}^{(t)}, \omega) \to X'(t, \omega)$, so that again $X'(\cdot, \omega)$ is separable with respect to D at t. Clearly, (38.11) implies (38.12). ∎

Example 38.7. One must allow for the possibility of equality in (38.12). Suppose that $V(\omega) > 0$ for all ω and that V has continuous distribution. Define

$$f(t) = \begin{cases} e^{-|t|} & \text{if } t \neq 0, \\ 0 & \text{if } t = 0, \end{cases}$$

and put $X(t, \omega) = f(t - V(\omega))$. If $[X_t': t \geq 0]$ is any separable process with the same finite-dimensional distributions as $[X_t: t \geq 0]$, then $X'(\cdot, \omega)$ must with probability 1 assume the value 1 somewhere. In this case (38.11) holds for $a < 0$ and $b = 1$, and equality in (38.12) cannot be avoided. ∎

If

$$(38.14) \qquad\qquad \sup_{t, \omega} |X(t, \omega)| < \infty,$$

then (38.11) holds for some a and b. To treat the case in which (38.14) fails, it is necessary to allow for the possibility of infinite values. If $x(t)$ is ∞ or $-\infty$, replace the third condition in (38.2) by $x(t_n) \to \infty$ or $x(t_n) \to -\infty$. This extends the definition of separability to functions x that may assume infinite values and to processes $[X_t: t \geq 0]$ for which $X(t, \omega) = \pm\infty$ is a possibility.

Theorem 38.1. *If $[X_t: t \geq 0]$ is a finite-valued process on $(\Omega, \mathcal{F}, P)$, there exists on the same space a separable process $[X_t': t \geq 0]$ such that $P[X_t' = X_t]$ $= 1$ for each t.*

It is for convenience assumed here that $X(t, \omega)$ is finite for all t and ω, although this is not really necessary. But in some cases infinite values for certain $X'(t, \omega)$ cannot be avoided—see Example 38.8.

PROOF. If (38.14) holds, the result is an immediate consequence of Lemma 2. The definition of separability allows an exceptional set N of probability 0; in the construction of Lemma 2 this set is actually empty, but it is clear from the definition that this could be arranged anyway.

The case in which (38.14) may fail could be treated by tracing through the preceding proofs, making slight changes to allow for infinite values. A simple argument makes this unnecessary. Let g be a continuous, strictly increasing mapping of R^1 onto $(0, 1)$. Let $Y(t, \omega) = g(X(t, \omega))$. Lemma 2 applies to $[Y_t: t \geq 0]$; there exists a separable process $[Y_t': t \geq 0]$ such that $P[Y_t' = Y_t] = 1$. Since $0 < Y(t, \omega) < 1$, Lemma 2 ensures $0 \leq Y'(t, \omega) \leq 1$. Define

$$X'(t, \omega) = \begin{cases} -\infty & \text{if } Y'(t, \omega) = 0, \\ g^{-1}(Y'(t, \omega) & \text{if } 0 < Y'(t, \omega) < 1, \\ +\infty & \text{if } Y'(t, \omega) = 1. \end{cases}$$

Then $[X_t': t \geq 0]$ satisfies the requirements. Note that $P[X_t' = \pm\infty] = 0$ for each t. ∎

Example 38.8. Suppose that $V(\omega) > 0$ for all ω and V has a continuous distribution. Define

$$h(t) = \begin{cases} |t|^{-1} & \text{if } t \neq 0, \\ 0 & \text{if } t = 0, \end{cases}$$

and put $X(t, \omega) = h(t - V(\omega))$. This is analogous to Example 38.7. If $[X_t': t \geq 0]$ is separable and has the finite-dimensional distributions of $[X_t: t \geq 0]$, then $X'(\cdot, \omega)$ must with probability 1 assume the value ∞ for some t. ∎

Combining Theorem 38.1 with Kolmogorov's existence theorem shows that *for any consistent system of finite-dimensional distributions $\mu_{t_1 \ldots t_k}$ there exists a separable process with the $\mu_{t_1 \ldots t_k}$ as finite-dimensional distributions.* As shown in Example 38.4, this leads to another construction of Brownian motion with continuous paths.

Consequences of Separability

The next theorem implies in effect that, if the finite-dimensional distributions of a process are such that it "should" have continuous paths, then it will in fact have continuous paths if it is separable. Example 38.4 illustrates this. The same thing holds for properties other than continuity.

Let $\overline{R}^T$ be the set of functions on $T = [0, \infty)$ with values that are ordinary reals or else ∞ or $-\infty$. Thus $\overline{R}^T$ is an enlargement of the R^T of Section 36, an enlargement necessary because separability forces infinite values. Define the function Z_t on $\overline{R}^T$ by $Z_t(x) = Z(t, x) = x(t)$. This is just an extension of the coordinate function (36.6). Let $\overline{\mathcal{R}}^T$ be the σ-field in $\overline{R}^T$ generated by the Z_t, $t \geq 0$.

Suppose that A is a subset of $\overline{R}^T$, not necessarily in $\overline{\mathcal{R}}^T$. For $D \subset T = [0, \infty)$, let A_D consist of those elements x of $\overline{R}^T$ that agree on D with some element y of A:

$$(38.15) \qquad A_D = \bigcup_{y \in A} \bigcap_{t \in D} [x \in \overline{R}^T : x(t) = y(t)].$$

Of course, $A \subset A_D$. Let S_D denote the set of x in $\overline{R}^T$ that are separable with respect to D.

In the following theorem, $[X_t : t \geq 0]$ and $[X_t' : t \geq 0]$ are processes on spaces $(\Omega, \mathcal{F}, P)$ and $(\Omega', \mathcal{F}', P')$ which may be distinct; the path functions are $X(\cdot, \omega)$ and $X'(\cdot, \omega')$.

Theorem 38.2. *Suppose of A that for each countable, dense subset D of $T = [0, \infty)$, the set (38.15) satisfies*

$$(38.16) \qquad A_D \in \overline{\mathcal{R}}^T, \qquad A_D \cap S_D \subset A.$$

If $[X_t : t \geq 0]$ and $[X_t' : t \geq 0]$ have the same finite-dimensional distributions, if $[\omega : X(\cdot, \omega) \in A]$ lies in $\mathcal{F}$ and has P-measure 1, and if $[X_t' : t \geq 0]$ is separable, then $[\omega' : X'(\cdot, \omega') \in A]$ contains an $\mathcal{F}'$-set of P'-measure 1.

If $(\Omega', \mathcal{F}', P')$ is complete, then of course $[\omega' : X'(\cdot, \omega') \in A]$ is itself an $\mathcal{F}'$-set of P'-measure 1.

PROOF. Suppose that $[X_t' : t \geq 0]$ is separable with respect to D. The difference $[\omega' : X'(\cdot, \omega') \in A_D] - [\omega' : X'(\cdot, \omega') \in A]$ is by (38.16) a subset of $[\omega' : X'(\cdot, \omega') \in \overline{R}^T - S_D]$, which is contained in an $\mathcal{F}'$-set N' of P'-measure 0. Since the two processes have the same finite-dimensional distributions and hence induce the same distribution on $(\overline{R}^T, \overline{\mathcal{R}}^T)$, and since A_D lies in $\overline{\mathcal{R}}^T$, $P'[\omega' : X'(\cdot, \omega') \in A_D] = P[\omega : X(\cdot, \omega) \in A_D] \geq P[\omega : X(\cdot, \omega) \in A] = 1$. Thus the subset $[\omega' : X'(\cdot, \omega') \in A_D] - N'$ of $[\omega' : X'(\cdot, \omega') \in A]$ lies in $\mathcal{F}'$ and has P'-measure 1. ∎

Example 38.9. Consider the set C of finite-valued, continuous functions on T. If $x \in S_D$ and $y \in C$, and if x and y agree on a dense D, then x and y agree everywhere: $x = y$. Therefore, $C_D \cap S_D \subset C$. Further,

$$C_D = \bigcap_{\epsilon, t} \bigcup_{\delta} \bigcap_s [x \in \overline{R}^T : |x(s)| < \infty, |x(t)| < \infty, |x(s) - x(t)| < \epsilon],$$

where ϵ and δ range over the positive rationals, t ranges over D, and the inner intersection extends over the s in D satisfying $|s - t| < \delta$. Hence $C_D \in \overline{\mathscr{R}}^T$. Thus C satisfies the condition (38.16).

Theorem 38.2 now implies that if a process has continuous paths with probability 1, then any separable process having the same finite-dimensional distributions has continuous paths outside a set of probability 0. In particular, a Brownian motion with continuous paths was constructed in the preceding section, and so any separable process with the finite-dimensional distributions of Brownian motion has continuous paths outside a set of probability 0. The argument in Example 38.4 now becomes supererogatory. ∎

Example 38.10. There is a somewhat similar argument for the step functions of the Poisson process. Let Z^+ be the set of nonnegative integers, let E consist of the nondecreasing functions x in $\overline{R}^T$ such that $x(t) \in Z^+$ for all t and such that for every $n \in Z^+$ there exists a nonempty interval I such that $x(t) = n$ for $t \in I$. Then

$$E_D = \bigcap_{t \in D} [x: x(t) \in Z^+] \cap \bigcap_{s,t \in D, s<t} [x: x(s) \le x(t)]$$

$$\cap \bigcap_{n=0}^{\infty} \bigcup_{I} \bigcap_{t \in D \cap I} [x: x(t) = n],$$

where I ranges over the open intervals with rational endpoints. Thus $E_D \in \overline{\mathscr{R}}^T$. Clearly, $E_D \cap S_D \subset E$, and so Theorem 38.2 applies.

In Section 23 was constructed a Poisson process with paths in E, and therefore any separable process with the same finite-dimensional distributions will have paths in E except for a set of probability 0. ∎

Example 38.11. For E as in Example 38.10, let E_0 consist of the elements of E that are right-continuous; a function in E need not lie in E_0, although at each t it must be continuous from one side or the other. The Poisson process as defined in Section 23 by $N_t = \max[n: S_n \le t]$ (see (23.5)) has paths in E_0. But if $N_t' = \max[n: S_n < t]$, then $[N_t': t \ge 0]$ is separable and has the same finite-dimensional distributions, but its paths are not in E_0. Thus E_0 does not satisfy the hypotheses of Theorem 38.2. Separability does not help distinguish between continuity from the right and continuity from the left. ∎

Example 38.12. The class of sets A satisfying (38.16) is closed under the formation of countable unions and intersections but is not closed under complementation. Define X_t and Y_t as in Example 38.1 and let C be the set of continuous paths. Then $[Y_t: t \ge 0]$ and $[X_t: t \ge 0]$ have the same finite-dimensional distributions and the latter is separable; $Y(\cdot, \omega)$ is in $\overline{R}^T - C$ for each ω, and $X(\cdot, \omega)$ is in $\overline{R}^T - C$ for no ω. ∎

Example 38.13. As a final example, consider the set J of functions with discontinuities of at most the first kind: x is in J if it is finite-valued, if $x(t+) = \lim_{s \downarrow t} x(s)$ exists (finite) for $t \geq 0$ and $x(t-) = \lim_{s \uparrow t} s(s)$ exists (finite) for $t > 0$, and if $x(t)$ lies between $x(t+)$ and $x(t-)$ for $t > 0$. Continuous and right-continuous functions are special cases.

Let V denote the general system

$$(38.17) \qquad V: \; k; r_1, \ldots, r_k; s_1, \ldots, s_k; \alpha_1, \ldots, \alpha_k,$$

where k is an integer, where the r_i, s_i, and α_i are rational, and where

$$0 = r_1 < s_1 < r_2 < s_2 < \cdots < r_k < s_k.$$

Define

$$J(D, V, \epsilon) = \bigcap_{i=1}^{k} [x : \alpha_i \leq x(t) \leq \alpha_i + \epsilon, t \in (r_i, s_i) \cap D]$$

$$\cap \bigcap_{i=2}^{k} [x : \min \{\alpha_{i-1}, \alpha_i\} \leq x(t) \leq \max \{\alpha_{i-1}, \alpha_i\} + \epsilon, t \in (s_{i-1}, r_i) \cap D].$$

Let $\mathcal{U}_{m,k,\delta}$ be the class of systems (38.17) that have a fixed value for k and satisfy $r_i - s_{i-1} < \delta, i = 2, \ldots, k$ and $s_k > m$. It will be shown that

$$(38.18) \qquad J_D = \bigcap_{m=1}^{\infty} \bigcap_{\epsilon} \bigcup_{k=1}^{\infty} \bigcap_{\delta} \bigcup_{V \in \mathcal{U}_{m,k,\delta}} J(D, V, \epsilon),$$

where ϵ and δ range over the positive rationals. From this it will follow that $J_D \in \overline{\mathcal{R}}^T$. It will also be shown that $J_D \cap S_D \subset J$, so that J satisfies the hypothesis of Theorem 38.2.

Suppose that $y \in J$. For fixed ϵ, let H be the set of nonnegative h for which there exist finitely many points t_i such that $0 = t_0 \leq t_1 \leq \cdots \leq t_r = h$ and $|y(t) - y(t')| < \epsilon$ for t and t' in the same interval (t_{i-1}, t_i). If $h_n \in H$ and $h_n \uparrow h$, then from the existence of $y(h-)$ follows $h \in H$. Hence H is closed. If $h \in H$, from the existence of $y(h+)$ it follows that H contains points to the right of h. Therefore, $H = [0, \infty)$. From this it follows that the right side of (38.18) contains J_D.

Suppose that x is a member of the right side of (38.18). It is not hard to deduce that for each t the limits

$$(38.19) \qquad \lim_{s \downarrow t, s \in D} x(s), \qquad \lim_{s \uparrow t, s \in D} x(s)$$

exist and that $x(t)$ lies between them if $t \in D$. For $t \in D$ take $y(t) = x(t)$, and for $t \notin D$ take $y(t)$ to be the first limit in (38.19). Then $y \in J$ and hence $x \in J_D$. This argument also shows that $J_D \cap S_D \subset J$. ∎

Separability in Product Space

Kolmogorov's existence theorem applies in $(\overline{R}^T, \overline{\mathcal{R}}^T)$: Construct on $(R^T, \mathcal{R}^T)$ a probability measure P_0 with the specified finite-dimensional distributions. The map $\rho: R^T \to \overline{R}^T$ defined by $\rho(x) = x$ is measurable $\mathcal{R}^T / \overline{\mathcal{R}}^T$; take $P = P_0 \rho^{-1}$. The process $[Z_t : t \geq 0]$ on $(\overline{R}^T, \overline{\mathcal{R}}^T, P)$ then has the required finite-dimensional distributions.

Apply Theorem 38.1 to $[X_t : t \geq 0] = [Z_t : t \geq 0]$ and $(\Omega, \mathcal{F}, P) = (\overline{R}^T, \overline{\mathcal{R}}^T, P)$. Specifically, apply the construction in Lemma 2, but without the restriction (38.11), so that infinite values are allowed. Since the set (38.10) lies in $\mathcal{F} = \overline{\mathcal{R}}^T$, the sets $N(t)$ of Lemma

1 can be taken to be $\mathcal{F}$-sets. Thus each $N(t)$ can be taken to be exactly the ω-set where $X(\cdot, \omega)$ is not separable with respect to D at t.

Let Ω' be the set where the new path $X'(\cdot, \omega)$ is the same as the old path $X(\cdot, \omega)$:

$$(38.20) \qquad \Omega' = \bigcap_{t \geq 0} [\omega: X'(t, \omega) = X(t, \omega)].$$

The proof of Lemma 2 shows that $X'(\cdot, \omega)$ is separable with respect to D for every ω. Therefore, $X(\cdot, \omega)$ is separable with respect to D if $\omega \in \Omega'$. It will be shown that Ω' has outer measure 1: $\Omega' \subset A$ and $A \in \mathcal{F} = \overline{\mathcal{R}}^T$ together imply that $P(A) = 1$.

By Theorem 36.3(ii), $A \in \sigma[X_t: t \in S]$ for some countable set S in T. If

$$(38.21) \qquad \bigcap_{s \in S} [\omega: X'(t, \omega) = X(t, \omega)] \subset A,$$

then $P(A) = 1$ will follow because $P[X_t' = X_t] = 1$ for each t. Now $X'(\cdot, \omega)$ is an element of $\Omega = \overline{R}^T$; denote it $\psi(\omega)$, so that $X'(t, \omega) = X(t, \psi(\omega))$. Since $X'(\cdot, \omega)$ is by construction separable D for each ω, so is $X(\cdot, \psi(\omega)) = \psi(\omega)$. Thus $\psi(\omega) \notin N(t)$ for each t. Then by the construction (38.13), $X(t, \psi,(\omega)) = X'(t, \psi(\omega))$ for all t, so that $\psi(\omega) \in \Omega'$ and hence $\psi(\omega) \in A$. But $X(t, \omega) = X'(t, \omega)$ for all $t \in S$ implies that $X(t, \omega) = X(t, \psi(\omega))$ for all $t \in S$; hence (Theorem 36.3(i)) A contains ω as well as $\psi(\omega)$. This proves (38.21).

Let $\mathcal{F}'$ consist of the sets of the form $\Omega' \cap A$ for $A \in \mathcal{F}$. Then $\mathcal{F}'$ is a σ-field in Ω', and because Ω' has outer measure 1, setting $P'(\Omega' \cap A) = P(A)$ consistently defines a probability measure on $\mathcal{F}'$. The coordinate functions $X_t = Z_t$ restricted to Ω' give a stochastic process on $(\Omega', \mathcal{F}', P')$ for which all paths are separable and the finite-dimensional distributions are the same as before.

Notes on the Problems

These notes consist of hints, solutions, and references to the literature. As a rule a solution is complete in proportion to the frequency with which it is needed for the solution of subsequent problems.

Section 1

1.1. (a) Each point of the discrete space lies in one of the four sets $A_1 \cap A_2$, $A_1^c \cap A_2$, $A_1 \cap A_2^c$, $A_1^c \cap A_2^c$ and hence would have probability at most 2^{-2}; continue.

(b) If, for each i, B_i is A_i or A_i^c, then $B_1 \cap \cdots \cap B_n$ has probability at most $\prod_{i=1}^n (1 - \alpha_i) \leq \exp[-\Sigma_{i=1}^n \alpha_i]$.

1.5. (a) Cover $A_b(\beta)$ by $(b-1)^n$ intervals of length b^{-n}.

(c) Go to the base b^k. Identify the digits in the base b with the keys of the typewriter. The monkey is certain eventually to reproduce the eleventh edition of the *Britannica* and even, unhappily, the fifteenth.

1.8. The interchange of limit and integral is justified because the series Σ_k $r_k(\omega)2^{-k}$ converges uniformly in ω (integration to the limit is studied systematically in Section 16). There is a direct derivation of (1.36): $\sin t = 2^n \sin 2^{-n}t \prod_{k=1}^n \cos 2^{-k}t$, as follows inductively by the double-angle formula; use the fact that $x^{-1} \sin x \to 1$ as $x \to 0$.

1.9. Within each b-adic interval of rank $n-1$, $r_n(\omega)$ assumes the value $b-1$ on a subinterval of length b^{-n} and the value -1 on subintervals of total length $(b-1)b^{-n}$. The proofs therefore parallel those for the case $b = 2$, $\beta = 1$.

1.10. Choose q_0 so that $\Sigma_{q \geq q_0} qf(q) < \epsilon$; cover A_f by the intervals $(pq^{-1} - f(q), pq^{-1} + f(q))$ for $q \geq q_0$ and $0 \leq p \leq q$.

See HARDY & WRIGHT, page 158, for a proof that infinitely many pq^{-1} come within q^{-2} of an irrational ω. If $q^2f(q)$ is nonincreasing and $\Sigma_q qf(q) = \infty$, it can be shown that A_f has negligible complement, but this is harder; see page 69 of A. Ya. Khinchine: *Continued Fractions* (University of Chicago Press, Chicago, 1964). For example, except for a negligible set of ω, $|\omega - p/q| < 1/q^2 \log q$ has infinitely many solutions.

Section 2

2.3. (b) Four points suffice.

2.5. (b) The class in question is certainly contained in $f(\mathcal{A})$ and is easily seen to be closed under the formation of finite intersections. But $(\bigcup_{i=1}^{m} \bigcap_{j=1}^{n_i} A_{ij})^c$ $= \bigcap_{i=1}^{m} \bigcup_{j=1}^{n_i} A_{ij}^c$, and $\bigcup_{j=1}^{n_i} A_{ij}^c = \bigcup_{j=1}^{n_i} [A_{ij}^c \cap \bigcap_{k=1}^{j-1} A_{ik}]$ has the required form.

2.6. If $\mathcal{K}$ is the smallest class over $\mathcal{A}$ closed under the formation of countable unions and intersections, clearly $\mathcal{K} \subset \sigma(\mathcal{A})$. To prove the reverse inclusion, first show that the class of A such that $A^c \in \mathcal{K}$ is closed under the formation of countable unions and intersections and contains $\mathcal{A}$ and hence contains $\mathcal{K}$.

2.8. Note that $\bigcup_n B_n \in \sigma(\bigcup_n \mathcal{A}_{B_n})$.

2.11. Suppose that $A_1, A_2, \ldots$ is an infinite sequence of distinct sets in a σ-field $\mathcal{F}$, and let $\mathcal{G}$ consist of the nonempty sets of the form $\bigcap_{n=1}^{\infty} B_n$, where $B_n = A_n$ or $B_n = A_n^c$, $n = 1, 2, \ldots$. Each A_n is the union of the $\mathcal{G}$-sets it contains, and since the A_n are distinct, $\mathcal{G}$ must be infinite. But there are uncountably many distinct countable unions of $\mathcal{G}$-sets, and they all lie in $\mathcal{F}$.

2.13. If $B_n = A_n \cap A_1^c \cap \cdots \cap A_{n-1}^c$, then $A = \bigcup_n B_n$, the B_n are disjoint, and $P(B_n) = P(A_n)$ because $A_n \Delta B_n \subset \bigcup_{m=1}^{n-1} (A_m \cap A_n)$.

2.15. For this and the subsequent problems on applications of probability theory to arithmetic, the only number theory required is the fundamental theorem of arithmetic and its immediate consequences. The other problems on stochastic arithmetic are 2.15, 3.2, 4.21, 5.16, 5.17, 6.17, 18.14, 25.16, 30.7, 30.8, 30.9, and 30.10. See also Theorem 30.3.

(b) Let A consist of the even integers, let $C_k = [m: v_k < m \leq v_{k+1}]$, and let B consist of the even integers in $C_1 \cup C_2 \cup \cdots$ together with the odd integers in $C_2 \cup C_4 \cup \cdots$; take v_k to increase very rapidly with k and consider $A \cap B$.

(c) If c is the least common multiple of a and b, then $M_a \cap M_b = M_c$. From $M_a \in \mathcal{D}$ conclude in succession that $M_a \cap M_b \in \mathcal{D}$, $M_{a_1} \cap \cdots \cap M_{a_j} \cap M_{b_1}^c$ $\cap \cdots \cap M_{b_k}^c \in \mathcal{D}$, $f(\mathcal{M}) \subset \mathcal{D}$. By the same sequence of steps, show how D on $\mathcal{M}$ determines D on $f(\mathcal{M})$.

(d) If $B_l = M_a - \bigcup_{p \leq l} M_{ap}$, then $a \in B_l$ and (the inclusion-exclusion formula requires only finite additivity)

$$D(B_l) = \frac{1}{a} - \sum_{p \leq l} \frac{1}{ap} + \sum_{p < q \leq l} \frac{1}{apq} - \cdots$$

$$= \frac{1}{a} \prod_{p \leq l} \left(1 - \frac{1}{p}\right) \leq \frac{1}{a} \exp\left(-\sum_{p \leq l} \frac{1}{p}\right) \to 0.$$

Choose l_a so that, if $C_a = B_{l_a}$, then $D(C_a) < 2^{-a-1}$. If D were a probability measure on $f(\mathcal{M})$, $D(\Omega) \leq \frac{1}{2}$ would follow.

2.16. **(a)** Use the fact that, if $k < n$, a cylinder of rank k can be split into 2^{n-k} cylinders of rank n, and the P-values add.

(b) It suffices (Example 2.10) to prove that $A_n \in \mathcal{C}_0$, $A_1 \supset A_2 \supset \cdots$ and $P(A_n) \geq \epsilon > 0$ together imply that $\bigcap_n A_n$ is nonempty. But each A_n is nonempty, and the diagonal method (Theorem 25.13) can be used to produce a sequence common to the A_n. (As a topological product, sequence space is compact, and the cylinders are closed.)

(c) The sequence (1.7) cannot end in 0's, but the sequence $(a_1(\omega), a_2(\omega), \ldots)$ can. Dyadic intervals can contract to the empty set, but nonempty cylinders cannot. The diagonal method plays here the role the Heine-Borel theorem plays in the proof of Lemma 2 to Theorem 2.2.

2.17. **(a)** Apply the intermediate-value theorem to the function $f(x) = \lambda(A \cap (0,x])$. Note that this even proves part (c) for λ (under the assumption that λ exists).

(b) If $0 < P(B) < P(A)$, then either $0 < P(B) \leq \frac{1}{2}P(A)$ or $0 < P(B - A) \leq \frac{1}{2}P(A)$. Continue.

(c) If $P(\bigcup_k H_k) < x$, choose C so that $C \subset A - \bigcup_k H_k$ and $0 < P(C) < x - P(\bigcup_k H_k)$. If $n^{-1} < P(C)$, then $P(\bigcup_{k<n}H_k) + h_n < P(\bigcup_{k<n}H_k) + P(H_n) + P(C) \leq P(\bigcup_{k<n}H_k) + h_n$. Zorn's lemma leads to a simpler proof.

2.18. **(a)** Choose B so that $P_1(B) = \frac{1}{2}$; take $A = B$ or $A = B^c$ according as $P_2(B) > \frac{1}{2}$ or $P_2(B) \leq \frac{1}{2}$. This corresponds to the scheme for dividing a piece of cake between two children: one divides and the other chooses.

(b) If $P_1, \ldots, P_n$ are nonatomic, Ω partitions into sets $A_1, \ldots, A_n$ such that $P_i(A_i) \geq 1/n$. The cake version of the proof runs as follows. The n children line up. The first cuts off $1/n$ of the cake (according to his measure) and passes it down the line, and any child who thinks it too large shaves it down to $1/n$, returning the shavings (somehow) to the body of the cake. The last child to shave the piece down (or the first child if none do) keeps it, and the remaining $n - 1$ children proceed to divide the remaining cake.

It can also be shown, although this is much harder, that $[(P_1(A), \ldots, P_n(A)): A \in \mathcal{F}]$ is a closed, convex set in n-space; see Paul R. Halmos: The range of a vector measure, *Bull. Amer. Math. Soc.*, **54** (1948), 416–421.

2.21. Use the fact that, if $\alpha_1, \alpha_2, \ldots$ is a sequence of ordinals satisfying $\alpha_n < \Omega$, then there exists an ordinal α such that $\alpha < \Omega$ and $\alpha_n < \alpha$ for all n.

2.22. Suppose that $B_j \in \bigcup_{\beta<\alpha}\mathcal{I}_\beta$, $j = 1, 2, \ldots$. Choose odd integers n_j in such a way that $B_j \in \mathcal{I}_{\beta_\alpha(n_j)}$ and the n_j are all distinct; choose $\mathcal{I}_0$-sets such that

$$\Phi_{\beta_\alpha(n_j)}(C_{m_{n_j1}}, C_{m_{n_j2}}, \ldots) = B_j;$$

for n not of the form n_j, choose $\mathcal{J}_0$-sets for which $\Phi_{\beta_\alpha(n)}(C_{m_{n_1}}, C_{m_{n_2}}, \ldots)$ is 0 or $(0, 1]$ as n is odd or even. Then $\bigcup_{j=1}^\infty B_j = \Phi_\alpha(C_1, C_2, \ldots)$. Similarly, $B^c = \Phi_\alpha(C_1, C_2, \ldots)$ for $\mathcal{J}_0$-sets C_n if $B \in \bigcup_{\beta<\alpha}\mathcal{J}_\beta$. The rest of the proof is essentially the same as before.

2.23. If E is a finite set of integers, then $E \subset \{1, 2, \ldots, n\}$ for some n, so that by nonnegativity $\Sigma_{k \in E} x_k \le \Sigma_{k \le n} x_k$. The set of partial sums $\Sigma_{k \le n} x_k$ thus has the same supremum as the larger set of sums $\Sigma_{k \in E} x_k$ (E finite).

2.24. Suppose each $\Sigma_j x_{ij}$ converges and $\Sigma_i \Sigma_j x_{ij}$ converges. If E is a finite set of the pairs (i, j), there is an n for which $\Sigma_{(i,j) \in E} x_{ij} \le \Sigma_{i \le n} \Sigma_{j \le n} x_{ij} \le \Sigma_{i \le n} \Sigma_j x_{ij} \le \Sigma_i \Sigma_j x_{ij}$; hence $\Sigma_{ij} x_{ij}$ converges and has sum at most $\Sigma_i \Sigma_j x_{ij}$. On the other hand, if $\Sigma_{ij} x_{ij}$ converges, then $\Sigma_{i \le m} \Sigma_{j \le n} x_{ij} \le \Sigma_{ij} x_{ij}$; let $n \to \infty$ and then $m \to \infty$.

Section 3

3.1. The examples in which P and P^* fail to agree on $\mathcal{J}_0$ are exactly those in which P is not countably additive there.

3.2. Note that $\{a\} = M_a \cap M_{2a}^c \cap M_{3a}^c \cap \cdots$. Use Theorem 3.1.

3.3. (b) Suppose that $P_*(A) = P^*(A)$ and choose $\mathcal{J}$-sets A_1 and A_2 in such a way that $A_1 \subset A \subset A_2$ and $P(A_1) = P(A_2)$. Given E choose an $\mathcal{J}$-set B such that $E \subset B$ and $P^*(E) = P(B)$. Use $P(A_2 - A_1) = 0$ to check (3.5).

3.7. If $\mathcal{J}$ is separable, then (Problem 2.5 (c)) there is a countable field $\mathcal{J}_0$ such that $\mathcal{J} = \sigma(\mathcal{J}_0)$. Given an $\mathcal{J}$-set A and a positive ϵ, choose $\mathcal{J}_0$-sets A_k such that $A \subset \bigcup_k A_k$ and $\Sigma P(A_k) < P^*(A) + \epsilon = P(A) + \epsilon$. If n is large enough and $B = \bigcup_{k \le n} A_k$, then $d(A, B) < \epsilon$. There are only countably many such sets B.

3.8. (b) Four points suffice.

3.18. (a) Since the $A \oplus r$ are disjoint Borel sets, $\Sigma_r \lambda(A \oplus r) \le 1$, and so the common value $\lambda(A)$ of the $\lambda(A \oplus r)$ must be 0. Similarly, if A is a Borel set contained in some $H \oplus r$, then $\lambda(A) = 0$.

(b) If the $E \cap (H \oplus r)$ are all Borel sets, they all have Lebesgue measure 0, and so E is a Borel set of Lebesgue measure 0.

Section 4

4.7. If $\{A_n\}$ is a Cauchy sequence, ensure by passing to a subsequence that $P(A_n \Delta A_{n+1}) < 2^{-n}$, take $A = \limsup_n A_n$, and apply the first Borel-Cantelli lemma and Problem 4.5(e).

4.11. Define A_{ij} as in Problem 4.8 and put $\mathcal{A} = \{A_{12}, A_{13}\}$. Although A_{23} is independent of $\mathcal{A}$, it is not independent of $\sigma(\mathcal{A})$, which in fact contains it—this despite the fact that the partitions for $\mathcal{A}$ and $\sigma(\mathcal{A})$ coincide (see Example 4.4). It is as though an observer with the items of information in $\mathcal{A}$ is unable to combine them to get information about $A_{23} = (A_{12} \Delta A_{13})^c$. For this reason, the partition defined by (4.14) should be viewed as specifying the information in $\sigma(\mathcal{A})$ rather than the information in $\mathcal{A}$.

4.13. If $(H \cap G_1) \cup (H^c \cap G_2) = (H \cap G_1') \cup (H^c \cap G_2')$, then $G_1 \Delta G_1' \subset H^c$ and $G_2 \Delta G_2' \subset H$; consistency now follows because $\lambda_*(H) = \lambda_*(H^c) = 0$. If $A_n = (H \cap G_1^{(n)}) \cup (H^c \cap G_2^{(n)})$ are disjoint, then $G_1^{(m)} \cap G_1^{(n)} \subset H^c$ and $G_2^{(m)} \cap G_2^{(n)} \subset H$ for $m \neq n$, and therefore (see Problem 2.13) $P(\bigcup_n A_n) = \frac{1}{2} \lambda (\bigcup_n G_1^{(n)}) + \frac{1}{2} \lambda (\bigcup_n G_2^{(n)}) = \Sigma_n (\frac{1}{2} \lambda (G_1^{(n)}) + \frac{1}{2} \lambda (G_2^{(n)})) = \Sigma_n P(A_n)$. The intervals with rational endpoints generate $\mathcal{G}$.

4.14. **(c)** Consider disjoint events $A_1, A_2, \ldots$ for which $P(A_n) = 1/2^n$.

4.19. The problem being symmetric in A_n and A_n^c, suppose that $\alpha_n = p_n$ and define A_n as in Problem 4.18. Define $B = \lim \inf_n A_n^c$, apply the first Borel-Cantelli lemma, and pass from sequence space to B and from A_n to $B \cap A_n$.

4.20. Show as in Problem 1.1(b) that the maximum of $P(B_1 \cap \cdots \cap B_n)$, where B_i is A_i or A_i^c, goes to 0. Let $A_x = [\omega : \Sigma_n I_{A_n}(\omega) 2^{-n} \leq x]$, show that $P(A \cap A_x)$ is continuous in x, and proceed as in Problem 2.17(a).

4.21. Calculate $D(F_l)$ by (2.22) and the inclusion-exclusion formula, and estimate $P_n(F_l) - P_n(F)$ by subadditivity; now use $0 \leq P_n(F_l) - P_n(F) = P_n(F_l - F)$. For the calculation of the infinite product, see HARDY & WRIGHT, page 245.

Section 5

5.5 **(a)** If $m = 0$, $\alpha \geq 0$, and $x > 0$, then $P[X \geq \alpha] \leq P[(X + x)^2 \geq (\alpha + x)^2] \leq E[(X + x)^2]/(\alpha + x)^2 = (\sigma^2 + x^2)/(\alpha + x)^2$; minimize over x.

5.8. **(b)** It is enough to prove that $\varphi(t) = f(t(x', y') + (1 - t)(x, y))$ is convex in t ($0 \leq t \leq 1$) for (x, y) and (x', y') in C. If $\alpha = x' - x$ and $\beta = y' - y$, then (if $f_{11} > 0$)

$$\varphi'' = f_{11}\alpha^2 + 2f_{12}\alpha\beta + f_{22}\beta^2$$

$$= \frac{1}{f_{11}}(f_{11}\alpha + f_{12}\beta)^2 + \frac{1}{f_{11}}(f_1 f_{22} - f_{12}^2)\beta^2 \geq 0.$$

5.9. Check (5.35) for $f(x, y) = -x^{1/p} y^{1/q}$.

5.10. Check (5.35) for $f(x, y) = -(x^{1/p} + y^{1/p})^p$.

5.16. For (5.39) use (2.22) and the fundamental theorem of arithmetic: since the p_i are distinct, the $p_i^{k_i}$ individually divide m if and only if their product does. For (5.40) use inclusion-exclusion. For (5.43), use (5.25) (see Problem 5.14).

5.17. (a) By (5.43), $E_n[\alpha_p] \le \sum_{k=1}^{\infty} p^{-k} \le 2/p$. And, of course, $n^{-1} \log n! = E_n[\log] = \sum_p E_n[\alpha_p] \log p$. (See Problem 18.16 for one proof of Stirling's formula.)

(b) Use (5.44) and the fact that $E_n[\alpha_p - \delta_p] \le \sum_{k=2}^{\infty} p^{-k}$.

(c) By (5.45),

$$\sum_{n<p\le 2n} \log p = \sum_{n<p\le 2n} \left(\left[\frac{2n}{p} \right] - 2\left[\frac{n}{p} \right] \right) \log p$$

$$\le 2n(E_{2n}[\log*] - E_n[\log*]) = O(n).$$

Deduce (5.46) by splitting the range of summation by successive powers of 2. The difference between the left side of (5.47) and $E_{[x]}[\log*]$ is $O(1)$.

(d) If K bounds the $O(1)$ term in (5.47), then $\sum_{p\le x} \log x \ge \theta x \sum_{\theta x < p \le x} p^{-1} \log p \ge \theta x(\log \theta^{-1} - 2K)$.

(e) For (5.49) use

$$\sum_{p\le x} \frac{\log p}{\log x} \le \pi(x) \le \sum_{p\le x^{1/2}} 1 + \sum_{x^{1/2}<p\le x} \frac{\log p}{\log x^{1/2}}$$

$$\le x^{1/2} + \frac{2}{\log x} \sum_{p\le x} \log p.$$

By (5.49), $\pi(x) \ge x^{1/2}$ for large x, and hence $\log \pi(x) \asymp \log x$ and $\pi(x) \asymp x/\log \pi(x)$. Apply this with $x = p_r$ and note that $\pi(p_r) = r$.

Section 6

6.3. Since for given values of $X_{n1}(\omega), \ldots, X_{n,k-1}(\omega)$ there are for $X_{nk}(\omega)$ the k possible values $0, 1, \ldots, k-1$, the number of values of $(X_{n1}(\omega), \ldots, X_{nn}(\omega))$ is $n!$ Therefore, the map $\omega \rightarrow (X_{n1}(\omega), \ldots, X_{nn}(\omega))$ is one-to-one, and the $X_{nk}(\omega)$ determine ω. It follows that if $0 \le x_i < i$ for $1 \le i \le k$, then the number of permutations ω satisfying $X_{ni}(\omega) = x_i$, $1 \le i \le k$, is just $(k+1)(k+2)\cdots n$, so that $P[X_{ni} = x_i, 1 \le i \le k] = 1/k!$ It now follows by induction on k that $X_{n1}, \ldots, X_{nk}$ are independent and $P[X_{nk} = x] = k^{-1}$ $(0 \le x < k)$.
Now calculate

$$E[X_{nk}] = \frac{k-1}{2}, \qquad E[S_n] = \frac{0 + 1 + \cdots + (n-1)}{2} = \frac{n(n-1)}{4} \sim \frac{n^2}{4},$$

$$\text{Var}\,[X_{nk}] = \frac{0^2 + 1^2 + \cdots + (k-1)^2}{k} - \left(\frac{k-1}{2}\right)^2 = \frac{k^2-1}{12},$$

$$\text{Var}\,[S_n] = \frac{1}{12}\sum_{k=1}^{n}(k^2-1) = \frac{2n^3 + 3n^2 - 5n}{72} \sim \frac{n^3}{36}.$$

Apply Chebyshev's inequality.

6.7. If $k^2 \leq n < (k+1)^2$, let $a_n = k^2$; if M bounds the $|X_n|$, then

$$\left|\frac{1}{n}S_n - \frac{1}{a_n}S_{a_n}\right| \leq \left|\frac{1}{n} - \frac{1}{a_n}\right| \cdot nM + \frac{1}{a_n}(n - a_n)M = 2M\frac{n - a_n}{a_n} \to 0.$$

6.17. From (5.49) and (5.50) it follows that $a_n = \Sigma_p\,n^{-1}[n/p] \to \infty$. The left side of (6.12) is

$$\frac{1}{n}\left[\frac{n}{pq}\right] - \frac{1}{n}\left[\frac{n}{p}\right]\frac{1}{n}\left[\frac{n}{q}\right] \leq \frac{1}{pq} - \left(\frac{1}{p} - \frac{1}{n}\right)\left(\frac{1}{q} - \frac{1}{n}\right) \leq \frac{1}{np} + \frac{1}{nq}.$$

Section 7

7.3. If one grants that there are only countably many effective rules, the result is an immediate consequence of the mathematics of this and the preceding sections: C is a countable intersection of $\mathcal{F}$-sets of measure 1. The argument proves in particular the nontrivial fact that collectives exist. See Per Martin-Löf: The literature on von Mises' Kollektives revisited, *Theoria*, **1** (1969), 12–37.

7.7. If $n \leq \tau$, then $W_n = W_{n-1} - X_{n-1} = W_1 - S_{n-1}$, and τ is the smallest n for which $S_{n-1} = W_1$. Use (7.8) for the question of whether the game terminates. Now

$$F_\tau = F_0 + \sum_{k=1}^{\tau-1}(W_1 - S_{k-1})X_k = F_0 + W_1 S_{\tau-1} - \frac{1}{2}(S_{\tau-1}^2 - (\tau-1)).$$

7.8. Let $\Sigma_0 = x_1 + \cdots + x_k$, $\Sigma_n = \Sigma_{n-1} - W_n X_n$, $L_0 = k$, and $L_n = L_{n-1} - (3X_n + 1)/2$. Then τ is the smallest n such that $L_n \leq 0$, and τ is by the strong law finite with probability 1 if $E[3X_n + 1] = 6(p - \frac{1}{3}) > 0$. For $n \leq \tau$, Σ_n is the sum of the pattern used to determine W_{n+1}. Since $F_n - F_{n-1} = \Sigma_{n-1} - \Sigma_n$, $F_n = F_0 + \Sigma_0 - \Sigma_n$, and $F_\tau = F_0 + \Sigma_0$.

7.9. Observe that $E[F_n - F_\tau] = E[\Sigma_{k=1}^{n} X_k I_{[\tau<k]}] = \Sigma_{k=1}^{n} E[X_k]P[\tau<k]$.

Section 8

8.8. Fix on a state i and let S_ν consist of those j for which $p_{ij}^{(n)} > 0$ for some n congruent to ν modulo t. Choose k so that $p_{ji}^{(k)} > 0$; if $p_{ij}^{(m)}$ and $p_{ij}^{(n)}$ are positive, then t divides $m + k$ and $n + k$, so that m and n are congruent modulo t. The S_ν are thus well defined.

8.10. Let P be the set of i for which $\pi_i > 0$, let N be the set of i for which $\pi_i \leq 0$, and suppose that P and N are both nonempty. For $i_0 \in P$ and $j_0 \in N$ choose n so that $p_{i_0 j_0}^{(n)} > 0$. Then

$$0 < \sum_{j \in N} \sum_{i \in P} \pi_i p_{ij}^{(n)} = \sum_{j \in N} \pi_j - \sum_{j \in N} \sum_{i \in N} \pi_i p_{ij}^{(n)}$$

$$= \sum_{i \in N} \pi_i \sum_{j \in P} p_{ij}^{(n)} \leq 0.$$

Transfer from N to P any i for which $\pi_i = 0$ and use a similar argument.

8.12. Show that Theorem 8.6 applies to the chain with transition probabilities $p_{ij}^{(t)}$.

8.13. For the symmetric random walk on the line, $\sum_n (p_{ii}^{(n)})^2 = \infty$, and so the coupled chain is persistent. But there can be no stationary distribution because $p_{ij}^{(n)} \to 0$.

8.17. The transition probabilities are $p_{0r} = 1$ and $p_{i,r-i+1} = p$, $p_{i,r-i} = q$, $1 \leq i \leq r$; the stationary probabilities are $u_1 = \cdots = u_r = q^{-1} u_0 = (r + q)^{-1}$. The chance of getting wet is $u_0 p$, of which the maximum is $2r + 1 - 2\sqrt{r(r+1)}$. For $r = 5$ this is .046, the pessimal value of p being .523. Of course, $u_0 p \leq 1/4r$. In more reasonable climates fewer umbrellas suffice: if $p = .25$ and $r = 3$, then $u_0 p = .050$; if $p = .1$ and $r = 2$, then $u_0 p = .031$. At the other end of the scale, if $p = .8$ and $r = 3$, then $u_0 p = .050$; and if $p = .9$ and $r = 2$, then $u_0 p = .043$.

8.20. Show that $\alpha_n = \pi_i$ and $\beta_n - \alpha_n^2 = 2n^{-1}(n-1)^{-1} \sum_{k=1}^{n-1} \pi_i((n-k)p_{ii}^{(k)} - \pi_i) = O(n^{-2} \sum_{k=1}^{n} (n-k)\rho^k) = O(n^{-1})$—$\rho$ as in Theorem 8.7.

8.26. **(b)** With probability 1 the population either dies out or goes to infinity. If, for example, $p_{k0} = 1 - p_{k,k+1} = 1/k^2$, then extinction and explosion each have positive probability.

Section 9

9.9. Because of Theorem 9.6 there are for $P[M_n \geq \alpha]$ bounds of the same order as the ones for $P[S_n \geq \alpha]$ used in the proof of (9.36).

Section 10

10.4. If $\mathcal{P} = \{0\}$, then $\sigma(\mathcal{P}) = \{0, \Omega\}$. Since Ω is not a union of $\mathcal{P}$-sets, no measure on $\sigma(\mathcal{P})$—not even a finite one—can be σ-finite on $\mathcal{P}$.

For a second example, let μ_1 be counting measure on the σ-field of all subsets of a countably infinite Ω, let $\mu_2 = 2\mu_1$, and let P consist of the cofinite sets. Granted the existence of Lebesgue measure λ on $\mathcal{R}^1$, one can construct a third example: let $\mu_1 = \lambda$ and $\mu_2 = 2\lambda$, and let P consist of the half-infinite intervals $(-\infty, x]$.

There are similar examples with a field $\mathcal{F}_0$ in place of P. Let Ω consist of the rationals in $(0, 1]$, let μ_1 be counting measure, let $\mu_2 = 2\mu_1$, and let $\mathcal{F}_0$ consist of finite disjoint unions of "intervals" $[r \in \Omega: a < r \le b]$.

10.5. **(a)** Use the monotone class theorem.

Section 11

11.4. **(b)** See the special cases (12.16) and (12.17).

11.5. **(a)** Consider, for example, the finite sets in an infinite Ω and the countable sets in an uncountable Ω.

(d) Suppose that $A \in f(\mathcal{A})$; if A can be represented as a finite or countable disjoint union of $\mathcal{A}$-sets A_n, put $\mu(A) = \Sigma_n \mu(A_n)$ (prove consistency); otherwise, put $\mu(A) = \infty$. This extension need not be unique—suppose $\mathcal{A} = \{0\}$ for example.

11.6. **(c)** See HALMOS, page 31, for the proof that pairwise additivity implies finite additivity in strong semirings. For a counterexample in the case of semirings, take $\mathcal{A}$ to consist of $0, \{1\}, \{2\}, \{3\}$, and $\Omega = \{1, 2, 3\}$ and make sure that $\mu\{1\} + \mu\{2\} + \mu\{3\} < \mu(\Omega)$.

That pairwise additivity implies finite additivity in strong semirings can be used to simplify slightly the proof of Theorem 12.5. Accounts treating only strong semirings omit the qualifier "strong."

11.7. Take preliminary $\mathcal{F}_0$-sets B_k satisfying $\mu(B_k \triangle A_k) < \epsilon/n^3$; add $\bigcup_{j \ne k} (B_j \cap B_k)$ to one of these sets and subtract it from the others and let B be the union of the new B_k. If A lies in $\mathcal{F}_0$, subtract $B - A$ from all the B_k and add $A - B$ to one of them.

Section 12

12.1. Take $\alpha = \mu(0, 1]$; for A an interval with rational endpoints, compare $\mu(A)$ with α by exhibiting A and $(0, 1]$ as disjoint unions of translates of a single interval. Such A's form a π-system.

12.4. **(a)** If $k = m - n$, then $2k\theta - (N_{2m} - N_{2n}) = \theta_{2m} - \theta_{2n}$ lies in $(0, \epsilon)$ and hence coincides with $\theta_{2k} = 2k\theta - N_{2k}$. If $l_0 = [\theta_{2k}^{-1}]$, then the θ_{l2k}, $1 \le l \le l_0$, form a chain of points in $(0, 1]$ such that the distance from each to the next is less than ϵ, the first is to the left of ϵ, and the last is to the right of $1 - \epsilon$.

(c) Use again the fact that θ is irrational.

12.5. See Example 2.10 (which applies to any finite measure).

12.7. By Theorem 12.3 and Problem 2.17(b), A contains two disjoint compact sets of arbitrarily small measure. Construct inductively compact sets $K_{u_1 \ldots u_n}$ (each u_i is 0 or 1) such that $0 < \mu(K_{u_1 \ldots u_n}) < 3^{-n}$ and $K_{u_1 \ldots u_n 0}$ and $K_{u_1 \ldots u_n 1}$ are disjoint subsets of $K_{u_1 \ldots u_n}$. Take $K = \bigcap_n \bigcup_{u_1 \ldots u_n} K_{u_1 \ldots u_n}$. The Cantor set is a special case.

Section 13

13.6. If $f = \Sigma_i x_i I_{A_i}$ and $A_i \in T^{-1} \mathscr{F}'$, take A_i' in $\mathscr{F}'$ so that $A_i = T^{-1} A_i'$ and set $\varphi = \Sigma_i x_i I_{A_i'}$. For the general f measurable $T^{-1} \mathscr{F}'$, there exist simple functions f_n, measurable $T^{-1} \mathscr{F}'$, such that $f_n(\omega) \to f(\omega)$ for each ω. Choose φ_n, measurable $\mathscr{F}'$, so that $f_n = \varphi_n T$. Let C' be the set of ω' for which $\varphi_n(\omega')$ has a finite limit, and define $\varphi(\omega') = \lim_n \varphi_n(\omega')$ for $\omega' \in C'$ and $\varphi(\omega') = 0$ for $\omega' \notin C'$. Theorem 20.1(ii) is a special case.

13.10. The class of Borel functions contains the continuous functions and is closed under pointwise passages to the limit and hence contains $\mathscr{X}$.

By imitating the proof of the π-λ theorem, show that, if f and g lie in $\mathscr{X}$, then so do $f + g, fg, f - g, f \vee g$ (note that, for example, $[g : f + g \in \mathscr{X}]$ is closed under passages to the limit). Show that $[A : I_A \in \mathscr{X}]$ is a λ-system containing (Problem 13.9) the sets $(-\infty, x]$. Conclude that $\mathscr{X}$ contains all indicators of Borel sets, all simple Borel functions, all Borel functions.

13.11. The set $[x : f(x) < \alpha]$ is open.

13.14. On the integers with counting measure, consider the indicator of $\{n, n + 1, \ldots\}$.

Section 14

14.4. **(b)** Since $u \leq F(x)$ is equivalent to $\varphi(u) \leq x, u \leq F(\varphi(u))$. Since $F(x) < u$ is equivalent to $x < \varphi(u), F(\varphi(u) - \epsilon) < u$ for positive ϵ.

(d) Let C be the set of continuity points of F. If $0 < u < v < 1$, then $P[u \leq F(X) < v, X \in C] = P[\varphi(u) \leq X < \varphi(v), X \in C]$. If $\varphi(u) \in C$, this is at most $P[\varphi(u) \leq X < \varphi(v)] = F(\varphi(v)-) - F(\varphi(u)-) = F(\varphi(v)-) - F(\varphi(u)) \leq v - u$; if $\varphi(u) \notin C$, it is at most $P[\varphi(u) < X < \varphi(v)] = F(\varphi(v)-) - F(\varphi(u)) \leq v - u$. Thus $P[F(X) \in [u, v), X \in C] \leq \lambda[u, v)$ if $0 < u < v < 1$, and it follows easily also if $u = 0$ or if (let $v \uparrow 1$) $v = 1$. The finite disjoint unions of intervals $[u, v)$ in $[0, 1)$ form a field there, and by addition $P[F(X) \in A, X \in C] \leq \lambda(A)$ for A in this field. By the monotone class theorem, the inequality holds for all Borel sets in $[0, 1)$. It is easy to check that it holds for $A = \{1\}$.

14.8. The sufficiency is easy. To prove necessity, choose continuity points x_i of F in such a way that $x_0 < x_1 < \cdots < x_k$, $F(x_0) < \epsilon$, $F(x_k) > 1 - \epsilon$, and $x_i - x_{i-1} < \epsilon$. If n exceeds some n_0, $|F(x_i) - F_n(x_i)| < \epsilon/2$ for all i. Suppose that $x_{i-1} \leq x \leq x_i$. Then $F_n(x) \leq F_n(x_i) \leq F(x_i) + \epsilon/2 \leq F(x + \epsilon) + \epsilon/2$. Establish a similar inequality going the other direction, and give special arguments for the cases $x \leq x_0$ and $x \geq x_k$.

Section 17

17.6. Part (a) can be elaborated: If f is Lebesgue integrable over each $[t, 1]$ and (17.19) holds, then f has Kurzweil integral I over $[0, 1]$. This integral was introduced in J. Kurzweil: Generalized ordinary differential equations and continuous dependence on a parameter, *Czechoslovak Math. Jour.*, **7** (82) (1957), 418–446.

17.10. Assume that $f_n' \to g$ almost everywhere and that g is continuous. The argument will go through if $\{f_n'\}$ is uniformly integrable, for example, or if $\Sigma_n \int_a^b |f_{n+1}' - f_n'| \, dx < \infty$. It is not hard to construct examples of sequences to which these results apply but to which the standard one does not.

17.12. The binomial formula is

$$(1 + x)^\alpha = \sum_{n=0}^{\infty} \binom{\alpha}{n} x^n,$$

valid for arbitrary α and $|x| < 1$; here $\binom{\alpha}{n} = \alpha(\alpha - 1) \cdots (\alpha - n + 1)/n!$. A simple way to verify the formula without considering the analyticity of the left-hand side is to show by the ratio test that the series converges for $|x| < 1$ to $f_\alpha(x)$, say, and then by a binomial identity to derive $(1 + x)f_\alpha'(x) = \alpha f_\alpha(x)$, whence it follows that the derivative of $f_\alpha(x)/(1 + x)^\alpha$ vanishes. Note that the series in (17.20) is nonnegative.

Section 18

18.6. If $E \in \mathcal{X} \times \mathcal{Y}$, then E lies in the σ-field generated by some countable class of measurable rectangles, which by the definition of $\mathcal{X} = \mathcal{Y}$ may be taken as the singletons $A_{mn} = \{(x_m, x_n)\}$ for some sequence $x_1, x_2, \ldots$. But then E must be the union of certain of the A_{mn}, or such a union together with $(\cup_{mn} A_{mn})^c$; either case is impossible.

18.9. See DOOB, page 63.

18.14. Put $f_p = p^{-1} \log p$ and put $f_n = 0$ if n is not a prime. In the notation of (18.18), $F(x) = \log x + \varphi(x)$, where φ is bounded because of (5.47). If $G(x) = -1/\log x$, then

$$\sum_{p \le x} \frac{1}{p} = \frac{F(x)}{\log x} + \int_2^x \frac{F(t)\, dt}{t \log^2 t}$$

$$= 1 + \frac{\varphi(x)}{\log x} + \int_2^x \frac{dt}{t \log t} + \int_2^\infty \frac{\varphi(t)\, dt}{t \log^2 t}$$

$$- \int_x^\infty \frac{\varphi(t)\, dt}{t \log^2 t}$$

18.16. Take the point of view of Problem 18.3 and use the translation invariance of Lebesgue measure in the plane. The T_k do not overlap because the slope log $(1 + k^{-1})$ of the lower boundary of T_k is less than the slope k^{-1} of a tangent to the upper boundary of T_{k-1} at its right-hand endpoint.

Section 19

19.2. Choose sets B_n such that diam $B_n < \epsilon$ and $c_m \Sigma_n$ (diam $B_n)^m < h_{m,\epsilon}(A)$ $+ \epsilon$. Replace B_n by the open set $[x: \text{dist } (x, B_n) < \delta_n]$; if the δ_n are small enough, the inequalities are preserved. Now use compactness.

19.3. (a) Inserting the extra point $f(t)$ cannot make the polygon shorter.

(b) If $L(f) = \infty$, $l(\cdot)$ is not well defined, but in this case there is nothing to prove. Apply Lemma 4 to the maps $f: [a, b] \rightarrow R^k$ and $l: [a, b] \rightarrow R^1$.

(c) Make sure that the trace is swept out several times as t runs from a to b.

(d) Let t_n be the supremum of the t in $[a, b]$ for which $f(t) \in B_n^-$. Now $f(a)$ lies in some B_{n_0}, $f(t_{n_0})$ lies in some B_{n_1}, $f(t_{n_1})$ lies in some B_{n_2}, and so on. The process must terminate with a t_{n_c} of b, because $n_0, n_1, \dots$ are distinct until then. Therefore, $|f(b) - f(a)| \le \Sigma_{i=0}^{c-1} |f(t_{n_{i+1}}) - f(t_{n_i})| \le \Sigma_{n=1}^N$ diam $B_n < h_{1,\epsilon}(f[a, b]) + \epsilon$.

(e) If $a = t_0 < \cdots < t_n = b$, then, since f is one-to-one and h_1 is 0 for singletons, $\Sigma_{j=1}^n |f(t_j) - f(t_{j-1})| \le \Sigma_{j=1}^n h_1(f[t_{j-1}, t_j]) = h_1(f[a, b])$.

19.4. If the curve has no multiple points, the result follows from (19.14), but it is easy to derive it in the general case directly from the definition of $L(f)$.

19.6. (a) $2\pi\sqrt{2}$.

(b) $2\pi [1/\sqrt{2} + \frac{1}{2} \log (1 + \sqrt{2})]$.

19.7. In the diagram (page 490) the circle has radius 1. Take $\theta = \pi/n$ and $x = 1/2m$, calculate the lengths of the numbered segments in order, and show that the triangles abc and abd have areas $\sin \theta[(1 - \cos \theta)^2 + x^2]^{1/2}$ and $x[2 - 2 \cos \theta]^{1/2}$. Since the polyhedron has $2mn$ triangular faces of each kind, the formula for A_{mn} follows. Since $\sin t \sim t$ as $t \rightarrow 0$, if m and n go to infinity in such a way that $m^2/n^4 \rightarrow \rho$, then $A_{mn} \rightarrow \pi + \pi[1 + \rho\pi^4]^{1/2}$. See FEDERER and RADÓ for accounts of this unimaginably complicated subject.

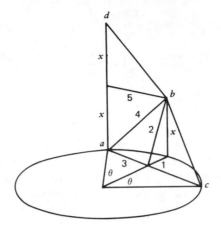

19.8. The rectangular coordinates are $(x, y, z) = (\cos \theta \cos \phi, \sin \theta \cos \phi, \sin \phi)$, and $|D_{\theta,\phi}|$ comes out to $\cos \phi$. The sector therefore has area $(\theta_2 - \theta_1)(\sin \phi_2 - \sin \phi_1)$.

19.10. **(a)** Choose sets B_n such that $A \subset \bigcup_n B_n$, diam $B_n < \epsilon$, and Σ_n (diam $B_n)^m \le h_m(A) + 1$, and note that $h_{m+\delta,\epsilon}(A) \le \epsilon^\delta(h_m(A) + 1)$.

(b) If T is a map from R^m to R^k to which Theorem 19.3 applies, and if $\int_A |D_x| \lambda_m(dx)$ is finite and positive, then dim $TA = m$. If T is not smooth, of course, anything can happen—consider a Peano curve.

19.11. This definition of dimension is due to Felix Hausdorff: Dimension und äusseres Mass, *Math. Ann.*, **79** (1919), 157–179. He showed that the Cantor set has dimension $\log 2/\log 3$. Eggleston showed that the set of points in the unit interval in whose dyadic expansions 1 appears with asymptotic relative frequency p has dimension $-p \log p - q \log q$; see BILLINGSLEY, Section 14.

For the connection between Hausdorff and topological dimension, see Withold Hurewicz and Henry Wallman: *Dimension Theory* (Princeton University Press, Princeton, New Jersey, 1948).

Section 20

20.8. By the argument following (8.16), $P_i[T_1 = n_1, \ldots, T_k = n_k] = f_{ij}^{(n_1)} f_{jj}^{(n_2-n_1)} \cdots f_{jj}^{(n_k-n_{k-1})}$. For the general initial distribution, average over i.

20.9. Integrate rather than differentiate.

20.10. Use the π-λ theorem.

20.11. **(a)** See Problem 20.10.

(b) Use part (a) and the fact that $Y_n = r$ if and only if $T_r^{(n)} = n$.

(c) If $k \leq n$, then $Y_k = r$ if and only if exactly $r - 1$ among the integers $1, \ldots,$ $k - 1$ precede k in the permutation $T^{(n)}$.

(d) Observe that $T^{(n)} = (t_1, \ldots, t_n)$ and $Y_{n+1} = r$ if and only if $T^{(n+1)} = (t_1,$ $\ldots, t_{r-1}, n + 1, t_r, \ldots, t_n)$, and conclude that $\sigma(Y_{n+1})$ is independent of $\sigma(T^{(n)})$ and hence of $\sigma(Y_1, \ldots, Y_n)$—see Problem 20.7.

20.16. If X and Y are independent, then $P[|(X + Y) - (x + y)| < \epsilon] \geq P[|X - x| < \frac{1}{2}\epsilon]P[|Y - y| < \frac{1}{2}\epsilon]$ and $P[X + Y = x + y] \geq P[X = x]P[Y = y]$.

20.20. The partial-fraction expansion gives

$$c_u(y - x)c_v(x) = \frac{uv}{\pi^2}\frac{1}{R}(A + B + C + D),$$

where $R = (u^2 - v^2)^2 + 2(u^2 + v^2)y^2 + y^4$ and

$$A = \frac{y^2 + v^2 - u^2}{u^2 + (y - x)^2}, \qquad B = \frac{2y(y - x)}{u^2 + (y - x)^2},$$

$$C = \frac{y^2 - v^2 + u^2}{v^2 + x^2}, \qquad D = \frac{2yx}{v^2 + x^2}.$$

After the fact this can of course be checked mechanically. Integrate over $[-t, t]$ and let $t \to \infty$: $\int_{-t}^{t} D \, dx = 0$, $\int_{-t}^{t} B \, dx \to 0$, and $\int_{-\infty}^{\infty} (A + C) \, dx = (y^2 + v^2 - u^2)u^{-1}\pi + (y^2 - v^2 + u^2)v^{-1}\pi = u^{-1}v^{-1}\pi^2 R c_{u+v}(y)$. There is a very simple proof by characteristic functions; see Problem 26.9.

20.22. See Example 20.1 for the case $n = 1$, prove by inductive convolution and a change of variable that the density must have the form $K_n x^{(n/2)-1}e^{-x/2}$, and then from the fact that the density must integrate to 1 deduce the form of K_n.

20.23. Show by (20.38) and a change of variable that the left side of (20.48) is some constant times the right side; then show that the constant must be 1.

20.27. **(a)** Given ϵ choose M so that $P[|X| > M] < \epsilon$ and $P[|Y| > M] < \epsilon$, and then choose δ so that $|x|, |y| \leq M$, $|x - x'| < \delta$, and $|y - y'| < \delta$ imply that $|f(x', y') - f(x, y)| < \epsilon$. Note that $P[|f(X_n, Y_n) - f(X, Y)| \geq \epsilon] \leq 2\epsilon + P[|X_n - X| \geq \delta] + P[|Y_n - Y| \geq \delta]$.

20.30. Take, for example, independent X_n assuming the values 0 and n with probabilities $1 - n^{-1}$ and n^{-1}. Estimate the probability that $X_k = k$ for some k in the range $n/2 < k \leq n$.

20.31. **(b)** For each m split A into 2^m sets A_{mk} of probability $P(A)/2^m$. Arrange all the A_{mk} in one infinite sequence and let X_n be the indicator of the nth set in it.

Section 21

21.6. Consider ΣI_{A_n}. A random variable is finite with probability 1 if (but not only if) it is integrable.

21.8. Calculate $\int_0^\infty x \, dF(x) = \int_0^\infty \int_0^x dy \, dF(x) = \int_0^\infty \int_y^\infty dF(x) \, dy$.

21.10. (a) Write $E[Y - X] = \int_{X<Y} \int_{X<t\leq Y} dt \, dP - \int_{Y<X} \int_{Y<t\leq X} dt \, dP$.

21.11. Fubini.

21.12. Show that $1 - F(t) \leq -\log F(t) \leq (1 - F(t))/F(\alpha)$ for $t \geq \alpha$ and apply (21.10).

21.13. Calculate

$$E[Y^r] = \int_0^\infty P[Y \geq s] r s^{r-1} \, ds \leq \int_0^\infty \int_{Y \geq s} X \, dP \, r s^{r-2} \, ds$$

$$= \frac{r}{r-1} E[XY^{r-1}] \leq \frac{r}{r-1} E^{1/r}[X^r] E^{1-1/r}[Y^r].$$

21.14. (a) Use (21.12).

21.15. (a) The most important dependent uncorrelated random variables are the trigonometric functions—the random variables $\sin 2\pi n\omega$ and $\cos 2\pi n\omega$ on the unit interval with Lebesgue measure.

21.19. Use Fubini's theorem; see (20.29) and (20.30).

21.20. Even if $X = -Y$ is not integrable, $X + Y = 0$ is. Since $|Y| \leq |x| + |x + Y|$, $E[|Y|] = \infty$ implies that $E[|x + Y|] = \infty$ for each x; use Problem 21.19. See also the lemma in Section 28.

21.21. Use Theorem 20.5.

21.25. (b) Suppose $\{X_n\}$ is d_p-fundamental. By Markov's inequality, the sequence is fundamental in probability and hence converges in probability to some X. By Fatou's lemma it follows from $X_n \in L^p$ that $X \in L^p$. Another application of Fatou's lemma gives $d_p(X, X_n) \leq \liminf_m d_p(X_m, X_n)$, and so $d_p(X, X_n) \to 0$.

21.28. If $x < \text{ess sup } |X|$, then $E^{1/p}[|X|^p] \geq x P^{1/p}[|X| \geq x] \to x$ as $p \to \infty$.

Section 22

22.10. (a) Let $r_n(t)$ be the Rademacher functions. The hypothesis implies that, for each t, $\Sigma_n r_n(t)X_n(\omega)$ converges for ω in a set of probability 1. Apply Fubini's theorem in $(0, 1] \times \Omega$.

(b) Note that $\Sigma_n \pm X_n(\omega)$ converges for all choices of the signs if and only if $\Sigma_n |X_n(\omega)| < \infty$.

22.14. (a) Put $U = \Sigma_k I_{[k \leq \tau]}X_k^+$ and $V = \Sigma_k I_{[k \leq \tau]}X_k^-$, so that $S_\tau = U - V$. Since $[\tau \geq k] = \Omega - [\tau \leq k - 1]$ lies in $\sigma(X_1, \ldots, X_{k-1})$, $E[I_{[\tau \geq k]}X_k^+] = E[I_{[\tau \geq k]}]E[X_k^+] = P[\tau \geq k]E[X_k^+]$. Hence $E[U] = \Sigma_{k=1}^\infty E[X_k^+]P[\tau \geq k] = E[X_1^+]E[\tau]$. Treat V the same way.

(b) By (7.7), S_τ is b with probability $(1 - \rho^a)/(1 - \rho^{a+b})$ and $-a$ with the opposite probability. Since $E[X_1] = p - q$,

$$E[\tau] = \frac{a}{q - p} - \frac{a + b}{q - p}\frac{1 - \rho^a}{1 - \rho^{a+b}}, \qquad \rho = \frac{q}{p} \neq 1.$$

22.19. For each θ, $\Sigma_n e^{iX_n}(e^{i\theta}z)^n$ has the same probabilistic behavior as the original series because the $X_n + n\theta$ reduced modulo 2π are independent and uniformly distributed. Therefore, the rotation idea in the proof of Theorem 22.6 carries over. See KAHANE for further results.

Section 23

23.3. Note that A_t cannot exceed t. If $0 \leq u \leq t$ and $v \geq 0$, then $P[A_t \geq u, B_t > v] = P[N_{t+v} - N_{t-u} = 0] = e^{-\alpha u}e^{-\alpha v}$.

23.4. (a) Use (20.37) and the distributions of A_t and B_t.

(b) A long interarrival interval has a better chance of covering t than a short one does.

23.6. The probability that $N'_{S_{n+k}} - N'_{S_n} = j$ is

$$\int_0^\infty e^{-\beta x}\frac{(\beta x)^j}{j!}\frac{\alpha^k}{\Gamma(k)}x^{k-1}e^{-\alpha x}\,dx = \frac{\alpha^k \beta^j}{(\alpha + \beta)^{k+j}}\frac{(j + k - 1)!}{j!(k - 1)!}.$$

23.7. The moment generating functions exist to the left of 0, and by the convolution law that of Y is $\exp\{(\beta - \alpha)(e^s - 1)\}$, which cannot be convex unless $\alpha \leq \beta$. Use the uniqueness theorem (see (22.8)) for moment generating functions of nonnegative random variables.

23.8. Restrict attention to subsets of a fixed interval $(0, a]$. Show that the sets with the desired property include the subintervals and form a λ-system of subsets of $(0, a]$. Note that, if $A_n \uparrow A$ and $N(A) < \infty$, then $N(A_n) = N(A)$ for sufficiently large n, and so $[N(A) = k] = \lim_n [N(A_n) = k]$ in the sense of (4.6).

23.10. (b) Suppose $\{N_t\}$ is a generalized Poisson process. Put $f(t) = E[N_t]$.

Since N_t has the form (23.5), the paths are right-continuous, and it follows by the dominated convergence theorem that f is right-continuous as well as non-negative and nondecreasing. Let $N_t^- = \sup_{s<t} N_s$. Show by a limiting argument that N_t^- has the Poisson distribution with mean $f(t-)$ and is independent of $N_t - N_t^-$; since the latter random variable assumes only the values 0 and 1, the distributions of N_t^- and $N_t - N_t^-$ cannot convolve properly unless $f(t-) = f(t)$ and $N_t^- = N_t$ with probability 1.

Let $\varphi(u) = \inf [t: u \le f(t)]$. Show that $M_u = N_{\varphi(u)}$ is an ordinary Poisson process and that $N_t = M_{f(t)}$.

23.11. Let $t \to \infty$ in

$$\frac{S_{N_t}}{N_t} \le \frac{t}{N_t} \le \frac{S_{N_t+1}}{N_t+1} \frac{N_t + 1}{N_t}.$$

23.13. Restrict t in Problem 23.12 to integers. The waiting times are the Z_n of Problem 20.8, and account must be taken of the fact that the distribution of Z_1 may differ from that of the other Z_n.

Section 25

25.1. (**d**) Let G be an open set that contains the rationals and satisfies $\lambda(G) < \frac{1}{2}$. For $k = 0, 1, \ldots, n - 1$, construct a triangle whose base contains k/n and is contained in G; make these bases so narrow that they do not overlap, and adjust the heights of the triangles so that each has area $1/n$. For the nth density, piece together these triangular functions, and for the limit density, use the function identically 1 over the unit interval.

25.2. By Problem 14.12 it suffices to prove that $F_n(\cdot, \omega) \Rightarrow F$ with probability 1, and for this it is enough that $F_n(x, \omega) \to F(x)$ with probability 1 for each rational x.

25.3. (**b**) It can be shown, for example, that (25.24) holds for $x_n = n!$ See Persi Diaconis: The distribution of leading digits and uniform distribution mod 1, *Ann. Prob.*, **5** (1977), 72–81.

(**c**) The first significant digits of numbers drawn at random from empirical compilations such as almanacs and engineering handbooks seem approximately to follow the limiting distribution in (25.25) rather than the uniform distribution over $1, 2, \ldots, 9$. This is sometimes called *Benford's law*. One explanation is that the distribution of the observation X and hence of $\log_{10} X$ will be spread over a large interval; if $\log_{10} X$ has a reasonably smooth density, it then seems plausible that $\{\log_{10} X\}$ should be approximately uniformly distributed. See FELLER, Volume 2, page 62, and R. Raimi: The first digit problem, *Amer. Math. Monthly*, **83** (1976), 521–538.

25.9. Use Scheffé's theorem.

25.10. Put $f_n(x) = P[X_n = k\delta_n]\,\delta_n^{-1}$ for $k\delta_n < x \le (k+1)\delta_n$. Construct random variables Y_n with densities f_n, and first prove $Y_n \Rightarrow X$. Show that $Z_n = [Y_n/\delta_n]\delta_n$ has the distribution of X_n and that $Y_n - Z_n \Rightarrow 0$.

25.11. For a proof of (25.26) see FELLER, Volume 1, Chapter 7.

25.13. (b) Follow the proof of Theorem 25.8, but approximate $I_{(x,y]}$ instead of $I_{(-\infty,x]}$.

Section 26

26.1. (b) Let μ be the distribution of X. If $|\varphi(t)| = 1$ and $t \ne 0$, then $\varphi(t) = e^{ita}$ for some a, and $0 = \int_{-\infty}^{\infty}(1 - e^{it(x-a)})\,\mu(dx) = \int_{-\infty}^{\infty}(1 - \cos t(x-a))\,\mu(dx)$. Since the integral vanishes, μ must confine its mass to the points where the nonnegative integrand vanishes, namely to the points x for which $t(x-a) = 2\pi n$ for some integer n.

(c) The mass of μ concentrates at points of the form $a + 2\pi n/t$ and also at points of the form $a' + 2\pi n/t'$. If μ is positive at two distinct points, it follows that t/t' is rational.

26.3. (a) Let $f_0(x) = \pi^{-1}x^{-2}(1 - \cos x)$ be the density corresponding to $\varphi_0(t)$. If $p_k = (s_k - s_{k+1})t_k$, then $\Sigma_{k=1}^{\infty} p_k = 1$; since $\Sigma_{k=1}^{\infty} p_k\varphi_0(t/t_k) = \varphi(t)$ (check the points $t = t_j$), $\varphi(t)$ is the characteristic function of the continuous density $\Sigma_{k=1}^{\infty} p_k t_k f_0(t_k x)$.

(b) For example, take $d_k = 1$ and $s_k = k^{-1}(k+1)^{-1}$.

(c) If $\lim_{t\to\infty}\varphi(t) = 0$, approximate φ by functions of the kind in part (a), pass to the limit, and use the first corollary to the continuity theorem. If φ does not vanish at infinity, mix in a unit mass at 0.

26.12. On the right in (26.23) replace $\varphi(t)$ by the integral defining it and apply Fubini's theorem; the integral average comes to

$$\mu\{a\} + \int_{x \ne a} \frac{\sin T(x-a)}{T(x-a)}\,\mu(dx).$$

Now use the bounded convergence theorem.

26.15. (a) Observe that $|\varphi_n(t+h) - \varphi_n(t)| \le \int_{-\infty}^{\infty}|e^{ihx} - 1|\,\mu_n(dx) \to \int_{-\infty}^{\infty}|e^{ihx} - 1|\,\mu(dx)$.

(b) Use part (a) or else integrate by parts.

26.17. (a) Use the second corollary to the continuity theorem.

26.19. For the Weierstrass approximation theorem, see RUDIN, Theorem 7.32.

26.26. (a) If a_n goes to 0 along a subsequence, then $|\psi(t)| \equiv 1$; use part (c) of Problem 26.1.

(c) Suppose two subsequences of $\{a_n\}$ converge to a_0 and to a, where $0 < a_0 < a$; put $\theta = a_0/a$ and show that $|\varphi(t)| = |\varphi(\theta^k t)|$.

Section 27

27.8. By the same reasoning as in Example 27.3, $(R_n - \log n)/\sqrt{\log n} \Rightarrow N$.

27.9. The Lindeberg theorem applies: $(S_n - n^2/4)/\sqrt{n^3/36} \Rightarrow N$.

27.16. Write $\int_x^\infty e^{-u^2/2}\, du = x^{-1}e^{-x^2/2} - \int_x^\infty u^{-2}e^{-u^2/2}\, du$.

27.17. For another approach to large deviation theory, see Mark Pinsky: An elementary derivation of Khintchine's estimate for large deviations, *Proc. Amer. Math. Soc.*, **22** (1969), 288–290.

27.18. See Chi Song Wong: A note on the central limit theorem, *Amer. Math. Monthly*, **84** (1977), 472.

27.19. (a) Everything comes from (4.7). If $A = [(l_1, \ldots, l_k) \in H]$ and $B \in \sigma(l_{k+n}, l_{k+n+1}, \ldots)$, then

$$|P(A \cap B) - P(A)P(B)|$$
$$\leq \Sigma\, |P([l_u = i_u, u \leq k] \cap B) - P[l_u = i_u, u \leq k]P(B)|,$$

where the sum extends over the k-tuples $(i_1, \ldots, i_k)$ of nonnegative integers in H. The summand vanishes if $u + i_u < k + n$ for $u \leq k$; the remaining terms add to at most $2\Sigma_{u=1}^k P[l_u \geq k + n - u] \leq 4/2^n$.

(b) To show that $\sigma^2 = 6$ (see (27.26)), show that l_1 has mean 1 and variance 2 and that

$$\int_{[l_1=i]} l_1 l_{1+n}\, dP = \begin{cases} P[l_1 = i]iE[l_{1+n}] & \text{if } i < n, \\ P[l_1 = i]i(i - n) & \text{if } i \geq n. \end{cases}$$

Section 28

28.4. (a) If all functions (28.12) are characteristic functions, they are all certainly infinitely divisible. Since (28.12) is continuous at 0, it need only be exhibited as a limit of characteristic functions. If μ_n has density $I_{[-n,n]}(1 + x^2)$ with respect to ν, then

$$\exp\left[i\gamma t + it \int_{-\infty}^\infty \frac{x}{1 + x^2} \mu_n(dx) + \int_{-\infty}^\infty (e^{itx} - 1 - itx) \frac{1}{x^2} \mu_n(dx) \right]$$

is a characteristic function and converges to (28.12). It can also be shown that every infinitely divisible distribution (no moments required) has characteristic function of the form (28.12); see GNEDENKO & KOLMOGOROV, page 76.

(b) Use (see Problem 18.18) $-|t| = \pi^{-1} \int_{-\infty}^{\infty} (\cos tx - 1) x^{-2} \, dx$.

28.14. If $X_1, X_2, \ldots$ are independent and have distribution function F, then $(X_1 + \cdots + X_n)/\sqrt{n}$ also has distribution function F. Apply the central limit theorem.

28.15. The characteristic function of Z_n is

$$\exp \frac{c}{n} \sum_k \frac{1}{(|k|/n)^{1+\alpha}} (e^{itk/n} - 1) \to \exp c \int_{-\infty}^{\infty} \frac{e^{itx} - 1}{|x|^{1+\alpha}} \, dx$$

$$= \exp \left[-c|t|^\alpha \int_{-\infty}^{\infty} \frac{1 - \cos x}{|x|^{1+\alpha}} \, dx \right].$$

Section 29

29.1. **(a)** If f is lower semicontinuous, $[x : f(x) > t]$ is open. If f is positive, which is no restriction, then $\int f \, d\mu = \int_0^\infty \mu[f > t] \, dt \le \int_0^\infty \liminf_n \mu_n[f > t] \, dt \le \liminf_n \int_0^\infty \mu_n[f > t] \, dt = \liminf_n \int f \, d\mu_n$.
(b) If G is open, then I_G is lower semicontinuous.

29.13. Let Σ be the covariance matrix. Let M be an orthogonal matrix such that the entries of $M\Sigma M'$ are 0 except for the first r diagonal entries, which are 1. If $Y = MX$, then Y has covariance matrix $M\Sigma M'$, and so $Y = (Y_1, \ldots, Y_r, 0, \ldots, 0)$, where $Y_1, \ldots, Y_r$ are independent and have the standard normal distribution. But $|X|^2 = \sum_{i=1}^k Y_i^2$.

29.14. By Theorem 29.5, X_n has asymptotically the centered normal distribution with covariances σ_{ij}. Put $x = (p_1^{1/2}, \ldots, p_k^{1/2})$ and show that $\Sigma x' = 0$, so that 0 is an eigenvalue of Σ. Show that $\Sigma y' = y'$ if y is perpendicular to x, so that Σ has 1 as an eigenvalue of multiplicity $k - 1$. Use Problem 29.13 together with Theorem 29.2 $(h(x) = |x|^2)$.

29.15. **(a)** Express the probability as a ratio of quantities (19.21), change variables $x_i = y_i \sqrt{n}$, show that the new integrands converge to $\exp(-\frac{1}{2} \sum_{i=1}^t y_i^2)$, and use the dominated convergence theorem.

Section 30

30.1. Rescale so that $s_n^2 = 1$, and put $L_n(\epsilon) = \sum_k \int_{|X_{nk}| \ge \epsilon} X_{nk}^2 \, dP$. Choose increasing n_u so that $L_n(u^{-1}) \le u^{-3}$ for $n \ge n_u$ and put $M_n = u^{-1}$ for $n_u \le n < n_{u+1}$. Then $M_n \to 0$ and $L_n(M_n) \le M_n^3$. Put $Y_{nk} = X_{nk} I_{[|X_{nk}| \le M_n]}$. Show that

$\Sigma_k E[Y_{nk}] \to 0$ and $\Sigma_k E[Y_{nk}^2] \to 1$, and apply to $\Sigma_k Y_{nk}$ the central limit theorem under (30.5). Show that $\Sigma_k P[X_{nk} \neq Y_{nk}] \to 0$.

30.3. (a) By Hölder's inequality $|\Sigma_{j=1}^k t_j x_j|^r \leq k^{r-1} \Sigma_{j=1}^k |t_j x_j|^r$, and so Σ_r $\theta^r \int |\Sigma_j t_j x_j|^r \mu(dx)/r!$ has positive radius of convergence. Now

$$\int_{R^k} \left(\sum_{j=1}^k t_j x_j \right)^r \mu(dx) = \Sigma t_1^{r_1} \ldots t_k^{r_k} \alpha(r_1, \ldots, r_k),$$

where the summation extends over k-tuples that add to r. Project μ to the line by the mapping $\Sigma_j t_j x_j$, apply Theorem 30.1, and use the fact that μ is determined by its values on half-spaces.

30.5. Use the Cramér-Wold idea.

30.6. Suppose that $k = 2$ in (30.29). Then

$M[(\cos \lambda_1 x)^{r_1}(\cos \lambda_2 x)^{r_2}]$

$$= M\left[\left(\frac{e^{i\lambda_1 x} + e^{-i\lambda_1 x}}{2} \right)^{r_1} \left(\frac{e^{i\lambda_2 x} + e^{-i\lambda_2 x}}{2} \right)^{r_2} \right]$$

$$= 2^{-r_1-r_2} \sum_{j_1=0}^{r_1} \sum_{j_2=0}^{r_2} \binom{r_1}{j_1}\binom{r_2}{j_2} M[\exp i(\lambda_1(2j_1 - r_1) + \lambda_2(2j_2 - r_2))x].$$

By (26.26) and the independence of λ_1 and λ_2, the last mean here is 1 if $2j_1 - r_1 = 2j_2 - r_2 = 0$ and is 0 otherwise. A similar calculation for $k = 1$ gives (30.27), and a similar calculation for general k gives (30.29). The actual form of the distribution in (30.28) is unimportant. For (30.30) use the multidimensional method of moments (Problem 30.5) and the mapping theorem. For (30.31) use the central limit theorem; by (30.27), X_1 has mean 0 and variance $\frac{1}{2}$.

30.8. If $n^{1/2} < m \leq n$ and the inequality in (30.32) holds, then $\log \log n^{1/2} < \log \log n - \epsilon (\log \log n)^{1/2}$, which implies $\log \log n < \epsilon^{-2} \log^2 2$. For large n the probability in (30.32) is thus at most $1/\sqrt{n}$.

Section 31

31.1. Consider the argument in Example 31.1. Suppose that F has a nonzero derivative at x and let I_n be the set of numbers whose base-r expansions agree in the first n places with that of x. The analogue of (31.16) is $P[X \in I_{n+1}]/P[X \in I_n] \to r^{-1}$, and the ratio here is one of $p_0, \ldots, p_r$. If $p_i \neq r^{-1}$ for some i, use the second Borel-Cantelli lemma to show that the ratio is p_i infinitely often except on a set of Lebesgue measure 0. (This last part of the argument is unnecessary if $r = 2$.)

The argument in Example 31.3 needs no essential change. The analogue of (31.17) is

$$F(x) = p_0 + \cdots + p_{i-1} + p_i F(rx - i), \qquad \frac{i}{r} \le x \le \frac{i+1}{r}, \quad 0 \le i < r - 1.$$

31.3. **(b)** Take $f_1 = I_{g^{-1}H_0}$ and $f_2 = F$; $(f_1 f_2)^{-1}\{1\} = H_0$ is not a Lebesgue set.

31.5. If $B = [(x, y): y - x \in A]$, then

$$\int \mu(A + x) \, dx = \int \mu[y: (x, y) \in B] \, dx = \int \lambda[x: (x, y) \in B] \, \mu(dy)$$

$$= \int \lambda(-(A - y)) \, \mu(dy) = 0.$$

31.7. Suppose that A is bounded, define μ by $\mu(B) = \lambda(B \cap A)$, and let F be the corresponding distribution function. It suffices to show that $F'(x) = 1$ for x in A, apart from a set of Lebesgue measure 0. Let C_ϵ be the set of x in A for which $F'(x) \le 1 - \epsilon$. From Theorem 31.4(i) deduce that $\lambda(C_\epsilon) = \mu(C_\epsilon) \le (1 - \epsilon) \lambda(C_\epsilon)$ and hence $\lambda(C_\epsilon) = 0$. Thus $F'(x) > 1 - \epsilon$ almost everywhere on A. Obviously, $F'(x) \le 1$.

31.9. Let A be the set of x in the unit interval for which $F'(x) = 0$, take $\alpha = 0$, and define A_n as in the first part of the proof of Theorem 31.4. Choose n so that $\lambda(A_n) \ge 1 - \epsilon$. Split $\{1, 2, \ldots, n\}$ into the set M of k for which $((k - 1)/n, k/n]$ meets A_n and the opposite set N. Prove successively that $\Sigma_{k \in M} [F(k/n) - F((k - 1)/n)] \le \epsilon, \Sigma_{k \in N} [F(k/n) - F((k - 1)/n)] \ge 1 - \epsilon, \Sigma_{k \in M} 1/n \ge \lambda(A_n) \ge 1 - \epsilon, \Sigma_{k=1}^n |f(k/n) - f((k - 1)/n)| \ge 2 - 2\epsilon$.

31.13. $\prod_{n=1}^\infty (\frac{1}{2} + \frac{1}{2} e^{2it/3^n})$.

31.15. **(a)** Since $(0, x) = \bigcup_n (\Sigma_{k=1}^{n-1} d_k(x) 2^{-k}, \Sigma_{k=1}^n d_k(x) 2^{-k}]$, $F(x)$ is the sum of the increments of F over these intervals. If $d_n(x) = 0$, the increment is 0; if $d_n(x) = 1$, it is the μ-measure of the set of points whose dyadic expansions begin with the digits $d_1(x), \ldots, d_{n-1}(x), 0$.

31.20. Let A be the x-set where (31.35) fails if f is replaced by $f\varphi$; then A has Lebesgue measure 0. Let G be the union of all open sets of μ-measure 0; represent G as a countable disjoint union of open intervals and let B be G together with any endpoints of zero μ-measure of these intervals. Let D be the set of discontinuity points of F. If $F(x) \notin A$, $x \notin B$, and $x \notin D$, then $F(x - h) < F(x) < F(x + h)$, $F(x \pm h) \to F(x)$, and

$$\frac{1}{F(x + h) - F(x - h)} \int_{F(x-h)}^{F(x+h)} f(\varphi(t)) \, dt \to f(\varphi(F(x))).$$

Now $x - \epsilon < \varphi(F(x)) \le x$ follows from $F(x - \epsilon) < F(x)$, and hence $\varphi(F(x)) = x$. If λ is Lebesgue measure restricted to $(0, 1)$, then $\mu = \lambda\varphi^{-1}$, and (31.36) follows by change of variable. But (36.36) is easy if $x \in D$, and hence it holds outside $B \cup (D^c \cap F^{-1}A)$. But $\mu(B) = 0$ by construction and $\mu(D^c \cap F^{-1}A) = 0$ by Problem 14.4(d).

Section 32

32.9. Define f and ν_s as in (32.7) and let f° and ν_s° be the corresponding function and measure for $\mathcal{F}^\circ$: $\nu(E) = \int_E f^\circ \, d\mu + \nu_s^\circ(E)$ for $E \in \mathcal{F}^\circ$ and there is an $\mathcal{F}^\circ$-set S° such that $\nu_s^\circ(\Omega - S^\circ) = 0$ and $\mu(S^\circ) = 0$. If $E \in \mathcal{F}^\circ$, then $\int_E f^\circ \, d\mu = \int_{E-S^\circ} f^\circ \, d\mu = \int_{E-S^\circ} f^\circ \, d\mu^\circ = \nu^\circ(E - S^\circ) = \nu(E - S^\circ) \geq \int_{E-S^\circ} f \, d\mu = \int_E f \, d\mu$. It is instructive to consider the extreme case $\mathcal{F}^\circ = \{0, \Omega\}$, in which ν° is absolutely continuous with respect to μ° (provided $\mu(\Omega) > 0$) and hence ν_s° vanishes.

Section 33

33.2. (a) To prove independence, check the covariance. Now use Example 33.7.

(b) Use the fact that R and Θ are independent (Example 20.2).

(c) As the single event $[X = Y] = [X - Y = 0] = [\Theta = \pi/4] \cup [\Theta = 5\pi/4]$ has probability 0, the conditional probabilities have no meaning, and strictly speaking there nothing to resolve. But whether it is natural to regard the degrees of freedom as one or as two depends on whether the 45° line through the origin is regarded as an element of the decomposition of the plane into 45° lines or whether it is regarded as an element of the decomposition of the plane into lines through the origin.

Borel's paradox can be explained the same way: The equator is an element of the decomposition of the sphere into lines of constant latitude; the Greenwich meridian is an element of the decomposition of the sphere into great circles with common poles. The decomposition matters, which is to say the σ-field matters.

33.3. (a) If the guard says, "1 is to be executed," then the conditional probability that 3 is also to be executed is $1/(1 + p)$. The "paradox" comes from assuming that p must be 1, in which case the conditional probability is indeed $\frac{1}{2}$. If $p = \frac{1}{2}$, then the guard gives no information.

(b) Here "one" and "other" are undefined, and the problem ignores the possibility that you have been introduced to a girl. Let the sample space be

$$bbo \quad \frac{\alpha}{4}, \qquad bby \quad \frac{1-\alpha}{4},$$

$$bgo \quad \frac{\beta}{4}, \qquad bgy \quad \frac{1-\beta}{4},$$

$$gbo \quad \frac{\gamma}{4}, \qquad gby \quad \frac{1-\gamma}{4},$$

$$ggo \quad \frac{\delta}{4}, \qquad ggy \quad \frac{1-\delta}{4}.$$

For example, bgo is the event (probability $\beta/4$) that the older child is a boy, the younger is a girl, and the child you have been introduced to is the older; and ggy is the event (probability $(1-\delta)/4$) that both children are girls and the one you have been introduced to is the younger. Note that the four sex distributions do have probability $\frac{1}{4}$. If the child you have been introduced to is a boy, then the conditional probability that the other child is also a boy is $p = 1/(2 + \beta - \gamma)$. If $\beta = 1$ and $\gamma = 0$ (the parents present a son if they have one), then $p = \frac{1}{3}$. If $\beta = \gamma$ (the parents are indifferent), then $p = \frac{1}{2}$. Any p between $\frac{1}{3}$ and 1 is possible.

This problem shows again that one must keep in mind the entire experiment the sub-σ-field $\mathscr{G}$ represents, not just one of the possible outcomes of the experiment.

33.18. If N is a standard normal variable, then

$$\frac{1}{\sqrt{n}} p_y \left(y + \frac{x}{\sqrt{n}} \right) = \frac{1}{\sqrt{2\pi}} e^{-x^2/2} f\left(y + \frac{x}{\sqrt{n}} \right) \Big/ E\left[f\left(y + \frac{N}{\sqrt{n}} \right) \right].$$

Section 34

34.3. If (X, Y) takes the values $(0, 0)$, $(1, -1)$, and $(1, 1)$ with probability $\frac{1}{3}$ each, then X and Y are dependent but $E[Y\|X] = E[Y] = 0$.

If (X, Y) takes the values $(-1, 1)$, $(0, -2)$, and $(1, 1)$ with probability $\frac{1}{3}$ each, then $E[X] = E[Y] = E[XY] = 0$ and so $E[XY] = E[X]E[Y]$, but $E[Y\|X] = Y \neq 0 = E[Y]$. Of course, this is another example of dependent but uncorrelated random variables.

34.4. First show that $P[B\|\mathscr{G}]$ is positive on B except for a set of P-measure 0 and hence is positive except on a set of P_1-measure 0. Now show that $P_1[A\|\mathscr{G}]P[B\|\mathscr{G}] = F[A\cap B\|\mathscr{G}]$ except on a set of P-measure 0. (First prove that $\int f\, dP_1 = \int_B f\, dP/P(B)$.)

34.9. All such results can be proved by imitating the proofs for the unconditional case or else by using Theorem 34.5 (for part (c), as generalized in Problem 34.7).

34.10. If $Y = X - E[X\|\mathscr{G}_1]$, then $X - E[X\|\mathscr{G}_2] = Y - E[Y\|\mathscr{G}_2]$, and $E[(Y - E[Y\|\mathscr{G}_2])^2\|\mathscr{G}_2] = E[Y^2\|\mathscr{G}_2] - E^2[Y\|\mathscr{G}_2] \leq E[Y^2\|\mathscr{G}_2]$. Take expected values.

34.11. First prove that

$$P[A_1 \cap A_3 \| \mathcal{G}_2] = E[I_{A_1} P[A_3 \| \mathcal{G}_{12}] \| \mathcal{G}_2].$$

From this and (i) deduce (ii). From

$$E[I_{A_1} P[A_3 \| \mathcal{G}_2] \| \mathcal{G}_2] = P[A_1 \| \mathcal{G}_2] P[A_3 \| \mathcal{G}_2],$$

(ii), and the preceding equation deduce $\int_{A_1 \cap A_2} P[A_3 \| \mathcal{G}_2] \, dP = \int_{A_1 \cap A_2} P[A_3 \| \mathcal{G}_{12}] \, dP$. The sets $A_1 \cap A_2$ form a π-system generating $\mathcal{G}_{12}$.

34.16. (a) Obviously (34.13) implies (34.12). If (34.12) holds, then clearly (34.13) holds for X simple. For the general X, choose simple X_k such that $\lim_k X_k = X$ and $|X_k| \leq |X|$. Note that

$$\left| \int_{A_n} X \, dP - \alpha \int X \, dP \right|$$

$$\leq \left| \int_{A_n} X_k \, dP - \alpha \int X_k \, dP \right| + (1 + |\alpha|) E[|X - X_k|];$$

let $n \to \infty$ and then let $k \to \infty$.

(b) If $\Omega \in \mathcal{P}$, then the class of E satisfying (34.12) is a λ-system, and so by the π-λ theorem and part (a), (34.13) holds if X is measurable $\sigma(\mathcal{P})$. Since $A_n \in \sigma(\mathcal{P})$, it follows that $\int_{A_n} X \, dP = \int_{A_n} E[X \| \sigma(\mathcal{P})] \, dP \to \alpha \int E[X \| \sigma(\mathcal{P})] \, dP = \alpha \int X \, dP$.

(c) Replace X by $X dP_0 / dP$ in (34.13).

34.17. (a) The Lindeberg-Lévy theorem.

(b) Chebyshev's inequality.

(c) Theorem 25.4.

(d) Independence of the X_n.

(e) Problem 34.16(b).

(f) Problem 34.16(c).

(g) Part (b) here and the ϵ-δ definition of absolute continuity.

(h) Theorem 25.4 again.

34.19. See S. M. Samuels: The Radon-Nikodym theorem as a theorem in probability, *Amer. Math. Monthly*, **85** (1978), 155–165, and SAKS, page 34.

Section 35

35.4. (b) Let S_n be the number of k such that $1 \leq k \leq n$ and $Y_k = \frac{3}{2}$. Then $X_n = 3^{S_n}/2^n$. Take logarithms and use the strong law of large numbers.

35.9. Let K bound $|X_1|$ and the $|X_n - X_{n-1}|$. Bound $|X_\tau|$ by $K\tau$. Write $\int_{\tau \leq k} X_\tau \, dP = \sum_{i=1}^k \int_{\tau = i} X_i \, dP = \sum_{i=1}^k \left(\int_{\tau \geq i} X_i \, dP - \int_{\tau \geq i+1} X_i \, dP \right)$. Transform the last integral by the martingale property and reduce the expression to $E[X_1]$ − $\int_{\tau > k} X_{k+1} \, dP$. Now

$$\left| \int_{\tau > k} X_{k+1} \, dP \right| \leq K(k+1)P[\tau > k] \leq K(k+1)k^{-1} \int_{\tau > k} dP \to 0.$$

35.13. If $A \in \mathcal{F}_n$, then $\int_A X_n \, dP \geq \int_A X_{n+1} \, dP$ by the result in Problem 32.9. Therefore, $X_1, X_2, \dots$ is a nonnegative supermartingale and by Theorem 35.4 converges with probability 1 to an integrable X.

If ν is absolutely continuous with respect to P on $\mathcal{F}_\infty$, then $X_n \to X_\infty = X$ with probability 1 by part (i) of Theorem 35.8.

Suppose P and ν are mutually singular on $\mathcal{F}_\infty$. If $A \in \mathcal{F}_k$ and $n \geq k$, then $\int_A X_n \, dP \leq \nu(A)$, and so $\int_A X \, dP \leq \nu(A)$ by Fatou's lemma. By the monotone class theorem this holds for all A in $\mathcal{F}_\infty$, and taking A so that $\nu(A) = 0$ and $P(A) = 1$ shows that $X = 0$ with probability 1.

Thus Theorem 35.8 is true without the assumption that P dominates ν on $\mathcal{F}_n$. To get the result called for in the problem split ν into its absolutely continuous and singular parts with respect to P on $\mathcal{F}_\infty$.

35.19. For a very general result, see L. J. Doob: Application of the theory of martingales, *Le Calcul des Probabilités et ses Applications* (Colloques Internationaux du Centre de la Recherche Scientifique, Paris, 1949).

35.22. (a) To prove that two sequences must have the same limit, interlace them. To prove the last assertion choose n and K so that $|E[X_\tau]| \leq K$ for $\tau \in T_n$, and note that $\tau \vee n \in T_n$ and $|E[X_\tau]| \leq \Sigma_{k=1}^{n} E[|X_k|] + |E[X_{\tau \vee n}]|$.

35.23. (a) Use Problem 13.16.

(b) If A_{mq} is the set where $|X_n - Y| < \epsilon$ for some n in the range $m \leq n \leq q$, then $P(\bigcup_q A_{mq}) = 1$ for each ϵ and m because Y is a cluster variable, and hence $P(A_{mq}) \uparrow 1$ as $q \uparrow \infty$.

(c) By parts (a) and (b) there is probability exceeding $1 - 2/k^2$ that $|X_n - Y_k| < 2/k$ for some n in the range $n_k \leq n \leq q_k$, in which case $|X_{\tau_k} - Y_k| < 2/k$. Thus $P[|X_{\tau_k} - Y| \geq 3/k] \leq 3/k^2$.

35.25. (a) Of course, $\{X_n^+\}$ is L_1-bounded. To show that it is an amart, it suffices (Problem 35.22 (b)) to show that $E[X_{\tau_n}]$ converges if $\tau_n \in T_n$ and $\tau_n \geq \tau_{n-1}$. Suppose that $n \leq q$ and put

$$\nu_{n,q} = \begin{cases} \tau_n & \text{if } X_{\tau_n} < 0, \\ \tau_q & \text{if } X_{\tau_n} \geq 0. \end{cases}$$

Check that $\nu_{n,q} \in T_n$ and

$$E[X_{\tau_n}^+] = \int_{X_{\tau_n} \geq 0} X_{\tau_q} \, dP + E[X_{\tau_n}] - E[X_{\nu_{n,q}}]$$

$$\leq E[X_{\tau_q}^+] + E[X_{\tau_n}] - E[X_{\nu_{n,q}}].$$

Now deduce (Problem 35.22(a)) that $\lim \sup_n E[X^+_{\tau_n}] \le \lim \inf_q E[X^+_{\tau_q}]$. Thus $\lim_n E[X^+_{\tau_n}]$ exists, although it might be infinite. Choose integers p_n such that $\tau_n \le p_n$ and put

$$\nu_n = \begin{cases} \tau_n & \text{if } X_{\tau_n} \ge 0, \\ p_n & \text{if } X_{\tau_n} < 0. \end{cases}$$

Check that $\nu_n \in T$ and use the estimate $E[X^+_{\tau_n}] \le E[X_{\nu_n}] + E[|X_{p_n}|]$ along with Problem 35.22(a) and the assumption that $\{X_n\}$ is L_1-bounded.

(b) Obviously, sums and differences of L_1-bounded amarts are again L_1-bounded amarts; note that $X_n \vee \alpha = X_n + (\alpha - X_n)^+$ and $X_n \wedge \alpha = X_n - (X_n - \alpha)^+$.

35.26. If $\nu_{\alpha,p} = p \wedge \inf[n: |X_n| > \alpha]$, then $\nu_{\alpha,p} \in T$ and $\alpha P[\max_{n \le p} |X_n| > \alpha] \le \alpha P[|X_{\nu_{\alpha,p}}| > \alpha] \le E[|X_{\nu_{\alpha,p}}|]$.

35.27. (a) Use Problem 35.25(b) and the fact that $X_n^{(\alpha)} = -\alpha \vee (X_n \wedge \alpha)$.

(b) Use Problem 35.26.

(c) Combine parts (a) and (b) with Problem 35.24.

35.28. (a) Since $E[X_n]$ increases to some limit, it suffices to show that $n \le \tau \le p$ implies that $E[X_n] \le E[X_\tau] \le E[X_p]$. To prove the left-hand inequality, note that

$$E[X_n] = \int_{\tau \ge n} X_n \, dP = \int_{\tau = n} X_\tau \, dP + \int_{\tau > n} X_n \, dP$$

$$\le \int_{\tau = n} X_\tau \, dP + \int_{\tau \ge n+1} X_{n+1} \, dP.$$

(b) For example, let X be a bounded, positive random variable and put $X_n = (-1)^n X/n$.

Section 36

36.6. To start from the beginning, assign cylinders* probabilities $P[\omega: a_i(\omega) = u_i, i \le n] = p_n(u_1, \ldots, u_n)$. Show that, if a cylinder of rank n is split into two cylinders of rank $n + 1$, the probabilities add. Next suppose that $C_1, \ldots, C_l$ are disjoint cylinders whose union is a cylinder. Suppose that $C_1 = [\omega: a_i(\omega) = u_i, i \le n]$, say, has maximal rank among them, and show that $[\omega: a_i(\omega) = u_i, i < n, a_n(\omega) = 1 - u_n]$ must also be one of the C_j, say C_2. But $C_1 \cup C_2$ is a cylinder of rank $n - 1$, and $\Sigma^l_{j=1} P(C_j) = P(C_1 \cup C_2) + \Sigma^l_{j=3} P(C_j)$. Apply the same argument to $C_1 \cup C_2, C_3, \ldots, C_l$, and proceed inductively.

---

* The analogue in sequence space of a cylinder in R^T is $[\omega: (a_{t_1}(\omega), \ldots, a_{t_k}(\omega)) \in H]$, where H is a set of k-long sequences of 0's and 1's. The special cylinders considered here might then be called "thin" cylinders.

Now by addition define P on the field $\mathcal{C}_0$ of finite disjoint unions of cylinders, and prove consistency and finite additivity. Prove countable additivity by the diagonal argument, and extend P from $\mathcal{C}_0$ to $\mathcal{C} = \sigma(\mathcal{C}_0)$.

This proof parallels that of Kolmogorov's theorem but is somewhat simpler, in part because regularity of measures is irrelevant and one can proceed directly to the compactness argument.

36.9. **(b)** Show by part (a) and Problem 34.18 that f_n is the conditional expected value of f with respect to the σ-field $\mathcal{T}_{n+1}$ generated by the coordinates $x_{n+1}, x_{n+2}, \ldots$. By Theorem 35.7 (see Problem 35.17), (36.21) will follow if each set in $\bigcap_n \mathcal{T}_n$ has π-measure either 0 or 1, and here the zero-one law applies.

(c) Show that g_n is the conditional expected value of f with respect to the σ-field generated by the coordinates $x_1, \ldots, x_n$, and apply Theorem 35.5.

36.11. Let $\mathcal{L}$ be the countable set of simple functions $\Sigma_i \, \alpha_i I_{A_i}$ for α_i rational and $\{A_i\}$ a finite decomposition of the unit interval into subintervals with rational endpoints. Suppose that the X_t exist and choose (Theorem 17.1) Y_t in $\mathcal{L}$ so that $E[|X_t - Y_t|] < \frac{1}{4}$. From $E[|X_s - X_t|] = \frac{1}{2}$, conclude that $E[|Y_s - Y_t|] > 0$ for $s \neq t$. But there are only countably many of the Y_t. It does no good to replace Lebesgue measure by some other measure on the unit interval.

36.12. That over D the path function for ω is integer-valued and nondecreasing with probability 1 is easy to prove. Suppose that ω has this property and that the path restricted to D has somewhere a jump exceeding 1. Then for some integer t_0 and for every sufficiently large n there is a $k \leq 2^n$ such that $N_{t_0 k/2^n}(\omega) - N_{t_0(k-1)/2^n}(\omega) \geq 2$. To prove that this has probability 0, use the fact that $P[N_{t+h} - N_t \geq 2] \leq Kh^2$, where K is independent of t and h.

36.13. Put $\mu_1 = \nu$ and inductively define $\mu_n(M \times H) = \int_M \nu_n(H; x_1, \ldots, x_{n-1}) \, d\mu_{n-1}(x_1, \ldots, x_{n-1})$ for $M \in \mathcal{R}^{n-1}$ and $H \in \mathcal{R}^1$; see Problem 18.19. Prove consistency and let $\{X_1, X_2, \ldots\}$ have finite-dimensional distributions $\mu_1, \mu_2, \ldots$. Let $\nu'_n(H, \omega) = \nu_n(H; X_1(\omega), \ldots, X_{n-1}(\omega))$ and prove by a change of variable that $\int_G \nu'_n(H, \omega) P(d\omega) = P(G \cap [X_n \in H])$ for $G = [(X_1, \ldots, X_{n-1}) \in M]$. In the sense of Theorem 33.3, $\nu'_n(H, \omega)$ is thus a conditional distribution for X_n given $\sigma[X_1, \ldots, X_{n-1}]$. For the part concerning integrals use Theorem 34.5.

If A supports ν and supports $\mu_n(\cdot; x_1, \ldots, x_{n-1})$ for $(x_1, \ldots, x_{n-1}) \in A \times \cdots \times A$, then $P[X_1 \in A, \ldots, X_n \in A] = \mu_n(A \times \cdots \times A) = 1$ for all n. In such a case it makes no difference how $\nu_n(H; x_1, \ldots, x_{n-1})$ is defined outside $A \times \cdots \times A$.

Section 37

37.2. **(a)** Use Problem 36.10(b).

(b) Let $[W_t: t \geq 0]$ be a Brownian motion on $(\Omega, \mathcal{F}, P_0)$, where $W(\cdot, \omega) \in C$ for every ω. Define $\xi: \Omega \to R^T$ by $Z_t(\xi(\omega)) = W_t(\omega)$. Show that ξ is measurable $\mathcal{F}/\mathcal{R}^T$ and $P = P_0 \xi^{-1}$. If $C \subset A \in \mathcal{R}^T$, then $P(A) = P_0(\xi^{-1}A) = P_0(\Omega) = 1$.

37.3. Consider $W(1) = \sum_{k=1}^{n} (W(k/n) - W((k-1)/n))$ for notational convenience. Since

$$n \int_{|W(1/n)| \geq \epsilon} W^2\left(\frac{1}{n}\right) dP = \int_{|W(1)| \geq \epsilon \sqrt{n}} W^2(1) \, dP \to 0,$$

the Lindeberg theorem applies.

37.12. By symmetry,

$$\rho(s, t) = 2P\left[W_s > 0, \inf_{s \leq u \leq t} (W_u - W_s) \leq -W_s \right];$$

W_s and the infimum here are independent because of the Markov property, and so by (20.30) (and symmetry again)

$$\rho(s, t) = 2 \int_0^\infty P[\tau_x \leq t - s] \frac{1}{\sqrt{2\pi s}} e^{-x^2/2s} \, dx$$

$$= 2 \int_0^\infty \int_0^{t-s} \frac{x}{\sqrt{2\pi} \, u^{3/2}} e^{-x^2/2u} \frac{1}{\sqrt{2\pi s}} e^{-x^2/2s} \, du \, dx.$$

Reverse the integral, use $\int_0^\infty x e^{-x^2 r/2} dx = 1/r$, and put $v = (s/(s + u))^{1/2}$:

$$\rho(s, t) = \frac{1}{\pi} \int_0^{t-s} \frac{1}{u + s} \frac{s^{1/2}}{u^{1/2}} \, du$$

$$= \frac{2}{\pi} \int_{\sqrt{s/t}}^1 \frac{dv}{\sqrt{1 - v^2}}.$$

Bibliography

HALMOS and SAKS have been the strongest measure-theoretic and DOOB and FELLER the strongest probabilistic influences on this book, and the spirit of KAC's small volume has been very important.

BILLINGSLEY: *Ergodic Theory and Information,* Patrick Billingsley. Wiley, New York, 1965.

BILLINGSLEY': *Convergence of Probability Measures,* Patrick Billingsley. Wiley, New York, 1968.

BIRKHOFF: *Lattice Theory,* rev. ed., Garrett Birkhoff. American Mathematical Society, Providence, Rhode Island, 1961.

BIRKHOFF & MAC LANE: *A Survey of Modern Algebra,* 4th ed., Garrett Birkhoff and Saunders Mac Lane. Macmillan, New York, 1977.

BREIMAN: *Probability,* Leo Breiman. Addison-Wesley, Reading, Massachusetts, 1968.

CHUNG: *A Course in Probability Theory,* 2nd ed., Kai Lai Chung. Academic, New York, 1974.

CHUNG': *Markov Chains with Stationary Transition Probabilities,* 2nd ed., Kai Lai Chung. Springer-Verlag, New York, 1967.

ÇINLAR: *Introduction to Stochastic Processes,* Erhan Çinlar. Prentice-Hall, Englewood Cliffs, New Jersey, 1975.

CRAMÉR: *Mathematical Methods of Statistics,* Harald Cramér. Princeton University Press, Princeton, New Jersey, 1946.

DOOB: *Stochastic Processes,* J. L. Doob. Wiley, New York, 1953.

DUBINS & SAVAGE: *How to Gamble If You Must,* Lester E. Dubins and Leonard J. Savage. McGraw-Hill, New York, 1965.

DYNKIN & YUSHKEVICH: *Markov Processes,* English ed., Evgenii B. Dynkin and Aleksandr A. Yushkevich. Plenum Press, New York, 1969.

FEDERER: *Geometric Measure Theory,* Herbert Federer. Springer-Verlag, New York, 1969.

FELLER: *An Introduction to Probability Theory and Its Applications,* Vol. I, 3rd ed., Vol. II, 2nd ed., William Feller. Wiley, New York, 1968, 1971.

GALAMBOS: *The Asymptotic Theory of Extreme Order Statistics,* Janos Galambos. Wiley, New York, 1978.

GNEDENKO & KOLMOGOROV: *Limit Distributions for Sums of Independent Random Variables,* English ed., B. V. Gnedenko and A. N. Kolmogorov. Addison-Wesley, Reading, Massachusetts, 1954.

507

HALMOS: *Measure Theory*, Paul R. Halmos. Van Nostrand, New York, 1950.

HARDY: *A Course of Pure Mathematics*, 9th ed., G. H. Hardy. Macmillan, New York, 1946.

HARDY & WRIGHT: *An Introduction to the Theory of Numbers*, 4th ed., G. H. Hardy and E. M. Wright. Clarendon, Oxford, 1959.

HAUSDORFF: *Set Theory*, 2nd English ed., Felix Hausdorff. Chelsea, New York, 1962.

KAC: *Statistical Independence in Probability, Analysis and Number Theory*, Carus Math. Monogr. 12, Marc Kac. Wiley, New York, 1959.

KAHANE: *Some Random Series of Functions*, Jean-Pierre Kahane. Heath, Lexington, Massachusetts, 1968.

KARLIN & TAYLOR: *A First Course in Stochastic Processes*, 2nd ed., Samuel Karlin and Howard M. Taylor, Academic, New York, 1975.

KINGMAN & TAYLOR: *Introduction to Measure and Probability*, J. F. C. Kingman and S. J. Taylor. Cambridge University Press, Cambridge, 1966.

KOLMOGOROV: *Grundbegriffe der Wahrscheinlichkeitsrechnung*, Erg. Math., Vol. 2, No. 3, A. N. Kolmogorov. Springer-Verlag, Berlin, 1933.

LÉVY: *Théorie de l'Addition des Variables Aléatoires*, Paul Lévy. Gauthier-Villars, Paris, 1937.

LOÈVE: *Probability Theory I*, 4th ed., M. Loève. Springer-Verlag, New York, 1977.

LUKACS: *Characteristic Functions*, Eugene Lukacs. Griffin, London, 1960.

NEVEU: *Mathematical Foundations of the Calculus of Probability*, English ed., J. Neveu. Holden-Day, San Francisco, 1965.

RADÓ: *Length and Area*, Tibor Radó. American Mathematical Society, Providence, Rhode Island, 1948.

RÉNYI: *Probability Theory*, A. Rényi. North-Holland, Amsterdam, 1970.

RÉNYI': *Foundations of Probability*, Alfred Rényi. Holden-Day, San Francisco, 1970.

RIESZ & SZ.-NAGY: *Functional Analysis*, English ed., Frigyes Riesz and Bela Sz.-Nagy. Unger, New-York, 1955.

ROYDEN: *Real Analysis*, 2nd ed., H. I. Royden. Macmillan, New York, 1968.

RUDIN: *Principles of Mathematical Analysis*, 3rd ed., Walter Rudin. McGraw-Hill, New York, 1976.

RUDIN': *Real and Complex Analysis*, 2nd ed., Walter Rudin. McGraw-Hill, New York, 1974.

SAKS: *Theory of the Integral*, 2nd rev. ed., Stanislaw Saks. Hafner, New York, 1937.

SKOROKHOD: *Studies in the Theory of Random Processes*, English ed., A. V. Skorokhod. Addison-Wesley, Reading, Massachusetts, 1965.

SPITZER: *Principles of Random Walk*, Frank Spitzer, Van Nostrand, Princeton, New Jersey, 1964.

SPIVAK: *Calculus on Manifolds*, Michael Spivak. W. A. Benjamin, New York, 1965.

List of Symbols

Addendum to Section 8: Markov Chains

The theory can be rounded out by an application of the coupling argument to the *persistent null* case, the case in which the chain is persistent but has no stationary distribution.

Theorem. *If an irreducible, aperiodic, persistent chain has no stationary distribution, then* $\lim_n p_{ij}^{(n)} = 0$ *for all i and j.*

PROOF. By the argument in the proof of Theorem 8.6, the coupled chain is irreducible. If it is transient, then $\Sigma_n (p_{ij}^{(n)})^2$ converges by Theorem 8.2, and the conclusion follows.

Suppose, on the other hand, that the coupled chain is persistent. Then the stopping-time argument leading to (8.32) goes through as before. If the $p_{ij}^{(n)}$ do not all go to 0, then there is an increasing sequence $\{n_u\}$ of integers along which some $p_{ij}^{(n)}$ is bounded away from 0. By the diagonal method (Theorem 25.13), it is possible by passing to a subsequence of $\{n_u\}$ to ensure that each $p_{ij}^{(n_u)}$ converges to a limit, which by (8.32) must be independent of i. Therefore, there is a sequence $\{n_u\}$ such that $\lim_u p_{ij}^{(n_u)} = \alpha_j$ exists for all i and j, where α_j is nonnegative for all j and positive for some j. If M is a finite set of states, then $\Sigma_{j \in M} \alpha_j = \lim_u \Sigma_{j \in M} p_{ij}^{(n_u)} \leq 1$, and hence $0 < \alpha = \Sigma_j \alpha_j \leq 1$. Now $\Sigma_{k \in M} p_{ik}^{(n_u)} p_{kj} \leq p_{ij}^{(n_u+1)} = \Sigma_k p_{ik} p_{kj}^{(n_u)}$; it is possible to pass to the limit ($u \to \infty$) inside the first sum (if M is finite) and inside the second sum (by the M-test), and hence $\Sigma_{k \in M} \alpha_k p_{kj} \leq \Sigma_k p_{ik} \alpha_j = \alpha_j$. Therefore, $\Sigma_k \alpha_k p_{kj} \leq \alpha_j$; if one of these inequalities were strict, it would follow that $\Sigma_k \alpha_k = \Sigma_j \Sigma_k \alpha_k p_{kj} < \Sigma_j \alpha_j$, which is impossible. Therefore $\Sigma_k \alpha_k p_{kj} = \alpha_j$ for all j, and the ratios $\pi_j = \alpha_j/\alpha$ give a stationary distribution, contrary to the hypothesis. ∎

Index